SURFACE MODIFICATIONS AND COATINGS

Proceedings of an International Conference held in conjunction with ASM's Materials Week '85

Toronto, Ontario, Canada
14-17 October 1985

Edited by
Richard D. Sisson, Jr.
Worcester Institute
Worcester, Massachusetts, USA

Sponsored by
ASM's Surface Treating and Coating Division

Published by

Library of Congress Catalog Card Number: 86-070859
ISBN: 0-87170-267-3
SAN: 204-7586

Printed in the United States of America

FOREWORD

This book contains the proceedings of the "1985 International Conference on Surface Modifications and Coatings" that was held during Materials Week '85 October 12-17, 1985 in Toronto, Ontario, Canada. This conference was sponsored by the Surface Treating and Coating Division Council of the American Society for Metals.

It was the goal of this conference to provide a forum for the presentation and discussion of a wide variety of surface modification and coating processes for selected applications. Both mature and emerging technologies were included. The applications included thermal, corrosion, oxidation and wear protection. The authors of the papers represented a wide variety of Industries as well as Laboratories and Universities. In addition, people from all over the world participated in the conference with representatives from China, Japan, England, Switzerland, France, Canada and the United States.

The Planning and organization of the conference took several years. The efforts of the session organizers are greatly appreciated. These people include:

- Robert Hochman, Georgia Institute of Technology —Surface Modifications by Ion Processes
- Alan Krauss, Argonne National Laboratory —Self-Renewing Coatings
- P. F. Woerner and Depak Bhat, GTE Valeron Corporation —Hard Coatings
- R. A. Miller, NASA Lewis Laboratory —Thermal Barrier Coatings
- Gordon H. Hansen, Ford Motor Company —Coatings for the Automotive Industry
- Sara A. Dillich, Worcester Polytechnic Institute —Coatings for Tribological Applications
- John Smeggil, United Technology Research Center —Coatings for High Temperature Oxidation Resistance
- Terence F. J. Quinn, Georgia Institute of Technology —Friction Materials Coatings
- J. W. Patten Cummins Engine Company —Coatings for Diesel Engine Applications

In addition, the efforts of Mike Rigdon, Christopher Brown, Ron Biederman and Ward Winer as session chairmen are gratefully acknowledged. The hard work and dedicated efforts of Ms. Shari Gerstenberger and Ms. Joan Bellian of the ASM staff is deeply appreciated.

The success of the conference is mainly due to the authors, presenters, and participants. Many thanks are due to these contributors whose work, ideas and enthusiasm made the conference a success.

Richard D. Sisson, Jr.
Conference Organizer
Worcester Polytechnic Institute
Worcester, Massachusetts

TABLE OF CONTENTS

Coatings for the Automotive Industry

Corrosion Resistance of Zn-Fe Alloy Electroplated Steel 1
T. Adaniya, T. Hara, M. Sagiyama, T. Watanabe, T. Honma

Undervehicle Corrosion Testing of Zinc & Zinc Alloy Coated Steels 11
R. J. Neville, K. M. de Souza

Cosmetic Corrosion Performance of Zinc Alloy Steel Sheets in Automotive Body . . . 23
S. Wakano, A. Sakoda, M. Nishihara, T. Kurimoto, Y. Hoboh

Zinc-Iron Alloy Coated Steel Sheets for Automotive Corrosion Protection 29
T. Hada, T. Fujiwara, T. Kanamaru, M. Nakayama, T. Horita

Enhanced Aluminum Auto Trim 37
J. H. Powers, J. A. Burke, M. Wiener

Coatings and Surface Modifications for Energy Systems

Ceramic Coating on Engine Parts 45
Y. Suzuki, S. Tanoue, H. Yamaguchi

Surface Modifications of Pistons and Cylinders Liners 49
Y. Suzuki

Production Based Ceramic Coatings for Diesel Engine Components 59
A. R. Nicoll, S. Keller, G. Wuest

Characterization of ZRO_2-Y_2O_3 Thermal Spray Powder Systems 65
M. J. Froning, N. Jayaraman, T. E. Mantkowski, D. V. Rigney

Rolls-Royce Experience with Thermal Barrier Coatings
Including High Time Service Evaluation 81
F. C. Toriz, A. Bennett, S. K. Gupta

Practical Aspects of Ultra-Thick Thermal Barrier Coatings 91
D. H. Harris

Advanced Thermal Barrier Coating Systems 97
J. D. Reardon, M. R. Dorfman

Optothermal NDE of Thermal Barrier Coatings 107
P. Cielo, S. Dallaire

Microstructure of Coatings Formed on Pack Aluminized Ni-Cr Alloys 117
N. Delanerolle, L. L. Seigle

Cyclic Oxidation of Chromized Steels and Competitive Materials at 650 to 815°C . . . 131
E. A. Loria

Effects of Microstructural Morphology on the Performance of Platinum Aluminide Coatings 143
P. Deb, D. H. Boone, R. Streiff

Na_2SO_4 Induced Corrosion of Nickel at High Temperature 161
A. K. Misra

Wear Resistant Coatings and Surface Treatments

Relationship Between Metal Cutting Performance and Material Properties of TiC Coated Cemented Carbide Cutting Tools 179
T. Cho, D. G. Bhat, P. F. Woerner

Machining Ferrous Materials with CVD Coated Carbides Part I: Interfacial Conditions 183
P. A. Dearnley, V. Thompson, A. N. Grearson

Machining Ferrous Materials with CVD Coated Carbides Part 2: Wear Mechanisms 199
P. A. Dearnley, V. Thompson, A. N. Grearson

Wear Behavior of TiB_2 and Boron Coated Cemented Carbides 217
D. E. Graham

Friction and Surface Chemistry of Implantation Modified Metal/Carbide Composites 229
S. A. Dillich, I. L. Singer

Low Friction and Antifretting Coating Inside Borings by Magnetically Enhanced Sputtering 239
B. Zega

Analytical Simulation of Friction Resistance Modeled by Ridge Mobility 245
Y. Nakamura, Y. Zhu, B. Avitzer

Plasma-Sprayed Coatings for Piston Rings—State of Development and Application Potential **255**
U. Buran, H.-C. Mader, M. Morsbach, B. A. Newman

Metallurgical Aspects of the Wheel/Rail Surface Contact **265**
M. N. Bassim, D. Shum, S. Yannacopoulos, M. R. Bayoumi, M. C. Chaturvedi

Characteristics of Coatings Produced Using a New High Velocity Thermal Spray Technique **275**
K. V. Rao

Flame Sprayed Ceramic Coatings as an Economic Alternative to Spray and Fuse Coatings for Frictional and Corrosive Use **283**
F. T. Jaeger

Process and Engineering Benefits of Sputter Ion Plated Titanium Nitride Coatings **291**
M. H. Jacobs

Tribocor 532N—A New Material for Environments Involving Wear and Corrosion **301**
P. F. Ziegler, J. J. Rausch

Diffusion Carbide Coatings Formed in Molten Borax Systems (Reaction in Borax Bath and Properties of Carbide Coated Steel) **311**
T. Arai, H. Fujita, Y. Sugimoto, Y. Ohta

Low Temperature Deposition of Tribological Coatings **319**
J. T. Prater

Surface Segregation and Self-Renewing Coatings

Equilibrium Surface Segregation—Recent Advances in Theory and Experiment **327**
P. Wynblatt, D. A. Steigerwald

Segregation and Ordering on Gold Alloy Surfaces Studied by Low Energy Ion Scattering and Low Energy Electron Diffraction **337**
T. M. Buck, E. G. McRae

Surface Precipitation and Its Application to Coatings **353**
K. Nii, K. Yoshihara

Applications of Self-Renewing Coatings to Improved Vacuum Materials, Hydrogen Permeation Barriers and Sputter-Resistant Materials **361**
A. R. Krauss

Preparation and Some Properties of Cu-Li Alloys Containing Up to 20 at .% Li 369
M. H. Mendelsohn, A. R. Krauss, D. M. Gruen

Effect of the Surface Segregation on the Ion Damages of Alloys During Sputtering at Elevated Temperatures 375
H. Shimizu, S. Ichumura, K. Morii, H. Murakami

Criterion for Self-Sustaining Coating in Metal- or Carbide-Coated Graphite 383
K. Morita, H. Ohno

Dependence of Sputtered Atom Angular Distribution on the Depth of Origin 391
T. D. Andreadis, M. L. Roush, G. Chambers, O. F. Goktepe

Surface Modifications and Coatings—General Sessions

Fundamentals of Ion Implantation 399
G. Dearnaley

Formation of Hard Surfaces by the Laser Melt-Particle Injection Process 409
K. P. Cooper

Influence of Machining on the Microstructure, Microhardness and Precipitation in the Aluminum Alloy 7075 421
C. A. Brown, F. Dupont, B. Ilschner

An Assessment of the Energy Conservation Potential of Surface Modification Technologies 427
H. K. Le

Surface Stress Modification for Repair of Stress-Relieved Carbon Steel Fabrications 439
C. F. Jenkins

The Microstructure of 55 w/o Al-Zn-Si (Galvaume) Hot-Dip Coatings 447
J. H. Selverian, M. R. Notis, A. R. Marder

The Wear of Stainless Steel and its Relation to Surface Modification 455
H. Hong, R. F. Hochman, T. F. J. Quinn

The following papers were presented at the International Conference on Surface Modifications and Coatings but were unavailable in manuscript form for publication in these proceedings.

Surface Modification by Ion Processes—A Rapidly Emerging Technology
R. F. Hochman

The Industrial Application of Ion Implantation Processes
J. K. Hirvonen

Refractory Coatings Formed at Room Temperature by Simultaneous Vapor Deposition and Ion Implantation
B. D. Sartwell, R. A. Kant

Effect of Structure Modification by Ion Implantation on Fatigue Behavior
S. D. Antolovich, J. G. Han, R. F. Hochman

Characterization of Surface Modified Layers
W. B. Carter

Chemistry and Structure of the Interface in Metal-Metal and Metal-Ceramics Ion Plated Materials
M. Rigsbee

An Overview of Cathodic Arc Technology
P. A. Lindfors

The Effect of Coating Thickness and Substrate Properties of Ion Plated Steel in Rolling Contact Fatigue
A. Erdimer, R. F. Hochman, F. J. Dolan, R. L. Thom

Stoichiometric Deposition of Alloys by Cathodic Arc Technology
P. A. Lindfors

Steered Arc Coating for the Deposition of Metal and Ceramic Layers
S. Ramalingam, R. F. Cheng

Calculated Equilibrium Surface Segregation of Impurity Cations at the Surfaces of Oxides
W. C. Mackrodt

Surface Segregation During Irradiation
L. E. Rehn, N. Q. Lam, P. R. Okamoto

CNTD Hard Coatings—Their Properties and Techniques for Characterization
R. A. Holzl, P. Dyer

The Role of Hard Coatings in Carbide Milling Tools
G. J. Wolfe, C. T. Petrosky, D. T. Quinto

Effects of Nitrogen Implantation on Cemented Carbides
D. W. Oblas, V. K. Sarin

Plasma D-Gun Coatings
M. K. Keshavan, R. C. Tucker

Determination of Phase Content of Multiphase Hard Coatings by X-Ray Defraction
D. Bhat, Y. DeKay

The Effects of Ion Implantation on the Surface Topography of WC-Co
D. W. Oblas, J. Hefter, T. A. Emma

Phase Stability in the $ZrO_2Y_2O_3$ System
T. E. Mitchell, D. S. Suhr

Phase Transformations in Partially Stabilized Zirconia During Thermal Fatigue
I. Sone, R. R. Biederman, R. D. Sisson, Jr.

Residual Stress Field in Plasma Sprayed Thermal Bearing Coatings
R. L. Mullen, R. C. Hendricks

Cyclic Stress Analysis of Ceramic Coated Gas Turbine Seals
J. Padovan, D. Dougherty, B. Hendricks

Performance of Ceramic Thermal Barrier Coatings in Diesel Engine Service
A. V. Levy, S. MacAdam

Automation of the Thermal Spray Process
F. J. Hermanek

Effect of Forming on Corrosion Performance for Galvanized Steel—Hot Dipped and One Side Electro
S. Denner

The Effects of Zinc Coating, Die Materials and Lubricants on Sheet Metal Formability
W. G. Brazier, R. W. Thompson

A New "Chrome-Like" Finish for Aluminum
D. Fern, W. Kellermann

On the Microstructure of Laser Surface Alloyed Fe-Cr-C Systems
J. Mazumder, S. Das, I. Dumier

Refractory Coatings Formed at Room Termperature by a Simultaneous Vapor Deposition and Ion Implantation
R. A. Kant, B. D. Sartwell

Laser Surface Treatment of Overlay Coatings
R. Streiff, P. Mazars, M. Caillet

The Utilization of Ion Implantation Techniques to Modify the Oxidation Characteristics of Palladium Alloys
J. G. Smeggil, C. O. Hulse, A. J. Shuskus, F. D. Lemkey

The Role of Active Elements in the Cyclic Oxidation Behavior of MCrAL Alloys
R. R. Biederman, R. D. Sisson, Jr.

Tribological Characteristics of Arc Coated Zirconium Nitride Films
S. Ramalingam, K. Kim, S. Kim

Friction and Wear Characteristics of Ion Implanted M-2 Steel in Lubricated Sliding Contact
S. Ramalingam, K. Kim, F. A. Smidt

Control of Coating Processes Through Analytical Instrumentation
T. R. Gallant

COATINGS FOR THE AUTOMOTIVE INDUSTRY

CORROSION RESISTANCE OF Zn-Fe ALLOY ELECTROPLATED STEEL

Takeshi Adaniya, Tomihiro Hara, Masaru Sagiyama, Tsutomu Watanabe, Toshiyuki Honma
Technical Research Center, Nippon Kokan
K. K. Kawasaki, Japan

Abstract

Zn-Fe alloy electroplated steel of which coatings contain 10 - 20%Fe exhibited excellent perforation corrosion resistance, scab corrosion resistance and blistering resistance. The corrosion resistance is equivalent to that of galvannealed steel and better than that of electrogalvanized steel. The enhanced corrosion resistance of Zn-Fe alloy electroplated steel may result from the suppressed activity and long duration of sacrificial protection action of the alloys.

THE IMPROVEMENT in corrosion protection of automobile bodies has been intensively promoted by automobile industries, since social attention had been focussed on corrosion damage to automobiles in the snow belt of North America and Nordic Europe. To achieve the improvement, automobile industries have been requesting precoated steel of high corrosion resistance. In response to the request, steel industries have developed various kinds of precoated steels.[1-7)]

Galvannealed steel is known to be a corrosion-resistant precoated steel.[8,9)] However, manufacturing process of galvannealed steel involves heating process which deteriorates mechanical properties of the products. For the reason, applications of galvannealed steel to car bodies were restricted. To overcome the shortcomings of galvannealed steel, authors have successfully developed Zn-Fe alloy electroplated steel and reported properties of the product.[10,11)]

In the present report, corrosion resistance, mainly perforation corrosion resistance and scab corrosion resistance, and corrosion behavior of Zn-Fe alloy electroplated steel are described.

EXPERIMENT

Zn-Fe alloy electroplated test panels were produced using laboratory flow cell. Zn-Fe alloy was electroplated on cold rolled steel from sulfate bath of which composition is shown in table 1. Fe content of the alloy was varied by adjusting $Fe^{2+}/Fe^{2+}+Zn^{2+}$ of the bath and plating conditions such as flow rate. As reference material for corrosion test, the followings were used: electrogalvanized steel Zn(E), galvannealed steel (GA), electro-galvanized and subsequently galvannealed steel (EGA), Zn-Ni alloy electroplated steel (Zn-Ni (E)), zinc rich primer coated steel (ZRP) and cold rolled steel (CRS).

Test panels for investigation on corrosion resistance and corrosion behavior were phosphated and painted as shown in table 2.

To evaluate corrosion resistance, the following corrosion tests were conducted: cyclic corrosion test, salt spray test, outdoor exposure with spraying 5% NaCl solution and outdoor exposure at beach. The test conditions are shown in table 3.

To investigate corrosion behavior, measured were phase composition, corrosion potential, change in rest potential during cyclic corrosion test and amount of dissolved coatings in alkaline solution.

X-ray diffraction was conducted to estimate phase composition. As a parameter of phase composition, we introduced peak intensity ratio difined as follows[11)]:

$$\text{peak intensity of the phase } j = \frac{\sum_i Iij}{\sum_j \sum_i Iij}$$

where, $\sum_i Iij$: the sum of intensity of peaks attributed to a phase j
$\sum_j \sum_i Iij$: the sum of intensity of all peaks observed

The conditions of X-ray diffraction are listed in table 4.

Corrosion potential of electrodeposited Zn-Fe alloy was measured in 5% NaCl solution at 25°C.

Potential measurement of panels exposed to cyclic corrosion test was conducted in 5% NaCl solution at room temperature after salt spray.

Dissolved amount of coatings in alkaline solution was determined by analyzing the immersion solution of 10^{-2}M NaOH after 66 Hr immersion using atomic absorption spectroscopy.

RESULTS and DISCUSSION

PERFORATION CORROSION- Cyclic corrosion test was carried out to study perforation corrosion of Zn-Fe alloy electroplated steel.

Relation between corrosion depth and the number of cycle is shown in Figs. 1 and 2. Steel substrate corroded from the beginning of the test, when the Zn-Fe alloy coatings contain 40% Fe or more. In contrast, corrosion of steel substrates initiated after some duration, when the Zn-Fe alloy coatings contain 19% Fe or less. The length of the duration in which corrosion of steel substrate is prevented at the initial stage predominantly determined the perforation corrosion resistance of Zn-Fe alloy electroplated steel.

The duration of initial suppression of corrosion relates to electrochemical properties of electrodeposited Zn-Fe alloy. The results of rest potential measurement during cyclic corrosion test are shown in Fig. 3. The test panels were E-coated and scribed before testing. Rest potential of Zn-Fe alloys containing 40% Fe or more approached to that of cold rolled steel in a short duration, while rest potential of Zn-Fe alloys containing 19% Fe or less and pure Zn shifted toward that of cold rolled steel at the start of the test, but was kept constant, where the potential was still more base than that of cold rolled steel. Rest potential of Zn-Fe alloys containing 19% Fe or less and pure Zn finally approached to that of cold rolled steel.

The duration in which rest potential is more base than that of cold rolled steel is almost equal to the duration in which corrosion of steel substrate is suppressed. The correlation between rest potential and corrosion depth indicates that sacrificial protection action of the coatings prevents corrosion of exposed steel substrate at the initial stage and dominates perforation corrosion resistance.

Relation between corrosion depth and Fe content in Zn-Fe alloy deposit is shown in Fig. 4. Corrosion depth remains almost constant in the Fe content range below 30%, while corrosion depth increases with increasing Fe content in the range above 30%. Comparing the influence of Fe content on corrosion depth with rest potential, we can expect effective sacrifice action of electrodeposited Zn-Fe alloy containing 30% Fe or less for the protection of steel substrate.

Relation between corrosion depth of unpainted panels and Fe content in deposit is shown in Fig. 5. Unpainted Zn-Fe alloy electroplated steel exhibited good perforation corrosion resistance in the Fe content range below 30%, especially in the Fe content range from 10% to 20%, where the minimum of corrosion depth is observed. The results are similar to those of E-coated Zn-Fe alloy electroplated steel, but slightly different in the Fe content range of 10-20% where the minimum of corrosion depth was not observed for E-coated Zn-Fe alloy electroplated steel.

Effect of coating weight of pure Zn and Zn-Fe (20% Fe) on corrosion depth of E-coated panels is shown in Fig. 6. Corrosion depth decreased with increasing coating weight for both pure Zn an Zn-Fe (20% Fe), and no difference between pure Zn and Zn-Fe (20%) was found. For the evaluation of perforation corrosion resistance, corrosion depth is indeed an important measure, but also corrosion area (width) of steel substrate is considered to be improtant for the detailed evaluation. From this point of view, corrosion width was also measured. Fig. 7 shows relation between corrosion width of steel substrate and Fe content in deposit. Corrosion depth exhibited the minimum in the Fe content range from 10% to 30%. Accordingly, the results of corrosion depth and corrosion width indicate that Zn-Fe alloy electroplated steel has the optimum Fe content range from 10% to 30% for perforation corrosion resistance, and has better performance than electrogalvanized steel.

To recognize the rank of perforation corrosion resistance of Zn-Fe alloy electroplated steel, corrosion depth of various precoated steels including Zn-Fe alloy electroplated steel was compared. The results are shown in Fig. 8. Perforation corrosion resistance of Zn-Fe alloy electroplated steel is equivalent to that of hot dip galvannealed steel, and better than that of Zn-Ni alloy electroplated steel and zinc-rich primer coated steel. Relatively poor perforation corrosion resistance of Zn-Ni alloy electroplated steel and zinc-rich primer coated steel may result from poor sacrificial protection action of their coatings.

BLISERING - Effect of Fe content in deposit on blistering was investigated in salt spray test. Fig. 9 shows the results. Blister formed remarkably on pure Zn coated steel, but decreased significantly with increasing Fe content in deposit. Blistering resistance is improved in the Fe content range above 10% especially in the range of 20-50% Fe. In the range above 50% Fe, blister increased again.

In corrosion in salt spray test, local cell is formed between exposed area at scribe (anode) and painted area near scribe (cathode).[12] Oxygen is reduced to hydroxylions at cathode, which increases pH value under paint. Consequently, environment under paint film at cathode becomes alkaline. The alkaline medium dissolves the phosphate film and plated layer, thus extending area of blister. Therefore, it is expected that the dissolved amount of deposit correlates with blistering. Relation between dissolved amount of deposit in 10^{-2}M NaOH and Fe content in deposit is shown in Fig. 10. Dissolved deposit was exclusively composed of zinc, which reflects a difference in solubility between zinc and iron. The dissolved amount decreased with increasing Fe content in deposit, and showed correlation with blistering in the Fe content range below 50%. In the Fe content range above 50%, the correlation was not recognized. This may result from poor corrosion resistance of Zn-Fe alloys containing 50% Fe or more, which extends corrosion area of coatings.

SCAB CORROSION - Scab corrosion resistance was evaluated in outdoor exposure with spraying 5% NaCl solution. Relation between Fe content and creepage width after 7 month exposure is shown in Fig. 11. Zn-Fe alloys exhibited narrow creepage width in the Fe content range below 30%, and the minimum width in the range of 10 to 25% Fe. Further exposure resulted in a slight shift of the optimum Fe content, which ranged from 10% to 20%. Accordingly, good scab corrosion resistance of Zn-Fe alloy electroplated steel is obtained in the range of 10-20% Fe.

Effect of coating weight on creepage width is shown in Fig. 12. Increase in coating weight reduced the creepage width, thus enhancing scab corrosion resistance.

Scab corrosion resistance of Zn-Fe alloy electroplated steel was evaluated in comparison with other precoated steels to recognize the rank of Zn-Fe alloy electroplated steel. The results are shown in Fig. 13. All the precoated steels showed better scab corrosion resistance than cold rolled steel, especially Zn-Fe and Zn-Ni alloy electroplated steels and hot dip galvannealed steel exhibited excellent performance.

Outdoor exposure test at beach showed the results similar to ones obtained in outdoor exposure test with spraying 5% solution with respect to an effect on Fe content on scab corrosion resistance.

PHASE COMPOSITION and CORROSION - Relation between phase composition and Fe content in deposit is shown in Fig. 14. η, δ_1, Γ and α phases appeared in electrodeposited Zn-Fe alloys, depending on Fe content. Fe content range where each phase of electrodeposited Zn-Fe alloy exists, differed from one shown in equilibrium diagram. It is known that alloys in non-equilibrium state are formed by electrodeposition.[12]

Corrosion potential measured in 5% Nacl solution is shown in Fig. 15. Corrosion potential of electrodeposited Zn-Fe alloys is divided into three regions which mainly correspond to η phase, δ_1 or Γ phase and α phase. In the region where η phase is a main phase, corrosion potential shifts toward more noble potential, as Fe content increases. In the region where δ_1 or Γ phase is a main phase, corrosion potential remains nearly constant. In the region where α phase is a main phase, corrosion potential shifts toward that of cold rolled steel with increasing Fe content.

When considering the results of corrosion tests, 10 to 20% Fe is an optimum composition of electrodeposited Zn-Fe alloy coatings for corrosion protection. Corrosion potential of electrodeposited Zn-Fe alloys containing 10 to 20% Fe is more noble than that of pure zinc by more than 100 mV, which indicates that the activity of pure zinc is suppressed by alloying. However, corrosion potential of Zn-Fe alloys remains nearly constant in the range of 13 to 45% Fe, which indicates that corrosion resistance of Zn-Fe alloys is enhanced not only by the suppressed activity of the alloys. As seen in the paragraph of perforation corrosion, both sacrificial protection action and its duration are necessitated for corrosion protection of steel. It is considered that the duration of sacrificial protection action tends to decrease faster in the range over about 20% Fe than in the range of 10 to 20% Fe, though corrosion potential at the initial stage of corrosion is nearly constant in the wide range of Fe content. Consequently, corrosion resistance may differ between the Zn-Fe alloys of 10 to 20% Fe and ones of over 20% Fe.

Electrodeposited Zn-Fe alloys of 10 to 20% Fe which exhibits excellent corrosion resistance is composed of η phase or $\eta + \delta_1$ phase. Γ phase exhibits less corrosion resistance than η and δ_1 phase, and α phase shows the least one.

CONCLUSIONS

(1) An optimum composition of electrodeposited Zn-Fe alloys for corrosion protection is 10 to 20% Fe.

(2) Zn-Fe alloy electroplated steel exhibits good perforation corrosion resistance, scab corrosion resistance and blistering resistance which are equivalent to those of galvannealed steel, and better than those of electrogalvanized steel.

(3) The enhanced corrosion resistance of Zn-Fe alloy electroplated steel may result from the suppressed activity and long duration of sacrificial protection action of the alloys.

REFERENCES

1. Watanabe, T., Ohmura, M., Honma, T., and Adaniya, T., SAE Technical Paper 820424 (1982)
2. Miura, N., Sato, T., Kanamaru, T., Shindo, Y., and Kitazawa, Y., Trans ISIJ 23, No.11, 913-922 (1983)
3. Shibuya, A., Kurimoto, T., Hoboh, Y., and Usuki, N., Trans. ISIJ 23, No.11, 923-929 (1983)
4. Nomura, S., Sakai, H., Nishimoto, H., Uegaki, T., Sakaguchi, M., Iwai, M., and Kokubo, I., Trans. ISIJ 23, No.11, 930-938 (1983)
5. Shindo, Y., Oka, J., Yoneno, M., Ejima, M., and Yamada, A., Seitetsu Kenkyu No.315, 25-33 (1984)
6. Okuda, H., Iino, S., Matsui, K., Noumi, R., and Matsuo, S., Tetsu-to-Haganē 69, No.5, S403 (1983)
7. Yamashita, M., Enatsu, A., Adaniya, T., Hara, T., Tetsu-to-Haganē 69, No.13, S1103 (1983)
8. Takahashi, H., Shirogane, Y., Matsudo, K., Adaniya, T., and Omura, M., SAE Technical Paper 800145 (1980)
9. Miyoshi, Y., Yoshida, K., Azami, T., Kanamaru, T., and Kado, S., Tetsu-to-Haganē 66, 858-867 (1980)
10. Adaniya, T., Hara, T., Sagiyama, M., Honma, T., and Watanabe, T., Plat. & Surf. Fin. 72, No.8, 52-56 (1985)
11. Hara, T., Adaniya, T., Sagiyama, M., Honma, T., Tonouchi, A., Watanabe, T., and Ohmura, M., Trans ISIJ 23, No.11, 954-958 (1983)
12. Uchida, K., and Deguchi, T., Nisshin Steel Technical Report 48, 32-43 (1983)
13. Raub, E., and Müller, K., Fundamentals of Metal Deposition (1967, Elsevier)

Table 1. Bath composition of Zn-Fe alloy electroplating

Component salts	Concentration (g/ℓ)
$FeSO_4 \cdot 7H_2O$ $ZnSO_4 \cdot 7H_2O$	} 500
Na_2SO_4	30
$CH_3COONa \cdot 3H_2O$	20
$C_6H_8O_7$	5

Table 2. Preparation of coated test panels

Test panel	Coating procedure
E-coated panel	Immersion type zinc phosphate coating (Bt-3030) ↓ Cathodic electropainting (El-9000) : 20 μm
3-coat painted panel	Immersion type zinc phosphate coating (Bt-3030) ↓ Cathodic electropainting (El-9000) : 20 μm ↓ Seal coat : 35 μm ↓ Top coat : 35 μm

Table 3. Corrosion test conditions

Corrosion test	Conditions	Evaluation	Test panel
Cyclic corrosion test (CCT)	→Salt spray (5% Nacl) : 50°C ↓ Drying : 70°C ↓ Immersion in 5% Nacl sol. : 50°C ↓ Drying : RT 1 cycle = 24 Hrs	Max. corrosion depth Max. corrosion width	E-coated panel Unpainted panel
Salt spray test	5% Nacl, 35°C (JIS Z 2371)	Blistering grade	E-coated panel
Outdoor exposure with spraying 5% Nacl sol.	Outdoor exposure with spraying 5% Nacl sol. once a weekday at NKK Technical Research Center, Kawasaki	Creepage width	E-coated panel
Outddoor expsoure at beach	Outdoor exposure at beach of Chikura, Chiba prefecture	Creepage width	3-coat painted panel

E-coated and 3-coat painted panels were scribed before testing.

Table 4. Conditions of X-ray diffraction

Target	Cu
Filter	- (monochromater)
Voltage	40 kV
Current	20 mA
Time constant	1 sec
Scanning speed	2 deg/min
Chart speed	2 cm/min

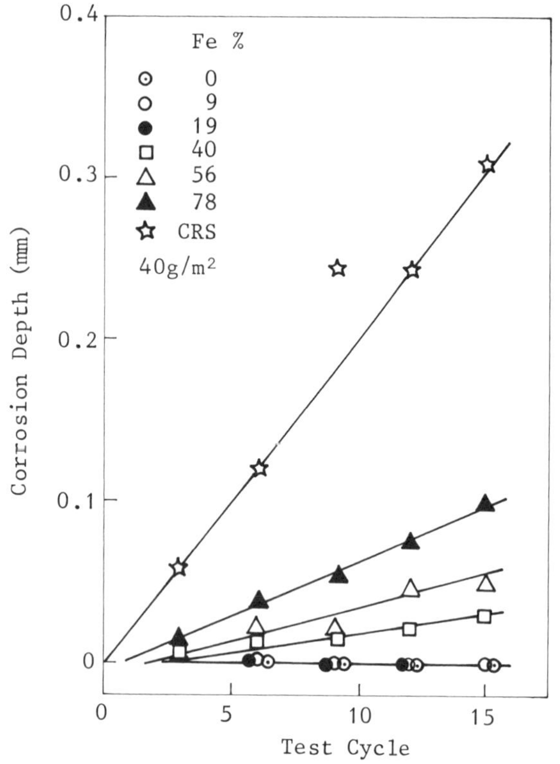

Fig. 1 Corrosion depth of E-coated Zn-Fe alloy electroplated steel exposed to CCT

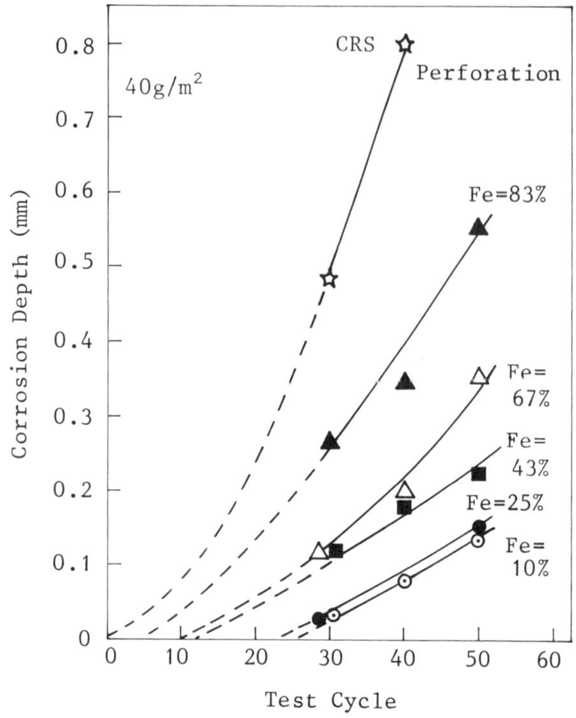

Fig. 2 Corrosion depth of E-coated Zn-Fe alloy electroplated steel exposed to CCT

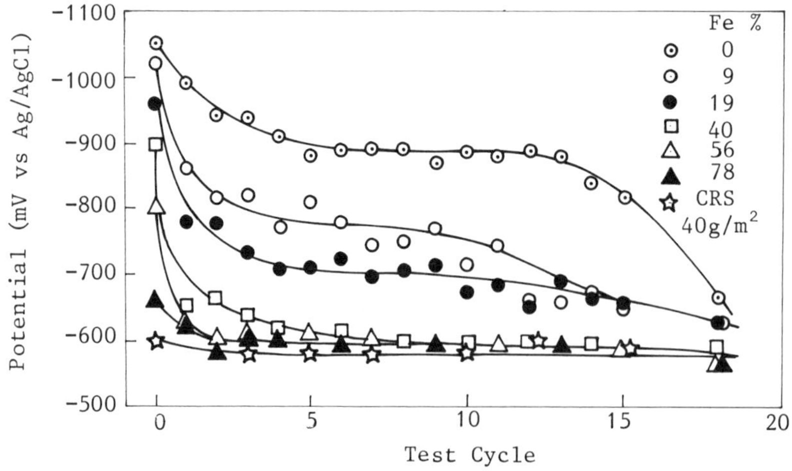

Fig. 3 Rest potential of E-coated Zn-Fe alloy electroplated steel exposed to CCT

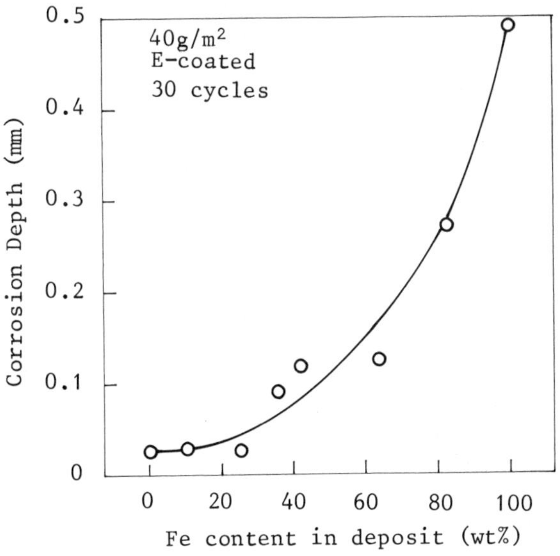

Fig. 4 Relation between corrosion depth and Fe content in deposit

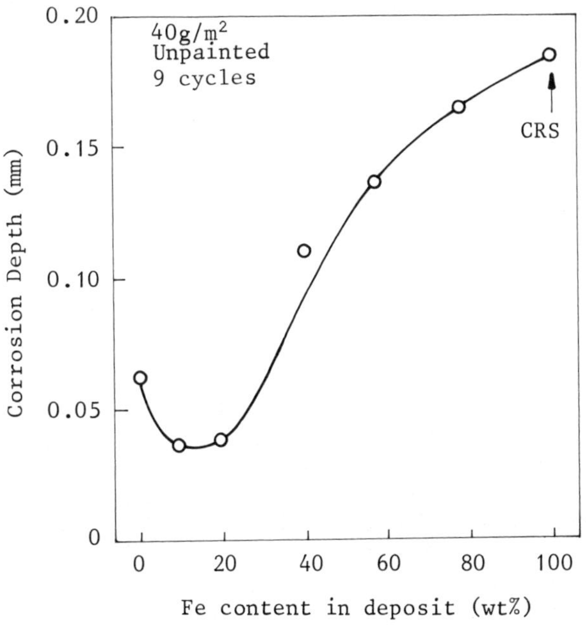

Fig. 5 Relation between corrosion depth and Fe content in deposit

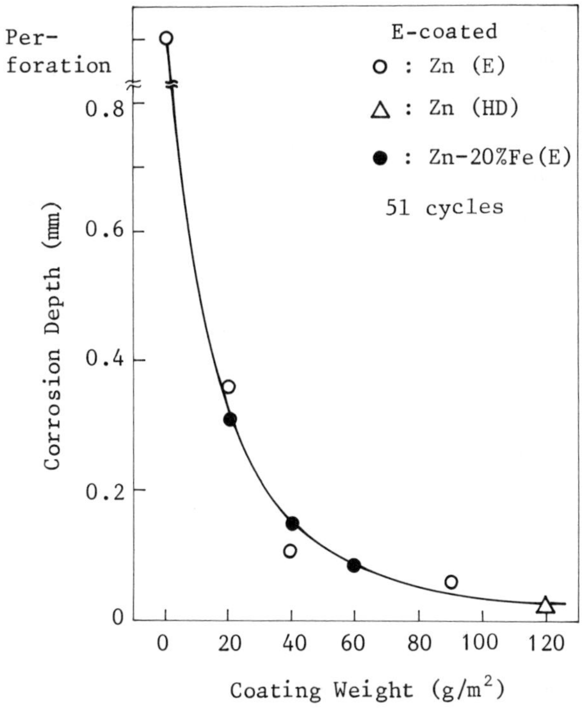

Fig. 6 Effect of coating weight on corrosion depth

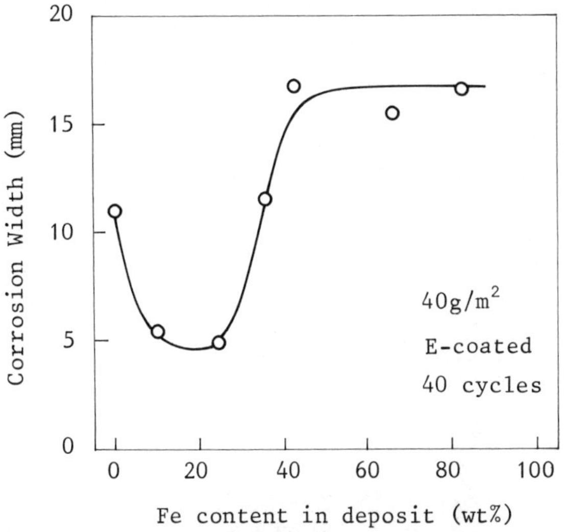

Fig. 7 Relation between corrosion width and Fe content in deposit

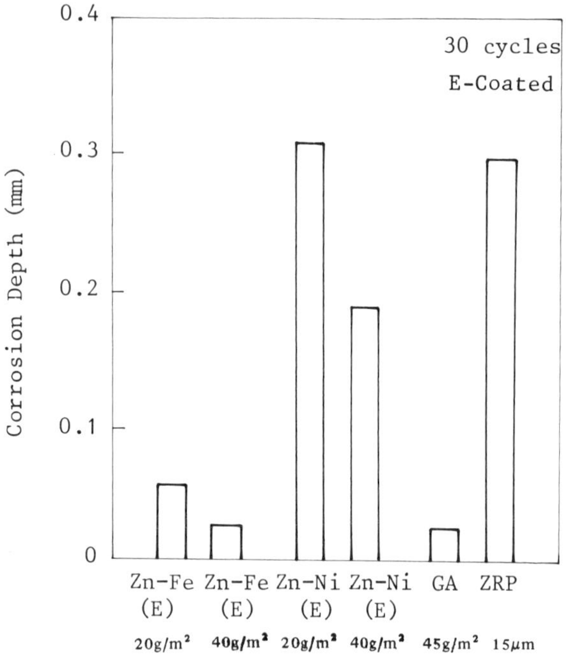

Fig. 8 Corrosion depth of precoated steel

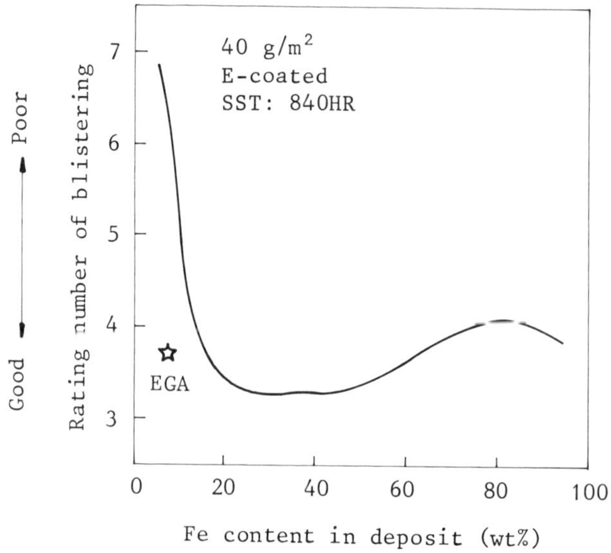

Fig. 9 Effect of Fe content in deposit on blistering

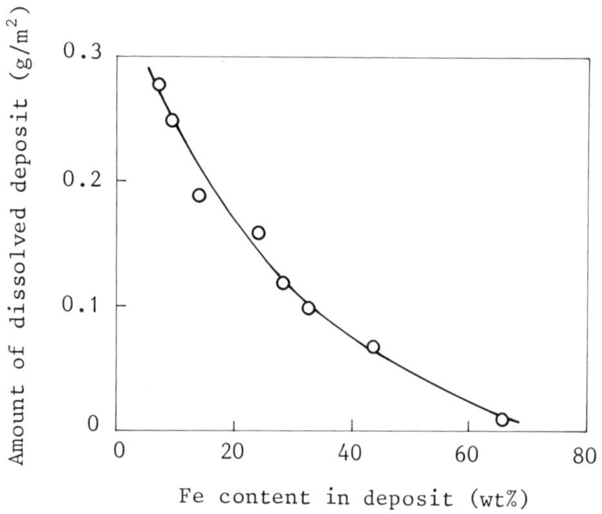

Fig. 10 Relation between amount of dissolved deposit and Fe content in deposit

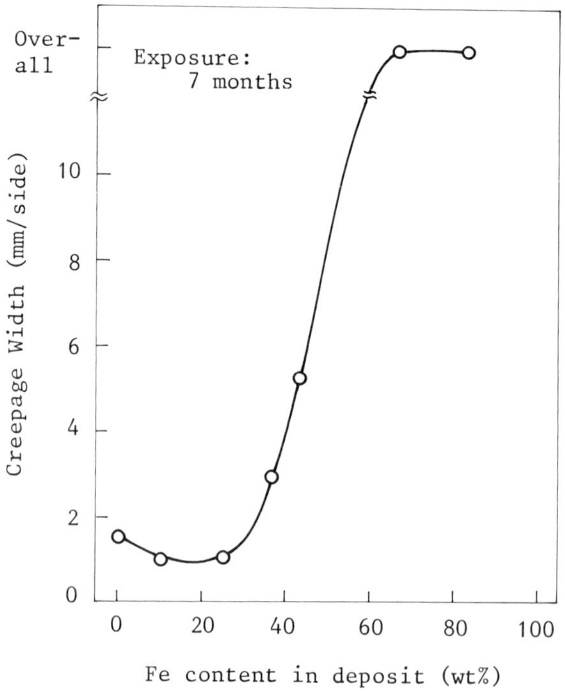

Fig. 11 Relation between creepage width and Fe content in deposit

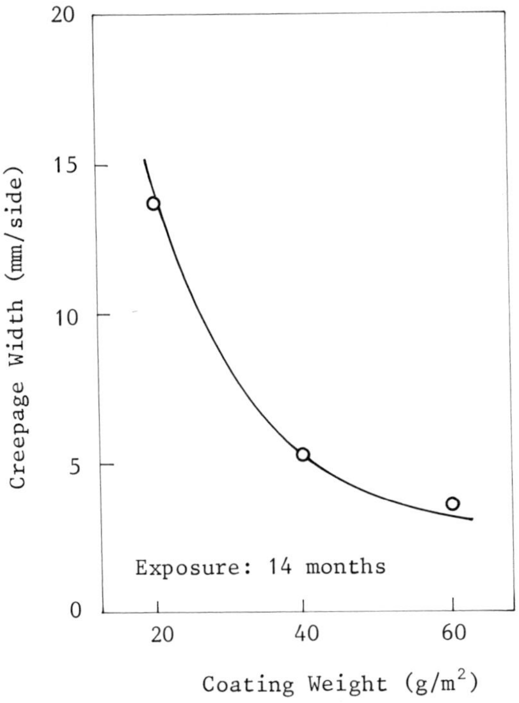

Fig. 12 Effect of coating weight on creepage width

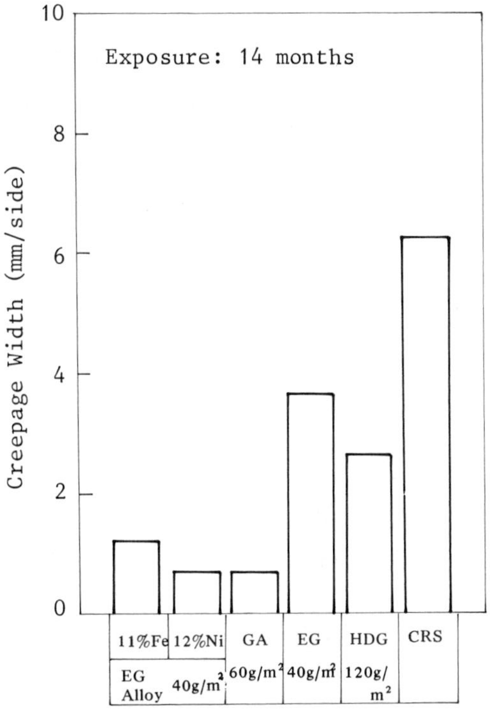

Fig. 13 Creepage width of precoated steels

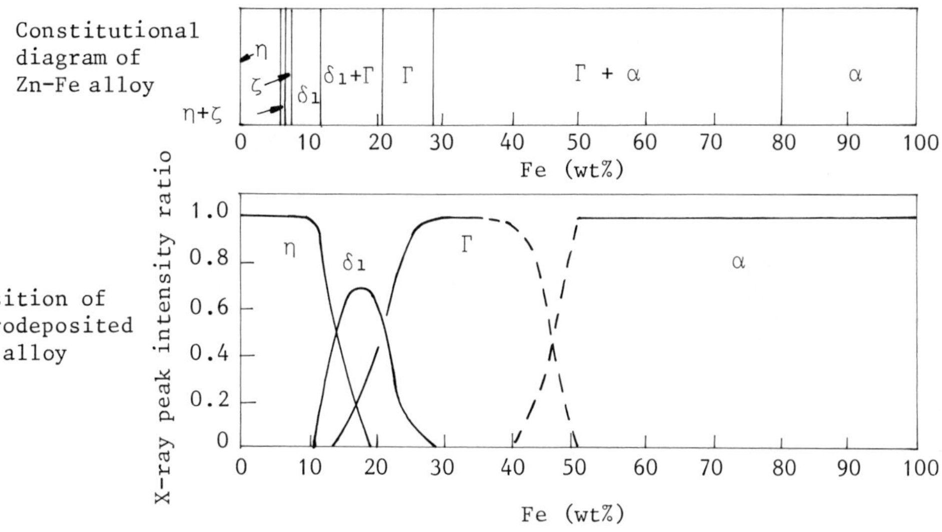

Fig. 14 Phase composition of electrodeposited Zn-Fe alloys

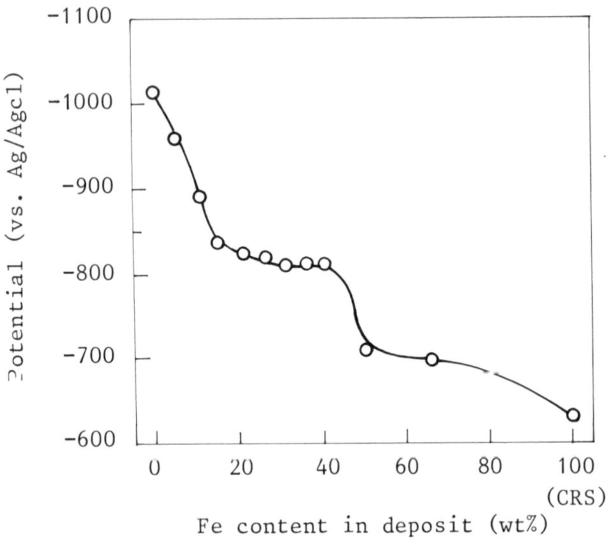

Fig. 15 Corrosion potential of electrodeposited Zn-Fe alloys

UNDERVEHICLE CORROSION TESTING OF ZINC & ZINC ALLOY COATED STEELS

Robert J. Neville, Kenneth M. de Souza
Dofasco Inc.
Hamilton, Ontario, Canada

ABSTRACT

Undervehicle and on-vehicle coupon corrosion test programs were initiated by Dofasco Inc. in 1981, using two commercial trucks operated in the deicing salt/snow belt area of Southern Ontario, Canada.

The purpose was to investigate the relative corrosion performance of numerous zinc and zinc alloy coated steels.

Seventeen coated steels were tested.

Results to date indicate that the hot dip coated steels with the thicker coatings are outperforming the electrolytic coated steels in both the unpainted and phosphated/cathodic primed conditions. Fully painted on-vehicle test coupons show minimal corrosion and little difference to date.

INTRODUCTION

IN THE LAST FEW YEARS, there has been an explosion of new coated steel products for automotive applications, mainly from electrogalvanizing lines. Corrosion resistance claims for most of these new electrodeposited zinc alloy coated steels, based on salt spray testing, have ranged from two to ten times the corrosion resistance of an equivalent thickness of zinc (1-9)*. This implies that these new coated steels with low coating weights can be substituted for conventional hot dip zinc coated steels and offer similar corrosion resistance.

This investigation is focused on a direct comparison of the corrosion resistance of the new electrolytic zinc and zinc alloy coated steels with the conventional hot dip coated steels of higher coating weight using coupons mounted on vehicles operated in the deicing salt/snow belt area of Southern Ontario, Canada. This location is considered a severe corrosive environment for vehicles by the automotive industry (10).

Our test program began in December, 1981, using two transport trucks. It consists of both undervehicle and on-vehicle exposure of 17 coated materials in each of three conditions:

(a) Unpainted.
(b) Zinc phosphated and cathodic ELPO primed.
(c) Full automotive paint system.

The test materials were obtained from various suppliers and are listed in Table 1. All were commercially produced except for item numbers 14 and 15 which were produced on a pilot line. Three of the coated materials joined the program one year after the start. These are materials with item numbers 4, 10, and 16, as shown in Table 1.

The two trucks were each fitted with one 2.15 metre long undervehicle test rack, and one 2.15 metre long front bumper (on-vehicle) rack (Figures 1 and 2).

A total of 492 coupons were installed:

- 232 coupons were installed in the unpainted condition and all have been removed and evaluated (one and two year exposures).

---

* Numbers in parentheses designate references at the end of the paper.

- 232 coupons were installed in the phosphated and primed condition and approximately one-half of these have been removed and evaluated (three and four year exposures).

- 28 coupons were installed in the fully painted condition on the front bumpers and all are still under test, having been exposed since December, 1981.

TABLE 1 - TEST MATERIALS

ITEM	MATERIAL TYPE	COATING WT./SIDE* (g/m^2)	STEEL THICKNESS (mm)
	HOT DIP		
1	Galvanized 1	120/150	0.71
2	Galvanized 2	100/120	0.90
3	Galvanized 3	55/90	0.45
4	Zn/5 Al	90/120	0.63
5	Galvannealed 1	80/120	1.42
6	Galvannealed 2	75/85	0.89
7	One Side Galvannealed	66	0.66
8	55 Al/Zn	75/100	0.64
	ZINC RICH PRIMER		
9	One Side Zincrometal	40	0.92
	ONE SIDE ELECTRODEPOSITED		
10	Electro Zn/15 Ni/0.4 Co	20	0.70
11	Electro Zn/15 Ni/0.4 Co	37	0.70
12	Electro Zn/16 Ni	20	0.68
13	Electro Zn/16 Ni	40	0.68
14	Electro Zn/16 Al	25	0.68
15	Electro Zn/22 Al	40	0.68
16	Electro Zn + Cr/CrO_x	56	0.77
17	Electro Zn	90	0.88
	UNCOATED		
18	Cold Rolled Steel	0	0.51

* The steel industry has traditionally followed the ASTM coating weight designation (e.g. G90, G60). However, most automotive companies are now revising and specifying their coating weight limits in grams per square metre per side.

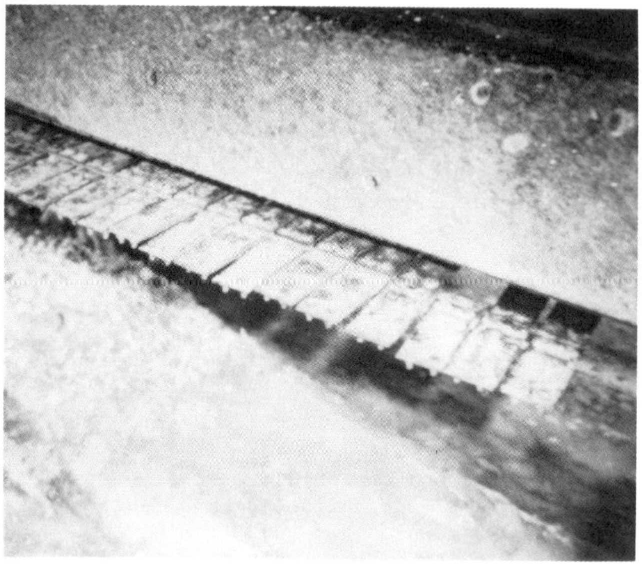

FIGURE 1B - CLOSE-UP OF "UNDER" VEHICLE TEST RACK

FIGURE 1A - LOCATION OF "UNDER" VEHICLE TEST RACK

FIGURE 2 - "ON" VEHICLE TEST RACK

TEST METHOD

COUPON PREPARATION

(1) Undervehicle Test: Test coupons 50 mm x 125 mm x thickness of each material were cut from larger sheets, identified and degreased using methyl ethyl ketone. Zincrometal® was not degreased and was tested as produced (unoiled).

Coating weights were measured for each material using weigh-strip-weigh techniques. Each coupon was weighed prior to assembly into the test configurations shown in Figure 3 for one side coated steels, and in Figure 4 for two side coated steels. The coupons were neither formed nor scribed in this test.

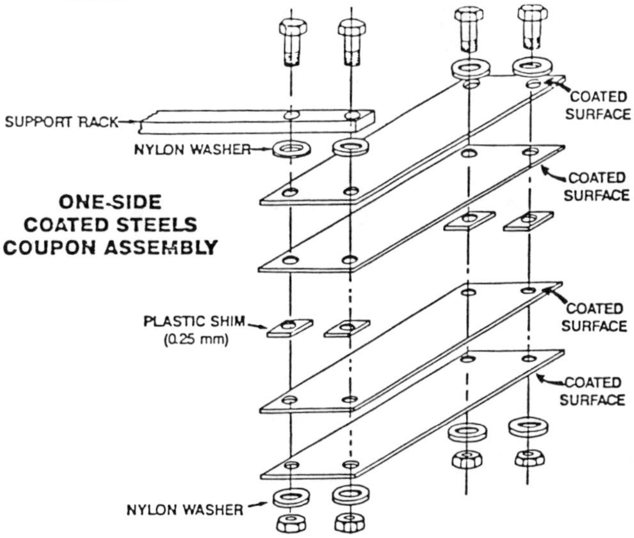

Figure 3
Test assembly, one side coated steels

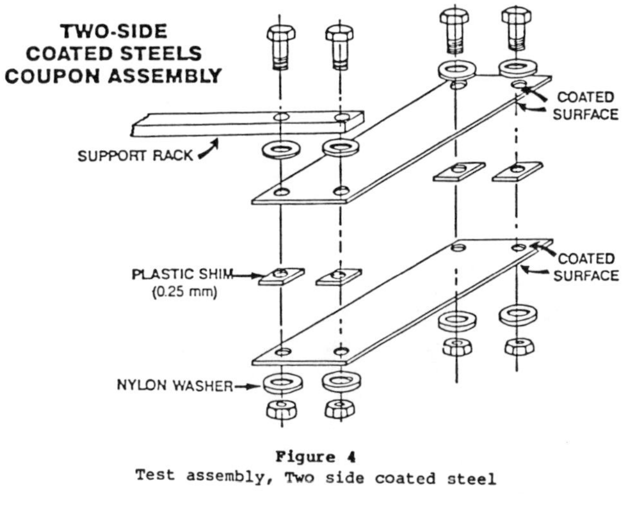

Figure 4
Test assembly, Two side coated steel

The assembly method follows the SAE J1293 "Undervehicle Corrosion Test" Recommended Practice. The only modification was the inclusion of four, rather than two coupons in the test assembly for one side coated steels. This allowed an evaluation of the coated steel in both exposed and crevice conditions.

All the materials in the undervehicle test were tested in two conditions:

(a) unpainted,
(b) zinc phosphated and cathodic ELPO primed.

(2) On-Vehicle Test: Flat panels measuring 75 mm x 150 mm were used for this test which involves exposing duplicate painted and scribed samples on the front bumper of the test vehicles. No unpainted samples were included in this test.

(3) Zinc Phosphate, Cathodic ELPO Primer and Top Coat: The "undervehicle" test assemblies to be ELPO primed, as well as all the "on-vehicle" samples, were laboratory pretreated with EP-1 (aim coating weight 150 - 300 mg/ft^2) by Parker Laboratories, Detroit.

The black ELPO paint primer was applied at the GM Technical Centre, Fisher Body Laboratories. Cathodic ELPO ED-3002V (ultra-filtered) was deposited to a dry film thickness of 15 µm at 300 Volts per GM 9984017 specification.

For the undervehicle test, the pretreatment and cathodic primer were applied on the fully assembled coupons to simulate actual production conditions. In general, the ELPO coating penetrated the 0.25 mm crevice, although on occasion, the primer was noted to have less than the required 15 µm thickness.

The ELPO primed "on-vehicle" coupons were sealed and top coated at Dofasco with a white solution acrylic lacquer in accordance with GM test method TM 55-34. A vertical scribe 110 mm long was made in the centre of the panel per GM specification 9102 P.

Zincrometal® is a registered trademark of Metal Coatings International Inc.

EXPOSURE PERIOD

Two transport trucks were used in this test program. The test coupons have been exposed for up to four winters (1225 days), with evaluations after one, two, three, and four winters. Both trucks were driven in the snow belt region of Southern Ontario where deicing salts are used extensively.

EVALUATION PROCEDURE

The detailed evaluation procedure is described fully in SAE Recommended Practice J1293, "Undervehicle Corrosion Test" (11).

Briefly, the test assemblies were removed from the vehicle test rack, disassembled and the individual coupons washed in warm water using a non-metallic scrub brush.

(1) Unpainted Materials: Three different criteria were used to evaluate the corrosion performance of the test materials in the unpainted condition. The evaluation for each test material included two exposed surfaces and two crevice surfaces for each nested assembly on each truck:

(a) Uniform Thickness Loss (Micrometres): This was determined by first removing all corrosion products according to ASTM G1-81, then determining the weight loss of each coupon in the assembly, and finally calculating coating thickness loss assuming a nominal exposed area of 50 mm x 125 mm x two surfaces and appropriate coating density.

This criterion was applicable only to the two side metallic coated steels and was used only if no steel substrate attack had occurred.

(b) Percent Surface Area Showing Base Metal Attack (%): This was determined by measuring the extent of attack on the steel substrate after stripping the remaining metallic coating according to ASTM A90-81. A low power (12x) binocular microscope and a clear plastic grid were used to estimate the percent surface area showing base metal attack. Evidence of steel surface roughening indicated base metal attack.

The corrosion evaluation was restricted to the surface area between the bolt holes at either end of the coupon - an area measuring 50 mm x 80 mm.

(c) Pitting Attack (Micrometres): Again, the corrosion evaluation was restricted to the surface area between the bolt holes.

Pitting was determined by locating the five deepest pits on each steel surface (after stripping the metallic coating) using a low power (12x) binocular microscope, and then measuring pit depth at 200x using a portable depth measuring microscope. The depth of each pit was determined by focusing first on the upper edge of the pit, then on the bottom, with the distance of travel measured on a vernier being the pit depth.

The five deepest pits were measured and an average pit depth calculated for each surface. Results are reported for exposed surfaces, crevice surfaces and the average pit depth for all four surfaces of the material in the test assembly.

(2) ELPO Primed Materials: Prior to evaluating the organic coated steels, the electrodeposited ELPO primer was removed by immersion for approximately one minute in a 1:1 mixture of tetrahydrofuran and dimethylformamide at 60°C (appropriate safety precautions are required when handling these two strong solvents). The metallic coatings were then removed with inhibited hydrochloric acid while Zincrometal was removed with methyl ethyl ketone.

Only two criteria, (b) and (c), were used for evaluating the corrosion performance of the ELPO primed materials.

(3) Fully Painted Materials: The "on" vehicle test coupons were evaluated after washing the painted and scribed surfaces with a mild soap and water solution. Paint creepback from the scribe was then measured. The criterion for failure has been set at $\geq$3 mm either side of the scribe.

RESULTS AND DISCUSSION

UNPAINTED MATERIALS

(1) Uniform Thickness Loss: This criterion of evaluation failed to provide meaningful results, and therefore, no results are reported.

The technique was limited to the two side metal coated steels which did not show steel substrate corrosion attack after exposure. The few results obtained were variable. Overall, the corrosion attack was seen as non-uniform, rather than uniform.

(2) Percent Surface Area Showing Base Metal Attack: Table 2 gives the results of percent surface area showing base metal attack of unpainted samples removed after one winter's exposure. The results are given separately for each truck, as each is considered an individual test.

We have grouped the materials in the categories of hot dip, zinc rich primer, electrodeposited and uncoated, and this grouping is consistent for all the tables in the paper.

The one winter exposure results generally indicate the hot dip coatings are still providing either complete corrosion protection to the steel substrate or are just beginning to show steel substrate attack. The electrodeposited coated steels, on the other hand, generally show a substantial area of attack of the steel substrate.

The results of coupons removed after two winters' exposure are shown in Table 3. The results show the same trend as the one winter exposure results and generally show more corrosion has occurred on each of the materials.

The electrodeposited coatings after two winters' exposure are near complete failure in the unpainted condition.

The results shown in Tables 2 and 3 indicate that the hot dip coated coupons out-performed the generally lighter coating weight electrodeposited coated coupons, particularly in the exposed condition.

(3) Pitting Attack: Tables 4 and 5 show the pitting results of the 18 materials (17 coated steels and 1 bare cold rolled steel for reference) after undervehicle exposures of one and two winters in the unpainted condition. Here, too, there are indications that the hot dip coated coupons experienced less pitting than the coupons with the electrodeposited coatings. Once again, the exposed surfaces show this trend more clearly than the crevice surfaces.

In summary, the results of both surface area of attack and pitting obtained on unpainted materials exposed for one and two winters undervehicles, do not substantiate the claims of superior corrosion resistance (based on salt spray) for electrodeposited zinc alloy coatings.

ZINC PHOSPHATED AND CATHODIC ELPO PRIMED MATERIALS

The expected improvement in corrosion protection of these steels by phosphating and the application of a cathodic electrodeposited primer was observed in our undervehicle coupon corrosion test.

Consequently, the first removal of phosphated and primed materials was delayed until after three winters' exposure, and only one set of test materials from one truck (#781) was removed.

Table 6 shows the results of percent surface area showing base metal attack, and Table 7 shows the results of the pitting evaluation for this set of materials. Generally, the superior corrosion resistance of the hot dip coated coupons versus the coupons with the electrodeposited coatings in the phosphated and primed condition, as in the unpainted condition, is indicated in these two tables.

This set of coupons also provided the first observation that coatings containing higher levels of aluminum appear to have a compatibility problem with the paint system as evidenced by primer blistering following exposure.

TABLE 2 - UNDERVEHICLE CORROSION OF UNPAINTED COUPONS
AFTER 1 WINTER'S EXPOSURE (1981 - 1982)**

PERCENT SURFACE AREA SHOWING BASE METAL ATTACK

MATERIAL	COATING WT/SIDE (g/m^2)	TRUCK NO. 781 EXPOSED SURFACES	TRUCK NO. 781 CREVICE SURFACES	TRUCK NO. 781 AVERAGE (4 SURFACES)	TRUCK NO. 782 EXPOSED SURFACES	TRUCK NO. 782 CREVICE SURFACES	TRUCK NO. 782 AVERAGE (4 SURFACES)
HOT DIP							
Galvanized 1	120/150	0.5	0.5	0.5	6.5	21.0	14.0
*Galvanized 1	120/150	0	0	0	0.5	0	0.3
Galvanized 2	100/120	0	0.5	0.3	0.5	18.5	9.5
Galvanized 3	55/90	1.0	2.0	1.5	0	0.5	0.3
*Zn/5 Al	90/120	0	0	0	0	0	0
Galvannealed 1	80/120	0	5.5	2.8	0	0	0
Galvannealed 2	75/85	0.5	0.5	0.5	0	5.0	2.5
One Side Galvannealed	66	12.0	1.0	6.5	72.5	10.0	41.3
55 Al/Zn	75/100	0.5	5.0	2.8	0	0.5	0.3
ZINC RICH PRIMER							
One Side Zincrometal	40	0.5	0.5	0.5	0.5	0.5	0.5
ONE SIDE ELECTRODEPOSITED							
*Electro Zn/15 Ni/0.4 Co	20	59.4	4.0	31.7	31.0	5.3	18.0
Electro Zn/15 Ni/0.4 Co	37	4.0	9.0	6.5	42.0	1.0	22.0
Electro Zn/16 Ni	20	80.0	32.0	56.0	55.0	44.0	50.0
Electro Zn/16 Ni	40	100.0	19.0	60.0	66.0	12.0	39.0
Electro Zn/16 Al	25	64.0	13.0	39.0	98.0	10.0	54.0
Electro Zn/22 Al	40	100.0	6.0	53.0	100.0	40.0	70.0
*Electro Zn + Cr/CrO_x	56	1.3	0	0.6	16.0	0.0	8.0
Electro Zn	90	100.0	80.0	90.0	93.0	16.0	55.0
UNCOATED							
Cold Rolled Steel	0	100.0	100.0	100.0	100.0	100.0	100.0

** 1981 - 1982 Exposure, Truck No. 781 - 192 days, 20500 km; Truck No. 782 - 192 days, 19500 km.
* 1982 - 1983 Exposure, Truck No. 781 - 304 days, 14310 km; Truck No. 782 - 304 days, 14500 km.

TABLE 3 - UNDERVEHICLE CORROSION OF UNPAINTED COUPONS
AFTER 2 WINTER'S EXPOSURE (1981 - 1983)**

PERCENT SURFACE AREA SHOWING BASE METAL ATTACK

MATERIAL	COATING WT/SIDE (g/m^2)	TRUCK NO. 781 EXPOSED SURFACES	TRUCK NO. 781 CREVICE SURFACES	TRUCK NO. 781 AVERAGE (4 SURFACES)	TRUCK NO. 782 EXPOSED SURFACES	TRUCK NO. 782 CREVICE SURFACES	TRUCK NO. 782 AVERAGE (4 SURFACES)
HOT DIP							
Galvanized 1	120/150	0.8	0.3	0.6	13.0	15.0	14.0
*Galvanized 1	120/150	8.5	0	4.2	0	0	0
Galvanized 2	100/120	0	0.7	0.3	0.5	54.0	27.3
Galvanized 3	55/90	0	1.0	0.5	0	10.0	5.0
*Zn/5 Al	90/120	1.5	1.5	1.5	0	0	0
Galvannealed 1	80/120	0	0	0	1.0	1.0	1.0
Galvannealed 2	75/85	0.5	0	0.3	10.5	55.0	32.8
One Side Galvannealed	66	48.2	1.9	25.0	100.0	13.0	56.5
55 Al/Zn	75/100	0.5	0	0.3	15.5	8.5	12.0
ZINC RICH PRIMER							
One Side Zincrometal	40	10.5	11.0	10.8	21.5	13.0	17.3
ONE SIDE ELECTRODEPOSITED							
*Electro Zn/15 Ni/0.4 Co	20	100.0	77.0	88.5	59.5	14.0	36.8
Electro Zn/15 Ni/0.4 Co	37	69.0	23.0	46.0	100.0	35.0	67.5
Electro Zn/16 Ni	20	100.0	69.0	85.0	100.0	87.0	93.5
Electro Zn/16 Ni	40	70.5	5.5	38.0	100.0	58.5	79.3
Electro Zn/16 Al	25	100.0	17.5	59.0	100.0	68.5	84.3
Electro Zn/22 Al	40	99.0	9.5	54.0	100.0	53.0	76.5
*Electro Zn + Cr/CrO_x	56	100.0	19.0	59.5	100.0	8.0	54.0
Electro Zn	90	97.0	25.0	61.0	100.0	72.0	86.0
UNCOATED							
Cold Rolled Steel	0	100.0	100.0	100.0	100.0	100.0	100.0

** 1981 - 1983 Exposure, Truck No. 781 - 660 days, 51000 km; Truck No. 782 - 660 days, 53500 km.
* 1982 - 1984 Exposure, Truck No. 781 - 603 days, 32580 km; Truck No. 782 - 603 days, 44900 km.

TABLE 4 - UNDERVEHICLE CORROSION OF UNPAINTED COUPONS AFTER 1 WINTER'S EXPOSURE (1981 - 1982)**

PITTING ATTACK (MICROMETRES)

MATERIAL	COATING WT/SIDE (g/m^2)	TRUCK NO. 781 EXPOSED SURFACES	TRUCK NO. 781 CREVICE SURFACES	TRUCK NO. 781 AVERAGE (4 SURFACES)	TRUCK NO. 782 EXPOSED SURFACES	TRUCK NO. 782 CREVICE SURFACES	TRUCK NO. 782 AVERAGE (4 SURFACES)
HOT DIP							
Galvanized 1	120/150	0	0	0	11	61	36
*Galvanized 1	120/150	0	0	0	0	0	0
Galvanized 2	100/120	0	27	13	4	29	17
Galvanized 3	55/90	0	0	0	0	11	6
*Zn/5 Al	90/120	0	0	0	0	0	0
Galvannealed 1	80/120	0	66	33	0	7	4
Galvannealed 2	75/85	0	22	11	20	9	15
One Side Galvannealed	66	37	20	29	67	51	59
55 Al/Zn	75/100	37	55	46	7	25	16
ZINC RICH PRIMER							
One Side Zincrometal	40	14	14	14	47	71	59
ONE SIDE ELECTRODEPOSITED							
*Electro Zn/15 Ni/0.4 Co	20	53	32	42	53	22	38
Electro Zn/15 Ni/0.4 Co	37	31	36	33	40	26	33
Electro Zn/16 Ni	20	53	68	61	61	64	63
Electro Zn/16 Ni	40	49	51	50	64	57	61
Electro Zn/16 Al	25	43	43	43	51	49	50
Electro Zn/22 Al	40	41	38	40	53	50	52
*Electro Zn + Cr/CrO_x	56	22	0	11	52	0	26
Electro Zn	90	46	72	59	49	33	41
UNCOATED							
Cold Rolled Steel	0	87	189	138	91	110	101

** 1981 - 1982 Exposure, Truck No. 781 - 192 days, 20500 km; Truck No. 782 192 days, 19500 km.

* 1982 - 1983 Exposure, Truck No. 781 - 304 days, 14310 km; Truck No. 782 304 days, 14500 km.

TABLE 5 - UNDERVEHICLE CORROSION OF UNPAINTED COUPONS AFTER 2 WINTER'S EXPOSURE (1981 - 1983)**

PITTING ATTACK (MICROMETRES)

MATERIAL	COATING WT/SIDE (g/m^2)	TRUCK NO. 781 EXPOSED SURFACES	TRUCK NO. 781 CREVICE SURFACES	TRUCK NO. 781 AVERAGE (4 SURFACES)	TRUCK NO. 782 EXPOSED SURFACES	TRUCK NO. 782 CREVICE SURFACES	TRUCK NO. 782 AVERAGE (4 SURFACES)
HOT DIP							
Galvanized 1	120/150	0	0	0	9	21	15
*Galvanized 1	120/150	27	0	14	0	0	0
Galvanized 2	100/120	0	21	11	0	111	56
Galvanized 3	55/90	0	0	0	0	30	15
*Zn/5 Al	90/120	25	0	13	0	0	0
Galvannealed 1	80/120	0	0	0	12	32	22
Galvannealed 2	75/85	21	0	11	45	128	86
One Side Galvannealed	66	58	37	48	77	56	67
55 Al/Zn	75/100	0	0	0	39	58	48
ZINC RICH PRIMER							
One Side Zincrometal	40	28	79	53	76	70	73
ONE SIDE ELECTRODEPOSITED							
*Electro Zn/15 Ni/0.4 Co	20	154	170	162	95	95	95
Electro Zn/15 Ni/0.4 Co	37	72	77	75	81	81	81
Electro Zn/16 Ni	20	77	90	83	92	107	100
Electro Zn/16 Ni	40	79	67	73	114	141	128
Electro Zn/16 Al	25	77	51	64	117	77	97
Electro Zn/22 Al	40	72	57	64	86	94	90
*Electro Zn + Cr/CrO_x	56	115	112	114	138	96	117
Electro Zn	90	64	64	64	105	134	120
UNCOATED							
Cold Rolled Steel	0	>250	>250	>250	>250	>250	>250

** 1981 - 1983 Exposure, Truck No. 781 - 660 days, 51000 km; Truck No. 782 - 660 days, 53500 km.

* 1982 - 1984 Exposure, Truck No. 781 - 603 days, 32580 km; Truck No. 782 - 603 days, 44900 km.

TABLE 6 - UNDERVEHICLE CORROSION OF ZINC PHOSPHATED AND ELPO PRIMED COUPONS AFTER 3 WINTER'S EXPOSURE (1981 - 1984)**

PERCENT SURFACE AREA SHOWING BASE METAL ATTACK

MATERIAL	COATING WT/SIDE (g/m^2)	TRUCK NO. 781: ONE EXPOSED[1] SURFACE	TRUCK NO. 781: TWO CREVICE SURFACES	TRUCK NO. 781: AVERAGE (3 SURFACES)
HOT DIP				
Galvanized 1	120/150	1.0	1.0	1.0
*Galvanized 1	120/150	0	0	0
Galvanized 2	100/120	0	0	0
Galvanized 3	55/90	1.0	1.0	1.0
*Zn /5 Al	90/120	0	0	0
Galvannealed 1	80/120	0	0	0
Galvannealed 2	75/85	0	0	0
One Side Galvannealed	66	1.0	0	0.3
55 Al/Zn	75/100	33.0	21.0	25.0
ZINC RICH PRIMER				
One Side Zincrometal	40	1.0	0	0.3
ONE SIDE ELECTRODEPOSITED				
*Electro Zn/15 Ni/0.4 Co	20	3.0	12.0	9.0
Electro Zn/15 Ni/0.4 Co	37	90.0	33.0	52.0
Electro Zn/16 Ni	20	54.0	10.5	25.0
Electro Zn/16 Ni	40	9.0	7.0	8.0
Electro Zn/16 Al	25	36.0	23.5	27.7
Electro Zn/22 Al	40	20.0	15.0	16.7
*Electro Zn + Cr/CrO_x	56	2.0	4.5	3.7
Electro Zn	90	13.0	13.0	13.0
UNCOATED				
Cold Rolled Steel	0	1.0	20.0	13.7

** 1981 - 1984 Exposure, 959 days, 71000 km.
* 1982 - 1985 Exposure, 1195 days, 67067 km.
[1] Top surface was accidentally shot blasted During repainting of truck. Results shown refer to bottom exposed surface only.

TABLE 7 - UNDERVEHICLE CORROSION OF ZINC PHOSPHATED AND ELPO PRIMED COUPONS AFTER 3 WINTER'S EXPOSURE (1981 - 1984)**

PITTING ATTACK (MICROMETRES)

MATERIAL	COATING WT/SIDE (g/m^2)	TRUCK NO. 781: ONE EXPOSED[1] SURFACE	TRUCK NO. 781: TWO CREVICE SURFACES	TRUCK NO. 781: AVERAGE (3 SURFACES)
HOT DIP				
Galvanized 1	120/150	0	0	0
*Galvanized 1	120/150	0	0	0
Galvanized 2	100/120	0	0	0
Galvanized 3	55/90	22***	30****	27
*Zn/5 Al	90/120	0	0	0
Galvannealed 1	80/120	0	0	0
Galvannealed 2	75/85	0	0	0
One Side Galvannealed	66	39	0	13
55 Al/Zn	75/100	103	72	82
ZINC RICH PRIMER				
One Side Zincrometal	40	0	0	0
ONE SIDE ELECTRODEPOSITED				
*Electro Zn/15 Ni/0.4 Co	20	73	57	62
Electro Zn/15 Ni/0.4 Co	37	33	79	63
Electro Zn/16 Ni	20	128	149	142
Electro Zn/16 Ni	40	69	37	47
Electro Zn/16 Al	25	130	176	160
Electro Zn/22 Al	40	122	97	105
*Electro Zn + Cr/CrO_x	56	53	39	43
Electro Zn	90	124	57	79
UNCOATED				
Cold Rolled Steel	0	57	189	145

** 1981 - 1984 Exposure, 959 days, 71000 km.
* 1982 - 1985 Exposure, 1195 days, 67067 km.
*** All pitting limited to a 4 mm^2 area.
**** All pitting limited to a 7 mm^2 area.
[1] Top surface was accidentally shot blasted during repainting of truck. Results shown refer to bottom exposed surface only.

The following year, another set of phosphated and primed test materials were removed from the other truck (#782). These materials were exposed for four winters and the results are shown in Tables 8 and 9. Again, the results agreed with the findings from earlier removals. Two additional sets of materials remain for future removals.

Most of the three year and four year test coupons exhibited some blistering of the cathodic primer predominantly on the exposed surfaces. The degree of "cratering" of the cured primed surface was evaluated quantitatively before exposure, but no relationship was found between cratering and blistering after exposure.

Some materials that exhibited a dense cratered surface (eg. Zincrometal), were not blistered after exposure, and some materials that showed little or no cratering (e.g. 55 Al/Zn) were blistered after exposure.

FULL AUTOMOTIVE PAINTED MATERIALS

All the fully painted test materials exposed on the front bumper for four years (December 5, 1981 to April 13, 1985) are peforming well with no discernible scribe creep. The samples were reinstalled for additional exposure.

FUTURE WORK

A new vehicle test program was initiated in November, 1984, and involves the detailed "on" and "under" vehicle testing of equivalent coating thicknesses (100 g/m^2 per side) of electrozinc and hot dip galvanized. Hot dip galvannealed (60 g/m^2 per side) is also included in this study.

Coupons in both the unpainted and in the phosphated and primed (EP-3/Uniprime 3150A) condition are on test.

CONCLUSIONS

Because of the large number of test materials examined and the limited amount of testing space available on a truck, it was not possible to incorporate enough samples to determine statistically significant conclusions. However, there are indications from the test coupons examined that:

(1) The hot dip coated steels are generally outperforming the lighter electrodeposited coated steels in the unpainted condition.

(2) Compared to the unpainted materials, the pretreated and cathodically ELPO primed test materials show improvement in corrosion performance. However, once again, the hot dip coated steels were generally observed to have less corrosion than the electrodeposited coated steels.

TABLE 8 - UNDERVEHICLE CORROSION OF ZINC PHOSPHATED AND ELPO PRIMED COUPONS AFTER 4 WINTER'S EXPOSURE (1981 - 1985)**

PERCENT SURFACE AREA SHOWING BASE METAL ATTACK

MATERIAL	COATING WT/SIDE (g/m^2)	TRUCK NO. 782 EXPOSED SURFACES	CREVICE SURFACES	AVERAGE (4 SURFACES)
HOT DIP				
Galvanized 1	120/150	0	0	0
*Galvanized 1	120/150	--	--	--
Galvanized 2	100/120	0	0	0
Galvanized 3	55/90	0.5	0	0.3
*Zn/5 Al	90/120	--	--	--
Galvannealed 1	80/120	0	0	0
Galvannealed 2	75/85	0.5	0.5	0.5
One Side Galvannealed	66	7.5	1.0	4.3
55 Al/Zn	75/100	37.5	7.8	22.7
ZINC RICH PRIMER				
One Side Zincrometal	40	0	0	0
ONE SIDE ELECTRODEPOSITED				
*Electro Zn/15 Ni/0.4 Co	20	--	--	--
Electro Zn/15 Ni/0.4 Co	37	40.0	15.0	27.5
Electro Zn/16 Ni	20	53.8	7.5	30.6
Electro Zn/16 Ni	40	9.4	8.8	9.1
Electro Zn/16 Al	25	48.8	25.6	37.2
Electro Zn/22 Al	40	71.3	25.9	48.6
*Electro Zn + Cr/CrO_x	56	--	--	--
Electro Zn	90	8.4	5.0	6.7
UNCOATED				
Cold Rolled Steel	0	3	5	4

** 1981 - 1985 Exposure, 1225 days, 107000 km.
* 1982 - 1986 Exposure, data available in 1986.

TABLE 9 - UNDERVEHICLE CORROSION OF ZINC PHOSPHATED AND ELPO PRIMED COUPONS AFTER 4 WINTER'S EXPOSURE (1981 - 1985)**

PITTING ATTACK (MICROMETRES)

MATERIAL	COATING WT/SIDE (g/m^2)	TRUCK NO. 782 EXPOSED SURFACES	CREVICE SURFACES	AVERAGE (4 SURFACES)
HOT DIP				
Galvanized 1	120/150	0	0	0
*Galvanized 1	120/150	--	--	--
Galvanized 2	100/120	0	0	0
Galvanized 3	55/90	0	0	0
*Zn/5 Al	90/120	--	--	--
Galvannealed 1	80/120	0	0	0
Galvannealed 2	75/85	11	13	12
One Side Galvannealed	66	102	16	59
55 Al/Zn	75/100	90	74	82
ZINC RICH PRIMER				
One Side Zincrometal	40	0	0	0
ONE SIDE ELECTRODEPOSITED				
*Electro Zn/15 Ni/0.4 Co	20	--	--	--
Electro Zn/15 Ni/0.4 Co	37	36	41	38
Electro Zn/16 Ni	20	101	61	81
Electro Zn/16 Ni	40	74	43	59
Electro Zn/16 Al	25	117	88	102
Electro Zn/22 Al	40	142	109	125
*Electro Zn + Cr/CrO^x	56	--	--	--
Electro Zn	90	82	67	75
UNCOATED				
Cold Rolled Steel	0	73	92	82

** 1981 - 1985 Exposure, 1225 days, 107000 km.
* 1982 - 1986 Exposure, data available in 1986.

ACKNOWLEDGEMENTS

The authors thank Dofasco Inc. for permission to publish this paper.

The contribution of associate Mr. Arden Cornford in preparing this paper is also appreciated.

Finally, the cooperation of Mr. Charles Bates and the staff of Bates Transport formerly Walmer Transport), Mississauga, Ontario, for the use of their trucks is gratefully acknowledged.

REFERENCES

(1) Kimura, H., Yoshihara, T., Harada, S., "Development of Zinc-Aluminum Electrodeposited Composite Coating For Improving Corrosion Resistance Of Electrogalvanized Steel Sheet"; Interfinish 80 (Proc.Conf.); 228-233.

(2) Kawasaki Steel Corporation. Kawasaki Newsletter; 4; (1); (1982).

(3) Noumi, R., Nagasaki, H., Foboh, Y., Shibuya, A., "Newly Developed Ni-Zn Alloy Plated Steel Sheet SZ and Its Application To Automotive Usages", Society of Automotive Engineers, Inc.; Paper No. 820332; 1982.

(4) Nippon Steel Corporation. "Corrosion Resistance of Unpainted ZINKLITE"; ZINKLITE; 1984; 19.

(5) Memmi, M., Bruno, R., Palladino, M. "Cr-CrO_x Electrodeposition for Corrosion Performance Improvement of Electrogalvanized Steel Sheet in Automotive Applications"; Corrosion/82 (Proc.Conf.), Houston, Texas; March 22-26; 1982; NACE; Paper No. 267.

(6) Hall, D. E., "Electrodeposited Zinc-Nickel Alloy Coatings - A Review"; Plat. Surf. Finish; 70; (11); (1983); 59-65.

(7) Shibuya, A., Kurimoto, T., Nishihara, M., Wakano, S., Kurashige, T., "Properties of Ni-Zn Alloy Plated Steel Sheet"; Sumitomo Metal Industries; Central Research Laboratories; unpublished report.

(8) Adaniya, T., Ohkubo, Y., Ohmura, M., Honma, T., Watanbe, T., Anan, T., "Electrodeposited Iron-Zinc Alloy Steel Sheet"; Nippon Kokan Tech. Rep. (Overseas); (34); (1982); 41-48.

(9) Dini, J. W., Johnson, H. R., "Electrodeposition of Zinc-Nickel Alloy Coatings"; Met. Finish; Aug.; (1979); 31-33.

(10) American Iron and Steel Institute. Cracking Down on Corrosion; AISI Publication SG 731-0285-10M-R1; 1985.

(11) Society of Automotive Engineers, Inc., "Undervehicle Corrosion Test"; 1985 SAE Handbook, Vol. 1, Materials; SAE J1293; 1980; 3.37-3.40

COSMETIC CORROSION PERFORMANCE OF ZINC ALLOY STEEL SHEETS IN AUTOMOTIVE BODY

Shigeru Wakano, Akito Sakoda, Minoru Nishihara
Sumitomo Metal Industries, LtD
Amagasaki, Japan

Tatsuo Kurimoto, Yoshihiko Hoboh
Sumitomo Metal Industries, LtD
Wakayama, Japan

ABSTRACT

In the application of precoated steel sheets for use as exterior body parts, many facters should be considered. This paper looks at cratering threshold susceptibility, paint adhesion and corrosion resistance of three kinds of electroplated steel sheet.

Fe-rich Fe-Zn alloy layer electroplated on Zn alloy plated steel sheets efficiently prevents occurrence of crater defects in cathodic electropainting and paint adhesion loss after warm water immersion. But in practice other means may be applied in the automotive paint process.

Cosmetic corrosion resistance depends on primarily on coating weight of the precoated layer. Alloy electroplated steel sheets with coating weight of 30 g/m² to 50 g/m² exhibits narrower paint creepage than Zn electroplated steel sheet with heavy coating weight and high red rust resistance in a modified Volvo test.

THE DEMANDS for protecting automotive bodies from corrosion have become increasingly severe, since the Canadian Anti-Corrosion Code was announced 8 years ago.

Recently automotive manufactures have tried to realize 10-5-2 year protection as an anti-corrosion goal. For 10-year protection against inside-to-out perforation damage, Japanese manufactures have adopted Zn alloy electroplated steel sheets with light coating weight and weldable paint coated steel sheets. The substrate for this is Ni-Zn alloy electroplated steel sheet. In the United States, Zn electroplated steel sheet with heavy coating weight is currently used and its application is expected to markedly increase in the future.

At present, 5-year protection against cosmetic damage is the major target in Japan and the United States. Many studies have been carried out to determine what kinds of precoated steel sheets are suitable for exposed parts.[1)-6)] The following factors are considered in the application of precoated steel sheets for use as exposed parts:

(1) press formability (the powdering off and the flaking off of the precoated layer during press forming)
(2) spot weldability
(3) cratering susceptibility
(4) paint adhesion
(5) corrosion resistance after painting

This paper looks at cratering susceptibility, paint adhesion and corrosion resistance of three kinds of electroplated steel sheet.

CRATERING SUSCEPTIBILITY IN CATHODIC ELECTROPAINTING

Zn alloy plated steel sheets exhibit a tendency for cathodic electropaint film to form "craters" when applied at high voltage.[7),8)] However when used for unexposed parts, these materials display few crater failures, because the deposition voltage is gradually applied to Zn alloy surfaces and no exposure to high deposition voltage. On the contrary, as high voltage is applied from the initial stage of paint deposition on precoated steel sheets for use as exposed parts, crater defects can occur on the surface. Therefore it may be necessary to improve the precoated layer. The cathodic electropaint used in this research was a conventional type U-30 (Nippon Paint Co.). Specimens were electropainted at a certain voltage for 3 minutes after immersion zinc phosphating. Table.1 shows the results of cratering threshold voltage. Both Ni-Zn alloy electroplated steel sheet and Fe-Zn alloy electroplated steel sheet showed lower cratering threshold voltages than cold rolled steel sheet and Zn elctroplated steel sheet. As crater-free film can be deposited on Zn electroplated steel sheet at deposition voltage above 300 volts, Zn alloy electroplated steel sheet which has Zn metal as an outer layer up to about 10g/m^2 was evaluated. Crater defects on these duplex steel sheets occurred at slightly higher voltage.

However Fe-Zn alloy as an outer layer greatly decreases the cratering susceptibility of alloy plated steels, bringing it to the level as cold rolled steel sheet. The outer layer consisted of Fe content at about 80 % and coating weight of about 3 g/m^2.

Recently high film build cathodic primer "Uni-Prime" has been adopted to the United States auto-industry. The cratering defect of "Uni-Prime" may occur at higher voltage, usually above 50 volts, than that of the conventional primer.

As mentioned above, Fe-rich Fe-Zn alloy layer can prevent crater defects on Zn alloy plated steel sheet. However in practice other means such as arrangements of counter electrodes in the paint bath and modes of applied voltage may also be applied in the automotive paint process for preventing craters.

PAINT ADHESION AFTER WATER IMMERSION

Paint adhesion failure occurs at the initial stage of underfilm corrosion. Better paint adhesion eventually provides better corrosion resistance. This is particularly important for precoated steel sheet used for exterior body panels.

In this research, two kinds of zinc phosphate solution were applied to precoated steel sheets for improving paint adhesion. Solution A was a conventional immersion zinc phosphate solution which has been adopted in Japanese automotive lines. This solution contains chlorate and Ni ion as accelerators. Solution B is a new type and contains fluoride instead of chlorate, Ni ion and Mn ion. Phosphated specimens were cathodically electropainted, seal-coated and top-coated with total film thickness of about 90 μm. Specimens were immediately scribed in block pattern of 2mm width (11x11 lines) after immersion in warm de-ionized water (40°C) for 240 hours. Then an adhesive tape peeling test was performed, and the area of remaining paint was evaluated.

All precoated steel sheets which were phosphated in solution A exhibited much poorer paint adhesion than cold rolled steel sheet as shown in Fig.1. Phosphating in solution B provided excellent paint adhesion for all specimens.

A modified Volvo test was also applied. The test consists of a standard atmospheric exposure with spraying twice weekly on the Central Research Laboratories corrosion test racks with a 5% NaCl solution. After 1 year of the test, paint adhesion of precoated steel sheets phosphated in solution B did not deteriorate.

According to X-ray diffraction analysis, phosphate crystal deposited from solution B consists only of Hopeite. The crystal deposited from solution A was the same. However

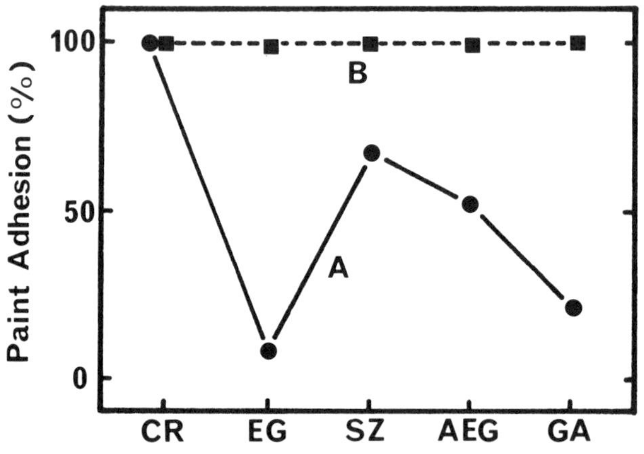

Fig.1 Influence of phosphate treatment on paint adhesion after immersion in de-ionized water at 40°C for 240 hours.
CR: cold rolled steel
EG: Zn electroplated steel
SZ: Ni-Zn alloy electroplated steel
AEG: Fe-Zn alloy electroplated steel
GA: galvannealed steel

Table. 1 Cratering Threshold Potentials in a Convetional Primer

Specimens	Condition		
	As recieved	Duplex 1 [1)]	Duplex 2 [2)]
Cold Rolled Steel	> 300	---	---
Zn Electroplated Steel	> 300	---	---
Ni-Zn Alloy Plated Steel	260	280	> 300
Fe-Zn Alloy Plated Steel	220	240	> 300

1) Zn layer, coating weight of about 10 g/m^2, is electroplated on the substrate as an outer layer.
2) Fe-Zn alloy layer, coating weight of about 3 g/m^2, is electroplated on the substrate as an outer layer.

the morphology of zinc phosphate crystal greatly varies as shown in Photo.1. Phosphate crystals deposited from solution B were fine and dense. Also it exhibits better alkali resistance as shown in Fig.2.

Furthermore the permeated water content in the paint film and the phosphate layer was calculated from weight decrease measurements after immersion in de-ionized water at 40°C for 20 days. Specimens used were three-coat painted Zn electroplated steel sheets. The specimen phosphated in solution B had a lower water content than that phosphated in solution A as shown in Fig.3. The fine and dense morphology of the phosphate crystal causes a

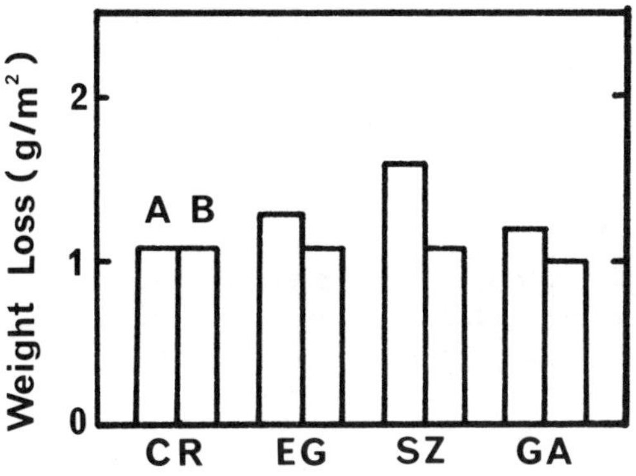

Fig.2 Weight loss of phosphate layer on precoated steel sheets in 0.1 N NaOH solution at 30°C for 5 min..

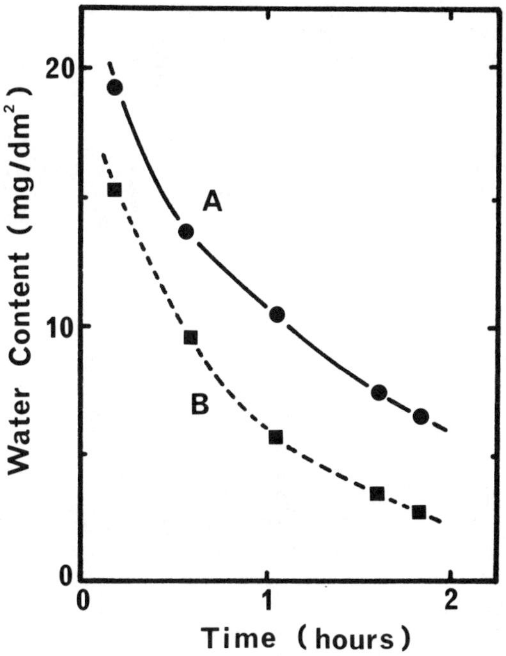

Fig.3 Influence of phosphate treatment on water content of 3-coat painted Zn electroplated steel sheet after 20 day immersion in de-ionized water at 40°C.

Photo.1 Morphology of phosphate crystal on precoated steel sheets

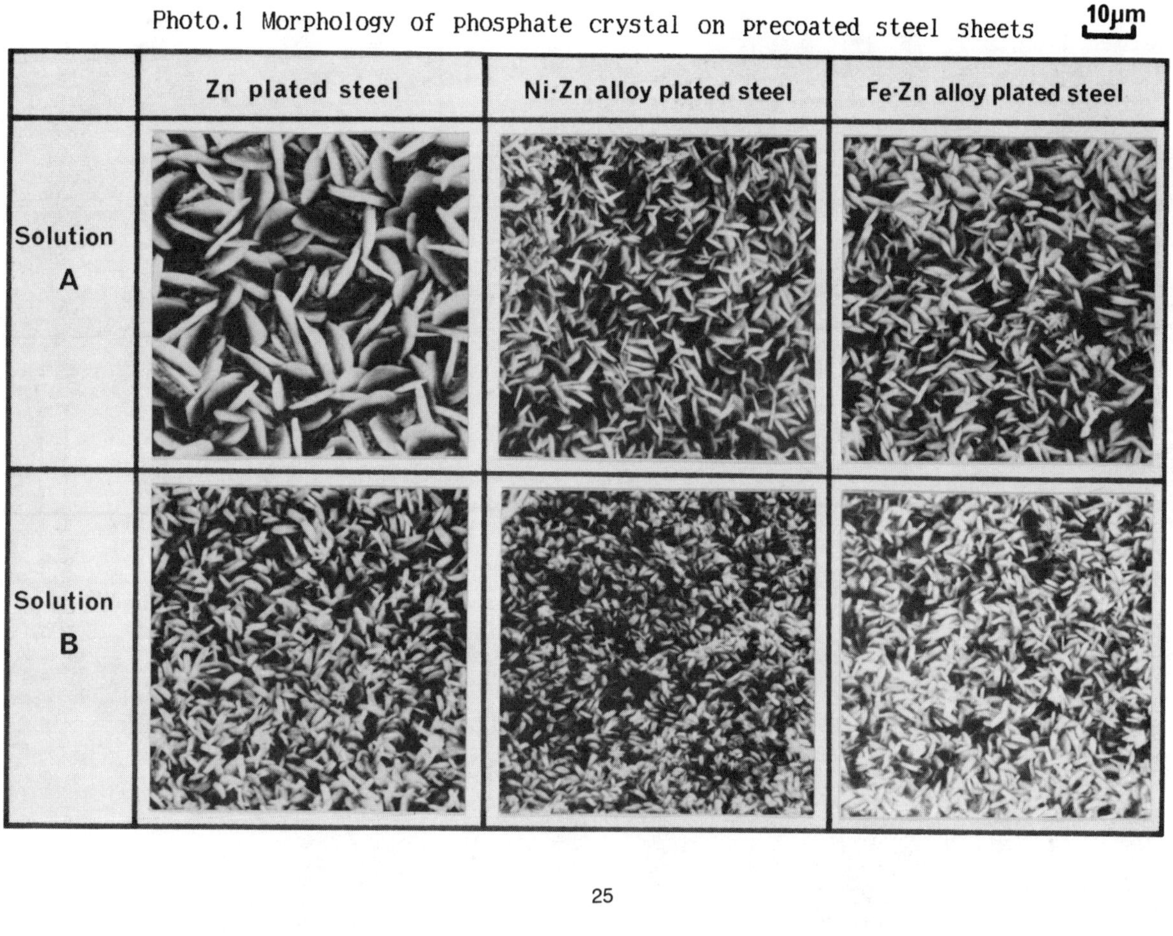

smaller space for permeated water and less porosity, also it may allow a large contact area between paint film and phosphate crystal as well as a stronger binding force between phosphate crystal and substrate.

In summary these results show that paint adhesion loss on precoated steel sheets will result from the following :

(1) Permeated water accumulating in the space of the phosphate layer and slightly changing to alkali solution at porosity.

(2) The alkali solution bringing about osmotic pressure in lifting up the paint film, while weakening the binding force between paint film and phosphate crystal.

Therefore paint adhesion is most inferior just after specimens emerge from de-ionized water.

COSMETIC CORROSION RESISTANCE

Specimens with several coating weights were tested mainly in the modified Volvo test, after they were painted with two paint systems as shown in Table.2. System 1 is representative of Japanese automotive paint process. System 2 is representative of the United States paint process. To facilitate corrosion testing, the test panels were scribed to the base metal. Furthermore specimens in system 1 were inflicted with two kinds of paint defect: chipping by diamond shot, 10 mg in weight at 170 Km/H, and chipping by arrow drop, 1g in weight from a height of 150 cm. On the other hand, specimens in system 2 had stone-chipped defect by gravelometer.

The results after 9 months are shown in Fig.4 and Fig.5. The precoated steel sheets exhibited better cosmetic corrosion performance than cold rolled steel sheet. In paint system 1, all precoated steel sheets with coating of 20 g/m^2 showed red rust at scribes after 6 months exposure. After 9 months exposure, Fe-Zn alloy electroplated steel sheet with 45 g/m^2 slightly exhibited yellow rust. And Ni-Zn alloy electroplated steel sheet and Zn electroplated steel sheet with 45 g/m^2 exhibited no red rust. Zn electroplated steel sheets showed much white rust even after 3 months exposure. Also the blister width of Zn electroplated steel sheet was very wide, especially at the coating weight of 20 g/m^2. Ni-Zn alloy electroplated steel sheet with 30 g/m^2 coating and Fe-Zn alloy electroplated steel sheet with 45 g/m^2 coating showed blister widths one third of the width of Zn electroplated steel sheet with coating over 45 g/m^2. Corrosion at arrow-chipped defects did not occur on the test panels except cold rolled steel sheet. Also paint creepage at diamond-shot defects did not occur on alloy plated steel sheets. But on Zn electroplated steel sheet, there was paint creepage and white rust.

Table. 2 Panel Preparation for Corrosion Test

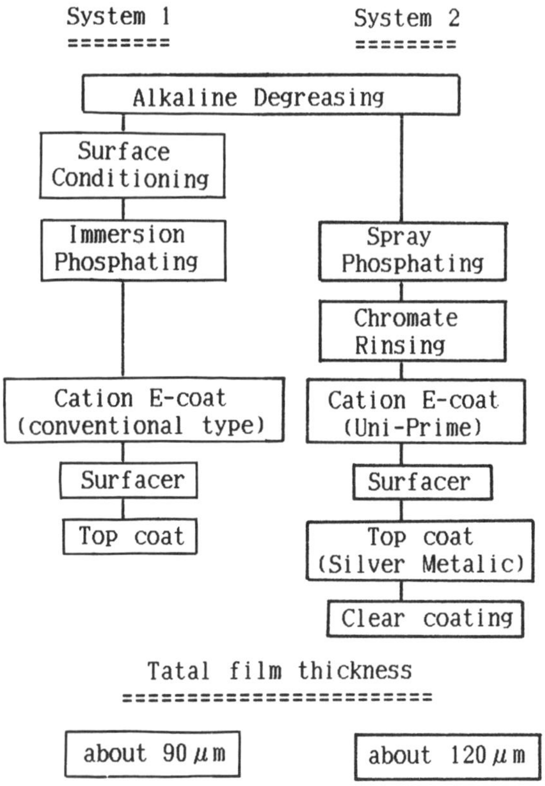

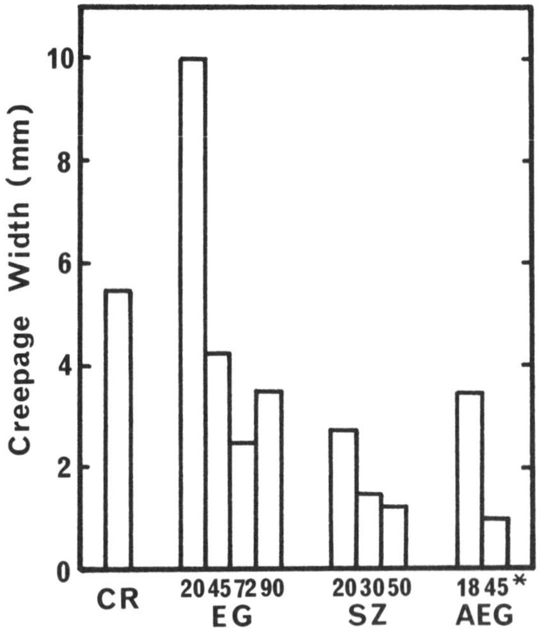

Fig.4 Corrosion behavior at scribed line of precoated steel sheets with paint system 1 after 9 months of the modified Volvo test.
*: coating weight (g/m^2)

In paint system 2, precoated steel sheets with coating weight of 20 g/m^2 exhibited red rust at scribes after 9 months exposure. No red rust occurred on precoated steel sheets with coating over 20 g/m^2. Paint system 2 may retard the occurrence of red rust on precoated steel sheets for 3 months longer than paint system 1 , due to chromate rinsing after phosphating and a thicker paint film. Ni-Zn alloy electroplated steel sheets with 30 g/m^2 and 50 g/m^2 and Fe-Zn alloy electroplated steel sheet with 45 g/m^2 exhibited smaller blister width than Zn electroplated sheel sheet with 90 g/m^2. But red rust at stone-chipped defects occurred on Zn electroplated steel sheet with 45 g/m^2 and on all alloy plated steel sheets. The red rust rate of Ni-Zn alloy plated steel sheets was better than that of Fe-Zn alloy plated steel sheets. There was much white rust and wider paint creepage on Zn plated steel sheets with 72 g/m^2 and 90 g/m^2 coatings.

In an additional corrosion test, the relationship between coating weight and paint creepage were evaluated on alloy electroplated steel sheets. Specimens were painted in a similar paint system to system 1. The corrosion test consisted of 1 hour of salt spraying at 35°C, 15 hours of drying at 50°C and 8 hours of humidifying at 50°C. The results shown in Fig.6 after 45 cycles were similar to the previous results. There was a

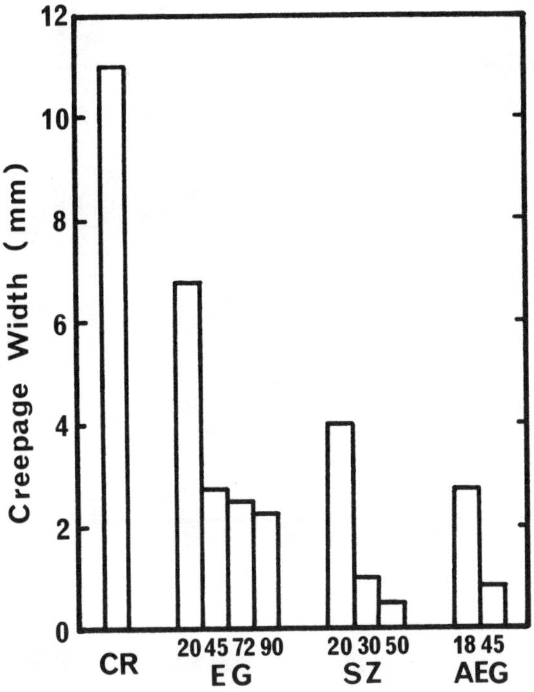

Fig.5 Corrosion behavior at scrived line of precoated steel sheets with paint system 2 after 9 months of the modified Volvo test.

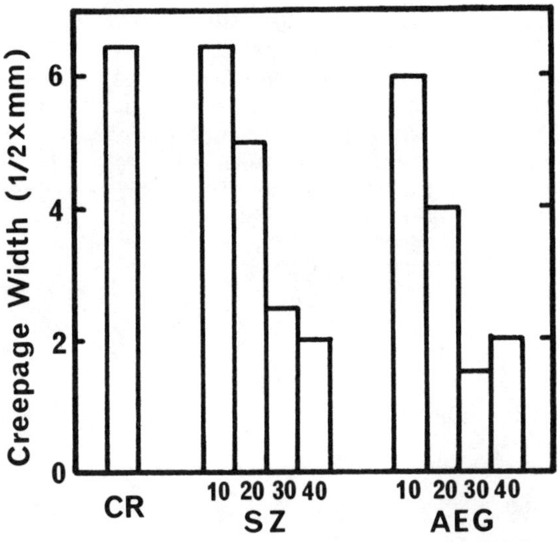

Fig.6 Effect of coating weight on creepage width after 45 cycles of scab corrosion test.

substantial difference in corrosion resistance between 20 g/m^2 and 30 g/m^2. This difference may result from the barrier effect of corrosion products and area ratio of anode to cathode at the scribe.

CONCLUSION

(1) Zn alloy electroplated steel sheet exhibits higher susceptibility to cathodic electropaint cratering than cold rolled steel sheet and Zn electroplated steel sheet. Thin Fe-Zn alloy layer, of which Fe content is about 80 %, electroplated on Zn-rich alloy plated steel sheets efficiently prevents occurrence of cratering. But a pure Zn layer is ineffective, even when it is thickly plated.

(2) The paint adhesion of precoated steel sheet after water immersion is inferior to that of cold rolled steel sheet. The addition of fluoride instead of chlorate and Mn ion to conventional immersion zinc phosphate solution greatly improves paint adhesion. This improvement results from the high alkali resistance of the phosphate layer and the small space of permeated water due to the fine and dense morphology of the phosphate crystals.

(3) Thin Fe-rich Fe-Zn alloy layer improves paint adhesion as well as cratering defect. However in practice other means such as improvement in electropainting process and application of new type phosphate solution may be applied.

(4) Cosmetic corrosion resistance of precoated steel sheets is better than that of cold rolled steel sheet. The coating weight of 20 g/m^2 is insufficient to protect substrates from corrosion. Zn plated steel sheet exhibits good red rust resistance, but

shows much white rust and wider paint creepage. Ni-Zn alloy electroplated steel sheet and Fe-Zn alloy electroplated steel sheet with coating weight of 30 g/m^2 to 50 g/m^2 exhibits narrower paint creepage than Zn electroplated steel sheet with 90 g/m^2 and high red rust resistance in the modified Volvo test.

REFFERENCES

(1) S. Wakano, A. Shibuya, Y. Hoboh and M. Nishihara : Trans. I. S. I. J. 23, 967-73 (1983)

(2) W. R. Johnson and J. B. Vrable : SAE Tech. Paper No.840213 (1984)

(3) J. B. Vrable : SAE Tech. Paper No.850002 (1985)

(4) Y. Miyoshi, M. Kitayama and S. Naito : SAE Tech. Paper No.850007 (1985)

(5) T. Kurimoto, Y. Hoboh, H. Oishi, K. Yanagawa and R. Noumi : SAE Tech. Paper No.831837

(6) L. L. Franks and E. T. Nowak : SAE Tech. Papar No.831819

(7) G. W. Froman and L. L. Franks : "Technological Impact of Surfaces", p.315 ASM, Metals Parks, Ohio (1981)

(8) R. G. Hart and H. E. Townsend : SAE Tech. Paper No.831818

ZINC-IRON ALLOY COATED STEEL SHEETS FOR AUTOMOTIVE CORROSION PROTECTION

Takashi Hada
R & D Laboratories-II
Nippon Steel Corp.
Sagamihara, Japan

Toshiro Fujiwara
Nagoya Works Nippon Steel Corp.
Nagoya, Japan

Tatsuya Kanamaru, Motohiro Nakayama, and Takashi Horita
Nagoya R&D Laboratories
Nippon Steel Corp.
Tokai, Japan

ABSTRACT

Galvannealed steel and zinc-iron alloy electroplated steel have coating layers composed of zinc-iron alloy. Based on the view that physical property of zinc-iron alloy common to the steels mentioned above determines an adequacy of their quality as corrosion-resistant steels for automotive body, quality characteristics of both steels were evaluated in a way of comparing them with a zinc coated steel. As a result of such research carried out, it was found that superior corrosion resistance, press formability and weldability could be achieved by zinc-iron alloy coated steels.

LONG LASTING LIFE is one of critical requirements for automobiles at present. To improve the present situation, various corrosion-resistant steels such as galvanized, zinc alloy coated, organic coated, organic composite-coated and others, are used for automotive body and their development is also in progress. Among those, development of zinc-iron alloy electroplated steel and its application to automotive body have attracted public attention recently[1~4].

On the background of that highlighted zinc-iron electroplated product, in addition to its coating layer characteristics, it is generally known that electrogavanizing process is more suitable to the material property of steel sheet than hot-dip galvanizing. Zinc-iron alloy coating has, however, been used for automotive body as it was known as galvannealed steel produced by hot-dip process in earlier days[5].

Hot-dip gavanizing technology has also made a remarkable progress in these days. Although its quality and variation thereof were previously considered inferior than those achieved by elctrogalvanizing, they have been improved by the better press stamping performance which ultra-low carbon steel can provide[6], development of one-side galvanizing technique[7], progresses in uniform coating[8] and alloy making techniques[9], etc. Nowadays, application of electrogalvanizing or hot-dip process does not introduce substantial quality difference in corrosion-resistant steel for automobile use. Therefore, it may be said that energy cost factor to coating quantity has now become a major interest.

For galvannealed steel or zinc-iron alloy electroplated steel, features of zinc-iron coating layer itself are reported that;-

(1) Suitable to subsequent protective painting[1,2,5,10];
(2) In press stamping process, its fragments do not build up on a die face since the layer is pulverized rather than flaking off[11]; and
(3) Life of welding electrode can be comparatively long[12].

Whether the process is electrolytic or hot-dip, coating is formed with zinc-iron alloy. Galvannealed steel is mainly composed of δ_1 phase and does not comprise η phase substantially. Zinc-iron alloy electroplated steel is mainly composed of δ_1 phase, metastable zinc-iron alloy phase with iron content ranging from 10% to 20%, and is produced under the plating condition that η phase is not comprised substantially[1].

Table 1 is to compare factors such as electrochemical potential and coating layer hardness of galvannealed steel and zinc-iron alloy electroplated steel to those of electrogalvanized and hot-dip galvanized steel. It is evident that the former two indicate almost same physical properties and these figures are different from those of the latter two.

From aforementioned viewpoint, research was instituted to evaluate an adequacy of zinc-iron coating quality as a corrosion-resistant steel for automobiles, in a way of comparing it with conventional zinc coating products like electrogalvanized and/or hot-dip galvanized steels. Then, corrosion resistance, press formability, weldability were investigated to clarify specific features of the zinc-iron coating.

Table 1 - Phisical Data of Zn-Fe Alloy Coatings and Zn Coatings

Coated Steel	Potential	Hardness
	(V vs. S.C.E. in 5% NaCℓ)	(Hv-10g)
Galvannealed	-0.9 (1)	355 for Compact δ_1
		263~326 for Palisade δ_1 (3)
Zn-Fe Alloy Electroplated	-0.9 (2)	320 (4)
Electrogalvanized	-1.1 (2)	35 for η (3)
Hot-Dip Galvanized	-1.1 (1)	50 (4)

(1) Described in Reference (6)
(2) Described in Reference (1)
(3) Described in Reference (13), where unit is measured by Hv-0.05 KP/mm^2
(4) Measured by authors of this report

EXPERIMENTAL

Corrosion-resistant steels used in the experiment were; galvannealed steel (GA), zinc-iron alloy electroplated steel (EL), electro-galvanized steel (EG) and hot-dip galvanized steel (GI) which are produced and supplied by Nippon Steel Co. for automobile use. Coating weight is identified together with respective data. Zinc-iron alloy electroplated steel used was a duplex type which had ferrous-rich coating in its upper layer (trade name EXCELITE)[2]. The upper layer had 75-85% of iron composition with coating weight of 3g/m^2, and the lower layer had 10-20% iron composition with coating weight of 20g/m^2.

Painting system used in corrosion resistant test was dip phosphate and primed with a 20 µm cathodic electro-deposition. In scab corrosion test, an intermediate coat, sanding and top coat were applied in addition, to prepare specimens with a 80 µm overall film thickness.

CORROSION RESISTANCE

Corrosion factors vary at each location of vehicle's body. The following items were used as a typical way to assess corrosion resistance:

(1) Blister resistance obtained by salt spray test.
(2) Scab corrosion resistance.
(3) Perforation resistance.
(4) Galvanic perforation resistance contacted with cold rolled steel.

BLISTER RESISTANCE - Figure 1 indicates blister resistance of cathodic electro-deposition primed plate obtained by salt spray test. Blister resistance was assessed by the specific time of salt spray required for a maximum creepage developed from a crosscut at one side to be 3 mm. Whether the process is hot-dip or electrolytic, blister resistance of zinc coating is determined only by its coating weight. Thus, the more coating weight, the better blister resistance. It is considered that thicker coating makes local cell interaction between zinc coating and steel substrate scarcely to reach out an interface between paint film and zinc coating. Galvannealed steel can achieve the blister resistance equivalent to that of zinc coated steel which has twice or more coating

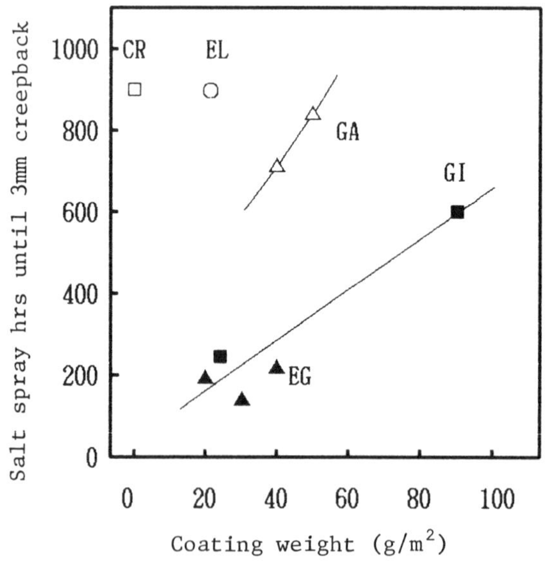

Fig. 1 - Blister resistance of 20µm cathodic electro-deposition primed sheets by salt spray test
(CR: cold rolled, EL: duplex zinc-iron alloy electroplated, GA: galvannealed, EG: electro-galvanized, GI: Hot-dip galvanized)

weight. This excellent blister resistance may be accomplished by less potential difference between galvannealed layer and steel substrate. Another possible reasoning is that superior alkali resistance of galvannealed steel inhibits further creepage development[10]. Duplex type zinc-iron alloy electroplated steel can ensure favorable blister resistance with lesser coating weight. This is considered due to a considerable effect given by the upper layer's ferrous-rich coating, in addition to an interaction provided by zinc-rich alloy coating of the lower layer. Such effect may be achieved by high alkali resistant phosphophyllite formed during the phosphating process and better compatibility of ferrous-rich coating with cathodic electro-deposition paint property as it makes the portion below the paint film to be relatively cathodic[1].

SCAB CORROSION RESISTANCE - To assess scab corrosion resistance, wet corrosive condition (Test A) and dry condition (Test B) were selected. Cyclic corrosive condition used in test A is specified below:

1)	Salt spray	35°C	1 day
2)	Humidifying	40°C, RH 85%	5 days
3)	Indoor		1 day

Cylic corrosive condition used in test B is specified below:

1)	Salt water immersion	5% NaCl	1 day
2)	Outdoor exposure		6 days

In both tests, specimens with 80 µm coating were scratched by a cutter, and a result was assessed by the length of longest rust or blister from the scratch. Figure 2 indicates the result obtained from test A. Under this corrosive condition, filiform rust was formed on cold rolled steel. No filiform rust was formed on zinc and zinc-iron alloy coated steels. Since filiform rusts are sometimes observed along scab edges on the vehicle's body in service, use of galvanized gives considerable effect under such corrosive environment that growth of filiform rust induces further scab corrosion. However, electrogalvanized steel tends to have blister, and galvannealed steel, zinc-iron alloy electroplated steel show better scab corrosion resistance.

Figure 3 shows the result obtained from test B. Under this corrosive condition, reddish rust formed a swelling around a scratch on cold rolled steel, and blisters were observed on zinc and zinc-iron alloy coated steels with no reddish rust. However, electrogalvanized steel presented quick growth of blister, and its size has exceeded that of cold rolled steel after seven weeks. Sizes of blisters appeared on galvannealed steel, zinc-iron alloy electroplated were substantially smaller than that made on cold rolled steel.

Similar to the case of salt spray test, coating weight might have given a certain influence in this scab corrosion test. Although used materials did not have an identical coating weight, these test results indicated that zinc-iron alloy coated steel had better scab corrosion resistance than zinc coated steel.

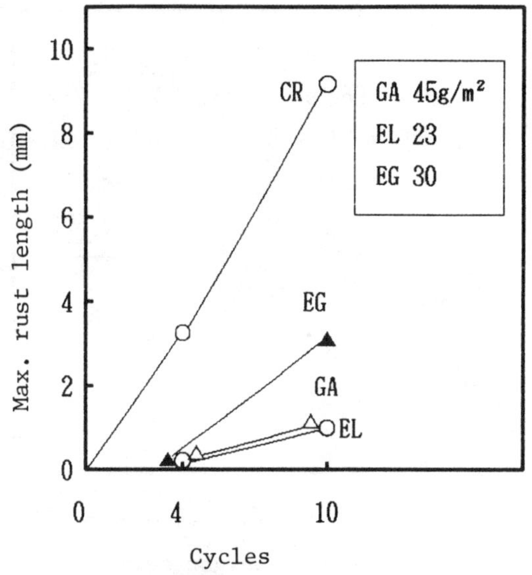

Fig. 2 - Scab corrosion test result by Test A (GA: galvannealed, EL: duplex zinc-iron alloy electroplated, EG: electrogalvanized, CR: cold rolled)

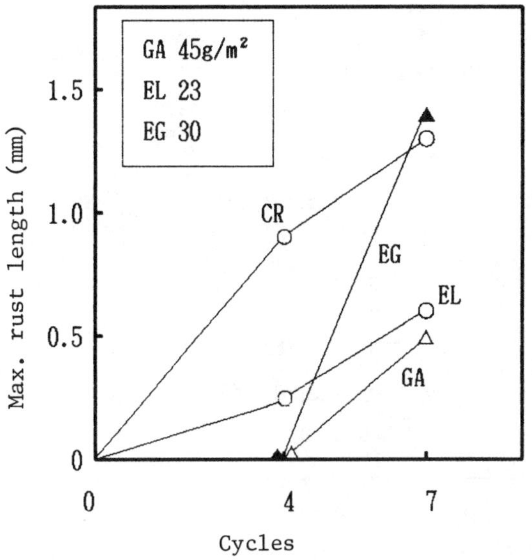

Fig. 3 - Scab corrosion test result by Test B (GA: galvannealed, EL: duplex zinc-iron alloy electroplated, EG: electrogalvanized, CR: cold rolled)

PERFORATION RESISTANCE - Figure 4 indicates results obtained from perforation resistance test. Cathodic electro-deposition primed specimen with a crosscut was used to provide a particular cyclic corrosive condition composed of salt spray and drying[14]. Perforation resistance is dependent on coating weight. It is, however, evident that zinc-iron alloy coated steel gave better result when comparing it to

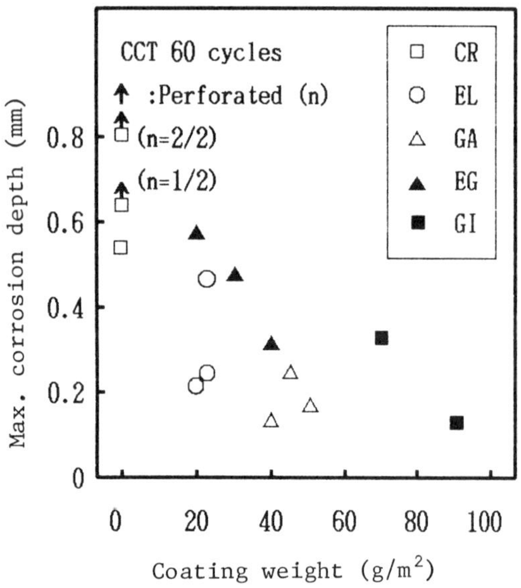

Fig. 4 - Perforation resistance by cyclic corrosion test
(CR: cold rolled, EL: duplex zinc-iron alloy electroplated, GA: galvannealed, EG: electrogalvanized, GI:, hot-dip galvanized)

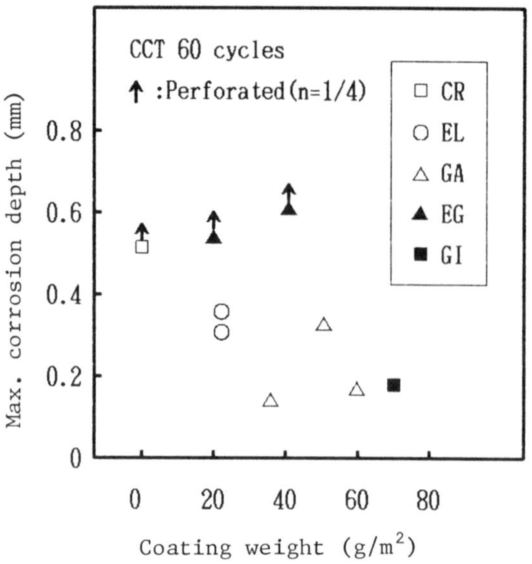

Fig. 6 - Galvanic perforation resistance contacted with cold rolled steel by cylic corrosion test
(CR: cold rolled, EL: duplex zinc-iron alloy electroplated, GA: galvannealed, EG: electrogalvanized, GI: hot-dip galvanized)

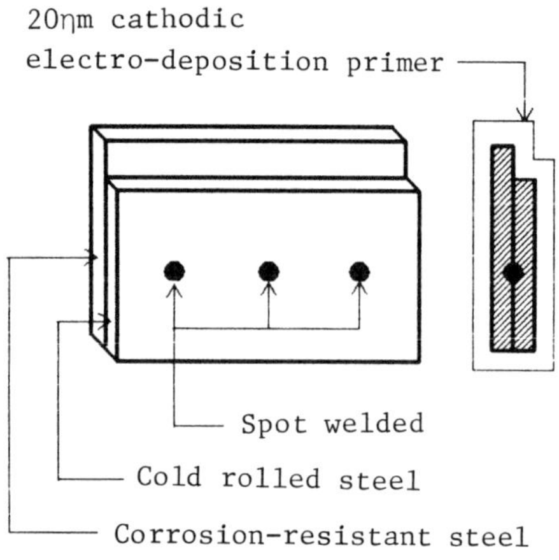

Fig. 5 - A coupled specimen with cold rolled and corrosion-resistant steels

zinc coated steel which had an identical coating weight. By salt spray test which provided more humid corrosive condition, perforation resistance of zinc coated steel was markedly inferior than that of zinc-iron alloy coated steel(2). This fact suggests ferrous composition existing in a coating layer inhibits flow-out of zinc corrosion products.

GALVANIC PERFORATION RESISTANCE CONTACTED WITH COLD ROLLED STEEL - Figure 5 indicates schematic of a coupled specimen with cold rolled and corrosion-resistant steels. With this specimen, cyclic corrosion test was carried out in the same manner as specified in figure 4. Figure 6 shows result thereof. Cathodic electro-deposition primer could scarcely goes into the internal surfaces of coupled plates, and corrosion was observed on overall internal surfaces of both, cold rolled steel and corrosion-resistant steel. Corresponding to a deep corrosion spot on the cold rolled steel, a significant corrosion depth was also found on corrosion-resistant steel. As both, cold rolled steel and corrosion-resistant steel presented similar levels of maximum corrosion depth, a mean of measurements taken from two couples, i.e. four plates, was used to represent the maximum depth.

Galvanic perforation resistance contacted with a cold rolled steel may also be affected by the coating weight in principle. Comparison of data given in figures 6 and 4 certainly gives clear distinction between zinc-iron alloy coated steel and zinc coated steel. Possible difficulty in drying internal surfaces between coupled plates is considered to be a particular reason for this.

Although the results obtained from aforementioned corrosion resistance tests may not perfectly represent all corrosions taken place on vehicle's body, they have respectively suggest reasonable correlations with some typical corrosions. Corrosion resistance is certainly dependent on coating weight. However, summarizing the results of four tests, it can be said that zinc-iron alloy coated steel has better corrosion resistance than zinc coated steel when they are compared at an identical coating weight.

PRESS FORMABILITY

Formability of corrosion-resistant steel is quite dependent on mechanical property of its base metal. Critical problem coating layer involves is falling-off which takes place in some cases during press stamping process.

Performances of galvannealed steel, zinc-iron alloy electroplated steel, electrogalvanized steel and hot-dip galvanized steel during continuous stamping process have already been reported[11]. While the former two (zinc-iron coated group) are reported that they can assure stable operation, the latter two (zinc coated group) are criticized as fallen-off coating particles build up on a die face which may subsequently form dents on a stamped panel.

Therefore, in this report, examples of such build-up formed by fallen-off coating are shown in photographs 1 and 2 to make a supplement to above-mentioned report. With galvannealed steel (photograph 1) and zinc-iron alloy electroplated steel, size of such particle did not exceed 50μm, which is equivalent to size of the particle fallen-off by forming. This situation suggests that particles made no secondary cohesion on a die face. In contrast, with galvanized steel (photograph 2), far greater size of particles are observed in the build-up formed on a die face. These large particles are obviously formed by the secondary cohesion, and assumed to result dents on a stamped panel. It is considered that secondary cohesion is correlated with hardness of the coating layer.

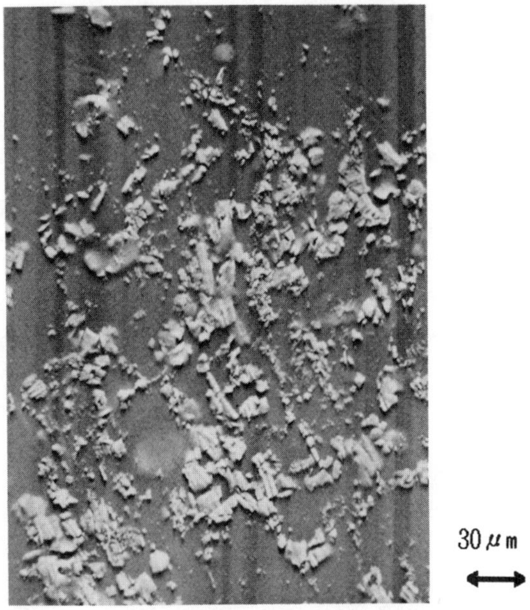

Photo 1 - Powder of galvannealed steel on a die face

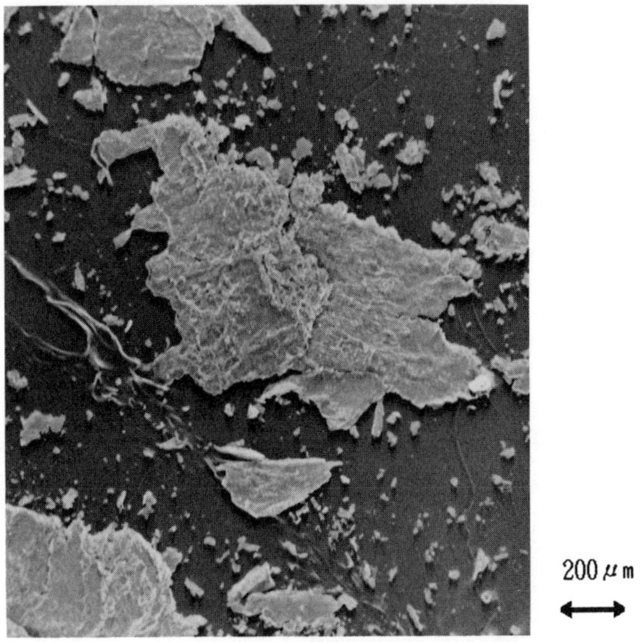

Photo 2 - Build-up of hot-dip galvanized steel on a die face

WEDLABILITY

Spot welding on zinc or zinc alloy coated steel generally gives tendencies such as higher weld current than that required to weld cold rolled steel, limited range of welding conditions and shorter life of electrode when used for continuous weld operation[12]. Although heavy coating weight ensures better corrosion resistance, weldability turns more difficult as coating weight becomes heavier[12]. Therefore, a balance between corrosion resistance and weldability must be considered when using corrosion-resistant steel for automobiles.

Corrosion-resistant steels have their optimum range of welding condition. Figures 7 and 8 indicate continuous spot weldability plotted on the coordinates of weld current and electrode load, for galvannealed steel and electrogalvanized steel, respectively. Comparing figures 7 and 8, it is evident that galvannealed steel gave the continuous spot weldability much better than that electrogalvanized steel could accomplish.

This is considered because the least current of nugget forming and the expulsion current are particularly low in a range of low electrode load with galvannealed steel. As a common matter applied to zinc or zinc alloy coated steel, ohmic resistance between steels declines as the melted coating layer has taken a certain extent of heat away and conductive area has been enlarged. Accordingly, contact resistance before the coating layer has molten gives remarkable effect for the coated steels. Since a coating layer of galvannealed steel is composed of zinc-

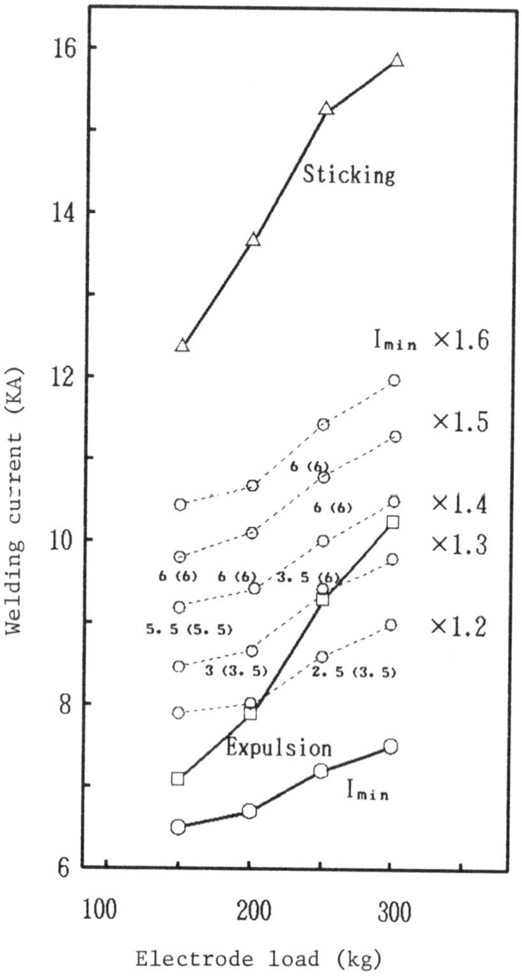

Fig. 7 - Weld lobe and continuous spot weldability of galvannealed steel (both side $45g/m^2$)
(t=0.8mm, Electrode: 6mmϕ dome type tip. I_{min} indicates the least current to form nugget dia more than $4\sqrt{t}$, where t is sheet thickness. Numeral is the number (x1000) of continuous weldable spots, assessed by 85% current. Numeral in parentheses is that assessed by the current showed.)

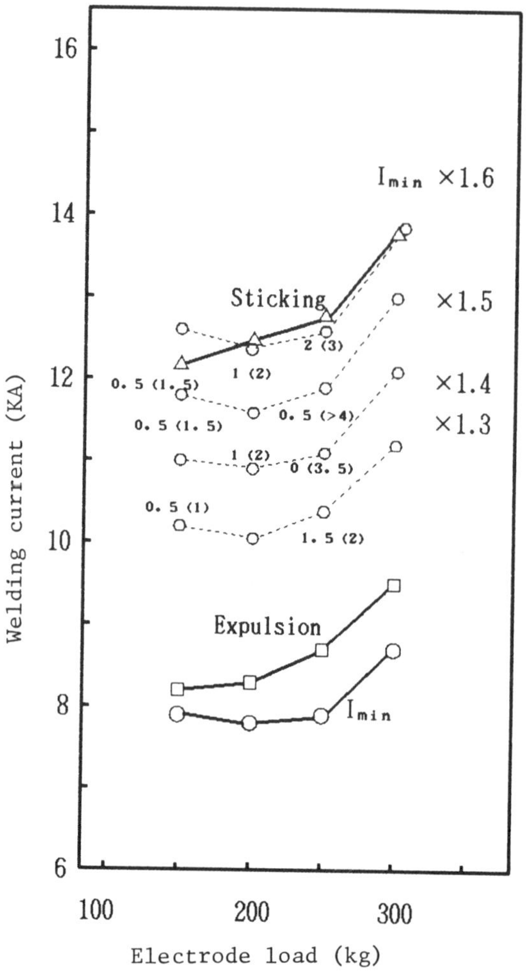

Fig. 8 - Weld lobe and continuous spot weldability of electrogalvanized steel (both side 20 g/m)

(Test condition is the same as Fig. 7.)

iron alloy which has a higher melting point, hardness of zinc-iron alloy is considered effective to keep the high-level of contact resistance at initial welding period, in addition to its lesser interactive resistance between steels decline caused by melting. When electrode is pressed down to a steel, a coating layer scarcely makes plastic deformation within the low-load range, and contact area with electrode tip is, therefore, not enlarged. This subsequently develops local expulsion and incurrs self-dressing of electrode during the continuous weld operation, thereby a shape of electrode tip remains fairly intact.

As mentioned above, logical analysis suggests physical property of zinc-iron alloy facilitates expansion of the specific weld condition range and improves continuous weldability.

Zinc-iron alloy electroplated steel basically presents behavior similar to that of galvannealed steel. As electrogalvanized steel has a softer coating layer, nugget-forming least current is relatively high due to the reason inverse to the above and its continuous weldability is inferior.

CONCLUSION

As a result of research carried out to evaluate adequacies for automobile use of galvannealed steel and zinc-iron alloy electroplated steel, both coated with zinc-iron alloy, in a way of comparing them with electrogalvanized steel and/or hot-dip galvanized steel, both with zinc, it was found that ;-

1) CORROSION RESISTANCE - If compared with an identical coating weight, zinc-iron alloy coated steels have better corrosion resistance than those offered by zinc coated steels;

2) PRESS FORMABILITY - Size of particles fallen off from the zinc-iron alloy coated steels is 50 μm or less, which scarecely results dents. The particles flaked off from the zinc coated steels tend to build up on die face due to the secondary cohesion, which subsequently forms dents on a stamped panel; and

3) WELDABILITY - Zinc-iron alloy coated steels offer larger range of available weld condition and superior continuous weldability. Hardness of zinc-iron alloy contributes for this.

REFERENCES

(1) Kojima, H., Yamamoto, T., Ito, K., Fujiwara, T. and Kanamaru T., SAE paper 840214 (1984)

(2) Toda, M., Kojima, H., Morishita, T., Kanamaru, T. and Arai, K., SAE paper 840212 (1984)

(3) Johnson, W.R. and Vrable, J.B., SAE paper 840213 (1984)

(4) Vrable, J.B., SAE paper 850002 (1985)

(5) Meguro, A., Yamamoto, T. and Aikawa, J., The Toyota Engineering 27, 553 (1978)

(6) Itoh, K., Tokunaga, Y., Sekiya, T. and Yamada, M., Bulletin of the Japan Institute of Metals 24, 316 (1975)

(7) Kitazawa, Y., Higuchi, S., Yamato, T., Kanamaru, T. and Sekiya, T., Tetsu-to-Hagané 68, A53 (1982)

(8) Kasugai, M., Tomari, S., Suhara, M., Miki, T., Ohkouchi, T. and Hasegawa, T., Tetsu-to-Hagané 71, S466 (1985)

(9) Tamoto, N., Miki, T., Sekiya, T. and Tsukamoto, S., Tetsu-to-Hagané 70, S466 (1984)

(10) Miyoshi, Y., Yoshida, K., Azami, T., Kanamaru, T. and Kado, S., Tetsu-to-Hagané 66, 858 (1980)

(11) Ejima, M., Tokunaga, Y. and Honda, T., Proc. 13th Biennial Congr. IDDRG, 317 (1984)

(12) Saito, T. and Takahashi, Y., Welding Technique ISSN 31, (4), 27 (1983)

(13) Neith, F. and Wiegand, H., Metalloberfläche 20, 393 (1966)

(14) Miyoshi, Y., Oie, Y., Amano, M. and Koyahara, H., SAE paper 820334 (1982)

ENHANCED ALUMINUM AUTO TRIM

John H. Powers
Alcoa
Alcoa Center, Pennsylvania, USA

John A. Burke
Ford
Dearborn, Michigan, USA

Max Wiener
H. Friedl Mfg. Ltd.
Richmond Hill, Ontario, Canada

ABSTRACT

This paper briefly summarizes process variables reported at the Society of Automotive Engineers (SAE) meetings in February 1983 to enhance the appearance of aluminum automotive trim. New data is presented comparing the relative influence of finishing, alloy, and metal purity. The results of a Ford/Magna/Alcoa production evaluation of alloy and process variations are reported. Finally, a new method for correlating optical measurements with appearance is presented.

BRIGHT DIPPED AND ANODIZED (BDA) aluminum auto trim has become the most attractive and cost-effective bright decorative material used on passenger automobiles, light trucks and vans. During the last 30 years, the aluminum industry has continually improved the alloy metallurgy and fabricating practices required for superior appearance. Responding to the automakers' desire for improved appearance, Alcoa Laboratories embarked on an R&D effort 3-1/2 years ago to focus on finishing process improvements for aluminum automotive trim.

An optical technique for quantifying appearance using a goniophotometer was developed (Appendix A). Optical properties that were correlated with appearance were image clarity (D/I) and color (2°). High image clarity and low color values appeared to correlate with the best appearance.

This optical technique was used to quantify the effect of variations in material, alloy, buffing, brightening, and anodizing on appearance. Two key findings were that 5657H25 alloy responded better to BDA than 5252H25 alloy, and that thinner anodic coatings produced at a lower anodizing voltage improved the appearance of auto trim significantly. The degree of improvement obtained with this modified anodizing was found to vary with variations in material within an alloy and between alloys.

These data were reported in a paper* presented at the Society of Automotive Engineers (SAE) meetings in February 1984. A summary of the findings that were reported is provided in Figure 1.

Since this paper was presented, additional studies have been conducted to develop methods for upgrading the appearance of aluminum automotive trim. These studies involved better definition and interpretation of optical properties, the effect of alloy purity on response and a production evaluation of the improved anodizing process described in the earlier paper. The production evaluation was conducted cooperatively by Ford, Magna (H. Friedl Mfg. Ltd.), and Alcoa. This paper reports the results of these R&D studies as well as the production evaluation.

OPTICAL PROPERTY ANALYSIS

Additional studies were made on flat panels with the same goniophotometer used in the previous studies and described in Appendix A. A total of 150 lots of production 5252 alloy were bright dipped and anodized (BDA) in the laboratory to a nominal coating thickness of 0.3 mil per present standards in the automotive industry. Optical properties were measured both with and across the rolling direction of the aluminum. In previous studies, the with and across values were averaged. In these studies, each measurement was examined independently to determine the best correlation with appearance.

It was found that color measurements were not independent of image clarity. In fact, there was a linear correlation within the range

*John H. Powers, "Enhanced Aluminum Auto Trim with Improved Luster, Color, and Texture", SAE, February 27-March 02, 1984

Figure 1 – Summary of Various Approaches to Improve Appearance of Aluminum Automotive Trim

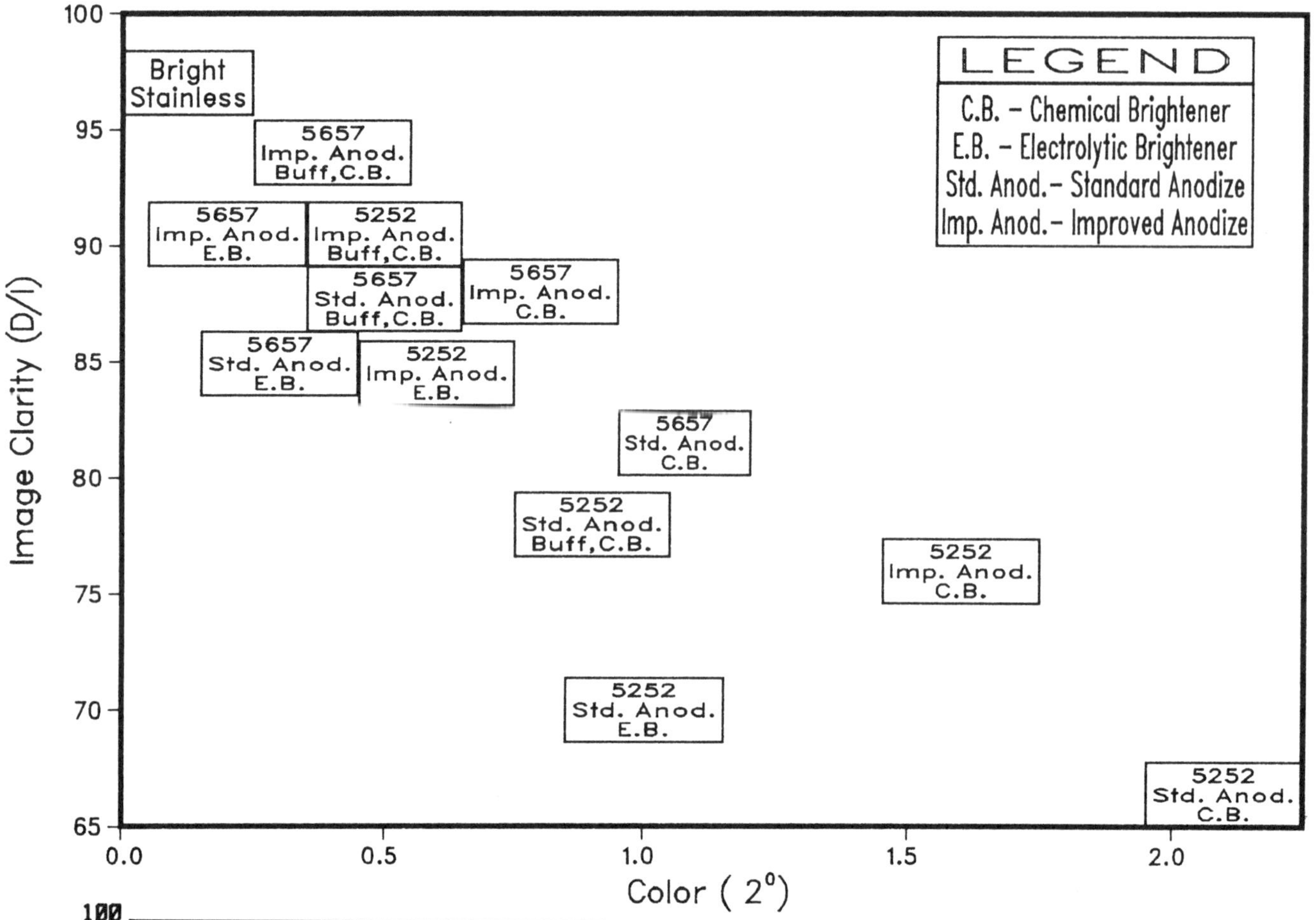

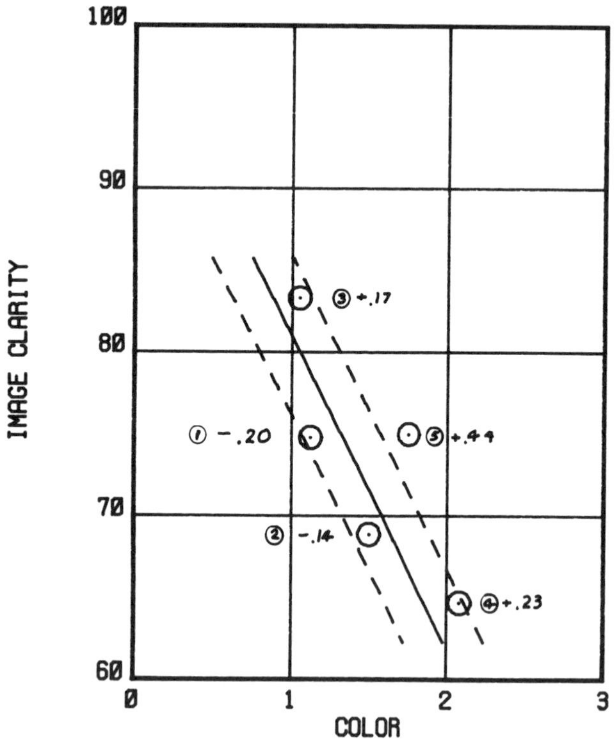

FIG.2 - 95% CONFIDENCE LIMITS FOR 150 LOTS OF 5252 ALLOY

studied, and the with-rolling readings correlated better than the average. This relationship (mean and 95% confidence limits) between color and image clarity is shown in Figure 2. These data could lead one to conclude that color cannot be evaluated independently since the number appears to be merely a reflection of the image clarity of the surface.

However, it was postulated that true color might indeed be quantified by measuring its deviation from the least squares predicted regression line. This idea was tested by visually examining high and low image clarity and color specimens from the 150 lots. These specimens were visually ranked from best to worst color and then the deviation from the regression line was computed. The data shown in Table I illustrate that there is indeed a correlation between visually observed color and deviation from the regression line. Negative deviations in color from the regression line have a better color than the more positive values. These data were superimposed on Figure 2 showing appearance ranking (1 thru 5) and color deviation from the regression line.

It is not surprising that the with-rolling optical properties show a better correlation with appearance than the average of with and across

Table I

Visual Color Ranking Versus Deviation From Least Squares Regression Line

Visual Rank	Image Clarity	Color	Deviation in Color from Regression Line
Best 1	74.8	1.12	- 0.20
2	68.9	1.49	- 0.14
3	83.3	1.05	+ 0.17
4	64.7	2.08	+ 0.23
Worst 5	75.0	1.75	+ 0.44

rolling optical properties. Factors affecting image clarity and color are on a microscopic scale and deal with light scattering within the aluminum oxide coating and at microdiscontinues at the metal oxide interface. Larger scale aberrations on the surface caused by the directionality of the rolled surface are indeed visually noticeable and do adversely affect the optical data. However, they do not influence image clarity and color provided the effect of directionality is not too pronounced. This is the case with most high quality automotive trim.

As a result of these findings, throughout the rest of the testing program, with-rolling optical properties were used.

EFFECT OF ALLOY AND PURITY ON FINISHING RESPONSE

In the previous paper, it was reported that 5657 alloy had a better appearance and optical properties than 5252 alloy after BDA. Additional studies were conducted on 14 lots of 5252 alloy and 27 lots of 5657 alloy. Also the effect of alloy purity on 5252 alloy was studied. Impurity levels (iron plus silicon) ranging from .01 to .08% by weight were produced using a 99.99% aluminum plus magnesium base material in combination with various quantities of 5252 alloy. Figure 3 shows the linear decrease in image clarity as impurities are increased. As part of these studies, four lots of high purity 5252 alloy with an iron plus silicon level of .014% were produced and optical properties were measured after standard BDA. Optical properties of this material along with the 5657 and 5252 alloys noted above are plotted in Figures 4 thru 6. These data show the improved BDA response of 5657 and high purity 5252. All three alloys show approximately the same slope regression line; however, the 5657 regression line is offset slightly to the left, suggesting a little better color than 5252 alloy.

FINISHING PROCESS

In the previous paper an improved anodizing process was recommended. Essentially this consisted of a reduced anodic coating thickness, 3.0 µm (.12 mil), and a lower anodizing voltage (-4 volts) obtained by elevating the anodizing electrolyte temperature. This improved anodizing

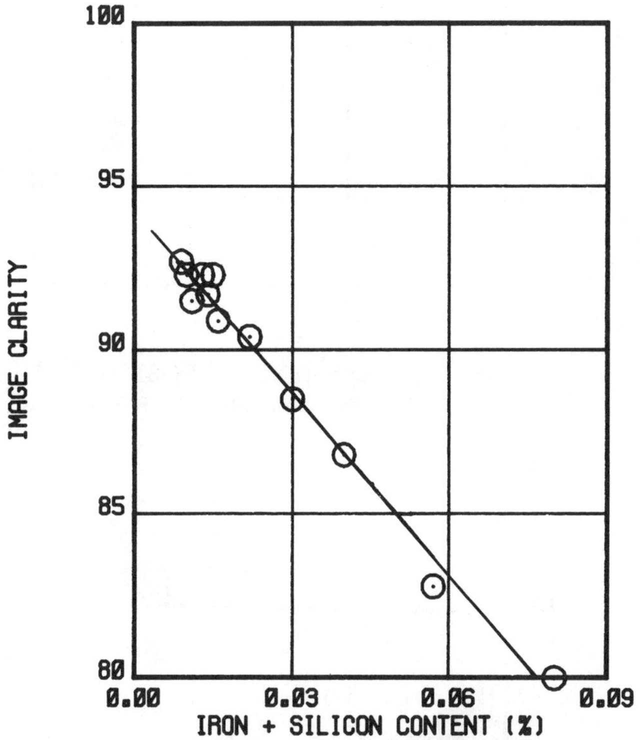

FIG.3 - EFFECT OF ALLOY PURITY ON IMAGE CLARITY

process was applied to the three alloy/purity materials reported above. A comparison of the optical properties resulting from the improved and standard anodizing process on the three materials is shown in Figures 4 thru 6.

The improved anodizing process significantly improves image clarity. Compared to the standard process, alloy 5252 reproduces 5657 levels of image clarity; and 5657 alloy is even better than high purity 5252. Color, judged visually, was also substantially better with the improved process. However, color data obtained with the improved process were offset to the right of the regression line. It was suggested earlier in the paper that this offset from the regression line should produce a poorer color.

When this anomalous result was studied further, it was found that the predicted regression lines for the improved and standard anodizing processes were definitely different.

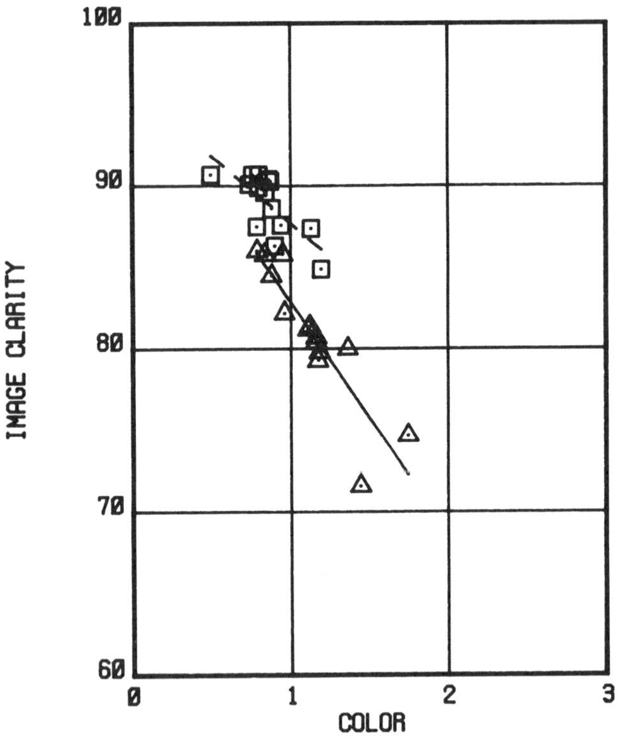

FIG.4 - 5252 ALLOY WITH STANDARD (——) AND IMPROVED (---) BDA.

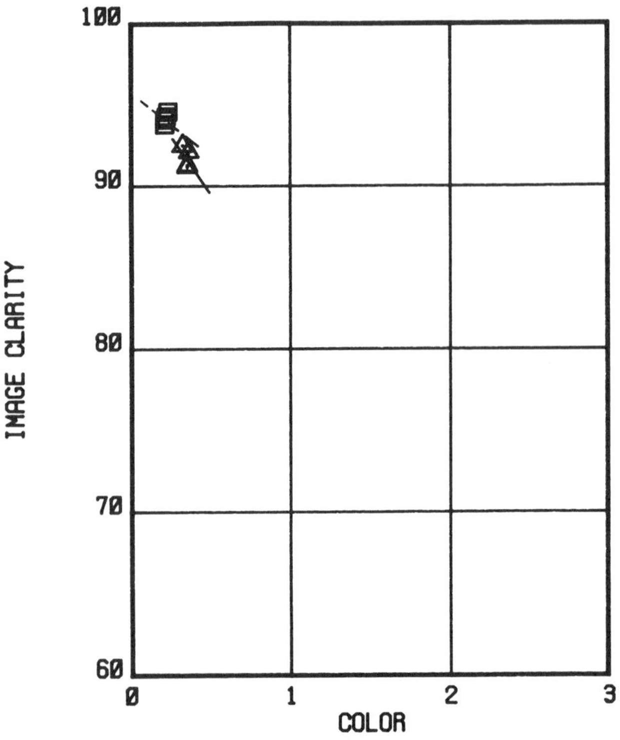

FIG.6 - HIGH PURITY 5252 ALLOY WITH STANDARD (——) AND IMPROVED (---) BDA.

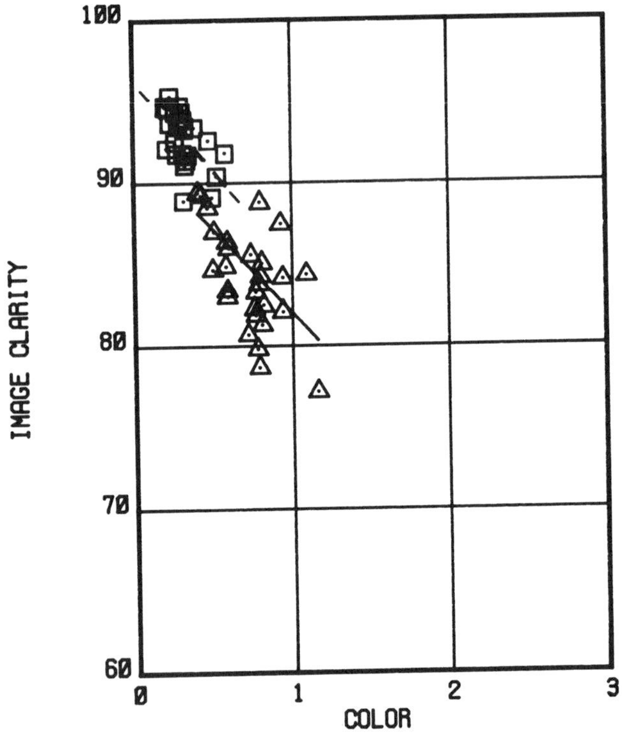

FIG.5 - 5657 ALLOY WITH STANDARD (——) AND IMPROVED (---) BDA.

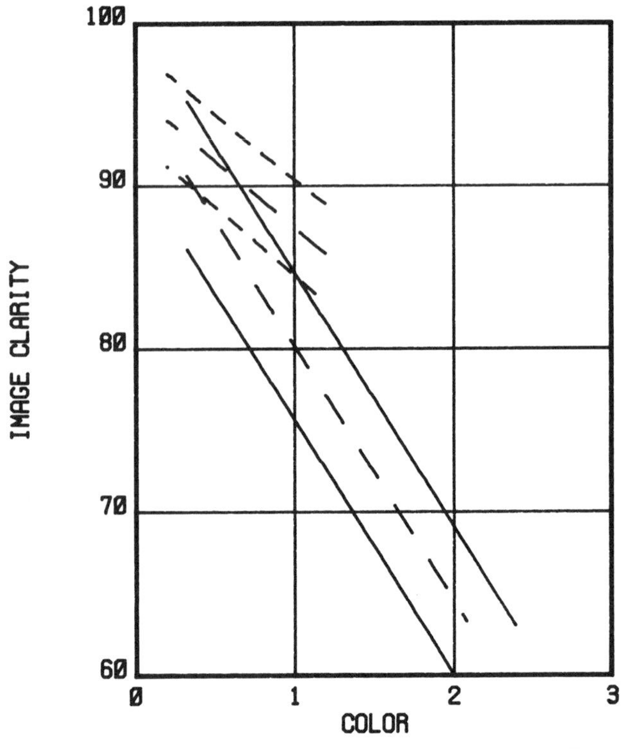

FIG.7 - 95% CONFIDENCE LIMITS FOR STANDARD (——) AND IMPROVED (---) BDA ON ALLOYS 5252, 5657, AND HIGH PURITY 5252.

Table II

Production Evaluation

Ct. Thickness μm (mils)	Optical Property	5252 Min.	5252 Max.	5252 Aver.	High Purity 5252 Min.	High Purity 5252 Max.	High Purity 5252 Aver.
7.5 (0.3)	D/I	76.5	88.9	81.7	74.7	83.9	79.3
	Color	0.50	0.88	0.67	0.19	0.25	0.22
5.0 (0.2)	D/I	67.4	85.8	76.2	78.3	81.8	80.0
	Color	0.41	0.83	0.66	0.22	0.22	0.22
2.5 (0.1)	D/I	86.9	90.7	88.8	87.7	90.1	88.9
	Color	0.38	0.58	0.48	0.19	0.21	0.20

It was found that all three alloys fit the same regression equations once it was determined that a separate equation was required for the two anodizing processes (Figure 7). This finding suggests that the proposed method for evaluating color by deviation from the D/I-color regression line has very serious limitations. Variations in oxide coating thickness have a significant effect and perhaps other process or material variations will change the relationship. At this point, it appears that the major utility in the method is for use in the laboratory where variables are under good control and held constant.

PRODUCTION EVALUATION

A production evaluation of the improved anodizing process and the high purity 5252 alloy was conducted by Ford Motor Co., H. Friedl Mfg. Ltd. (Magna), and Alcoa. The program included auto trim parts and 6 inch x 4 inch panels which were used to measure optical properties and conduct accelerated testing and outdoor exposure. Three random lots of 5252 alloy and one of high purity 5252 were processed in a fully automatic anodizing line with four computerized hoists. Three production-sized loads were anodized to produce nominal coating thicknesses of 7.5, 5, and 2.5 μm (.3, .2, .1 mil). All three loads were anodized for 30 minutes using standard production practices except for the 5 and 2.5 μm coating thicknesses. In these two cases, the current density was reduced 33% and 67% respectively compared to the 7.5 μm coating thickness.

It has been reported*(**) that anodizing voltage has an effect on the image clarity of bright trim alloys. One theory is that higher anodizing voltages significantly increase the reaction rate of the microconstituents in the alloy. This leads to larger irregularities in the oxide and at the metal/oxide interface which scatter light and reduce appearance. Laboratory studies have shown that either elevated electrolyte temperatures or reduced current densities provide the same improvement in optical properties due to the lower anodizing voltage.

The optical properties of replicate panels from each lot and each anodic coating thickness were measured and are listed in Table II. On 5252 alloy, 2.5 μm (.1 mil) coating thickness showed an improvement in both image clarity and color. The 5.0 μm (.2 mil) coating thickness did not show an improvement in this production trial in contrast to results of laboratory tests conducted earlier. The high purity 5252 had excellent color and reductions in coating thickness improved the image clarity.

*W. C. Cochran and F. Keller, Trans. Amer. Electroplaters Society, 1961, Vol. 48, p. 82

**W. E. Cooke, "Factors Affecting Loss of Brightness and Image Clarity During Anodizing of Bright-Trim Aluminum Alloys in Sulfuric Acid Electrolyte," Plating, Vol. 49, Nov. 1962, p. 1157

ARGUMENTS AGAINST NEW PROCESS

- Thin coatings will not pass certain accelerated corrosion tests.
- Satisfactory performance of thin coatings on the car has not yet been proven.

ARGUMENTS FOR NEW PROCESS

- Accelerated tests predict pitting corrosion; but field failures on BDA trim have been mainly blushing, which is influenced by seal quality rather than coating thickness.
- Seal quality with the improved process is as good as with the standard process.
- Thin coatings, 2.5-5.0 μm (.1-.2 mil), are being successfully used in Europe on auto trim.
- The improved process gives a better appearance at no cost premium with no equipment change.
- The improved BDA process reduces thermal crazing which has caused pitting problems in the field when using the standard BDA process.

CONCLUSIONS

Optical testing has proved to be a very useful method to assess BDA response of aluminum auto trim. It eliminates subjective evaluations and provides continuity in the evaluation. A key factor that remains to be determined is the relative importance of image clarity and color when one is viewing a trim part on a car. The relative importance of each optical property may vary with the type, position, and design of the part as well as individual preferences.

Various options are available to improve the appearance of BDA aluminum trim. Alloy 5657 is clearly the best choice where high strength is not required. If the mechanical properties of 5252 (~25% higher than 5657) are necessary, either high purity 5252 alloy with the conventional BDA process or standard 5252 alloy with the improved BDA process will upgrade the appearance. High purity 5252 involves an increase in cost whereas the improved BDA process does not.

The U.S. automotive industry is still investigating whether thinner anodic coatings will perform adequately on the car. These tests coupled with the longer term exposure programs should provide the data necessary to decide on the suitability of the improved BDA process.

APPENDIX A

Quantifying Appearance - Many different terms such as brightness, luster and color, which mean different things to different people, are used to describe the appearance of BDA trim. Before proceeding with research studies, we needed to define what appearance characteristics were critical and attempt to quantify them.

A literature review suggested that the two most important appearance characteristics are image clarity and color. Image clarity refers to the distinctness of an image reflected off the surface. Color is subjective and refers to the chrome-like bluish color or, with reference to a BDA surface, freedom from or lack of milkiness or bloom.

We chose a measuring technique reported at SAE in 1965.* A goniophotometer is used to measure distinctness of image (D/I) and bloom or, as we shall call it in this report, color. A commercial instrument** is available to make these measurements on flat panels.

The method consists of directing a beam of light onto the surface at an incident angle of 30° and measuring the relative amount of light reflected at various angles compared to the reflectance angle. Image clarity is measured at 10 minutes from the specular angle and color at 2° from the specular angle. This method is based on the light scattering characteristics imposed by surface heterogenity on the incident light beam.

Since anodized aluminum surfaces exhibit directionality, the values reported represent the average of measurements along and across the rolling direction of the aluminum.

*W. H. Tingle and D. J. George, "Measuring Appearance Characteristics of Anodixzed Aluminum Automotive Trim," SAE, May 17-21, 1965.
**Dorigon D47 Glossmeter, Hunterlab Assoc.

REFERENCES

John H. Powers, "Enhanced Aluminum Auto Trim with Improved Luster, Color, and Texture", SAE, February 27-March 02, 1984.

W. C. Cochran and F. Keller, Trans. Amer. Electroplaters Society, 1961, Vol. 48, p. 82.

W. E. Cooke, "Factors Affecting Loss of Brightness and Image Clarity During Anodizing of Bright-Trim Aluminum Alloys in Sulfuric Acid Electrolyte," Plating, Vol. 49, Nov. 1962, p. 1157.

W. H. Tingle and D. J. George, "Measuring Appearance Characteristics of Anodized Aluminum Automotive Trim," SAE, May 17-21, 1965.

Dorigon D47 Glossmeter, Hunterlab Assoc.

COATINGS
AND SURFACE MODIFICATIONS
FOR ENERGY SYSTEMS

CERAMIC COATING ON ENGINE PARTS

Yoshihiro Suzuki, Shigeru Tanoue, Hiroshi Yamaguchi
Izumi Automotive Industry Co., Ltd.
620, Onakai Kawagoe Saitama, Japan

ABSTRACT

Ceramic coatings on diesel engine parts can be regarded as promissing to raise the reliability of engine parts, because the advantages of both metals and ceramics can be combined in the coating on metals. The post densification process developed by Kaman Sciences Corporation offers one of such coating possibility for engine parts. Some characteristics of the process,such as wear resistance, bond strength and corrosion property, are investigated. Some results of engine test are also reported.

THE IMPROVEMENT OF ADVANCED CERAMIC MATERIAL is recently remarkable and many attempts are already made to use ceramics as engine parts to raise the thermal and mechanical efficiency and to increase its durabilty. Although ceramics has many advantageous properties, such as high wear and corrosion resistance, high thermal strength and low heat conductivity etc, application of monolithic ceramics to engine parts is at present not always successful, because of the large variation of mechanical strength and difficulty of inspecting inherent defects in the ceramic material.

Ceramic coating by plasma spray on metal parts is another approach, in which the reliability of metal and the inherent excellent properties of ceramics can be combined.

The post densification process developed by Kaman Sciences Corporation offers unique advantages which could be attractive for engine parts.(1)※ Both porous monolithic ceramics and sprayed layer can be densified by the process to increase its hardness and mechanical strength. It is necessary to investigate the properties of the new process in order to evaluate the possibility of applying it to engine parts.

VARIOUS PROPERTIES OF POST DENSIFIED COATINGS

POST DENSIFYING PROCESS-- K-Ramic process, trade name of Kaman Sciences Corporation, can be applied to three cases.

The Silica, Chromia, Alumina system (SCA) utilizes chromium oxide to make a fine and extremely hard ceramic coating on the metal substrate. The SCA coating can be applied on the substrate by spraying, dipping or draining. The Coating thickness ranges from 30 to 130μm. The coating is then heated in the furnace at 510℃, to convert CrO_3 to Cr_2O_3. Densifying process,immersing the coating in the chromic acid solution and firing it, is repeated until all the pores are densified by Cr_2O_3. This increases the hardness and strength of the coating.

※ Numbers in parentheses designate references at the end of paper.

In Chromia Combined Alumina system (CCA), impregnation process to immerse the fired refractories and then to fire it is repeated until all the voids in the refractory are filled with Cr_2O_3. This will increase the wear resistance and strength of the refractory with maximum dimensional accuracy.

In Densifying Hard Chrome plating system (DHC), the impregnation process is repeated to fill the crazed network of electrochemically plated hard chrom layer to increase corrosion property.

In the following test, the SCA coating is applied on the test pieces to the thickness of about 100µm. The impregnation process is repeated 10 times to fill the voids in the coating.

BONDING STRENGTH---Any kind of coatings can prove its capability, so long as they remain attached to the metal substrate. Therefore, bondings strength was the first object for investigation. Diesel engine parts which constitute combustion chamber must endure not only mechanical but also thermal load. From the above reason, bonding strength test at room temperature and thermal shock test were conducted.

Fig.1 shows the bonding test. After applying SCA coat on the test piece, the test piece was bonded to the other test piece without SCA coat by the epoxy adhesive, and the tensile strength was measured. As shown in Table 1, all the test pieces were broken at the point of epoxy bonding layer.

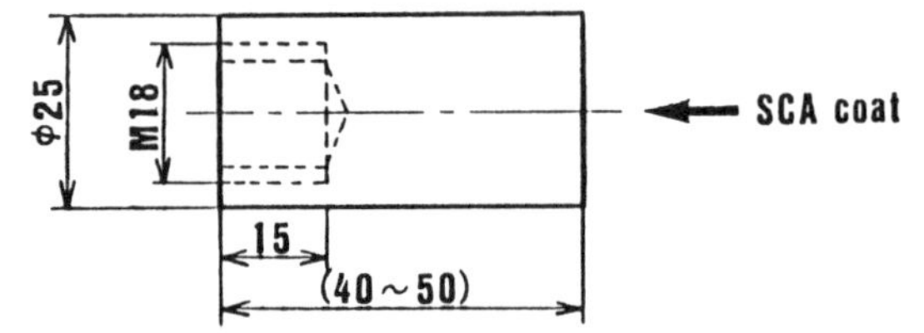

SCA test piece for bond strength

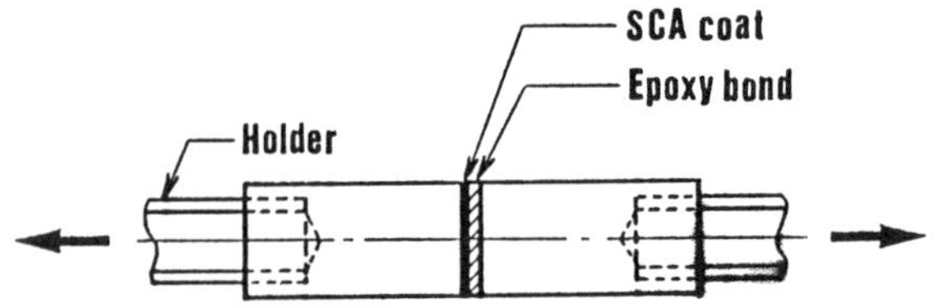

Fig.1 Bond strength test

As stronger adhesives than the epoxy were not available, the real bonding strength was not measured. But it can be concluded from the test that the bonding strength of SCA coat is higher than 70N/mm².

Table1 SCA bond strength

No.	Test piece	Load	Break position	Bond strength
1	SCA	2960 kg	Epoxy bond	>59.1 N/mm²
2	↑	3310	↑	66.2 ↑
3	↑	3160	↑	63.1 ↑
4	↑	2730	↑	54.6 ↑
5	↑	3320	↑	66.3 ↑
6	Without SCA	3075	↑	61.3 N/mm²
7	↑	2900	↑	57.9 ↑
8	↑	3300	↑	66.0 ↑
9	↑	2530	↑	50.6 ↑
10	↑	3585	↑	71.7 ↑

Table 2 SCA thermal shock test

O:No crack
X:Cracked

Thermal shock \ Test piece	Ring[1)]	Steel pipe[2)]
600℃ → Water	O	O
700℃ → Water	O	X
800℃ → Water	O	
900℃ → Water	X	

Note:
1) Material:Stainless steel,OD:78 mm,ID:63, SCA:50µm,inside, length:35mm
2) OD:107mm,ID:105,length:204, SCA:50um,inside

Heat shock test was conducted by applying SCA coat on two types of test pieces and suddenly immersing them in the water from various temperatures. Table 2. Test pieces are; thick ring from stainless steeel and thin steel pipe. The SCA coat was applied in the bores of the both test pieces. After immersing the test piece in the water, the bore was checked if the cracks were produced in the SCA coat. The SCA coat applied in the ring could stand the heat shock from 800℃ to 20℃ water, whereas SCA coat in steel pipe only from 600℃. The lower heat shock resistance of SCA in steel pipe can be attributed to the deformation of the pipe.

WEAR AND CORROSION RESISTANCE---Through Cr_2O_3 impregnation, the hardness of porous ceramics or SCA coating can be extremely high up to 1300HV. The resulting wear reisitance is expected to be very high. To evaluate the wear property, wear test was conducted by pressing the rotor of cast iron against test pieces (See Table 3). Fig. 2 shows the test result. The SCA coating has much better wear resistance than the Cr-plating.

Table 3 Conditions for wear test with oil lubrication

Test piece	SCA (80 μm)
↑	Hard chrome (100 μm)
Rotor	Special cast iron
Rub distance	570 m
Rub speed	0.121 m/sec.
Max. load	18.9 Kg
Lubricating oil	SAE C-C grade #30, 2.5ml/min. drop

Fig. 2 Wear test result under oil lubrication

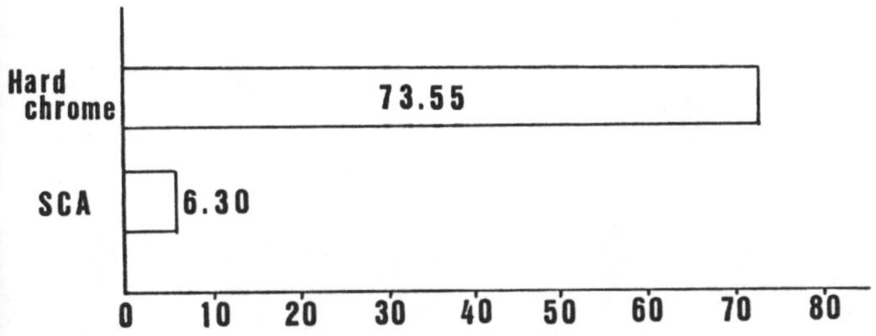

Corrosion property of the system against various chemicals is already reported in other paper(1)※. In following test the SCA coating was exposed to the corrosive atmosphere in which the salt solution was sprayed on the coating and the corrosion attack was observed. The test was conducted 4 times with differrent thickness levels. See Table 4 and 5. It is concluded that the SCA coating does not protect the metal substrates from humidity and salt, because the coating is mostly peeled off due to the extreme rust on the substrate. The other wear test has revealed that the wear resistance of the SCA coating is remarkably reduced, if water is added as a lubricant. The SCA coating of about 80μm thickness is completely worn off at the same test condition as shown in Table 3, but with water lubrication.

Table 4 SCA test piece for salt spray test

Test	1st	2nd
Size	150×70×1 (mm)	150×70×4.5 (mm)
Substrate	Steel (ASTM 40)	←
Blasting	Alumina grit	←
Coating	Spray	←
Coat thickness	100 μm 250 μm	150 μm 250 μm
Dip-bake Cycle	Impregnation: 10 times 510℃-1Hr	←

Table 5 Salt spray test condition

Salt sol. concentration	5 ±1%
Air pressure	0.7~1.8Bar
Ambient temp.	35±2℃
Spray temp.	33-35℃
Spray feed	0.5-3.0ml/80cm²/Hr
Test hours	500Hr

ENGINE TEST RESULTS

Some engine tests were conducted to see if Cr_2O_3 impregnation process offers some improvements.

The SCA coating of about 0.1mm thickness was applied on the cylinder bore and tested in various engines. All the engine tests revealed that the wear resistance of the cylinder bore was remarkably increased and breaking off or peeling off of the SCA coating was not observed. But the wear of the Cr-plated piston ring was, on the contrary, increased. For applying the process for cylinder liners, the mating partner, the piston ring material, should carefully be selected. Roy Kamo et al(2) suggests the promissing material in this direction.

The precombustion chamber cup is sometimes eroded by corrosive combustion gas. To protect the cup from the corrosive attack, the SCA coating was applied and tested in the engine (Fig.3).

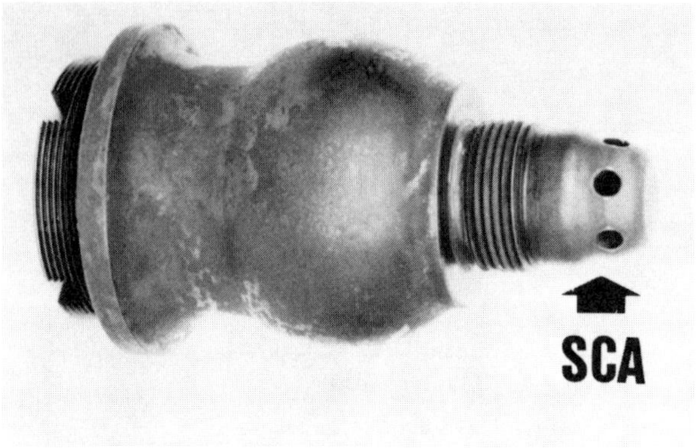

Fig.3 Precombustion chamber cup with SCA coating

The protection of the coating was proved excellent and the coating is applied for serial production.

SUMMARY

1. The post densified Cr_2O_3 coating has excellent anti-wear property with high bonding strength to the substrate.
2. Regarding corrosion resistance, the application of the coating should be checked case by case.

References

1. J.Car,J.Jones, "Post Desified Cr_2O_3 Coatings for Adiabatic Engine," SAE Paper 840432, 1984.
2. P.Moorhouse, B.Mortimer, R.Kamo, "Solid lubrication Studies for Adiabatic Diesel Engines, SAE Paper 850508, 1985.

SURFACE MODIFICATIONS OF PISTONS AND CYLINDER LINERS

Yoshihiro Suzuki
Izumi Automotive Industry Co., Ltd.
620, Onakai Kawagoe Saitama, Japan

ABSTRACT

With the higher BMEP of a diesel engine, pistons and cylinder liners suffer from increasing mechanical and thermal loading, which causes several problems on these vital engine parts. The main critical problems are thermally induced cracking on piston head and scuffing on cyliner bore. Hard anodizing the piston head which is currently the most effective counter-measure against heat cracking is described. Another promissing method to reinforce the piston head by SiC-whisker is also reported. The new process to improve the surface lubrication of the cylinder liner was developed. The bore has numerous finely distributed micro pits which act as very good oil reservoir. This improves antiscuffing property of the cylinder liner remarkably.

TURBO CHARGING THE DIESEL ENGINE is a very effective way to raise the engine rating and reduce fuel consumption. The resulting higher mechanical and thermal loading will create several ploblems e.g. piston pin bore cracking, thermal crack on the head, higher wear in the ring groove etc. of the pistons, or higher wear and scuffing of cylinder bore.

Thermal cracking at the combustion bowl edge of the DI pistons which is caused mainly by cyclic thermal loading is a troublesome obstacle which must be overcome for most of the modern highly rated diesel engines. Limit of aluminium piston alloys is already reached and the normal countermeasure e.g. large radius at the combustion bowl edge cannot always be applied because of the deteriorated thermal efficiency. Lowering the piston temperature by introducing the cooling oil in the cast-in cooling cavity in the piston head is often effective, because the strength of the aluminium alloy will be increased in proportion to the reduced temperature. But resulting steeper temperature gradient will produce higer thermal stress which may in some cases increases the danger of thermal cracking. Hard anodizing the piston head can generally be recommended as a solution against the thermal cacking with the competitive production cost. It is therefore worthwhile to consider the process if one has thermal cracking problems. But the best solution of today cannot always be efficient enough for future engines. Some possible alternatives are therefore reported here e.g. squeeze casting and fiber reinforced material.

The increased maximum combustion pressure and heat release by turbo charging are making the lubricating condition between piston ring and cylinder worse and worse, resulting in scuffing and increased wear of

both piston rings and cylinder liners.
Honing the cylinder liner with cross hatch micro grooves has been the most popular and effective surface treatment to improve the running in characteristic and to keep the lubricating oil consumption within the acceptable range. But scuffing problem cannot always be completely solved, especially at heavy duty diesel engines. Plateau honing or lubriting would be regarded as a better countermeasure against scuffing, especially during running in phase of an engine. In some cases, however, there is a possibility where scuffing tends to be caused after honing structure has been disappeared. The scuffing in this case cannot always be completely avoided by plateau honing or lubriting.
The cylinder liner with finely distributed micro pits on the bore proved to be very effective against scuffing and would be a potential alternative surface treatment for cyliner liners of heavy duty diesel engines.

HARD ANODIZING

MECHANISM OF THERMAL CRACKING ON PISTON HEAD---Many factors have influence on the formation of thermal cracking on the combustion bowl edge of highly loaded diesel pistons. The following are the main factors.

Material strength at higher temperature---The combustion gas flowing out from the combustion bowl in the piston head at the TDC of the piston has its maximum speed around the combustion bowl edge, resulting in the highest heat transfer into the piston at the bowl edge. Therefore the highest temperature in the piston will appear at the bowl edge, ranging normally from about 300℃ to 370℃, depending on the engine and piston types. At this temperature range, the aluminium alloys lose its strength considerably (see Fig.7 P.6). Softening of the piston material promotes the formation of thermal cracking.

Steady state thermal cracking---The temperature distribution in DI pistons at full load is schematically shown in Fig.1. The expansion of the material at the hotter inner zone is restrained by the cooler outer zone, which causes thermally induced compressive stress at the hotter zone. If the stress is higher than the yielding point of the material, the tensile stress will be produced at the inner zone when the piston is cooled down. This cycle overstrains the material and the cracking will finally be produced.

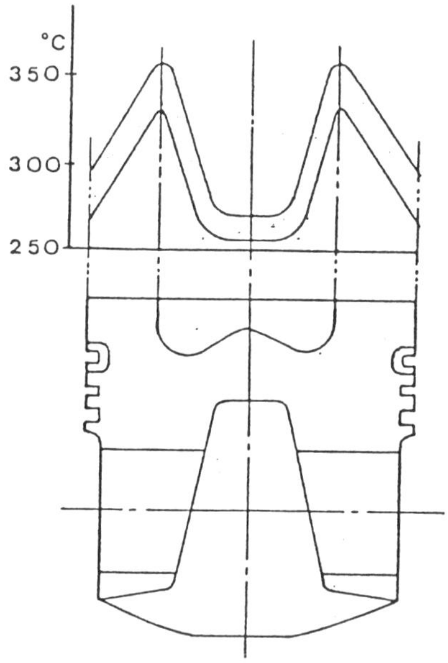

Fig.1 Temperature distribution on the piston head of direct injection type (1)*

Transient thermal stress---As the load condition of an engine changes, the temperature distribution in the piston will also be changed, which creates unsteady state thermal stress. The transient thermal stress is reported to be rather bigger than the steady state thermal stress(2).

Mechanical stress---Under the combustion gas pressure, the stress distribution shown in Fig.2 is produced at the combustion bowl edge of the piston. The compressive stress in thrust- and antithrust direction is overlapped with the compressive stress

*: Numbers in parentheses designate References at end of paper

under thermal loading, which promotes the formation of thermal cracking in this direction. This coincides with the practical experience of preferred occurrence of thermal cracking in this direction.

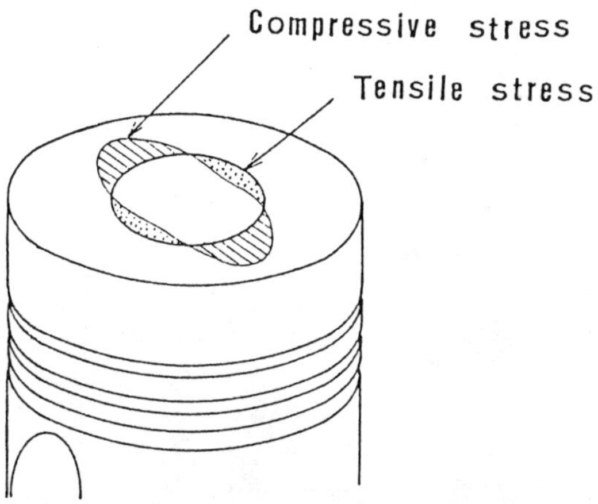

Fig.2 Stress distribution at the combustion bowl edge of the DI piston under combustion gas pressure

HEAT SHOCK TEST---Final evaluation of the thermal cracking on the piston head should be made in the engine, as many factors influence the formation of the crack as described above. However, the appropriate test rig which can simulate the thermal loading on the piston in the engine should be chosen, in order to develop effective countermeasures against the thermal cracking.

Fig.3 shows the test rig to heat up the combustion bowl edge of the piston by induction heating and to cool down it by air blowing.
On the test rig, thermal cracking can be reproduced just like the one in the engine. Fig.4 shows some test results obtained on the test rig. By improving the mechanical strength of the material, such as by squeeze casting or forging, heat shock resistance can be increased. Hard anodizing is most effective to reduce thermal cracks. In the actual engine operation, it has been proven that the life of the piston can be improved by 3 to 7 times by hard anodizing.

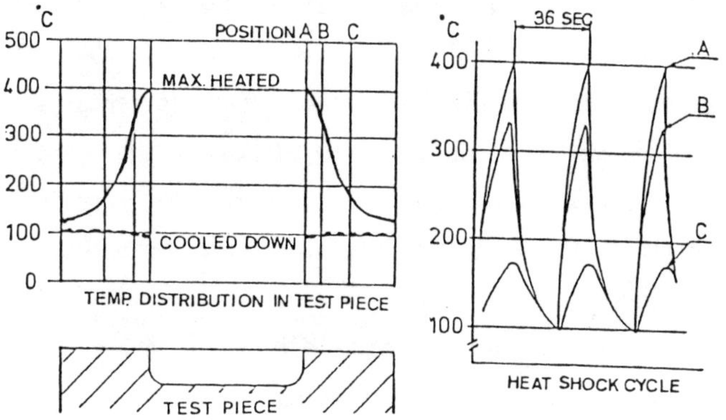

Test condition

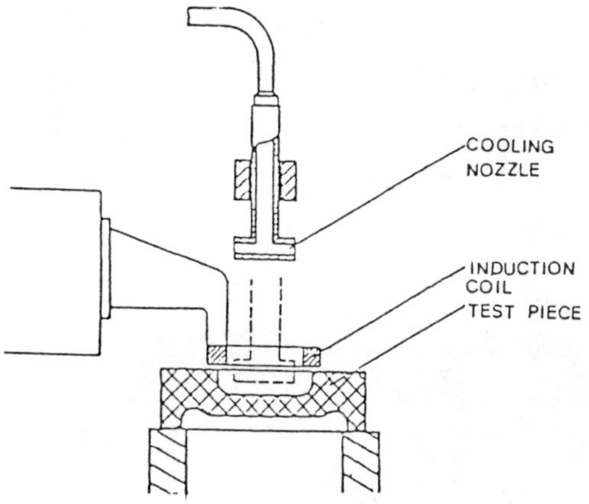

Test apparatus

Fig.3 Heat shock test of piston head by induction heating

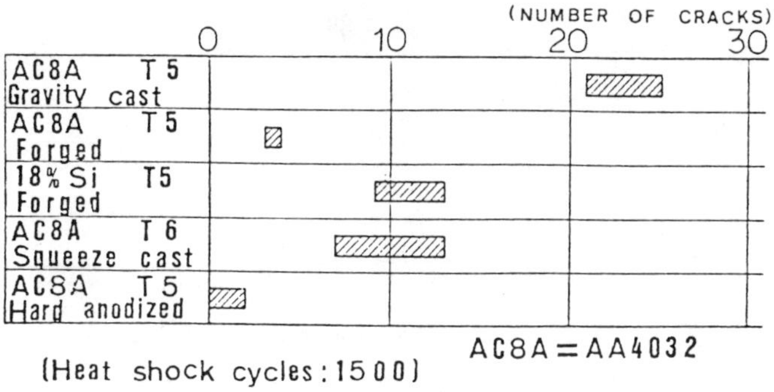

Fig.4 Heat shock test results
---effect of processing and hard anodizing---

HARD ANODIZING---Hard anodized layer on aluminium parts is produced in the electrolyte of either organic or inorganic acid. The process is normally applied on the aluminium parts to improve the wear and corrosion properties. For applying the process on the piston head, the following aspects should be considered.

Adhesion to the ground material---In case of sprayed ceramic film or electrically plated layer on aluminium, the layers are deposited on the substrate and the bonding force is produced mechanically or chemically. Contrary to these layers, the anodized oxide layer is produced in the aluminium itself, which explaines excellent durability of the anodized layer in the engine. The layer grows in both directions; into and above the aluminium.(Photo.1)

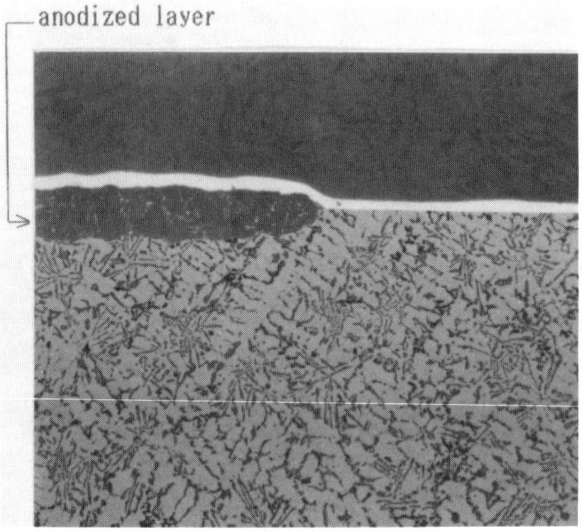

×100×0.85

Photo.1 Growth of anodized layer on Al-Si alloy

The phenomenon that half of the anodized layer thickness swells above the aluminium should be considered for applying the process to pistons i.e. increased compression height.
The smaller bowl edge radius of the DI piston is preferred to make best use of the micro swirl around the combustion bowl edge for better fuel economy and emission. But the anodized layer tends to have cracking(see Photo.2) during processing, if the radius of the bowl is too small. To avoid such cracking during hard anodizing process, the radius should be bigger than 0.4mm.

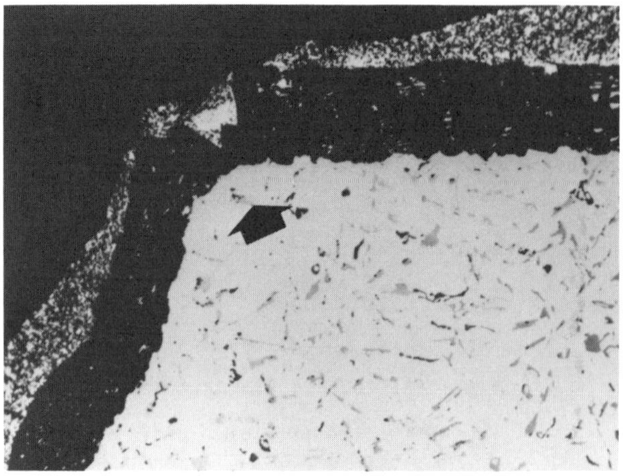

×200×0.85

Photo 2 Crack in anodized layer caused during anodizing process

Even if the radius is bigger than 0.4mm, there is a possibility that the cracking in the anodized layer should occur during engine operation, because the anodized layer expands far less than the aluminium at higher temperature. Photo 3 shows such example after extremely severe engine test for 500hrs under abruptly changing load conditions.
The test was so severe that the edge temperature was higher than 400℃ at full load and the anodized layer could not prevent from the thermal cracking at the bowl edge of the piston in thrust- and anti-thrust direction. The cracking in the anodized layer itself was also produced in circumferential direction of the bowl due to the big difference of

expansion coefficient between anodized layer and the substrate, but peeling-off of the layer is not observed, which explaines excellent bonding of the layer to the substrate. The cracking in the anodized layer propagated into the aluminium of about 1mm depth. But it is assumed that the crack in the anodized layer did not induce the thermal crack in the piston in thrust- and antithrust direction, because the directions of the two cracks are not the same.

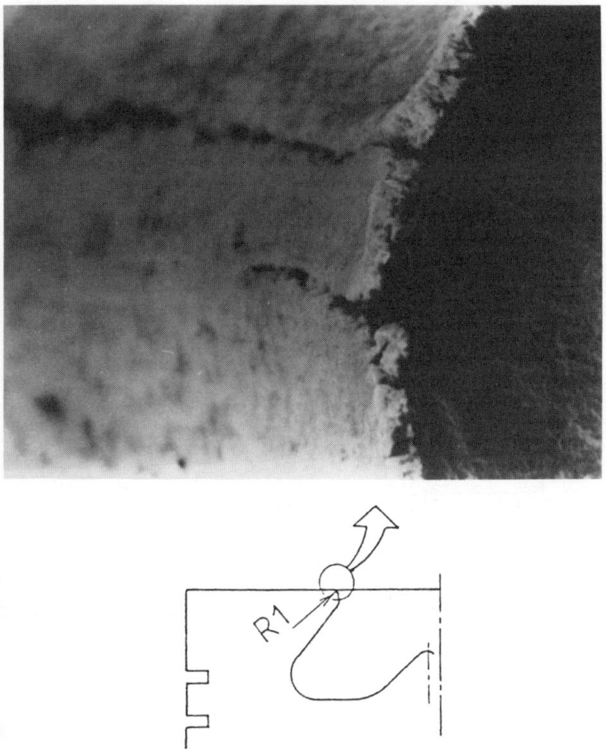

Photo.3 Cracks in anodized layer caused during engine operation

Layer quality and thickness---Through inadiquate anodizing process, the quality of anodized layer can be deteriorated. Photo 4 shows such example. During normal anodizing process, the silicon crystals in the aluminium matrix are not oxided and embedded in the anodized layer. But in this case, the eutectic silicon crystals and anodized layer around silicon crystals are dissolved in the electrolyte and the porous anodized layer without silicon crystals is produced.

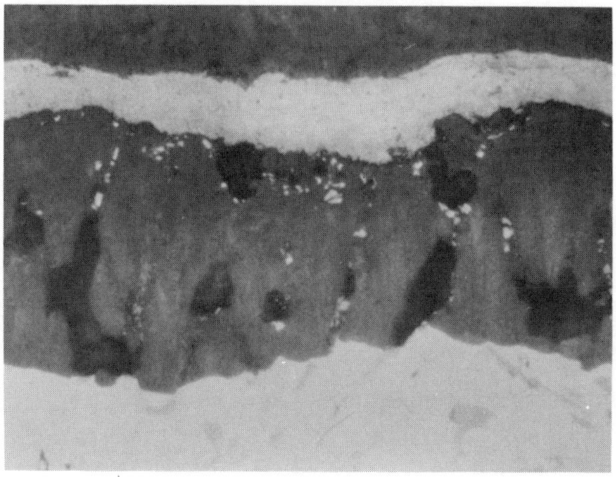

Photo.4 Porous anodized layer

This ill-produced anodized layer is not so effective as normal anodized layer in preventing thermal cracking. See Fig.5

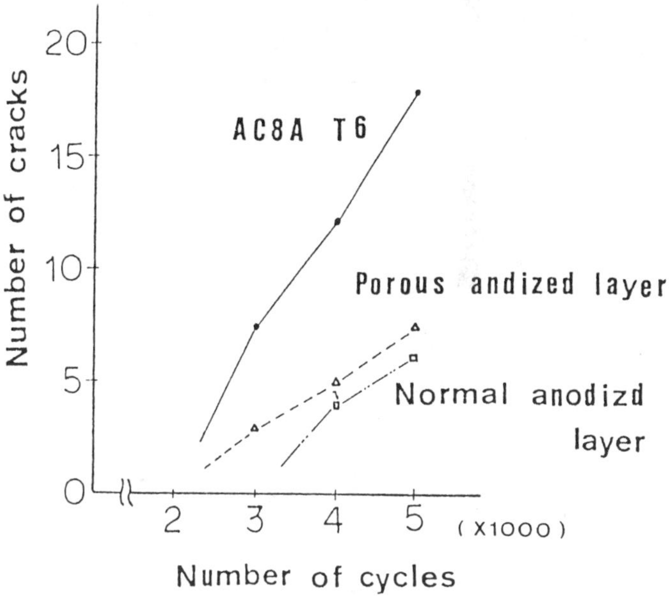

Fig.5 Heat shock test for anodized layer quality

The thickness of the anodized layer has an influence on the effectiveness against thermal cracking. The thickness of about 90μm will give the best protection to the piston head. See Fig.6

Further improvement of the heat shock resistance for the piston head can be achieved by anodizing the squeeze cast aluminium. During the squeeze casting process, the very high pressure of about 1000bar is applied to the aluminium melt. Through the accelerated solidification speed of the melt by squeeze casting, the grain size of the Al-Si eutectic will considerably be smaller than that of gravity die-cast eutectic.

The anodized layer of such fine matrix will also be dense and hard. This is the reason for considerably improved heat shock resistance of anodized layer on the squeeze-cast aluminium. It must be noted here, however, that the squeeze-cast bare aluminium may not always be more effective than anodized gravity die-cast aluminium as shown here in Fig.6. The test results are obtained on the test piece of ϕ 110×35mm. With such small test piece, solidification speed of the melt during squeeze-casting is higher than the actual piston. In case of squeeze-casting for actual piston, the solidification speed may considerably be slower than with the small test piece, depending on the size and gating system for the sqeeze

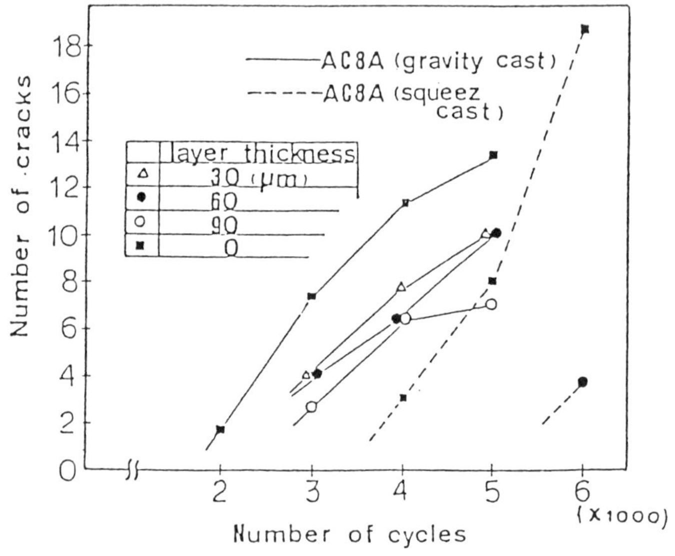

Fig.6 Heat shock test

---Effect of anodized layer thickness and ---

squeeze cast

casting. Therefore the effectiveness of the squeeze casting shoud be checked case by case.

FIBER REINFORCING---By selecting proper reinforcing material, the mechanical strength of the aluminium alloys can remarkably be improved. SiC whisker is one of such promissing material. Fig.7 shows how mechanical strength of the aluminium alloy can be improved by SiC-whisker especially at higher temperatures. Heat shock resistance of aluminium alloys can also be increased by SiC reinforcing. See Fig.8.

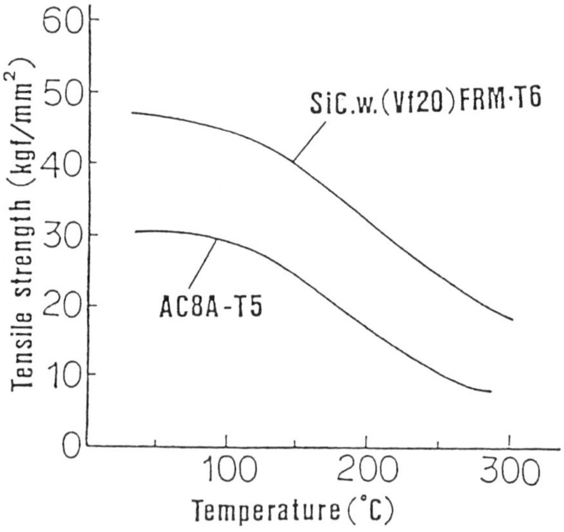

Fig.7 Tensile strength of aluminium alloy and FRM at higher temperature

For evaluating heat shock resistance of the aluminium piston, the thermal conductivity of the material should also be considered. If the thermal conductivity of the material is worse, the temperature gradient in the piston will be bigger at a given loading of an engine, resulting in higher thermal stress. SiC is one of the best ceramics with good thermal conductivity. Fig.9 shows that the SiC reinforced material with higher volume fraction shows as good a thermal conductivity as the alumina fiber reinforced material with considerably lower volume fraction. This means also that the final decision to select the best solution against

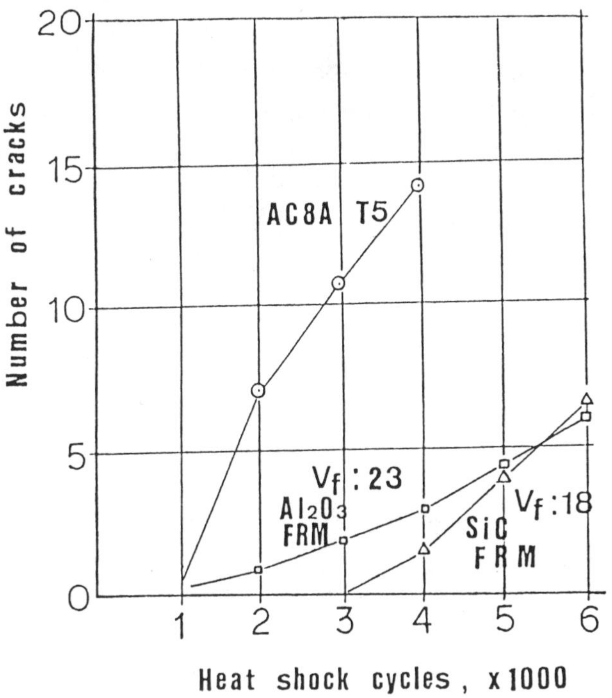

Fig.8 Heat shock test results of FRM

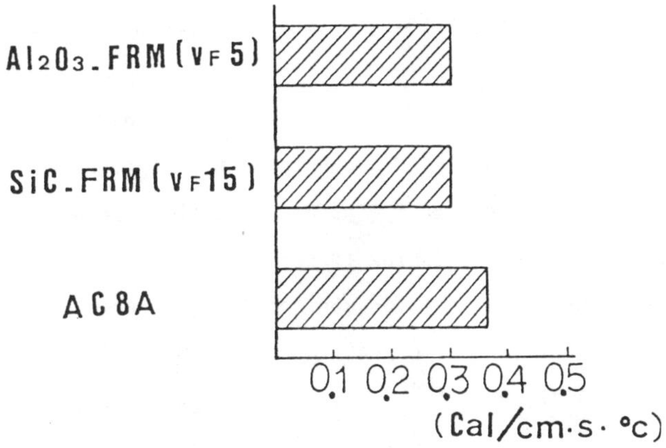

Fig.9 Heat conductivity of FRM

thermal cracking should be made in the engine. In one severe engine test, it is already proven that the pisotn reinforced by SiC whisker could endure 700 hours without thermal cracking, whereas the anodized piston had cracking after 400 hours. Considering that the reinforcing the aluminium improves the mechanical property at higher temperature, the FRM piston may offer a good solution for future heavy duty engine with higher maximum combustion pressure.

CYLINDER LINER WITH MICRO PITS

SCUFFING OF CYLINDER BORE---With increasing combustion pressure and thermal loading of the diesel engine, many efforts have been concentrated for better design and material of piston rings and cylinder liners to leave minimum oil film between these two mating parts, so that the possibility of scuffing and increased wear should be avoided, keeping the lubricating oil consumption as low as possible.

The scuffing of the cylinder bore can be classified into three phases according to the time of occurrence. The first phase is during running-in period. The second phase is the phase after running-in. In the second phase the bore has no honing structure more which retains lubricating oil to avoid scuffing. Therefore it is important that bore material has a good metallurgical structure; good shaped graphite for good lubrication, tough matrix which avoids adhesion and deformation and hard itermetallic compound e.g. steadite to reduce wear. The scuffing in the second phase may often occur, when the graphite is smeared by the deformed matrix and do not appear on the bore surface. In the third phase scuffing is often promoted by increased blow-by when the wear of piston ring groove, piston ring and cylinder bore is considerably increased.

Among the three phases, it is of utmost importance to avoid scuffing during running-in, because even the slightest scuffing in the first phase will deteriorate the bore and ring surfaces, which increases the wear and may shorten the engine life.

The most common methods for better running-in are, honing with cross hatch micro grooves, plateau honing and/or lubriting of the cylinder bore.

SIC IMPREGNATION PROCESS---The cylinder bore with impregnated SiC particles has excellent wear resistance property.(3) This process can be applied to almost all the metal parts to increase wear resistance. Fig.10 shows how the wear of cr-plating can further be reduced by SiC impregnation. For impregnating SiC, the cylinder bore is lapped by using lapping oil with suspended SiC particles. In the pursuit of applying this process to many metals, it was found that even mild steel can be a good cylinder liner material. For diesel engines, cast iron is at present the only suitable material for cylinder liner. The flake graphite acts as a good solid lubricant. The matrix consists of fine pearlite; slightest appearance of ferrite is not preferred because ferrite induces adhesion to mating material. However, the impregnation process converts the most unlikely material, mild steel of ferrite and without graphite, to very promissing material for cylinder liners. Fig. 11 shows the engine test results after 250 hours comparing cylinder liner of mild steel impregnated by SiC particles with cast iron cylinder liner which is also impregnated by SiC.

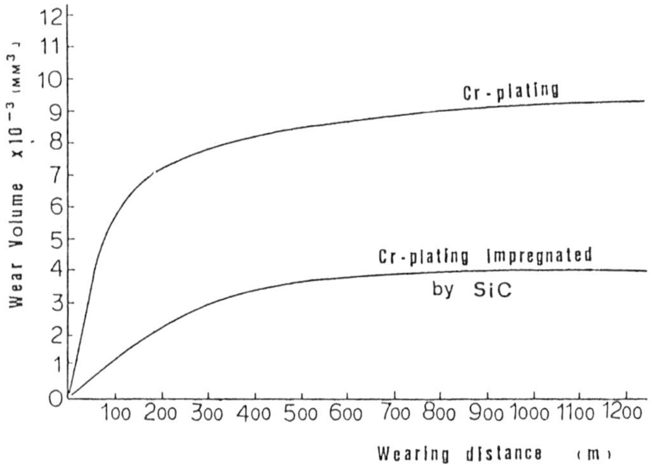

Fig.10 Wear test of Cr-plating and Cr-plating with SiC impregnation

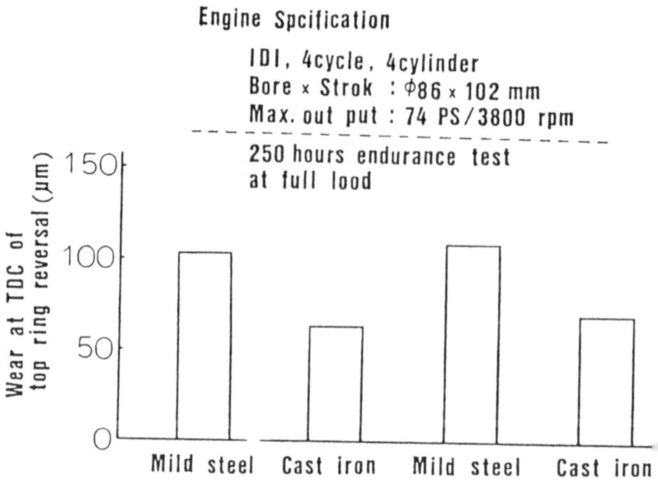

Fig.11 Wear of SiC impregnated cylinder liner in the engine

The wear of mild steel liner is a little bit higher than that of cast iron liner. But considering that the wear of cast iron liner is already reduced by SiC impregnation, it can be said that the wear of mild steel liner is rather good.

The mild steel cylinder liner impreganted by SiC could stand another severe scuff test in the engine with higher cooling water temperature.

CYLINDER LINER WITH MICRO PITS---Encouraged by the excellent anti-scuffing property of SiC impregnated mild steel, the reason for the good antiscuffing property was investigated. During impregnation process, the rugged surface is created, which seems to act as a good oil reservoir.The impregnation process with other ceramic particles offerred also good anti-scuffing property. Then the efforts have been concentrated to distribute the micro pockets of the same size. It is expected that very high oil pressure would be created in the separately distributed micro pits, when the piston ring travels on the bore. Because the oil retained in the pits cannot flow out. The high oil pressure will avoid scuffing. This was achieved by choosing the appropriate lapping pressure with proper particle size.

Photo.5 shows the surface structure of cylinder liner with finely distributed micropits.
The black polygons are the micro pits of about 0.03 to 0.1 mm. Fig.12 shows surface roughness which looks like the one of plateau honed structure.

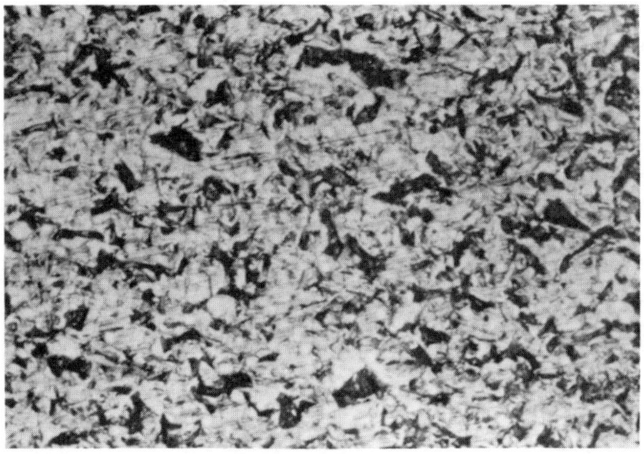

×100×0.85

Photo.5 Surface structure of micro pits cylinder liner

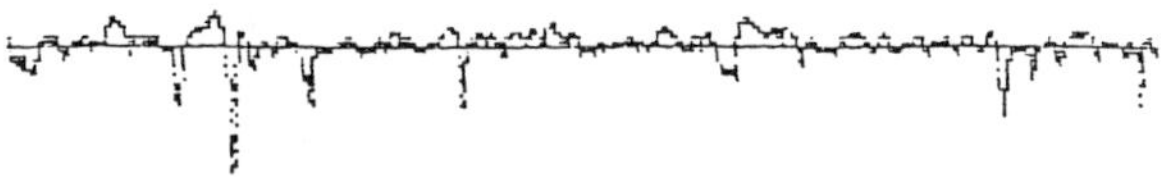

Magnification Horizontal: ×60×0.85
Vertical: ×2000×0.85

Fig.12 Surface roughness of micro pits cylinder liner

The cylinder liner was tested in the heavy duty diesel engine. Bore dia.160mm, 105ps/cyl. The engine had the problem that the scuffing of cylinder bore occurrs after 100 to 150 hours operation at full load, when honed structure just disappears at the top ring reversal. Lubriting the cylinder liner, which is conventionally the most effective process against scuffing during running-in period could not be the solution for the scuffing. The micro pits cylinder liner could stand the severe endurance test for more than 500 hours without problem. The reason would be the pits remaining even after 300 hours operation.

SUMMARY

1. Hard anodizing is the very effective current process to avoid thermal cracking in the piston for heavy duty diesel engines.
2. Hard anodizing of the squeeze cast piston can further improve the heat shock resistance of the piston head.
3. Reinforcing the piston by SiC-whisker is a potential candidate for future heavy duty diesel engines.
4. The cylinder with finely distributed micro pits offers excellent anti-scuffing property.

Reference

1) M. Röhrle : MAHLE Kolbenkolloquium (1969), p.18
2) R. Munro, W.J.Griffiths: CIMAC 1979) D46
3) J. E. Tanner; SAE Paper 830063(1983)

PRODUCTION BASED CERAMIC COATINGS FOR DIESEL ENGINE COMPONENTS

A. R. Nicoll, S. Keller
Plasma-Technik AG
Switzerland

G. Wuest
Plasma Technics Inc.
USA

ABSTRACT

The present trends in engine technology are to increase engine ratings, reduce fuel consumption and allow use of lower cost fuels. Ceramics have been investigated to reduce the heat losses from adiabatic and low heat loss diesel engines to increase fuel efficiency and reduce cooling requirements. The majority of ceramic thermal barrier coatings being investigated are well known in the aero engine industry. Experience is being gained at the present time with production coatings of this type in the field of diesel engines. This paper discusses the problems associated with prototype production schemes in this field and describes the production of a 3 mm ceramic thermal barrier coating on a cast iron cylinder head.

INTRODUCTION

The principle function of a thermal barrier coating is to insulate the underlying substrate metal from high temperature conditions (1,2). The heat conductivities of ceramics lie between 0.7 and 5 W/MK (0.7 - 1.4 W/MK for ZrO_2) and compare very favourably with metal surfaces which have values as high as 25 W/MK. Ceramic coatings with the thermal conductivity given above can significantly lower the temperature of a cooled metal component operating in a high temperature environment. This effect can be exploited in several ways. Firstly the flow of cooling medium (air or water) can be kept constant, thereby decreasing metal temperature and increasing the mechanical and oxidation-corrosion life of the component. Alternatively, the cooling medium supply can be appropriately reduced, whereby the mechanical and oxidation-corrosion life of the component remain the same, but the efficiency, in terms of specific fuel consumption is improved. A further benefit is that thermally induced strains are reduced by the insulating effects of the thermal barrier coating which results in a significant improvement in thermal fatigue resistance.

Thermal barrier coatings are generally plasma sprayed systems comprising an inner metallic coatime layer intended to provide environmental oxidation protection to the component substrate alloy and an outer ceramic or metal-ceramic coating layer with low thermal conductivities intended to provide thermal insulation of the substrate.

The metallic coating layer or bond coat comprises a NiCr, NiAl or MCrAlY (where M = Fe, Ni, Co, CoNi or NiCo) depending on the application and is applied generally using air (APS) or vacuum plasma spraying (VPS) if particularly good environmental protection is necessary. The bond coat is followed by either a cermet (oxide + metal), a graded ceramic-metal or a ceramic layer.

The use of a cermet can give rise to problems with the metering of powder of this type (i.e. differing densities and powder separation in the powder hopper). The grading of 2 powders requires individual powder hoppers and accurate metering systems to produce the correct material distribution in the coating. Figs. 1 and 2 show examples of graded coatings using NiCrAlY and ZrO_2 8 % Y_2O_2 on aluminium and cast iron.

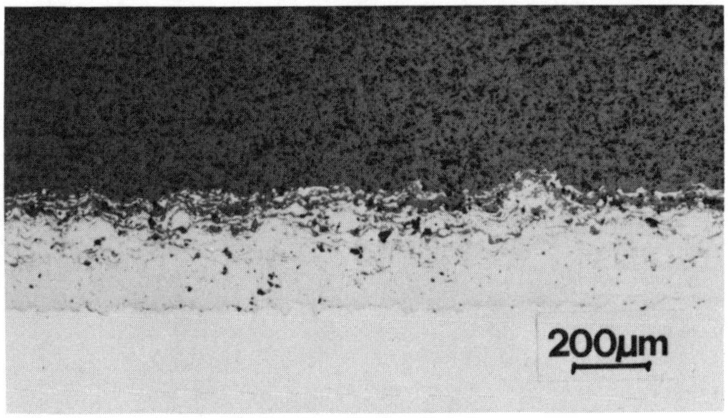

Fig. 1 : Graded coating on an aluminium substrate.

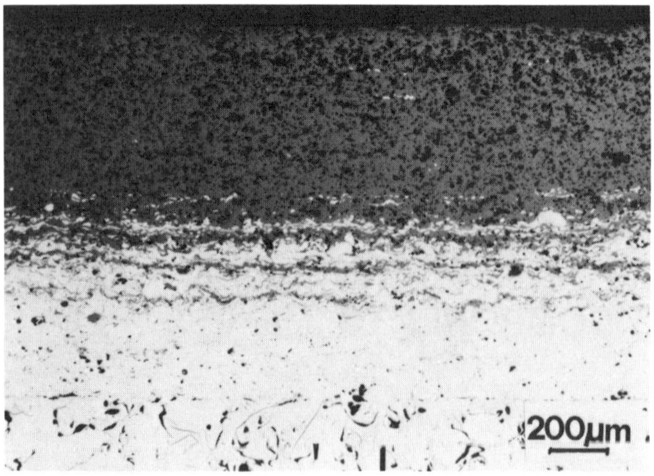

Fig. 2 : Graded coating on a cast-iron substrate.

An example of a triplex system is to be found in the combustion chamber of the Rolls-Royce RB 211 engine (3). This engine has utilized magnesia stabilized zirconia thermal barrier coatings for the last decade. Metal temperatures have been reduced by 50^{0}C or more and the combustor life has been at least doubled. This three layer coating is produced with a nickel chromium bond coat, a nickel chromium-magnesia zirconia inter layer and an outer ceramic layer consisting of zirconia with 24 weight % magnesia. This type of material becomes unstable at temperatures above 950^{0}C as magnesia precipitates from the zirconia, thus increasing the thermal conductivity and leading to the formation of monoclinic zirconia on cooling. Subsequent phase transformation on thermal cycling causes coating failure.

Techniques such as controlled porosity, segmentation, microcracking and residual stress control have been developed in connection with the plasma spray process to increase the thermal shock resistance of thermal barriers. These techniques are based on controlling the powder particle distribution, power levels, torch substrate stand-off, substrate temperature control and the use of a post-coating heat treatment. An excellent review of thermal barriers has been given by Goward (4). Of particular importance is the correlation between melting particles in the plasma and quenching them on impact with air cooling jets thus creating microcracking via rapid solidification (Fig. 3). This can only be achieved with thick coatings using an absolutely constant relation between plasma torch and substrate.

The use of aero engine developed thermal barriers for marine, truck or automobile applications also requires a certain level of thermal barrier corrosion resistance. Large scale programmes (5) have demonstrated that thermal barriers such as ZrO_2 stabilized with MgO and Y_2O_3 and $CaOSiO_2$ are attacked by impurities in the fuel and air such as vanadium, sodium and sulphur. V_2O_5 for example can attack the Y_2O_3 in ZrO_2 causing destabilization.

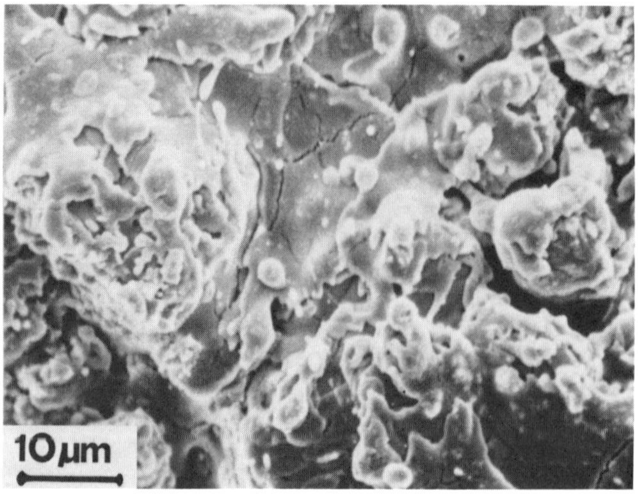

Fig. 3 : Microcracking in plasma sprayed ZrO_2-13%Y_2O_3.

This inturn causes mechanical break down and spalling of the coating. In contrast to the aircraft industry where a single component is normally discussed these engine industries as mentioned previously deal with batch systems involving cylinder heads, valves, and pistons. These components vary in composition and therefore the mechanical adherence has to be considered on these different substrates, examples of which are aluminium, cast iron, steel or nimonic 80 A.

PRODUCTION ASPECTS

As in the fabrication of coatings on aeroengine components, fabrication of coated engine components puts emphasis on large batch or mass production systems. In such cases, priority has to be given to the quality specifications provided for each part of the production process. A process for example normally comprises degreasing, masking, grit blasting, grit and powder control, plasma parameter and part-torch relative movement control and some of these points are discussed below.

In order to gain the advantages provided by ceramic coatings in engine fabrication, consideration has to be given to whether parts are produced in a batch or continuous production mode. A break down of the operations involved in the complete process can be given as follows, degreasing 5 %, masking 5 %, grit blasting 20 %, spraying - bond coat 15 - 20 %, top coat 45 %, mask removal, cleaning 5 %. This indicates that decreasing, masking and grit blasting have some importance in their own right as compared to the actual spraying of the component. Masking involves not only protection of surrounding areas, but also for example with cylinder heads the protection of the inlet and outlet ports and the fuel injection nozzle. Cylinder heads have 4 masking parts which have to be individually placed in position and centered. Obviously this is an area where much attention has to be paid to automatic placement of these types of individual

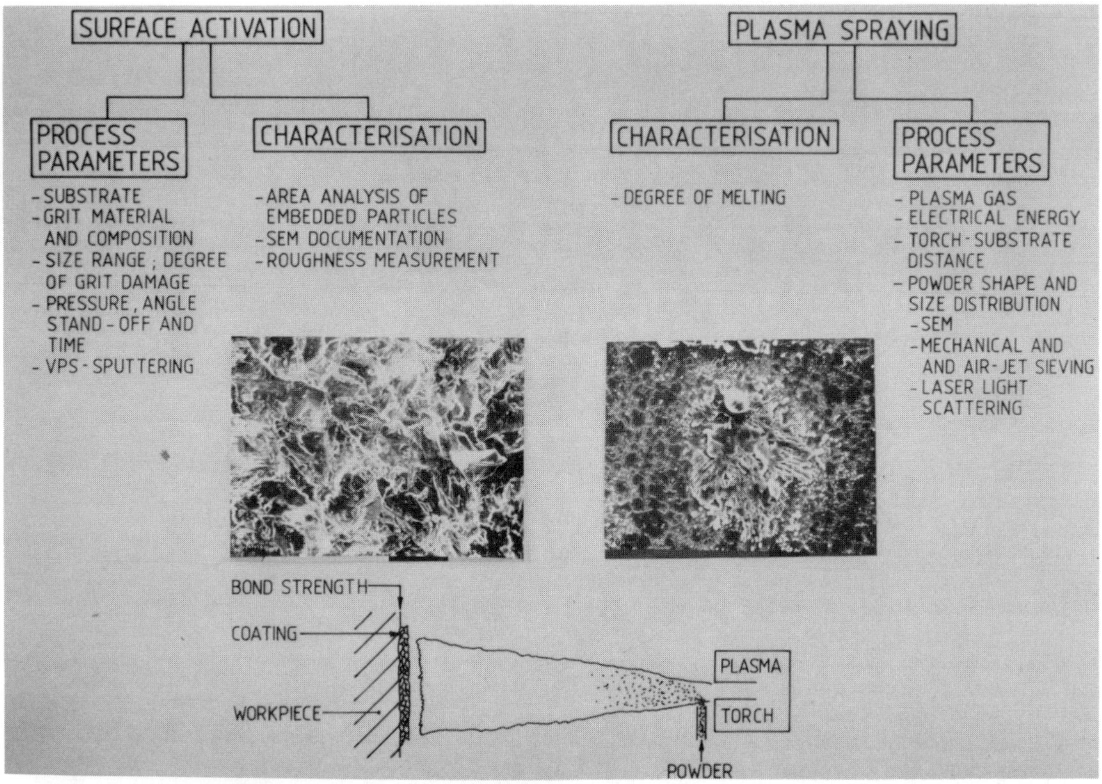

Fig. 4 : Interactions between surface activation and plasma spraying.

masks. A further important point is the treatment of these masks that they can be used repeatedly.

Following masking, parts are grit blasted in order to roughen the surface and create sufficient mechanical adherence between the substrate and the coating. Fig. 4 shows the interaction between the concepts of surface activation and plasma spraying. The process parameters are shown together with the methods of characterization for both surface activation and plasma spraying. It is of great importance that areas to be coated are satisfactorily grit blasted. Control of grit blasting demands that a suitable automatic machine is used which controls air pressures, blast angles, nozzle stand-offs and time of grit blasting. Further quality control standards have to be employed to control the grit size range, the degree of nozzle damage, and the degree of grit damage with a particular substrate.

In the plasma spray process powder is introduced into the plasma at a constant rate where it softens, melts and is accelerated towards the substrate. Therefore quality levels are maintained by controlling the powder feed rate, powder size and distribution, plasma gas flow and plasma arc power level. Constant powder rates can be guaranteed using a volumetric feed system (6). An example of this for NiAl is shown in Fig. 5 over a period of 40 minutes. At no time does the feed rate exceed the +/- 1 % limit. Use of a powder feeder of this type with this level of accuracy guarantees very good consistency in production applications, whether the powder container is full or nearly empty.

The shape of ceramic powder can have a great influence on the spray deposition efficiency. Also the particle size distribution can affect the level of porosity and coating structure. A quality system has to be enforced to ensure that the powder shape factors and distributions are accurately controlled. The laser light particle analysers appear to provide the best production quality assurance pf powder distributions.

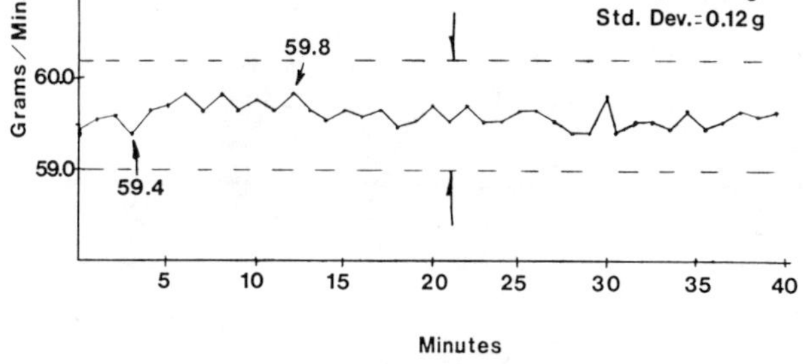

Fig. 5 : Twin 10 powder feed unit at a feed rate of 59,5 g/min with NiAl Powder.

Following grit blasting, components are cleaned of dust using clean compressed air and transferred to the spraying area. In the aircraft industry, units of the type shown in Fig. 6

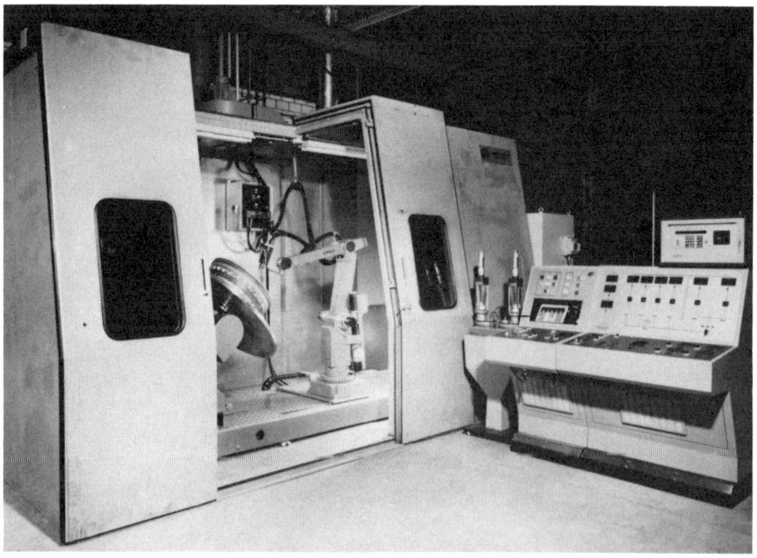

Fig. 6 : Automated plasma spray system.

are commonly used. The full sound enclosure with air inlets, air exhaust systems and anti-dazzle windows ensures environmental control for the operator. The plasma torch is mounted on a programmable robot that can also be positioned relative to the working table. The micro-processor controlled plasma console is seen outside of the sound enclosure. This type of unit with full parameter storage capacity can be coupled to robot programmes and allows fully automatic schedules to be achieved for individual batches of parts.

A robot system has the necessary flexibility with regard to ease of programming and usage with widely differing parts. Fig. 7 shows valves prior to spraying.

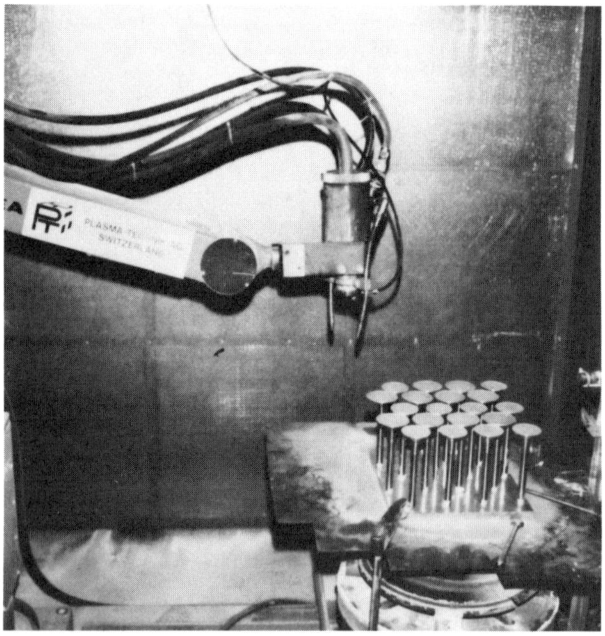

Fig. 7 : Robot spraying of engine valves.

Fig. 8 : Typical piston head for robot spraying.

Fig. 8 shows a piston head, which requires complex programming. Coating thickness on these parts can exceed 1,6 mm of ceramic. Diskette storage of programmes provides for allocation of parts to different machines or to factories at other locations.

Cooling of parts is carried out using air cooling nozzles mounted on the torch. The advantage of the robot arm is that the stand-off is always held constant which means constant coating build-up and cooling conditions. Coating temperatures can be monitored and held constant using infra-red cameras with feedback control to the plasma control console.

As mentioned above, quality control is one of the major factors involved in producing thick coatings. Using a batch process, steel coupons are sprayed with the parts to obtain metallurgical samples. However, evaluation of these parts must be carried out with a standard automatic method of metallographic preparation (6).

Otherwise, defects and over average porosity levels are easily introduced into the metallurgical mounted sections.

The result of such considerations is shown in Fig. 9. This sample was taken from a batch of 20 parts and shows a triplex coating on a cast iron cylinder head. The bond coat is Ni - 16 % Al - 6 % Al - 0,5 % Y and the ceramic coating is ZrO_2 13 % Y_2O_3. The top ceramic layer has a thickness of 1.6 mm. The cermet layer was fabricated using two powder feeders running at different speeds and two powder injectors.

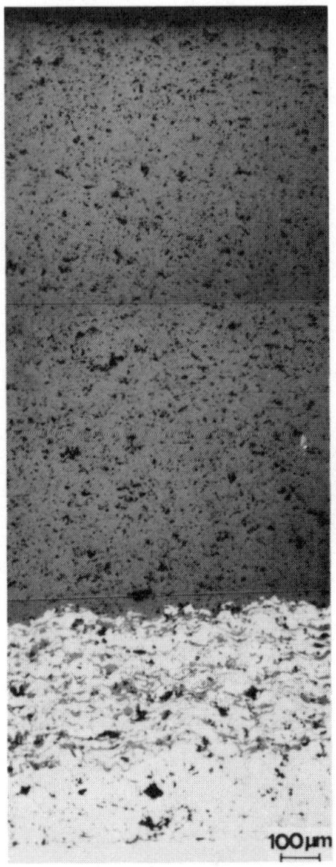

Fig. 9 : Triplex plasma sprayed coating with 1,6 mm ceramic on cast iron.

Following removal of masks, the surface of the coating was ground down flush with the original surface and run in a diesel engine for over 30 hours under extreme test conditions without spalling of the coating. Microprocessor controlled plasma parameters and robot movement provided absolute repeatability for all components.

FUTURE ASPECTS

While confidence in the use of sprayed coatings has increased based on service experience and carefull process control exercised during spraying, the need for component inspection is apparent. It is of importance to be able to detect any lack of adhesion between coating and substrate, excessive porosity or lack of coherence in the coating. In the field of non-destructive testing (7) various methods (ultrasonic C-scan (8), acoustic emission monitoring, optical imaging (9) and thermal wave interferometry) continue to make progress.

Obviously in a market such as engine manufacture, a need will arise for materials that are less expensive yet exhibit the same coating properties. Some ideas being investigated here address ZrO_2/CaO and Al_2O_3 base materials (10). Both provide economic advantages compared to ZrO_2 - Y_2O_3. Al_2O_3 has favourable corrosion properties in dirty fuels, but transforms during plasma spraying (11). Research work is continuing on the stabilization of Al_2O_3.

A further requirement is the ability to spray parts to their final dimensions thus reducing or totally removing the need for machining the ceramic which can be an expensive process.

CONCLUSIONS

The production aspects of fabricating thick (1.5 - 2 mm) ceramic coatings for thermal barrier applications have been discussed. The use of robots is shown to be important for the absolute control of substrate-torch stand-off and therefore, constant coating structure. Examples have been given for different types of coatings on various substrate materials.

Importance is attached to the use of automated equipment for grit blasting and spraying and their associated quality control procedures for batch or volume production. Future material advances will see a trend away from ZrO_2 - Y_2O_3 to ZrO_2 - CaO and Al_2O_3 type materials.

REFERENCES

1. R. Kamo and W. Bryzik, Ceramics for Adiabatic Turbocompound Diesel Engines. Proc. Int. Symposium on Ceramic Components for Engines, Japan, 1983.

2. J. W. Patten, Opprotunities for coatings in adiabatic Diesel Engines. Nato Advanced Workshop on Coatings for Heat Engines, Acquafredda di Maratea, Italy, April 1984.

3. G. Meetham, Service Experience with Protective Coatings in Industrial, Marine and Aero Gas Turbines. ibid.

4. G. W. Goward, Thermal Barrier Coatings for Gas Turbines, a Review of Development and Production Capability. ibid.

5. I. Kvernes, Coating of Diesel Engine Components, Ed. E.Lang, p. 361 in "Coatings for High Temperature Applications". Applied Science Publishers, London, 1983.

6. G. Wuest, S. Keller, A. R. Nicoll, A. Donnelly, Plasma Spray Deposition Efficiencies. JVST, Nov/Dec A, 1985.

7. D. F. Almond and H. Reiter, Novel ways of looking at thermal spray coatings. The Surfacing Journal, Vol. 16, No. 1, 1985, p. 4.

8. P. Boch, J. F. Coudert and C. Grault, Study of the thermal fatigue resistance of plasma sprayed zirconia coatings for diesel components. Proc. of Int. Symp. on Ceramic Components for Engines, Japan, 1983.

9. H. A. Crostack, A. Krüger, H. D. Steffens, Non-Destructive Testing of Arc and Plasma Sprayed Coatings by Ultrasonic Imaging Technique. ICMC, Los Angeles, 1985, to be published in JVST, 1985.

10. H. Haas and A. R. Nicoll, Thermal Shock Resistant ZrO_2 and Al_2O_3 Base Plasma Spray Coatings. DVS-Berichte Band 98, p. 103, Deutscher Verlag für Schweisstechnik (DVS) GmbH, Düsseldorf/FRG, 1985.

CHARACTERIZATION OF ZrO_2-Y_2O_3 THERMAL SPRAY POWDER SYSTEMS

Marc J. Froning, N. Jayaraman
Department of Materials Science & Engineering
University of Cincinnati
Cincinnati, Ohio, USA

Thomas E. Mantkowski, David V. Rigney
Aircraft Engine Business Group
General Electric Company
Evendale, Ohio, USA

Abstract

Powder samples of ZrO_2-8%Y_2O_3 manufactured by different processes such as (i) spray drying, (ii) spray drying and sintering, (iii) sintering and crushing, (iv) casting and crushing, and (v) casting, crushing and fusing were initially characterized for particle size and shape, microstructure and morphology, surface area, density, flow characteristics, and qualitative phase distribution by x-ray diffraction. The powder manufacturing process did appear to have significant effects on all the above characteristics.

The powders were air plasma sprayed on Hastalloy-X substrate with a Vacuum Plasma Sprayed (VPS) NiCrAlY bond coating. The coated samples were characterized for microstructure, erosion resistance, and qualitative phase distribution. The coated samples were subjected to thermal cycling at 1093°C, and similar characterizations were carried out on the exposed samples. In general, as observed by other investigators, the highly porous coatings appeared to withstand a greater number of thermal cycles. Also the coatings containing small fractions of monoclinic phase appeared to withstand less number of thermal cycles than those containing only tetragonal or cubic phases. In addition, the thermal cycling appears to transform all the available phases to tetragonal form. Further work is in progress to confirm the preliminary thermal cycling test results, and to characterize the coated and thermal cycled samples.

THE DESIRE TO IMPROVE the performance and efficiency of aircraft turbine engines has led to higher operating temperatures in the turbine sections of the engines. To accommodate these higher temperatures, new superalloys and material systems have been developed, but the practical temperature limits of these materials still limits engine performance. The use of thermal barrier coatings (TBCs) on turbine components has the potential to provide a dramatic increase in engine performance by allowing higher gas temperatures in the turbine and by reducing the need for cooling air in the turbine components. The desire to expand TBC use from combustors and other low-risk applications to more critical turbine applications has created the necessity for an improved, more reliable thermal barrier coating.

Earlier work on TBCs has focused on the use of stabilized or partially-stabilized forms of zirconia as the insulating material. These materials can be plasma sprayed onto the engine part over a plasma sprayed oxidation resistant bond coat (1). The system currently being developed for employment on turbine hardware is zirconia partially stabilized with 8% yttria. This material provides coatings which have good thermal shock performance, hardness, and erosion characteristics (2,3,4).

Past work on plasma sprayed TBCs has demonstrated wide variations in the performance of the coatings (5,6). Increased automation of the thermal spray process has demonstrated that significant differences exist in plasma spray performance. The recent work by R. J. Keller (2) demonstrated that the initial powder size distribution and chemical homogeneity had a dramatic effect on TBC performance. In addition, work on other thermal spray systems within General Electric has demonstrated that an understanding of the plasma spray raw material is critical to the development of reliable, reproducible coatings.

The purpose of this program is to analyze the various zirconia-yttria powders currently available, to characterize those powders based upon their methods of manufacture, to gain insight into the powder-plasma flame interactions, and, ultimately, to determine the effects of the raw material on the resulting thermal barrier coating.

This paper reports the preliminary results of this effort. Work on this program is continuing to more accurately define and explain these preliminary results.

MATERIALS AND EXPERIMENTAL METHODS

The nine different thermal spray powders used in this program were supplied by seven different vendors. Each material in nominally zirconia-8% yttria (by weight). The actual compositions were taken from the vendors specification sheets. As a rough test of the particle-to-particle consistency, five particles of each powder were examined by KEVEX. In addition, a full field (120X) KEVEX chemistry was taken. A comparison of these analyses is shown in Table 1. The vendor-supplied analyses are shown in the wet chemistry column. The KEVEX column gives the full field chemistry and then lists (in parentheses) the particle-to-particle range. In both cases, the balance of the material is Zr.

Prior to spraying, physical data was obtained on each of the 9 powders. The particle size distribution of each material was obtained with a Leeds and Northrup Microtrac (Model 7995-12, extended range). The samples were run in water using a trace of T-X100 surfactant when needed. The specific surface area of each powder was measured by the Branaur, Emitt, Turnel (BET) method using a Quantachrome Model QS-8 Quantasorb. A three-point BET was run for each powder using nitrogen as the adsorbant in a helium carrier-gas stream. The true powder particle density was measured in a Quantachrome PY-5 Null Pycnometer employing pure helium gas. The average of three runs is reported. Powder flowability was measured using a Hall Flow Cup (7). Rather than reporting the time for 50 grams of material to flow, the data was inverted to provide a flow rate (in grams/sec.). In addition, the powders were examined by Scanning Electron Microscopy (SEM) for overall shape and structure, and they were evaluated metallographically for microstructure.

The powders were air plasma sprayed using identical spray parameters onto Hastalloy-X buttons (1" diameter X 0.125" thick) with an 0.008 inch VPS Ni-22Cr-10Al-1Y bond coating. The spray parameters for both operations are shown in Table 2. The typical roughness of the bond coats was 600 u in AA. The air plasma work was performed on an X-Y traverse to insure the reproducibility of the process throughout this test and future iterations.

The as-sprayed coatings were evaluated for erosion resistance by impinging 50 micron Al_2O_3 at a 20° angle onto the surface. The depth of penetration was measured, and the results reduced to an erosion reading in seconds per mil. All of the values were normalized against a Lexan® standard to allow future comparison of the data. The coatings were also evaluated metallographically for structure and porosity.

To test the TBCs' thermal cycling properties, the coatings were cycled in a C-M, Inc., Rapid Temp Furnace (Model 1700S-DBL). This furnace is bottom loading, microprocessor controlled, and capable of automatically testing 15 run conditions simultaneously. The samples were run in an air environment through cycles consisting of a 10 minute heat up to 1093°C, a 40 minute soak at 1093°C, and then 10 minutes of forced air cooling to approximately 204°C. A specimen failed when 10% of the coating had been lost, but the exposure was continued until 20% of the coating was lost in order to fully observe the transformations which took place during cycling. The failed samples were then evaluated metallographically to determine coating changes during thermal exposure.

In order to evaluate the effects of phase structure on TBCs, x-ray diffraction (XRD) analyses were performed on the as-received powders, the as-sprayed coatings, and the coatings after thermal cycling. A Philips-Norelco diffractometer, automated by a microcomputer (8) for data acquisition, storage and analysis was used. A line source of Chromium radiation was used at operating conditions of 35 kV and 20 mA. A Soller slit assembly and a 1° divergent slit were used on the incident beam side. A receiving slit, a Soller slit assembly and a diffracted beam monochrometer with a graphite crystal were used on the diffracted beam side. The diffracted beam intensity was detected by a scintillation detector. The diffractometer was calibrated periodically with Si (311) diffraction peak from an NBS standard powder sample. The procedure for obtaining a diffractogram consisted of initially step scanning by intervals of 0.25° of 2θ with 40 seconds of counting time at each step. Initial scanning of samples was done in the 2θ angle range of 20°-150°. Subsequently, all samples were analyzed by slow step scanning by intervals of 0.05° of 2θ in angular ranges closer to strong diffraction peaks. More attention was paid to angular ranges of 40°-50° and 120°-135° of 2θ, where the resolution of (111) and $(11\bar{1})$ monoclinic peaks in the former angle range and (400) and (004) tetragonal peaks in the latter range becomes possible.

RESULTS AND DISCUSSION

POWDER CHARACTERISTICS - The results of the various powder analyses are shown in Table 3. While each material is unique, clearly the powders can be classified by their method of manufacture. In general, spray dried TBC powders have the lowest true density, the highest specific surface area, and the lowest mass flow rates of all powders. Sintered materials have intermediate values of density and flow, and surface areas which are an order of magnitude lower than those for spray dried. Finally, fused materials were dense, had high mass flow rates, and very low surface areas. The size distributions provided by the various vendors fit into a very loose specification, and the different distributions are a function of the vendors' processes and/or of the size distributions which the vendors chose to supply.

Metallographic evaluations of the powders (see Figure 1) demonstrated the wide differences in powder shape, density, and homogeneity. The spray dried materials were composites of discrete ZrO_2 and Y_2O_3. These materials have a wide size distribution of Y_2O_3 particles, which cause chemistry variations between spray dried particles in addition to the microscopic inhomogeneity inherent in spray dried powders. The spherical nature of the powders caused them to flow smoothly, but the low particle density lowered the material's mass flow rate. Sintered materials were chemically homogeneous at 200X and tended to have a rough, rounded spherical shape. These powders tend to be relatively porous, although their true densities are higher than those of spray dried powders. These powders flowed fairly well, but the powder porosity lowered the mass flow rates. Fused materials are dense, angular powders with low surface areas. When the powders flow they tended to pulse, but the dense nature of the powders led to high mass flows.

XRD analysis of the raw powders demonstrated the differences inherent in the various manufacturing methods (see Figures 2, 3, 4). In these figures the tetragonal ZrO_2 peaks were labeled. The spray dried materials were typically tetragonal, which would be expected from a homogeneous ZrO_2-8%Y_2O_3 material. The materials which were sintered tended to have a substantial amount of cubic phase, along with some tetragonal phase. These observations were consistent with the phases predicted by the ZrO_2-Y_2O_3 phase diagram, which is shown in Figure 5 (9). The spray dried materials are ZrO_2, with only a small amount of Y_2O_3 dissolved in the ZrO_2; therefore, the room temperature phase is monoclinic. In the case of fused ZrO_2 powder, all of the Y_2O_3 was in solution and, upon cooling (quenching) to room temperature, this material forms the tetragonal phase as predicted by the phase diagram. In the case of the sintered ZrO_2 powders, the materials reflected the high temperature equilibrium phases which are tetragonal and cubic. The Y_2O_3 content in these phases was controlled by the sintering temperature and, depending on this temperature and the cooling rate after the sintering process, these powders were likely to retain the tetragonal phase in addition to the stable cubic phase at room temperature.

COATING CHARACTERISTICS - The characteristics of the as-sprayed TBCs provide a definite reflection of the raw materials employed. It appears that the major powder factors affecting TBC performance are the powder size distribution and the powder's chemical homogeneity and phase structure.

The size distribution of the thermal spray powder is critical because it largely controls the density of the as-sprayed structure. Both the mean particle size and standard deviation of the distribution were found to be significant and of the same order of magnitude in affecting the coating. Figure 6 shows the structures of three coatings sprayed with different size distributions ranging from an average size (by Microtrac) of about 20 μm to an average size of 74 μm. The change in coating porosity with mean powder size is quite obvious. It is well recognized (2,4) that the porosity of the coating has a distinct effect on TBC thermal cycle life and erosion resistance. These data confirmed that the most dense coatings (made from W's spray dried powder and all of those made from fused materials) had the highest erosion resistance and the lowest thermal cycle lives (see Table 4), while the most porous coatings (vendors A, B, and C's powders) had lower erosion resistances and good thermal cycle lives. Further work is needed to improve upon this relationship and to attempt to increase erosion resistance without adversely affecting thermal cyclic life.

The starting powder's chemical homogeneity and phase structure have a major effect on the as-sprayed TBC structure. Coatings made from spray dried powders tend to be predominantly tetragonal and retain some of the monoclinic phase of the powder (Figure 2); coatings made from fully tetragonal fused materials remained tetragonal (Figure 3), and coatings sprayed with sintered materials retain a substantial amount of the original cubic phase, although some shift towards the tetragonal phase is evident (Figure 4). These observations are consistent with the phase diagram (Figure 5) in that quenching of the liquid powder particles during spraying would produce the tetragonal phase. Retention of the monoclinic or cubic phase may be explained by incomplete melting of the powders during the plasma spray operations. While the amount of phase transformation upon spraying varied from manufacturer to manufacturer, it may not be possible to produce superior quality TBCs from certain types of powders, regardless of the spray parameters employed. Further work is needed to study the effect of particle size distribution and spray parameters on the phase transformations which occur during spraying.

THERMALLY CYCLED COATING CHARACTERISTICS - All of the samples sprayed for this test were subjected to thermal cyclic testing until 20% of the TBC has spalled. The metallographic evaluations produced after cyclic exposure are shown in Figure 7, XRD analyses are shown in Figures 2-4, and the number of cycles to failure are given in Table 4. In general, the coatings with the longest life were very porous and had some degree of cubic phase in the as-sprayed coating. The intermediate life TBCs had intermediate porosity levels and were fully tetragonal in the as-sprayed condition. Finally, the lowest life samples were dense and had a relatively high concentration of monoclinic phase in the coating.

It can be seen that all of the coatings tended to become tetragonal upon thermal cycling and that, after 1000 cycles, this transformation was fairly complete even for the least- tetragonal as-sprayed coatings (Figure 4). This observation is very different from the observations made by Herman et al. (10) and Miller et al. (6,11). They observed that continuous exposure of the coatings at a temperature greater than 1100°C for periods of time up to 100 hours tended to produce varying amounts of monoclinic and cubic phases with decreasing tetragonal phase. This difference between the observations of the previous investigations and our present work may be explained by the differences in the types of thermal exposures involved in the investigations. For example, the coatings of Herman et al. and Miller et al. were heated to a constant temperature continuously for an extended period of time. From Figure 5 one can see that at their heat treatment temperatures, the equilibrium phases are the cubic and the transformable tetragonal phases. The time involved in their exposures is sufficient to complete the diffusion controlled processes involved in formation of the equilibrium phases; therefore, upon cooling the stable monoclinic and cubic phases were formed. Opposed to this, in the present investigation, the coatings were thermally cycled between 1100°C and 200°C for periods of 40 minutes at the high temperature. This repeated heating and cooling of the coatings may be compared to the repeated heating and cooling of martensite in steel. Such cycling was found to significantly retain the high temperature austenitic phase at lower temperatures (12). It appears that the result of thermal cycling of ZrO_2-Y_2O_3 coatings is similar to this. Thus after cyclic testing a greater fraction of the high temperature phases was retained.

CONCLUSIONS AND FUTURE WORK

The examinations of the powders used in this program and the coatings which they produced indicated significant differences in powders both from different methods of manufacture and from vendor to vendor for the same generic type of powder. These differences are somewhat removed during the plasma spray process, but they are still transmitted to the resultant coating, ultimately affecting TBC performance. While the results of our original tests showed that sintered ZrO_2-8%Y_2O_3 powders gave the longest life, the sintered materials employed in this program also had the most narrow, coarsest size distribution of any of the powders used. Future work on this program will involve reforming the original powders into comparable size distribution, spraying these "new" powders, and retesting them. This will remove one of the major variables from the comparison and give a clearer picture of the relative merits of each type of powder.

Upon cyclic exposure, ZrO_2-8%Y_2O_3 deposits transform from this as-sprayed condition to the tetragonal phase. This differs from the transformations which occur in constant - exposure conditions, and demonstrates a need to make testing procedures closely simulate the engine conditions to which the coatings will be exposed.

This preliminary work does indicate that certain types of ZrO_2-8%Y_2O_3 powders will not be suitable for producing consistent, highly reliable thermal barrier coatings. While it is not yet clear what constitutes an ideal raw material, it is quite likely that plasma spray TBC powders which contain a great degree of chemical inhomogeneity will not be capable of producing a superior quality thermal barrier coating.

ACKNOWLEDGMENTS

The authors would like to thank the General Electric Company's Metals Manufacturing Applications Center for its support in this endeavor.

REFERENCES

(1) Stecura, S., "Optimization of the NiCrAl-Y/$ZrO_2Y_2O_3$ Thermal Barrier System," NASA Technical Memorandum 86905, April 1985.

(2) Keller, R. J., "Research and Development for Improved Thermal Barrier Coatings," AFWAL-TR-84-4008, April 1984.

(3) Ramadass, N., Mohan, S. C., and Ravindra Reddy, S., "Studies on the Metastable Phase Retention and Hardness in Zirconia Ceramics," Materials Science and Engineering 60, 65-72 (1983).

(4) Anderson, C. A., "Advanced Ceramics For Industrial/Utility Gas Turbine Applications," DOE/NASA/0110-1; NASA CR-165619: February 1982.

(5) Stecura, S., "Effects of Compositional Changes on the Performance of Thermal Barrier Coatings System," NASA-TM 78976, 1978.

(6) Miller, R. A., Garlick, R. G., and Smialek, J. L., "Phase Distribution in Plasma Sprayed Zirconia-Yttria," Ceramic Bulletin 62(12), 1355-1358 (1983).

(7) ASTM B213-77.

(8) Jayaraman, N. and Verrilli, M. J., "Oxidation Stress in Polycrystalline Ni 200," Accepted for publication in Met. Trans.

(9) Scott, H. G., "Phase Relationships in Zirconia-Yttria Systems," Journal of Materials Science 10, 1527-1535 (1975).

(10) Shankar, N. R., Berndt, C. C., and Herman, H. "Phase Analysis of Plasma-Sprayed Zirconia-Yttria Coatings."

(11) Miller, R. A., Smialek, J. L., and Garlick, R. G., "Phase Stability in Plasma-Sprayed Partially Stabilized Zirconia-Yttria," in "Advances in Ceramics: Vol. 3," Science Technology of Zirconia," edited by A. H. Hena and L. W. Hobbs, ACES Publishing.

(12) Goldberg, A., "Effects of Repeated Thermal Cycling on the Microstructures of 300-Grade Maraging Steel," Trans. ASM 61, 26-36 (1968).

Table 1

Chemical Composition of Experimental Powders

Vendor Codes	Powder Types	Composition [by KEVEX]	Composition [wet chemistry]
W	Spray Dried	2.60% Hf [0.0- 3.0] 3.40% Y [2.1- 4.6]	0.6% Si 0.9% Hf 6.8% Y
J	Spray Dried	0.24% Ca [0.0- 0.6] 0.00% Ti [0.0- 0.5] 2.60% Hf [0.0- 1.2] 8.60% Y [8.4-35.6]	0.3% Ti 0.4% Si 1.3% Hf 7.3% Y
A	Spray Dried	0.00% Ti [0.0- 0.3] 0.00% Ca [0.0- 0.3] 0.31% Fe [0.0- 0.3] 2.19% Hf [0.0- 2.3] 12.20% Y [5.1-15.3]	0.2% Ti 0.7% Si 8.4% Y
C	Spray Dried & Sintered	2.41% Hf [1.5- 3.1] 9.10% Y [8.7-11.8]	0.2% Ti 0.7% Si 7.6% Y
B	Sintered & Crushed	0.00% K [0.0- 0.2] 0.25% Ni [0.0- 0.0] 1.49% Hf [0.9- 3.3] 11.10% Y [7.0-13.0]	0.2% Ti 0.3% Si 8.2% Y
Y	Sintered & Crushed	0.00% Ca [0.0- 0.3] 0.00% Ti [0.0- 0.3] 0.00% Fe [0.0- 1.1] 1.35% Hf [0.0- 2.5] 7.81% Y [4.7- 7.0]	0.1% Ti 0.2% Si 1.5% Hf 7.4% Y
J	Cast, Crushed, & Fused	0.00% Ti [0.0- 0.3] 2.11% Hf [0.6- 1.4] 8.16% Y [4.3-12.4]	0.1% Si 0.2% Ti 7.0% Y
H	Fused, Cast, & Crushed	0.00% Ti [0.0- 0.1] 0.00% Mg [0.0- 0.2] 1.21% Hf [0.0- 1.5] 10.20% Y [6.4- 8.0]	0.2% Mg 0.1% Ti 0.3% Si 8.6% Y
J	Fused, Cast, & Crushed	0.00% Mn [0.0- 0.2] 0.14% Ti [0.0- 0.0] 1.95% Hf [0.0- 1.3] 9.24% Y [4.9- 8.5]	0.1% Si 0.2% Ti 6.4% Y

Table 2

Plasma Spray Parameters

	Air	Vacuum
Plasma Gun	Metco 7MB	Metco 7MB
Primary/Secondary Gas	N_2/H_2	Ar/H_2
Gun Power	35 KW	50 KW
Traverse Rate	100 sfpm	100 sfpm
Powder Feed Rate	6 Lbs./Hr.	10 Lbs./Hr.
Preheat	None	1800°F
Spray Distance	5 In.	12 In.
Other	90° air impingement cooling	40 torr Abs. Pressure, RTA cleaned

Table 3

Physical Properties of the As-Received Experimental Powders

Vendor Codes	Powder Types	Shape [by SEM]	Color	Density [g/mil]	Flowability [hall, g/s]	Size [Microtrac]	Surface Area [sq. m/g]
W	Spray Dried	Porous Spheres	White	5.54	5.2	*	3.60
J	Spray Dried	Spheres	White	5.55	6.4	MV = 46.7 SD = 23.0	1.60
A	Spray Dried	Spheres	Pale Yellow	5.60	7.2	*	1.39
C	Sintered	Rough Spheres	Pale Yellow	5.82	7.6	MV = 67.7 SD = 30.0	0.187
B	Sintered & Crushed	Rough Spheres	Light Yellow	5.96	7.0	MV = 74.4 SD = 25.4	0.256
Y	Sintered & Crushed	Irregular	Light Yellow	5.96	7.0	MV = 53.2 SD = 31.5	0.39
J	Cast, Crushed and Fused	Spheres/ Angular/ Acicular	Dark Yellow	5.67	9.6	MV = 50.0 SD = 23.4	0.073
H	Fused Cast, and Crushed	Irregular	Dark Yellow	6.03	8.7	MV = 65.4 SD = 35.6	0.00656
J	Fused Cast, and Crushed	Angular	Dark Yellow	6.05	8.3	MV = 51.1 SD = 22.0	0.081

* Powders Dissolve in Water.

Table 4

TBC Life and Erosion Resistance

Vendor Codes	Powder Types	Erosion Resistance [sec/mil]	Thermal Cycles to Failure
W	Spray Dried	9.5	40
J	Spray Dried	7.4	285
A	Spray Dried	4.1	700
C	Sintered & Crushed	6.1	660
B	Sintered & Crushed	7.1	1000
Y	Sintered & Crushed	6.6	410
J	Cast, Crushed & Fused	10.5	380
H	Fused, Cast & Crushed	9.6	310
J	Fused, Cast & Crushed	11.6	390

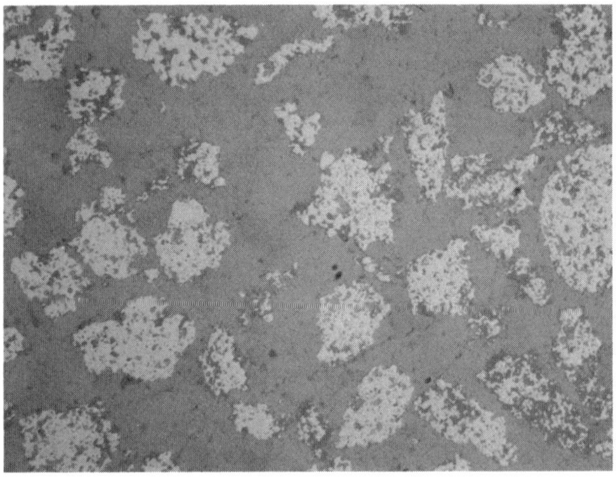

Vendor B - Sintered

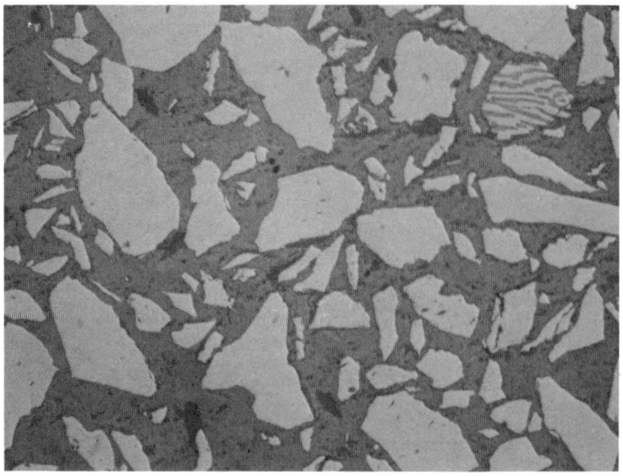

Vendor H - Cast and Crushed

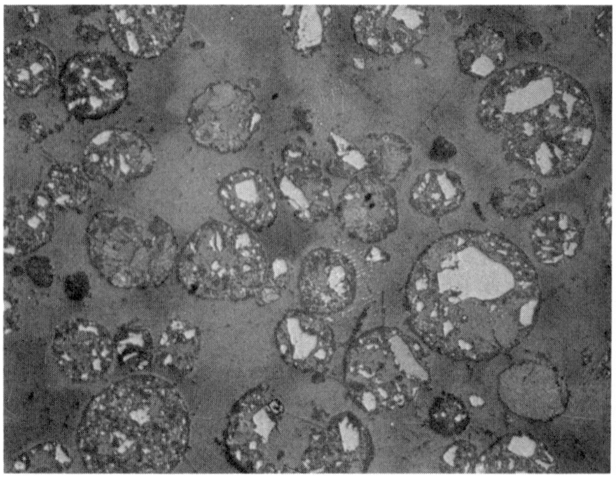

Vendor J - Spray-Dried

Figure 1. Metallographic Evaluation of ZrO_2-8%Y_2O_3 Powders.

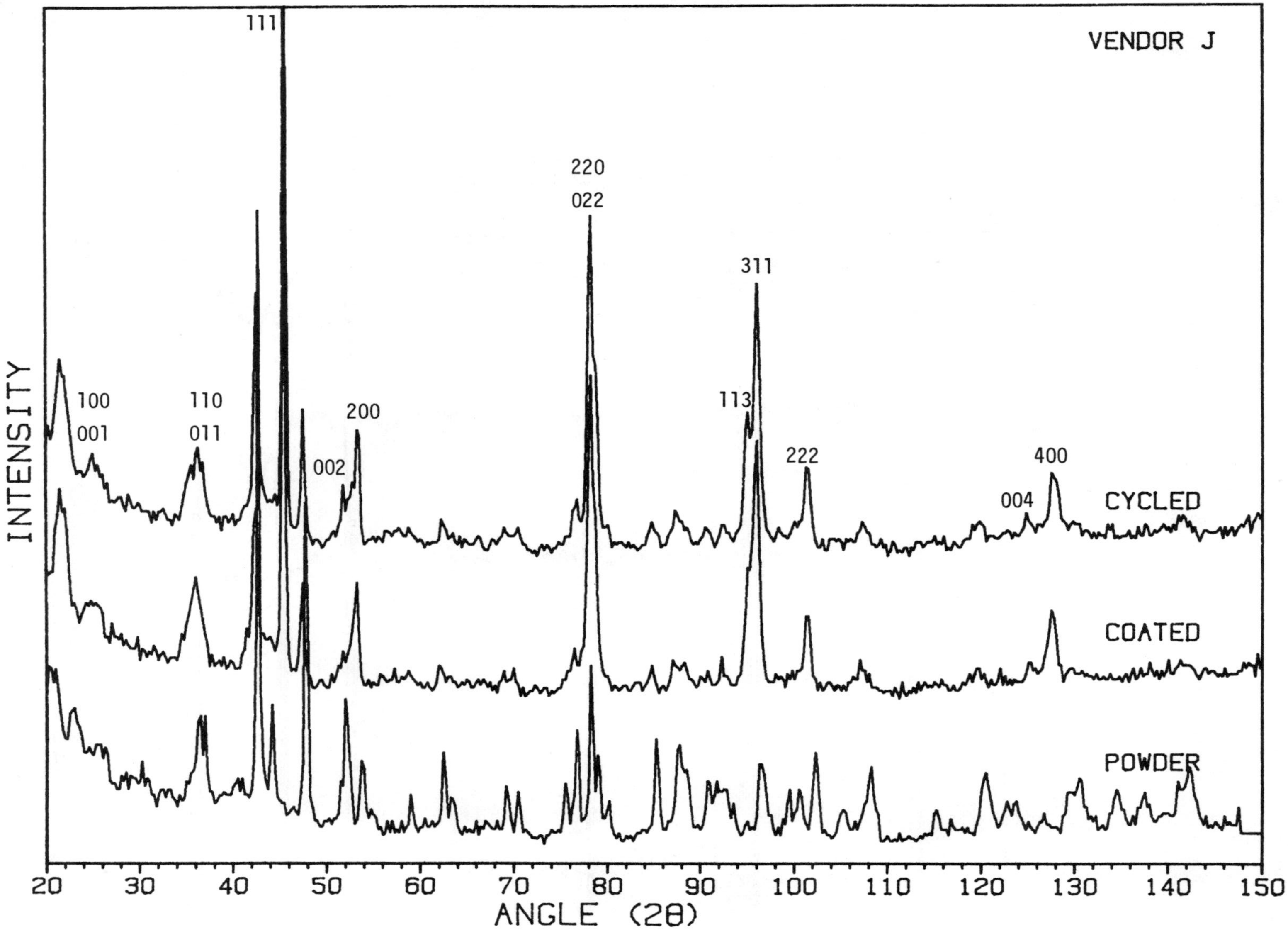

Figure 2. XRD Analysis of Vendor J's TBC System.

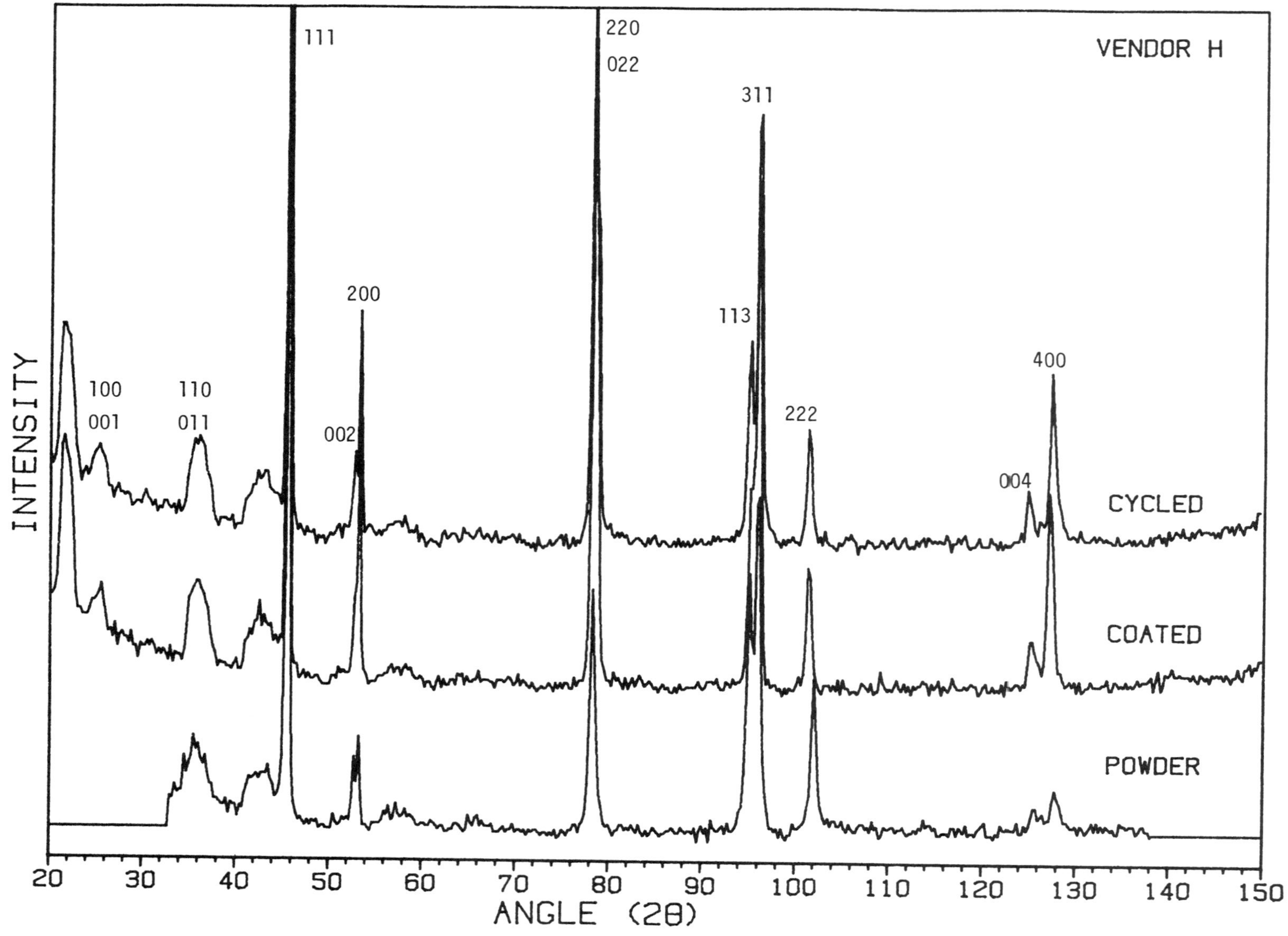

Figure 3. XRD Analysis of Vendor H's TBC System.

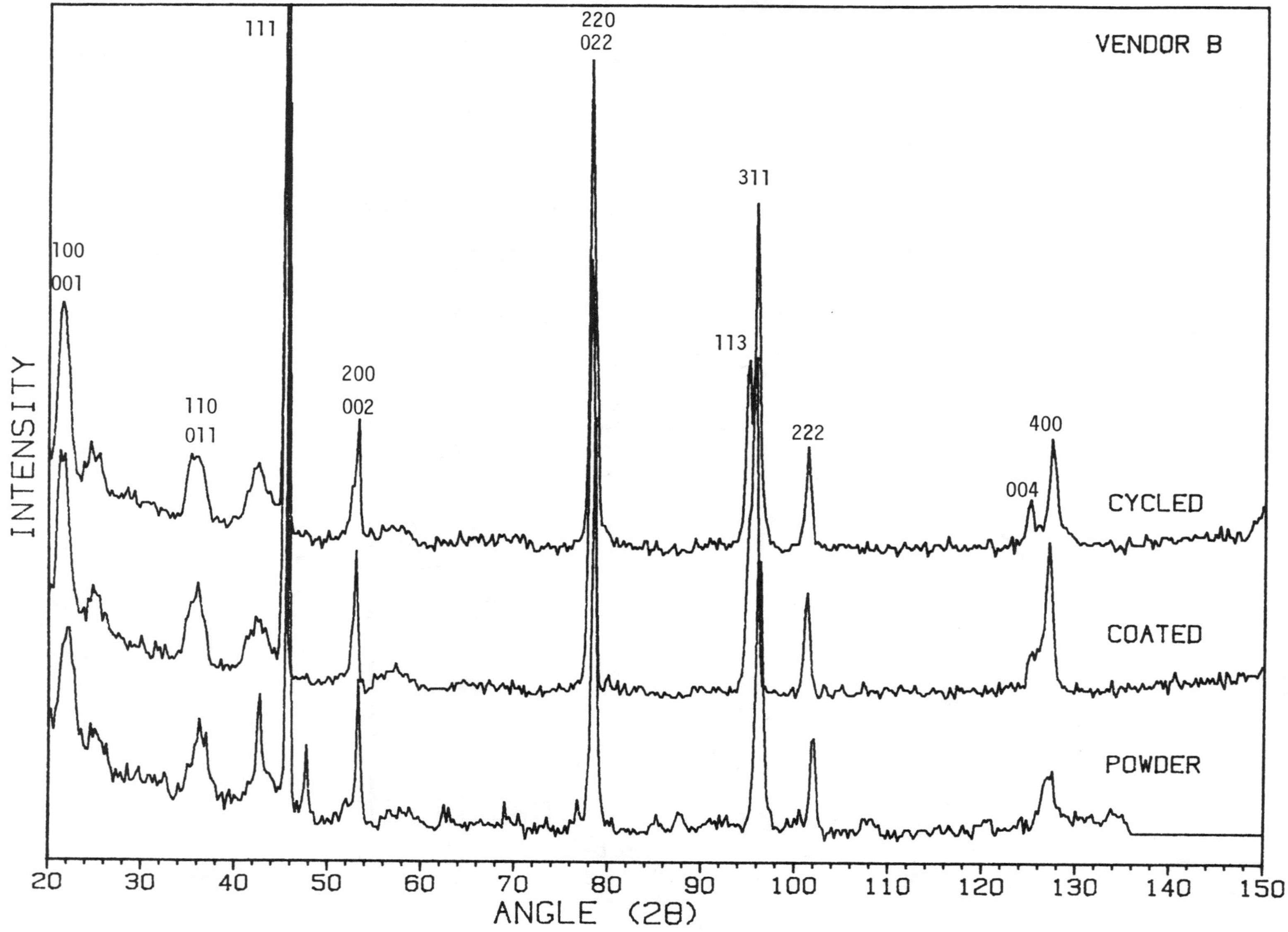

Figure 4. XRD Analysis of Vendor B's TBC System.

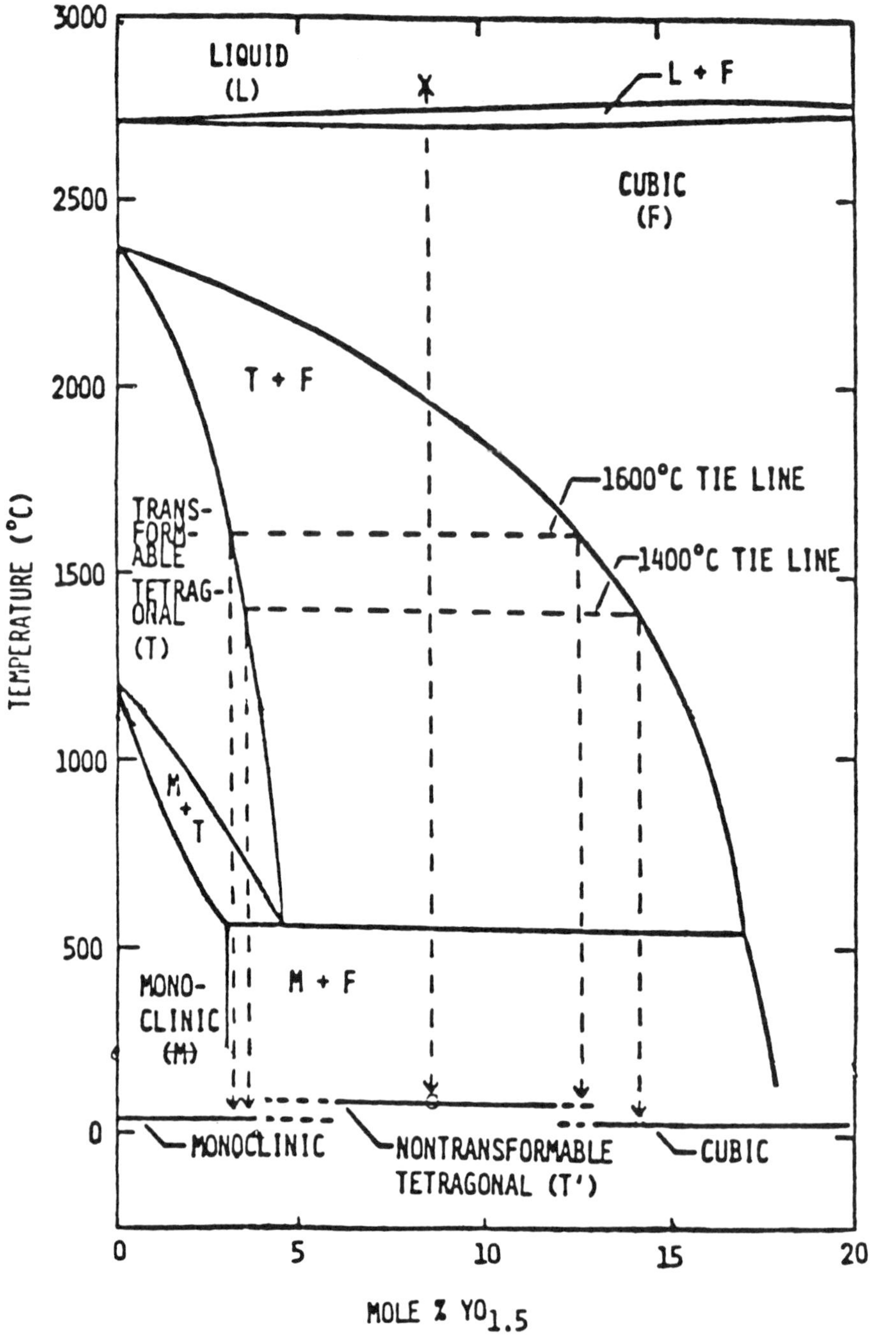

Figure 5. Low Yttria Region of the ZrO_2-Y_2O_3 Phase Diagram (9).

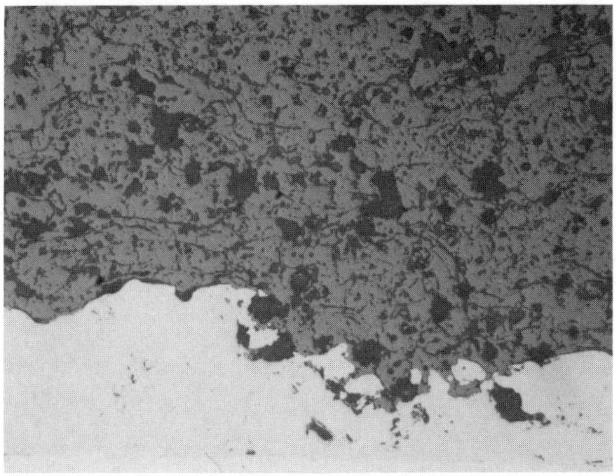

Vendor B

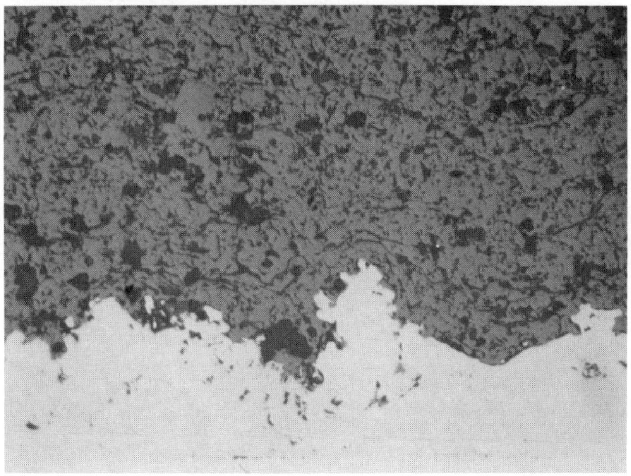

Vendor H

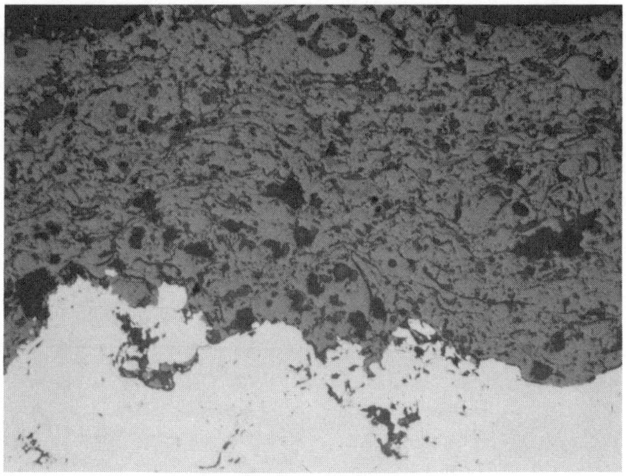

Vendor J

Figure 6. Metallographic Evaluation of As-Sprayed TBCs.

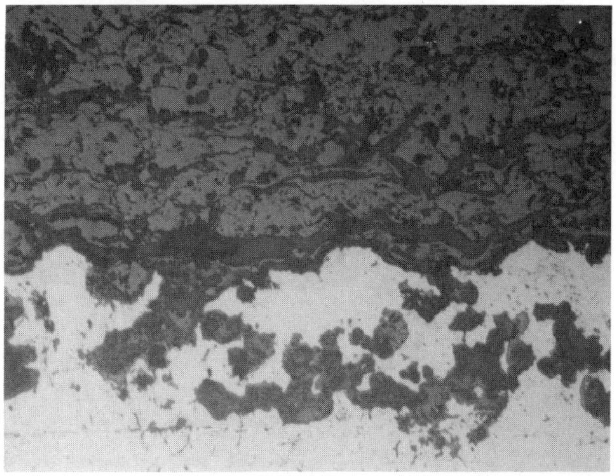

Vendor B - 1000 Cycles

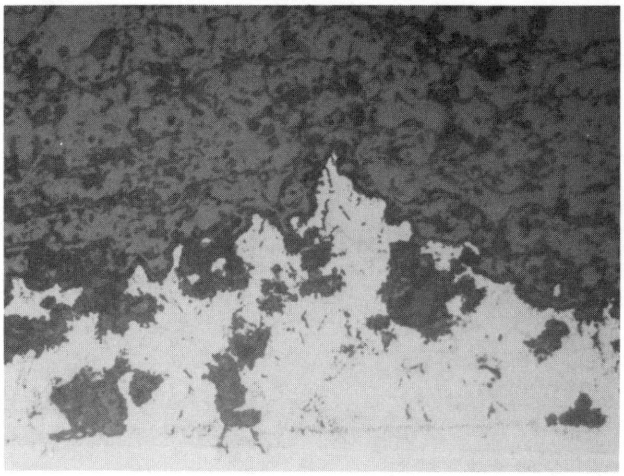

Vendor H - 310 Cycles

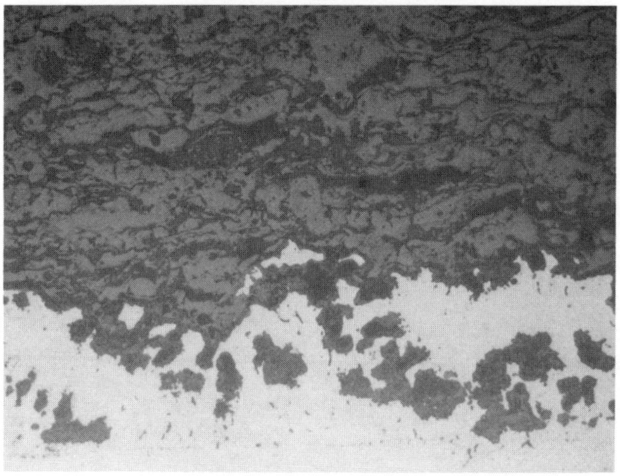

Vendor J - 285 Cycles

Figure 7. Metallographic Evaluation of Cycled TBCs.

ROLLS-ROYCE EXPERIENCE WITH THERMAL BARRIER COATINGS INCLUDING HIGH TIME SERVICE EVALUATION

F. Chris Toriz, Suresh K. Gupta
Rolls-Royce Inc.
Atlanta, Georgia, USA

A. Bennett
Rolls-Royce Limited
Derby, England

ABSTRACT

Rolls-Royce has been actively involved in the development, evaluation and the use of thermal barrier coating systems for well over a decade. This paper briefly reviews this effort with special emphasis on a current in-service evaluation program of several Zirconia/MCrAlY based coating systems. In this program some coated turbine nozzle guide vane sets have accumulated over 6000 hrs and 3000 cycles. Prospective coating systems for a quantum jump in life improvement, "Second Generation TBC's" are identified for near term evaluation.

Rolls-Royce experience extends from carousel rig testing of coated coupons, to engine rig evaluation, and engine development testing of coated turbine blades, and finally to Rolls-Royce service engine experience on coated combustors, and turbine nozzle guide vanes. Several different coating systems with a range of variables such as bond coat composition and application techniques have been evaluated, and are briefly discussed in this paper.

ROLLS-ROYCE EXPERIENCE

BACKGROUND - Rolls-Royce Ltd. has used thermal barrier coatings in the field for over ten years, beginning with the RB211-22B engine, for increasing combustor life, up to todays use on the RB211-535E4 high pressure turbine nozzle guide vanes (NGV's). (See Figure 1) During that time they have proven to be reliable within certain constraints, and of considerable value. They have always suffered problems of one form or another however, and these fall into two categories: those associated with extremes of temperature, and those associated with small radii. Thus certain areas of combustion ware have traditionally been difficult to coat, although the large majority of the coated areas remain intact for thousands of hours.

These two problems have been accentuated by the requirement to exploit the TBC benefits in the turbine, on both static and rotating members, on airfoils and on platforms. In these applications, the temperatures can be several tens of degrees higher, and there are critical small radius areas which must remain coated if the system is to be exploited adequately. High substrate strains for rotating members,

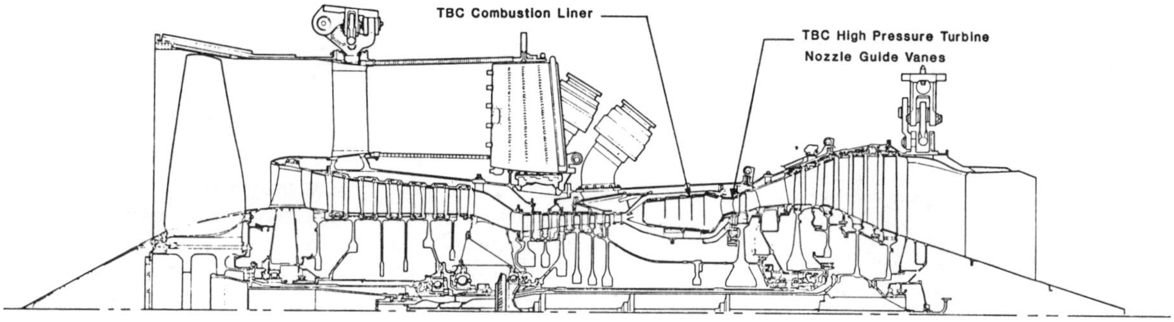

Figure 1 Use of TBC's on the Rolls-Royce 535-E4

and the possibility of ballistic impact of semi-molten debris which passes through the engine add to the general problems. The former gives rise to even higher strain cycles than imposed by purely thermal means, and the latter can give impact damage, erosion damage, or if the debris sticks and builds up, it can modify the TBC properties and lead to surface spallation.

Aerodynamic effects lead to the requirement for these coatings to be as smooth as possible, which for plasma sprayed deposits means extensive surface polishing. Once the benefits of a material such as TBC's are demonstrated there is a demand for improved properties of that material, such as improved insulating capability and coefficient of expansion matching. The requirements for development of new improved TBC systems were thus clearly stated in the late 1970's, and materials and processing developments have been pursued steadily since that time.

MATERIALS

Mag Zirc (MZ) - The traditional system employed by Rolls-Royce is known as the "three layer Mag-Zirc" (3LMZ) coating. It is comprised of 1) air plasma sprayed nickel-chromium alloy bond coat to which free aluminum is added as an aid to adhesion, 2) a physical mixture of this material with the top coat ceramic material to produce a cermet layer, and finally 3) the top coat which is a 50/50 molar mixture of zirconium and magnesium oxides. These are deposited to give 0.003-0.005" bond coat, 0.003-0.005" cermet, 0.006-0.008" ceramic.

Failure Mechanisms - The 3LMZ has three major problems. The bond coat is poor in oxidation and above 950°C it will rapidly oxidize undermining the stability of the coating. Secondly, the ceramic is not stable and undergoes phase changes, again at around 950°C and above, which affect its mechanical stability, but more importantly can triple its thermal conductivity. Thus at hot spots it degrades fastest and most completely, to give least protection. Finally, the cermet, which theoretically should ameliorate the expansion coefficient differential between the bond coat and the ceramic, proves to be the weak link in this system. The metallic constituent can be at temperatures up to 100°C (200°F) above the bond coat and exists as tiny irregular shaped particles in a ceramic matrix known to transmit oxygen, and in any case, to be highly porous. The metal fragments oxidize rapidly, bloat, and cause stresses which tear the cermet apart. See Figure 2.

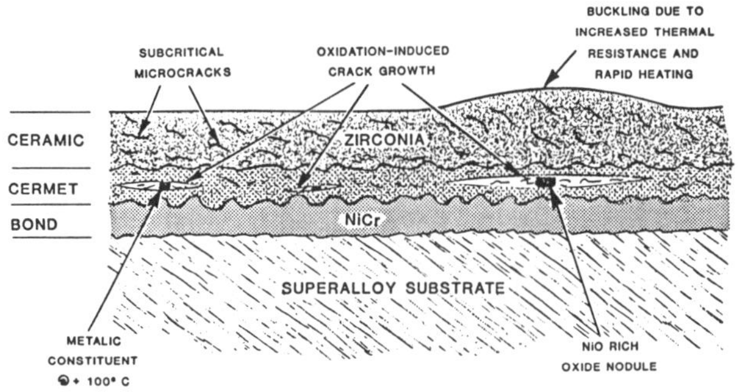

Figure 2 Spalling Due to Oxidation in Plasma Sprayed TBC's

To address these problems a range of actions have been taken. Simply deleting the cermet layer for instance gives a 30% cheaper coating with around five times the thermal cycle durability (Figure 3) in a standard Rolls-Royce (RRL) cyclic test to above 1100°C. (See appendix A for test definition)

For relatively moderate temperature operation this simple modification can prove to be most cost effective, but it still incorporates the magnesia stabilized zirconia, and thus experiences phase changes and conductivity increases. The two other common stabilizing agents for zirconia are calcia and yttria, but when these are added at levels which give full cubic phase stabilization their thermal cycle resistance is very poor. For some while it was recognized that metastable materials may give high levels of durability, but poor insulation, where as fully stable systems gave good insulation, but poor cyclic capability. Metallographic sections of these materials traced the origin of the poor cyclic durability of fully stable systems to their inherent crack networks in the as sprayed state. The magnesia stabilized system had few cracks near the interface, and a random dispersion of microcracks through its thickness. The stable materials by contrast had heavy crack networks close to and parallel to the interface. On thermal cycling these coalesce rapidly, and failure occurs through this region. What was needed was some system which did not have the crack network of the fully stable systems, or the phase instability of the metastable system.

Yttria Stabilized Zirconia (YSZ) - At around this time, much literature was being published about the partially stabilized zirconia systems, (i.e. References 1 and 2) and on spraying these were found to have the right sort of crack networks. On cycling, they also give lives comparable with or above the magnesia system. The theory went that with inadequate yttria to give full stability of the cubic phase, the coating should exist as a mixture of cubic and monoclinic, and the volume changes attendant on the formation and transformation of the monoclinic phase would produce crack networks

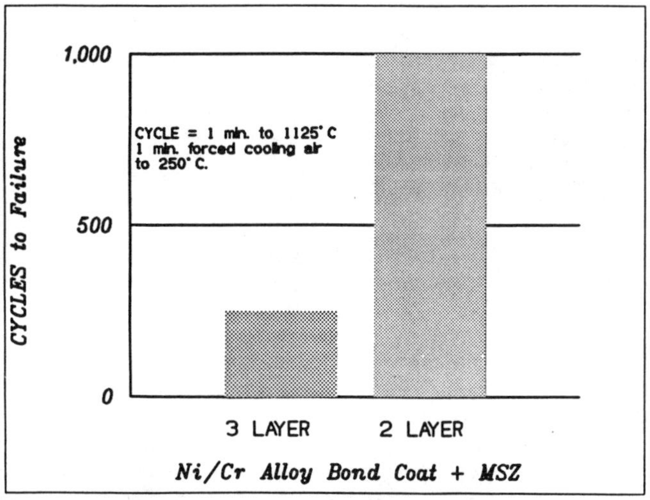

Figure 3 Thermal Cycle Life of MZ Coatings

which would be beneficial. X-ray diffraction work however proved the major phase to be a tetragonal (T') form of zirconia, which had not decayed to monoclinic as it should have, according to the phase diagram of the time, and more over, it was highly supersaturated with yttria. To compound matters further, at temperatures up to 1200°C (2200°F) or so, (its typical use temperature in a TBC application) it was phase stable for hundreds of hours, and had stable thermal insulation.

Phase Effect - Several conferences have since been held to discuss the phase chemistry of the low yttria zirconias both in their applications as bulk ceramics, (i.e. References 3 and 4) and with reference to their use as TBC's, and the non-transformable T' phase is a well-established fact. The mechanisms for its formation and its stability are becoming clearer, but there remains much work to be done to fully understand this most peculiar and beneficial of phases. The effect of this phase is evident from carousel rig testing as shown in Figure 4. 20% stabilized YSZ is predominantly cubic phase and has a short thermal cyclic life. 8% YSZ with no T' phase has an even shorter thermal cycle life, but increasing the T' to 85% increases the thermal life 8 times and going to 100% T' phase increases the thermal cyclic life 22 times.

Bond Coat - Having found a ceramic capable of extended use at high temperature, it became evident, that although eventual spallation was an entirely cohesive failure in the ceramic adjacent to the bond coat, the bond coat itself played a major role in the failure mechanisms, and that bond coat oxidation was of great importance. This has been demonstrated by several workers who have shown the effect of dwell times at the peak condition in cyclic tests, (References 5 and 6) and also within Rolls-Royce in two sets of experiments. One involved isothermal soaking trials to study bond coat oxidation, and showed that after certain soak periods the coating spontaneously spalled as the sample cooled down. The second was learned accidently when a combustion rig flame was allowed to run under reducing instead of oxidizing conditions, and the coating on test lasted several times it standard life. When it was finally removed from the rig, unfailed, it was found to have sustained minimal damage.

These experiments and confirmation of similar effects led us to the conclusion that the role of bond coat oxidation is to continuously alter and amplify the stresses imposed on the ceramic as each thermal cycle goes by. Failure is thus due to a continuously increasing load per cycle, not a repeated constant load per cycle.

High Density Bond Coats - The problem remained, of how to combat this effect, and as nobody wanted to run the engines under reducing condition, we had to turn to a more

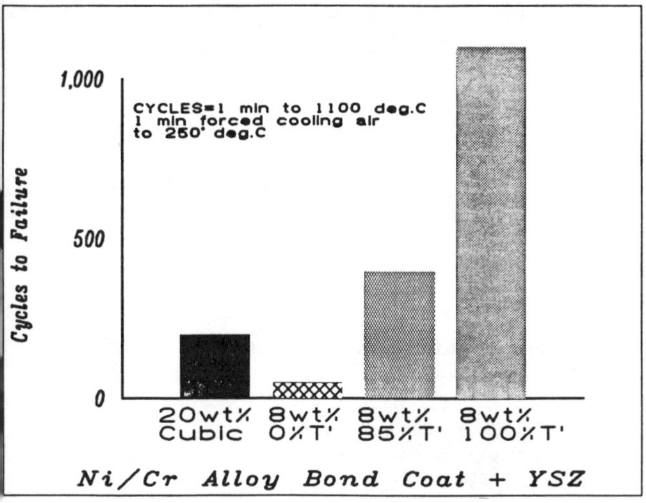

Figure 4 Effects of Zirconia Phase on YSZ Thermal Cycle Life

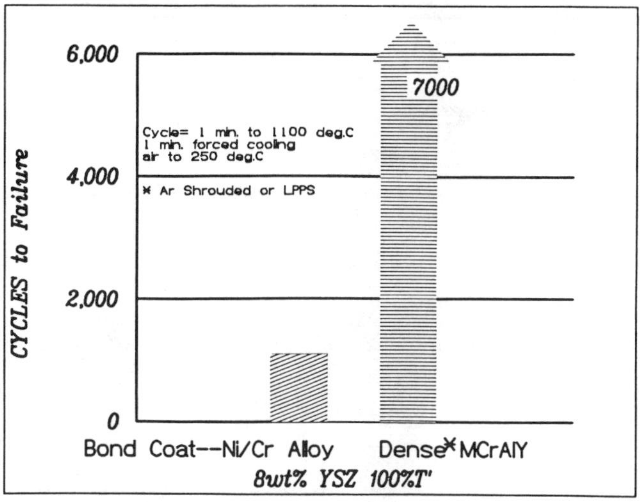

Figure 5 Effects of Bond Coat Phase on YSZ Thermal Cycle Life

oxidation resistant bond coat, and to the MCrAlY series of alloys with which we had several years research experience on the oxidation protection front. Two layer plasma sprayed systems experience similar oxidation induced failures as discussed earlier on the 3LMZ system (Figure 2). Oxidation induced crack growth causes failure in the ceramic top coat adjacent to the bond coat. Simply by replacing our original bond coat by a high density MCrAlY applied by Argon Shrouded plasma (SPS) or low pressure plasma (VPS) techniques, we were able to increase thermal cycle life by a factor of around seven times relative to the previous two layer systems or around thirty-five times relative to the three layer Mag Zirc material! (Figure 5)

Other Ceramic Materials - This system is now our current datum system, and is in service on the RB211-535 E4 HP NGV platform.(Figure 1) It does not represent an ultimate or even optimum system, and we recognize there is a long way to go before we have such a system ready for service. One much reported aspect of TBC use which we in Rolls-Royce have not yet had significant first-hand experience of is acid leaching of yttria from the zirconia by agents such as vanadium, phosphorous and sulphur, being compounds typically found in lower grade fuels. A very interesting new material said to be equally as durable as the zirconias, of similar thermal insulation capability, but not prone to acid attack, is calcium titanate, first brought to our attention by Vogan and Stetson of Solar Turbine Inc. (Reference 7) Coated samples supplied by Solar did very well in thermal cycle durability in the standard RRL cyclic test to 1125°C. (Figure 6) The samples were at least as good as the 8% YSZ samples, having surpassed 6000 cycles without failure. Testing continues. This is one potential material which, with others, we are keeping a close watch on for future potential use.

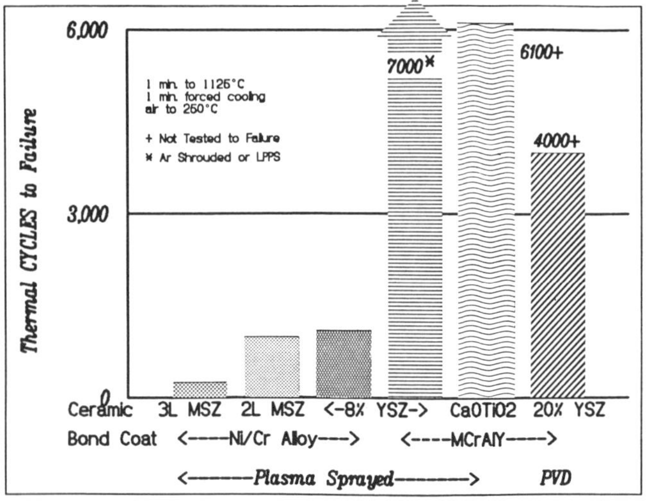

Figure 6 Effects of Material and Process Variants on Thermal Cycle Life

The other great area of activity, besides materials development, is in processing research.

PROCESSING

Plasma Spray Coatings - The traditional method of TBC application is to plasma spray in air. This tends to cause oxidation and loss of the reactive elements such as aluminum, chromium, etc. in the bond coat alloy, reducing its oxidation resistance. It also gives a degree of porosity in the bond coat which may be undesirable. Argon shrouding (SPS), or the use of low pressures of argon known as either low pressure plasma spraying (LPPS) or vacuum plasma spraying (VPS) can produce dense clean bond coat layers, and much research is being directed into these areas to produce the next generation of coatings. These are already available as commercial systems, and one of the combustion chamber components in the military Adour engine (in service in 14 countries throughout the world) has run such a coating for over fifteen times the average life of the three layer "mag zirc" coating, with no sign of deterioration whatsoever. Internal research is concentrating on the development of manipulative systems for coating complex bodies, and on the continued use of plasma sprayed MCrAlY's for oxidation protection as much as for TBC bond coats.

Optimizing Spray Parameters - Regarding plasma spraying of ceramics, there are probably in excess of twenty process, chemistry and interface geometry parameters which can be varied, largely independently, all of which may yield significant effects on the life of the TBC as a system. Several of these have recently been outlined by Stecura (Reference 8) who has generated an equation with twenty-four variables in it, for evaluating coating life. This is the most thorough, quantitative study we have seen published to date, but still leaves many of the vital parameters untouched. Although the general principles of the effects of these remaining parametric variables are understood in a qualitative way, the task of preparing a full quantitative mapping of their effects remains a daunting exercise. By educated guesswork we believe we have at least doubled the thermal cycle capability of our coatings, but there can be no shortcuts to find the processing technology which will yield a consistent, reproducible optimized coating. One of the early pieces of work which has contributed to our quest for reproducibility concerned the selection of a supply source for the 8% Y_2O_3 in ZrO_2 we currently use. Some fifteen powders were sprayed up and tested, and two orders of magnitude separated the best from the worst. On x-ray diffraction of the as sprayed deposites, it was seen that the intensity of the monoclinic reflections increased as coating life decreased. The presence of monoclinic zirconia was taken to indicate a less than fully homogenized coating composition, and it was thus determined that only suppliers whose powders gave fully homogenized sprayed coatings could be included in the specification. It was noted however that this did not bar some powders which contained high levels of monoclinic in the powder itself, but did exclude some powders with substantially less monoclinic material in the powder form. Being produced by different processing routes, the interactions of these powders with the plasma lead to different degrees of mixing, and thus to different degrees of homogeneity in the sprayed coating. By restricting the source of supply in this manner, higher reproducibility can be achieved, but this is still on the order of ±20%. (Figure 6)

Other Deposition Processes - Alternatives to the plasma system exist, of which PVD is perhaps the best known. In this system an ingot is melted in a vacuum chamber, a vapor cloud is formed, and a coating condenses on the work piece from the vapor cloud. The process has severe constraints on the cleanliness and in the details of process control, and these can be even more strict when the ceramic is to be processed. To deposit the ceramic it is possible to either melt a metal and use an oxygen bleed into the chamber, or to coat directly from a ceramic "ingot" with zirconia. There is a partial dissociation at high temperature and low oxygen partial pressures, which cause the oxide to turn grey, and this can be recovered to a creamy white by air heat treatment after coating. Given the strict control necessary over chemistry, cleanliness and general processing, early samples tended to be highly variable. In one batch of material Rolls-Royce obtained, there were samples with capabilities as good as any plasma coating system we had seen at that time, and others which were already spalled off the backing plate on receipt. PVD samples thermally cycled were at least as good as the 8% YSZ samples having surpassed 4000 cycles without failure. (Figure 6) Testing continues. Much work has since gone into the development of this process and if it can be guaranteed reproducible, it should have significant potential, particularly as it produces extremely smooth coating surfaces.

Future Work - Having demonstrated some of the factors which control reproducibility and mean capability, we are now engaged in work to progress well beyond our current datum coating towards systems which are sufficiently reliable to consider designs of components which depend on the TBC remaining intact in order for the component to survive more than one or two flight cycles. Such critically cooled components are still a long way off, but must be the ultimate goal if we are to fully exploit the TBC effect.

SERVICE EXPERIENCE AND ENGINE TRIALS

OBJECTIVES - One of our current programs to evalute different TBC coating systems in a turbine environment is a flight service evaluation program. While cycles play an important role in the life of a TBC, time at temperature is also an important factor for bond coat oxidation failure, as discussed earlier. Many airline operations fly long routes and therefore have high hours to cycles ratios. Rolls-Royce therefore put together a flight service evaluation program to evaluate the long term effect on TBC's.

The purpose of this service evaluation program is four-fold

1. To compare various TBC systems in the turbine environment as compared to 3 layer mag zirc (3LMZ) with which we have much experience. And of course if 3LMZ would hold up in the intermediate or high pressure turbine, so much the better.
2. Determine the effects of long-time and cycle exposure, try for 10,000 hrs/5000 cycles.
3. Compare flight data to carousel cycle data to see if a simple consistant correlation exists.
4. Identify new coating systems worth testing in this program.

TEST VEHICLE - For a flight service evaluation, the component chosen to coat must not be dependent on the protective coating, so that if the coating failed, the hardware would not suffer significantly. At the same time the heat flux of the component must be high to give the TBC a good test. The component chosen to meet these requirements was the intermediate pressure turbine (IPT) NGV. See Figure 7. It was applied only to the inner diameter pressure side platform to minimize the flow area loss. The cooled platform runs in excess of 925°C while the scrubbing temperature is in the neighborhood

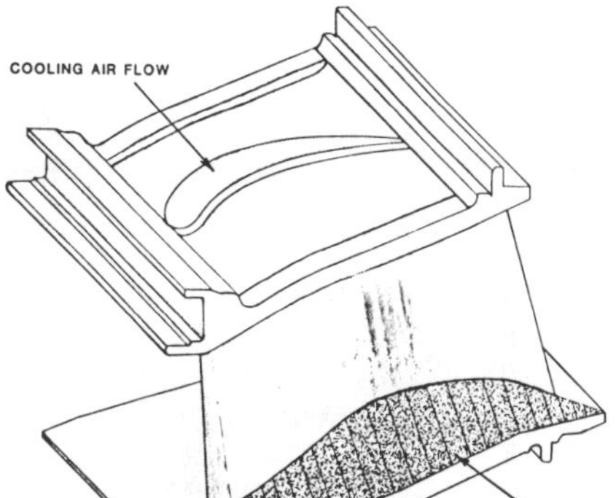

Figure 7 RB-211 Intermediate Pressure Turbine Nozzle Guide Vane with TBC

of 1000°C. The vane is nominally 8" long, 3" wide and 4" tall, is made of a nickel base alloy (C1023), and is normally uncoated.

COATING SYSTEMS EVALUATED - Several different coating combinations or systems were tested. See Figure 8. The base line 3LMZ was given an identifying "A" designation. The coatings were tested in rainbow fashion, typically 5 coatings per engine, 5 vanes per coating on 4 engines. Coating A was installed in all four engines for a total of 21 vanes. The vanes were clocked for a different location in each engine. Coating B, 2LMZ, was only installed in engines 1 and 2 for a total of 11 vanes. Coating C used the same 443 bond coat but with a 6-8% Yttria stabilized zirconia top coat, hereafter called 7% YSZ. It was installed and clocked in all 4 engines. Coating D has a CoNiCrAlY bond coat applied by VPS plus a 7% YSZ top coat. This coating was not originally in engine build 1, but was installed at the first teardown (2500 hrs), now designated build 1a. Coating E has the same CoNiCrAlY bond coat as "D", but it was applied by SPS . The ceramic is 7% YSZ. Vanes were installed on all 4 engines for a total of 21 vanes. Coating G is another variation of "D" and "E" using the same (CoNiCrAlY) bond coat composition, but plasma sprayed in air, so we will have a comparison of the bond coat applied by 3 significantly different techniques. 5 vanes were exchanged when build 2 was inspected, but they were replaced with the like coating. Five vanes were uncoated, typically one per engine. That brings the total to 112 vanes.

OTHER VARIABLES - There were a few other minor variables, such as coating thickness, type and source of ceramic powder. Builds 1 and 2 vanes, and all coating E vanes were new hardware. The remainder were cleaned up used vanes of unknown age, but probably had between 2000 and 4000 hrs on them.

FLIGHT STATUS - For the 4 engines flying, the status as of August 3, 1985 is given in Figure 9. The time shown is the time on the TBC. Two engines have over 6000 hours, one over 5000 and the other over 4000 hrs. To get an approximation of the associated cycles divide the hours by 2. For example when build 1 was pulled with 4160 hours it had 2271 cycles. Or roughly 2 hours per flight cycle. To date, 3 of these engines have been in the shop. However, only on builds 1 and 2 was the 05 module (where our TBC vanes are) torn down so that we could closely examine and positively identify the vanes.

STATUS AS OF AUGUST 1985			
ENGINE BUILD	ENGINE NO.	FLIGHT HOURS	INSPECTED AT-HOURS
1	10676	6188	4160
2	10575	6537	2539
3	10650	4275	--
4	10398	5522	3871

Figure 9 TBC Service Evaluation Status as of August 1985

RESULTS

Build No. 2 - Engine build 2 was removed in December 1983 for high oil consumption. At that time the TBC vanes had 2539 hours and 1353 cycles. This engine had 5 different coatings systems (Fig. 10), 5 vanes each, plus one uncoated vane (the borescope vane). The vanes were grouped together starting at the borescope vane going counter clockwise when looking aft. (Fig. 11) The borescope vane is roughly at 7:30 so that the first group is centered at approximately the 6:00 position.

IDENT	BOND	CERAMIC	BOND COAT METHOD	ON ENGINES	TOTAL VANES
A	443+441	210 MZ	Plasma	1234	22
B	443	210 MZ	Plasma	12	12
C	443	YSZ	Plasma	1234	22
D	CoNiCrAlY	YSZ	VPS	1a234	20
E	CoNiCrAlY	YSZ	SPS	1234	21
G	CoNiCrAlY	YSZ	Air Sprayed	34	10
Uncoated Vanes					5
	TOTALS				112

Figure 8 Coatings used on RB-211 Service Evaluation IPT NGV's

BUILD 2

	COATING	POSITION
A	M443+441+210MZ	3:30
B	M443 + 210 MZ	1:30
C	M443 + YSZ	11:00
D	CoNiCrAlY[2] + YSZ	9:00
E	CoNiCrAlY[1] + YSZ	6:00

M=Metco, MZ=Mag Zirc
1 = SPS
2 = VPS

Figure 10 Engine Positioning of TBC's for Flight Service Evaluation

For build 2 at 2539 hours, the vanes with "A" had already lost a moderate amount of coating. Figure 12 shows how the typical 3LMZ vane looked. It has loss of coating at the leading edge (LE) and along the platform edge. The bond coat was loose in the fillet radius, about 2 inches from the trailing edge (TE). Other vane photographs can be found in Reference 10.

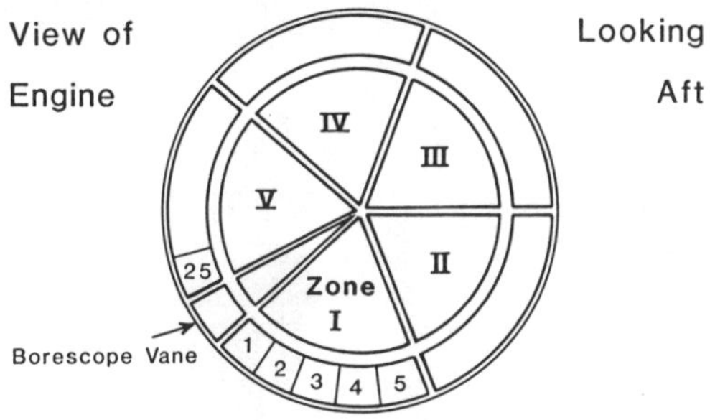

ENGINE ZONE NUMBER	I	II	III	IV	V
		COATINGS			
1	A	B	C		E
2	E	A	B	C	D
3	D	E	A	G	C
4	C	D	E	A	G

Figure 11 Location of Coatings used in Service Evaluation

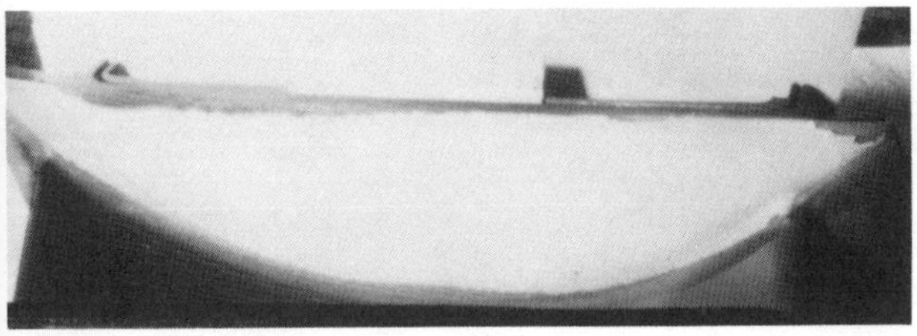

Figure 12 3LMZ @ 2539 hrs/1353 Cycles

The 2LMZ, coating B looked good. The worst of the 5 coated vanes is shown in Figure 13. There is a small loss of coating at the airfoil LE and along the platform edge. These could however be due to handling. Note that the bond coat is not adhering well in the fillet radius (FR) and that some is missing. One other vane had bond coat missing. The other three vanes had essentially no coating loss.

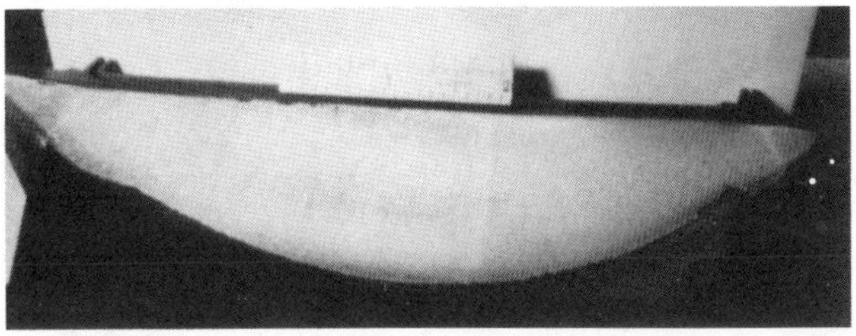

Figure 13 2LMZ @ 2539 hrs/1353 cycles

Coating C with the Metco 443 bond coat and the 7% YSZ also looked good at this point as can be seen in Figure 14. No coating loss or loose bond coat was noted on any vane although the bond coat was quite rough on 3 vanes at the coating FR interface.

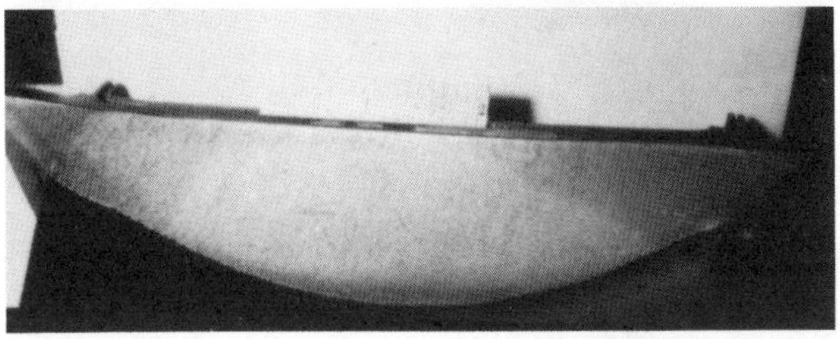

Figure 14 Metco 443 + YSZ @ 2539 hrs

Coating D, the VPS applied CoNiCrAlY bond coat + 7% YSZ is seen in Figure 15. There was some coating loss near the airfoil LE. Three of the vanes had small coating loss along the platform edge . The bond coat in the FR was good on all five vanes.

Figure 15 VPS MCrAlY + YSZ at 2539 hrs

Coating E uses the same CoNiCrAlY bond coat as "D", but it was applied by SPS + 7% YSZ. All 5 vanes were in excellent condition (see Figure 16) with no coating spallation or bond coat problems.

Figure 16 SPS MCrAlY + YSZ @ 2539 hrs/1353 cycles

BUILD 1

	COATING	POSITION
A	M443+441+210MZ	6:00
B	M443 + 210MZ	3:00
C	M443 + YSZ	12:00
E	CoNiCrAlY[1] + YSZ	9:00

1=SPS

Figure 17 TBC's Used and Positioning for Flight Service Evaluation

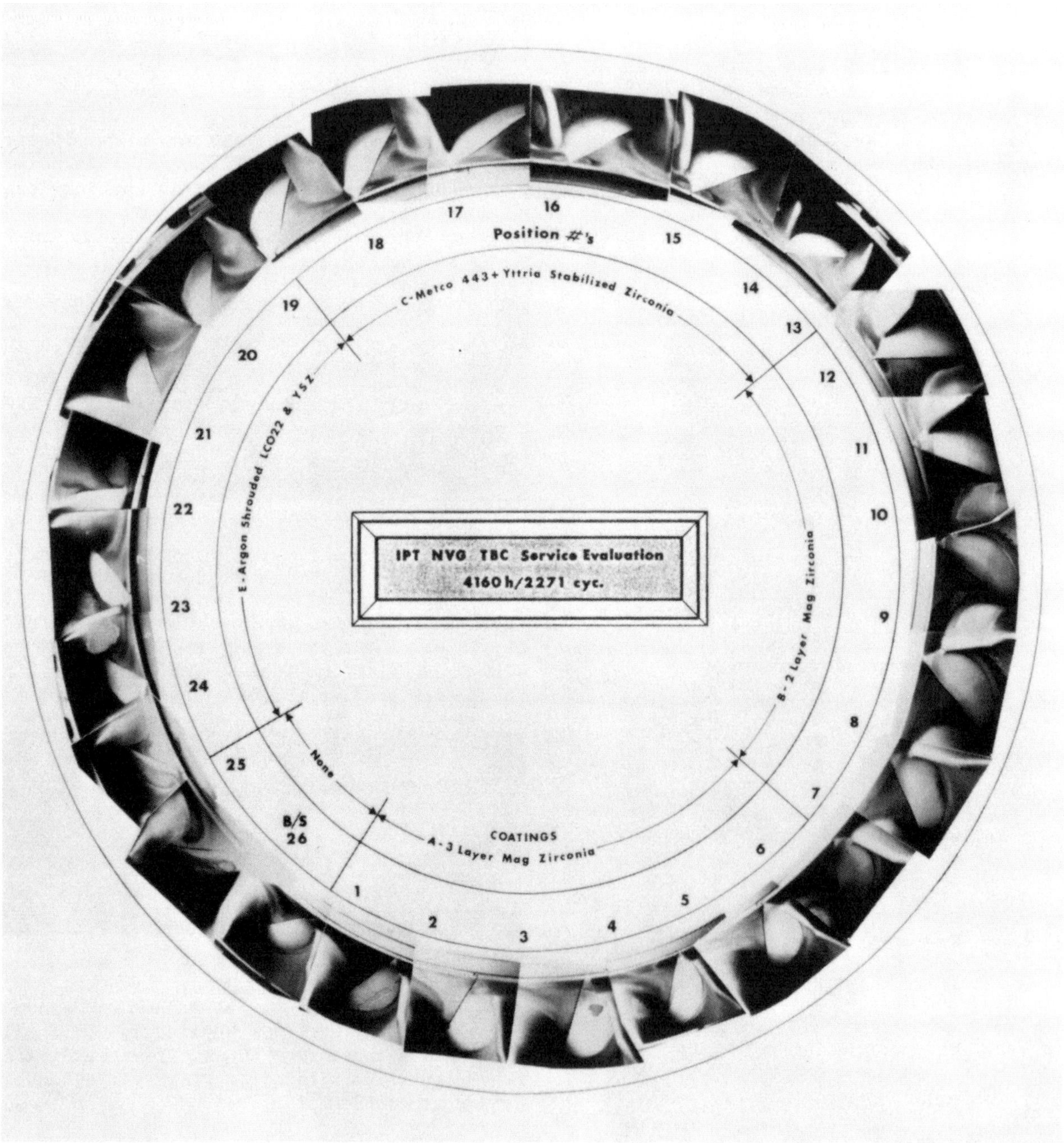

Figure 18 Overall View of TBC's After 4160 in Build 1

Build No. 1 - Engine build 1 was removed in July 1984 for a time expired compressor shaft. At that time the TBC vanes had accumulated 4106 hours and 2271 cycles. This engine had only 4 coating systems, 6 vanes per coating plus two uncoated vanes (Figure 16). Coating A vanes were grouped at the 6:00 location, "B" at the 3:00 position, "C" at the 12:00 position and "E" at roughly the 9:00 position when looking aft. Refer to figure 17.

Figure 18 gives a feeling for how the IP turbine vanes looked in situ for build 1. At just over 4000 hours, the 3LMZ vanes (6:00) can be seen to have a lot of spallation, the 2LMZ now has a moderate amount of coating loss, but the YSZ coatings by and large look very good. The figure also gives a feel for where the hot spots were. Vanes 9, 10 and 23 were obviously hot. Note that the "A" coatings were not in a particularly hot segment of the engine but that "B" and "E" were.

Coating A, the 3LMZ looks worse at 4100 hrs, than the set with 2500 hrs did. In Figure 19 the coating has failed between the ceramic and the cermet. On some of the vanes the failure occurs at the cermet/bond coat interface resulting in significant coating loss, as in vane 4 (Figure 18). Two vanes had loose bond coat in the FR.

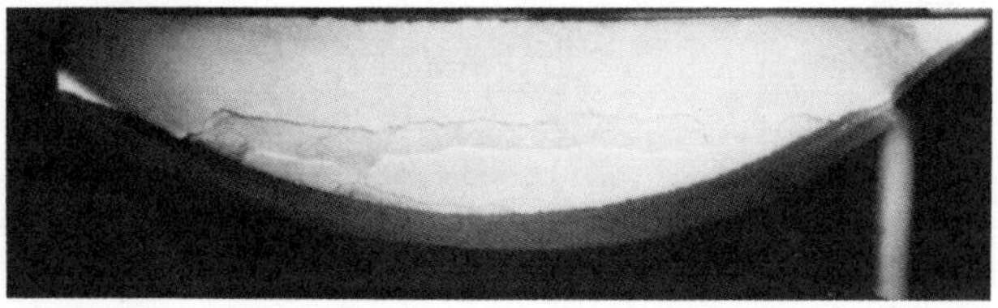

Figure 19 3 Layer Mag Zirc after 4160 hrs/2271 cycles

The 2LMZ ("B") also had more spallation at 4100 hrs. In Figure 20 a large piece is missing at the platform LE. Failure occurred predominately within the ceramic near the bond coat interface, with some within the bond coat itself. The typical vane however had only small losses at the LE and along the platform edge. The bond coat in the FR looked suspicious on two vanes including this one. While worse than the earlier set the 2LMZ still looks a lot better than the 3LMZ.

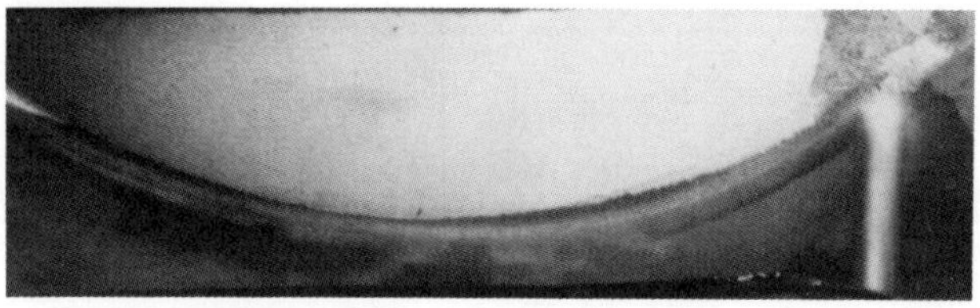

Figure 20 2LMZ after 4160 hrs

Coating C looked very good at 2500 hrs and still looks good at 4100 hrs (Figure 21) with only very minor coating loss. However, the bond coat in the fillet radius is loose and some has fallen off. Two vanes were like this which probably indicates that they were approaching the effective life of the 443 coating.

Figure 21 Metco 443 + YSZ after 4160 hrs

Coating E (SPS MCrAlY) again looked very good, only one vane showing any sign of coating loss at all, (Figure 22), and this could have been from new. The bond coat was in excellent condition.

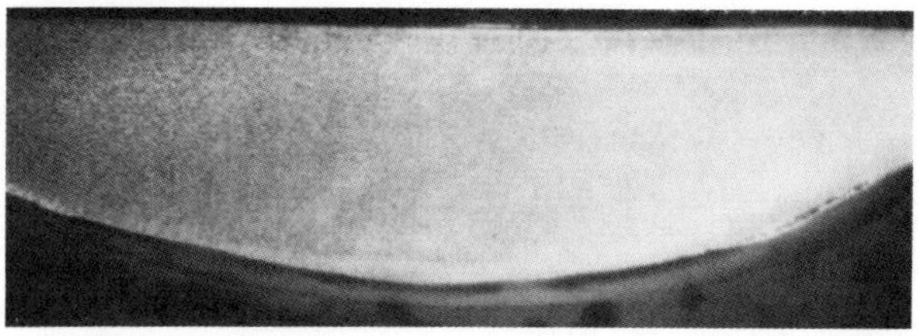

Figure 22 SPS MCrAlY + YSZ @ 4160 hrs

COATING LOSS WITH TIME - By measuring the area of coating missing (adjusted to 5 vanes) and plotting it against time we get the chart of Figure 23. It shows that spallation is progressive with time for both MZ coatings. The 3LMZ starts very early, probably around 100 hrs where the 2LMZ does not see significant spalling until about 3000 hrs, it may however catch up to the 3LMZ by the next shop visit.

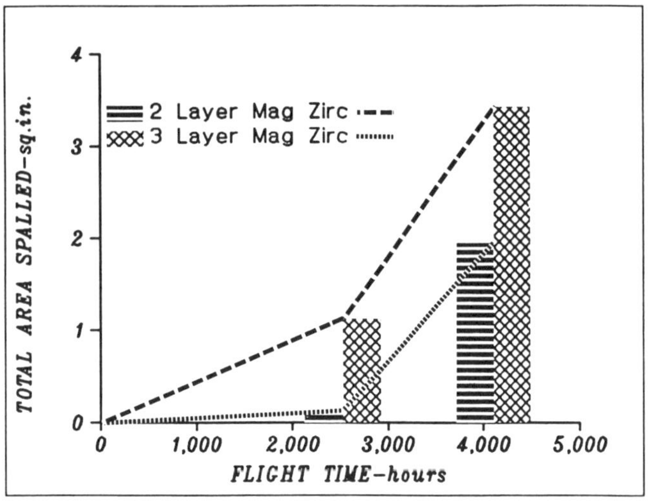

Figure 23 Comparison of Spallation with Time Between 2 and 3 Layer MZ.

SUMMARY - In conclusion then the 3LMZ is not adequate in the turbine. Its failure at low hours does however indicate that it was subjected to a higher heat load than it experiences in the combustor where the coating typically lasts 5000 to 10,000 hrs. 2LMZ, while significantly better, is not really suitable for the turbine applications either, except perhaps for a 3000 hrs or less application. An improved bond coat would probably add a few thousand hours to that.

YSZ was superior in every instance and with the proper bond coat tailored to the component it should give the required coating lives.

As for bond coats, Metco 443 has demonstrated a limited oxidation life and is probably not advisable for use on airfoils or rotating tubine parts. The VPS MCrAlY was somewhat suspect on build 2, but preliminary indications are that it looked very good on build 3 at 3500 hours. The best bond coat tested to date is the argon shrouded CoNiCrAlY.

Future work in this program will include service evaluation of other promising TBC systems. Much work has been done by the coating community on the standard plasma MCrAlY + 8% YSZ system. However to see the quantum jump that is required for the next generation of engines and to take full advantages of the property gains of new material process such as single crystal castings, other coating materials and processes must be explored. Several promising coating systems have been identified during this program, two of which have done extremely well in cyclic rig testing. Coating systems to be evaluated next are:

1. Plasma (NiCrAlY + CaO TiO_2)
2. PVD (NiCoCrAlY + 20% YSZ)
3. PVD (NiCoCrAlY + 6-8% YSZ)

We also plan to move on the airfoil with some of the more promising TBC systems.

APPENDIX A

The standard Rolls-Royce thermal cyclic durability test consists of alternately rotating two inch square coated coupons between 1125°C and 250°C continuously and is known as "carousel rig testing". The rig consists of an 8 arm spider, each holding one tested specimen. The spider is pneumatically turned clocking the specimens alternately into a torch flame for 1 minute than into a cooling jet for 1 minute.

REFERENCES

1. Stecura, S., "Two Layer Thermal Barrier Coating for High Temperature Components", Am. Ceramic Society Bulletin, Vol. 56, No. 12, December 1977, p.1082-1085.

2. Levine, S.R., Miller, R. A., Hodge, P. E., "Thermal Barrier Coatings for Heat Engine Components", SAMPE Quarterly, Vol. 12, No. 1, October 1980, p. 20-26.

3. Heuer, A. H., editor, "Science and Technology of Zirconia", American Ceramic Society, Advances in Ceramics Series, Vol. 3, 1981.

4. Heuer, A. H., editor, "Science and Technology of Zirconia II", American Ceramic Society, Advances in Ceramics Series, Vol. 12, 1983.

5. McDonald, G., Hendricks, R. C., "Effects of Thermal Cycling on ZrO_2-Y_2O_3 Thermal Barrier Coatings", NASA-TM-81480, 1980

6. Miller, R. A., Lowell, C. E., "Failure Mechanisms of Thermal Barrier Coating Exposed to Elevated Temperatures", NASA-TM-82905, 1982.

8. Stecura, S., "Optimization of the NiCRAlY/ZrO_2-Y_2O_3 TBC's", NASA-TM 86905, May, 1985.

9. Vogan, J.W., Stetson, A. R., Solar Turbines, Inc., "Advanced Ceramic Coating Development for Industrial/Utility Gas Turbines", NASA-CR 169852, January 1982.

10. Toriz, F. C., "High Time Evaluation of Thermal Barrier Coatings on the Rolls-Royce RB211 Engine", NASA TBC Workshop Abstracts and Figures, p.187-196, May 1985.

PRACTICAL ASPECTS OF ULTRA-THICK THERMAL BARRIER COATINGS

Doug H. Harris
APS-Materials, Inc.
Dayton, Ohio, USA

Abstract
Ultra-thick [>2.5mm (0.1 in.)] thermal barrier coatings can be produced using arc-plasma deposits of stabilized zirconia. Success requires attention to geometric details, effects of substrate material, and spraying conditions.
spraying conditions.

THE STATE OF THE ART for thermal barrier coating (TBC) thickness has been expressed as less than 1mm (.039in.)[*] although several major projects have demonstrated thicknesses above 1.5mm (.059in.).

The production of a thick TBC's using arc-plasma spraying (APS) methods has been demonstrated using fully and partially stabilized zirconia. For our applications, many of the TBC overlays have optimum thermal insulative properties at thicknesses of 7mm (0.25in.) but coatings up to 19mm (0.75in.) thick have been successfully constructed.

The need has come from various process industries where the coating is important to the operation of critical components. Applications have included bearing mounts, press platens (TBC used to protect hydraulics), brake pistons, and compressor pistons.

We have had opportunity to practice our skills on piston engines; primarily diesels. There is a growing interest in applications to reciprocal engines of all types with motives being the retarding of hot-corrosion, improved power and torque performance, and reduced fuel consumption. Ultra-thick TBC's are frequently required and it seems appropriate to discuss some of the more practical aspects of depositing such materials.

*TBC's have been largly explored by the turbine industry for aerospace applications. This figure reflects industry opinion.

SOME PRACTICAL CONSIDERATIONS

The success of a thick TBC is related to the internal stresses of the coating and how they respond to the external thermal and mechanical are applied. The evidence points to the use of partially stabilized, 6-8$\%_{mole}$ yttria zirconia with a porosity of 12% as the material of choice for optimum resistance to these stresses and the maintenance of long life under cyclic conditions.

In such material the phenomena of transformation toughening results in a well distributed series of microcracks that serve to relieve the intrinsic strain of the zirconia coating. It has been demonstrated that through proper control of the coating morphology, these cracks remain "micro" and do not combine significantly enough to produce coating failures. It should be noted that the porosity of 12% is achieved by adjusting the deposition parameters.

It must be remembered that this evidence regarding thermal shock, corrosion, thermal cycling, and lifetimes by virtue of closely controlled porosity and percent stabilization is the result of the airborne turbine engine industry and government agency studies related thereunto. When considering other applications, i.e. highly stressed diesel engines, such data may be only a starting point for additional study.

The diesel brings to the TBC system fresh demands such as wider temperature fluctuations (cyclic), fuels with higher corrosive content, and cyclic pressure changes that can range from 1.38 to 13.8 MPa (200 to 2000 psi).

Another factor that has to be faced when considering diesels is the lifetime of the TBC system. Aircraft engines are subject to regulatory inspection and overhaul where replacement of such coatings is usual practice. This is not the norm for industries where the diesel predominates!

PLASMA SPRAYING CONSIDERATIONS

The transformation toughening properties and the microstructure of the zirconia is very dependent upon the plasma spraying and material parameters. Unfortunately, some of these parameters are usually compromised by virtue of substrate geometry or system control over the period required for deposition of the ceramic layer.

MATERIAL CONSIDERATIONS, limited to the plasma spray powders, should concern the ability to deposit homogeneous, stoichiometric product.

We have found in our studies that this is best done with fully fired or fused ceramic powder; but it is not enough to assume that $7\%_{mole}$ yttria stabilized zirconia will arrive at the substrate as the same composition! Process optimization always should include a study of the effects of altered composition on the powder of choice.

PROCESS CONSIDERATIONS for ultra-thick TBC must include the long-term control of the deposition parameters. It is probable that many of the failures observed thus far in the application of such coatings were derived from the use of hand-spraying methods where operator fatigue, and perhaps lack of skill, caused poor morphology.

Process Interruptions are of equal concern to the spraying parameters. We have found that failures of ultra-thick TBC can often be traced to an interruption in the deposition process of the zirconia. Not always has the mechanism been clearly defined. Some of the interruptions are caused by the need to refill powder feeders, install fresh electrodes, or by a simple operator break.

Under microscopic examination at higher magnifications, the deposit still appears homogeneous across the break interface, yet the coating will spall at that location under later testing. Figure 1. shows a low-magnification cross-section of an interrupted coating which does reveal an interface.

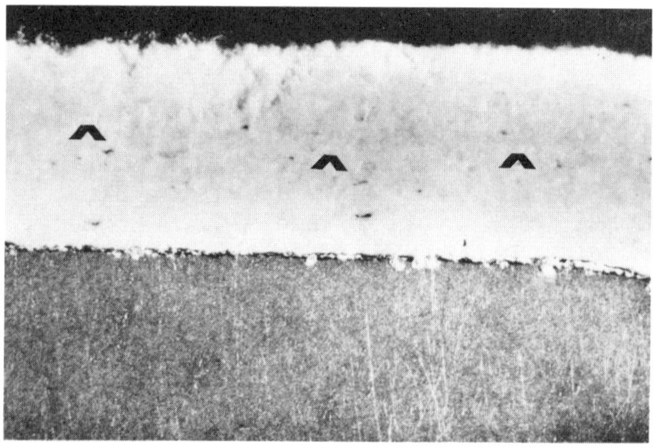

Fig. 1.- 10X cross section showing an interface as the result of an interrupted plasma process.

In order to avoid such 'hidden' interfaces, we make a practice of using a powder feeder and a quantity of powder sufficient enough for the job, and new or nearly new electrode sets for the plasma torch.

GEOMETRY EFFECTS are more subtle in that what we usually ignore on thinner deposits becomes very troublesome to the morphology of ultra-thick deposits.

Perpendicular Walls One example relative to the diesel engine would be the requirement of depositing the TBC within the confines of a machined cavity such as the fire deck of the head. In this particular example there are several vertical walls which frame the periphery of the cavity and the valve seats. In addition, there may be several more vertical walls created by masking of the threaded holes used for the fuel injector, a glow-plug, or laboratory instrumentation. Furthermore, it may not be practical to provide large radii at the intersection of these surfaces that are normal to each other.

The object of the plasma process in this example, is to provide and TBC which is an inlay in the provided cavity. The coating has to be deposited at some thickness which is greater than the cavity depth and machined flush with the overall head surface. In the performance of this inlay, two problems are present: (1) it is most difficult to apply the deposit to all subject surfaces at the proper angle of incidence which has been determined through earlier optimization, and (2) there exists zones of higher turbulence in the cavity which alter the flow of the plasma gasses, the molten droplets, and the cooling gasses.

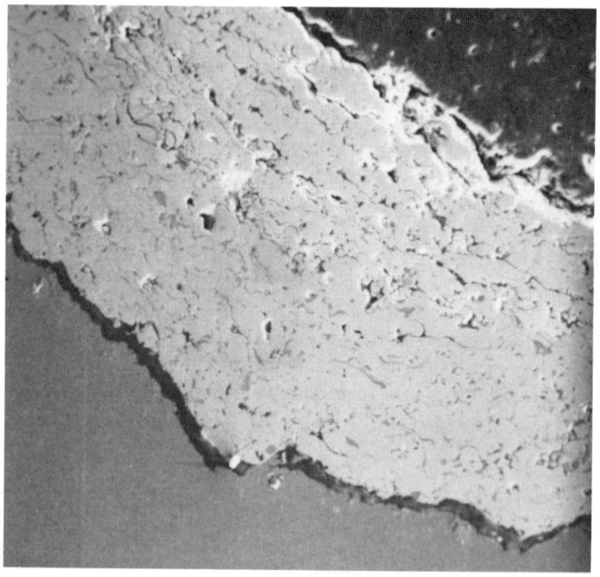

Fig. 2 - SEM (100X) of improperly deposited zirconia on L-shaped substrate showing unmelted particles, voids, and cracks which are effects of low impingement angle and turbulence.

Both problems may alter the morphology of the TBC by simple reason of compromised parameters; but a greater threat is imposed by the presence of rebounding molten particles, some of them not fully melted, which may attach to other surfaces to become part of the overall zirconia structure.

Figure 2 partially illustrates the problem. The portion of the coating shown is that located on the radius of an L-shaped geometry representative of the 'eyebrow' formed by valve relief cutouts on a piston. Even at this low magnification, the SEM shows the larger rounded unmelted particles lodged in the coating. The excessive pull-out is likely due to the entrapping of over-spray dust which was not properly blown away by cooling streams. The overall effect of macro-sized cracks following the curvature is also evident.

The use of metal shadow masks can aid in solving this problem, but care must be taken not to embed masking in the coating or that the mask itself becomes a source of additional turbulence.

Raised Edges are another geometric source of concern. The best example relevant to a diesel is the top of a piston. An ultra-thick TBC brought to the very peripheral edge of the top is usually structurally weak. Masking 'tricks' must be used to create a wall of TBC which is a vertical extension of the piston skirt. These efforts are again the source of turbulence and morphology problems discussed previously.

We have noted in thermal cycling studies of such edges that the corner portion of ceramic which is not in direct contact with any other metal surface, is prone to higher temperatures than the major portion of the surrounding surface, and yet because it is protruding into the gas stream which is the thermal energy source, it will respond with less thermal inertia. The effect is an enhanced cyclic fluctuation of this corner.

We feel that many of the failures that occur in this geometry are caused by this greater thermal stress and the likelihood of impaired zirconia morphology.

COMPONENT DESIGN FOR ULTRA-THICK TBC

Current work with ultra-thick TBC's usually employs adaptation to already designed and existing components. It will be obvious from the above that the difficulties encountered by so doing may deny us valid data points, or at least limit the experiment.

Any consideration given to the use of ultra-thick TBC on production hardware must include new designs to simplify coating procedures.

Relevant to the diesel, other failure modes not yet encountered must also be contended with.

SUBSTRATE-TBC INTERFACES are needed to bring about optimum bonding. To some extent the choice of substrate material will effect the method and perhaps even the ultimate thickness of the TBC. The thermal expansion properties of aluminum alloy, for example, make it a poor choice for ultra-thick TBC. In general, we are going to have to contend with ferrous castings (because of cost) though alloys of lower expansion would be more attractive.

The difference of thermal expansion between the ferrous materials and the zirconia ceramic (10.6 to 14.9 x 10^{-6} m/m/K versus 5.6 x 10^{-6} m/m/K) is a major cause for concern. Most seek an 'adjustment' of some kind at the interface so that external mechanical stresses to the TBC will be minimized. The use of compliant layers is such an approach. Without such compensation, the zirconia becomes its own 'complaint' layer and hope rests on the retarding of enough heat flux from the substrate to prevent excessive stress from forming microcracks into macrocracks and debonding.

Single Bond Coats of MCrAlY, where M=Ni,Fe, or Co are used throughout the industry for TBC's. As applied to ultra-thick TBC, they have been effective at times for overlays up to 7mm (0.275in.) thick, though they offer no adjustment to the thermal expansion mismatch.

The experience of the turbine engine industry shows that vacuum plasma applied bond coats are far more resistant to corrosion attack than air-spray deposits. We have not, as yet, used vacuum plasma on diesel engine components. At present technology levels, it is doubtful that testing will get far enough for a component to suffer corrosive failure. But should vacuum plasma become a necessity, the complexity of that process deems it necessary that significant design changes take place for many of the combustion chamber components.

Graded Coatings which employ interface mixtures of MCrAlY and zirconia continuously varying in ceramic content from 0 - 100% over a short distance have been tried as another approach to interface adjustment.

We have pursued this approach with vigor using zirconia and other ceramics on behalf of several coal-fired magneto-hydrodynamic generator projects with operational wall temperatures in excess of 2200°C

Our findings were always the same; ceramics which operated with transformation toughening characteristics fractured just above the interface within the graded mix, delaminating from the substrate within a few hours of operation. Those who did not work with this principle survived very well by comparison.

Analysis has shown that the metal inclusions, having greater expansion than the surrounding zirconia may have served as a kind of 'micro-wedge' which widened the microcracks of the partially stabilized material to failure point.

Subsequent testing on diesel components at lower temperatures has shown similar failures and the approach is not considered promising at this time.

Feltmetal Complaint Layers have also been used extensively by ourselves and others [1,2] with various degrees of success. These are formed of sintered, interlocking metal alloy fibers where alloy selection is made based on the thermal conditions encountered and the required mechanical properties. The pad is brazed to the substrate and forms the surface of deposition. Variation of the density of the feltmetal is another method of altering the thermal characteristics.

We have seen the material used both with and without MCrAlY bondcoats. Part of the reasoning behind the use of feltmetal is that the open porosity presents an ideal surface for the penetration of molten particles thus providing enhanced bonding. Since the feltmetal itself has poor heat transfer, its use generally means that thinner deposits of zirconia are required which makes it very effective.

When used with ultra-thick TBC, we find empirical evidence that the penetration of the zirconia into the pad may be a weak point. Like the graded coatings mentioned above, the presence of a higher expansion material surrounded by the zirconia enables that round cross-sectional metal to act as a wedge.

IMPROVED TBC RESISTANCE to wear and corrosion is necessary if applications are going to be made to diesel engine combustion chambers. Some of the area, at least a portion of the cylinder liner, may require a thermal barrier which will see frictional wear from piston rings.

Sealing the Zirconia is one way to improve wear and perhaps the overall protection of the substrate from corrosive media. We have worked with ceramic slurry coatings** that contained chrome oxides, alumina, and silica dispersed in a chromic acid media. With several applications and firings at 540°C, the slurry coating could, on occasion penetrate and seal off the TBC.

This sealing process was used by Cummins Engine on the adiabatic project[2] and was found to yield an excellent wear surface with low friction coefficients. But some failures during application and in later use caused us to examine the the process further.

We were embarrased to find that chromic acid dissolves yttria--even when fired with zirconia. Coatings which were fully stabilized became partially, with unknown percentages and almost certain inhomogeneity.

It is now known that diesel fuels are apt to be more corrosive with some of the salts directly attacking the TBC by removal of the yttria thus reducing stabilization.

Top Coats are of greater interest now that we are aware of the corrosion problem. We have used two materials primarily on ultra-thick TBC; (1) chrome oxide, and (2) a $20\%_{mole}$ chromia-alumina solid solution ceramic. Both are used in highly corrosive industrial applications with the chromia-alumina showing a great deal of success operating at temperatures up to 1200°C. They have been tried as top coats on diesel components but tests were not sufficiently long to determine corrosion resistance.

Hollow Microspheres have been reported used in TBC[1] for the purpose of enhancing corrosion resistance. Both zirconia and alumina silicate ceramics (as hollow spheres) have been dispersed in the zirconia coatings and tested in actual engines.

Figures 3 and 4 show two such coatings using the alumina silicate spheres which we believe will toughen the microstructure of the TBC against high temperature flame erosion. These photos show, respectively, a $20\%_{vol.}$ loading of spheres into zirconia and a coating comprised of 100% spheres.

It is known that flame speeds increase significantly in TBC-lined combustion chambers and evidence of such damage already exists in several applications to high performance gasoline or racing fuel engines. We applied the principles illustrated here to a set of aluminum pistons for the formula 1 March-Cosworth engine. Bench tests thus far at 11,000 rpm have shown no deterioration for these coatings where as the 'standard' TBC (at <0.5mm thick) would likely be damaged under the same conditions.

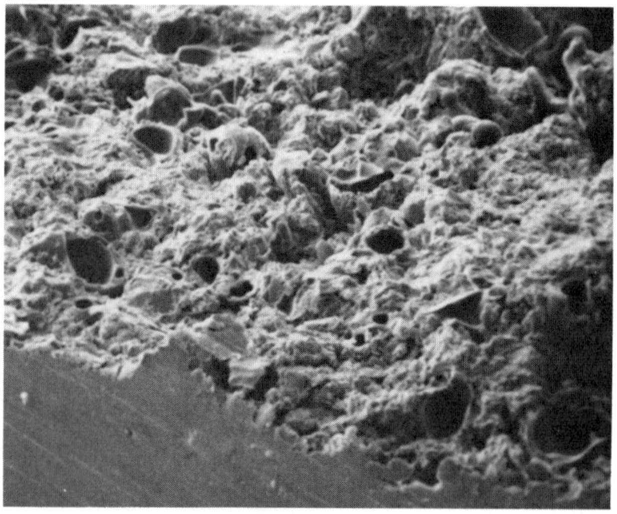

Fig. 3 - Structure of $20\%_{vol.}$ alumina silicate microspheres added to zirconia TBC. SEM at 100X of sawed substrate plus fresh break fracture of ceramic.

**K-ramic™ - Kaman Sciences Corporation

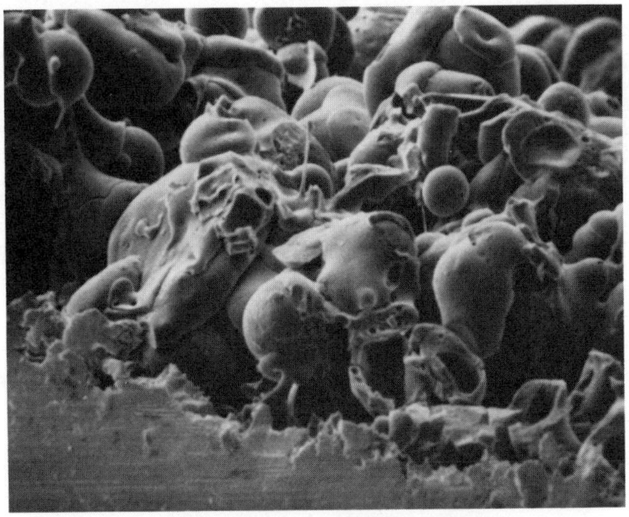

Fig. 4 - Structure of 100% alumina silicate microsphere coating. SEM at 100X of sawed substrate plus fresh break fracture.

THERMAL CONDUCTIVITY of these and other TBC's is always of interest to the component designer. Figure 5 portrays a collection of data taken from the work at Cummins and Detroit Allison. It shows that the 'standard' ultra-thick TBC using NiCrAlY bondcoat as supplied by us to Cummins offers the best thermal resistance.

The Allison HYSZ designation is an attempt at measuring the thermal conductivity of the hollow zirconia spheres in coating form. Lacking appropriate samples, a cemented sample was used to approximate a plasma-sprayed version.

The Allison 80/20 YSZ is similar to our $20\%_{vol}$. alumina silicate hollow sphere version. Although it is not shown on this figure, their plasma-sprayed partially stabilized zirconia tests almost overlap these data.

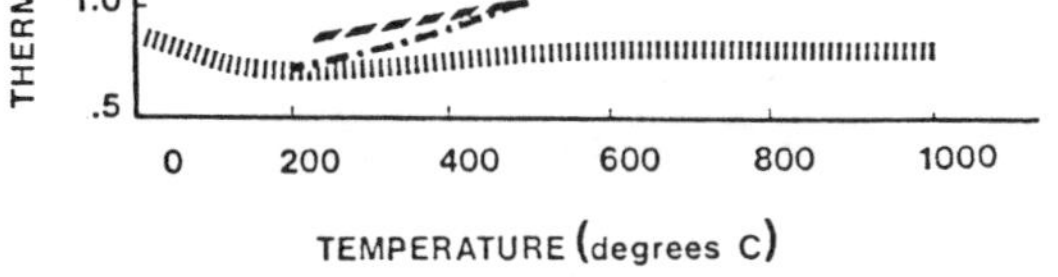

Fig. 5 - Thermal conductivity data for various TBC's and solid zirconia ceramics. Heavier data lines are from Cummins and the lighter lines are Allison data.

A DEMONSTRATION ENGINE

A "classic" illustration of the use of ultra-thick TBC was the approach-to-adiabatic experimental diesel which resulted from the joint development program by the U.S. Army-Tank Automotive Command and Cummins Engine.*** This engine was based on the currently used NHC-250 engine; a 6-cylinder in-line turbocharged unit with a 5.5" bore and a 6" stroke, 855 C.I.D.

The combustion chamber of the engine was insulated using a fully stabilized (yttria) zirconia sprayed to a thickness of 6.3mm (0.25in.) on the heads, approximately 2mm (0.08in.) on the pistons, and 4mm (0.16in) on the I.D. of the cylinder liners. The cylinder liners were ground to final dimension and treated with the slurry sealer coating described previously to provide a better wear surface than the "raw" zirconia. Other than the sealer, no modification was made to the zirconia or the NiCrAlY bond-base coating that was used on the ferrous alloy parts.

The demo-engine was mounted in a 5-ton Army truck (M813) where the elimination of the water cooling system resulted in a net +15 horsepower gain. According to Cummins Engine, a lifetime fuel savings of $7,089 could be expected based on 2,500 hours of operation at $1.25/gal. fuel.

TBC APPLICATION PROBLEMS related to this engine were confined to the head and pistons. These components demonstrated virtually all the difficulties discussed earlier.

Heads - The use of inlay methods raises the question as to which surfaces of the cavity shall have priority for bonding? For example, if the cavity is 6-7mm deep, there are substantial surfaces that are normal to the head plane. These include the cavity periphery, the valve seats, and injector port. With equally strong attachment on all these surfaces, radial cracks about the periphery and the internal circumferences can be expected to develop do to the tensile stresses in the TBC.

The use of compliant layers may add to the problem. If the feltmetal is used, its stiffness must be high enough to support the chamber pressure under maximum load. In some instances, a deflection >0.1mm has been noted and will result in TBC cracking as well.

Furthermore, the feltmetal pad must be sealed at its periphery if hot gasses are not to penetrate.

Finally, a suspicion exists that the pressures at certain periods in the chamber may not be uniform. If this is the case and the coatings are not sealed, gas surges through the cracks and open porosity may occur which would cause spalling.

***APS-Materials was the vendor for the TBC's used on this demonstration engine.

Pistons were bowl-shaped with rather sharp radii at the top. The coatings extended to the periphery with full thickness. Multiple masking techniques were used to attempt to maintain desirable microstructure throughout. No attempt was made to use a compliant layer.

IN SUMMARY

The designing of components for the use of ultra-thick TBC requires a full understanding of the difficulties that are encountered in producing the plasma sprayed coatings.

Relative to the diesel engine, We believe that the current approach of taking an existing engine and modifying the dimensions to accept a coating is only a research tool; that real progress will come only when we design for TBC from the very beginning.

Designs must take into consideration how the coatings are to be sprayed. I must be remembered that any ultra-thick TBC which is interrupted in its application is automatically a candidate for rejection. This fact will have significant effect on how such coatings are applied in production.

Recent programs involving the use of TBC's on diesels have reported some results as being geometry dependent. The reality is that the geometry seriously effects the morphology of the coating. New concepts are needed simplifying the geometry so as to eliminate the difficulty.

There are several problems that need emphasis. Among them are new attachment methods for TBC which permit the expansion differences without compressive movement, and a sealer and topcoat which resists the corrosion and provides an improved wear surface.

References:

1. Holtman, R., Layne,J., and B. Schechter, "An Investigation of Enhanced Capability Thermal Barrier Coating Systems for Diesel Engine Components" NASA CR-174820 and DOE/NASA/0326-1, Aug. 1984
2. Kamo, R., and C. Cheng, reports on Joint Development Program; U.S. Army-Tank Automotive Command and Cummins Engine Co. DAA07-80-C-9125
3. Kvernes, I. and B. Oberlander, "Studies of Microstructure and Properties of Thermal Barrier Coatings Required for Diesel Engine Applications" Presented Int. Conf. Met. Coatings 1983
4. H. Herman, "Failure During Thermal Cycling of Plasma-Sprayed Thermal Barrier Coatings." Proc.Int. Conf. Met. Coatings 1983, p427+
5. Watson, J. and S. Levine, "Deposition Stress Effects on Thermal Barrier Coating Life." Proc. Int. Conf. Met. Coatings 1984

ADVANCED THERMAL BARRIER COATING SYSTEMS

J. D. Reardon, M. R. Dorfman
METCO Inc.
1101 Prospect Avenue
Westbury, New York, USA

Abstract

Current state-of-the-art thermal barrier coating (TBC) systems consist of partially stabilized zirconia coatings plasma sprayed over a MCrAlY bond coat. Although these systems have excellent thermal shock properties, they have shown themselves to be deficient for a number of severe diesel and aircraft applications.

New ternary ceramic plasma coatings are discussed with respect to their possible use in TBC systems. Ceria-yttria stabilized zirconia (CYZ) coatings have been developed with low thermal conductivities, excellent thermal shock resistance and erosion resistance comparable to baseline yttria stabilized zirconia (YSZ) coatings. In addition, dense zirconia-titania-yttria (ZTY) and chromia-titania-silica (CTS) coatings have shown particle erosion resistance properties exceeding conventional stabilized zirconia coatings.

Coatings have been evaluated in conjunction with a NiCr-Al-Co-Y_2O_3 bond coat. Also, multilayer - hybrid coatings consisting of a bond coat with subsequent coatings of ceria-yttria stabilized zirconia (CYZ) or yttria stabilized zirconia (YSZ) and top coats of zirconia-titania-yttria (ZTY) or chromia titania silica (CTS) have been evaluated. These coating systems combine the enhanced thermal shock performance characteristics of CYZ or YSZ with the improved erosion resistance of ZTY or CTS coatings.

THERMAL BARRIER COATINGS are employed in the manufacture of aircraft gas turbine and diesel engines to protect metal components from high temperature operating environments. These coatings consist of an oxidation resistant bond coat and a thermally insulating ceramic top coat, both applied using the plasma spray process. Initially developed by NASA for aircraft gas turbine engines, a present state-of-the-art TBC consists of a dense, oxide free NiCrAlY bond coating and a porous, finely microcracked 6-8% yttria stabilized zirconia (YSZ) top coat (Fig. 1). These coatings show excellent thermal shock and insulating properties. Thicknesses are typically 0.13 mm and 0.39 mm for the NiCrAlY bond coat and YSZ top coat, respectively.

The primary purpose of these thermal barrier coating systems is to reduce the component metal temperature [Ref. 1,2]. Lower metal temperatures result in reduced creep stress, lower oxidation rates, and ultimately greater component durability. Large temperature drops may allow engine manufacturers to reduce cost by permitting the use of cheaper, less exotic materials in design of engine components or to increase the fuel efficiency of the engine by operating at higher temperatures with reduced cooling flows. The magnitude of the temperature gradient is determined by the heat flux, coating thickness, and thermal conductivity of the ceramic. Estimated savings of 250,000 $/yr. per engine may be realized by improving the fuel efficiency of large aircraft engines by 6-12% [Ref. 3].

In addition to improving fuel efficiency, diesel manufacturers also are particularly interested in the cost savings associated with burning residual fuel [Ref. 4]. Unfortunately, residual fuels are corrosive and erosive to diesel components. Plasma sprayed yttria stabilized zirconia coatings afford some protection but are not optimum for diesel applications. In addition to thermal shock resistance, a requirement common to all engine applications, thermal barrier coatings for diesel engines require greater environmental stability (hot corrosion resistance) and erosion/abrasion resistance than required for aircraft engine applications [Ref. 5].

Recent NASA studies have shown TBC thermal shock life of YSZ coatings to be reduced because of residual fuel by-products. These by-products, including vanadium, sodium, and sulfur salts, chemically attack and erode YSZ coatings. Vanadium salts attack yttria causing destabilization of the ceramic [Ref. 1,3]. Molten sodium sulfate, although not reactive with YSZ, penetrates the coating pores and condenses. The combined effect of these two contaminates is to reduce the thermal shock life of the YSZ coating by creating a thermal expansion mismatch between the stabilized zirconia and the condensed salts or the partially stabilized phases. Thus, there is a paradoxical situation, coating porosity is needed for thermal shock life but also contributes to the reduction of thermal shock resistance. New coatings are required which will successfully withstand the diesel environment.

This paper will discuss development of alternate coating systems to replace traditional YSZ thermal barrier systems. The new coating systems have been designed to exhibit improved erosion resistance, compared to YSZ coatings, without sacrificing thermal shock resistance. As an additional criterion, candidate materials were selected for their potentially superior hot corrosion resistance.

COATING DEVELOPMENT

New multilayer TBC systems are being evaluated for diesel applications. These systems incorporate:

*a nickel chromium-aluminum-cobalt-yttria bond coat

*an intermediate thermal barrier of either yttria stabilized zirconia (YSZ) or ceria-yttria stabilized zirconia (CYZ).

*a thin erosion corrosion resistant top coat of chromia titania silica (CTS) or zirconia titania yttria (ZTY).

Chemical composition and physical property data on the above ceramic powders are presented in Table 1. All four ceramic powders are spherical with a particle distribution ranging between 10-120 microns. CTS and ZTY powders are manufactured as composite powders, defined as a mixture of two or more raw material components held together by an organic binder. YSZ and CYZ are prealloyed powders which insures that stabilization (critical to TBC performance) is complete and not dependent upon the somewhat unreliable alloying during spraying. CTS and ZTY powders are sized slightly finer ($<$89 microns) than YSZ or CYZ to insure higher coating density.

All four powders were plasma sprayed using a METCO 9MB plasma spray system. Figure 2 is a photomicrograph of a typical three layer coating microstructure under development. Similar spray parameters produce equivalent CYZ and YSZ coating microstructures. For both coatings, 15-17 volume percent porosity levels are typical with Rc 26-28 macrohardness, DPH 400-500 microhardness, and 0.7-1.1 w/m K estimated thermal conductivity values. Expansion coefficient for CYZ and YSZ coatings are 8.5 x 10^{-6} in/in °C and 10.5 x 10^{-6} in/in °C respectively. Dense CYZ coating with 5-7 volume percent porosity and Rc 41 macro-hardness readings was also evaluated. CYZ and YSZ coatings sprayed denser than this resulted in severe microcracking.

CTS and ZTY powders produced denser coating microstructure without cracking and exhibited higher thermal conductivity and hardness values than CYZ and YSZ coatings. Thermal expansion coefficients were lower than YSZ and CYZ coatings. Porosity levels for CTS and ZTY measured under 3 volume percent. The physical property data for all coatings are presented in Table 2.

EXPERIMENTAL

Thermal shock and hot blast erosion tests were used to evaluate thermal barrier coating performance. All coatings were plasma sprayed using automated part handling equipment. Metallic bond coat thickness of 0.18 mm was used in conjunction with a 0.31-0.39 mm ceramic top coat. A duplex ceramic layer consists of a porous 0.26-0.31 mm intermediate and a dense 0.08-0.13 mm top coat.

THERMAL SHOCK TEST - The mechanical durability of various coating systems were tested using a thermal cycling test. The measure of coating performance was the total number of thermal cycles before a coating cracked or delaminated. A schematic diagram of the test apparatus is shown in Figure 3. In this procedure various coating systems were sprayed onto solid Inconel 718 shafts ½" dia. x 5" long. Samples were then rotated at 400 RPM and subjected to 30 seconds of heating (up to 980°C) followed by 30 seconds of forced air cooling.

HOT PARTICLE EROSION - Figure 3 schematically illustrates the particle erosion apparatus. The coated substrate was heated to 870°C using an oxy-propane flame while the coating was eroded by a controlled stream of fine aluminum oxide particles. Coating erosion was reported as the volume loss of coating per gram of abrasive per second.

RESULTS AND DISCUSSION

BLAST EROSION - Results of the hot particle

blast testing are presented in Table 3. These data indicate that the dense ceria stabilized zirconia (CYZ) and the yttria stabilized zirconia (YSZ) coatings have similar erosion rates at both high (60°) and low (30°) particle impingement angles. The erosion rates for the CTS and ZTY coatings are also similar to each other but are both substantially lower than the rates observed for the dense CYZ and YSZ coatings. It should be noted that the porosity levels of the ZTY and CTS coatings are less than the CYZ and YSZ coatings. As would be expected for brittle materials, all of the ceramic coatings demonstrated a higher erosion rate at the higher incident angle.

Erosion resistance of these coatings is dependent upon high coating density and high macrohardness. Use of high velocity plasma parameters, fine particle size, and judicious choice of chemistry all promote the formation of the hard, dense coatings which are necessary for superior erosion resistance. Titania, a common constituent in the CTS and ZTY coatings, depresses the melting point of these ceramics. This results in better fluidity during melting which in turn results in greater interparticle bonding and coating density. Interestingly, CTS and ZTY can be sprayed to a greater density than the CYZ and YSZ coatings. CYZ and YSZ coatings are prone to cracking if sprayed with less than 5-7 volume percent porosity.

THERMAL SHOCK - The number of thermal cycles to coating failure are presented in Table 3. These data suggest that a coating's thermal shock properties are a combination of its intrinsic material and coating properties. High porosity levels, low thermal conductivity, high thermal expansion coefficients and high temperature phase stability are critical for thermal shock resistance. Porosity acts to stress relieve the coatings. During cycling, cracks that may have been formed are arrested at pores. Expansion differences between the base material and top coat create stress during cyclic conditions. Stress may be minimized by designing TBC systems with expansion coefficients more compliant with the base material. Stress may also be minimized by using ceramics that have high temperature phase stability.

Chromia-titania-silica has the highest thermal conductivity and the lowest expansion coefficient of the various coating systems tested. These properties along with a dense structure results in CTS coatings having the poorest thermal shock performance. ZTY has better properties than CTS, despite an equivalent coating microstructure, a phenomenon which is attributed to a higher expansion coefficient and lower thermal conductivity of the ZTY. Porous CYZ coatings perform better than CYZ coating with dense microstructures. YSZ and CYZ coatings show the greatest thermal shock resistant properties because of its lowest thermal conductivity and high expansion coefficient.

Analyzing the thermal shock and erosion data together, it is apparent that the coatings with the highest erosion resistance (ZTY, CTS) also have the worst thermal shock performance. Likewise, coatings with the best thermal shock resistance (YSZ, CYZ) have the worst erosion performance. To combine both erosion resistance and thermal durability, hybrid coatings consisting of a NiCrAlY-Y_2O_3 bond coat, a porous CYZ (15-17% porous) intermediate coat (0.52 mm) and a thin (0.13 mm) top coat of either CTS or ZTY have been developed. The hybrid coatings were based upon hypotheses that the intermediate CYZ layer would withstand the thermally induced stresses while the erosion resistant top coat would armor the system and protect it from particle erosion. Both hybrid coating systems were subjected to thermal shock evaluation. The CTS and ZTY systems showed 55 and 120 cycles to failure, respectively. Compared to the single ceramic coatings (See Table 3), both CTS and ZTY hybrid systems showed improved shock resistance. The ZTY hybrid system, in fact, showed thermal shock performance equivalent to the baseline YSZ coating.

CONCLUSIONS

1. The new ceria-yttria-zirconia (CYZ) thermal barrier coatings, when compared to the baseline yttria stabilized zirconia (YSZ) coatings, exhibit superior thermal shock resistance without sacrificing erosion resistance.

2. The dense, tough ceramics zirconia-titania-yttria (ZTY) and chromia-titania-silica (CTS) exhibited superior erosion resistance to yttria stabilized zirconia (YSZ) and ceria-yttria-zirconia (CYZ) coatings.

3. Hybrid multilayer ceramic coating systems (CYZ/ZTY and CYZ/CTS) combine improved thermal shock and particle erosion resistance.

FUTURE WORK

The initial laboratory evaluation of CYZ and the hybrid multilayer coatings looked encouraging. Additional work is necessary to determine the long term stability of the CYZ with respect to YSZ and to establish the hot corrosion resistance of CYZ, CTS, and ZTY in the presence of vanadium, sodium, sulfur, and chlorine containing salts.

The hypothesis has been advanced that the dense ZTY and CTS ceramics will provide an impervious surface which would prevent corrosive salts from penetrating the coating and causing accelerated failure. This theory will be evaluated.

Most importantly, engine testing will be undertaken for specific diesel and gas turbine applications.

REFERENCES

1. Levine, S. R., Miller, R. A., and Hodge, P. E., "Thermal Barrier Coatings for Heat Engine Components, " SAMPE Quarterly, December 1, 1980, p. 20.

2. Sheffler, K. D. and Anderson, N. P., "Development of Strain Tolerant Thermal Barrier Coating Systems," United Technology Corp., PWA Engineering Division, prepared for NASA, Contact NAS3-22548 [September 1983].

3. Stevens, R., "An Introduction to Zirconia," Sherwin River Ltd., U.K., (1983).

4. Holtman, R., Layne, J., Schechter, B., "An Investigation of Enhanced Capability Thermal Barrier Coating Systems for Diesel Engine Component," Allison Gas Turbine Division General Motors Corporation, Indianapolis, IN, prepared for NASA, contract DEN3-326 [August 1984].

5. Davis, A. G., Boone, D. H., and Levy, A. V., "Erosion of Ceramic Thermal Barrier Coatings," paper presented at Corrosion 84 - International Corrosion Forum - April 2-6, 1984, New Orleans, Louisiana - New Orleans Hilton.

Figure 1

State-of-the-Art Thermal Barrier Coating
Developed by NASA

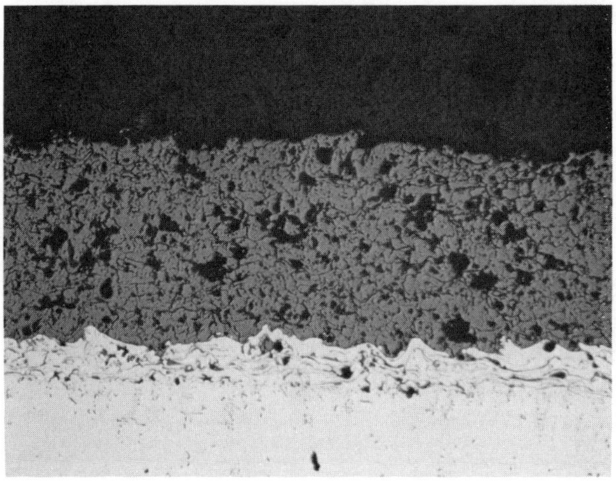

100X Magnification

Figure 2

Hybrid Multilayer Ceramic Coating System

50X Magnification

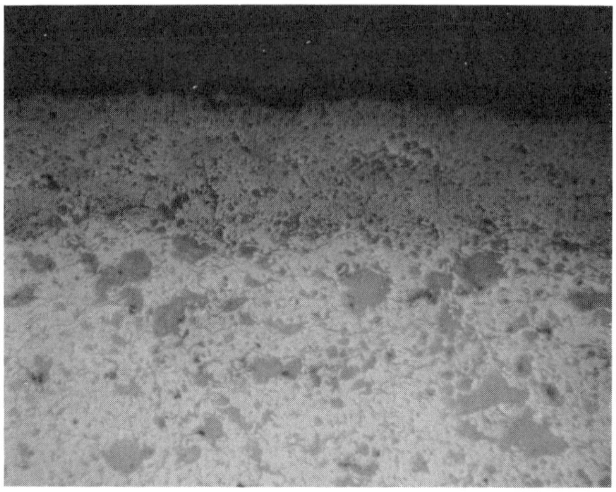

100X Magnification

FIGURE 3

Experimental Test Procedure

*HIGH TEMPERATURE PARTICLE EROSION

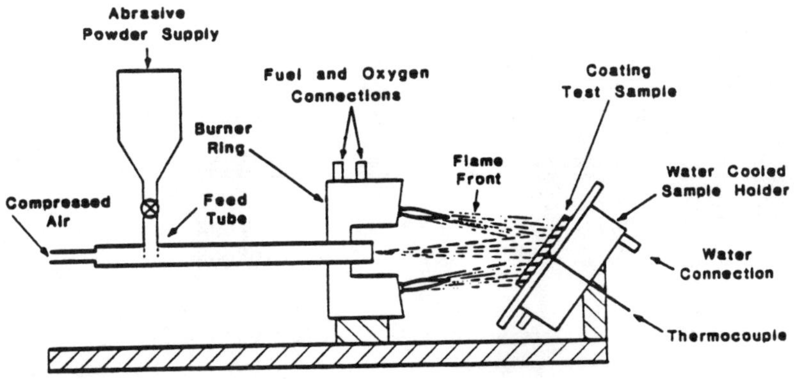

*CYCLIC THERMAL SHOCK

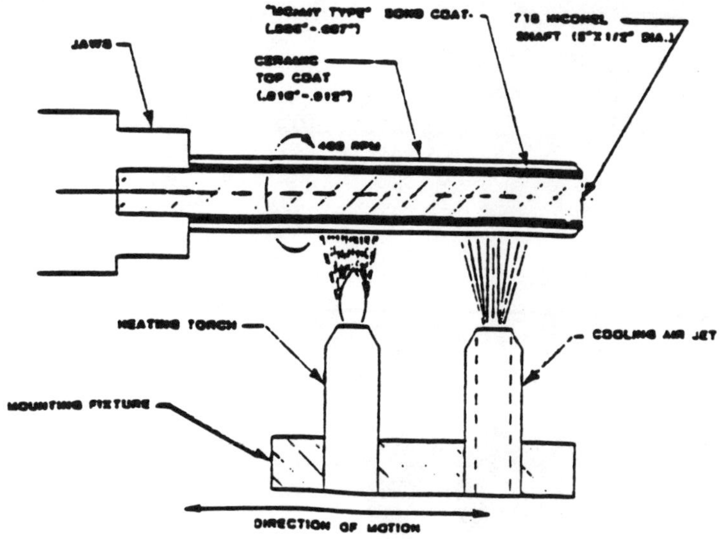

Table 1

POWDER CHARACTERISTICS

	Yttria Zirconia	Ceria Yttria Zirconia	Zirconia Titania-Composite Powder	Chromia-Silica-Powder
Nominal Chemistry(wt%)	Zirconia Base	Zirconia Base	Zirconia Base	Chromia Base
	Yttria-8%	Ceria-26%	Titania-18%	Silica-5%
		Yttria-2.5%	Yttria-10%	Titania-3%
Powder Size	-117 microns +10 microns	-89 microns +10 microns	-74 microns +10 microns	-55 microns +15 microns
Powder Morphology	Fused Spherical Powder	Fused Spherical Powder	Spherical Spray Dried Composite	Composite Powders

Table 2

COATING PROPERTIES

Coating System	Cross Sectional Hardness (DPH_{300})	Macro-Hardness Rc	Porosity	Thermal Conductivity w/m K	Thermal Expansion Coefficient in/in °C x 10^{-6}
YSZ	400-500	28	15-17	0.7-1.1	10.5
CYZ	400-500	26	15-17	0.7-1.1	8.5
CYZ	400-500	41	5-7	0.8-1.2	8.5
ZTY	600-650	55	1-3	1.2-1.6	6.5
CTS	900-1100	63	1-3	2-3	5

Table 3

EROSION RESISTANCE AND THERMAL SHOCK DATA

Coating System	Porosity	Volume Loss Rate ($cm^3 \times 10^4$) 30°	60°	Number of Cycles to Thermal Shock Failure
YSZ (dense)	5-7	1.35	2.12	60
CYZ (dense)	5-7	1.43	2.05	70
CYZ	15-17	2.06	4.64	180
YSZ (baseline)	15-17	2.11	4.59	120
ZTY	1-3	0.29	0.83	60
CTS	1-3	0.25	0.80	35

OPTOTHERMAL NDE OF THERMAL-BARRIER COATINGS

P. Cielo, S. Dallaire
National Research Council Canada
Industrial Materials Research Institute
Boucherville, Quebec, Canada

"Abstract"

An optothermal technique is described for the nondestructive evaluation of thermal barriers, such as ceramic plasma-sprayed coatings. This method is noncontact and requires access from one side only of the sample, thereby making it easier to inspect awkward-shaped or difficult to reach pieces. A laser beam irradiates the coated material, raising its surface temperature to some degrees above ambient. The surface temperature is monitored during or after irradiation by an infrared detector to obtain a characteristic temperature-time curve (thermogram). After appropriate processing, the thermogram provides information on the thermal properties of the coating, if its thickness is known. Alternatively, the thickness of the coating can be estimated if its thermal properties are known. In either case, the optothermal analysis allows to evaluate the degree of thermal insulation provided by the coating. Each measurement is performed nondestructively and within a time period of less than 0.5 seconds/point using the single-shot method.

CERAMIC THERMAL-BARRIER COATINGS[1], are being increasingly used to protect the base metal and allow higher operating temperatures in gas-turbine and diesel-engine components[2]. Plasma-sprayed stabilized ZrO_2 coatings can reduce the heat flow and thus the cooling requirements by up to 80%[2], depending on the thickness and on the thermal conductivity of the coating material[3].

The measurement of the thermal properties of plasma-sprayed ceramic coatings at different temperatures has been the object of several recent investigations[4-6]. Such properties are typically measured by pulse-heating one side of a spalled coating sample and temperature-monitoring on the other side of the coating. Typically, the thermal conductivity of plasma-sprayed ceramics is lower by a factor of 3 or 4 as compared to the conductivity of the sintered ceramic material. The thermal conductivity of the coating shows however irreversible variations, approaching the values for the sintered material, after sustained thermal cycling at temperatures of the order of 1000°C. Such a behaviour is believed to be produced by sintering and phase-transformation effects at high temperatures.

A nondestructive technique making it possible to evaluate in-situ the degree of thermal insulation provided by the thermal-barrier coating before and during service would be quite convenient. In-service inspection is particularly valuable to follow the evolution of the coating thermal properties which may vary because of irreversible thermal-conductivity variations or hot-corrosion of the coating. Such a technique should be fast and deployable directly on the coated engine components in the field. In this paper we analyse some optothermal techniques as possible candidates for such measurements. Different heating cycles and signal processing methods are described and compared.

OVERVIEW OF AVAILABLE TECHNIQUES FOR NDE OF COATINGS

Several methods are available for the inspection of coatings[7-10]. Some techniques are destructive, such as sample sectioning followed by microscopic inspection, scratching, gravimetric analysis before and after coating for an average thickness reading, and coulometric techniques by which the coating thickness is evaluated from the time required to chemically dissolve the coating by an electrolyte. Such methods are not applicable to in-service 100% inspection.

A number of nondestructive testing techniques have been employed to determine the coating thickness or the presence of interface disbonds. The most currently used are the

EXPERIMENTAL RESULTS - MgO-stabilized ZrO_2 coatings on steel were tested with the experimental apparatus shown in fig. 1. Four coatings of thickness 0.16, 0.35, 0.45 and 0.6 mm respectively were plasma-sprayed following the manufacturer's recommended parameters using a vertical-horizontal automatic transverse system. The spraying procedure was slightly modified at the end of the operation for the 0.16 mm and the 0.6 mm-thick samples resulting in a greyish surface appearance. Fig. 6 shows the metallographic sections of a 0.45 mm-thick coating on steel.

Fig. 7 shows the superposed raw signals obtained for the four coated samples with an irradiation time of 3s. There is a clear difference in the saturation levels reached with samples of different coating thicknesses, however such a difference could also be caused by unpredictable parameters such as a different laser absorptivity or IR emissivity due to surface contamination. The shape of the curve is more reliable than the absolute temperature to quantitatively evaluate the coating properties.

The four raw signals shown in fig. 7 were computer-processed (logarithmically smoothed, converted to temperature using the built-in detector-output-to-temperature calibration curve, and displayed logarithmically) to obtain the four curves of fig. 8 (only the first 1 s heating-period is displayed). As we can see, the curves start at different levels, corresponding to different absorbed power levels or emissivity values for the different samples. Indeed, the samples corresponding to curves (a) and (d) had a darker surface appearance than the others, as mentioned above, because of their different surface morphology and colour which resulted from incompletely molten particles laying on the surface as observed by SEM. However, the shape of the different curves, e.g. the time period required to reach the saturation level is clearly correlated to the coating thickness.

This is shown clearly in fig. 9, which gives a plot of the slopes of the thermograms shown in fig. 8, i.e. the logarithmic derivatives $\frac{d\log T}{d\log t} = \frac{t}{T}\frac{dT}{dt}$ obtained for the four samples versus the coating thickness after 0.2 s, 0.3 s and 0.5 s after starting heating. If we choose 0.3 s as the observation time, the correlation between the slope of the thermogram at such a time and the coating thickness is quite good, providing a thickness-evaluation capability with a precision which can be estimated in the 50 μm-range. An alternative approach would be to record for each curve of fig. 8 the time delay at which the curve has reached a given value of the slope, e.g. 0.3. The thickness of the coating is proportional to the square-root of such a time delay.

The inaccuracy in the coating-thickness evaluation by such a method depends on a number of factors. The intrinsic signal reproducibility is better than 5%, as was verified by multiple tests performed at conveniently-spaced time intervals over the same position on the coated sample. Slightly different values were obtained over different positions, but this was related to the slightly tapered profile of the thickness as was verified by metallographic inspection. The coating roughness or local porosity hardly affected the recorded values because the measurement was made in each case over a 2 mm-diameter area. Variations in the optical penetration depth and in the diameter of the heated area may affect the measured slope values. For best results the laser and IR wavelengths should thus be chosen in spectral regions of high absorbance or scattering by the coating material, while the heated area should be large with respect to the coating thickness.

As said above, the coating thickness may be evaluated if the thermal properties of the coating are known, or vice versa. If the coating thickness is known, the arguments given in the previous section show that slope measurements of the kind described here are related both to the thermal diffusivity $k/\rho c$ and to the thermal effusivity $(k\rho c)^{\frac{1}{2}}$ of the coating. As the values of the mass density and specific heat have a relatively small variability with the spraying parameters and thermal cycling, as compared to the thermal conductivity, (they vary typically by 30% as compared with variations by a factor of 2 or 3 for the conductivity), we can conclude that our measurements are mainly related to the thermal conductivity. If, as it is usually the case, neither the coating thickness nor the conductivity are known, we can interpret a delayed signal (e.g. of curve (c) as compared to curve (a) in fig. 8) as either a larger thickness or a smaller thermal conductivity of the coating. Our measurements are thus essentially related to the ratio of the coating thickness and the thermal conductivity, i.e. to the thermal resistivity of the coating which is the important parameter to evaluate in a thermal-barrier coating.

MODULATED-BEAM TECHNIQUE

An alternative to the transient heating scheme described in the last section is the modulated-heating approach. In this case the heating laser beam is amplitude modulated and the stationary AC signal from the detector is processed to obtain the required information on the plasma-sprayed coating[14-16]. Two examples of the signal obtained with the apparatus of fig. 1 when the laser beam is square-wave modulated at frequencies of 0.17 Hz and 10 Hz, respectively, are shown in fig. 10.

electromagnetic and the radiographic techniques.

Electromagnetic methods are either of the capacitive or of the eddy-current type. In both cases an electrode is put in contact with the coating surface and the metallic substrate acts as the opposite electrode. The DC voltage or the AC impedance read by the instrument depends on the thickness and on the dielectric properties of the nonconductive coating. Such methods require a flat and smooth coating surface to match the surface of the electrode, while variations in the dielectric properties of the coating do not correspond necessarily to variations in its thermal insulation properties.

Radiographic methods including beta-backscattering and X-ray fluorescence methods use a nuclear-radiation source to irradiate the coated material and detect the backscattered radiation which is dependent on the coating properties. Problems with such methods are the need for dangerous nuclear radiation as well as the detection of coating properties, such as density variations, which do not correspond necessarily to variations of its thermal conductivity.

Ultrasonic techniques have been extensively used for the detection of disbond defects in plasma-sprayed coatings[11,12]. One problem with this approach is the need for a good liquid-coupling between the ultrasonic transducer and the coating surface. Another problem is the large attenuation of the ultrasonic beam in the porous and stratified coating, which is of the order of 500 dB/cm.[13] This problem has induced some researchers having previously worked with ultrasonic methods[11,13] to shift to the thermal approach[14,15].

Thermographic techniques[12] have been shown to detect lack-of-adhesion flaws with a diameter of at least some millimeters. Main problems of the IR camera approach are the high cost and lack of uniformity of the temperature distribution over the heated area. On the other hand, an IR camera is better suited for the detection of subsurface flaw than for the evaluation of the thickness of a uniform sample.

Single-spot photothermal methods using as the heating source either a modulated laser[14-16] or a pulsed thermal source[17-18] are less expensive than the IR camera technique and are better suited to a quantitative evaluation of the coating thickness or thermal properties. Although such methods are slower than the IR camera approach to provide a full image of a given area (except for the method described in ref. 17, which however gives up much time-resolved information) a spot-by-spot scanning approach has at least the advantage of avoiding the distribution inhomogeneity of an extended heating source. Different spot-heating configurations will be described and compared in this paper, with particular emphasis on a step-heating technique[19] which has not been previously applied to the analysis of stratified materials.

TRANSIENT-HEATING METHODS

THEORETICAL CONSIDERATIONS - Figure 1 shows the basic principle of the method. A 2W Argon laser beam is directed toward the coated-sample surface, where it heats a surface of the order of 2 mm in diameter. A HgCdTe infrared detector is pointed to the center of the irradiated area, either directly or through a laser-reflecting, IR-transmitting dichroic beam-splitter. A chopper in front of the detector (not shown) is used in conjunction with a lock-in amplifier to perform synchronous detection at a typical chopping frequency of 800 Hz, to avoid low-frequency detector noise in the 0 to 50 Hz signal-frequency band. The time-resolved signal output from the detector is computer-processed to provide the experimental thermogram.

Single-shot, or transient, thermograms are obtained by recording the transient surface temperature before the stationary configuration is reached. Such signals thus provide information on the thermal diffusivity of the coated sample surface. Two basic heating procedures may be conceived, namely the pulse-heating and step-heating methods, although all the intermediate cases are obviously possible.

In the pulse-heating approach the surface is heated by a short laser pulse and the temperature-time history is analyzed during the cooling period. An example of a finite-difference computed thermogram of this kind is shown in fig. 2. A 15 ms surface-heating extended-area laser pulse irradiates (a) a soft-steel material, (b) a 0.2 mm-thick ZrO_2 coating on steel, or (c) a 0.4 mm-thick ZrO_2 coating on steel. The assumed absorbed energy density was of 10 W/cm^2, while the thermal conductivity, specific heat and mass density were, for steel[20], $k = 0.7$ W/cm K, $c = 0.4$ J/g K and $\rho = 7.8$ g/cm^3 respectively. The respective properties for the zirconia plasma-sprayed material were taken as $k = 0.01$ W/cm K, $c = 0.75$ J/g K and $\rho = 5$ g/cm^3 on the basis of the published data[4-6].

All of the curves shown in fig. 2 have a decay rate of slope near to -0.5 in the log-log scale, in agreement with the one-dimensional analytical model[18]. The slope of curves (b) and (c) markedly departs from such a uniform value when the thermal front reaches the coating-substrate interface, attaining a maximum value of nearly 1.7 in absolute value at the inflexion point. The time-period required to reach the inflexion point is of the order of Z^2/α, where Z is the coating thickness and $\alpha = k/\rho c$ is the thermal diffusivity of the coating, while the vertical displacement between the initial portion of the curves, and thus the value of the slope at the inflexion point, are related to the effusivity $(k\rho c)^{\frac{1}{2}}$ of the coating[18]. The thickness or thermal properties of the coating can thus be inferred from an analysis of the experimentally-recorded thermogram.

Fig. 3 shows a similar computation in the case of step-heating, i.e. when the heating power is suddenly switched on and remains at a constant value of 1.3 W/cm^2 during the observation time. The initial slope of all curves is equal to +0.5 in agreement with the analytical one-dimensional model[21], while decreasing to values of the order of 0.15 at the inflexion point for the coated-sample curves (b) and (c). The information which can be obtained from such step-heating thermograms is thus similar to the information which can be obtained from the pulsed thermograms shown in fig. 2.

The choice between a pulsed-heating and a step-heating configuration must be made as a function of the material properties, i.e.:

1) The pulsed-heating approach provides a quicker time-response and larger signal visibility for coatings of different thicknesses[18], as can be seen by comparing curves (b) and (c) of figs. 2 and 3. It must however be noted that because of the logarithmic scale, the absolute temperature differential between the two curves is larger in the step-heating case.

2) Curves of the kind shown in fig. 2 have the advantage of allowing more time for averaging during the last portion of the curve when more precision is required because of the lower temperature signal. For example, the low-temperature portion of curve (b) in fig. 2 is acquired during 2 seconds between t=1 s and t= 3 s, while the low-temperature portion of curve (b) in fig. 3 is acquired between t = 0.01 s and t = 0.03 s. This advantage is however mainly effective with thin coatings for which the signal-frequency band is in the range of 1 KHz or more, where the detectivity is limited by the electronic white noise. For thick or low-thermal-conductivity coatings such as those analyzed in this paper, low-frequency environmental thermal noise[22] nullify the advantage to average over longer periods.

3) If the coating has a thermal effusivity smaller than the substrate, as is usually the case for thermal-barrier coatings, the step-heating approach has the advantage of requiring a smaller dynamic range. Curve (b) in fig. 3 spans less than one decade of the temperature range, while the same curve in fig. 2 extends over more than two decades. To avoid surface overheating, temperatures of the order of 0.05 K above ambient must be detected in the pulsed case, to be compared with minimum temperatures of the order of 1 K in the step-heating case.

Because of arguments (2) and (3), a step-heating approach was chosen for the single-shot thermal-barrier analysis reported in this paper. Such an investigation also presents certain interest from the technical point of view, because the step-index approach has never been applied before to the inspection of coated materials.

The computations of figs. 2 and 3 have been performed under rather idealized conditions, i.e. one dimensional heat propagation and surface absorptance. Some more realistic curves are shown in fig. 4 for the step-heating approach, to allow a more meaningful comparison with the experimental results which are described in the next section. In fig. 4, curve (a) corresponds to the ideal case of an extended heating source, giving one-dimensional thermal propagation within the material, and of a sample which is opaque to both the laser and the IR wavelength of detection. Curve (b) shows the effect of radial thermal propagation when the heating source has an r = 1 mm radius (the attainment of a saturation level after a time of the order of r^2/α), while curve (c) shows the effect of a finite optical penetration depth. The latter curve was obtained by assuming an exponential absorption profile with 1/e point at 0.1 mm of depth within the coating for both the laser and IR-detector wavelengths. It should be noted that the slope of curve (c) is nearly equal to 1 in the first portion of the curve (i.e. for a time period shorter than the thermal-propagation time through the optical-penetration depth). This effect must be taken into account when the slope of the experimental thermogram is used to evaluate the coating properties.

The optical penetration depth within the ceramic coating is strongly dependent on the wavelength. The penetration depth in ZrO_2 ceramics is particularly large, several hundreds of microns, in the 2 to 4 μm spectral range[4]. The penetration depth in the visible is much shorter, mainly because of scattering at the particle-to-particle interfaces. The optical penetration in the 10 to 20 μm region is again reduced because of molecular absorption by the Zirconium oxyde, as fig. 5 shows. This figure shows the reflectance spectrum of a mixture of 2.4% ZrO_2-powder used for our coatings in a KBr-powder matrix sample 3 mm thick. Because the KBr matrix is neutrally backscattering over the whole IR spectrum, this spectrum is equivalent to a transmission spectrum from a nearly 50 μm-thick ZrO_2 sample.

To minimize the optical penetration effects in our experiments, we used a visible-light laser (λ = 514 nm) as the heating source, and a 9 to 13 μm HgCdTe as the IR detector which was available to us. Better results could probably be obtained with a 14 to 20 μm HgCdTe detector which is commercially available on special order.

Each cycle of the 0.17 Hz signal is very similar to the signals shown in fig. 7: the signal saturates after nearly 1 s because of both thermal quenching by the conductive substrate and of three-dimensional heat spread in the sample. The 10 Hz signal in fig. 10 corresponds to the first 0.1 s portion of the signals shown in figs. 7 or 8: the average temperature over the optical-penetration volume within the coating increases almost proportionally with the injected laser energy and thus with time.

The complete plot of the signal characteristics for this sample over the 0.17 to 10 Hz frequency range is shown in fig. 11. In the saturation region below 0.5 Hz the signal amplitude hardly changes with frequency, while in the 5 to 10 Hz region the slope of the amplitude plot tends to 1. The phase of the detected signal with respect to the laser input signal changes from nearly zero (fig. 10, upper picture: almost square wave matching the square-wave chopped input) to nearly 90° (fig. 10, lower picture: almost triangular wave corresponding to the time-integral of the laser-input square wave). The curves in fig. 11 thus contain an information equivalent to that of the thermograms shown in fig. 8.

To obtain a coating-evaluation curve similar to that of fig. 9, we can thus plot either the slope of the amplitude-curve of fig. 11, or the value of the phase, at a convenient modulation frequency, obtained with the samples of different coating thicknesses. Such a plot is shown in fig. 12 for three different values of the modulation frequency. If a convenient value of the modulation frequency is chosen, the measured value of the phase shift is well correlated to the coating thickness independently of the level of the absorbed power or of the surface emissivity.

The choice between the transient-heating or the modulated-heating methods again depends on the kind of coating to be analyzed:

1) If the coating is relatively thin or highly conductive, the relatively high beam modulation frequencies which can be used in the modulated approach provide good signal averaging capabilities within a relatively short measurement time. For thermal-barrier coatings of the kind considered in this paper, however, it may take typically 30 seconds to reach a phase reading within a two degree precision to the final value, as fig. 13 shows. Such a time scale is hardly compatible with rapid industrial scanning needs.

2) If the coating is thin or if time is not at a premium, the modulated approach provides a better signal-to-noise ratio than the single-shot transient technique. This makes possible to use a lower-power heating laser, an important consideration from the eye-safety point of view in an industrial environment. Consideration should however be given in this case to the possibility to repetitively perform the transient measurement with the low-power laser and to improve the signal-to-noise level of the transient thermogram by superposing the recorded thermograms to obtain an averaged transient curve.

3) A weak point of the modulated approach is the difficulty to differentiate between coating thickness and the presence of defects. For example, if the recorded value of the phase at a given frequency is high, one cannot say whether such a value is due to a thicker coating or to a disbond defect at the coating-substrate interface. A glance at the time-resolved thermogram in the case of the transient approach easily permits to differentiate between such cases[18,21]. A similar capability in the modulated approach would require the recording of multiple-frequency curves of the kind of fig. 11, requiring an even longer observation time for each measurement. A possible improvement of the modulated method would be to average the AC signal (fig. 10) over several cycles to obtain a good-quality raw signal which could be processed using methods similar to those described in the previous section.

CONCLUSION

An optothermal technique has been described for the nondestructive characterization of plasma-sprayed ZrO_2 coatings of the kind used as thermal barriers in high-temperature applications. The technique appears to be performant and relatively precise, provided that the optical penetration depth within the coating is much smaller than the coating thickness and that the diameter of the heated area is much larger than the coating thickness. This method is particularly attractive because it is rapid, noncontact, and can be applied to remote or difficult-to-access coated components.

ACKNOWLEDGEMENTS

The precious technical assistance of M. Lamontagne of IMRI is gratefully acknowledged.

REFERENCES

1. C.H. Liebert and R.A. Miller, Ind. Eng. Chem. Prod. Res. Dev. 23, 344 (1984).
2. I. Kvernes "Coating of diesel-engine components" in: Coatings for High Temperature Applications, E. Lang editor, pp 361-394. Applied Science Publishers, London (1983).
3. C.H. Liebert and R.E. Gaugler, Thin Solid Films, 73, 471 (1980).
4. R. Brandt, High Temp. - High Pressures 13, 79 (1981).
5. R. McPherson, Thin Solid Films, 112, 89 (1984).
6. L. Pawlowski, D. Lombard, A. Mahlia, C. Martin and P. Fauchais, High Temp.-High Pressures 16, 347 (1984).

7. G.W. Scott "NDI of thin plasma-sprayed ceramic and cermet protective coatings for coal conversion and utilization equipment". Report TM-6210, Oak Ridge Nat. Lab., Oak Ridge, Tenn., (1978).
8. R.C. Stinebring, Materials Research and Standards, 12, 18 (1972).
9. T.D.T. Latter, Polymers Paint Colour Journ. 173, 119 (1983).
10. G.A. DiBari, Metal Finishing, p. 29 (March 1984).
11. R.L. Cox, D.P. Almond and H. Reiter, NDT Int. 13, 291 (1980).
12. H.D. Steffens, H.A. Crostack and J. Beczkowiak "Testing and inspection of coatings" in: Coatings for High Temperature Applications, E. Lang editor, pp. 193-215. Applied Science Publishers, London (1983).
13. R.L. Cox, D.P. Almond and H. Reiter, Ultrasonics, 19, 17 (1981).
14. D.P. Almond, P..M. Patel, I.M. Pickup and H. Reiter, NDT Int. 18, 17 (1985).
15. P.M. Patel and D.P. Almond, J. Mater. Science 20, 955 (1985).
16. J. Jaarinen and M. Luukkala "The use of thermal waves for NDE of plasma-sprayed coatings" 3rd European Conf. NDT, Florence, Oct. 15-18, 1984.
17. D.R. Green, M.D. Schmeller and R.A. Sulit "Thermal NDE method for thermal-spray coatings" Proc. Oceans '82 Conf., Washington, D.C. Sept. 1982. Marine Technology Society.
18. P. Cielo, J. Appl. Phys. 56, 230 (1984).
19. A.W. Schultz, Appl. Opt. 7, 1845 (1968).
20. Y.S. Touloukian "Thermophysical properties of matter" IFI/Plenum, New York (1970).
21. P. Cielo "Analysis of pulsed thermal inspection" Proc. 14th Symp. on NDE, D.W. Moore and G.A. Matzkanin editors, pp. 418-428. SRI/NTIAC, San Antonio, TX. (1983).
22. P. Cielo, S. Dallaire, G. Lamonde and S. Johar "Measurement of thermal inertia by the reflective-cavity method". 4th Int. Top. Meet. Photoacoustic, Thermal and Related Science, L'Estérel, Aug. 4-8, 1985. To be published in Can. J. Phys. (1986).

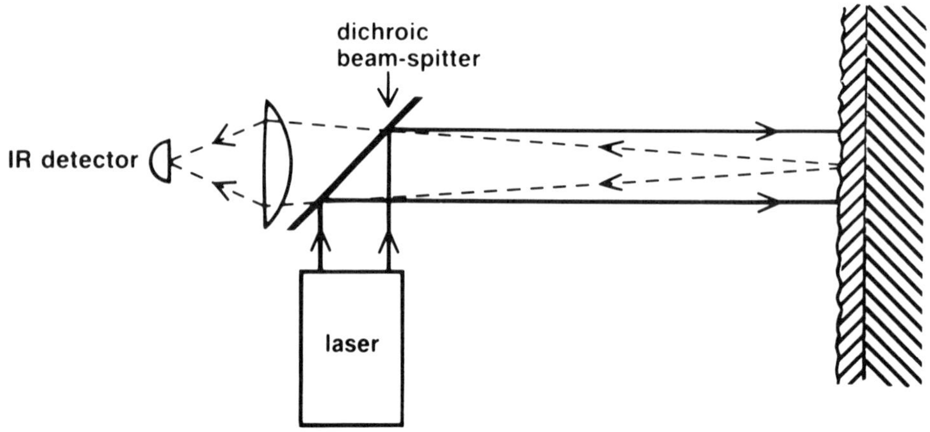

Fig. 1: Experimental Apparatus

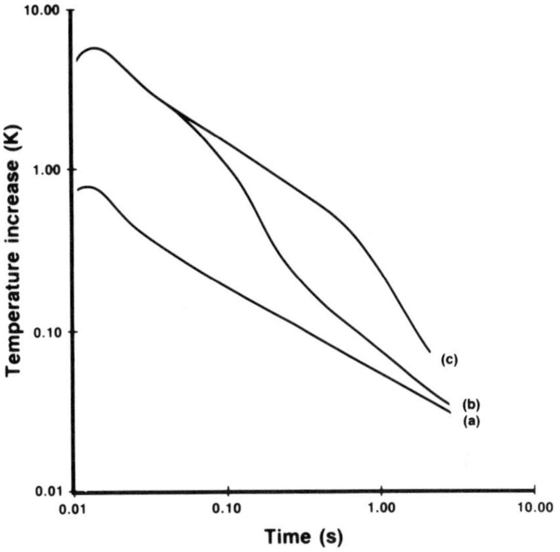

Fig. 2: Computed one-dimensional pulsed thermogram for an irradiated material which is (a) bare soft steel, (b) steel with a 0.2 mm-thick ZrO_2 coating and (c) steel with a 0.4 mm-thick ZrO_2 coating.

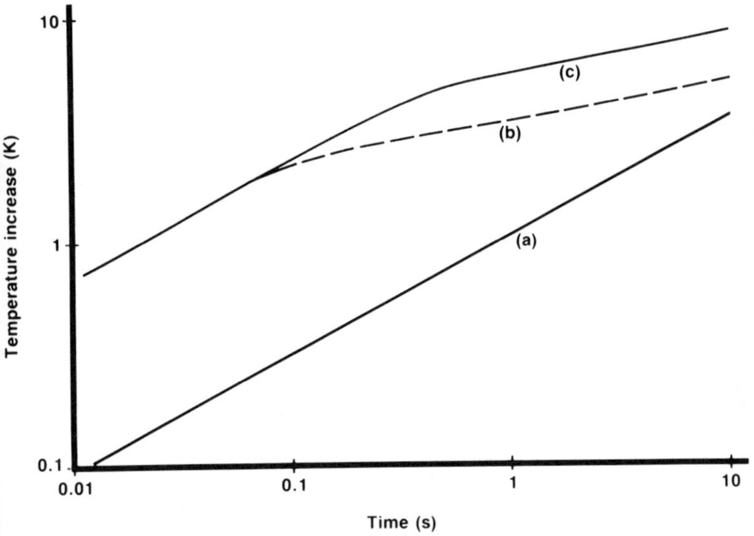

Fig. 3: Computed one-dimensional step-heating thermogram for the same materials as in fig. 2.

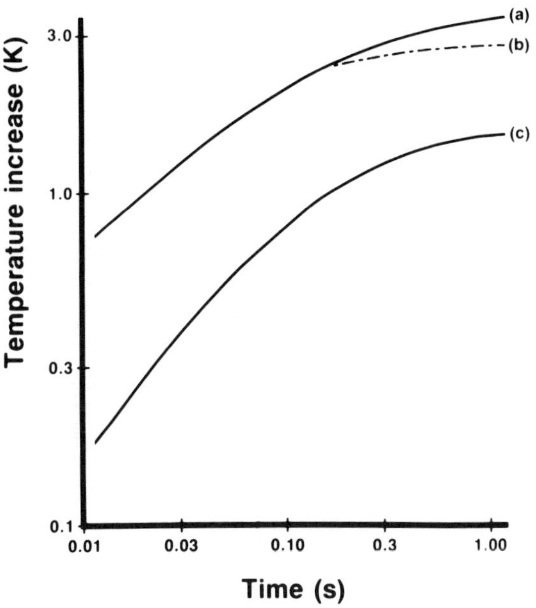

Fig. 4: Computed three-dimensional thermogram of a 0.2 mm-thick ZrO_2 coating on steel: (a) extended thermal source, surface absorption; (b) 2 mm-diameter source, surface absorption; (c) extended source, optical penetration depth 100 ım.

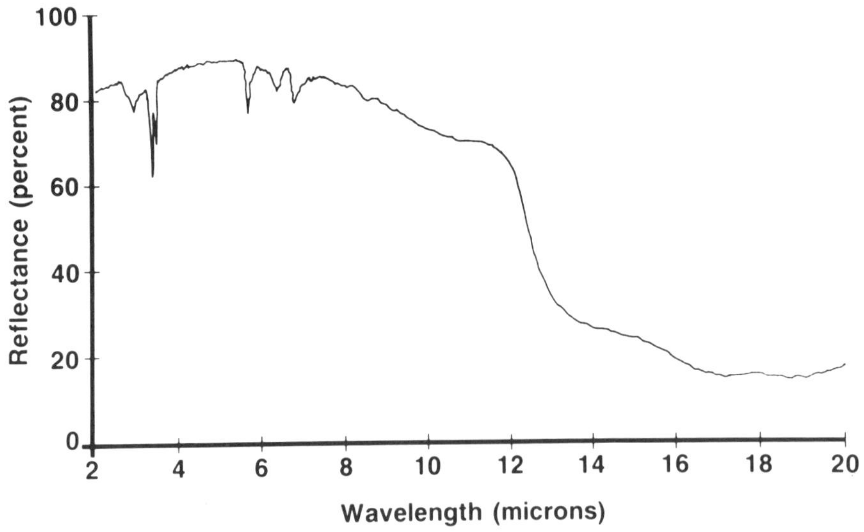

Fig. 5: Reflectance spectrum of a 2.4% ZrO_2-powder mixture in a neutrally reflecting, mm-thick KBr 3 matrix, equivalent to a transmission spectrum.

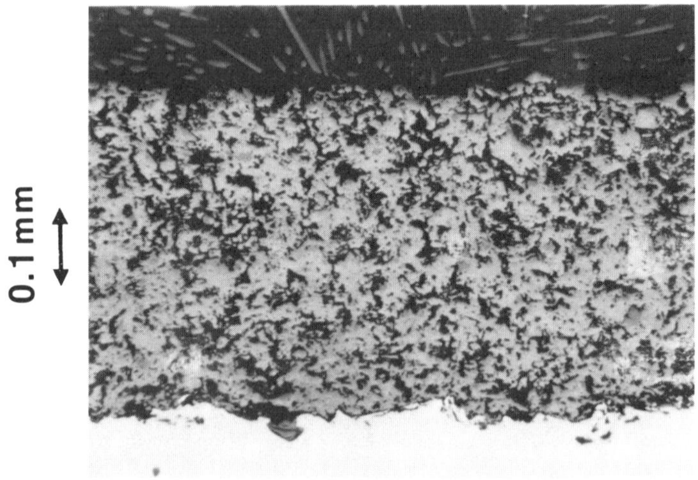

Fig. 6: Cross-section of the 0.45 mm-thick ZrO_2 coating.

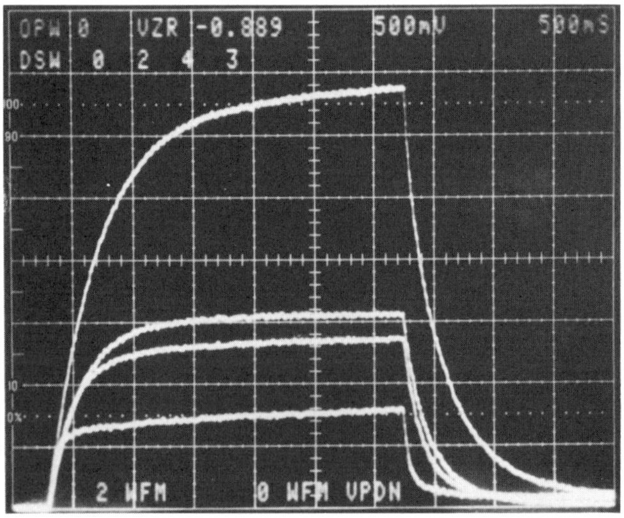

Fig. 7: Raw signals obtained for the four samples with different coating thicknesses. Vertical scale 0.5 V/div., horizontal scale 0.5 s/div.

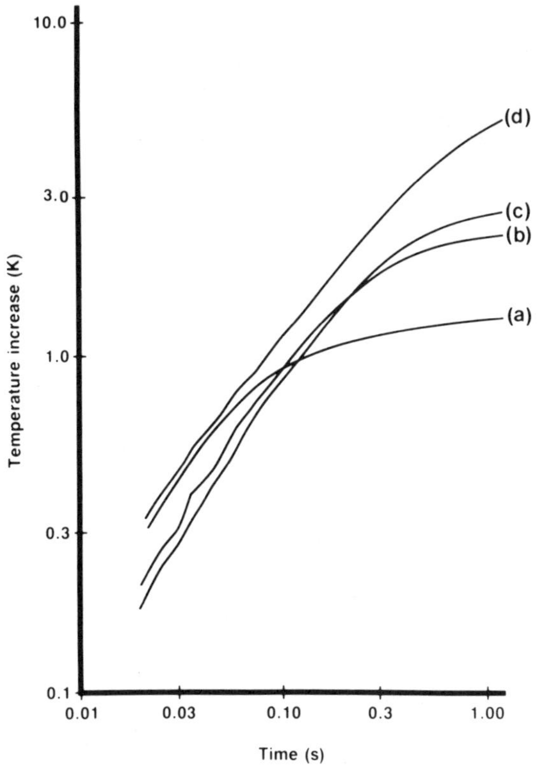

Fig. 8: Processed thermograms obtained for the samples with coating thickness equal to (a) 0.16 mm, (b) 0.35 mm, (c) 0.45 mm and (d) 0.6 mm.

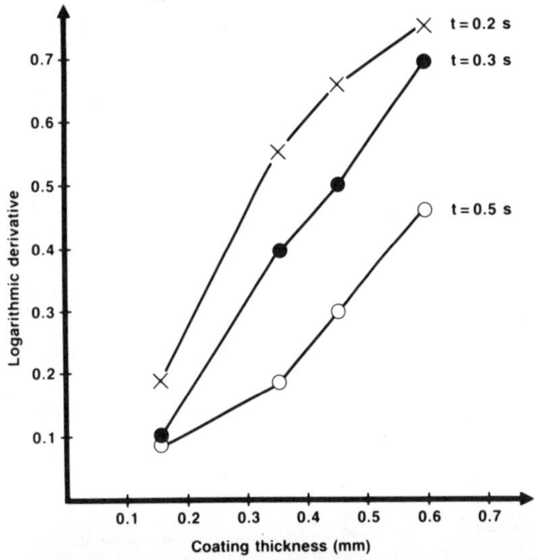

Fig. 9: Plot of the logarithmic derivative vs. coating thickness for the four samples at different times after starting heating.

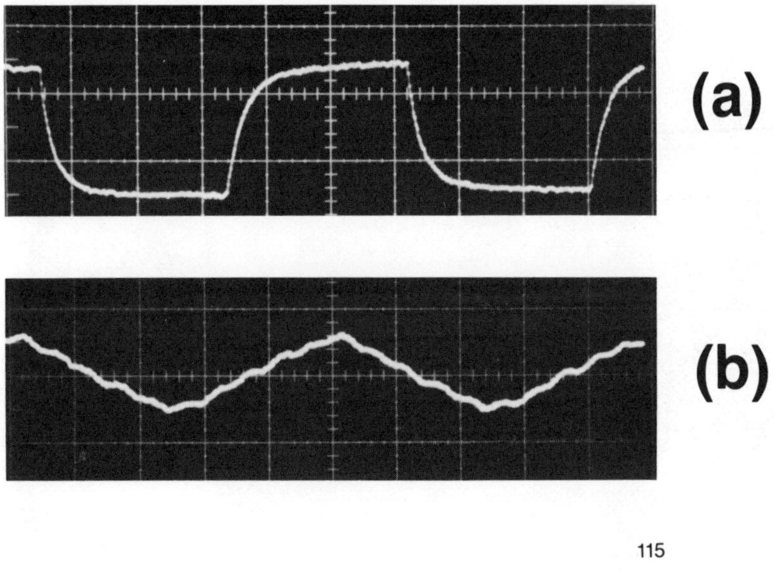

Fig. 10: Raw-signal AC output with a 0.45 mm-thick coating when the laser beam is square-wave modulated at a frequency of (a)0.17 Hz and (b) 10 Hz.

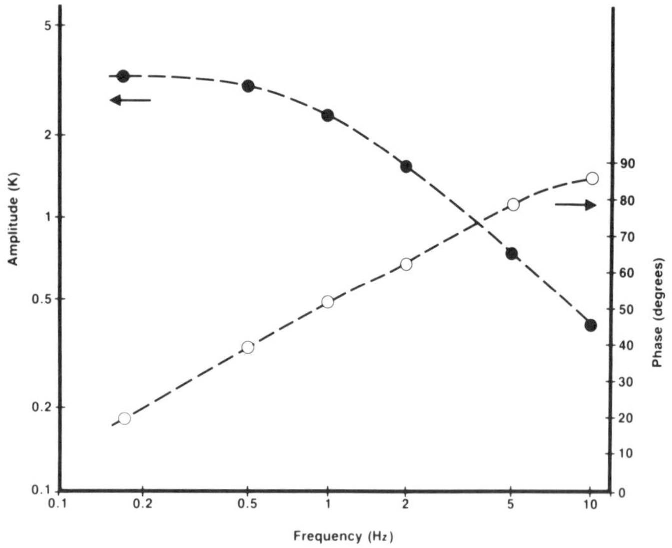

Fig. 11: Plots of the signal amplitude and phase with respect to the laser-irradiation signal vs. the modulation frequency in the case of the 0.45 mm-thick coating.

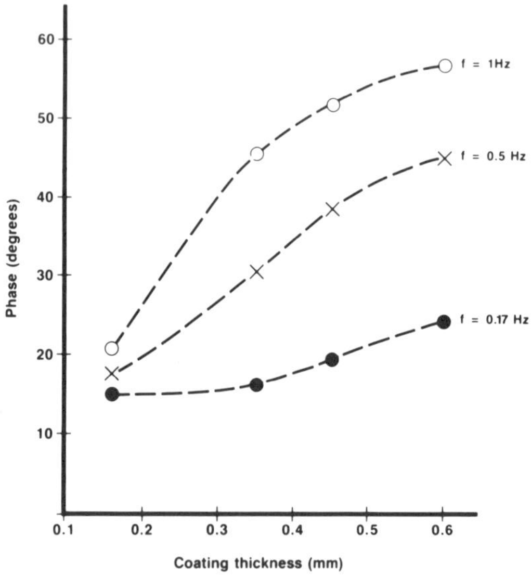

Fig. 12: Plot of the phase shift between the laser-input signal and the detected signal vs. the coating thickness for different modulation frequencies.

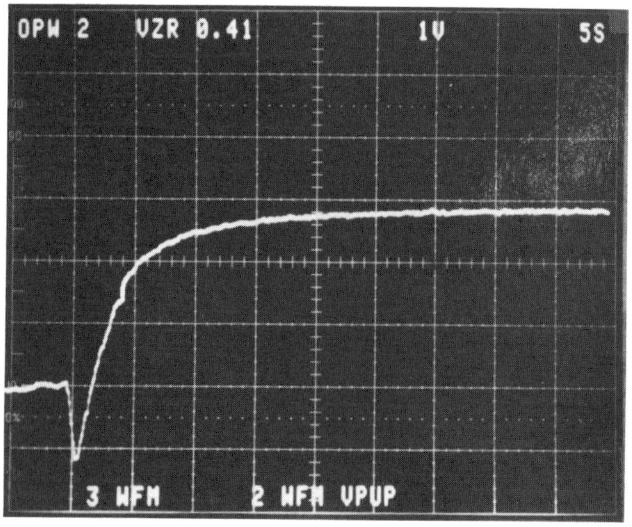

Fig. 13: Typical phase-signal output for the 0.45 mm-thick coating at a modulation frequency of 0.5 Hz. Vertical scale: 20 degrees/div; horizontal scale: 5 s/div.

MICROSTRUCTURE OF COATINGS FORMED ON PACK ALUMINIZED Ni-Cr ALLOYS

Nimal Delanerolle
Standard Microsystems Corporation
Hauppauge, New York, USA

L.L. Seigle
Department of Materials Science & Engineering
State University of New York
Stony Brook, New York, USA

ABSTRACT

The microstructures of coatings on several pack aluminized Ni-Cr alloys with varying Cr content were carefully studied using scanning electronic microscopy and optical microscopy. The morphology of the multiphase structures found in these coatings is described and correlated with the Cr content of the substrate and aluminization time.

THE MICROSTRUCTURES FORMED on aluminization of commercial Ni based superalloys are extremely complex. This stems partly from the fact that the superalloys themselves are metallurgically and chemically complicated. Although many studies have been made to elucidate the structures of aluminide coatings on superalloys 1-8, there are still unresolved problems. The current investigation attempts to make a contribution to the clarification of these structures. To avoid chemical complexity, the major components of the Ni-based superalloys - Ni and Cr - were selected as the components of the substrates used in this investigation.

Recent studies 9,10 using scanning electron microscopy with an energy dispersive system have verified the identity of the phases formed during pack aluminization of Ni-Cr alloys at 1025°C for 18 hours and has also been successful in profiling the variation in composition of the components in each phase of the coatings. The present investigation is a sequel to the above, and considers the formation and evolution of coatings formed by aluminization for different times.

Ni-Cr alloy specimen with compositions shown in Table 1 were aluminized at 1025°C in a 60 at % Al alloy pack containing 2 at % AlF_3 as activator, in a hydrogen atmosphere, for the times shown. Detailed description of the aluminization process is to be found in Reference 11. After aluminizing, the specimen were sectioned and prepared metallographically for SEM observation.

Resolution and adequate depth of field were important for the observation of microstructures. Hence, a working distance of 8 - 15 cm, a tilt angle of 15 - 45°, and accelerating voltage of 30 keV and an emmission current of 150 uA were employed. An increased working distance of 23-33 cm was used in some cases to obtain increased depth of field.

The study of the morphology of α-Cr precipitates required an etchant that dissolved away the β-phase completely from the coating region, leaving the α-Cr lamellae projecting into the region from the substrate. The etchant used by Cline [12] - 100 ml H_2O, 20 ml HCl, 5g CRO_3 - served the purpose. The method was to immerse the sample in the etchant for 1/2-2 hours. Repeated examination with an optical microscope within this period gave a quick indication of amount of phase remaining. Once all the β-phase had been removed, the samples were ready for the morphological examination in the SEM. Varying tilt angles enabled one to observe the actual shape of the precipitates.

OBSERVATIONS

The coatings studied in this investigation can be broadly classified into the two types shown sketched in Figure 1, and this figure also helps in identifying the different features observed. An overall picture of the coating microstructure can be obtained by studying the low magnification SEM micrographs in Figures 2-4. The most striking feature is that the coatings appear to be divided into distinct zones - an initial zone containing single phase β, an intermediate zone containing isolated α-Cr precipitate particles and a final zone containing non-isolated lamellar type α-Cr. The initial zone is a single phase region of NiAl (β) beginning from the surface of the coating. The intermediate zone begins with the first appearance of particles of α-Cr[9]. In this zone the minor α-Cr phase is distributed in an isolated fashion in the matrix β-phase. The maximum length of these precipitate particles was about 10 um. Their identity can be inferred from previous work [9-11]. One notices from the micrographs that the isolated precipitates are concentrated at the grain boundaries of the β-phase in this zone. The final zone also possesses a two phase $\alpha+\beta$ structure, but in this zone the α-Cr phase occurs in the form of a lamellar like structure growing from the interface with the γ-phase substrate. That is the α-phase is nonisolated. The lamellae have lengths in the range of 20-30 um, oriented in the direction of transport of Al to the bulk alloy. These lamellar-type precipitates are qualitatively similar to those found in aluminide coatings on superalloys [1-8] and to the Widmenstatten-type precipitation in ternary alloy diffusion couples [13-18]. All three zones are present in the higher Cr ($\geq$25 at% Cr) samples. Whereas, the low Cr (= 15 at% Cr) sample show only the initial and final zones (Figure 4).

Examination of the lower Cr samples with an optical microscope, showed that the final zone is of a distinctly different color than the initial zone. This is understandable as the final zone is of higher Ni content as shown by the previous analytical work [9]. A similar observation could not be made with the higher Cr samples as the etchant discolored the surface.

The low magnification study enabled thickness measurements of all the different zones. These are shown plotted in Figures 10-12. The conclusions can be made from them:

(i) Each zone appears to obey a parabolic growth equation, irrespective of the presence of two-phase regions and three-phase boundaries.

(ii) An incubation period for layer growth exists (assuming that extrapolation to time zero is valid).

The low magnification micrographs also show that the size of the grains in the single-phase zone is much larger than that in the two-phase $\alpha+\beta$ zones. One observes that these grains are dimensionally larger in the direction of transport of Al. A close examination of the micrographs in Reference 10 shows similar behavior in coatings on superalloys.

The shape of the non-isolated α-Cr precipitates is better visualized from the higher magnification micrographs in Figures 5-7. The higher Cr samples show a more lamellar type morphology. A rod-like morphology is exhibited by the low Cr samples (Figure 7). The effect of time of aluminizing on the shape of the rods is well illustrated from the series for the Ni-35 at % Cr substrate (Figure 5). Thickening as well as shape distortion are exhibited at longer times. Whereas the one hour structure comprises mainly of single lamellae, with increasing time more and more bridging between the lamellae occurs, to such an extent that a single lamella is difficult to come by at 18 hours.

The effect of substrate Cr composition on the microstructure can be seen by comparing the two sets of micrographs from the Ni-25 at % (Figure 6) and Ni-35 at % CR (Figure 5) samples. For the same time, the lamellar-type precipitates in Ni-25 at % CR samples are less distorted in shape than those in the Ni-35 at % Cr samples, and there is less joining between adjacent units.

Nonisolated α-Cr precipitates in the Ni-15 at % Cr samples also exhibit thickening and lengthening with time, as shown in Figure 7. These α-Cr precipitates appear to be mostly rod-like units. Joining of units is less frequent, though this increases with increasing time. Figure 7 (c) shows an interesting example of joining of rods by bridging across at their terminus. The growth of the rods need not terminate at this point, as in this example, but can continue from the position of the bridge as in Figure 7 (d).

Another effect of greater time of aluminization on the α-Cr phase is the increased breaking-up of the non-isolated precipitation, shown by the series of low magnification micrographs for Ni-35 at % Cr (Figure 2). This appears to be connected with the formation and growth of the intermediate zone. These observations agree well with those of Taylor [13].

The deviation from rod-like structure is clearly shown in the set of micrographs shown in Figure 8. Also manifesed is the relatively rod-like structure in the lower Cr samples (Figure 8 (a)). The mushroom-shaped cap at the end of the rods (in Figure 9a) is probably the beginning of a fault (which could be a break), in the otherwise smooth cylindrical morphology (as for example breaking-up or joining). Figure 8 (d) for Ni-35 at % Cr and 18 hours indicates that the structure is distorted from the rod-like shape shown in the earlier micrographs (Figure 8a). Thus, it is possible that in the beginning (i.e. at short times) the morphology is entirely rod-like, and with faulting and thickening becomes altered in shape. It is clear from these micrographs that the morphology show a distinct directionality. This directionality appears to coincide with the direction of transport of Al to the bulk alloy

Two-phase structures were not the only types of microstructures that were observed. Coatings on Ni-5 at% Cr alloys are comprised of only a single layer of β-phase. Optical micrographs show two differently colored zones in this single phase. Tu [11] concluded from his microprobe studies that they correspond to Al-rich and N-rich β.

SUMMARY AND CONCLUSIONS

This study of the microstructures of aluminide coatings formed on Ni-Cr alloys helps to clarify the influence of both Cr content and aluminization time on coating structure. The Cr content of all of the specimens investigated is high enough so that γ' does not form and the phases appearing in the coatings are limited to α-Cr, and β, in addition to the original γ substrate. Only β was observed in coatings formed on the 5 at% Cr substrate, with the particular alloy pack used for aluminization, but both α-Cr and β appeared in coatings in the 15,25 and 35 at% Cr substrates. This is explained, in part by the fact that the maximum solubility of Cr in β at 1025°C is 8 at%. The α-Cr precipitate in the 15 at% Cr sample was in a lamellar or rod-like form at all times, but in the 25 and 35 at% Cr samples it appeared in both lamellar and isolated particle morphologies. At short times only the lamellar morphology was observed in these substrates but a layer containing an isolated precipitates of α-Cr in β appeared at longer times. Since the zone of lamellar structure must move through the coatings as it grows, it is possible that the α-Cr particles in the zone of isolated precipitation are the remnants of partly dissolved lamellae. This dissolution of α-lamellae may be connected with the mode of growth of β in the coating. Since the surface is at 55 at% Al,[9] inward diffusion of Al is very likely occurring at this location. On the other hand Al content of β at the α-Cr +β interface is only 33 at%, and therefore, outward diffusion of Ni from this interface is a possibility [10]. Thus, new β-phase is possibly forming somewhere within the $\beta + \alpha$-Cr layer and this may be the cause of dissolution of the lamellar structure[2]. However, the diffusion processes in this ternary system are complex and the exact mechanisms of precipitation remain to be worked out.

It is interesting that the thickness vs. time data observed in this study suggest a parabolic layer growth law, even though the layers are complex structures with two intermingled phases. This observation is significant with respect to the possibility of computer modeling the process of coating formation in practical systems.

The structures in these aluminized Ni-Cr alloys duplicate many of the important microstructural features of both aluminide coatings on superalloys and ternary alloy diffusion couples.

ACKNOWLEDGEMENTS

This material is based upon work supported by the National Science Foundation under Grant number DMR 8027494.

REFERENCES

1. Goward, G.W. and D.H. Boone, J. Oxidation of Metals 3, p. 475 (1971
2. Goward, G.W., D.H. Boone and C.S. Giggins, Trans. Amer. Soc. Met. 60, p. 228 (1967)
3. Petit, F.S. and G.W. Goward, in Superalloys Source Book, ed. Donachie, M.J. 170, published by American Society of Metals, Metal Park, Ohio (1984)
4. Jackson, M.R. and J.R. Rairden, General Electric Co. Report No. 77CRD16, Schenectady, New York (1977)
5. Janssen, M.M.P. and G.D. Rieck, Trans. Met. Soc. of AIME 239, p. 1372 (1967)
6. Fleetwood, M.J., J. Inst. Metals 98, p. 1 (1970)
7. Lewis, S.R. and R.M. Caves, J. Electrochemical soc. 121, 1051-1063 (1974)
8. Redden, T.K., Trans. Met. Soc. AIME 242 1695-1702 (1969)
9. DeLanerolle, N. and L.L. Seigle, J. Vac. Sc. & Tech. A 24, Nov/Dec 1985
10. Delanerolle, N., Ph.D. Thesis, State University of New York, Stony Brook, New York (1985)
11. Tu, D., Ph.D. Thesis, State University of New York, Stony Brook, New York (1982)
12. Cline, H.E., Met. Trans 2, p. 189 (1971)
13. Taylor, C.W., Dayananda, M.A. and R.E. Grace, Met. Trans. 1, p. 127 (1970)
14. Cheng, G.W. and M.A. Dayananda, Met. Trans, 10A, p. 1407 (1979)
15. Sisson, R.D. and M.A. Dayananda, Met. Trans. 3A, p. 647 (1972)
16. Wirtz, C.E. and M.A. Dayananda, Met. Trans. 8A, p. 567 (1977)
17. Cheng, G.H. and M.A. Dayananda, Met. Trans. 10A, p. 1407 (1979)
18. Kirkaldy, J.S. and D.G. Fedach, Trans. Met. Soc. AIME 224, p. 490 (1962)

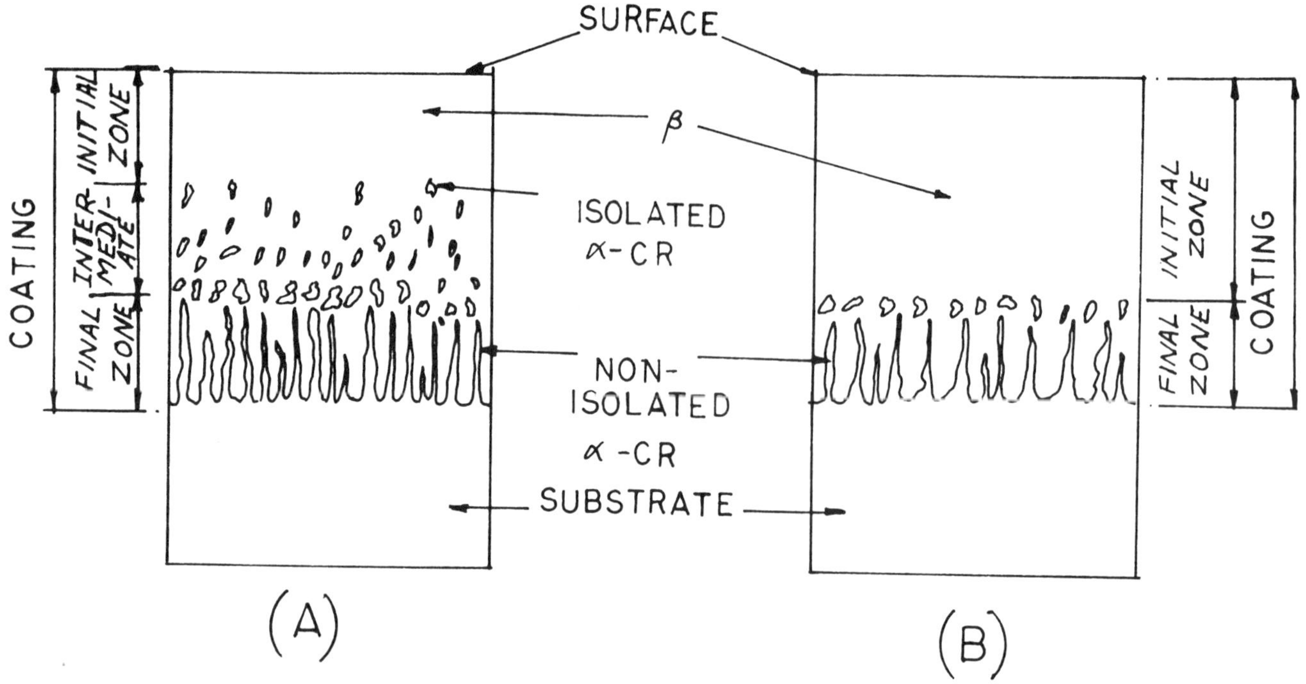

Figure 1 - Sketch showing the basic features of coatings on
(A) A high chromium (⩾ 25 at% Cr) substrate
(B) A low chromium (=15 at% Cr) substrate

Table 1 - Aluminization Parameters of Samples used in the microstructural investigation

Pack Composition	Time of Aluminization (hours)	Substrate
60 at% Al	1, 4, 10 and 18	Low Cr samples Ni-15 at% Cr
	1, 4, 10 and 18	High Cr samples Ni-25 at% Cr Ni-35 at% Cr

The temperature of aluminization was 1025°C
The Ni-5 at% Cr sample was aluminized for 18 hours with 60 at% Al pack.

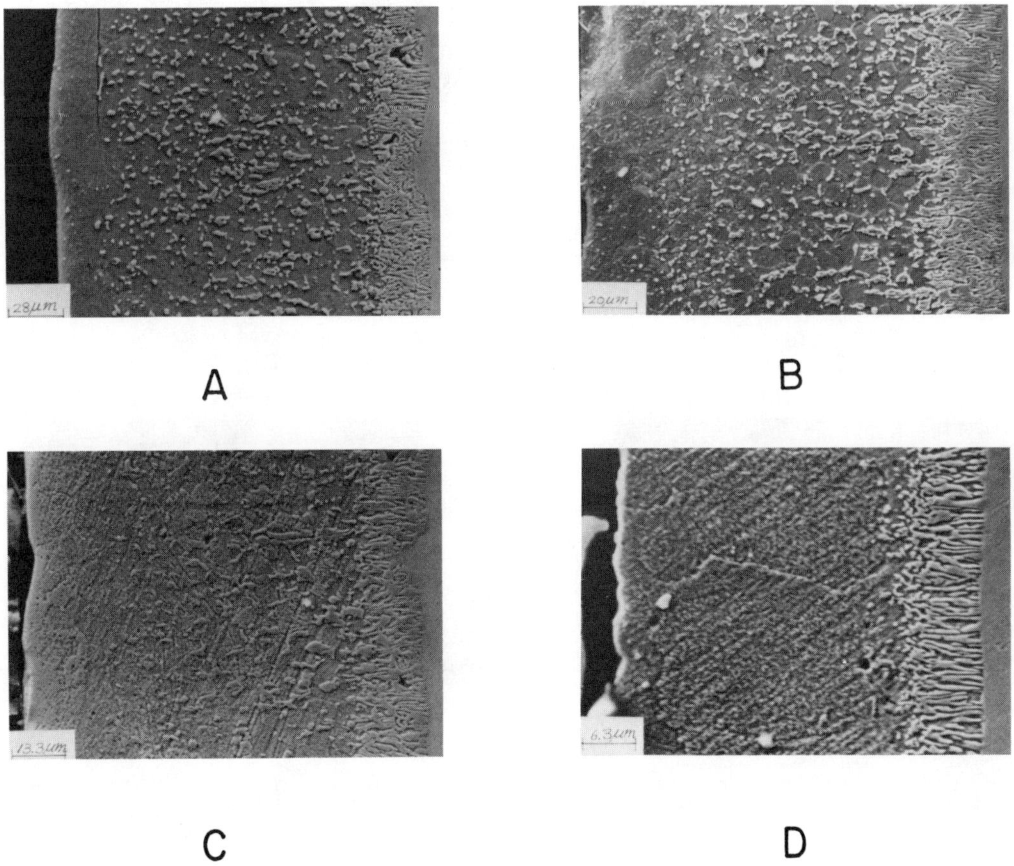

Figure 2 - Scanning electron micrographs of aluminized (60 at% Al, 2% AlF_3 pack) coatings on Ni-35 at% Cr samples at 1025 $^{\circ}C$ for
(A) 18 hours
(B) 10 hours
(C) 4 hours
(D) 1 hour

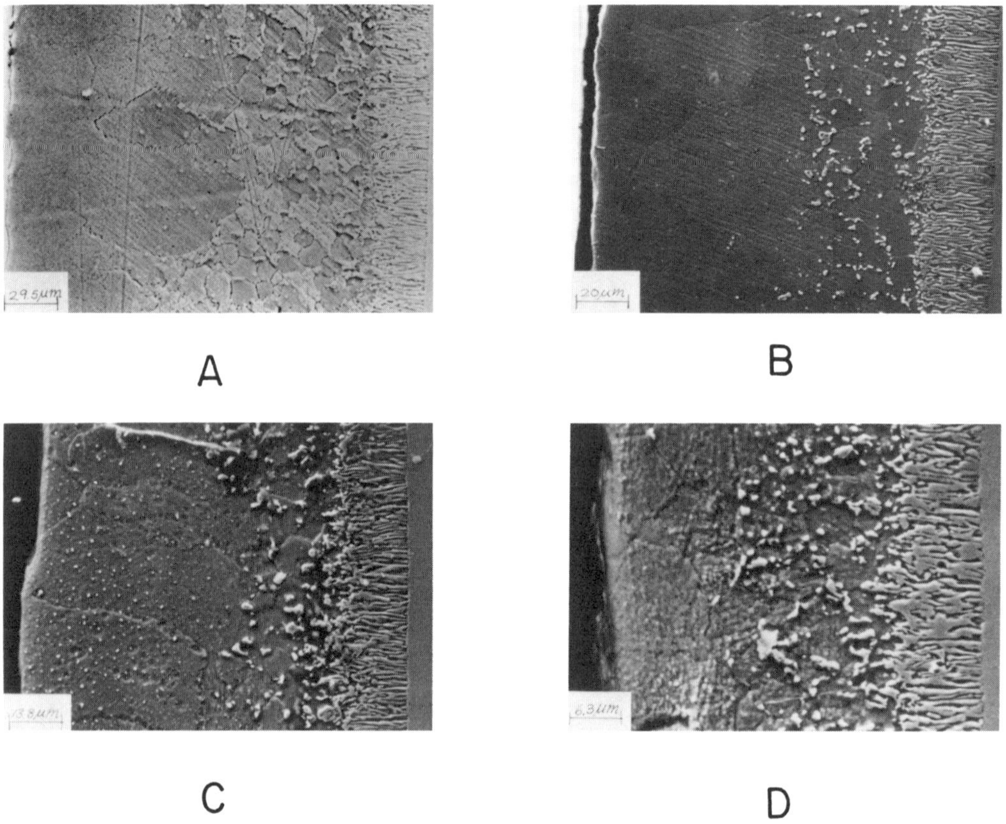

Figure 3 - Scanning electron micrographs of aluminized (60 at% Al, 2% AlF_3 pack) coatings on Ni-25 at% Cr samples at 1025°C for
(A) 18 hours
(B) 10 hours
(C) 4 hours
(D) 1 hour

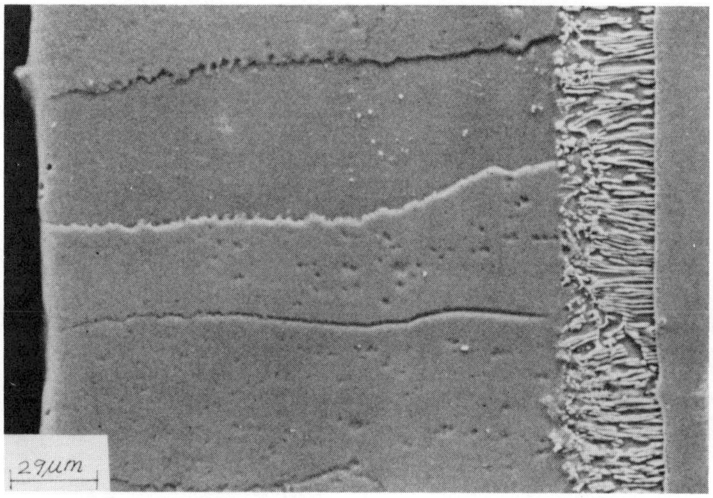

A

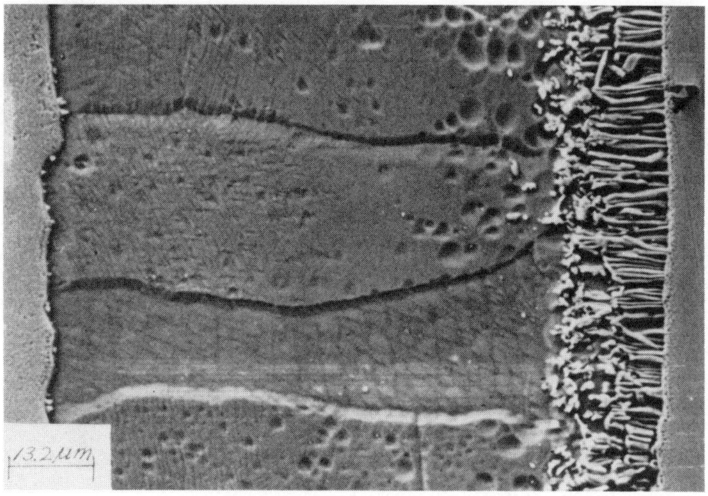

B

Figure 4 - Scanning electron micrographs of aluminized (60 at%, 2% AlF_3 pack) coatings on Ni-15 at% Cr samples at 1025 °C for
(A) 18 hours
(B) 4 hours

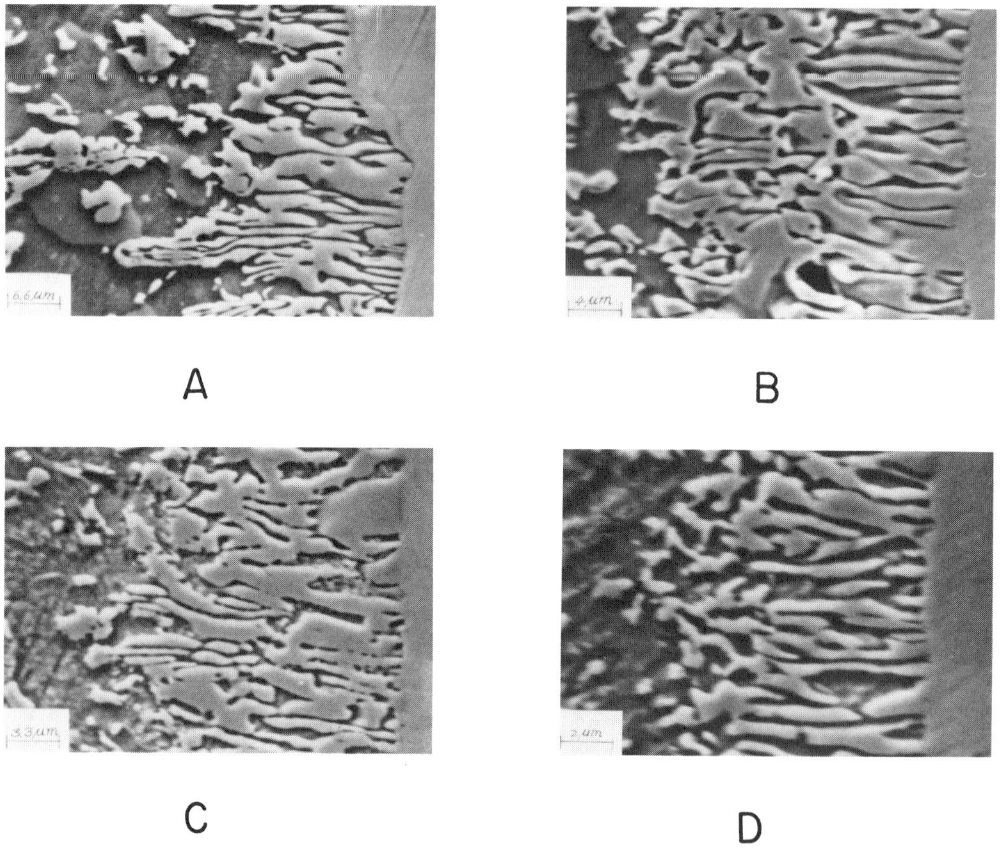

Figure 5 - Scanning electron micrographs of the $(\alpha+\beta)/\gamma$ region of aluminized (60 at% Al, 2 at% AlF_3 pack) coatings on Ni-35 at% Cr samples at 1025°C for
(A) 18 hours
(B) 10 hours
(C) 4 hours
(D) 1 hour

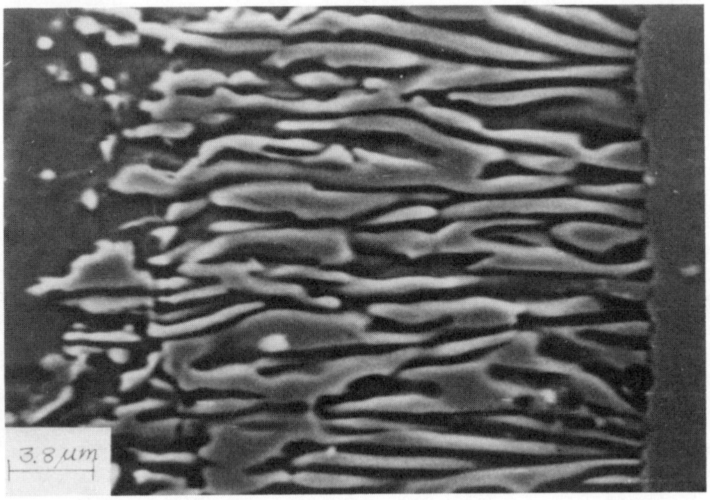

A

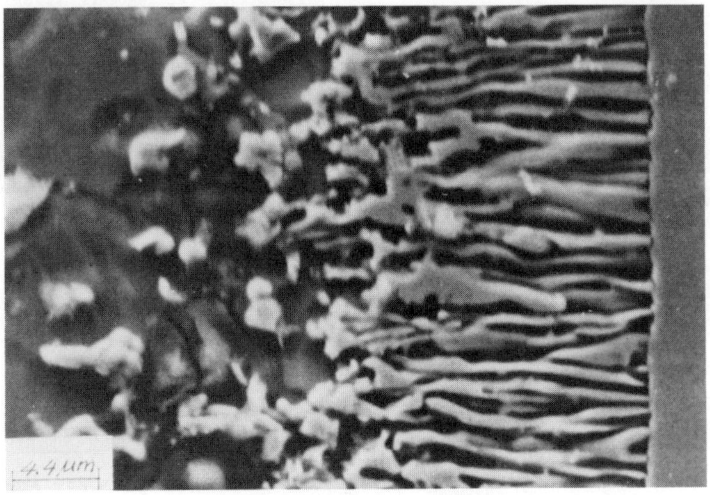

B

Figure 6 - Scanning electron micrographs of the $(\alpha + \beta)/\gamma$ region of aluminized (60 at% Al, 2 at% AlF_3 pack) coatings on Ni-25 at% Cr samples at 1025 °C for
(A) 10 hours
(B) 4 hours

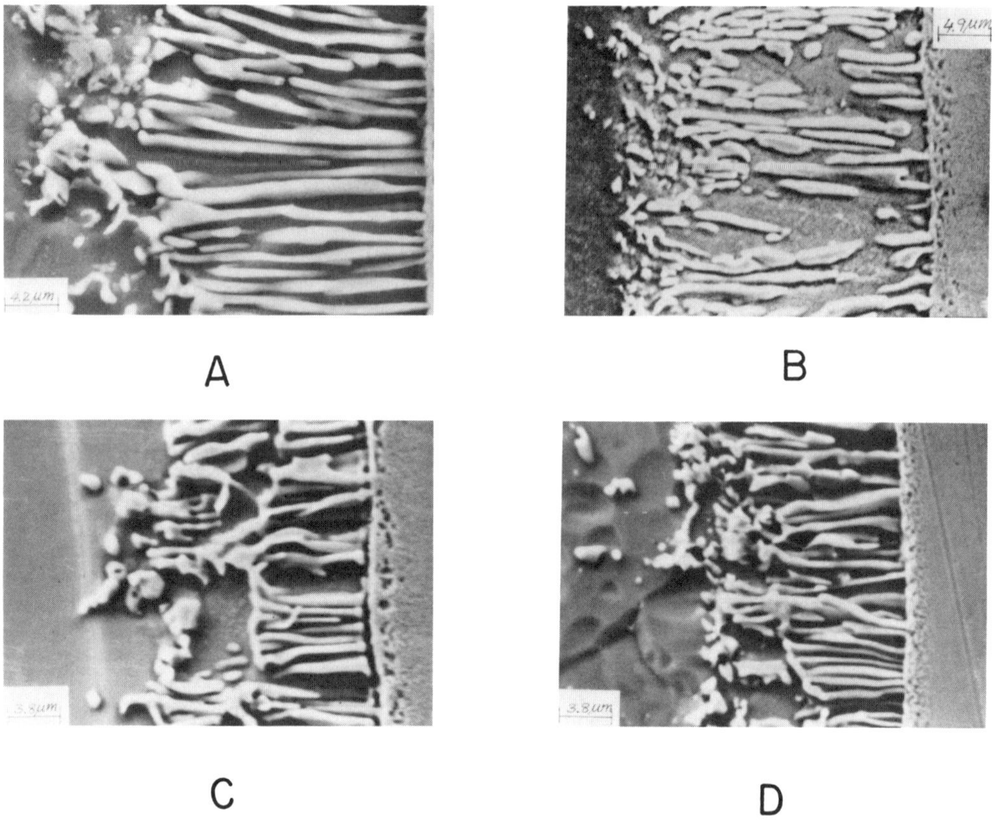

Figure 7 - Scanning electron micrographs of the
($\alpha + \beta$)/γ region of aluminized
(60 at% Al, 2 at% AlF_3 pack) coatings
on Ni-15 at% Cr samples at 1025°C for
(A) 18 hours
(B) 18 hours
(C) 10 hours
(D) 4 hours

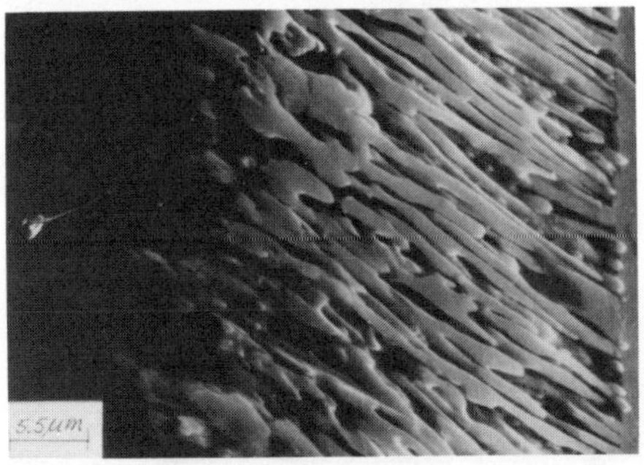

A

B

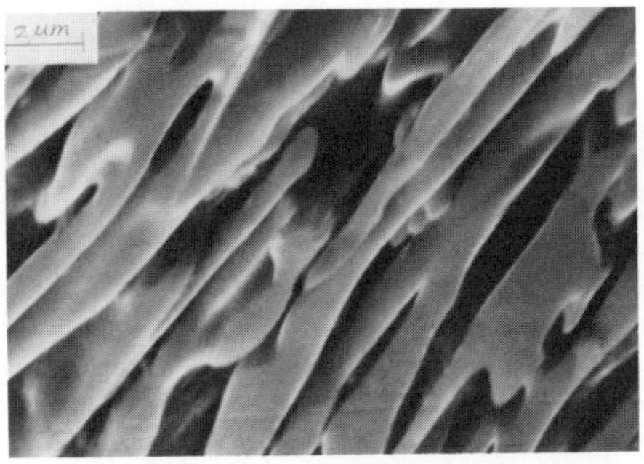

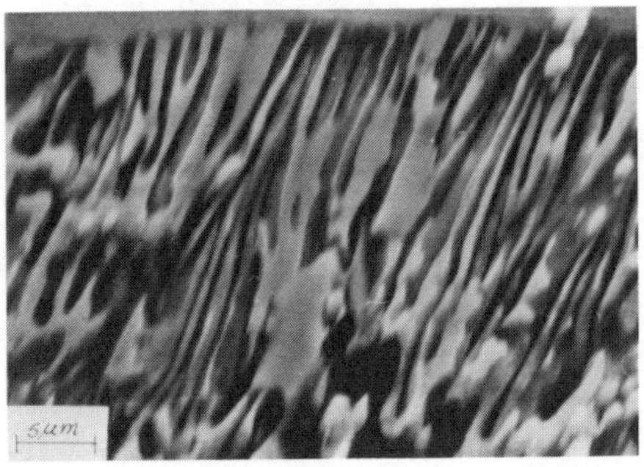

C

D

Figure 8 - Scanning electron micrographs of the morphology of α-Cr nonisolated precipitates in aluminized (60 at% Al, 2 at% AlF_3 pack) coatings on
(A) Ni-15 at% Cr at 1025°C for 1 hour
(B) Ni-25 at% Cr at 1025°C for 1 hour
(C) Ni-25 at% Cr at 1025°C for 1 hour
(D) Ni-35 at% Cr at 1025°C for 18 hours

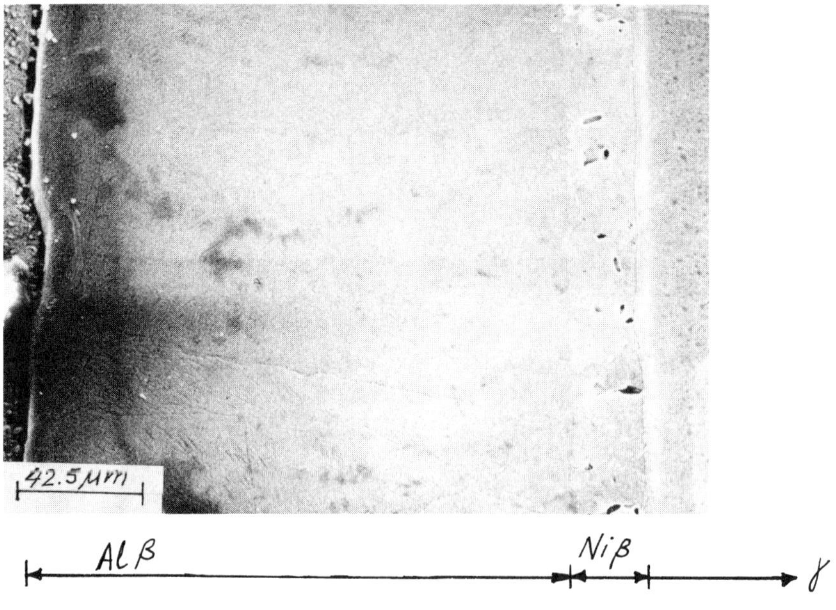

Figure 9 - Scanning electron micrograph of aluminized (60 at% Al, 2 at% AlF_3 pack) coating on a Ni-5 at% Cr sample at 1025°C for 18 hours

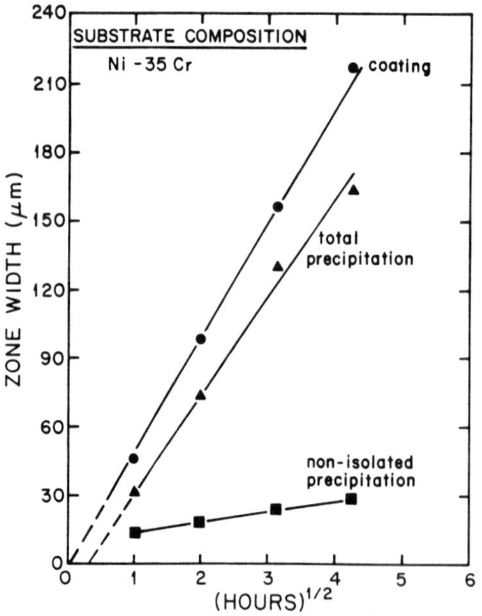

Figure 10 - Plot of layer width vs $(\text{time})^{\frac{1}{2}}$ from a coating on a Ni-35 at% Cr sample. Observe that there is an incubation time needed for the appearance of Cr precipitation. Temperature of aluminization was 1025°C.

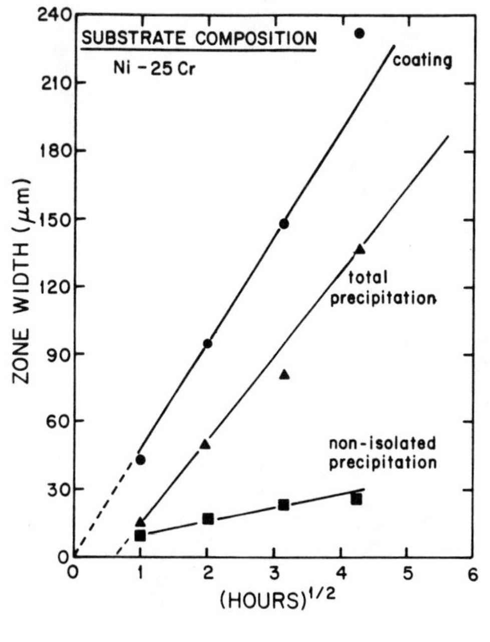

Figure 11 - Plot of layer width vs (time)$^{\frac{1}{2}}$ from a coating on a Ni-25 at% Cr sample. Observe that there is an incubation time needed for the appearance of Cr precipitation. Temperature of aluminization was 1025°C.

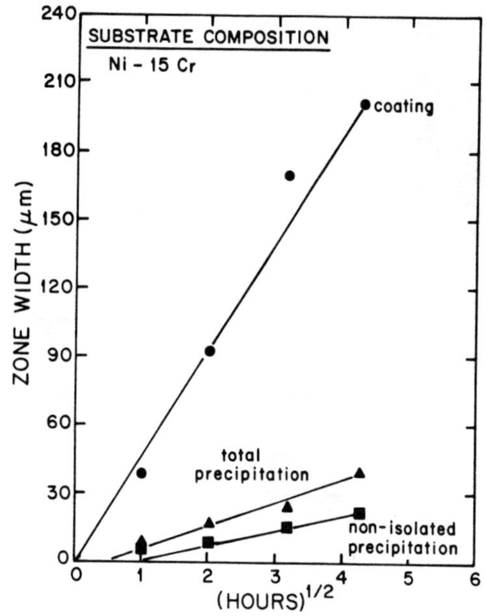

Figure 12 - Plot of layer width vs (time)$^{\frac{1}{2}}$ from a coating on a Ni-15 at% Cr sample. Observe that there is an incubation time needed for the appearance of Cr precipitation. Temperature of aluminization was 1025°C.

CYCLIC OXIDATION OF CHROMIZED STEELS AND COMPETITIVE MATERIALS AT 650 TO 815°C

Edward A. Loria
Metallurgical Consultant
Pittsburgh, Pennsylvania, USA

SOLID STAINLESS STEELS such as Type 409, 410 and 430 have been employed or considered for automotive mufflers, catalytic converters and exhaust manifolds but chromized steel and aluminized steel are alternate materials if chromium conservation and material cost become primary considerations. Since these devices are subject to thermal cycling conditions, a study of the oxidation of chromized steels and competitive materials at 650, 704 and 815°C (1200, 1300 and 1500°F) was initiated. Oxidation behavior will be discussed using data obtained from laboratory cyclic oxidation tests. Such tests cannot be used to predict how long the materials will be resistant to attack in a given application. However, these tests can be used to describe mechanistically how degradation occurs as well as the effects of different elements upon the degradation process. Hence, they do provide a comparison of the relative rates of oxidation in still air of chromized steel and aluminized steel with solid Type 409, 410 and 430 stainless steel.

There are numerous publications on chromizing and oxidation under isothermal conditions. (1,2), suffice it to say here that the chromized steels evaluated in this anisothermal study came from basically two processes that have been described in the literature. In the first series, product from 7 inch wide pilot plant coils of chromized steel produced by the first process (3,4), was compared with product from a commercial coil of Ti stabilized steel B-1 that was chromized by the second process (5,6) as well as product from commercial coils of aluminized steel and Type 430 stainless steel. The three chromized variations were on substrates of vacuum degassed steel N-43, Ti stabilized steel N-106 and low carbon rimmed steel N-111. In the second series, product from 30 to 36 inch wide commercial coils of chromized steel via the first process (3,4) were examined, namely low carbon rimmed steel W-10, Ti stabilized steel W-41 and partially Cb stabilized steel W-70. Also included was the product from a commercial coil of Ti stabilized steel B-2 produced by the second process (5,6) and another one identified as JL chromized, as well as product from coils of Type 409 and 410 stainless steel and Al-10 pct Si coated steel identified as JL Aludip. This steel has a protective layer of Fe-Al-Si alloy which is the basis for an improved aluminized steel (7) that can provide heat and corrosion resistance equal to Type 409 stainless steel.

PROCEDURE

Two specimens (panels) of each material measuring 2 inches by 4 inches were run under identical conditions. The specimens were cut from the center of each coil and hence had uncoated or unprotected edges. In the first series, the B-1 chromized and the aluminized specimens were around fifty percent thicker than the rest of the chromized steels and the solid Type 430 steel. In the second series, all of the specimens were of comparable thickness except for the B-2 chromized steel (taken from another coil) which again was thicker by about the same percentage, per Table I. The specimens were hung from nichrome hooks and exposed in still air in electrically heated box furnaces held continuously at 650°C, 704°C and 815°C. After 48 hours at temperature, they were removed and allowed to cool for one-half hour to room temperature. They were weighed and then placed back in the furnace for another 48 hour cycle. The tests were run for 480 hours or, in other words, for ten 48 hour cycles. Graphs of weight gain vs. time for each material were prepared and the relative rates of oxidation in still air at each temperature determined. Hence, they do show accumulative weight gained by the various

---

Retired Supervisory Metallurgist, Universal-Cyclops Specialty Steel Division. Cyclops Corporation, Pittsburgh. At the time of this investigation, Supervisor, Alloy Development, National Steel Research Center, Weirton, WV.

TABLE 1 TOTAL WEIGHT GAIN ON THERMAL CYCLING AT INDICATED TEMPERATURES

Material	Initial Gage	Total Weight Gain, mg/sq. in. 1200°F	1300°F
Type 409	.047	1.98	2.30
Type 410	.035	1.07	7.59
W-70	.038	3.56	1.78
W-41	.035	4.78	5.11
Bethlehem-2	.063	8.92	12.33
JL Aludip	.036	5.51	10.56
JL Chromized	.040	9.46	11.68
N-43	.024	4.6	11.0
N-106	.031	5.1	9.5
N-108	.036	7.4	17.5
N-111	.035	7.8	32.5
Bethlehem-1	.058	3.4	15.0
Aluminized	.050	9.0	22.5
Type 430	.022	-	2.5

samples. The value for the weight gain after 480 hours is the total weight gained by the sample over the entire testing cycle, and the amount of weight gained for each 48 hour period can also be determined from these graphs.

RESULTS

The weight gain vs. time curves for the first series of materials upon thermal cycling at 650°C are presented as Figure I. Uniformity in the rate of oxidation between specimens of 0.058 inch thick B-I chromized and 0.024 inch thick N-43 chromized steel at this temperature is noteworthy. The initial rapid period of oxidation is completed in the same time interval of around 144 hours in N-43 and N-106 and in B-I chromized product. The rate of oxidation afterwards is constant and quite low. Each curve has essentially the same slope and the rate constant could be estimated to be 0.010 milligrams per square inch per 48 hour cycle. The superiority of the two low-carbon chromized products vs. aluminized steel under thermal cycling is evident.

The weight gain vs. time curves upon thermal cycling at 704°C are shown in Figure 2. The better performance of the low-carbon coatings, N-43 and N-106, is quite apparent. The initial high rate of oxidation is completed between 96 and 144 hours, and the rate of oxidation thereafter is as low at 704°C as it was at 650°C. Chromizing vacuum-degassed, low-carbon steel N-43 produced the best overall result at 704°C. The uniformity of performance of the two specimens of this product is also noteworthy and all four specimens are better than the performance of two specimens of the heavier gage Ti-stabilized chromized steel B-I.

It is noteworthy that a variable weight gain curve can be obtained from two panels of the same product. This non-uniformity is evident

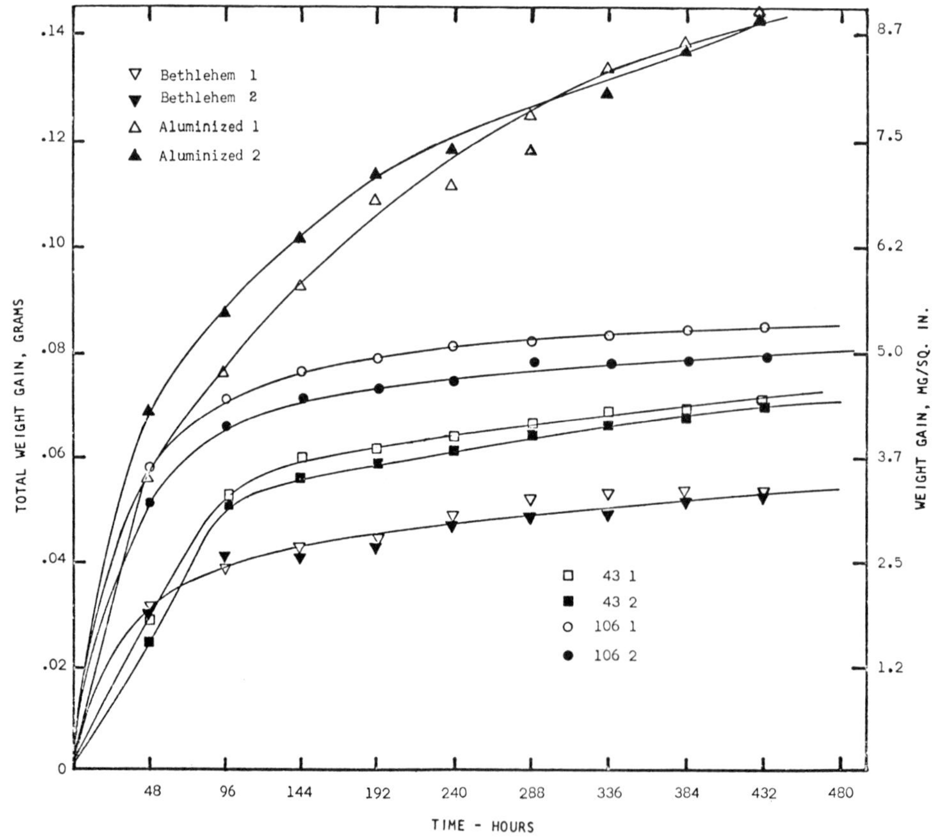

Fig. I - Weight gain vs. time at 650°C.

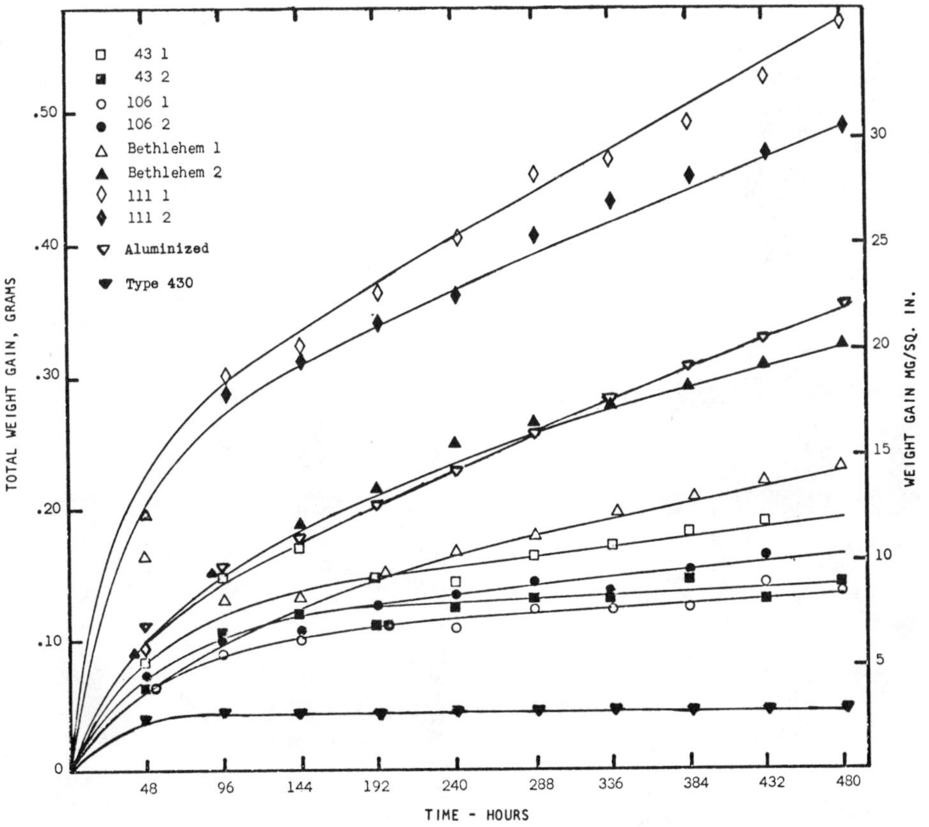

Fig. 2 - Weight gain vs. time at 704°C.

both in the case of the Ti-stabilized steel B-I and the rimmed steel N-111. On the other hand, this did not occur in vacuum decarburized steel N-43 nor the Ti stabilized steel N-106 at 650°C or 704°C. From the shape of the curves at 704°C, it appears that there is no appreciable oxidation after 144 hours and the rate of oxidation from that point onward, at least in N-43 is nearly equivalent to that of solid Type 430 stainless steel. The surface appearance of the respective panels after total exposure at 650°C is depicted in Figure 3. N-43 and N-106 compare favorably with the chromized B-I and the aluminized steel which were 50 pct thicker for better flatness and support under thermal cycling stresses. Figure 4 shows the surfaces of these low carbon coatings after total exposure at 704°C.

The weight gain vs. time curves upon thermal cycling the second series of materials at 650°C are presented in Figure 5. Again, the slopes of the curves indicate the degree of the initial rapid oxidation. For the better performing materials, this time interval is between 48 and 96 hours. The two solid stainless steels show very little, if any, gain in weight beyond 96 hours, in other words their rate of oxidation is nil. The W-70 chromized steel is the best of the coated steels followed by W-41 and JL Aludip, and it should be noted that the oxidation or rusting which accounts for the weight gain occurs along the unprotected sheared edges and not by attack of the coated surface itself. W-41 shows a higher rate of oxidation initially but after 96 hours exposure the oxide layer stabilizes and beyond 192 hours there is little or no more growth. On the other hand, less oxide formed during the first 96 hours on the Aludip coating but the layer continued to grow throughout the 480 hour test period and was greater than W-41 beyond 336 hours. The B-2, JL and W-10 chromized steels appear satisfactory, with the heavier gage 0.063 inch thick B-2 product showing good performance because the rate of oxidation beyond 144 hours is nil. The surface appearance of the respective test panels after total exposure at 650°C is depicted in Figure 6 and reveals the nature and degree of surface deterioration and edge rusting in the coated steels.

The weight gain vs time curves upon thermal cycling at 704°C are plotted as Figure 7. Again the solid stainless steels resist oxidation better than any of the coated steels. Better performance from the standpoint of an adhering oxide layer and low total weight gain was obtained on W-41, JL and B-2 chromized and JL Aludip steel. The JL chromized and Aludip steels

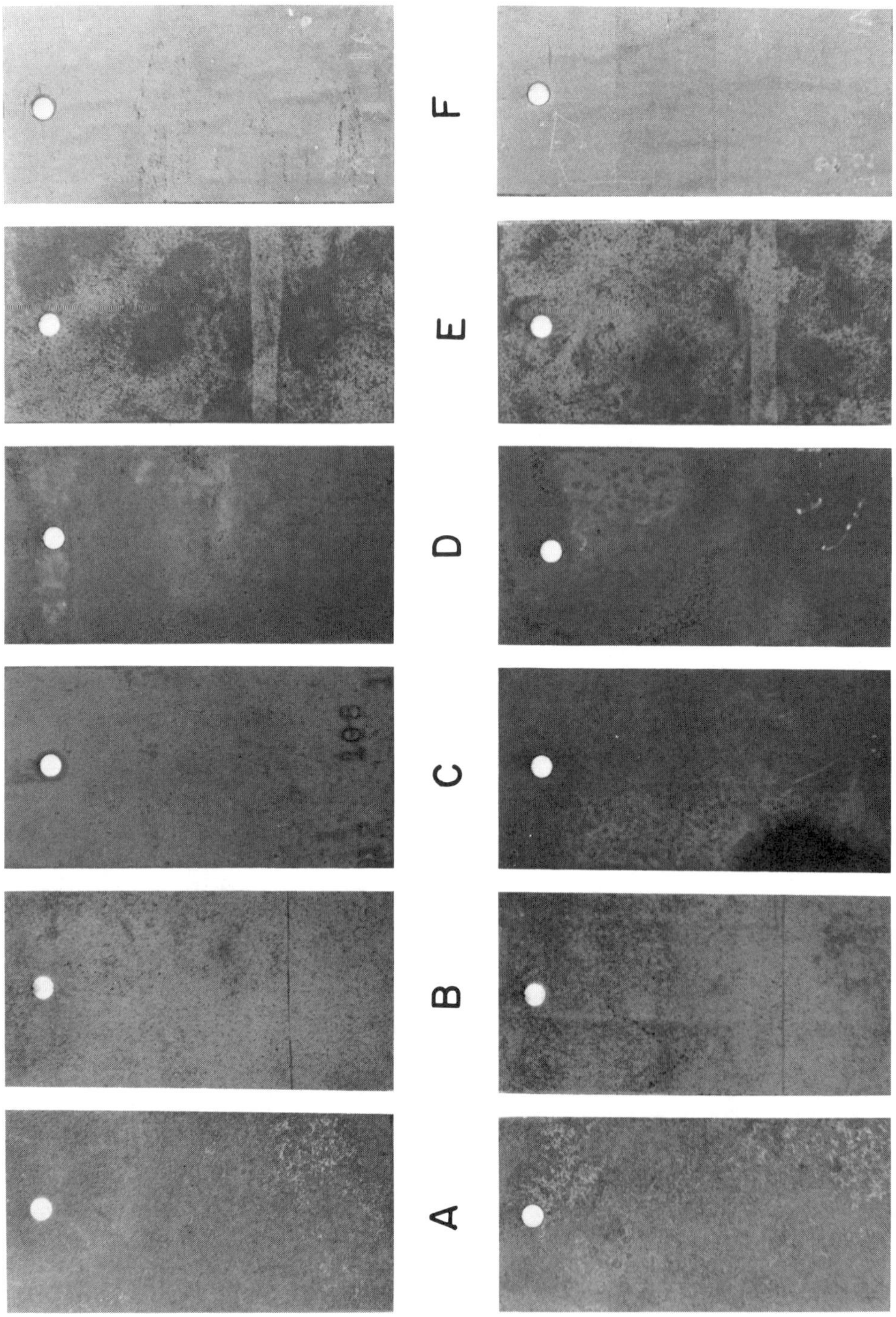

Fig. 3 - Surface appearance after 480 hours at 650°C. A = N-43, B = N-106, C = N-198, D = N-111, E = B-1, F = Aluminized.

had adherent scales and their parabolic growth curves closely simulate the behavior of the B-2 chromized steel in that some oxidation continues even after the faster initial oxidation period is over. On the other hand, the curves for the solid stainless steels are flat after the initial layer is formed which indicates that there is little or no more oxidation thereafter.

Surprisingly, the 0.035 inch thick W-41 performed better at 704°C than at 650°C in that the oxide layer remained mostly adherent throughout the test and total growth was lower. On the other hand, the 0.038 inch thick W-70 which performed well at 650°C lost part of its surface (and edges) between 48 and 96 hours exposure at 704°C and the characteristic parabolic rate of oxidation could not be obtained. These inconsistencies point out the importance of variations from one surface area to another in regard to coating morphology which will govern stability or failure on thermal cycling at high temperatures. Also, it should be recognized that there is a greater possiblity of rupture of an oxide film at the surface by the continual thermal cycling stresses in thinner gage material. In this regard, the heavier gago 0.063 inch thick chromized steel retained its oxide layer both at 650°C and 704°C.

The total weight gain on thermal cycling at 650°C and 704°C for the two series of material is enumerated in Table I. At 650°C, Type 409 and 410 showed the least amount of oxidation and exhibited true parabolic scale-growth and subsequent passivity behavior throughout the test period. W-70 chromized steel was the best of the coated steels followed by W-41, N-43, N-106 and Aludip. Throughout half of the test exposure, W-70 was nearly equivalent to Type 409 but, after 480 hours, its oxide layer was about 50 pct heavier than Type 409. Comparing W-41 and Aludip, although W-41 showed higher initial growth, it had a stable layer and very little growth beyond 192 hours and hence exhibited better oxidation resistance than Aludip after 480 hours. At 704°C, the solid stainless steels resisted oxidation better than any of the coated steels. Of the latter, better performance from the standpoint of an adhering oxide layer and low total weight gain was obtained in W-41, N-106, JL Aludip, N-43 JL and B-2 chromized. W-70 which performed well at 650°C lost part of its surface and edges before 96 hours of testing. Completely catastropic oxidation occurred in W-10 which precluded curve plotting.

Practically all of the coated steels could not withstand thermal cycling at 815°C and the progress of oxidation according to a parabolic law could not be followed. Unprotected edge rusting, which became more acute with rising temperature, caused a breakdown in the coating near the edges and a consequent weight loss in the test panel. Cracking and shredding of scale and the heavy attack along the unprotected edges led to complete destruction of the carbon steel

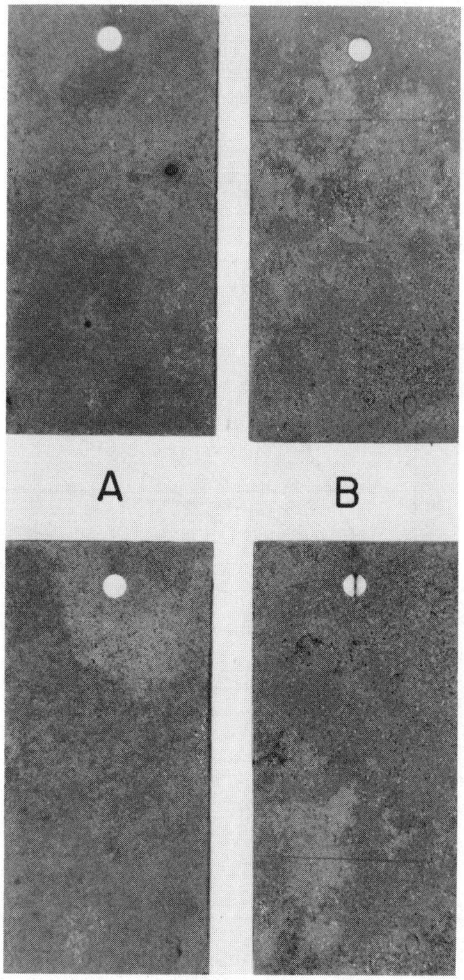

Fig. 4 - Surface appearance after 480 hours at 704°C. A = N-43, B = N-106.

substrate in some instances. Under these circumstances, all of the thinner gage chromized steel products must be considered as inadequate for this type of service. The two solid stianless steels showed a slight continual weight gain and then passivity over the entire 480 hour test period, the values being an insignificant difference between 1.07 mg/sq.in. for Type 409 and 1.72 mg/sq.in. for Type 410. Again, to point out the difference in surface morphology from one location to another that can exist even in solid stainless steel, these results at 815°C are lower than those obtained at 704°C which are listed in Table I.

As seen in Figure 8, the heavier gage B-2 chromized steel remained intact with pitting of the surface and loss of scale which prevented the plotting of a complete parabolic curve of scale growth. Also, the JL Aludip specimen survived the 480 hour test period at 815°C, with heavy

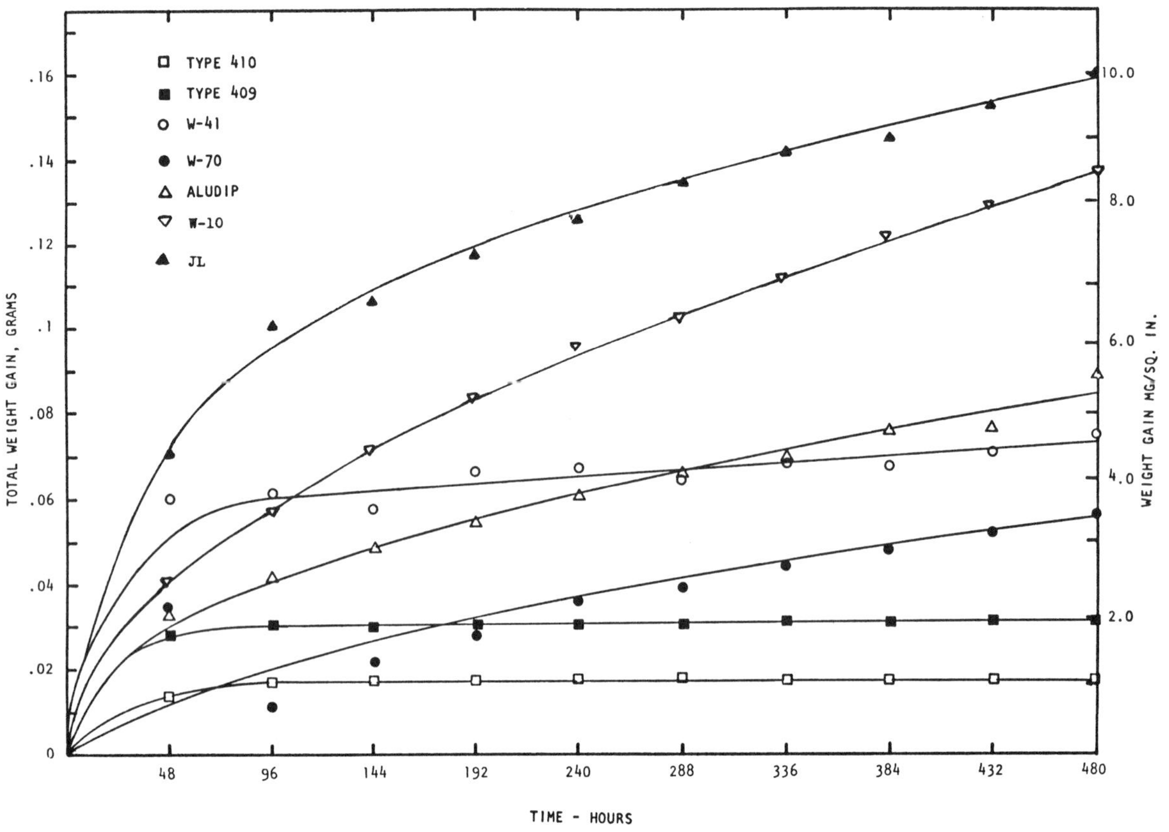

Fig. 5 - Weight gain vs. time at 650°C.

pitting and edge rusting, and produced an essentially sharp-rising parabolic curve with a total weight gain of 122.56 mg/sq.in. However, even these two products would not be considered satisfactory for thermal cycling service at 815°C, but this conclusion should be tempered by the understanding that the coated steel panels had unprotected edges which exposed the steel substrate to 815°C and part of the heavy scale growth (and also loss) is attributed to this condition rather than scale growth on the two flat coated surfaces alone.

The importance of edge oxidation on the interpretation of oxide weight gain measurements of aluminum coated steel has been demonstrated and 80 pct of the weight gain has been attributed to edge oxidation for the cyclic test conditions used (8,9). On this basis, the calculated value for the Aludip product would be 24 mg/sq.in. which is still considerably higher than the values obtained on Type 409 and 410. Also, early scale loss in the 0.058 inch thick B-1 chromized steel from the first series of tests prevented the plotting of a parabolic curve but 50 mg/sq.in. was measured after 480 hours at 815°C. If 80 pct of this value still represents edge oxidation, then the remaining 10 mg/sq.in. could be acceptable for scale growth on the two flat coated surfaces alone. However, the question of adherence and reproducibility from one coil to another or from different locations in the same coil remains. These questions can be directed, to a lesser degree, at Type 409 stainless steel which can lose weight (in relatively small amounts) in some of the test intervals of cyclic oxidation at 815°C and really means that the surface layer is not adherent even when the specimen is solid and devoid of edge oxidation.

DISCUSSION

In an effort to establish reasons for the relative performance of the materials studied, the respective coatings were measured and analyzed, per Table 2. Stripped foil representing the total coating was obtained directly by micrometer while metallographic thickness measurements represent solid coating only (below surface effects). The Cr analysis was made by direct spectrographic method on the stripped foils via a compaction-briquetting technique and C by standard wet chemistry procedure. All of the coatings, except for the JL chromized, had adequate stripped foil thickness and the

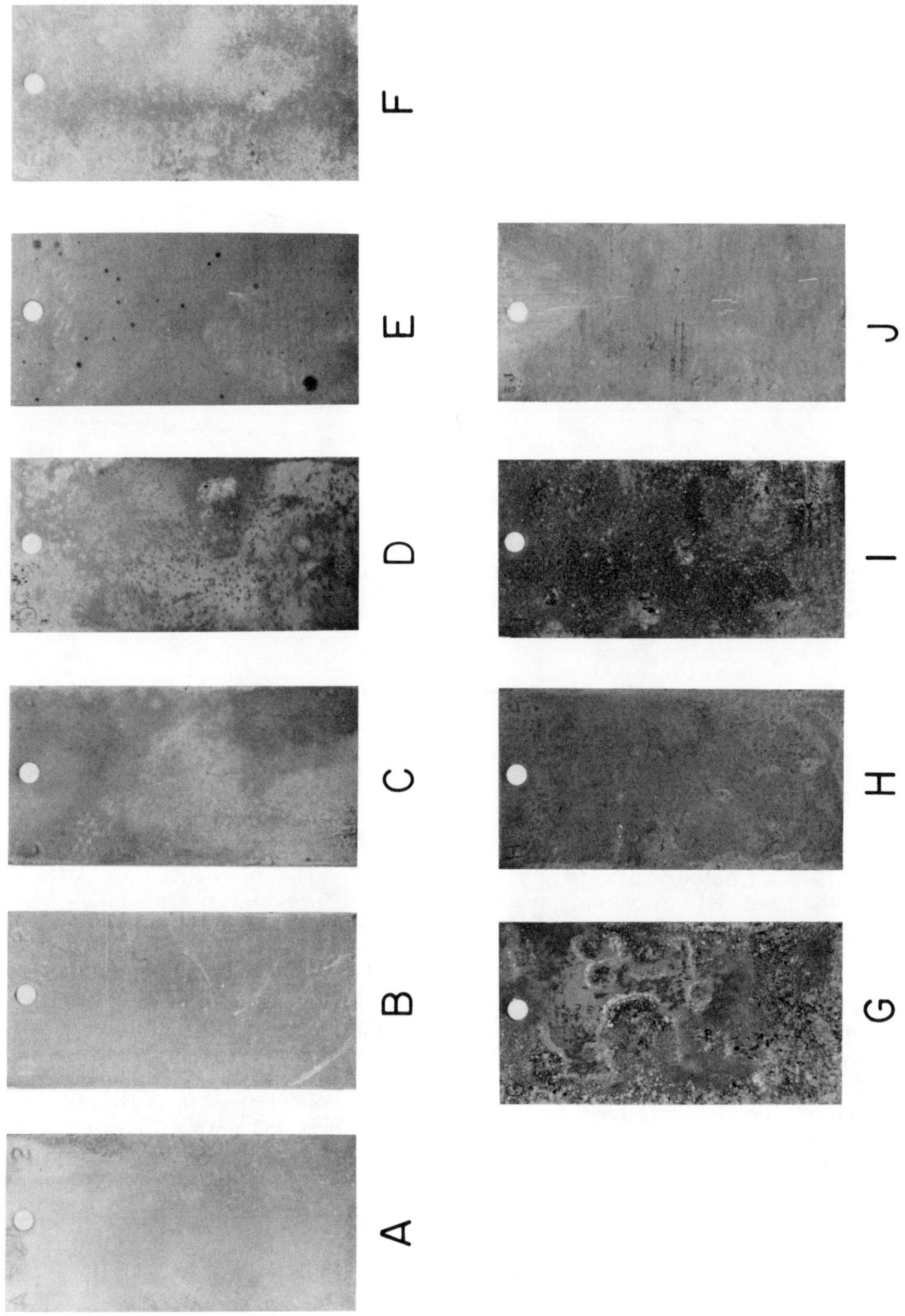

Fig. 6 - Surface appearance after 480 hours at 650°C. A = Type 410, B = Aludip, C = JL, D = W-10, E = B-2, F = W-41, G = W-69, H = W-70, I = W-71, J = Type 409.

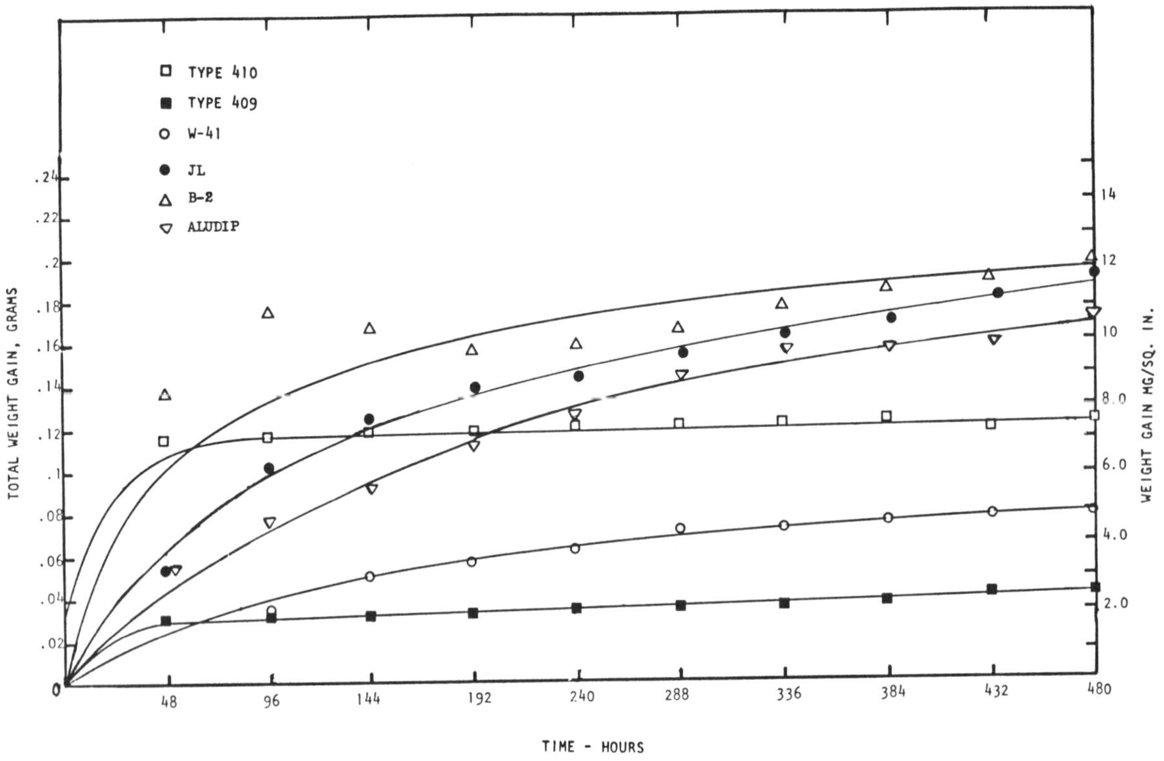

Fig. 7 - Weight gain vs. time at 704°C.

difference in the average chromium and carbon contents are evident. In this regard, it appears that the better performing coatings are those containing lower carbon, namely W-70 and W-41. W-70 was a columbium treated substrate steel while W-41 was stabilized with titanium. This view is substantiated by the complimentary result obtained on the B-1 and B-2 chromized and even the thin layer JL product, also, the 7-inch wide coils N-43 and N-106 were chromized low carbon and Ti-stabilized steel respectively and their coatings analyzed only 0.06 pct C while N-108 and N-111 analyzed 0.27 and 0.21 pct C respectively.

The merit of maintaining low carbon content in the diffusion layer as an important factor in achieving both corrosion and oxidation resistance has been stated in most investigations of chromized steel. The maximum carbon which can be tolerated has not been defined. On the basis of the work done to date, one would propose that a limit of 0.10 pct carbon in the coating would be appropriate for chromized steel. At the same time one would call attention to an intangible aspect of the chromizing procedure itself insofar as it affects the degree of homogenization of both the surface and the depth of the diffusion layer. In other words, one could speculate that variations in chromium content, segregated carbides and pitted areas could act in the same manner as chromium depleted areas which have been observed on the surface of improperly processed (solid) stainless steels, such as Type 430, and which have lowered both corrosion and oxidation resistance.

The microstructure of the coated materials also provides a correlation with the oxidation behavior under thermal cycling conditions. The cross section of the N-43 coating on the vacuum decarburized steel substrate reveals a smooth surface with little or no porosity nor migrated carbides therein, per Figure 9a, whereas the JL coating shows their presence in the outer layer, per Figure 9b. On the other hand, the B-2 chromized steel (also B-1) has a heavily creviced surface and porous layer over a solid layer, per Figure 9c. The Aludip coating, illustrated in Figure 9d, has a complex layer with a considerable amount of second phase, possibly resulting from the 10 pct Si addition, in the thicker outer layer which measures 1.2 mils.

The basic cause of coating spalling (or shredding) is the large thermal expension mismatch between the coating and substrate and the stresses developed upon cyclic cooling from the service properties. The most promising approach to improving spalling resistance has been based on the recognition that failure is caused by flaws which induce crack propogation in the plane of the coating. The stresses causing such propagation might be reduced by controlling

TABLE 2 COATING THICKNESS AND ANALYSIS

Material	Coating Thickness, mil Stripped Foil	Metallographic	Stripped Coating Avg. Analysis, % Cr	C
JL	2.0-1.0	1.0-0.5	23.7	.29
B-1	4.3	2.0	16.9	.04
B-2	4.0	1.3	22.6	.07
W-41	3.0	1.8	16.9	.08
N-43	3.8	2.5	21.6	.06
N-106	5.5	2.5	18.6	.06
W-10	3.5-3.0	2.4	20.0	.24
W-70	2.2	1.5	21.1	.14
N-111	4.0	2.4	25.2	.21
Type 409	-	-	(11.7)	(.06)
Type 410	-	-	(13.0)	(.12)

coating structure so that appropriate discontinuities cause a reduction in elastic modulus of the coating, thereby reducing local stresses. Such discontinuities include increased porosity, random microcracking and columnar structures with the columns normal to the substrate surface.

Accordingly, the ability of the B-2 chromized and the Aludip coating to withstand 480 hours of thermal cycling at 815°C can be explained by the porous outer structure with random microcracking within the coating in the case of the B-2 chromized steel, per Figure 9c, and the inner layer with columnar spikes extending into the outer layer which existed in the Aludip coating, per Figure 9d. Notwithstanding, the excellent performance and equivalence of N-43 and N-106 to B-2 chromized product at 650°C and 704°C should be noted in view of the fact that these steels were around 50 pct thinner in gage. The stability of the heavier substrate in withstanding thermal cycling stresses and which warp or curl thinner gauge specimens, thereby opening up the surface to further attack, should be recognized. Taking this factor into consideration and the excellent results obtained at 650°C and 704°C, it can be hypothesized that, on an equivalent gage basis, the chromizing process employed on either stabilized steel N-107 or vacuum carbon degassed steel N-43 should produce an equivalent oxidation rating to B-2 chromized product at 815°C.

The tensile test results on most of the materials are listed in Table 3. The strength parameters are a reflection of the fineness of the ferrite grain size after the coating process. As seen in Figure 9d, the substrate of the Aludip coating had a very fine grain size measured as ASTM 10-11. The JL and B-2 chromized steels, as well as N-106, had comparable ASTM 6-7 grain sizes which are illustrated in Figures 9b and 9c. Their carbon content was stabilized by the addition of Ti or Cb and their properties can be compared with those obtained on W-10 rimmed steel which obtained its strength from the residual carbon that did not form a carbide barrier under the surface, thereby hindering chromium diffusion. The importance of having a very low C content for diffusion chromizing is well known and can be attained either by lowering it to a sufficiently low level (such as in N-43) or by stabilizing it with carbide formers. A low C level in the substrate results in a coarse grain size and lower strength after chromizing. Consequently, for product applicability, it is preferable to stabilize the free C with strong carbide formers Ti, Cb and Zr, thereby providing adequate hardenability (or mechanical properties). Also, the substrate can contain the ferrite stabilizing element Cr in order to attain a certain coating thickness in a shorter time than a steel alloyed with austenite stabilizing elements like Mn or Ni. Finally, in the case of aluminized steel (7), Ti in the substrate also stabilizes C and N in the form of titanium carbonitrides, thereby avoiding the formation of a subsurface layer or barrier to coating diffusion that tends to restrict growth of the protective layer of Fe-Al-Si alloy.

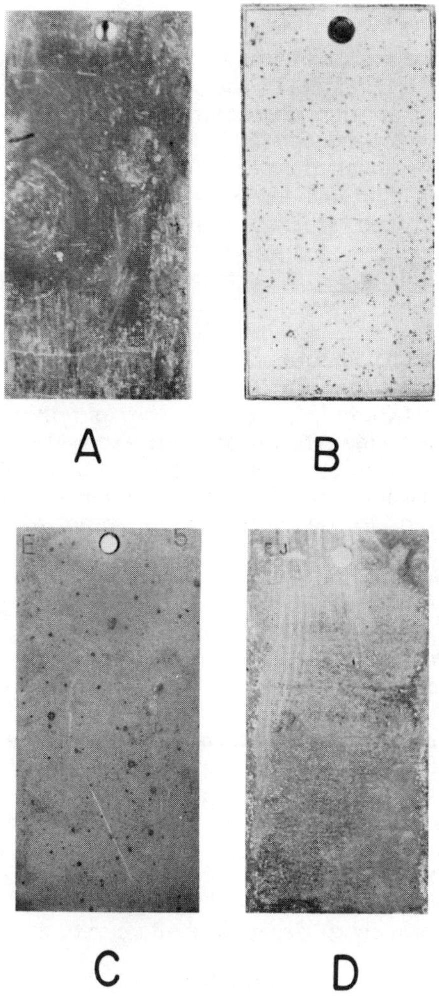

Fig. 8 - **Surface appearance after 480 hours at 815°C. A = Type 410, B = Aludip, C = B-2, D = Type 409.**

TABLE 3 TENSILE TEST RESULTS

Material	Gage Inch	Yield Stress, psi	Ultimate Stress, psi	Elongation Unif.	Elongation Tot.
Type 410	.035	47,000	70,800	17.0	21.5
Aludip	.038	42,800	57,200	15.8	21.5
JL Chrom.	.040	25,000	36,800	22.0	44.0
Bethlehem	.063	29,000	43,000	18.0	35.0
W-10	.040	31,000	36,900	23.4	43.5
W-41	.035	22,200	43,600	-	35.0
W-69	.037	22,500	39,600	-	35.0
W-70	.038	23,500	42,500	-	35.0
W-71	.038	22,100	40,800	-	33.0

CONCLUSIONS

On the basis of still air oxidation tests at 650, 704 and 815°C, under heating and cooling cycles every 48 hours up to 480 hours, the following conclusions are derived on various forms of chromized steel, aluminized steel and Type 409, 410 and 430 stainless steels.

1. Chromized steels can withstand long periods of thermal cycling at 650°C. Also, certain chromized steels can withstand thermal cycling at 704°C provided that sheared edges are protected. Better performing coatings are those containing lower carbon as the result of utilizing stabilized steel or low carbon steel as the substrate.
2. The oxidation characteristics of chromized steel depend upon the free carbon content of the substrate steel. When the chromized layer contains less than 0.10 pct carbon, chromized steel can probably be used in oxidizing atmospheres at temperatures approaching 815°C. However, if the coating contains more carbon, temperatures of about 704°C appear to be the limit.
3. Superior performance was obtained on the stainless steels at 650 and 704°C which can be attributed, in part, to the fact that the coated steels had base steel exposed at the sheared edges whereas the solid stainless steels did not. Hot dipped aluminized steel proved to be better than some of the chromized steels but inferior to others.
4. Type 409 and 410 stainless steels were the only materials to exhibit parabolic scale growth followed by passivity over the 480 hour test period at 815°C. Cracking and shedding of scale plus heavy attack along the unprotected edges and warping of the thinner gages of chromized steel led to destruction of the steel substrate and caused such products to be inadequate for this service. Heavier gage chromized or aluminized (Al-10Si) steel that had been stabilized remained intact but the former exhibited pitting and loss of scale and the latter a sharp rising parabolic curve.

REFERENCES

1. Menzies, I.A. and D. Mortimer, Observations on Chromizing of Iron and Plain Carbon Steels, Proceedings, Second International

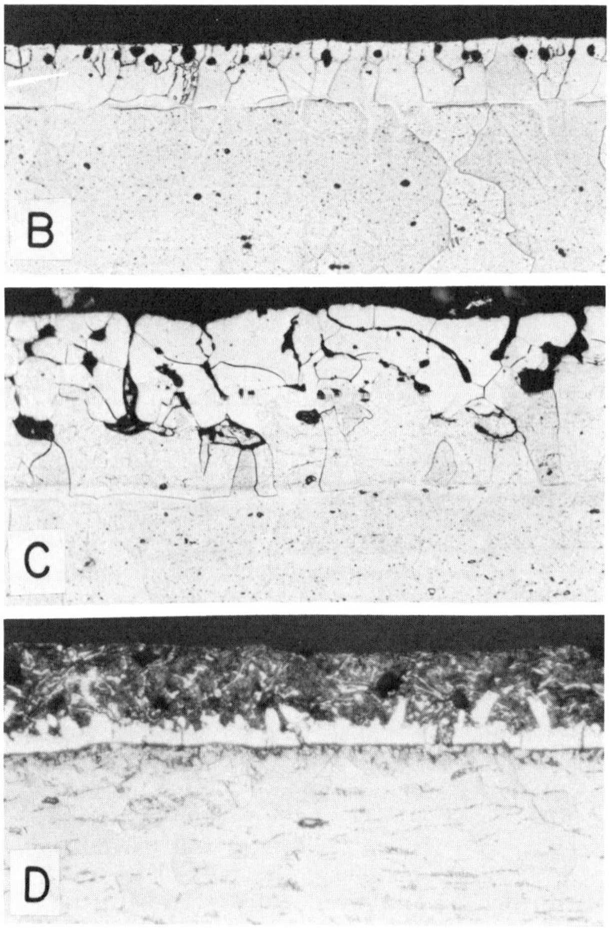

Fig. 9 - Cross sections of coated steel at X400. A = N-43, B = JL, C = B-2, D = Aludip.

Congress of Metallic Coatings, NACE, p. 733 (1963).

2. Sully, A.H. and E.A. Brandes, Oxidation Properties of Chromized Steel, Metallurgia, Vol. 49, p. 165 (1954).
3. Loria, E.A., Fabrication of Stainless Steel Foil Utilizing Chromized Steel Strip, Metall. Trans., Vol. 11A, p. 1755 (October 1980).
4. Beiser, C.A., National's Sheet Chromizing Process, U.S. Patent 3,969,550, 10 pp (July 13, 1976).
5. Mayer, E.H. and R.M. Willison, Bethlehem's Sheet Chromizing Process, Yearbook AISI, p. 251 (1967).
6. Willison, R.M. and J.D. Forstmann, Chromized Steel-Corrosion Resistance at Less Cost than Stainless Steel, Metal Progress, Vol. 91, p. 99 (January 1967).
7. Kim, Y.W. and R.A. Nickola, A Heat Resistant Aluminized Steel for High Temperature Applications, Paper 800316, presented at SAE Congress, Detroit (February 1980).

8. Gibson, A.F., F.M. Kilbane and F.C. Dunbar, Aluminum Coated Steels for Elevated Temperature Applications, Paper 810037, presented at SAE Congress, Detroit (February 1981).
9. Kilbane, F.M., Contribution of the Oxidation of Uncoated Edges to the Weight Gain of Metallic-Coated Steels Heated in the 600-800°C Range, ASTM Symposium on Indistrial Methods for Testing in High-Temperature Environments, Phoenix (May 1981).

EFFECTS OF MICROSTRUCTURAL MORPHOLOGY ON THE PERFORMANCE OF PLATINUM ALUMINIDE COATINGS

P. Deb and D. H. Boone
Materials Science Group
Naval Postgraduate School
Monterey, California, USA

R. Streiff
Universite De Provence
Marseille 13331 - France

ABSTRACT

The use of noble metal modified aluminide coatings has gained renewed interest as the requirements of improved protectivity for gas turbine engine components has increased. Increasing temperature, severity of operating environment and required lifetime extension are cited as reasons for their selection. While the literature documents examples of significant improvements in hot corrosion resistance at low (<800°C) and high (>800°C) temperatures and in cyclic oxidation with Pt additions, there are also examples of limited improvements and even detrimental effects. These controversies have been preliminarily ascribed to the considerable differences in microstructural features possible in these systems and to the variation in test procedures between laboratories, and between rigs and engines. A systematic study was therefore initiated to identify the microstructural variations obtained for the platinum-modified and chromium modified platinum aluminide coatings on the IN738 and IN100 nickel-base superalloys and to develop a fundamental understanding of their mechanism of formation. In this paper, the several classical structural types are defined together with the beginnings of their formation understanding. These microstructures were then subjected to a series of tests under conditions including hot corrosion at 700 and 900°C, and cyclic oxidation. In addition, measurements of change in mechanical properties such as DBTT were performed.

It was found that platinum addition to a standard aluminide is beneficial at 900°C, so far as hot corrosion attack is concerned. Hot corrosion resistance at 700°C is strongly structure dependent with most structural types offering little advantage over the unmodified aluminides. Pre-aluminizing heat treatments and their resulting effect on the coating structure and composition are important variables that seem to be the dominant controlling factors for the tensile ductility and corrosion resistance. The resulting surface roughness of these coatings and the selective interaction of the Pt with specific substrate elements may play an important role in the ultimate behavior of these coatings and their performance in cyclic oxidation.

I. INTRODUCTION

The development of new generation, higher performance, and cost-effective heavy duty gas turbine engines demands longer service life as compared to the engines used in the early seventies. These engine components, made of superalloys, are generally exposed to very high operating temperatures in conjunction with a wide variety of lower grade fuels, and often are used in marine environments. Therefore, high temperature hot corrosion and oxidation are common failure modes of these turbine engine components and/or other fuel-fired energy systems.

Diffusion aluminide coatings (1-3), commonly deposited by the pack cementation process, were first applied on superalloy substrates to improve high temperature oxidation and hot corrosion resistance and thus improve the service life of gas turbine engines and their components. It has been reported that the simple or unmodified aluminides suffer hot corrosion attack caused by the fluxing of the protective scale by molten salts and the impurities present in the combustion gases (4). Besides this, the protectivity and durability of the simple aluminide coatings depend to a major extent on the substrate alloy composition. This prompted the development of the MCrAlY class of overlay type coatings designed to provide a protection system essentially independent of the substrate and with a variable composition optimized to provide hot corrosion resistance and in some instances improved mechanical properties such as ductility and thermal fatigue resistance. The MCrAlY overlay coatings are generally produced by electron beam physical vapor deposition (EB-PVD) or advanced plasma spray techniques (5,6).

The aluminide coatings, modified by the additions of critical alloying elements, have gained renewed interest because they offer economical alternatives to the EB-PVD and plasma spray deposition techniques. A number of alloying elements have been tried to improve the coatings, including chromium, silicon, and the "active elements" like hafnium and, finally, the noble metals such as platinum (7,8).

Aluminide coatings modified by the additions of platinum, rhodium and palladium have been known for a number of years and used in production applications for the protection of airfoils in heavy duty gas turbine engines and various other aircraft and marine engines. However, there have been some controversies on the advantages of, and the reason for, the beneficial effects of platinum in the modified aluminide coatings corrosion and oxidation resistance (9-24). It has been suggested that these conflicting findings are mainly the results of differences in the structures reported for the various types of coatings investigated. Few data are available in the literature on the mechanical behavior of these systems.

In recent years there has been a renewed interest in the possible performance and economic benefits to be gained from the use of the platinum aluminide coatings. It is also seen in the literature that there has been an increase in research activity into the platinum modified aluminide coatings. The objective of this paper is to review the present state of knowledge in this area together with the latest test results developed in this program to provide a better understanding about these unique surface structures and their performance under hostile environments.

II. BACKGROUND

A. THE AVAILABLE Pt-ALUMINIDE COATINGS

In a review of the structures of "standard" Pt-Al structures (1-24), it was apparent that a wide range of variations were possible for the available commercial and developmental coatings, and that the existance of such differences might help to explain some of the apparent test anomalies reported. The first commercial Pt-Al system was produced by Lehnert, et al. (7), designated LDC-2, with a reported four-fold life increase in cyclic oxidation and a greater than two-fold improvement in hot corrosion resistance over a simple unmodified aluminide coating. The structure of the LDC-2 coating was found to have a single phase $PtAl_2$ surface structure containing some Ni. It was later learned that Cape (8) and others (9) had earlier disclosed the possible beneficial effects of Pt additions to coating systems.

A second coating structure described by Seelig, et al. (10), and designated RT-22, though formed by the same process, i.e., initial Pt electrodeposition followed by pack aluminizing, exhibited a two-phase surface structure. In comparison with the LDC-2 version, a more uniform, deeper gradient of a lower concentration of Pt dissolved in the cubic β (NiAl) phase is present in the RT-22 together with a dispersion of a $PtAl_2$ second phase. In this coating system as well as others, the density and distribution of this second phase appears to vary in a number of the samples examined and is hence a fruitful area for research.

A potential third version (although not necessarily a different type) of Pt-aluminide structure developed at Johnson Mathey, in collaboration with Rolls-Royce, designated JML-1, was perfected and described by Wing and McGill (11). A fused salt platinum electrodeposition technique was used followed by a high activity aluminizing treatment. This coating is reported to have a different structure and platinum distribution with the presence of a thick Pt_2Al_3 outer layer over a duplex $(Ni,Pt)_2Al_3$ + PtAl structured zone. The apparent absence of $PtAl_2$, the Pt-rich phase reported in many Pt-aluminide coatings, is not explained. (Note that this Pt_2Al_3 phase has not been reported in other studies, and a subsequent paper identified the phase as $PtAl_2$).

It was apparent from these and other observations that a range of possible Pt-aluminide structures could be produced within a fairly standard envelope of coating processing parameters. (Unfortunately, in most of the earlier studies, little information on actual processing parameters or the resulting structures was provided). The possbile variables included amount (thickness) and deposition mode of the initial Pt layer, the possibility of pre-aluminizing diffusion of the platinum with the substrate, the aluminizing cycle at activities varying from high to low, followed by post aluminizing coating diffusion and substrate solution heat treatments at varying temperatures for varying times.

B. PRESENT STATE OF KNOWLEDGE ON HIGH TEMPERATURE CORROSION AND OXIDATION BEHAVIOR

A variety of results have been reported on the hot corrosion and oxidation resistance of platinum modified aluminide coatings. The major conclusion of these studies are discussed in this section. In a comparative investigation on a series of aluminides containing noble metals, i.e., Pt-aluminide and Pt-Rh-aluminide, on two cobalt-base superalloys X-40 and MAR-M-509, tested along with a standard aluminide and a CoCrAlY coating in a low-velocity atmospheric-pressure burner rig, R. M. Clarke (12) has shown that in high temperature 900°C hot corrosion conditions, the Pt-aluminide was the most corrosion-resistant aluminide coating tested. It performed almost as well as the CoCrAlY composition specially designed to resist high temperature hot corrosion. The ranking of the tested coatings after 500 hours of attack indicated an increase in corrosion resistance from the modified aluminides to Pd, Pt-Rh, and Pt-aluminide coatings. The Pt-aluminide exhibits half as much oxide penetration as the standard aluminide. However, in low temperature (704°C) hot corrosion tests, the Pt-aluminide was found to be only slightly more resistant than the simple aluminide and performed more advantageously than CoCrAlY, while in high temperature burner-rig tests the Pt-aluminide on MAR-M-509 and the CoCrAlY coating had nearly identical corrosion rates. Both a structural effect and a substrate effect were observed for these Pt-aluminide coatings. The two-phase RT-22* coating on MAR-M-509 appeared to have slightly better hot corrosion resistance, in both temperature ranges, than a single CODEP B2 + Pt (experimental coating structure attempted for this study) on the X-40 substrate.

Bauer, Schneider and Grunling (13) confirmed the improvement with Pt additions for two common Pt-aluminide coatings, RT-22 and LDC-2 in a laboratory hot corrosion crucible test as well as in engine service operation. In the tests Bauer et al. found that, at the 850°C intermediate temperature, both Pt-aluminide coatings were able to prevent IN 738 base material corrosion attack for periods of about 1000 hours while an unmodified aluminide of similar thickness was degraded in less than 500 hours. Microstructural analyses revealed that one Pt-Al coating suffered extensive internal corrosion of the NiAl layer near the interdiffusion zone. This attack could be considered to be the predominant failure mechanism.

*It should be noted here and elsewhere that distinctions between coating types should not be made on the simple basis of the letter designation (i.e., LDC-2, RT22, etc.) but rather on the specific coating structure and Pt content of the specimens tested. Unfortunately in many studies this information is not provided.

In engine service testing on first stage blades, more pronounced internal attack of one structure than the other also occurred. At the higher termperature, 950°C, less severe corrosion conditions exist where the SO_3 partial pressure drops to a level where acidic fluxing becomes negligible compared to the rate of oxidation and the protection lifetime of the Pt-aluminide increases dramatically. The protection mechanism seems to be based on the formation of adherent scales and possibly the form and distribution of the Pt in the NiAl layer in juxtaposition. No more corrosion attack occurs and the LDC-2 coatings become much more protective providing twice as long a penetration life as RT-22.

A similar result is reported by Hanna, et al. (14) for the 1000°C oxidation of a series of Pt-aluminized IN 738 samples with various structures obtained by aluminization of electroplated platinum of various thickness. They observed that the initial oxidation rate (i.e., weight gain) for any of the Pt-aluminide coatings was more than one order of magnitude slower than for the unmodified aluminide, and that the performance of the (Ni,Pt)Al single-phase coating was slightly superior to the other platinum-aluminides.

The JML-1 coating, tested for a 500-hour period by Wing and McGill (11) on the two MAR-M-002 and IN 738 alloys in burner rig cyclic conditions shows apparent outstanding performance, slightly better for the 850°C intermediate temperature than for the 750°C low temperature hot corrosion attack. A comparison with a JML-2 system (a modification of JML-1 with a different platinum distribution provides fair protection at the higher temperatures but has an even greater limitation for low temperature hot corrosion) enabled them to conclude that the single phase $PtAl_2$ layer, JML-1 system, is a better corrosion resistant coating.

Chigasaki, et al. (15) have reported that in burner rig testing at 850°C and below in sulfur bearing fuel, the hot corrosion resistance of a two zone aluminide coating structure (i.e., outer layer is a high Al NiAl and the inner layer consists mainly of low Al NiAl), can be improved. This is accomplished by increasing the thickness of the outer layer of the coating to >80 μm. This system then becomes even more resistant than the platinum aluminide coatings where they found preferential attack of the $PtAl_2$ phase dispersed in the outer layer of the modified coating. Preferential attack of $PtAl_2$ in a two phase structure has also been reported by Bauer, et al. (13), as previously noted.

In a study of the composition and morphology of the platinum aluminide coatings produced by one aluminizing process, Jackson & Rairden (16) have found that the morphology of the resulting coating is highly dependent upon the thickness of the platinum pre-deposit. With thick platinum layers, the refractory metals from the substrate alloy are excluded from the outer surface of the coating. Herein may reside, at least in part, the reason of the improvement of hot corrosion resistance of superalloys when protected by platinum aluminide coatings. If potential deleterious elements from the alloy, such as Mo, W, Ti, etc. are "buried" beneath or somehow tied up near the substrate interface, they cannot contribute to accelerated hot corrosion at the surface.

In electrochemical measurements of a platinum modified (designated LDC-2) and a platinum-free aluminide coating in a $(Na,K)_2SO_4$ melt at 900°C, Rahmel (17) observed that the Pt-rich surface layer exhibits a broader passive range than the Pt-free surface. The Pt-free coating corrodes significantly faster in the passive range. The resistance to acidic fluxing is similar for both coating types, but the Pt-rich surface has a significantly higher resistance to basic fluxing. Finally, in an earlier investigation and later rerun as part of a study of Cr effects, Boone, et al. (18) found little difference between any of the commercial structural variation of platinum-aluminide coatings in furnace hot corrosion testing in the low temperature range.

This survey of the present understanding of the corrosion behavior of platinum modified aluminide coatings reveals that the performance of these systems is strongly dependent on structure and composition. In these studies there is some controversy about the optimum Pt level required for these coatings, with 40-60% often being mentioned. When a $PtAl_2$ phase is sampled, clearly a "high" Pt level is reported. However, volume fraction of $PtAl_2$ and the composition of the matrix NiAl phase may be even more critical in the resulting protectivity such that reliance on Pt level alone may not be sufficient. Because these structures vary with the source and the production time frame (21), it is not possible today to have a clear view on critical factors affecting protectiveness of these systems in the various hot corrosion conditions.

For these reasons studies by Boone, et al. (18-21) were initiated under Office of Naval Research sponsorship aimed at defining the relationship between the platinum modified aluminide coating structure and their protective properties. A preliminary investigation (19) of the thermal stability by heat treatment in air at 1080°C of a series of Pt-aluminide coatings from two commercial sources, i.e., with the two archetype structures, (1) single phase $PtAl_2$ and (2) two-phase $PtAl_2$ + NiAl has pointed out that the oxidation resistance may also be markedly dependent on the structure. A sequence of specific structures of the "A" series of the platinum modified coatings, which will be outlined below, related to the processing parameters, i.e., the platinum deposition treatment and the aluminizing conditions, has been established (19). This series of structural types were evaluated in furnace hot corrosion tests under both low temperature acidic fluxing conditions at 700°C, and high temperature basic fluxing conditions at 900°C (20). Results are reported in the following sections.

III. PRESENT STATUS OF PLATINUM MODIFIED ALUMINIDE COATINGS UNDERSTANDING

A. STRUCTURE AND FORMATION MECHANISMS OF PLATINUM MODIFIED ALUMINIDE COATINGS

The structure and formation mechanisms of diffusion aluminide coatings are well understood (1,2). However, their performance under harsh environments, i.e., especially under hot corrosion, is limited and they are highly substrate composition dependent. As mentioned earlier, the addition of noble metals such as platinum to the aluminide coatings may be beneficial under hot corrosion and oxidation environments. It is reported that the addition of Pt to the aluminide coatings improves hot corrosion resistance at high temperature but that is not as beneficial at low temperatures. In the following paragraphs, the structure and formation mechanisms of Pt modified aluminide coatings as they are presently understood are discussed in greater detail. It should be noted that during the course of this research, significant progress has been made in the development of the understanding of these mechanisms. However, additional research is needed particularly in the areas of phase diagrams and diffusional mechanisms and is in progress.

Several application techniques are commercially available for the deposition of Pt on superalloy substrates (22). They include (i) electroplating, (ii) ion plating, and (iii) sputtering. Of these application techniques, the electro deposition technique (i.e., electroplating) is the most commonly used for the production of Pt-Al coatings. The amount of Pt deposition can be varied, but in these studies, was kept constant at 8-10 μm. It is known that these process variables can produce different surface structures, but at present, there is no indication of any benefit of one over the other except in the possible area of cost.

In these studies to be reported here, the electroplated specimens* are given selective pre-aluminizing diffusion heat treatements listed in Table I. Pre-aluminizing diffusion heat treatments facilitate Pt diffusion and thus change the surface Pt content and structure which result in selective interaction with substrate elements. The heat treated specimens are then aluminized by pack cementation and vapor phase processes (20,23). Two aluminizing conditions (1) high activity, aluminum rich packs at low temperatures (about 700-900°C) known as LTHA and (2) low activity, lean aluminum packs at high temperatures (1000-1100°C) known as HTLA are generally used for the deposition of aluminum on the pre-diffused platinum enriched layer. These aluminizing conditions control the as-deposited aluminum level, surface phase structure and hence, the nature of diffusion processes occurring in the Pt enriched layer. A complete understanding of the effect of Pt level on Al and Ni diffusion processes in high Pt alloys is not known but some preliminary thoughts will be presented separately. The amount of Al deposited is mainly controlled by time in the aluminizing process. Preliminary studies however, revealed that the presence of Pt increases the rate of Al pick up over the unmodified aluminide coatings for both types of aluminizing processes (24).

In a manner similar to the processes found for the diffusion aluminide coatings, two major types of structures, i.e., (1) inward and (2) outward can also be defined for the Pt modified aluminide coatings. Inward type Pt-Al coatings are formed by aluminizing pre-diffused electroplated platinized layer in LTHA packs+. The aluminum diffuses through the pre-diffused Pt enriched surface layer and thus results in a high aluminum gradient across the Ni-Al coating. After the aluminizing treatment, the specimens are given a further heat treatment referred to as the post-aluminizing heat treatment which is mandatory for the Pt-Al coatings produced by the LTHA process and for many substrates as well.

---

*All data reported herein are based on IN-738 superalloy.

+While designated here as an Inward Type, selective Al diffusion through the pure Pt has not yet been clearly established. However, once aluminum penetrates through the Pt-enriched zone, the coating is of sufficient aluminum content (i.e.; hyperstoichiometric β(NiAl) to result in inward aluminum diffusion.

The magnitude of the Pt level and the extent of diffusion distance are strongly dependent on the pre-aluminizing diffusion heat treatments. A very high level of Pt in a limited zone near the surface can be achieved with a minimum pre-aluminizing diffusion treatment. This high level of Pt near the surface interacts with the incoming aluminum atoms during the aluminizing process and thus produces a thin continuous single phase layer. Streiff, et al. (20) have identified this single phase layer as $PtAl_2$. However, there is some ambiguity regarding the identification of this single phase structure (11). In this paper, this continuous single phase structure will be designated as $PtAl_2$. In these studies it became very apparent that the aluminum activity level in the coating process was extremely critical in controlling the final structure of the coatings. Unfortunately, it was not always possible to control this variable to the degree required such that some coating structural variability was seen even for samples produced under the so-called "same condition".

Figure 1 illustrates the typical optical photomicrographs of an inward Pt-Al coating with various pre-aluminizing diffusion heat treatments. The inward type Pt-Al coating exhibits the four zone structure shown in Figure 1(a). The surface zone consists of a high platinum content single phase $PtAl_2$. It should also be noted that random grit blast particles are also present at the interface of the initial substrate surface and the Pt over layer. These particles act as excellent markers for the identification of subsequent diffusion processes. The outer intermediate zone can consist of either fine platinum-rich precipitates in an NiAl* matrix enriched in Al or an Al-rich NiAl precipitate in continuous $PtAl_2$ phase. The matrix and morphology of these precipitates depend on several factors including the Al activity level during the aluminizing treatment and possibly the structure of the platinum interdiffusion layer. The inner intermediate zone is a single phase NiAl matrix (Ni-rich) denuded of any other phases or substrate elements. The innermost zone is the so-called interdiffusion zone which consists of substrate phases and elements insoluble in (NiAl) which formed when Ni was withdrawn by the Al-rich surface as previously established for simple aluminides. Any Pt pre-diffused into these outer zones would be expected to be precipitated as $PtAl_2$ (at sufficiently high Al levels as discussed below).

---

*There is some indication of two types of NiAl phase structure (20). Besides this, a range of single phase NiAl structures can be formed by varying the relative amounts of Ni and Al.

Figures 1(b) and 1(c) exhibit the typical three zone structure. The outer zone is platinum rich $PtAl_2$ phase in an NiAl matrix rich in Al; while the intermediate zone is a single phase NiAl matrix and the innermost zone is the so-called interdiffusion zone. Streiff, et al. (20) have reported the structural changes of Pt-Al coatings produced by the LTHA process, as determined by x-ray diffraction study. The x-ray data of Pt-Al coatings are listed in Table II. It is interesting to note that the surface structure changes from a single phase to a two phase and then followed by a single phase βNiAl with prolonged pre-aluminizing heat treatments. When these specimens were examined under the Scanning Electron Microscope (SEM), it was found that much better microstructural detail was obtained than with optical metallography. A typical SEM photomicrograph of an inward Pt-Al coating is shown in Figure 2 which exhibits the four zone structure. The surface zone is Pt-rich single $PtAl_2$ phase. As mentioned earlier, the outer intermediate zone consists of fine platinum-rich $PtAl_2$ and refractory metal enriched precipitates in an NiAl matrix rich in Al. It should be noted that the fine $PtAl_2$ precipitates are present in the upper region while refractory metal enriched precipitates are present in the lower portion. However, the volume fraction of these (refractory metal enriched) precipitates in an NiAl matrix is increased substantially as Ni content increases in the NiAl phase. The inner intermediate zone is a single phase NiAl matrix containing high nickel and the innermost zone is the interdiffusion zone.

As pointed out earlier, outward type Pt-Al coating structures are produced by a HTLA process. Like the inward type structure, the diffusional interactions among the elements during the low activity aluminizing process are also very complicated. Aluminum apparently diffuses inward until it forms a single NiAl phase containing low Pt and Al through which essentially only Ni (not Al) diffuses outward from the substrate to the pre-diffused Pt enriched layer. This results in a lower Al gradient across the Ni-Al coating. Furthermore, platinum diffusion continues during the aluminizing process, and it is limited near the substrate at much higher temperature for longer times presumably by the outward flux of Ni. The actual Al and possible Ni level were found to be extremely important in influencing the relative movement of the elements and the resulting phase distribution. However, the observation of diffusion of platinum through the NiAl (containing low Al) is still considered significant and is an area of active investigation. Present understanding and speculation in this area will be presented later.

Figure 3 illustrates typical optical photomicrographs of the outward type Pt-Al structures given various pre-aluminizing diffusion heat treatments. Figure 3(a) exhibits a typical two-zone structure. The surface zone consists of platinum rich $PtAl_2$ phase and the inner zone is the interdiffusion zone. Figures 3(b) and 3(c) exhibit typical three-zone structures. The surface zone consists of Pt-rich precipitate (i.e., $PtAl_2$) in the Al rich NiAl matrix while the intermediate zone is a single phase NiAl matrix rich in nickel, denuded of any other phases or substrate elements. It should also be noted that the $PtAl_2$ in this structure is not a continuous phase although in some structures it appears to be and this difference in distribution is an area for further evaluation and clarification. The formation mechanisms of fine $PtAl_2$ precipitates in an Al rich NiAl matrix are not yet understood but may be related to the pre-aluminized diffusion structure. It is reported (20) that the structure of Pt-Al coatings produced by HTLA process changes from a single phase to a two phase structure with increasing pre-aluminizing diffusion time and temperature, as determined by x-ray diffraction analysis. The x-ray data are also listed in Table II.

Therefore, the microstructure of Pt-Al coatings produced either by the LTHA or HTLA process is strongly dependent on such factors as, (1) the amount of platinum (although not an intentional variable in this study), (2) the pre-aluminizing diffusion time, temperature and the resulting structure, (3) the aluminum activity level and the time and temperature conditions of the aluminizing process. As an example, Figures 4(a) and 4(b) exhibit structural changes as a result of variations of the pre-aluminizing time, temperature and aluminum activity level. Figure 4(a) exhibits a typical single phase $PtAl_2$ in the outer zone of an inward Pt-Al coating while Figure 4(b) depicts a typical two phase ($PtAl_2$ plus NiAl matrix) structure in the outer zone of an outward type Pt-Al coating.

Streiff, et al. (20) have also reported that the diffusional stability of Pt modified aluminide coatings is significantly dependent on the structure, i.e., the amount and the distribution of Pt (and possible Al level) in the coating. The effect of prolonged annealing in air at 1080°C of two commercial Pt-Al coatings is shown in Figures 5 and 6. It can be noted that the single phase inward Pt-Al coating (Fieure 5(a)) did not change structure during prolonged annealing (i.e., 24 hours) at 1080°C (Figure 5(b)) while the initial two phase structure shown in Figure 6(a) in the outer zone of the outward type Pt-Al coating rapidly changes to the single phase structure shown in Figure 6(b) during prolonged exposure (24 hours) at 1080°C. Although, in coatings produced by the high activity process, the structure did not change during the 24 hours treatment, the composition across the coating will not remain the same and a structural transition will occur, apparently at some critical Al level. Prolonged annealing produces a more uniform single phase structure and elemental distribution in the outer zone of low activity Pt-Al coatings. The Al level of the NiAl is now postulated to affect the movement of Pt as it does for Ni and Al and hence initial coating Al contents can have a significant effect on the system's structural and diffusional stability.

Summarizing the x-ray diffraction and electron probe data of two structural types of Pt-Al coatings, it is found that the diffusional interaction between the elements present in the coatings and in the substrate is very complex in nature. In the following paragraphs, the formation mechanisms of these coatings as they are presently understood will be discussed in greater detail. As presented earlier, pre-aluminizing treatments produce a higher concentration of Pt near the surface. The extent of Pt diffusion is strongly dependent on the amount of time and the pre-aluminizing temperature. Figure 7(a) presents an initial schematic representation of Pt diffusion during heat treatment. A typical elemental distribution across an inward type Pt-Al coating is shown in Figure 7(b). It is further important to note that as a result the Pt concentration profile is relatively flat to about 25 μm of the final coating thickness followed by a sharp drop in concentration level to zero. At these lower pre-aluminizing temperatures, platinum interdiffusion is not significant. Therefore, a high level of Pt is available near the surface to interact with the incoming aluminum atoms during the aluminizing treatment and thus produces a continuous $PtAl_2$ phase during the LTHA aluminizing process. With longer diffusion times at higher temperatures prior to aluminizing the Pt is distributed over a wider zone near the surface and the surface Pt content is lower.

Figure 7(c) illustrates the elemental distribution across the outward type Pt-Al coating. It is important to note here that the Pt concentration decreases gradually from the surface to a minimum near the substrate followed by a small peak in the interdiffusion zone near the interface between the substrate and the coating, while the Ni content maintains a fairly high level close to the surface (see Figure 7(c)). The Al concentration profile is also found to decrease (not as fast as Pt) from the surface to near the interface. The initial level of Al available in the HTLA process (see Figure 7(c)) is also less than that in the LTHA process (see Figure 7(b)), as expected from the aluminum activity level in the pack. However, the presence of a sufficiently high level of Pt near the surface can interact with the (presumed) incoming Al atoms and thus results in $PtAl_2$ phase (either as formed or precipitated during subsequent cooling). These fine discontinuous $PtAl_2$ phase precipitates are present in Al-rich NiAl phase. The formation of Al rich NiAl phase (as opposed to the continuous $PtAl_2$ layer) is mainly the result of the high level of Ni (i.e., about 40%) near the surface which thus interacts with the available Al atoms.

The simultaneous diffusion of Pt and Ni through the NiAl phase during the HTLA process is also very complex. Figure 8 is a schematic representation of possible platinum diffusion effects on the NiAl phase components. Elegant measurements of D_{Al}/D_{Ni} have been made in the Al-Ni system (25). It is reported that the Ni is the predominant diffusing constituent in hypo-stoichiometric βNiAl while in hyper-stoichiometric βNiAl, Al becomes the predominant diffusing species. Therefore, it is of interest to speculate as to the the effect of Pt additions to the Al-Ni system at 1150°C. The solid lines in Figure 8 show the interdiffusion coefficient ratio of D_{Al}/D_{Ni} in an Al-rich NiAl phase. Pt additions to the Al-Ni system are found experimentally in these studies to move these interdiffusion coefficient ratio lines toward the left (i.e., towards lower Al levels) as shown with the broken line. The extent of this shift has not yet been established but it is now postulated to be a strong function of Pt level. Pt addition to NiAl phase reduces the apparent Al level of NiAl phase at which Al is the dominating diffusional element. Then apparently once the β(NiAl) composition falls below a critical Al level (i.e., a new stoichiometry of the ternary NiAl phase containing Pt), Pt can diffuse very rapidly (as does Ni in pure NiAl phase) and the $PtAl_2$ phase is not stable.

B. HOT CORROSION DATA

1. High Temperature

Recently Deb, et al. (26) reported that coatings produced with the low aluminum activity process, in general, appear to be less sensitive to surface attack at 900°C as compared to samples produced by the high aluminum activity process (Note: in this study the HTLA coating was applied in the vapor phase, the so-called out-of-the pack and the effect of this variable has yet to be assessed). However, with extended pre-aluminizing diffusion treatment at 1080°C for 4 hours the protectivity provided by the Pt decreased and little difference between the two aluminizing processes were seen (see Figure 9). Even with this diffusion treatment, however, specimens given either LTHA or HTLA aluminizing treatments were attacked at a rate of only 0.2 μm/hour, at least a factor of six times more corrosion resistance than the unmodified aluminides (1.5 μm/hour) exposed in the same test as control samples. The increased surface degradation behavior of both the HTLA and LTHA type coatings after prolonged diffusion is not understood.

2. Low Temperature

Streiff, et al. (21) and Deb, et al. (26,27) have studied the hot corrosion behavior of platinum modified aluminide coatings on the IN-738 superalloy substrate at 700 and 900°C respectively. In these studies (21,26,27) it is observed that the addition of platinum to a standard aluminide is found to have only minimal benefit at 700°C (i.e., about a two-fold increase) as compared to hot corrosion behavior of platinum additions at 900°C (i.e., a six-fold increase). However, they have reported that the three zone inward type coating showed much less attack than the standard two zone outward type coatings. They have also noted that a continuous, pure, $PtAl_2$ phase, shown in Figure 10(a), is present in the more resistant coatings. In low temperature hot corrosion environments, this continuous $PtAl_2$ layer apparently prevents (i.e., acts as a barrier to) hot corrosion attack. Figure 10(b) exhibits a typical low temperature hot corrosion attack.

Recently Dust, et al. (28) have studied the low temperature hot corrosion behavior of platinum-aluminide and chromium modified platinum-aluminide coatings at 700°C on the IN-100 superalloy substrate. They have reported that the LTHA platinum-aluminide coating did not display particularly good resistance to LTHC (i.e., about 50 m minimum penetration in 100 hours) as compared to HTLA platinum-aluminide coatings. However, the LTHA platinum-aluminide coating showed (i.e., 0.5 μm/hour surface attack) little improved LTHC resistance (i.e., two-fold) over the unmodified-aluminide coatings (1.0 μm/hour) at 700°C reported by Deb and Boone (27). This is also consistent with the above results (21,26) reported here. Although this sample had a surface layer of $PtAl_2$ as shown in Figure 11(a), there was little improved LTHC resistance (see Figure 11(b)) over the HTLA platinum-aluminides. The two HTLA platinum-aluminide coatings did provide a good resistance to LTHC attack (i.e., about 40 m minimum penetration) and it is interesting to note that a thick two-phase structure of $PtAl_2$ and β-NiAl (Al-rich) was present underneath the thin surface layer of $PtAl_2$. Even after this surface layer was penetrated, the Pt-rich $PtAl_2$ precipitates provided some hot corrosion resistance. In other testing (11,27,30,31) the continuous $PtAl_2$ layer has been reported as the LTHC resistant structure with the two-phase $PtAl_2$/ β-NiAl structure performing little better than the unmodified-aluminide. Recently Dust, et al. (28) and Deb, et al. (29) noted that the Pt diffusion through the hypo-stoichiometric β-NiAl composite is not yet well understood but it is apparently a strong function of Al level. It appears that once the β-NiAl composition falls below a critical Al level (i.e., a new stoichiometry of the ternary β-NiAl containing Pt), Pt can diffuse very rapidly (as does Ni in hypo-stoichiometric β-NiAl) and thus $PtAl_2$ becomes unstable. This could be one of the reasons why the LTHA platinum-aluminide coating, even though it had a thick (i.e., about 20 μm) continuous layer of $PtAl_2$ phase at the surface, did not show good resistance to LTHC attack as compared to the HTLA platinum-aluminide coatings. In these tests, the coating's lifetime may be dependent on the optimum thickness and stability of this $PtAl_2$ phase surface layer.

The chromium-modified platinum-aluminides also differed greatly in LTHC resistance. Process D (Cr-Pt-Al) which exhibited a good LTHC resistance (about 34 m penetration), had a high $PtAl_2$ surface content with little Ni or Cr, again confirming the observation that a continuous single $PtAl_2$ phase at the surface is beneficial. Process B (Pt-Cr-Al), which exhibited poor resistance to LTHC attack (about 66 m penetration), had less continuous $PtAl_2$ phase with high amounts of Ni (probably NiAl) and some chromium. In both coatings, the most important factor for the LTHC resistance appeared to be the amount of $PtAl_2$ content at the surface layer because they both had a relatively high level of chromium concentration near the surface. Applying chromium first and then platinum prior to aluminizing allows an effective layer of $PtAl_2$ to form at the surface, while reversing the alloying addition sequence (i.e., Pt-Cr-Al) disperses the Pt in the intermediate zone. As a result, a continuous single $PtAl_2$ phase cannot form at the surface during subsequent aluminizing and post-aluminizing heat treatment. It is therefore important to emphasize that the structure of these chromium-modified platinum-aluminides is strongly dependent on the modifying element which is first applied.

C. CYCLIC OXIDATION

It is known that pre-aluminizing diffusion heat treatments produce an exaggerated micro-surface roughness on the platinum modified aluminide coatings, as compared to the unmodified coatings. Recently, the effects of surface structure under cyclic oxidation at

1100°C of the two structural groups of unmodified and platinum modified aluminide coatings were studied (24). In that investigation, variously heat treated platinum free and platinum modified aluminide coatings were exposed to cyclic oxidation environment at 1100°C for times up to 250 hours. The specimens were cycled for 60 minutes at 1100°C and then cooled 10 minutes outside the furnace prior to reheating. It was found that surface plastic instability occurred, called "rumpling", shown in Figures 12(a) and 12(b), and that the degree of rumpling was observed more in the platinum modified aluminide specimens. Figure 12(c) shows SEM photomicrograph of the corresponding cross-section. Surface upheavals on the order of 50 µm were found on both systems. The amount of surface rumpling during cyclic oxidation testing can be attributed to several factors such as coefficient of thermal expansion mismatch between coating and substrate, thermal gradients across the specimen (i.e., thermal strains) produced during heating and cooling, coating strength which depends on the microconstituents and composition present in the coating, and finally, the coating thickness (which also affects the coating composition and structure). Therefore, the resulting performance of the coatings is a complex interplay of the above mentioned variables. At present, there is no mention in the literature of surface rumpling of the platinum aluminide coatings or even of the unmodified aluminides. However, similar surface rumpling has been reported on some overlay MCrAlY coating system.

In this study, it was established that the rumpling observed could not be attributed to simple coating affected substrate melting. It was further observed that the platinum modified aluminide coatings did not show oxide spallation and that oxide spallation of unmodified coatings appeared with a secession of further rumpling. Finally it was found that the thicker coatings on the IN-738 substrate showed a lower propensity to rumple as compared to the thinner coatings. However, for the thick coating samples which showed minimum rumpling, transverse coating cracks were found while no such cracks were observed in the thinner coatings which underwent significantly more plastic deformation. Further studies of this phenomena and the variables operating are underway.

D. MECHANICAL PROPERTIES

As previously detailed, a wide range of structures and compositions of Pt-Al coatings can be prepared by varying the pre-aluminizing and post-aluminizing diffusion heat treatments. There are however variables other than oxidation and hot corrosion which can directly or indirectly control the coating performance under severe environmental conditions. It is therefore expected that the mechanical properties of diffusion type coatings such as the Pt modified aluminides will depend on the type of structure and the corresponding elemental distribution within the coating as well. There are however, little data available on the mechanical properties of the platinum modified aluminides and none relating to specific structures. There are only limited data available on the unmodified aluminides (32-34) and the overlay type MCrAlY coatings (35).

It is reported (24) that in high temperature cyclic testing, the platinum modified aluminides showed surface rumpling which is mainly attributed to the coefficient of thermal expansion mismatch induces strains between the coating and the substrate. It is also known that the coating's ability to accommodate additional strain is its ductility. Therefore the ductile to brittle transition temperature (DBTT) of these coatings is very important as far as coating mechanical properties and the engineering design of coated components is concerned.

Recently Vogel (36) has studied the DBTT of several structural variations of the Pt-Al coatings. It was found that the DBTT is strongly structural dependent. It is also interesting to note that the DBTT of outward type Pt-modified aluminides (i.e., HTLA) is in the range of 640-700°C, while inward diffusion type Pt-modified aluminides did not show DBTT up to the maximum experimental temperature (i.e., 800°C) shown in Figure 13. As might be expected, the data reveal that structures with high Pt concentrations near the surface and high Al levels have higher DBTT's. Lower Al levels and two-phase structures (i.e., $PtAl_2$ plus NiAl) which are generally provided by the HTLA coating process are more ductile. Because it is difficult to separate out the platinum, aluminum and structural effects which are produced by the various aluminizing treatments, additional testing and more detailed structural characterization is underway.

A high DBTT of any given coating is undesirable since the coating remains in the brittle mode for a longer period of time as compared to a coating which has a lower DBTT. A low DBTT will place the coating substrate system into a ductile mode at lower operating temperatures. Therefore the probability of cracking type failure for a low DBTT coating is much less than in high DBTT coating when turbine engine experienced various spectrum loading over a wide range of temperatures under hostile environments. Optical metallographic examination of these tested Pt-Al specimens revealed wide and well spaced cracks in the coatings tested above the DBTT while the inward and the outward type coatings tested below the DBTT exhibited straight, very fine and closely spaced cracks. These observations are also in good agreement with the published crack morphology data of the unmodified aluminides (32-34). Even though the high Al and Pt content coatings exhibited a high DBTT, their room temperature ductility appeared greater than that in the "more ductile" coatings. Initial analysis has related this to a high level of residual compressive stress in the coatings at room temperature, the higher Al and Pt content coatings having the higher compressive stresses. The magnitude of these stresses and their effects on handling and rumpling properties are being investigated, but a strong correlation is postulated.

E. CONCLUSIONS

The structures of the Pt aluminide coatings are capable of being varied over a wide range. A number of processing variables can play an important role in the attainment of a specific structural type. While these various coating types are in the process of being more carefully defined and documented, some of the underlying diffusional and phase relations involved in the Ni-Al-Pt system need much additional study. The effects of some of these structural variations has resulted in the following additional information:

1. Hot Corrosion

The presence of platinum in an aluminide coating improves hot corrosion resistance at 900°C; the degree of improvement depends on the structures and possible processing parameters. Low aluminizing activity processes (i.e., HTLA) apparently produce coatings with improved hot corrosion resistance over those produced by the high aluminum activity process (i.e., LTHA). In low temperature hot corrosion only minimum benefit (about 2.5 times improvement) of platinum additions were found over the unmodified aluminide coatings.

2. Cyclic Oxidation

In an initial cyclic oxidation test, it is found that cycling to elevated temperature produces surface rumpling in both unmodified and platinum modified coatings. The degree of rumpling is a function of platinum level and structure and appears related to coefficient of thermal expansion mismatch induced strains and coating properties.

3. Mechanical Properties

At comparable aluminum levels, the presence of platinum increases the DBTT and room temperature residual compressive stress level. DBTT and residual stress can be varied by changing the composition and structure of the coating.

REFERENCES

1. G. W. Goward, D. H. Boone and C. S. Giggins, Trans. ASM, 60, 228, 1967.
2. G. W. Goward and D. H. Boone, Oxidation of Metals, v. 3, No. 5, pp. 475-495, 1971.
3. G. W. Goward, presented at Les Arcs NATO, July 1983 and published in the Proceedings.
4. F. H. Pettit and G. H. Meier, Superalloys 84, AIME, New York, p. 653, 1984.
5. D. H. Boone, Material Coating Techniques, AGARD lecture series No. 106, paper 8, Lisbon, 1980.
6. W. F. Shilling, Coatings for Heat Engines, (to be published), NATO Workshop, Aquafredda, 1984.
7. C. Lehnert and H. W. Meinhardt, Electrodep. Surface Treat., 189, 1972/73.
8. A. T. Cape-U.S. Patent 3,107,175, Coast Metals Inc., 1963.
9. U.S. Patent 3,494,748, 1966.
10. R. P. Seelig and R. J. Stuber, High Temp. High Pressures, 10, 207, 1978.
11. R. G. Wing and I. R. McGil, Platum Met., Rev., 25, 94, 1981.
12. R. L. Clark, DTNSRDC REPORT SME-80/81, 1981.
13. R. Bauer, K. Schneider and W. H. Grunling, in Proceedings 10th ICMC, San Diego, 1983.
14. M. D. Hanna, R. B. Argent and G. W. Haworth, Pers. Communication (to be published).
15. M. Chigasaki, H. Soeno, Y. Fukui and K. Otaka, Japan Inst. Iron & Steel, 69, 128, 1983.
16. M. R. Jackson and J. R. Rairden, Met. Trans, 8A, 1697, 1977.

17. A. Rahmel, Final Report COST 50-5 Round D/2, 1983.

18. D. H. Boone, R. Steiff and D. E. Peacock, Coatings for Heat Engines (to be published), NATO Workshop, Aquafredda, 1984.

19. R. Streiff, D. H. Boone and L. J. Purvis, Jour. Metal. d'Automne, Paris, 1983.

20. R. Streiff, D. H. Boone and L. J. Purvis, Surface Engineering, NATO ASI, Les Arcs, France, 1983.

21. R. Streiff and D. H. Boone, in Proceed, 9th International Congress on Metallic Corrosion, p. 409, Toronto, Canada, 1984.

22. D. H. Boone, P. Deb, L. J. Purvis and D. V. Rigney, in 12th ICMC, Los Angeles, CA, 1985 and published JVSTA, Nov-Dec 1985.

23. LT M. W. Dust, Hot Corrosion Resistance of Chromium Modified Platinum-Aluminide Coating, NPS Thesis, Monterey, California, Dec. 1985.

24. CAPT T. F. Manley, Plastic Instability of Aluminide and Platinum Modified Diffusion Coatings During 1100°C Cyclic Testing, NPS Thesis, Monterey, California, Dec 1985.

25. S. Shankar and L. L. Seigle, Met. Trans., v. 9A, pp. 1467-1476, 1978.

26. P. Deb, D. H. Boone and R. Streiff, in 12th ICMC, Los Angeles, CA, 1985 and published in JVSTA, Nov-Dec 1985.

27. P. Deb and D. H. Boone, to be submitted for publication, 1986.

28. M. W. Dust, P. Deb, D. H. Boone and S. Shankar, will be presented at the June 1986 Gas Turbine Conference, Dusseldorf, W. Germany and published in ASME as a technical paper.

29. P. Deb, D. H. Boone and R. Streiff, presented at the International Conference on Metallurgical Coating, Toronto, Canada, October 1985 and will be published in ASME technical paper.

30. R. Streiff and D. H. Boone, Reactivity of Solids, Eds. P. Barret and L. -C.Dufour, Elsevier Science Publisher B. V., Amsterdam, pp. 195-199, 1985.

31. W. T. Wu, A. Rahmel and M. Schorrm, Oxidation of Metals, v. 22, pp. 59-81, 1984.

32. G. W. Goward, Journal of Metals, Oct 1970, p. 31.

33. G. W. Goward, symposium on Properties of High Temperature Alloys, Las Vegas, USA, Oct 1976, p. 806.

34. D. H. Boone, Overlay Coatings for High Temperature Applications, Airco Temescal, Jan 1976.

35. D. H. Boone, Airco Temescal Data Sheets, Airco Inc., Berkeley, CA, USA, Jun 1977.

36. LCDR David Vogel, Determination of the Ductile to Brittle Transition Temperature of Platinum - Aluminide Gas Turbine Blade Coatings, NPS Thesis, Monterey, California, Sep 1985.

TABLE I. PRE-ALUMINIZING DIFFUSION HEAT TREATMENTS

HEAT TREATMENTS	TEMPERATURE, °C	DIFFUSION TIME, HOURS
I	870	0.5
II	980	2
III	1040	3
IV	1080	4

TABLE II. X-RAY ANALYSIS OF PT MODIFIED ALUMINIDE COATINGS

Aluminizing Processes:		Low Temperature High Activity, INWARD		High Temperature Low Activity, OUTWARD
Sample	Heat Treatment	As-coated	As-coated 4 hrs @ 1080° C	As-coated
IN-738		Ni_2Al_3	NiAl	NiAl
Platinum		$PtAl_2$+PtAl		$PtAl_2$
IN-738+Pt	1/2 hrs @ 870° C		$PtAl_2$	$PtAl_2$
IN-738+Pt	2 hrs @ 980° C		$PtAl_2$+ NiAl	$PtAl_2$+ NiAl
IN-738+Pt	3 hrs @ 1040° C		$PtAl_2$+ NiAl	$PtAl_2$+ NiAl
IN-738+Pt	4 hrs @ 1080° C		$PtAl_2$	$PtAl_2$+ NiAl

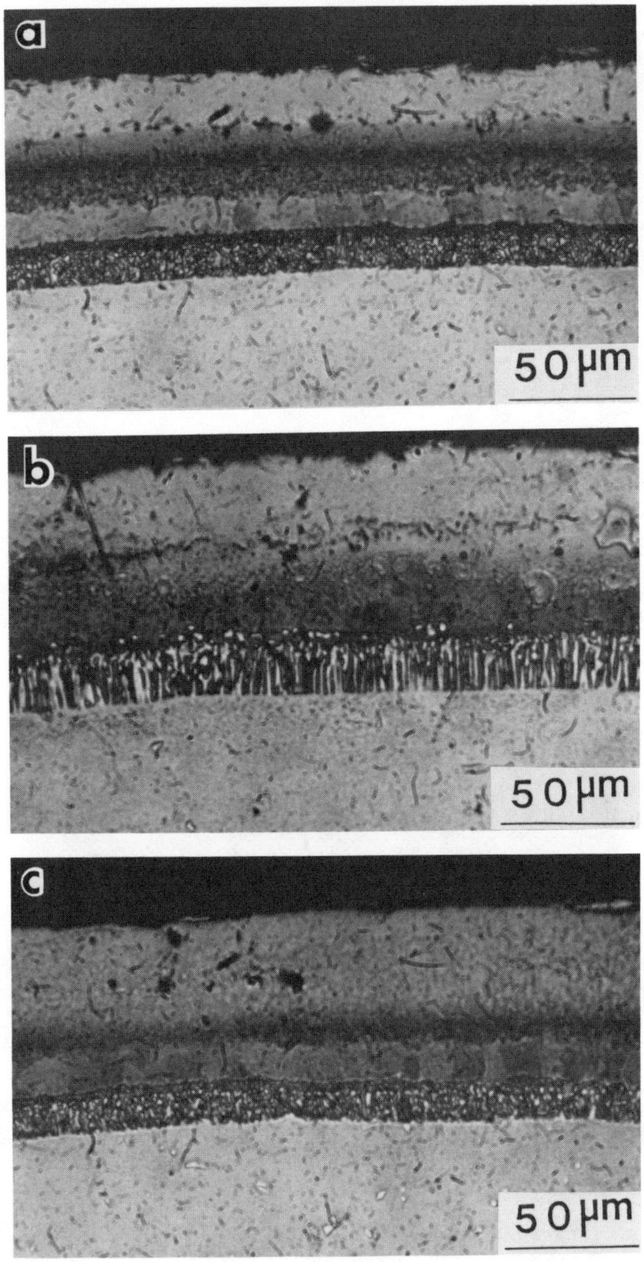

Figure 1. Typical inward type Pt-Al structures exhibit effect of pre-aluminizing heat treatments: (a) 1/2 hr. at 870°C, (b) 2 hrs. at 980°C, and (c) 1 hr. at 1050°C.

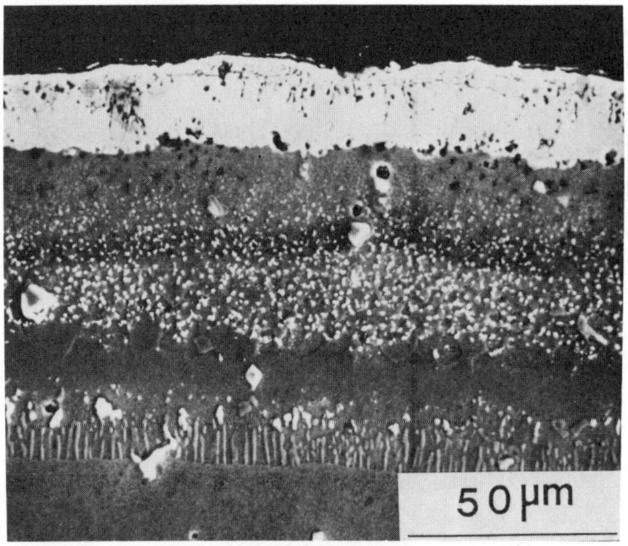

Figure 2. SEM photomicrograph of inward type Pt-Al structure contains four zones.

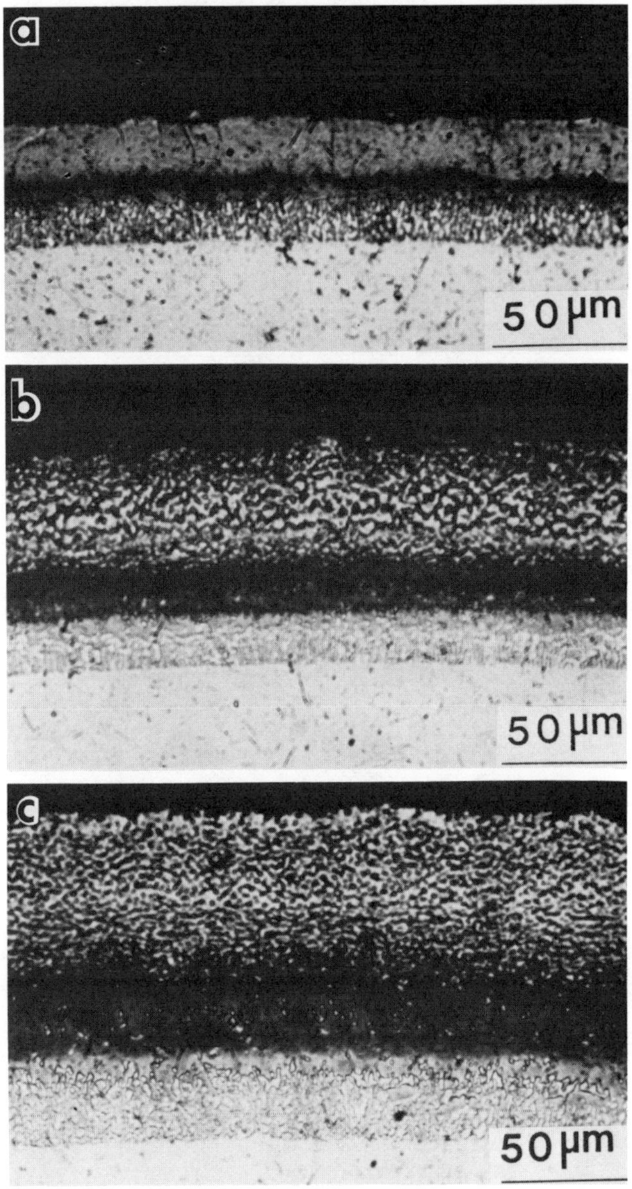

Figure 3. Typical outward type Pt-Al structures exhibit effect of pre-aluminizing heat treatments: (a) 1/2 hr. at 870°C, (b) 3 hrs. at 1040°C, and (c) 4 hrs. at 1080°C.

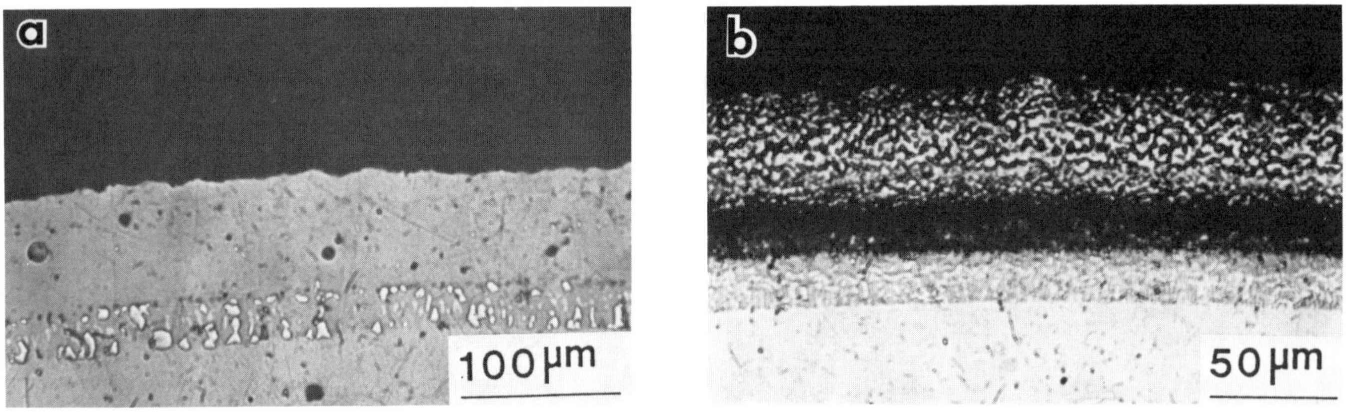

Figure 4. The outer-zone structure of Pt-Al coatings exhibit: (a) single phase, inward type structure and (b) two phase outward type structure.

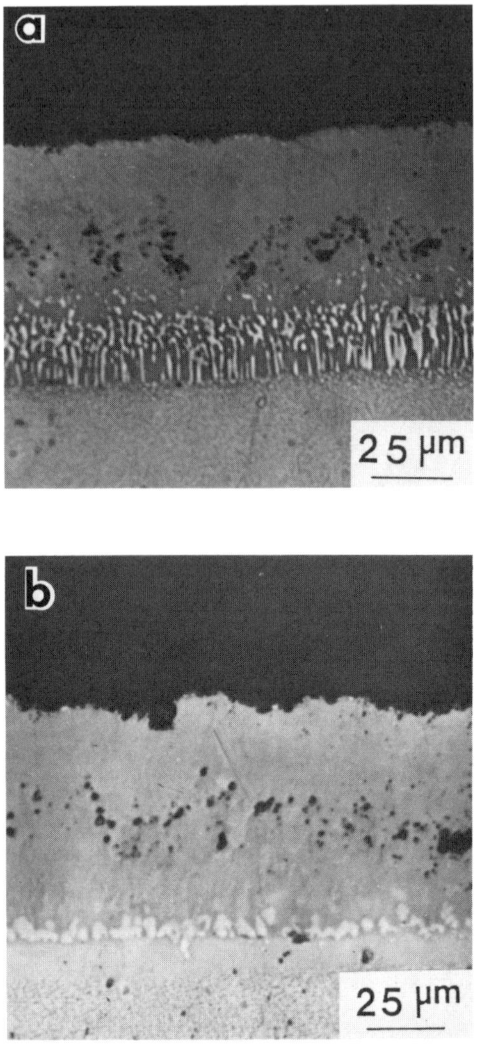

Figure 5. Effect of prolonged exposure of Pt-Al coatings in air at 1080°C exhibit: (a) single phase structure after 1 hr. heat treatment and (b) single phase structure after 24 hrs. heat treatment.

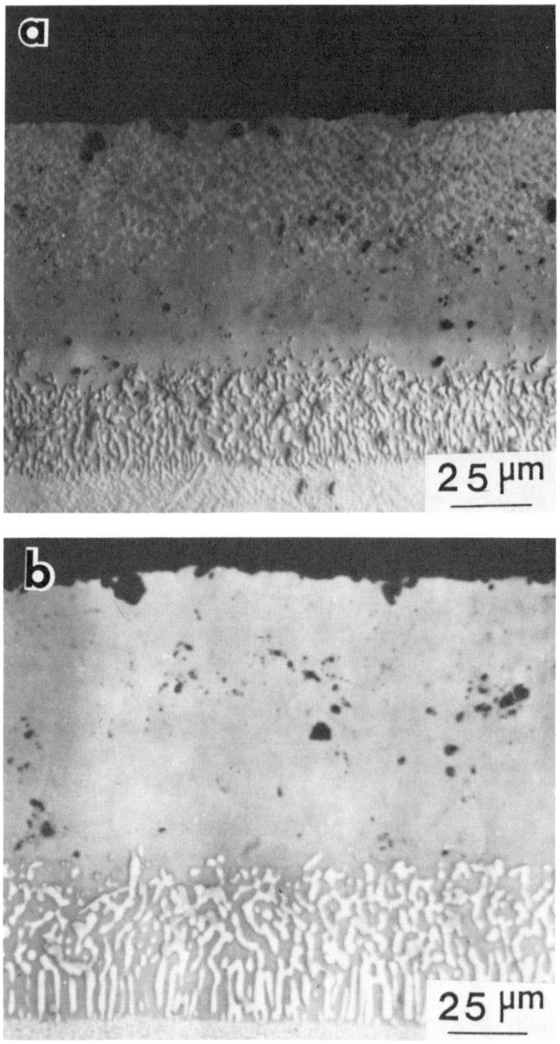

Figure 6. Effect of prolonged exposure of outward type Pt-Al coatings in air at 1080°C exhibit: (a) two phase structure after 1 hr. heat treatment and (b) single phase structure after 24 hours heat treatment.

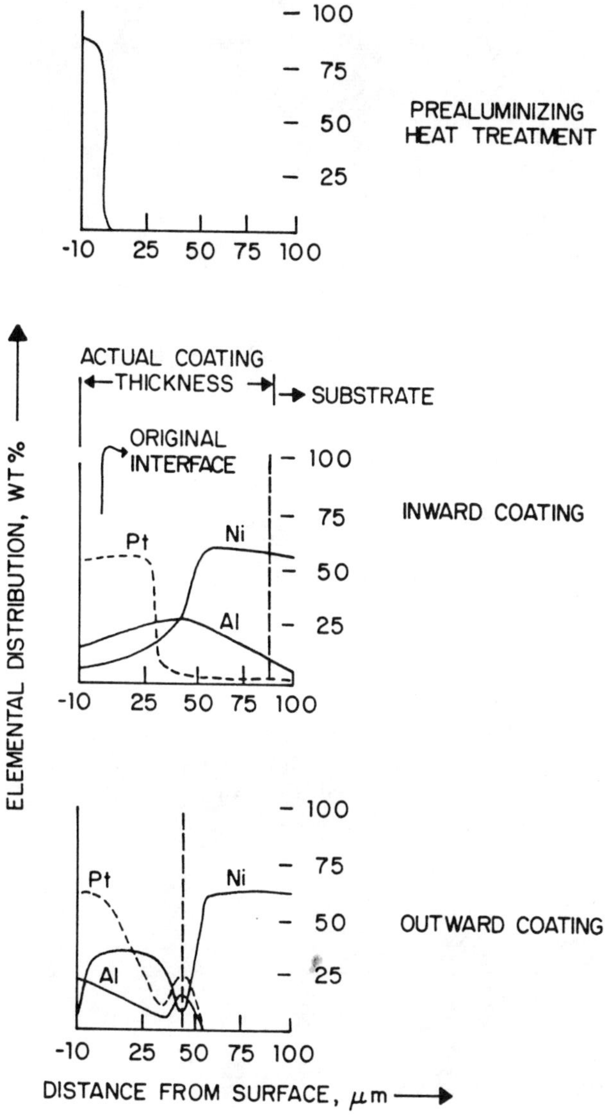

Figure 7. Elemental distribution across the Pt-Al coatings: (a) Pre-aluminizing heat treatment, 1/2 hr. at 870°C, (b) Inward type coating and (c) outward type coating.

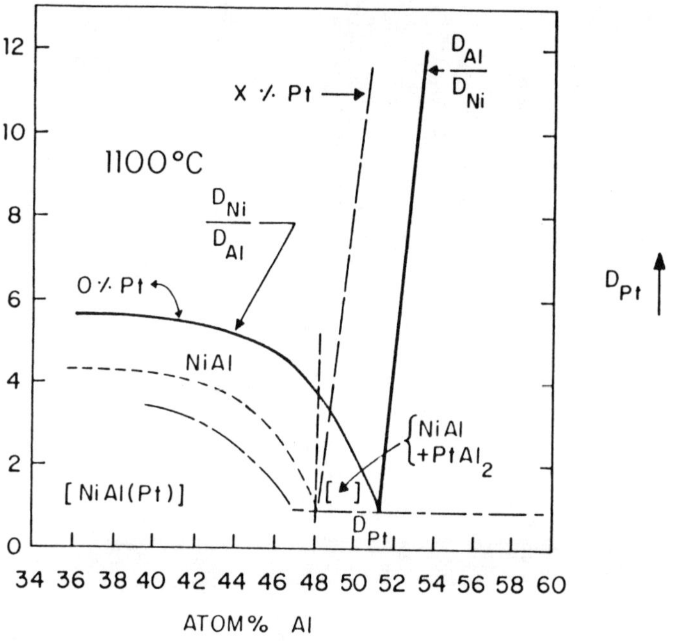

Figure 8. Possible modification of NiAl diffusion data (29) with the addition of Pt.

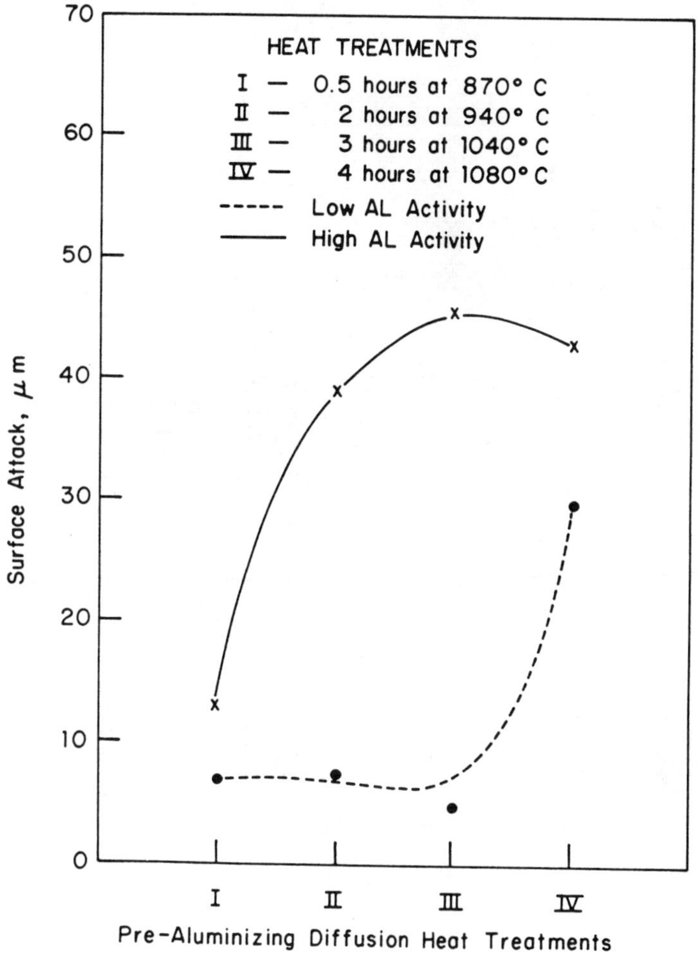

Figure 9. Surface attack of Pt modified aluminide coatings exposed for 200 hours hot corrosion condition at 900°C with various pre-aluminizing diffusion heat treatments.

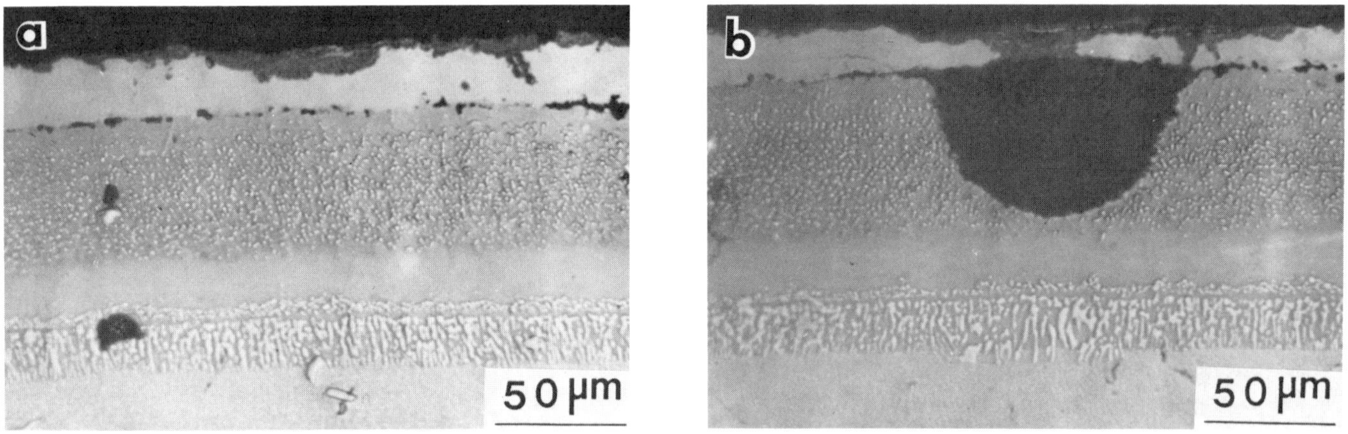

Figure 10. Typical low temperature hot corrosion attack at 700° C:
(a) specimen exhibits little attack on $PtAl_2$ phase and
(b) substantial hot corrosion attack beneath the $PtAl_2$ phase.

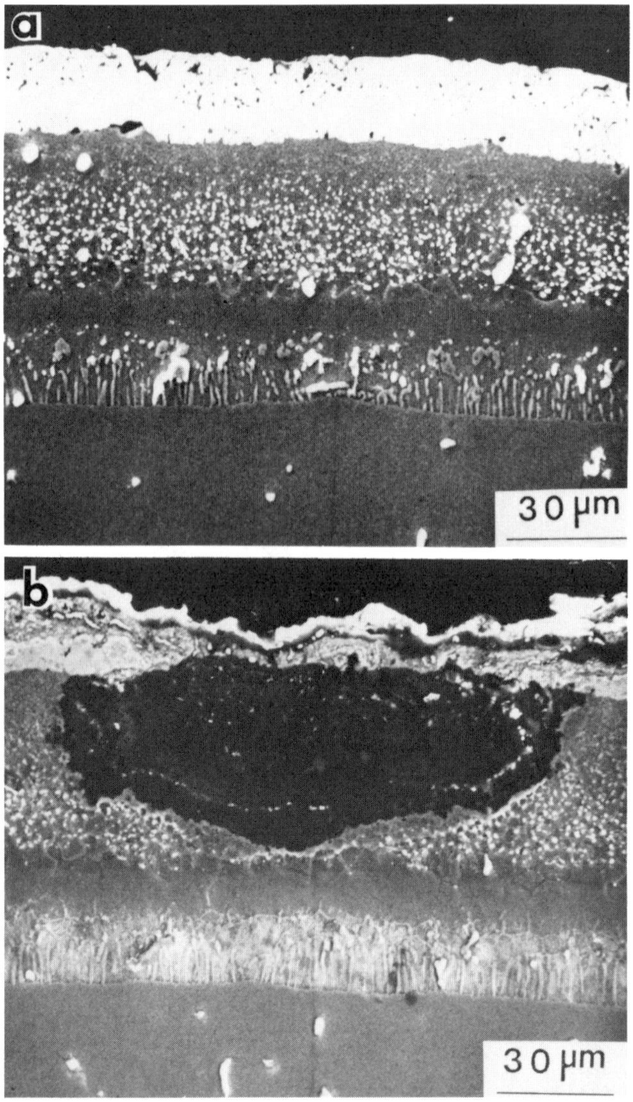

Figure 11. SEM photomicrographs exhibit: (a) inward type LTHA platinum-aluminide coating and (b) typical LTHC attack after 100 hours exposure at 700°C.

Figure 12. Typical example of surface rumpling of Pt-Al coatings: (a) sample exhibits macro surface roughness (8x), (b) enlarged photomicrograph (64x) and (c) cross-section of same specimen showing surface upheaval on the order of 50 µm.

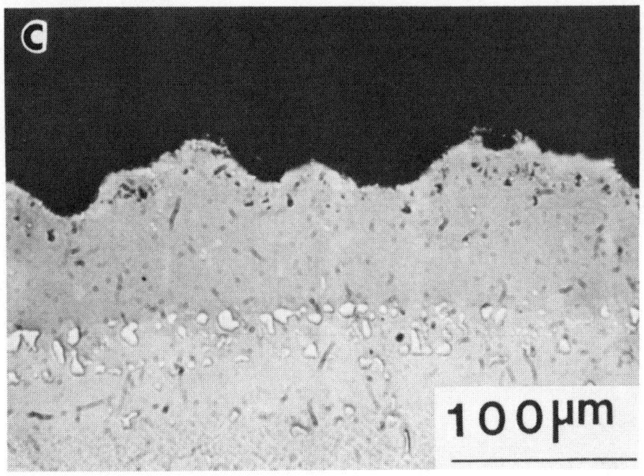

Figure 12. Typical example of surface rumpling of Pt-Al coatings: (a) sample exhibits macro surface roughness (8x), (b) enlarged photomicrograph (64x) and (c) cross-section of same specimen showing surface upheaval on the order of 50 µm.

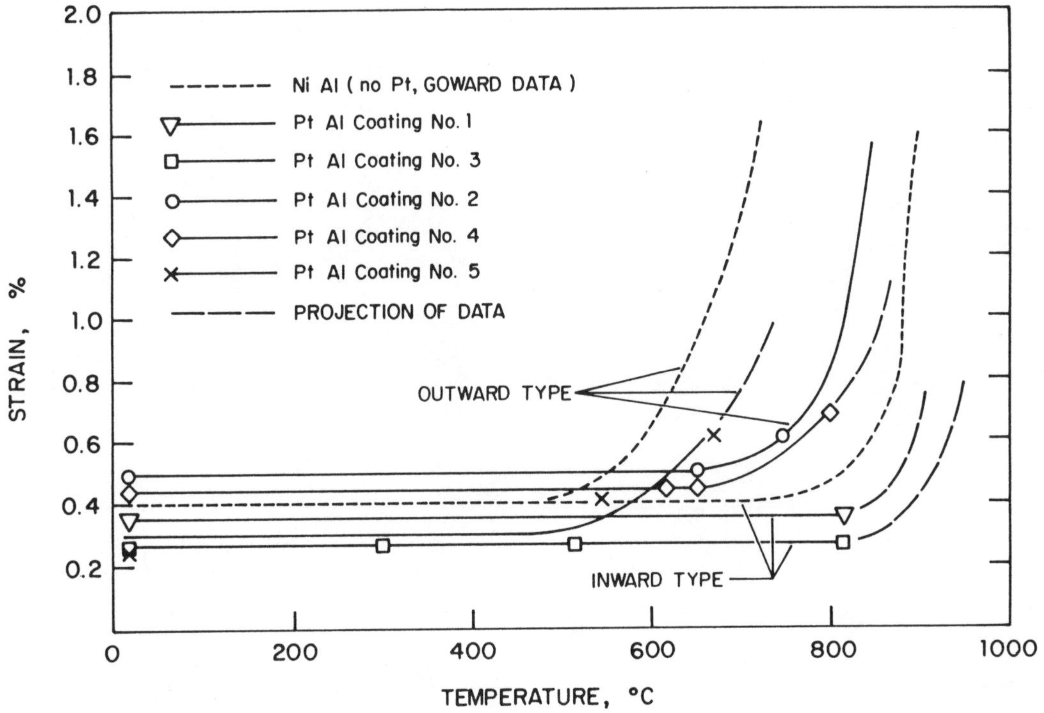

Figure 13. Ductile to brittle transition temperature (DBTT) and fracture strain of Pt-Al coatings as function of structure and Al-Pt content.

Na_2SO_4 INDUCED CORROSION OF NICKEL AT HIGH TEMPERATURE

A. K. Misra*
NASA Lewis Research Center
Cleveland, Ohio, USA
*Associated with
Dept. of Metallurgy and Materials Science,
Case Western Reserve University,
Cleveland, Ohio

ABSTRACT

Sodium sulfate induced corrosion of nickel was studied at 900°C as a function of oxygen partial pressure. For high O_2 partial pressures, accelerated corrosion during the first few minutes occurred by rapid penetration of the melt along the metal grain boundaries, and a mechanism is proposed to explain this phenomena. Repetitive scale metal detachment was observed for corrosion in lower O_2 partial pressures and during the later period of corrosion in higher O_2 partial pressures. The effect of preoxidation on the hot corrosion has also been studied. An induction period is observed before the onset of rapid corrosion for the preoxidized samples and the onset of rapid corrosion is associated with sudden cracking of the scale. The length of the induction period for the preoxidized samples is a function of the length of preoxidation and appears to be related to the structure of the oxide scale after the preoxidation treatment.

*Associated with Dept. of Metallurgy and Materials Science, Case Western Reserve University, Cleveland, OH

DEPOSITION OF MOLTEN Na_2SO_4 on the blades and vanes of combustion gas turbines causes corrosion attack of the turbine materials and this is conventionally known as hot corrosion. Nickel base superalloys and coating systems are extensively used as gas turbine blading material, and, therefore, considerable studies have been made to understand the hot corrosion mechanisms for pure nickel.[1-6] In the presence of a thin layer of molten Na_2SO_4, the oxidation of nickel generates a porous, friable, oxide scale, and accelerated corrosion is observed. Goebel and Pettit[1] have studied the hot corrosion of nickel at 1000°C in 1 atm. O_2 and postulated a mechanism in which sulfides are introduced in the metal, resulting in an increase in the Na_2O activity of the melt. The modified Na_2SO_4 melt then dissolves the NiO at the melt-oxide interface and the dissolved oxides are subsequently precipitated at a distance a way from the melt-oxide interface, thereby leading to the development of a porous oxide scale. The corrosion process described above is known as the basic fluxing mechanism. In a later study Goebel and Pettit[2] have modified the basic fluxing mechanism and postulated that the accelerated corrosion occurs by localized penetration of the scale by Na_2SO_4 followed by stripping of the oxide scale. The localized penetration was believed to occur by reaction of the oxide in the localized regions of the scale with the oxide ions in the melt. Reising[3] has described a mechanism in which O_2 diffuses through the sulfate melt to the melt-metal interface where oxidation of Ni takes place to form Ni^{++} and O^{--}. The Ni ions diffuse towards the melt-gas interface where NiO is formed. According to this mechanism, Na_2SO_4 acts as a fast diffusion medium for Ni^{++} and O_2. Lillerud and Kofstad[5] have studied the corrosion of nickel in O_2+4.2% SO_2 at 900°C and suggested a mechanism in which NiO, instead of being formed as a continuous layer, forms as separate grains. This results in a NiO scale with molten sulfate distributed as a network between the NiO grains, and the corrosion proceeds by inward transport of O_2 through the molten sulfate network. Kofstad and his coworkers[4,5] observed sulfides for corrosion in O_2+SO_2 atmospheres, however, no sulfides were observed for Na_2SO_4 induced corrosion in 1 atm. O_2. This is in contrast to the results of Goebel and Pettit.[1]

From the above discussion it becomes apparent that hot corrosion of nickel needs further study. All the corrosion studies on nickel so far have been conducted on the bare metal, although in many situations in actual service, a layer of oxide might be present on the metal surface before the onset of Na_2SO_4 deposition. Therefore, it would be instructive to study the hot corrosion of preoxidized nickel and examine how a protective scale breaks down under hot corrosion situations. The objective of the present study is two fold: (1) to examine the mechanisms for the accelerated corrosion of nickel in the presence of molten Na_2SO_4 and; (2) to study the effects of preoxidation treatment on the hot corrosion process.

EXPERIMENTAL PROCEDURE

High purity nickel (99.99%) was used in the present study. The nickel coupons were cut from 1.25 mm diameter rods into disks of 2 mm thickness. The coupons were polished through 600 grit and cleaned ultrasonically in trichloroethylene, detergent, distilled water, alcohol, and then dried in an oven at 120°C. The samples were coated with Na_2SO_4 by airbrushing a saturated solution of salt onto coupons heated on a hot plate to about 200°C. The corrosion experiments were performed in a vertical tube furnace and the weight gains were monitored continuously by a microbalance. A few experiments were conducted in Ar+O_2 mixtures in order to examine the effect of O_2 partial pressure on the corrosion process. The test samples corroded for different lengths of time were washed in hot distilled water, and the solutions were analyzed for various water soluble species like Na, SO_4^{--}, and Ni. The concentration of SO_2 in the exit gas was monitored continuously with a Thermoelectron pulsed SO_2 analyzer. The corroded samples were examined by conventional metallography and electron optical techniques. Few samples were dry polished in order to preserve the water soluble species like Na_2SO_4 in the scale.

RESULTS AND DISCUSSION

CORROSION OF Na_2SO_4 COATED Ni IN ATM. O_2 - There is considerable scatter in the observed corrosion kinetics, as illustrated in Figure 1. In spite of the scatter, the general trend in the kinetics is observed to be same for all the runs. There is rapid weight gain during the first 5-6 minutes of corrosion followed by slowing down of the rate of weight gain. The time at which the weight of the sample starts to increase at a rapid rate varies from run to run and normally lies in the range of 1.5-3 minutes.

It is not clear whether this can be considered as an induction period since it takes 1.5-2 minutes for the sample to reach the desired test temperature after the furnace is raised. From Figure 1 the rate of weight gain between 3-6 min. is observed to be of the order of 100-400 mg/cm^2-hr, which is much higher than that of pure oxidation. It must be realized that the weight gain is due to the pickup of O_2, and, therefore, it becomes apparent that the presence of a layer of molten Na_2SO_4 on the metal surface accelerates the oxidation process. After about 20-25 min. the rate of corrosion slows down considerably, and for longer corrosion periods, the corrosion rate was observed to be nearly the same as that of oxidation without a coating of Na_2SO_4

The top surface of nickel, coated with approximately 2.5 mg/cm^2 Na_2SO_4 and oxidized for different lengths of time between 0.5-4 minutes, is shown in Figure 2. Before the beginning of the rapid weight gain i.e., at 40 seconds, the top surface is observed to be completely covered with Na_2SO_4. At the beginning of rapid corrosion i.e., about 3 min. several cracks are observed on the surface and there is no melt on the surface, suggesting flow of the sulfate melt through these cracks. In some areas of the specimen surface severe bending and extrusion is observed in the cracked regions (figure 2c), suggesting the development of large stresses in the scale. The typical cross section during the first 3-4 min. of corrosion, illustrated in Figure 3, shows a thick oxide layer and several vertical channels in the oxide layer extending from the scale-gas to the scale-metal interface. From the electron probe micro analysis, the results of which are shown in Figure 4, the vertical channels in the oxide scale in Figure 3 are observed to be rich in Na and S; the X-ray diffraction of the scale showed NiO and Na_2SO_4. Obviously the channels in the scale are filled with Na_2SO_4. X-ray maps in Figure 4b also show Na_2SO_4 to be present at the scale-metal interface. The cross section in Figure 3 does not show any internal attack in the metal; however, in a few cases detailed examination of the cross section of the samples corroded between 1-3 min. shows internal penetration, as illustrated in Figure 5. The internal attack appears to be due to the formation of a thin layer of sulfide along the grain boundaries. The cross section of a sample that has been corroded till a weight gain of about 10 mg/cm^2 is shown in Figure 6. The scale can be divided into two distinct layers -- an outer and an inner scale. The outer part of the scale, which is about 200 μm thick, consists of several channels extending from the scale-gas interface, and this part of the scale has the same characteristics as that of the Figure 3. The inner part of the scale however, has a layered structure, and the oxide in each layer except for the bottom most layer near the scale-metal interface has several microfissures perpendicular to the scale-metal interface. There is a thin band of sulfide, of the order of 1-2 μm, at the scale-metal interface. The scale-metal interface is observed to be non-planar and has a wavy appearance. Further corrosion with time takes place by the growth of the inner layered type of scale, which is shown in Figure 7. Note that the oxide has been removed at several places probably during metallographic preparation. The scale-metal interface in Figure 7 has be same general features as that of Figure 6.

The results of the chemical analysis for samples corroded for different lengths of time are shown in Table I. The chemical analysis showed primarily Na and SO_4^{--} in the solution; only a small amounts of Ni, of the order of 4-5 μg, was detected in the solution, and the amount of Ni in the solution was observed to be nearly the same for all corrosion times. Rapid weight gain takes place between 2-6 minutes, and it is interesting to note that almost all of the Na_2SO_4 applied could be recovered from the solution for these times. For example, the amount of Na_2SO_4 recovered from the solution at 2 and 6 minutes of corrosion corresponds to 98.9 and 94 percent respectively of the amount originally applied. To examine if the reaction of NiO with reagant grade Na_2SO_4 occurs, NiO and Na_2SO_4 were mixed with a Pt crucible, and the resulting mixture was heated in 1 atm. O_2 at 900°C for 24 hours. The mixture after exposure was dissolved in distilled water and the solution was analyzed for nickel. No nickel could be detected, which is consistent with the results of Goebel and Pettit.[1] Thus, the nickel in the water washed solution for the corroding sample must be as a result of the basic dissolution of NiO in the Na_2SO_4 melt, as originally suggested by Goebel and Pettit.[1] The basic dissolution occurs due to the increase in the Na_2O activity of the melt because of the removal of sulfur from the melt by sulfide formation.

CORROSION OF Na_2SO_4 COATED Ni IN AR+O_2 MIXTURES - The corrosion kinetics for different O_2 contents in the Ar+O_2 mixture is shown in Figure 8. For 2.2 and 1 percent O_2 in the atmosphere, the corrosion kinetics are similar as that in 1 atm. O_2. However, the corrosion continues for longer periods of time and the rate of corrosion is lower than that of 1 atm. O_2. For O_2 contents $\lesssim$ 500 ppm, an induction period is observed before the onset of rapid corrosion, and the length of the induction period is of the order of 1-4 hours. After the induction period rapid corrosion continues for 6-10 hours, followed by a decrease in the rate of corrosion.

The scale morphology for corrosion in Ar+1% and Ar+2% O_2 is observed to be very similar to that of 1 atm. O_2. The top surface of a sample during the induction period for corrosion in Ar+233 ppm O_2, shown in Figure 9a, is observed to be completely covered with Na_2SO_4, and the cross section at this time (figure 9b) shows the formation of a mixed oxide and sulfide scale and a layer of sulfide underneath it. During the period of rapid corrosion, the edge[8] of the sample are observed to suffer from severe degradation, as illustrated in Figure 10. A stratified scale consisting of several layers of oxide is observed at the edges of the sample, whereas, a nearly compact scale is observed in the other regions. The cross-section during the period of rapid corrosion in Ar+500 ppm O_2, illustrated in Figure 11, shows localized corrosion, and it appears that the localized corrosion occurs by the detachment of the scale from the metal surface at localized regions. Also, similar to that of corrosion in Ar+233 ppm O_2, a layered scale is observed in the corroded regions. The chemical analysis of the water washed solution for a sample corroded for 1 hr. in Ar+233 ppm O_2 shows 6 µg of nickel in the solution, suggesting basic dissolution of the oxide in the melt.

SO_2 EVOLUTION DURING CORROSION IN O_2 AND AR+O_2 MIXTURES - The amount of SO_2 evolved as a function of time for different O_2 partial pressures in the atmosphere is shown in Figure 12. Large amounts of SO_2, of the order of 40-70 ppm , were evolved during the first few minutes of corrosion in atmospheres containing less than 500 ppm O_2. In general, the amount of SO_2 evolved decreases with an increase in the O_2 partial pressure; however, there is a sharp decrease in the amount of SO_2 evolved for O_2 contents higher than 500 ppm. Also, no SO_2 (within the detection limits the analyzer) could be detected for corrosion in 1 atm. O_2.

SO_2 evolution is possibly due to the decomposition of the sulfate melt, which can be described by the following reactions:

$$Na_2SO_4 = Na_2O + SO_3 \qquad [1]$$

$$SO_3 = SO_2 + 1/2O_2 \qquad [2]$$

It should be noted that the analyzer measures the SO_2 only; thus, the total amount of sulfurous gases (SO_2 plus SO_3) evolved due to the decomposition of the sulfate melt would be higher than that shown in Figure 12. Decomposition of the sulfate melt was observed to occur only in the presence of a corroding metal beneath the melt; for example, no SO_2 evolution occurred when a Pt coupon was coated with Na_2SO_4 and heated at 900°C in 1 atm. argon. Also, no SO_2 evolution occurred when a Pt crucible containing a mixture of Na_2SO_4 and NiO was heated in 1 atm. argon at 900°C. These results indicate that reaction of Na_2SO_4 with nickel results in the decomposition of the melt, the extent of reaction/decomposition increasing with the decrease in O_2 partial pressure.

CORROSION MECHANISM - In terms of the chemical interaction the role of the Na_2SO_4 melt is two fold: (1) to introduce sulfides into the scale, and (2) to dissolve the protective oxides. Both these effects are observed in the present study as evidenced by the formation of sulfides in the cross section and detection of Ni in the water washed solution of the corroded samples; however, the highest corrosion rates are not necessarily associated with the situations in which the interaction of Na_2SO_4 melt with the metal is the maximum. For example, the extent of sulfidation is the highest during the initial period of corrosion in Ar+233 ppm O_2, and the amount of nickel in the water washed solutions is of the same order of magnitude as that of atmospheres containing higher O_2 levels. However, the rate of corrosion is much lower than the higher oxygen containing atmospheres. Corrosion in Ar+233 ppm for longer times results in rapid corrosion only at the edges of the samples, and a layered type of scale is observed, which suggests repetitive detachment of scale from the metal surface. It is well known from the oxidation literature[7] that the corners and the edges of the samples are the regions of high stress concentration and are favorable sites for the scale metal detachment to occur. Although a layered type of scale in the edges was not developed for the oxidation of uncoated nickel in 1 atm. O_2 at 900°C; it's development is possible in the presence of Na_2SO_4 melt due to either one or a combination of the following reasons: (1) at oxygen contents below 500 ppm, a thick mixed oxide plus sulfide scale is formed, and it is well known from the oxidation literature[7] that it is difficult for a thick scale to plastically deform in order to maintain contact with the underlying metal, thus leading to the scale metal detachment. (2) corrosion in the presence of a Na_2SO_4 melt results in the formation of a thin layer of sulfide at the scale-metal interface. Sulfides in metals and alloys are known to cause embrittlement[8,9,10] under the action of mechanical stress. Thus it is likely that a thin layer of sulfide at the scale-metal interface can cause cracking at this interface under the action of the oxide growth stress.

Once scale metal detachment occurs, the sulfate melt can again come in contact with the metal and the sequence would be repeated, resulting in a layered type of scale. The existence of an induction period for corrosion in lower oxygen contents in the atmosphere is probably related to the time required for the initial scale-metal separation to occur in the localized regions.

A layered type of scale is obtained only during the later period of corrosion in 1 atm. O_2. The initial rapid corrosion is due to the formation of cracks and the flow of the melt through these cracks to the scale-metal interface, resulting in channels of sulfate melt in the oxide scale. A non uniform distribution of the salt is obtained and some areas of the sample are in contact with the sulfate melt, while oxygen has free access to the metal surface in other areas. The surface cracks and the melt filled channels in the scale are probably as a result of preferential attack along metal grain boundaries as explained below. Initially the sulfides are formed only along the metal grain boundaries and this increases the Na_2O activity of the melt in the localized regions just above the metal grain boundaries, resulting in the dissolution of oxide along metal grain boundaries. This process has the same effect of creating a crack in the grain boundary region. The diffusion of sulfur along the metal/alloy grain boundaries is known to cause rapid propagation of the crack along the grain boundaries under the action of mechanical loading.[8,9] The presence of large stresses in the scale was readily evident from the bending and extrusion observed in certain areas of the scale surface. Thus it is likely that once a crack is created along the grain boundaries because of localized basic dissolution, the crack can propagate rapidly by a combination of the growth stress due to rapid oxidation of the metal in between the grain boundaries and the presence of sulfur along the grain boundaries. The sulfate melt flows through these cracks while the metal in between the grain boundaries rapidly oxidizes, resulting in several melt filled channels in the oxide scale.

It is interesting to note that sulfidation exclusively along the grain boundaries was observed at high O_2 partial pressures in the atmosphere, whereas, no such preferential sulfidation along the grain boundaries was observed for the lower O_2 partial pressures; instead, a uniform sulfide layer was observed at the scale-metal interface. Therefore, the presence of a uniform layer of sulfide underneath the oxide scale at the scale-metal interface for longer corrosion periods in 1 atm. O_2 suggests that Po_2 beneath a thick porous scale containing several melt filled channels must be low. An atmosphere of low Po_2 can be attained beneath a thick porous scale if the O_2 diffusion through the pores becomes important and O_2 has to travel a tortuous path in the porous scale. Also, there would be a difference in the Po_2 in different regions beneath the porous scale; for example, the Po_2 beneath a long column of Na_2SO_4 would be lower than that below a column of porous oxide because of the low solubility of O_2 in the sulfate melt. These differences in the Po_2 in different regions of the same sample cause the scale growth rate to be different in different areas, thus resulting in the development of a wavy scale-metal interface as observed. The corrosion beneath the thick outer porous layer proceeds by the repetitive detachment of the scale from the metal surface, as was the case for low Po_2 atmospheres. It is known from oxidation literature[7] that scale-metal separation occurs with relative ease for the curved surfaces; thus the development of a wavy scale-metal interface clearly helps in the scale-metal detachment process. The microfissures in each oxide layer are probably due to either one or a combination of the following: (1) anisotropic disassociation of the oxide along certain sites after the detachment of the scale from the metal surface, as has been suggested by Mrowec,[7] and (2) preferential dissolution of oxide in the melt along the oxide grain boundaries. In either case, the development of microfissures in the oxides would lead to melt penetration through these fissures. The accelerated corrosion stops when all the Na_2SO_4 are distributed in the scale, presumably by sticking to the walls of the pores and fissures, and no more Na_2SO_4 is available at the reaction site.

EFFECT OF PREOXIDATION ON CORROSION IN 1 ATM. O_2 - Nickel coupons were preoxidized for various lengths of time at 980°C and the preoxidized samples were corroded in the presence of Na_2SO_4 at 900°C in 1 atm. O_2. Before describing the corrosion of the preoxidized samples, it is imperative to examine the oxide structure after the preoxidation treatment. Figure 13 shows the top surface of the oxide for different times of oxidation at 980°C. At 1 hour, the oxide surface shows several ridges, which is due to the preferential growth along the oxide grain boundaries. At 5 hours, the ridge like structure is not observed; instead, smooth oxide grains are observed. At 20 hours, large faceted oxide grains are observed, and a comparison of the oxide at five hours and 20 hours shows that oxide grain growth has occurred. From these observations it can be derived that initially the oxide growth takes place preferentially along the oxide grain boundaries, possibly by diffusion of oxygen along these boundaries. The evolution of oxide structure with time is also reflected in the oxidation kinetics, which is shown in Figure 14. Parabolic oxidation kinetics are observed only after 4-5 hours of oxidation, and this is probably due to the faster rate of oxidation during the first few hours because of the preferential growth along the oxide grain boundaries. These results are consistent with the observation of other investigators.[11]

The hot corrosion kinetics for the preoxidized samples, illustrated in Figure 15, shows an induction period before the onset of accelerated corrosion. The end of induction period is marked by a sudden rise in the weight of the sample, as evidenced by a sharp transition point in the corrosion kinetics. The length of the induction period is observed to be a function of the preoxidation time, and in general, increases with an increase in the length of the preoxidation time. The increase in the length of the induction period is dramatic for preoxidation times between 3-5 hours; for example, the length of the induction period increases to 40 hr. from 6 hr. by increasing the preoxidation time from 3 hr. to 4 hr., and no accelerated corrosion is observed even up to 200 hours for the sample preoxidized for five hours and longer. No such dramatic increase in the corrosion resistance was observed for increase in preoxidation time between 1-3 hours.

During the induction period, the surface of the preoxidized samples is covered with a layer of Na_2SO_4, and the oxide morphology beneath the sulfate melt, as observed after washing the top surface with water, appears to be very similar to that observed before the application of the sulfate coating. This is shown in Figure 16.

The examination of the top surface just at the beginning of rapid corrosion (figure 17) shows the hot corrosion to be initiated by cracking of a part of the outer oxide and the development of a porous oxide in this region. In other parts of the scale the surface shows similar characteristics as that during the induction period. The cross section during the period of rapid corrosion, illustrated in Figure 18, shows localized detachment of the oxide scale from the metal surface and the development of a porous scale beneath the detached oxide layer. The morphology of the porous scale beneath the stripped outer scale is very similar to that observed for the corrosion of bare nickel in the presence of the sulfate melt.

SUMMARY AND CONCLUSION

The Na_2SO_4 induced corrosion of nickel was studied at 900°C as a function of O_2 partial pressure in the atmosphere. For high Po_2 atmospheres, rapid corrosion occurred within the first few minutes of corrosion, which is attributed to the following sequence of events: (1) initially sulfides are formed only along the metal grain boundaries, which increases the Na_2O activity of the melt in the localized regions just above the metal grain boundaries, (2) basic dissolution of the oxide in the region above the metal grain boundaries occurs which has the effect of creating a crack at the grain boundaries, (3) the crack propagation occurs along the metal grain boundaries as a result of the combination of the stress generated due to the growth of the oxide in the region in between the grain boundaries and the formation of sulfides along the grain boundaries, (4) the sulfate melt rapidly flows through these cracks, and simultaneously the metal region in between the grain boundaries undergo rapid oxidation resulting in a porous oxide scale with several channels filled with the melt. For lower Po_2 atmospheres and during the later stages of corrosion in higher Po_2 atmospheres, accelerated corrosion is due to repetitive detachment of the scale from the metal surface resulting in the development of a stratified scale. An induction period is observed before the onset of rapid corrosion in lower Po_2 atmospheres and the length of the induction period is probably related to the time necessary for the first scale-metal detachment to occur.

Corrosion of the preoxidized samples show the existence of an induction period before the onset of accelerated corrosion, the length of the induction period increasing with an increase in the preoxidation time. The onset of accelerated corrosion is associated with the sudden cracking of the scale and melt penetration through it to the scale-metal interface. The length of the induction period is observed to be strongly related to the oxide structure at the end of the preoxidation treatment. Scale cracking appears to be easier and the length of the induction period is shorter if the oxide at the end of the preoxidation treatment shows preferential growth along the oxide grain boundaries. On the other hand, if the oxide shows uniform scale growth, and parabolic kinetics have been attained during oxidation, no scale cracking and melt penetration is observed even after 200 hours of corrosion beneath a sulfate melt.

REFERENCES

1. J. A. Goebel and F. S. Pettit, Metall. Trans., 1, 1943-1954, (1970).
2. J. A. Goebel and F. S. Pettit, "Metal-Slag-Gas Reactions and Processes," p. 693, Z. A. Foroulis and W. W. Smeltzer, eds., Electrochemical Society Publication, Pennington, New Jersey, (1975).
3. R. F. Reising, ibid, p. 747.
4. P. Kofstad and G. Akesson, Oxid. Met., 14, 301-323 (1980).
5. K. P. Lillerud and P. Kofstad, Oxid. Met., 21, 233-270 (1984).
6. A. K. Misra and D. P. Whittle, Oxid. Met., 22, 1-33 (1984).
7. S. Mrowec, Corr. Sci., 7, 563-579 (1967).
8. S. Floreen and R. H. Kane, Metall. Trans., 13A, 145-152, (1982).
9. S. Floreen and R. H. Kane, Metall. Trans., 15A, 5-10, (1984).
10. G. A. Whitlow, C. G. Beck, T. Viswanathan and E. A. Crombie, Metall. Trans., 15A, 23-28, (1984).
11. D. Caplan, M. J. Graham and M. Cohen, J. Electrochem. Soc., 119, 1205-1215 (1972).

TABLE I

RESULTS OF THE CHEMICAL ANALYSIS AT 900 °C

TIME, min	WEIGHT GAIN, mg/cm^2	AMOUNT OF Na_2SO_4 APPLIED, mg	AMOUNT OF Na_2SO_4 RECOVERED, mg	AMOUNT OF Ni IN THE SOLUTION, µg
2	4.5	7.8	7.72	4
6	15	2.5	2.35	4.5
24	22	5.5	4.39	4
180	25	5	3.33	5

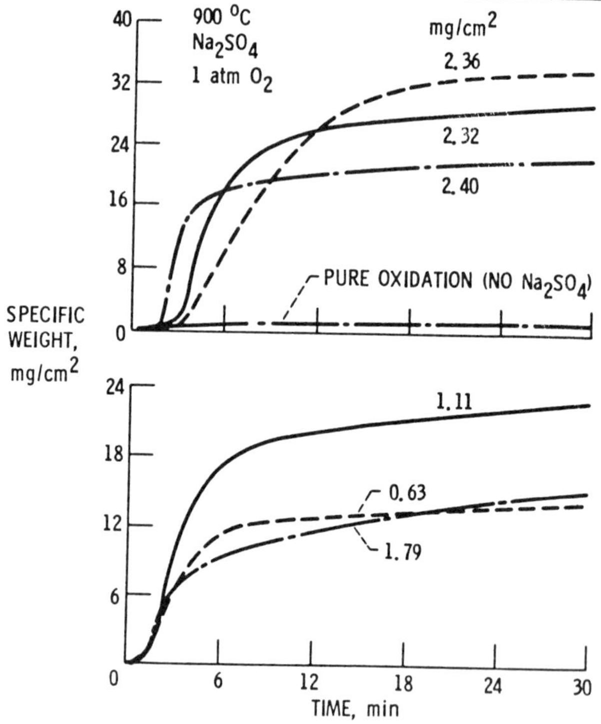

Fig. 1 Corrosion kinetics for nickel, coated with different amounts of Na_2SO_4, and oxidized at 900°C in 1 atm. O_2.

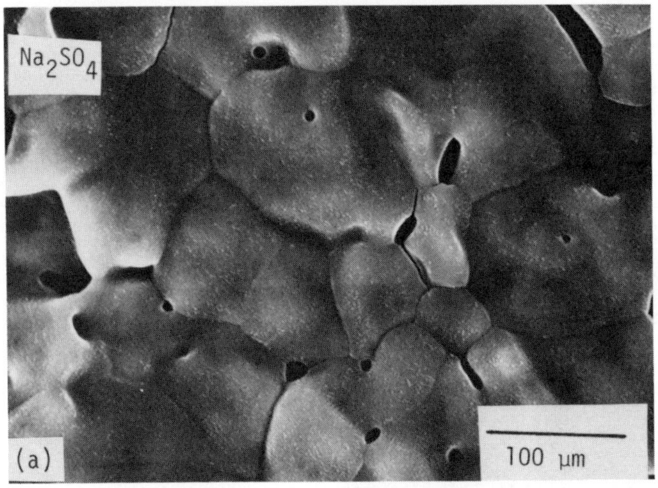

(b)

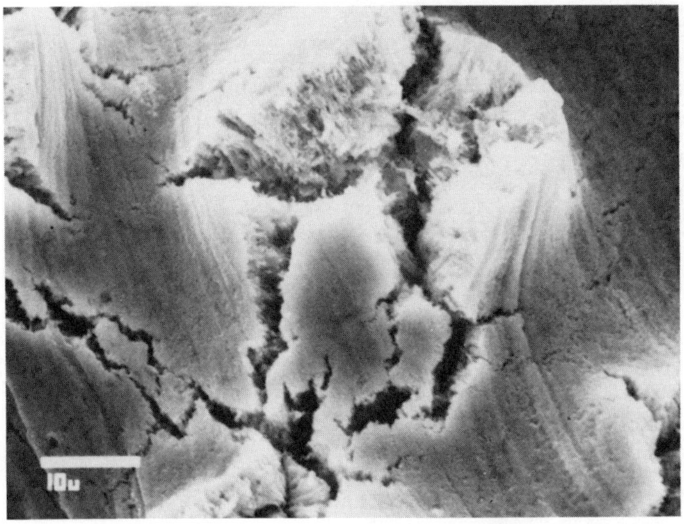

(c)

Fig. 2 Top surface of Ni, coated with 2.5 mg/cm^2 Na_2SO_4 and oxidized at 900°C in 1 atm. O_2.

a. 40 seconds - top surface covered with Na_2SO_4
b. 3 minutes
c. 3 minutes (one of the cracked regions at a higher magnification)

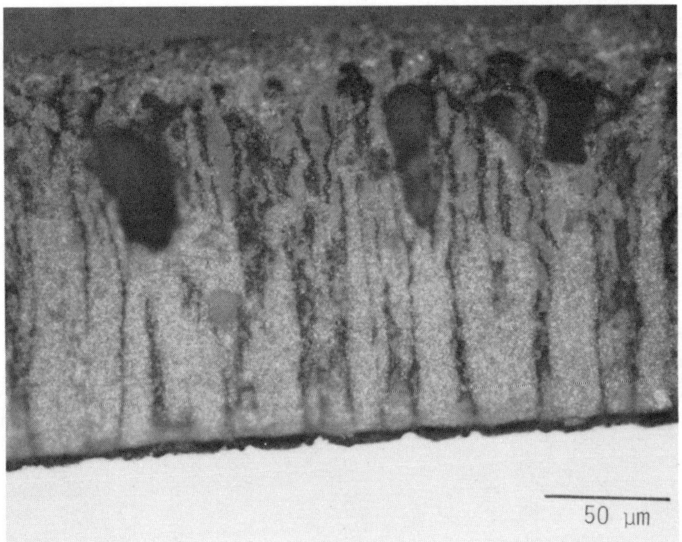

Fig. 3 Cross section of Ni, coated with 3 mg/cm^2 Na_2SO_4, and oxidized for 3.5 minutes at 900°C in 1 atm. O_2.

S. E. Image Sodium Sulfur

(a)

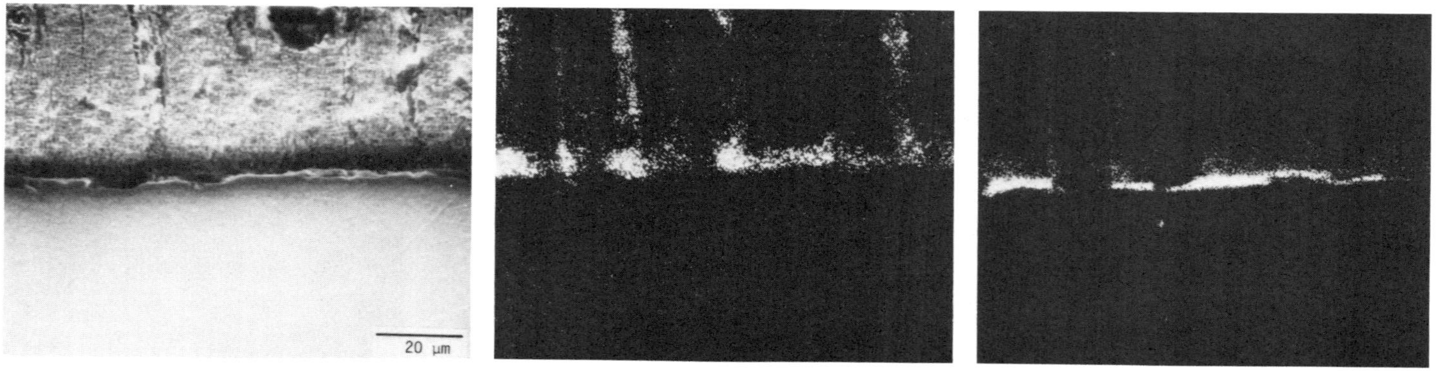

S. E. Image Sodium Sulfur

(b)

Fig. 4 X-ray maps for Na and S for the cross-section shown in figure 3.
a. outer part of the scale
b. inner part of the scale

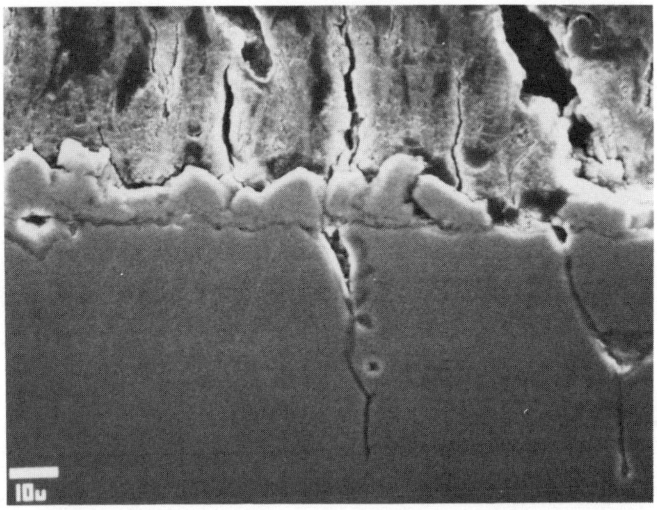

Fig. 5 Scale-metal interface showing internal attack along the grain boundaries for Ni, coated with Na_2SO_4, and oxidized at 900°C between 1.5-3 minutes.

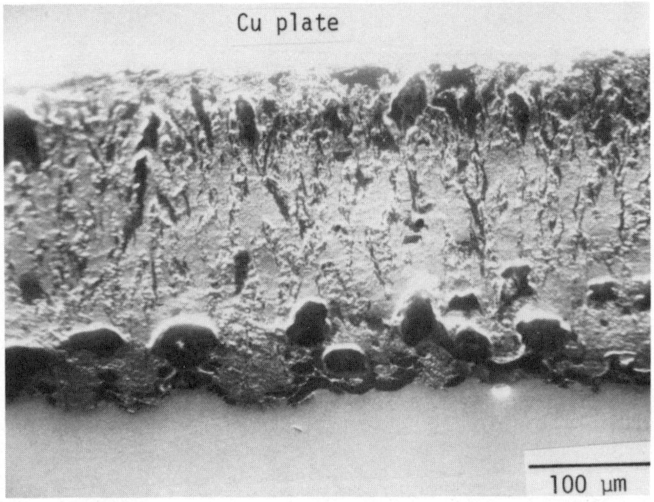

(a)

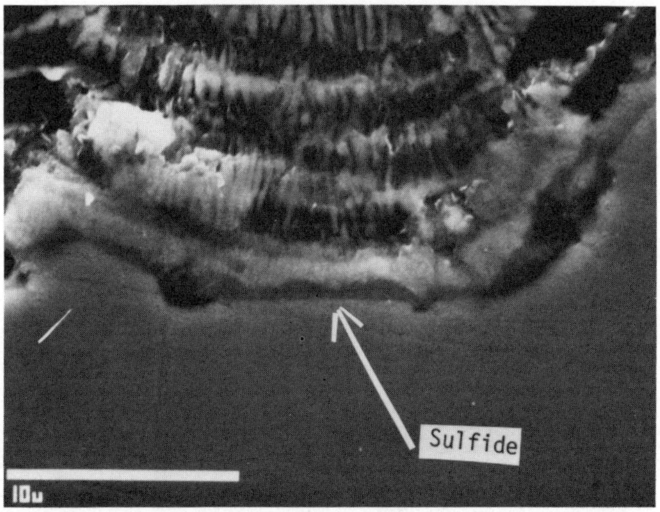

(b)

Fig. 6 Cross section of Ni, coated with 2.2 mg/cm^2 Na_2SO_4, and oxidized for 4 minutes at 900°C in 1 atm. O_2.
a. overall cross section
b. scale-metal interface

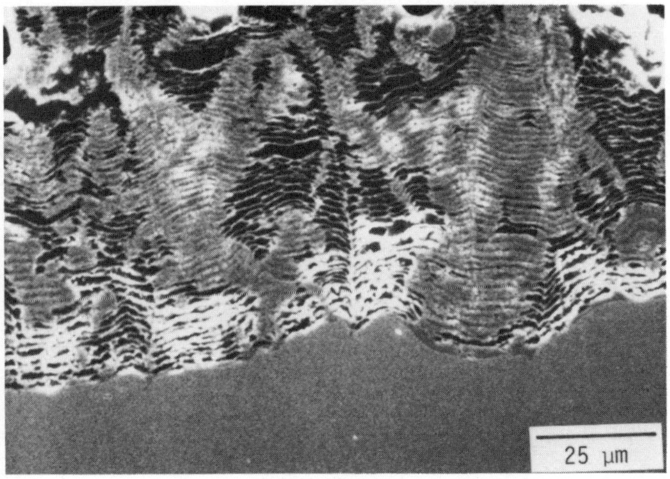

Fig. 7 Inner part of the scale for Ni, coated with 2 mg/cm^2 Na_2SO_4, and oxidized for 6 minutes in 1 atm. O_2 at 900°C.

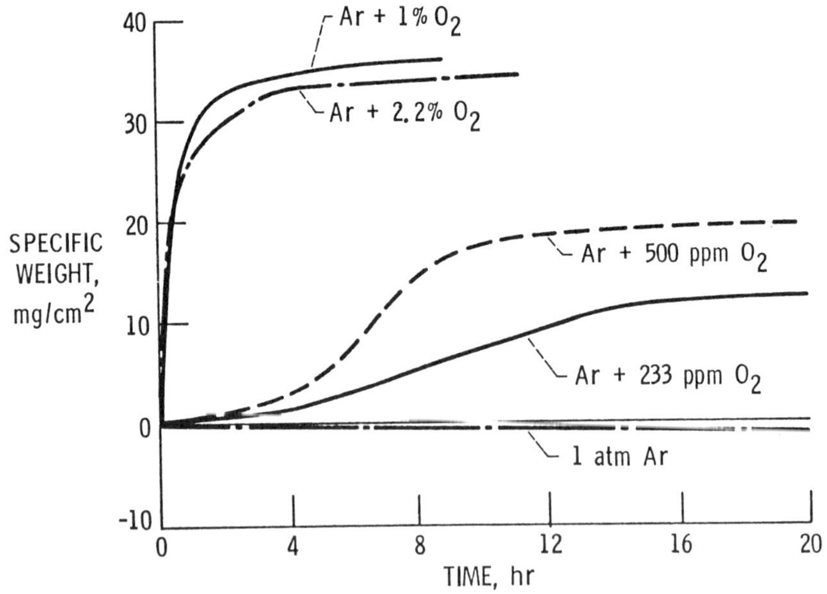

Fig. 8 Corrosion kinetics for nickel, coated with 2 mg/cm^2 Na_2SO_4, and oxidized at 900°C in Ar+O_2 mixture with different O_2 content.

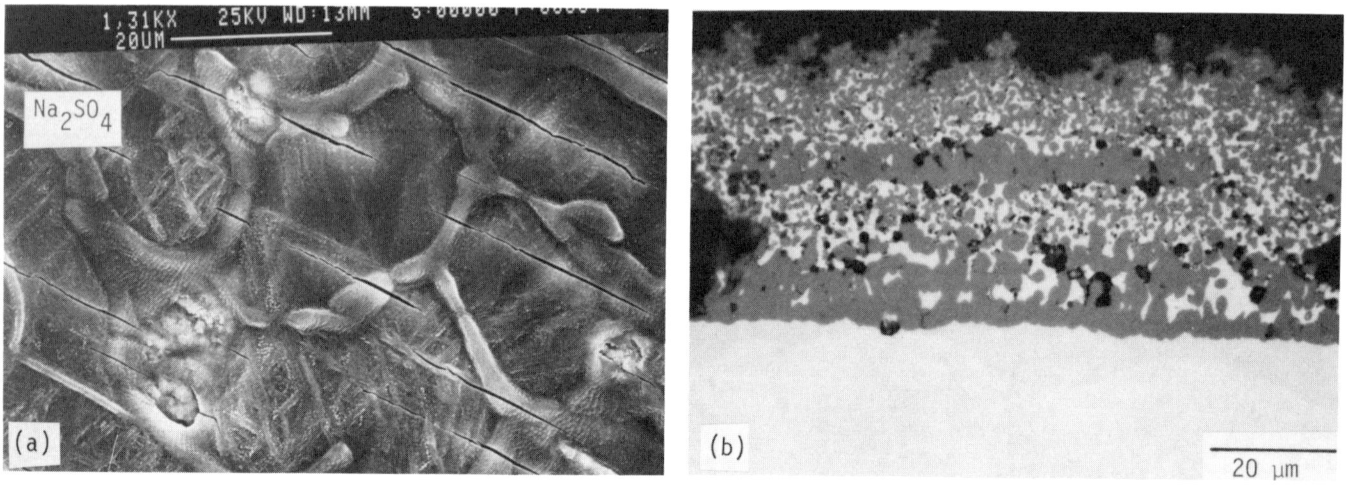

Fig. 9 Scale morphology for nickel, coated with 2 mg/cm^2 Na_2SO_4, and oxidized for 1 hour in Ar+233 ppm O_2 at 900°C.

a. top surface

b. cross section

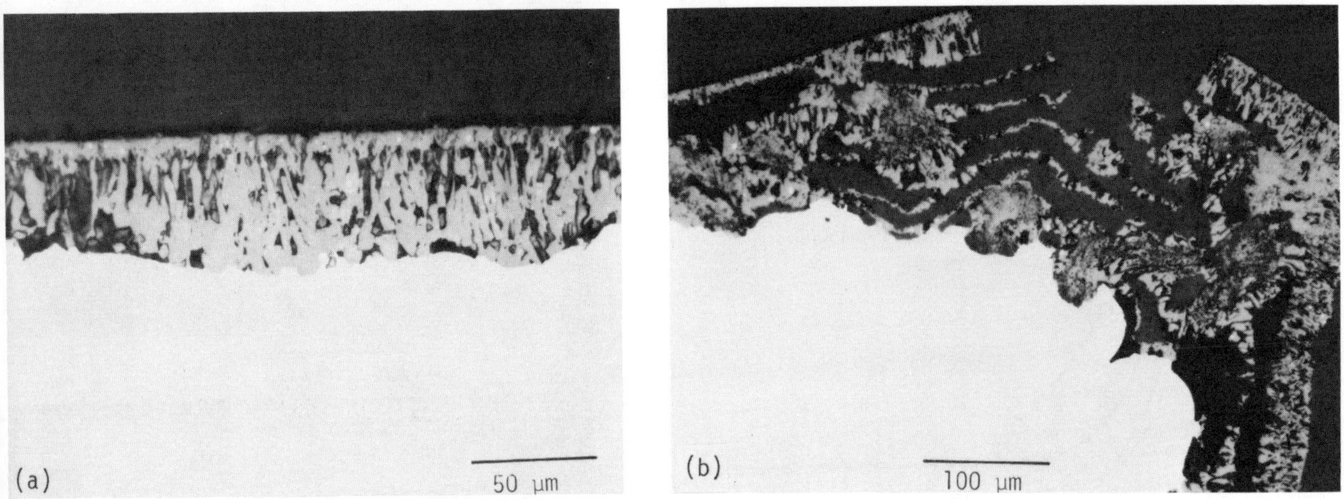

Fig. 10 Cross section of Ni, coated with 2 mg/cm^2 Na_2SO_4, and oxidized for 12 hours in Ar+233 ppm O_2 at 900°C.

a. cross section on flat surface

b. cross section near the edges

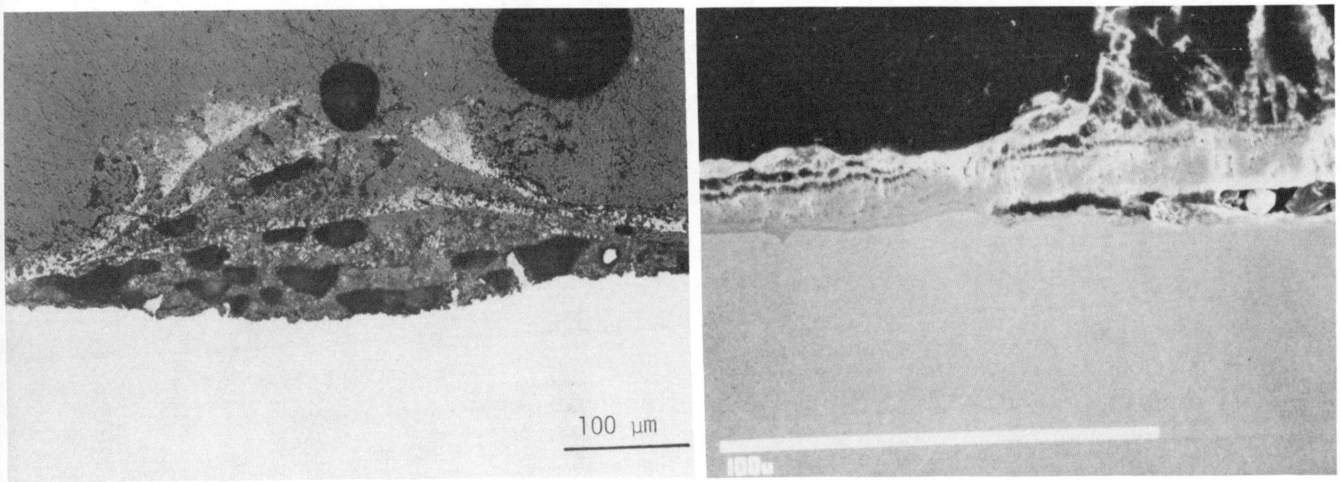

(a) (b)

Fig. 11 Cross section of Ni, coated with 2 mg.cm^2 Na_2SO_4, and oxidized for 7 hours in Ar+500 ppm O_2 at 900°C.

a. overall cross section showing localized corrosion

b. detachment of scale from the metal surface

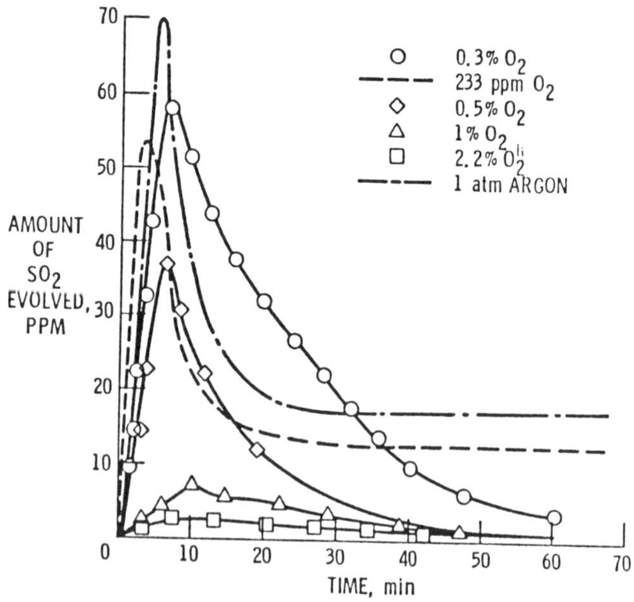

Fig. 12 SO_2 evolved as a function of time for corrosion of Na_2SO_4 coated Ni at 900°C as a function of O_2 partial pressure.

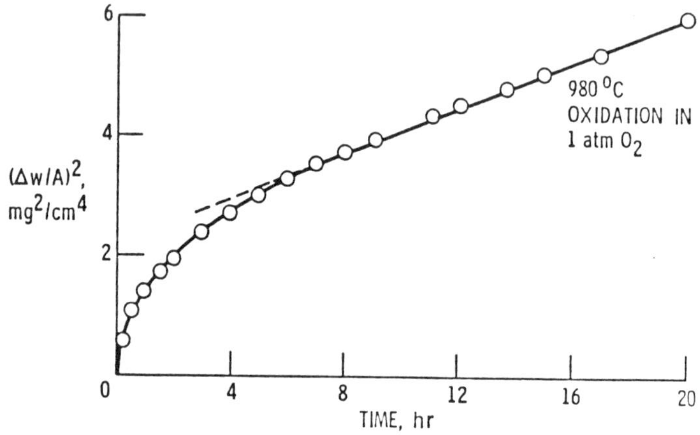

Fig. 14 Oxidation kinetics for uncoated nickel at 980°C in 1 atm. O_2.

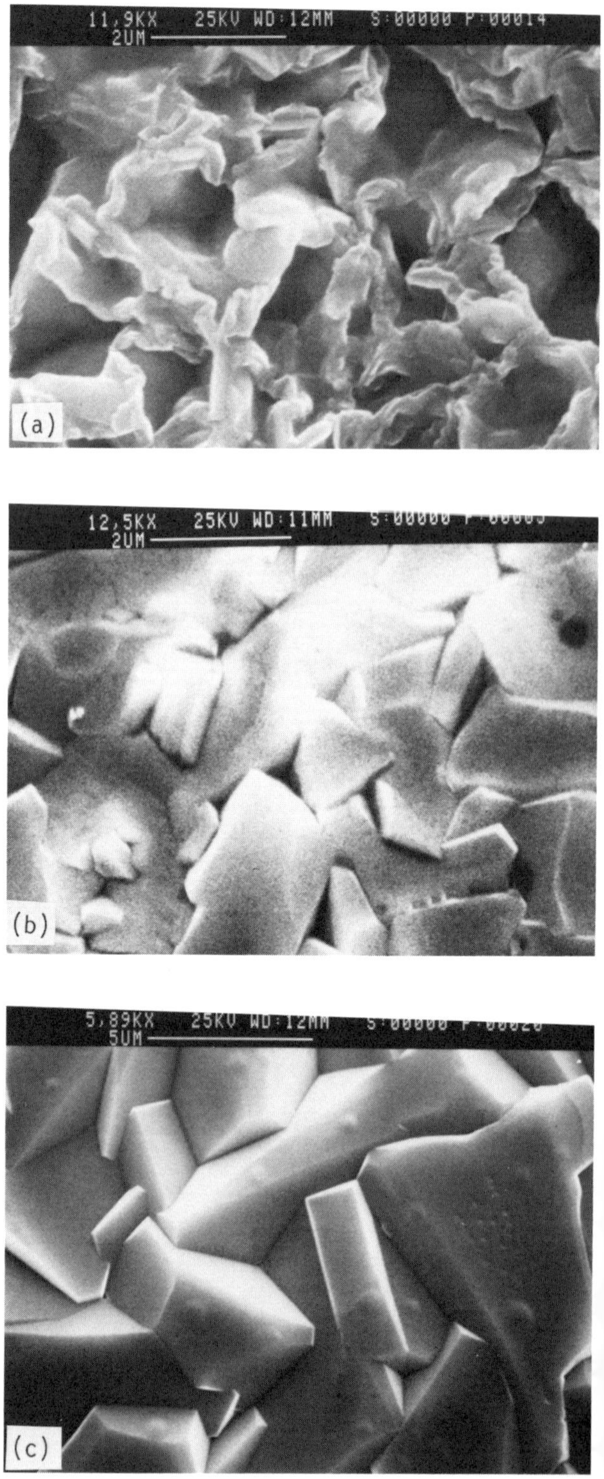

Fig. 13 Top surface of oxide after different times of oxidation of uncoated nickel at 980°C in 1 atm. O_2.

a. 1 hour

b. 5 hours

c. 25 hours

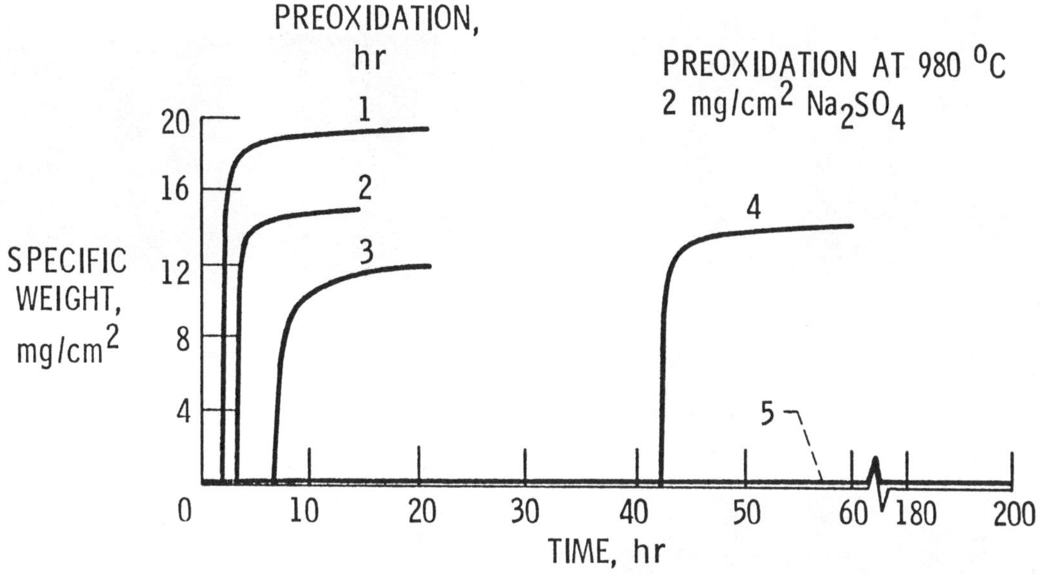

Fig. 15 Hot corrosion kinetics in 1 atm. O_2 as a function of the length of preoxidation.

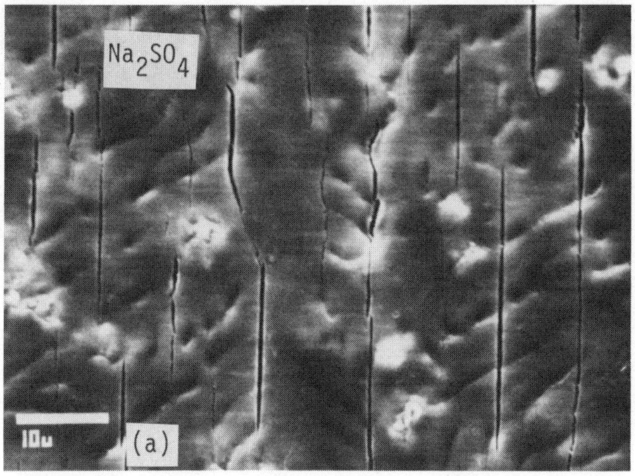

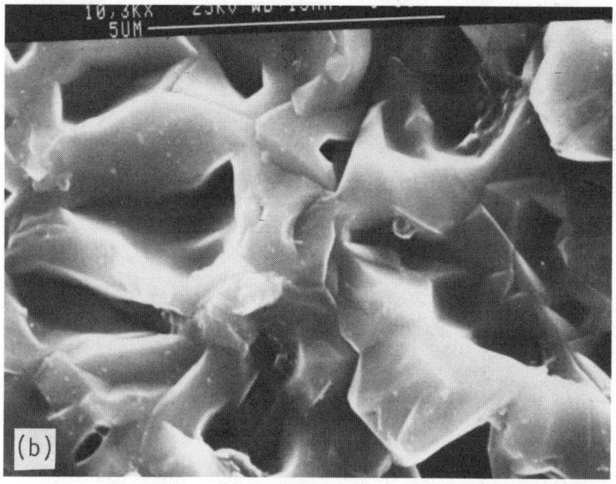

Fig. 16 Top surface during the induction period for the hot corrosion of preoxidized samples

a. top surface covered with Na_2SO_4
b. top surface after washing the sulfate layer with water for a sample that was preoxidized for 4 hours.
c. top surface after washing the sulfate layer in water for a sample that was preoxidized for 25 hours.

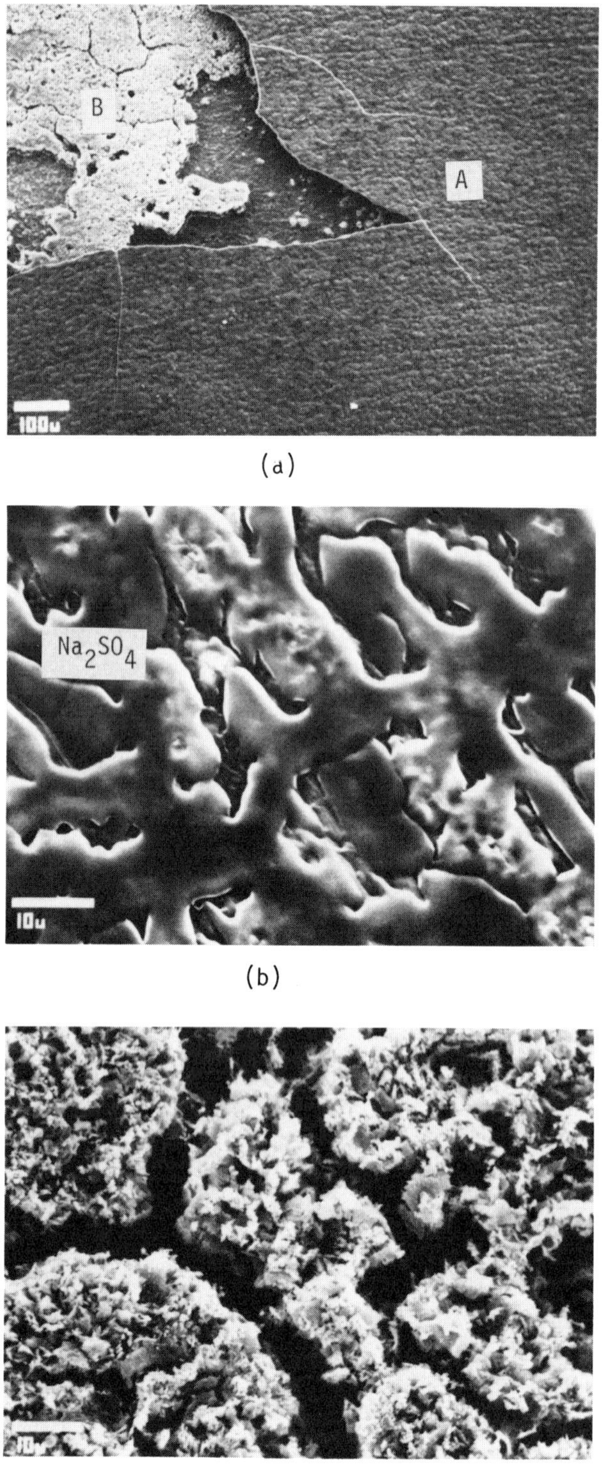

Fig. 17 Top surface of the preoxidized sample at the beginning of accelerated corrosion.

a. overall view of the surface
b. details of the region 'A'
c. details of the region 'B'

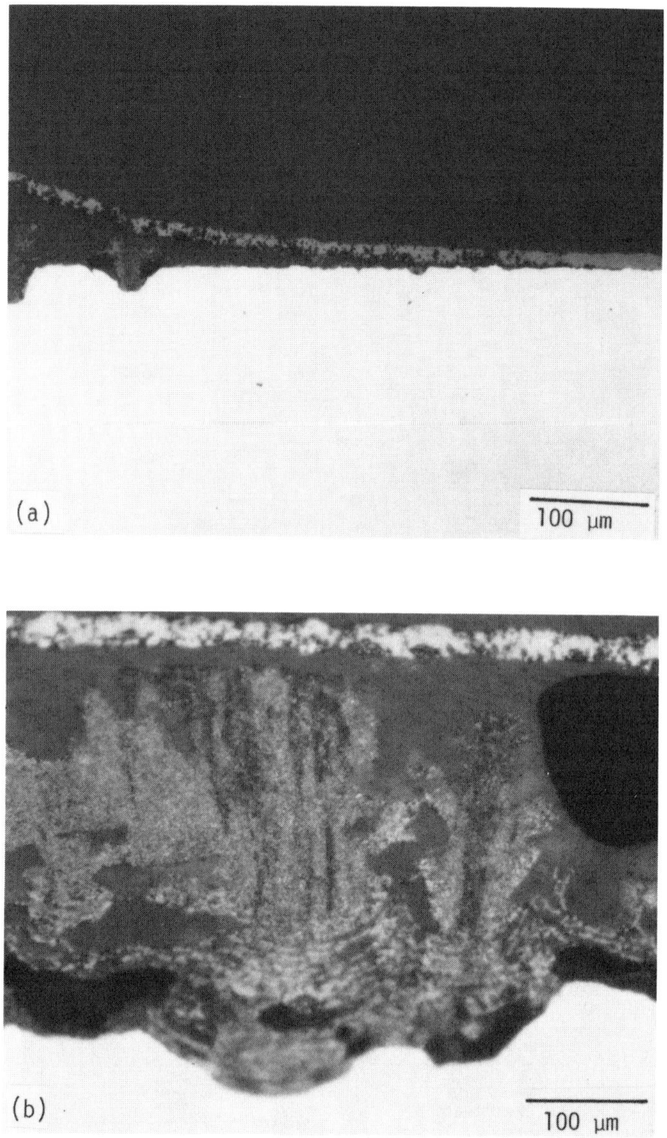

Fig. 18 Cross section during the period of rapid corrosion for the hot corrosion of preoxidized samples (a and b are at two different locations).

WEAR RESISTANT COATINGS AND SURFACE TREATMENTS

RELATIONSHIP BETWEEN METAL CUTTING PERFORMANCE AND MATERIAL PROPERTIES OF TiC COATED CEMENTED CARBIDE CUTTING TOOLS

Toshiaki Cho, Deepak G. Bhat, Paul F. Woerner
GTE Valeron Corporation
Troy, Michigan, USA

Abstract

The performance of TiC-coated cemented carbide cutting tool inserts was evaluated in steel turning and cast iron milling. The machining behavior was correlated with the properties of the coating and the chemical vapor deposition (CVD) parameters. It is shown that improved performance can be obtained by controlling the grain size and orientation of the coating, and by choosing appropriate parameters to minimize the formation of eta-phase in the substrate.

MORE THAN FIFTEEN YEARS have passed since the first TiC coated cemented carbide tools were introduced in the cutting tool market. Since then, significant improvements in coating techniques, adhesion and cutting performance have been achieved. At the same time, extensive studies of the properties of CVD coatings have been done, with the result that cemented carbides coated with TiN, Ti(C,N), Al_2O_3 and HfN have been put on the market. However, TiC is still one of the popular coatings for the cutting tool. This is due to the properties of TiC which has a higher hardness compared with other materials such as TiN, Al_2O_3 and HfN. Titanium carbide coating has also been used as an interlayer to improve adhesion between the Al_2O_3 coating and the cemented carbide. In spite of the importance of the CVD TiC coating, there is very little literature available concerning the relationship between the cutting performance and the metallurgical properties of CVD TiC coating on the cemented carbide. Therefore, the present study was undertaken to examine the interrelationship between the properties of TiC coating and the performance of the coated tools. The details of this relationship are discussed in this report.

EXPERIMENTAL PROCEDURE

CVD TiC coating was deposited in a graphite resistance furnace from $TiCl_4$-CH_4-H_2 mixtures under various conditions. The deposition process was performed at around 1050°C and atmospheric pressure.

The morphology and grain size of the coatings were examined by using the scanning electron microscope (SEM). The grain size is the size of crystallites on the surface of TiC. Coating thickness was measured by various techniques such as beta-ray backscatter, cross-section, and SEM. The porosity in the coating and thickness of eta-phase were measured on the polished cross-section of each sample. Knoop hardness measurements were taken on the cross-section of coatings. X-ray diffraction (XRD) was used for the phase indentification and measurement of orientation of TiC. The performance of TiC coating in machining was evaluated on TNG 333 or TNU 333 inserts of a steel cutting grade with an appropriate edge hone. All inserts for the machining tests were coated with a 5.0 micron (±1.0 micron) TiC coating. Two types of machining tests were performed: A facing test on alloy steel to examine the wear resistance, and a milling test to examine the adhesion between the TiC coating and the substrate.

The end point of tool life in the facing test was set at 0.015 inch flank or nose wear, and the number of cuts to reach the end point was counted. In the milling test, the spalled area of the coated layer was measured by the optical microscope after removing the adherent work material by a mild acid etch. The degree of spalling is represented as a percentage of the contact area on the rake face where coating is spalled. Typical rake faces of inserts after milling are shown in Figure 1.

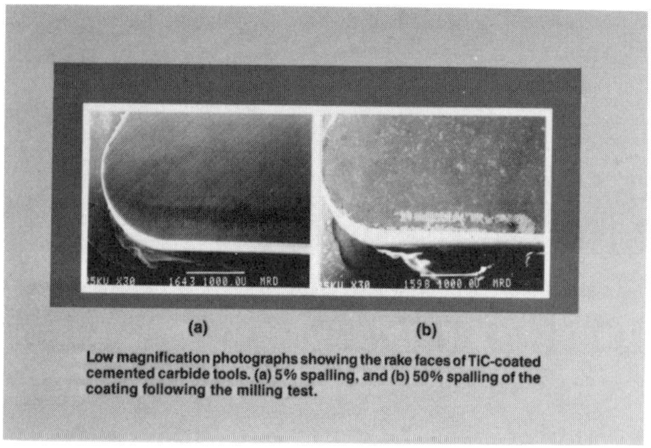

Fig. 1 - Low magnification photographs showing the rake faces of TiC-coated cemented carbide tools: (a) 5% spalling, and (b) 50% spalling of the coating following the milling test.

RESULTS AND DISCUSSION

Three major properties of the CVD TiC coating were found to have a pronounced effect on the cutting performance: (a) grain size, (b) thickness of eta-phase, and (c) orientation of TiC. These are discussed below.

EFFECT OF GRAIN SIZE - The grain size of deposited TiC was found to be the most important property related to wear resistance of CVD TiC coating. This relationship is shown in Figure 2.

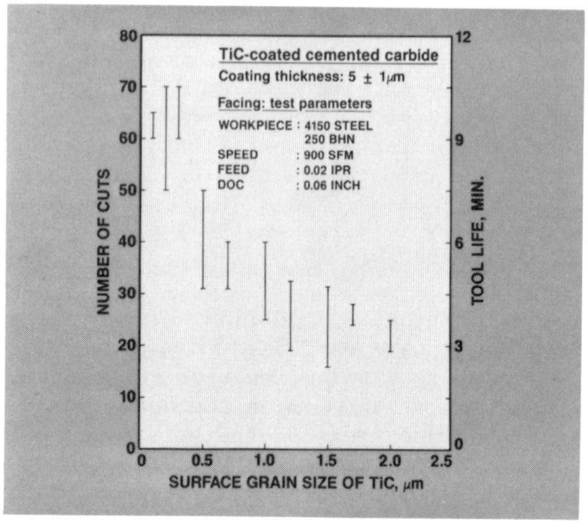

Fig. 2 - Effect of surface grain size of coating on the wear resistance of TiC-coated cemented carbide tool in machining of alloy steel.

It may be seen that the grain size of TiC has a strong effect on the wear resistance of coated inserts. Fine-grained TiC coated inserts showed better wear resistance than those having a relatively coarse grain size. However, it was observed that the improvements in wear resistance appeared to reach a maximum at about 0.3 micron grain size, and no further improvements were found at the lower values of grain size. Figure 3 shows the cross-section and surface of a fine-grained and a coarse-grained TiC coating.

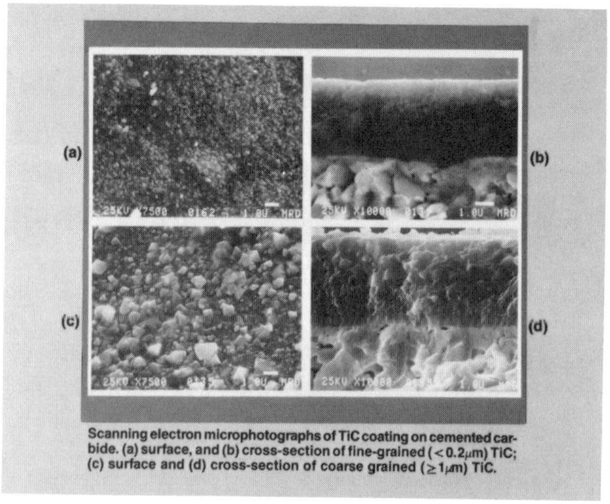

Fig. 3 - Scanning electron microphotographs of TiC coating on cemented carbide. (a) surface, and (b) cross section of fine-grained (<0.2 micron) TiC; (c) surface, and (d) cross section of coarse-grained (≥ 1 micron) TiC.

It was expected that a finer grain size would result in a higher microhardness of TiC. However, no major differences in microhardness were found between the fine and coarse-grained TiC coatings. The reasons for this are not clear at this time. We also did not find any correlation between the hardness and orientation of TiC coating. The microhardness of both coatings was measured to be around HKN_{10} = 2,600 kg/mm². This value is in good agreement with those obtained by other investigators [1].

EFFECT OF ETA-PHASE FORMATION - It is reported [2] that the existence of the eta-phase ($M_{12}C$ or M_6C) improves the wear resistance of coated inserts. However, no positive effect of the eta-phase on the wear resistance was observed in this experiment. It was observed that the TiC coated inserts with a heavy eta-phase (over 2.0 micron thickness) had a tendency to chip around the cutting edge during the facing test. The inserts with a heavy eta-phase were also found to have a higher spalling percentage in milling test on nodular iron, as shown in Figure 4. This indicates that the existence of the heavy eta-phase has a negative influence on the machining performance of TiC coated cutting tool.

The cross-sections of tested inserts in milling were examined to investigate the spalling mechanism. It was observed that the "spalled area" was fairly deep, and the weak point was not the interface between TiC coating and the eta-phase but the inter-region between the eta-phase and the cemented carbide substrate. The cross-section of a tested insert is shown in Figure 5.

It may be seen that the coated layer spalled at the interface between the eta-phase and the substrate, and not at the coating/substrate interface. This result indicates that the failure occurs by microchipping for the insert having a heavy eta-phase. This is probably because of the brittle nature of the eta-phase.

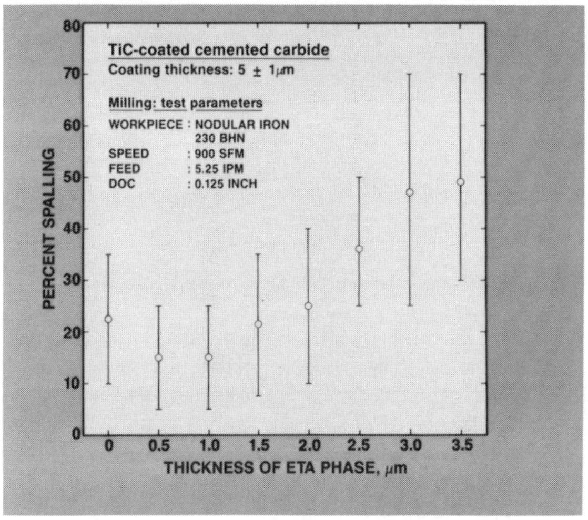

Fig. 4 - Effect of the thickness of eta-phase on the spalling resistance of TiC coating in the nodular iron milling test.

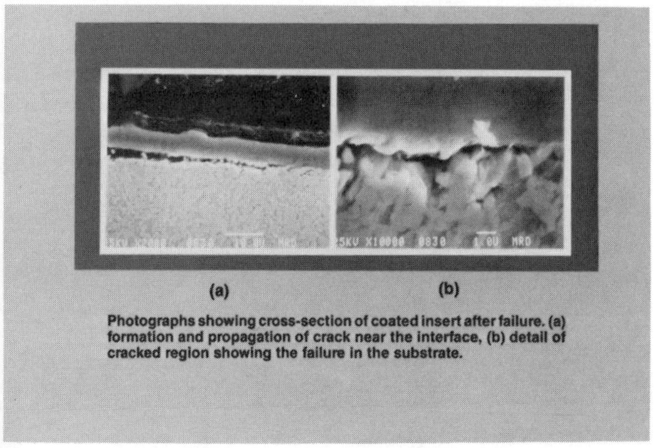

Fig. 5 - Photographs showing cross-section of coated insert after failure. (a) formation and propagation of crack near the interface, (b) detail of cracked region showing the failure in the substrate.

The existence of the eta-phase may suggest improved adhesion between the TiC coating and the cemented carbide [3], because the formation of eta-phase is caused by the diffusion of carbon from the substrate into the coating. However, for a coated cutting tool, the existence of a heavy eta-phase makes the tool life of the insert very short, due to chipping or breaking of the cutting edge. Similar observations are also reported by S. Murai, et al [4]. No significant differences in spalling in the milling test were found between TiC coated inserts with a medium eta-phase (1.0-2.0 microns) and those without the eta-phase formation. This indicates that the light eta-phase formation (less than 1.0 microns) has little influence on the cutting performance of TiC coated insert. This observation is in agreement with the findings of Sarin and Lindstrom [3].

EFFECT OF ORIENTATION - The effect of orientation on the machining performance of coated TiC was examined. No significant effect of orientation was observed for the fine-grained TiC coating in the facing test. This lack of correlation is not surprising, since the facing test primarily measures wear resistance which is related to hardness. As seen earlier, no differences were found in the hardness of TiC coating as a function of orientation or grain size. It was found that the orientation of TiC was closely related to the adhesion of TiC to cemented carbide. It is known that the degree of preferred orientation of CVD coating is dependent upon the coating conditions. The relationship between the orientation of CVD TiC coating on the cemented carbide and the cutting performance is not clearly established. In the present study, the orientation of coated TiC was found to be strongly related to the cutting performance in milling on nodular iron, which measures the adhesion of the coating. This relationship between the intensity of (111) and (200) orientations of coated TiC and the cutting performance in milling test is shown in Figure 6.

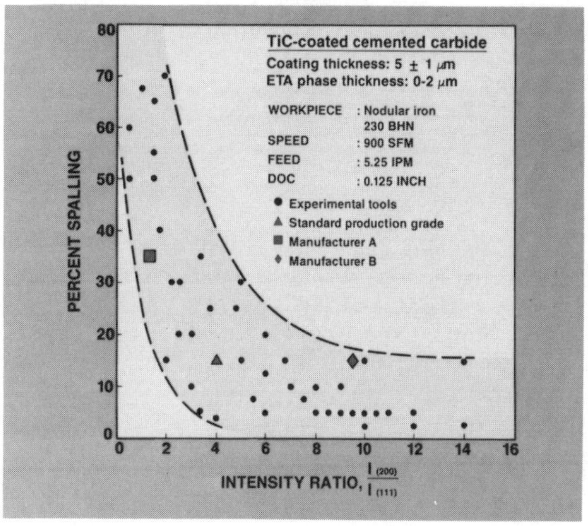

Fig. 6 - Effect of crystallographic orientation on the spalling resistance of TiC coating in the nodular iron milling test.

As far as we have been able to determine, this is the first published account linking the orientation of TiC coating directly to the performance of the coated tool. The results show that the TiC coating with a strong (200) preferred orientation shows stronger adhesion to the cemented carbide substrate, compared with a coating having no special orientation, or a weak (111) orientation. The reason for this apparent relationship is not clear. Further work is in progress to understand this phenomenon.

According to Lee, et al [1], (110) preferred orientation of CVD TiC can be obtained by using a higher deposition temperature. This effect was also observed in the present work, but no additional experiments were carried out in this direction because a fine-grained TiC is very difficult to deposit at high deposition temperatures. As mentioned earlier, the grain size of TiC is a very critical property for improving the wear resistance. The TiC coating with a strong (111) orientation was not tested, because the coating conditions to obtain a strong (111) orientation tended to cause a heavy eta-phase formation.

The spalling of coated inserts in milling is related to two factors: (a) presence of eta-phase, and (b) orientation of the coating. When a heavy eta-phase is present, the failure always occurs in the eta-phase/substrate region, irrespective of the orientation of TiC. When a light or no eta-phase is present, the failure is due to a loss of adhesion at the coating/substrate interface. It is in this situation that the orientation of TiC becomes very important. The results of the present tests indicate that a TiC coating with a strong (200) orientation gives a superior spalling resistance to the (111) orientation. Thus, even though interdiffusion at the interface may permit better bonding between the coating and the substrate, attempts to cause diffusion will frequently result in the formation of heavy eta-phase, along with the Kirkendall porosity [3]. Therefore, promoting a (200) preferred orientation by choosing appropriate CVD parameters is found to be preferable for improved performance.

CONCLUSIONS

1. It was observed that the grain size of CVD TiC coating has a strong effect on the wear resistance. TiC coated inserts consisting of fine, equi-axed grains showed excellent machining performance in the facing test on alloy steel.

2. The eta-phase formation and the preferred orientation of CVD TiC were found to be the key properties related to the adhesion of coating as evaluated in milling test on nodular iron.

3. It is suggested that a CVD TiC coating which consists of fine, equi-axed grains and has a strong (200) preferred orientation, with a minimum of eta-phase is most desirable for the cemented carbide cutting tool. This can be achieved by an appropriate selection of key CVD parameters.

ACKNOWLEDGEMENTS

The authors express their appreciation to Michael Shippell for CVD experiments, Mark Jagner and Dr. Joseph Keeley for their assistance in the microstructural evaluations, Prasad Boppana for conducting the machining tests, and Dennis Rich and Dhiraj Shah for helful discussions and suggestions.

REFERENCES

1. Lee, C. W., Nam, S. W. and J. S. Chun, J. Vac. Sci. Technol. 21(1), 42-46, (May/June 1982).

2. Sproul, W. D. and M. H. Richman, Met. Technol. 3, 489-493 (1976).

3. Sarin, V. K. and J. N. Lindstrom, pp.389-402, Proceedings of the Sixth International Conference on Chemical Vapor Deposition, L. F. Donaghey, P. Rai-Choudhury and R. N. Tauber (Editors), The Electrochemical Society, Princeton, New Jersey (1977).

4. Murai, S., Nishiyama, A. and Y. Sugisawa, p.62, Summary Report of the Fall Meeting of the Japan Society of Powder Metallurgy (1980).

MACHINING FERROUS MATERIALS WITH CVD COATED CARBIDES. PART 1: INTERFACIAL CONDITIONS

P. A. Dearnley
W.I.S.E., Dept. Metallurgy & Materials
University of Birmingham
Birmingham, England

V. Thompson, A. N. Grearson
Sandvik Hard Materials Ltd.
Research Centre
Coventry, England

ABSTRACT

In order to interpret the wear surfaces of CVD coated and uncoated cemented carbides, it is important to appreciate the nature of the interfaces formed between the tool, the forming chip and the newly created workpiece surface. To elucidate this a series of quick-stop experiments and tool temperature estimates were made with coated and uncoated cemented carbide tools. Three distinct interfacial zones were observed: a) the main chip/tool interface; b) the peripheral chip/tool contact at the outer edges of the chip underside; c) the flank/workpiece interface. These interfaces were essentially similar for CVD coated and uncoated carbides, but the strength of bonding across the main chip/tool interface was greater for the uncoated carbides, which led to a larger chip/tool contact length and higher rake face temperatures for these tools when cutting steel.

CEMENTED CARBIDES COATED WITH single (Fig. 1) or composite layers (~5 to 10 µm thick) of TiC, TiN and/or Al_2O_3 represent approximately 70% of all the coated carbides used in metal cutting. The coatings are applied by Chemical Vapour Deposition (CVD) at temperatures ~1000°C and are, in general, more adherent and reliable than coatings applied by sputtering or ion plating.

It is well known that the wear rates of coated cemented carbides are often substantially less than those of uncoated carbides, when used in many steel and cast-iron cutting operations. However, the extent of this improvement depends upon the cutting conditions selected and the precise workpiece being machined, and as shown in Part 2, there is no single coating material which resists all forms of wear better than others. In order to understand this fact and account for the observed wear rate rankings and wear patterns described in Part 2, it is essential to evaluate the conditions pertaining at the interface formed between the tool, the forming chip and the adjacent workpiece new surface. Hence, in Part 1, evidence is presented concerning the interfacial contact conditions and the distribution of temperature within coated and uncoated cemented carbide tools used for cutting steels and cast-irons.

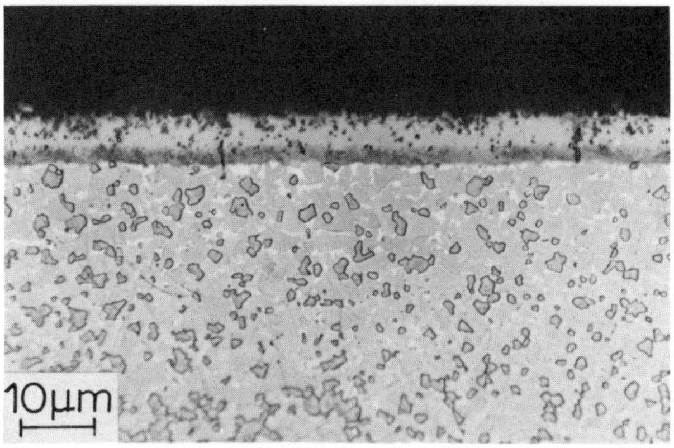

Fig. 1 - Metallographic section of TiC coated cemented carbide (type 1).

EXPERIMENTAL

TOOL MATERIALS - The composition and hardness of the uncoated carbides and the and the substrates of the coated carbides, used in this investigation, are given in Table 1, together with coating composition and thickness. For comparison purposes, a few Al_2O_3/ZrO_2 ceramics were also included in the investigations described in Parts 1

Table 1 Tool Designation, Composition and Hardness

Tool Type	Coating Composition (where applicable) or Tool Material	Nominal Coating Thickness, μm	Nominal Composition, wt% WC	Co	TiC	TaC	NbC	Mean Hardness (substrates of Coated Carbides), HV 30 kg
1	TiC	5-8	85.9	5.5	2.6	3.6	2.4	1550
2	TiN (exterior) TiC (interior)	1 5-8	85.9	5.5	2.6	3.6	2.4	1550
3	TiN (exterior) Ti(C,N) (inter mediate) TiC	5-6 2 1.5	84.0	6.0	5.0	5.0 (TaC + NbC)		1630
4	TiN	11-12	52.8	10.5	18.9	16.5	1.3	1448
5	$\alpha+\kappa Al_2O_3$ (exterior) TiC (interior)	1 5-8	85.9	5.5	2.6	3.6	2.4	1550
6	α-Al_2O_3 (exterior) Cubic Carbide (interior)	8 3-5	85.7	5.9	2.5	5.8	0.1	1550
7	TiN (exterior) Al_2O_3 (intermediate) TiC (interior)	<1 3-4 3-4	85.4	6.3	2.6	3.6	2.1	1591
8	Uncoated Carbide	...	94.0	6.0	...	...	...	1578
9	Uncoated Carbide	...	85.9	5.5	2.6	3.6	2.4	1509
10	Uncoated Carbide	...	55.5	9.5	19.0	12.2	3.8	1533
			Nominal Composition, wt% Al_2O_3	ZrO_2				
11	Al_2O_3/ZrO_2 Ceramic		85.0	15.0				1541

and 2.

The grain size and morphology of the CVD coatings was determined by SEM of fracture and etched surfaces, optical microscopy and TEM studies of foils (Refs 1 and 2). The TiC coatings of types 1 and 2 comprised equiaxed grains, mostly about 0.8 μm across, in the bulk of the coating, but towards the coat/substrate interface grains approximately 0.2 μm across were observed. The TiN coatings of tool types 2 and 3 were equiaxed (~0.5 μm) and columnar (~1.0 x 4.0 μm) respectively, while the Al_2O_3 coatings of tool types 5 and 6 were acicular and (0.1 x 0.6μm) and equiaxed (~0.5 μm), respectively.

Estimates of the preferred orientation of the coatings, attained from relative intensity values, during X-ray diffractometer measurement, showed that the TiC coating of tool types 1, 2 and 5 were near randomly oriented, whereas, the Al_2O_3 coat of type 6 and the TiN of tool type 3 had a majority of grains with {0001} and {220} planes, respectively, lying parallel to the substrate surface.

CVD coatings faithfully follow the topographical irregularites of the substrate, which in general constitute an appreciable proportion of the coating thickness. In the case of the coated carbides used in this investigation, the uneveness took the form of a three dimensional array of "hills" and "valleys". The hills were ~30-100 μm apart and the average vertical distance between each hill and valley was ~1-3 μm, Fig. 2. Superimposed upon these features was a finer network of surface asperities, which varied according to the type of coated carbide. For types 1, 2 and 5 the asperities were hemispherical (Fig. 2) and about 1 μm across, while for type 3, faceted asperities were observed and irregularly shaped spheres, also about 1 μm across were seen on tool types 4 and 6.

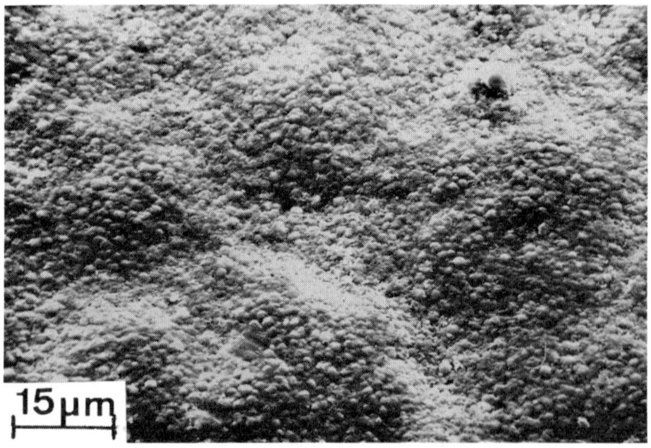

Fig. 2 - Topography of TiN/TiC coated carbide (type 2), SEM.

Table 2 Composition and Hardness of Workpiece Materials.

Work-Material	Mean Hardness HV (20 kg)	Composition, wt% C	Si	Mn	Ni	Cr	Mo	S	P	Mg	Ce
080M40	177	0.42	0.29	0.82	0.05	0.04	0.01	0.03	0.021	...	...
817M40	326	0.39	0.30	0.53	1.38	0.99	0.24	0.04	0.014	...	...
GA-iron	177	2.00	1.03	0.56	0.04	0.06	<0.001	0.065	0.035	<0.001	<0.001
SG-iron	235	2.00	1.43	0.37	0.02	0.04	<0.001	0.003	0.059	0.006	0.001

WORKPIECE MATERIALS - Tool temperature and quick-stop experiments were made on SG and GA-cast irons and 080M40 and 817M40 steels. The 080M40 steel was annealed and the 817M40 was hardened and tempered. The cast-iron workpieces were sand cast ingots, comprising pearlitic matrixes with spheroidal (SG) and flake (GA) graphite morphologies. The compositions and hardnesses of the workpieces are shown in Table 2. All of the workpieces were ~150 mm dia. and ~0.8 m long.

QUICK-STOP CUTTING TESTS - Quick-stop tests were made using DSG and VDF manual lathes. Initial experiments were made using a device described by Williams et al (Ref 3). Basically the device consists of a pivoted tool holder which rests upon a shear pin during the cutting operation. The cutting process is then arrested by firing the "captive bolt" of a humane killer gun, onto the top face of the tool holder, breaking the shear pin, and propelling the tool away from the cutting zone. To ensure clean separation of the tool from the chip and workpiece shoulder, initial experiments were conducted with the device mounted on a 10° block. However, later experiments were conducted with a redesigned device, having the tool holder pivot inclined at 45°, which had the same effect. High speed cine made of the separation event, showed that the device could be used up to 600 m/min, without the tool fouling the chip (Ref 4).

TOOL TEMPERATURE ESTIMATION TECHNIQUE - Tool temperature maps were derived from tools with silicon and iron-silicon binders (Ref. 5). Each tool material has a characteristic austenetisation temperature, which is determined from calibration experiments using a rapid quench furnace facility. During metal cutting sufficient heat is frequently generated to cause partial austenitisation of the binder, which upon withdrawal of the tool from the cut, forms a transformation product. The heat affected zone (HAZ) is deliniated by etching metallographic sections (taken normal to the rake face of the tool, in the position of greatest wear) in 0.25% nital. The boundary of the HAZ corresponds to the pre-determined austenitisation temperature, and, can be considered as an isotherm. Hence, by using several tool materials (at the same cutting condition) with differing austenitising temperatures, a tool temperature map can be constructed. The isotherms are within ± 13°C.

Some of the iron bonded tools were coated with TiN, by the "Sputter Ion Plating" technique (Ref. 6), so that the effect of coating on tool temperature could be determined. All the coatings remained adherent during the tests.

CUTTING CONDITIONS- Quick-stop tests were made using positive and negative rake angle tools, while tool temperaures were only estimated for negative rake angle tools (see Table 2 in Part 2).

For the cast-irons, quick-stops were made at 60 m/min and above, while for the steels, quick-stops were made at 180 m/min and above. A depth of cut of 2.0 and 1.3 mm, was used respectively and a constant feed rate of 0.25 mm/rev. Identical conditions were used for tool temperature estimates, although only one speed was selected when cutting the cast-irons (200 m/min).

RESULTS

QUICK-STOP EXPERIMENTS - Steels- During quick-stopping, the coated carbide and ceramic tools always separated cleanly along the chip/tool interface, such that most topographical features of the rake face contact zone was replicated in the chip underside (Fig. 3a,b). The cutting edge or edge chamfer of these tools were also replicated. Little, if any, work material, remained adherent to the rake and flank faces of the tool after quick-stopping. At the outer sides of the chip underside, replication of the tool rake face desisted and shallow grooves, parallel to the direction of chip travel were visible. This region, is shown in Fig. 4, zone E. The replicated tool surface, zone A, is shown on the right hand side of the micrograph.

No replication of the flank face was apparent in the workpiece surface, adjacent to the replicated cutting edge, Fig 3b, even after quick-stopping tools with a 0.10 mm flank wear land. Instead, shallow grooves, parallel to the direction of workpiece rotation, and small near circ-

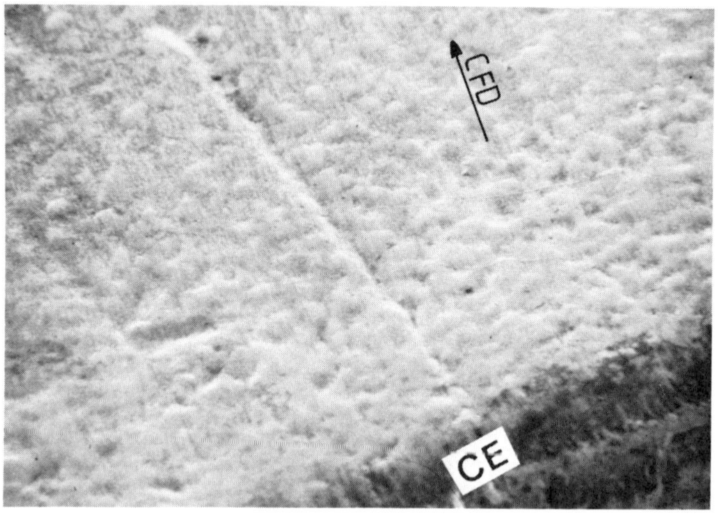

(a)

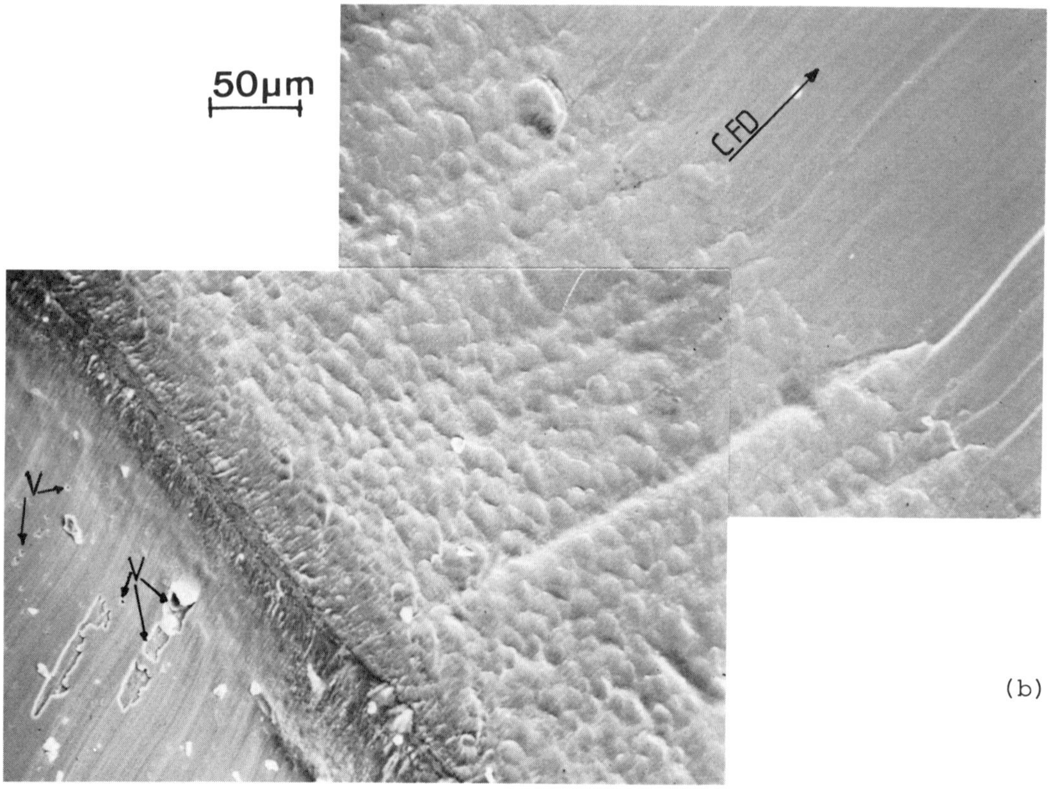

(b)

Fig. 3 Quick-stop surfaces after cutting 817M40 at 180 m/min for 30 secs with TiC coated carbide (type 1):
(a) Tool rake face and cutting edge, showing "hills" and "valleys" and a coated grinding mark. Note absence of work material debris.
(b) Corresponding chip underside, showing replication of tool rake face topography, and adjacent smooth workpiece surface containing voids (arrowed).
CFD = Chip Flow Direction; CE = Cutting Edge

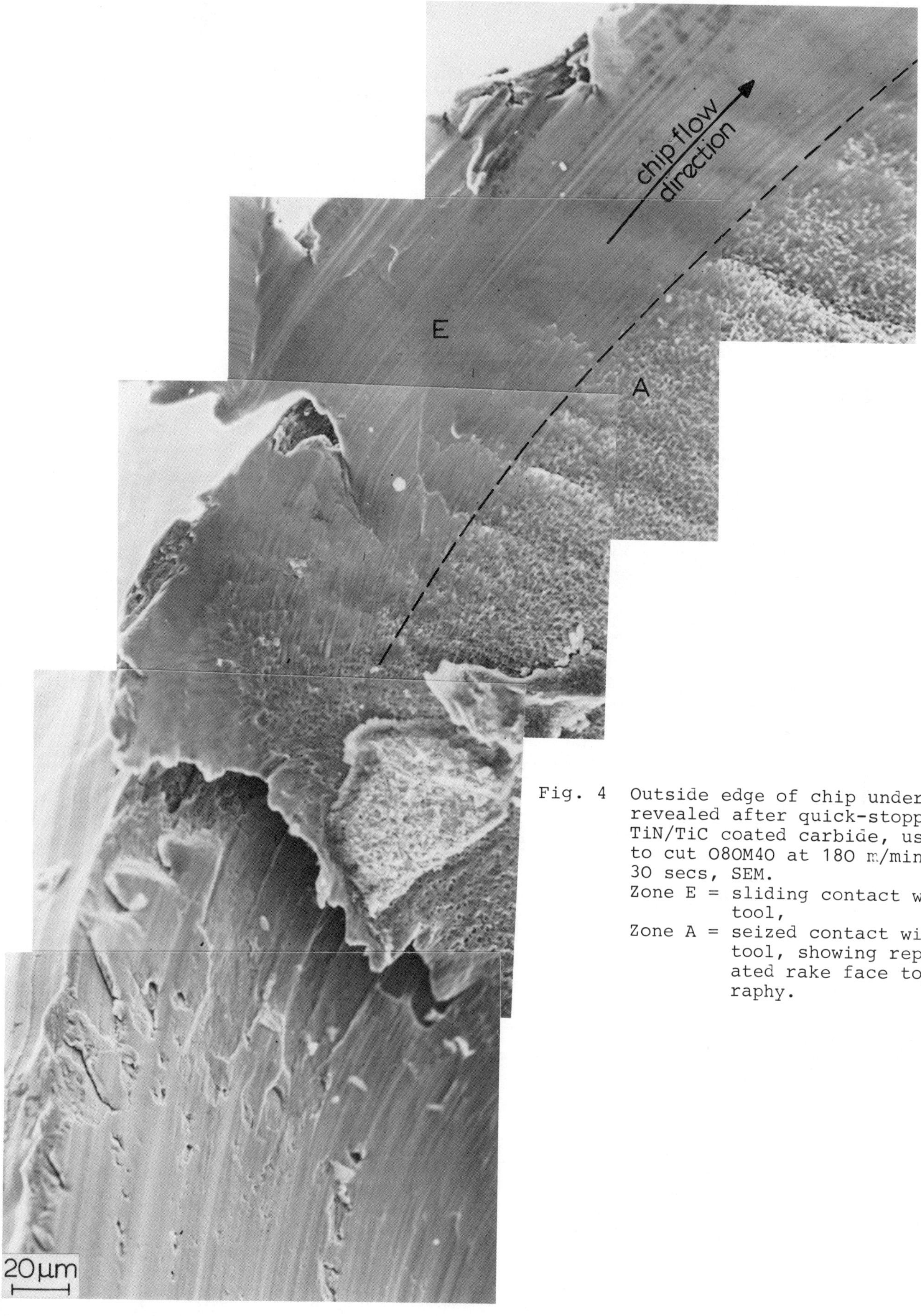

Fig. 4 Outside edge of chip underside revealed after quick-stopping TiN/TiC coated carbide, used to cut 080M40 at 180 m/min for 30 secs, SEM.
Zone E = sliding contact with tool,
Zone A = seized contact with tool, showing replicated rake face topography.

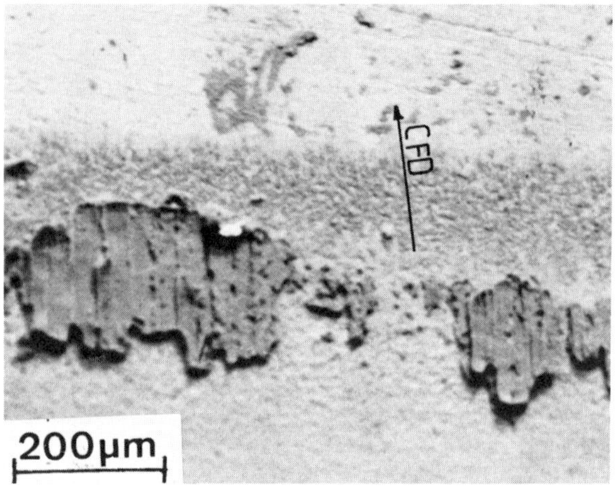

(a)

CFD = Chip Flow Direction

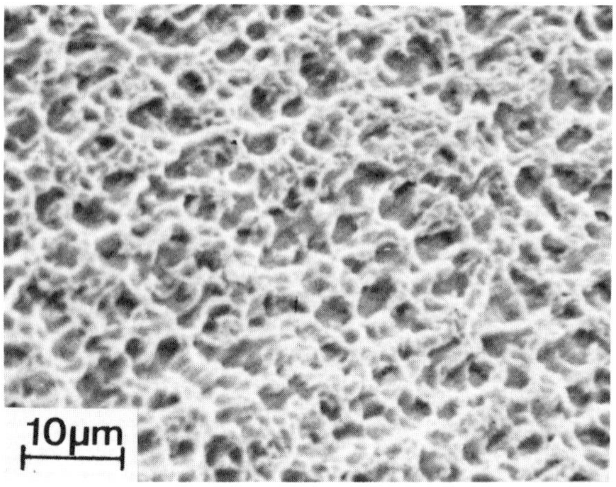

(b)

Fig. 5 SEM's of quick-stop surfaces after cutting 080M40 at 180 m/min with uncoated carbide tool (type 9):
(a) view of rake face, showing adherent work material debris
(b) view of chip underside, showing ductile "dimple-like" fracture surface.

ular voids, ~1-4 µm wide were visible. Also larger void-like features, ~1µm deep and 8 µm wide, extending up to 100 µm in length, in a direction parallel with the direction of rotation of the workpiece, were visible (arrowed in Fig. 3b).

For the uncoated carbides, separation during quick-stopping, took place in a plane above the chip/tool interface. Consequently, fragments of chip material remained adherent to the tool (Fig. 5a) and the chip underside had a dimpled appearance (Fig. 5b), indicitive of ductile fracture. The latter features extended to where the cutting edge had been during the cutting operation. When viewed in cross-section it was possible to see where fragments of material had been torn from the chip underside, and, in some cases, bonding was so strong at the chip/tool interface, that the tool fractured and lodged in the chip root (Fig. 6a). At the outer edges of the chip underside, the dimple-like features (Fig. 5b), gave way to smoother surfaces, lightly grooved in the direction of chip travel, and had an identical appearance to those observed after quick-stopping the coated carbide tools (i.e., zone E in Fig. 4).

Examination of the workpiece surface which had been in contact with the flank face of the uncoated carbides during the cutting process, showed no evidence of a definite contact zone.

Metallographic sections of the quick stops made with ceramic, coated and uncoated cemented carbide tools, revealed a continuous chip form (Figs 6 and 7), with a well defined secondary shear zone or flow zone, next to the chip underside, which merged with a thinner shear zone, lying adjacent to the workpiece new surface (Fig. 7b).

QUICK-STOP EXPERIMENTS - Cast-Irons- Quick-stops made at 60 and 200 m/min demonstrated that the general chip form was segmented or discontinuous, when cutting both GA- and SG-irons, irrespective of the tool material used, Fig. 8a,b. At 60 m/min, the chip was associated with a built-up-edge (BUE), Fig. 9. The size of the BUE was much reduced for the coated carbides, when cutting GA-iron (Fig. 10a,b) while there was no significant difference in BUE size for either type of tool when cutting SG-iron (Fig. 10c,d).

At 200 m/min, a small BUE was still visible on the rake faces of the ceramic and, coated and uncoated cemented carbide tools used to cut SG-iron (Fig. 11),while the same tool materials, quick-stopped after cutting GA-iron, showed no adherent BUE (Fig. 12a). However, the rake and flank faces of the coated carbides and ceramics in particular, were covered by a thin (~1 µm thick) adherent layer, which X-ray energy dispersive mapping revealed to be enriched in Si, Mn and Al (Fig. 12). Ca was also sometimes detected. Such layers were insoluble in 50% HCl suggesting they were of an oxide nature. Whether cutting GA- or SG-iron at 200 m/min, the pearlite lamallae of the matrix, although plastically deformed, could still be observed in many regions adjacent to the chip underside, Fig. 13, and bore no resemblance to the highly strained secondary shear zone, observed in the same region of the steel chips.

TOOL TEMPERATURES - When cutting steel the size of the heat affected zone (HAZ) was larger for the uncoated carbide than for the TiN coated carbide tools (Figs 14 and 15). The magnitude of the maximum rake face temperature and its po-

(a) (b)

Fig. 6 Metallographic sections of 817M40 chip/workpiece, obtained at 180 m/min after quick-stopping:
(a) uncoated carbide tool (type 9), (b) TiC coated carbide (type 1)

sition, relative to the cutting edge, was also greater for the uncoated carbides. However, when cutting GA- or SG-irons there was little difference in the size of the HAZ developed in the uncoated and TiN coated carbides, and the position of the highest isotherm was similar (Fig 16).

Whether cutting cast-iron or steel the size of the HAZ increased with increasing cutting speed.

Although no tests were made with Al_2O_3 or TiC coated carbide tools in the present test series, earlier tests (Refs 1 and 2), with CVD coated iron bonded carbides, did show that the size of the HAZ, when cutting 080M40 steel, was about the same for TiC and Al_2O_3/TiC coated tools. Hence, it appears probable that the temperature distributions for Al_2O_3 and TiC coated tools would not differ significantly from those shown for the TiN coated carbides (Figs 14b, 15b, 16b,d).

It is not yet possible to measure the temperature distribution in Al_2O_3/ZrO_2 ceramic cutting tools. However, experiments with edge chamfered uncoated carbides, used to cut 080M40 steel, have shown that the size of the HAZ and the magnitude of the maximum rake face temperature, is much larger than when using sharp edged tools (Ref. 7). Hence, the edge chamfer causes more heat to be created during chip formation. It is therefore probable that the maximum rake face temperature maybe higher for the Al_2O_3/ZrO_2 ceramic tools than for the sharp edged uncoated cemented carbides, even though the distribution of temperature within the ceramic tools will probably differ, in view of their relatively lower thermal diffusivity.

DISCUSSION

CHIP FORMATION - Cutting steels above 150 m/min, produces a continuous chip form (Figs 6 and 7), whereas cutting GA- and SG-irons at 60m/min produces a segmented chip (Fig. 8). Both situations are schematically depicted in Figs 17 and 18. At 60 m/min, the segemented chip form of the cast-irons was associated with a built-up-edge (BUE), Fig. 18a, whereas, at 200 m/min a small BUE was prevalent only when cutting SG-iron (Fig 18b) and not when cutting GA-iron (Fig. 18c).

Both steel and cast-iron chips are formed by plastic deformation in two shear zones: a) along the primary shear plane (A in Fig 17 and GH in Fig. 18), and, b) within the secondary shear zone adjacent to the chip underside (region B in Fig. 17). When cutting steel the latter zone is typically ~10-20 μm thick. Secondary shear is more complicated when cutting cast-iron at 60 m/min, since a large BUE is formed on the tool (Fig. 9). Such BUE's are commonly observed when cutting duplex phased materials at low cutting speeds (Ref. 8), and comprise highly strain hardened work material (Ref. 9). The occurance of the BUE at

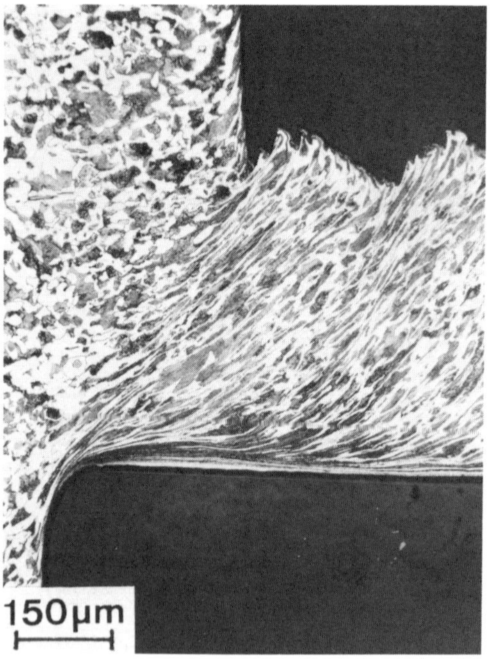

(a)

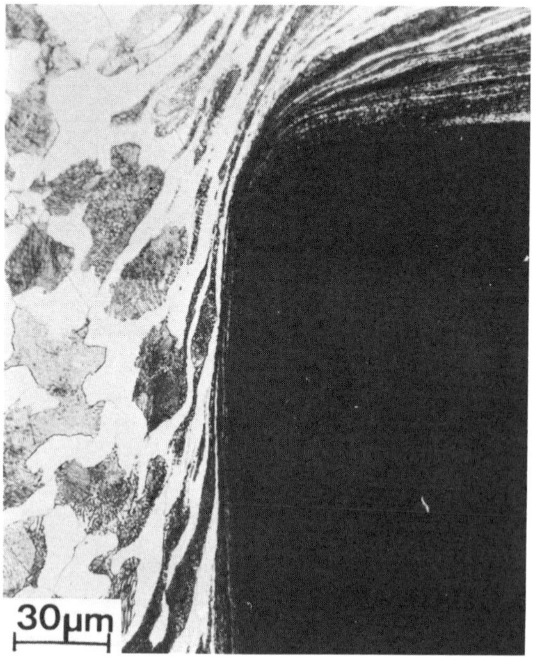

(b)

Fig. 7 Metallographic sections of 080M40 chip/workpiece, obtained at 180 m/min after quick-stopping:
a) TiN/TiC coated carbide (type 2), general view of chip form,
b) Al_2O_3/TiC coated carbide (type 5), detail showing blending of secondary shear zone of chip (top right) with shear zone adjacent to workpiece surface (oriented vertically).

200 m/min when cutting SG-iron, is not entirely unexpected, since others have observed BUE at relatively high speeds when cutting cast-irons (Refs 8 and 9).

Depending upon the size of the BUE secondary shear maybe continuous or discontinuous. When the BUE is large (Fig 18a) secondary shear is accomplished over the BUE (between positions G and E in Fig 18a) and then resumed, perhaps intermittently, between positions D and L. However, at higher cutting speeds, when the BUE is small (in the case of SG-iron - Fig. 18b), secondary shear is practically continuous between positions G, C and L, while when the BUE is absent (as for GA-iron at 200 m/min) secondary shear is continuous (along BL in Fig. 18c).

The segmentation of the chip when cutting the cast-irons, appears to be caused by fracture during primary shear (GH in Fig. 19), probably initiated in the vicinity of the graphite/pearlite matrix interface. This fracture process seems to happen after the pearlitic matrix has only been lightly deformed. Hence, primary shear is restricted. In the case of SG-iron, however, fracture is probably not so quickly initiated, since substantial deformation of the graphite nodules takes place prior to fracture (Fig. 8a). There was less evidence of deformation of the graphite flakes (Fig. 8b).

The reason for the reduced BUE size when cutting SG-iron at 200 m/min, and its absence when cutting GA-iron, is that the higher temperatures generated during chip formation at the higher speed cause the severely strain hardened BUE structure to undergo recovery and recrystallisation processes, softening the BUE and causing it to collapse, i.e., the BUE is no longer able to support the high compressive forces created during primary shear. It remains unclear why no small BUE was observed when cutting GA-iron at 200 m/min. This may have been partly due to the formation of non-metallic layers on the rake face of the tool, as shown in Fig. 12, i.e., they may have reduced the strength of bonding between the BUE and the rake face. It is also a possibility that recovery/recrystallisation processes were more rapid in the GA-iron BUE than in the SG-iron BUE (Ref. 10).

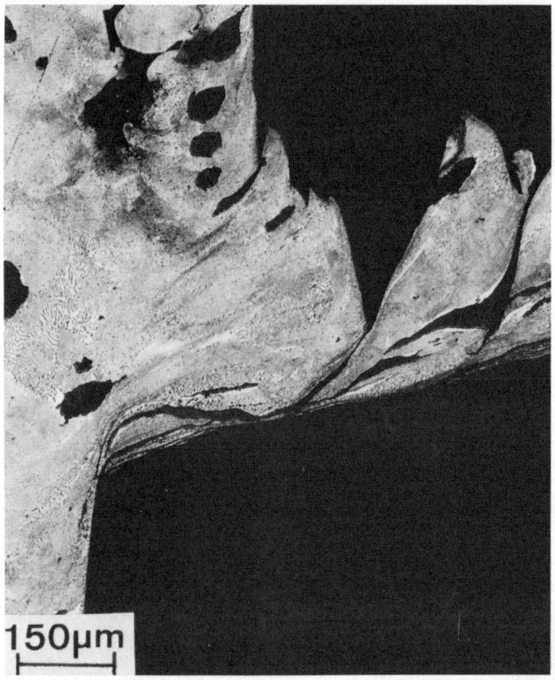

(a)

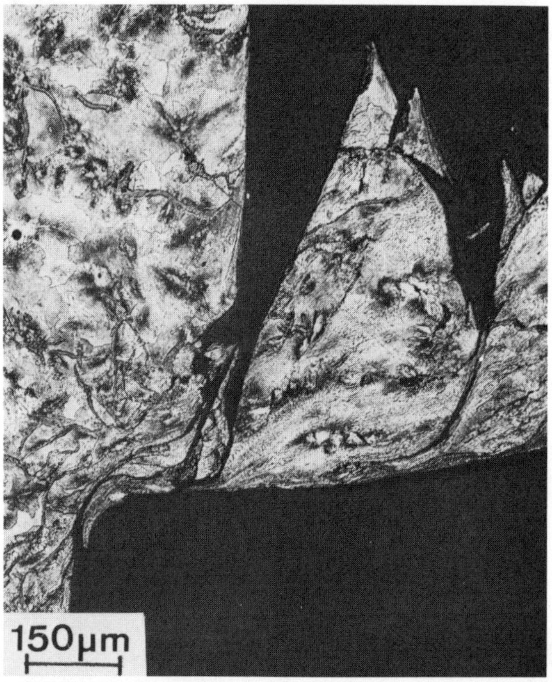

(b)

Fig. 8 Metallographic sections of chip/workpiece, obtained at 60 m/min after quick-stopping:
(a) uncoated carbide (type 8) used to cut SG-iron, note BUE at chip root,
(b) TiC coated carbide (type 1) used to cutGA-iron, note small BUE at chip root.

THE CHIP/TOOL INTERFACE AND THE ORIJIN OF TOOL TEMPERATURES- It has been claimed by some workers (Ref 11) that interfacial sliding may occur in the main part of the chip/tool interface, especially near the cutting edge. The quick-stop evidence presented here, however, does not support this hypothesis. Instead they show that the majority of the rake face contact area is in intimate contact with the chip underside, as shown by the replication of the coated carbide and ceramic tool rake face surfaces in the chip underside (Fig. 3). The cutting edge is also replicated, and, in the case of the uncoated carbide tools, bonding between the chip underside and the tool rake face is so strong, in the vicinity of the cutting edge,that the tool often fractures (Fig 6a). However, at the outer edges of the chip underside, interfacial contact is obviously very different, since the tool rake face is no longer replicated (zone E in Fig. 4) and this portion of the chip underside is relatively smooth, containing numerous shallow grooves (parallel to the direction of chip travel), which are suggestive of localised interfacial sliding. This region is very narrow, ~80 μm wide, and represents only a small part of the contact made by the chip with the tool. Sliding in this zone is also coroborated, by the fact that there is no observable secondary shear adjacent to the chip underside, as shown in Fig. 19.

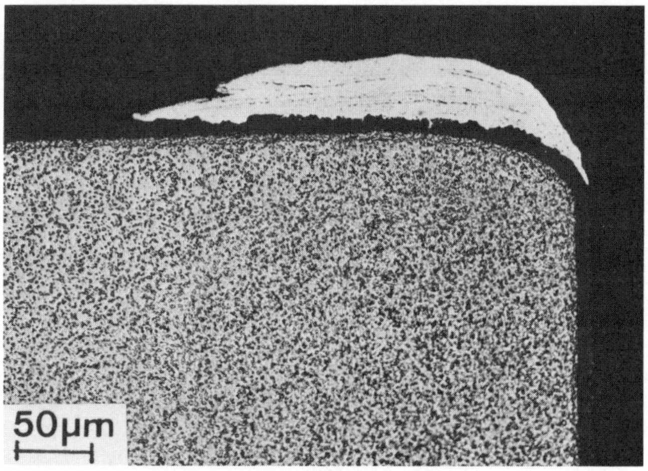

Fig. 9 Adherent BUE on Al_2O_3 coated carbide (type 6). Tool quick-stopped after cutting GA-iron at 60 m/min for 30 secs. Note Al_2O_3 coating appears black.

(a) (b) (c) (d)

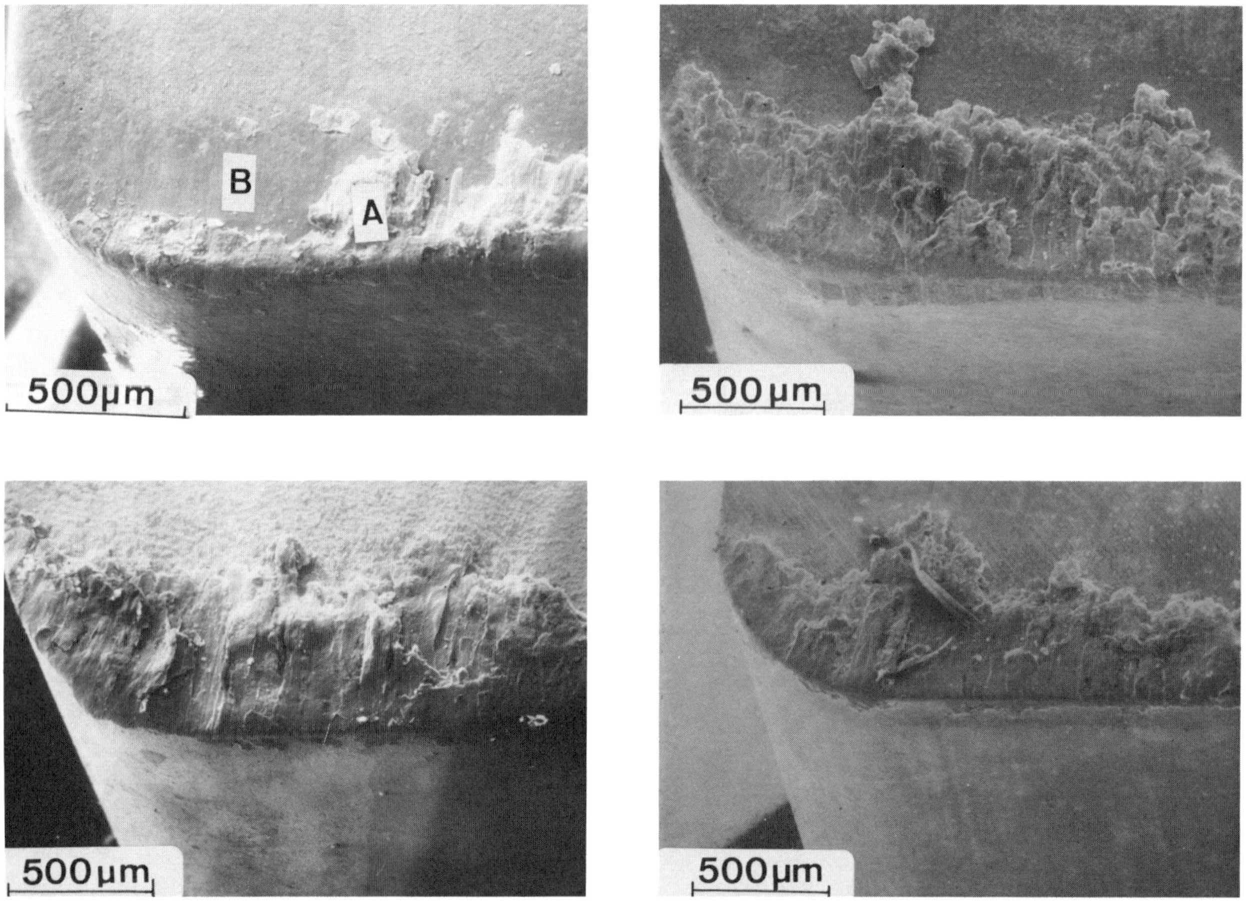

Fig. 10 Cutting tools, quick-stopped at 60 m/min after 30 secs, SEM's: (a) TiC coated carbide (type 1) after cutting GA-iron; A = adherent BUE B = lightly worn coating; (b) Uncoated carbide (type 8) after cutting GA-iron; (c) TiC coated carbide (type 1) after cutting SG-iron; (d) Uncoated carbide (type 8) after cutting SG-iron.

It is proposed therefore, that when cutting steel above 150 m/min, and GA- and SG-irons at 200 m/min, the majority of the chip underside is strongly bonded to the rake face of the tool, which forces the chip to shear internally, within the narrow secondary shear zone mentioned above. Since this zone is only ~10 to 20 µm thick, the shear strains and strain rates are very high, being in the order of 50 and 10^4-10^5 s^{-1} respectively (Ref 5). The heat created during this intense plastic deformation is so large that it frequently raises the temperature of the tools above 900°C (Figs 14 and 15).

The magnitude of the maximum rake face temperature and the size of the heat affected zone (HAZ) was greater for the uncoated carbides than for the TiN coated carbides, when cutting steel (Figs 14 and 15). There was, however, no appreciable difference in these parameters, between the two types of tool, when cutting either GA- or SG-irons (Fig. 16).

Fig. 11 Al_2O_3 coated carbide tool (type 6) quick-stopped after cutting SG-iron at 200 m/min. Note adherent BUE (white).

(a) (b) (c) (d)

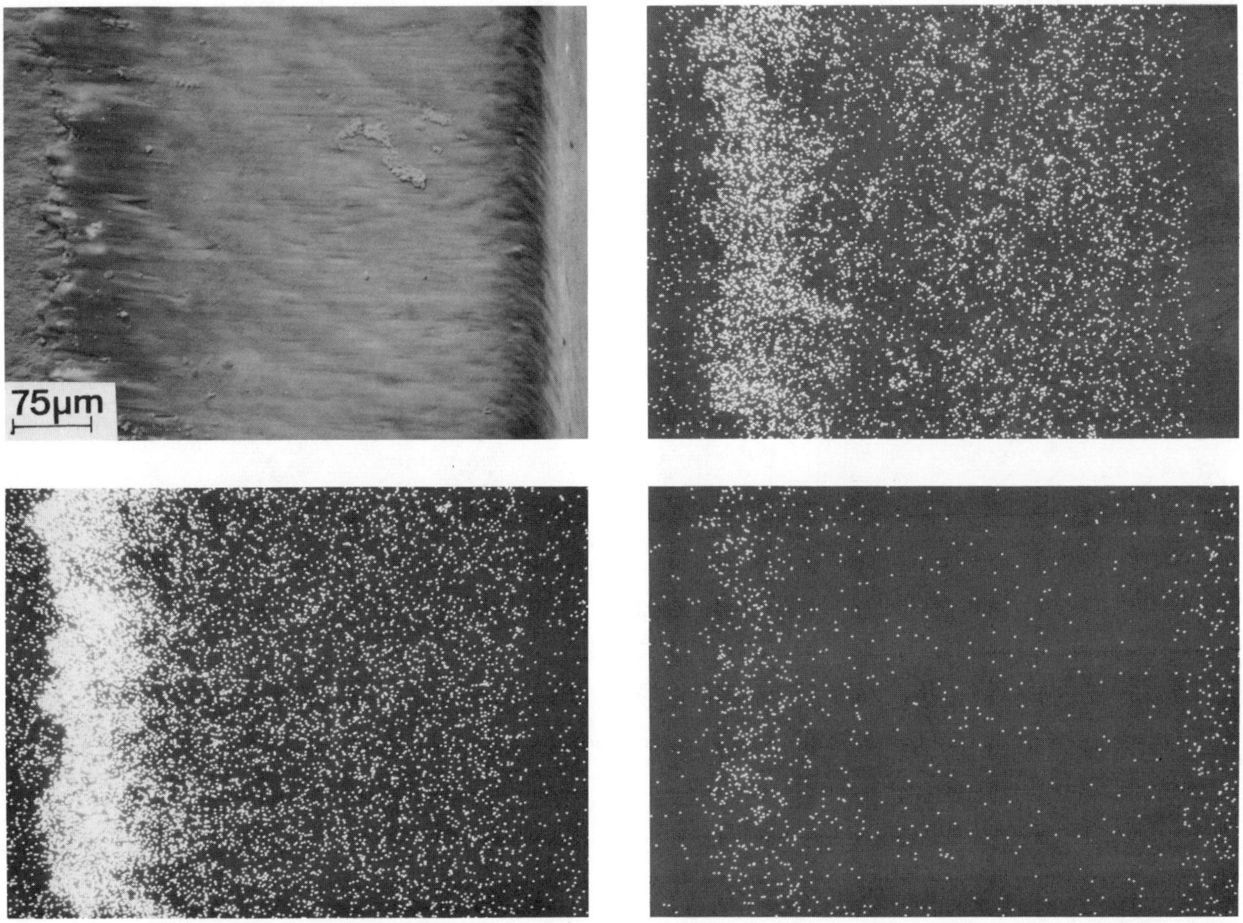

Fig. 12 TiC coated carbide (type 1), quick-stopped after cutting GA-iron at 200 m/min for 30 secs:
(a) secondary electron image, showing adherent layer on rake face;
(b) Si K_α X-ray map; (c) Mn K_α X-ray map; (d) Al K_α X-ray map

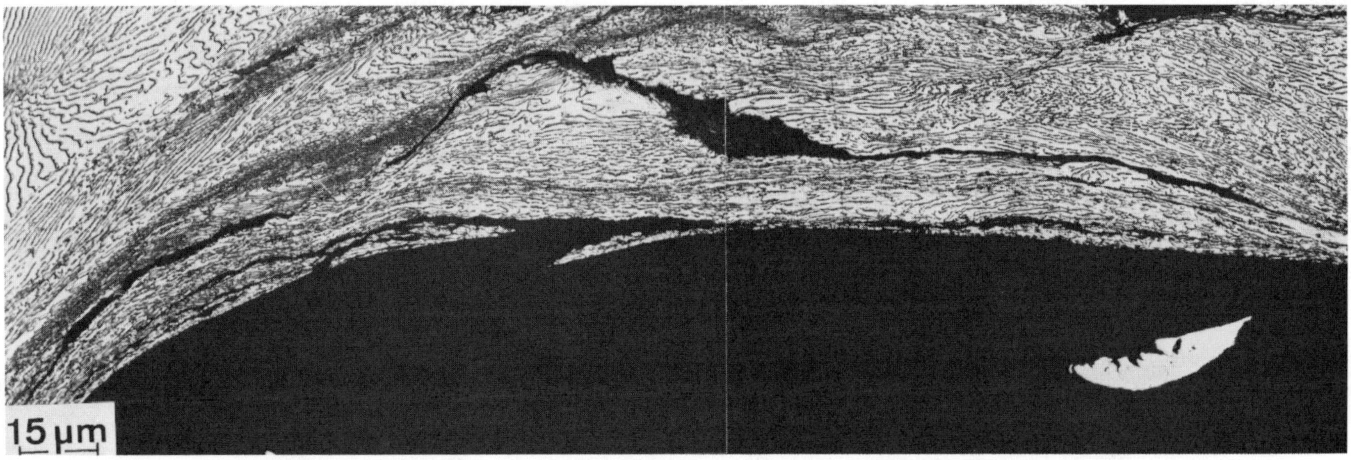

Fig. 13 Microsection of GA-iron chip, secondary shear zone, as formed adjacent to the rake face/cutting edge of an Al_2O_3 coated carbide tool (type 6). Revealed by quick-stopping tool, after being used at 200 m/min.

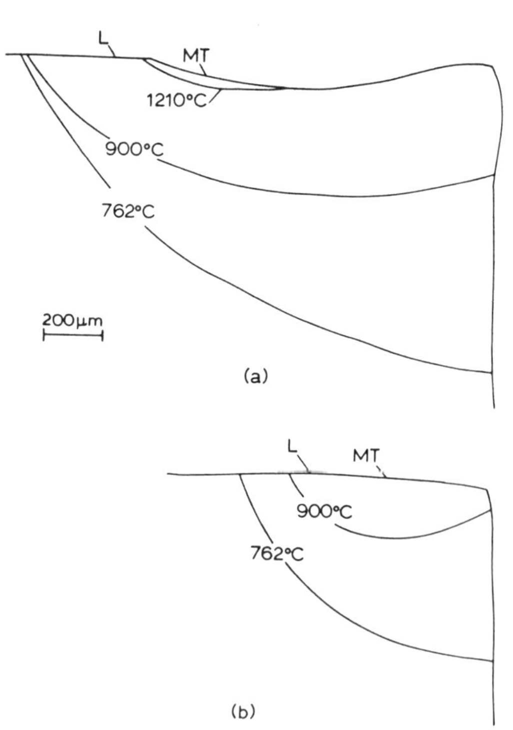

Fig. 14 Tool temperature maps derived after cutting 080M40 steel at 300 m/min for 30 secs. L = chip/tool contact length extremity. MT = Approx. position of max. rake face temperature.
(a) uncoated carbide; (b) TiN coated carbide.

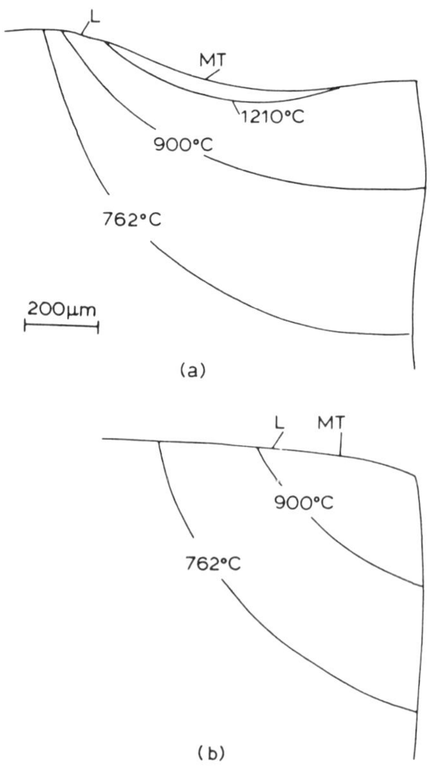

Fig. 15 Tool temperature maps derived after cutting 817M40 at 240 m/min for 10 secs. L = chip/tool contact length extremity, MT = Approx. position of max. rake face temperature.
(a) uncoated carbide; (b) TiN coated carbide.

This effect is probably related to the chip/tool contact length. When cutting the steels, the chip/tool contact length was greater for the uncoated carbides than for the coated carbides, whereas it was practically identical for both types of tool when cutting the cast irons, as shown in Table 3. There were no significant differences in contact length observed between cemented carbides with exterior coatings of TiC, TiN or Al_2O_3.

The larger chip/tool contact lengths observed for the uncoated carbides, used to cut steel, causes larger plastic strains to be produced within the secondary shear zone, than in the case of the coated tools. Hence, the work done and the heat created in the secondary shear zone is higher when using uncoated carbides, and consequently these tools become hotter than the coated carbides, as shown in Figs 14 and 15.

The factors which determine the chip tool contact length, are not clearly understood. However, the intrinsic strength of bonding between the chip underside and the tool rake face, is probably one important factor. Towards the back of the chip/tool contact length, the high compressive stress which results from the primary shear of the chip, diminishes markedly (Refs 12 and 13). Hence, the chip is no longer "forced" onto the rake face of the tool, but merely held in position by the intrinsic strength of bonding between the chip and tool. When the interfacial bond strength is weak (as demonstrated for the coated carbides and ceramics - tool separation during quick-stopping taking place at the chip/tool interface), the resulting chip/tool contact length is short. However, when interfacial bonding is strong (as for uncoated carbides - tool separation during quick-stopping taking place above the chip/tool interface), the chip/tool contact length is relatively large.

It is proposed that the chip/tool contact length is determined by other factors when cutting cast-irons. The secondary shear, adjacent to the chip underside (Fig. 13), appeared less severe when cutting both types of cast-iron, compared to cutting the steels, since the

Table 3 Chip/Tool Contact Length Data

Work Material	Cutting Speed (m/min)	Tool Material	Chip/Tool Contact Length (mm)
080M40	180	Uncoated Carbide (type 8)	0.94
080M40	180	$TiN/Al_2O_3/TiC$ Coated Carbide (type 7)	0.68
817M40	180	Uncoated Carbide (type 8)	0.61
817M40	180	$TiN/Al_2O_3/TiC$ Coated Carbide (type 7)	0.48
GA-iron	200	Uncoated Iron Bonded Carbide	0.46
GA-iron	200	TiN Coated Iron Bonded Carbide	0.42
SG-iron	200	Uncoated Iron Bonded Carbide	0.44
SG-iron	200	TiN Coated Iron Bonded Carbide	0.48

(a) (c)

(b) (d)

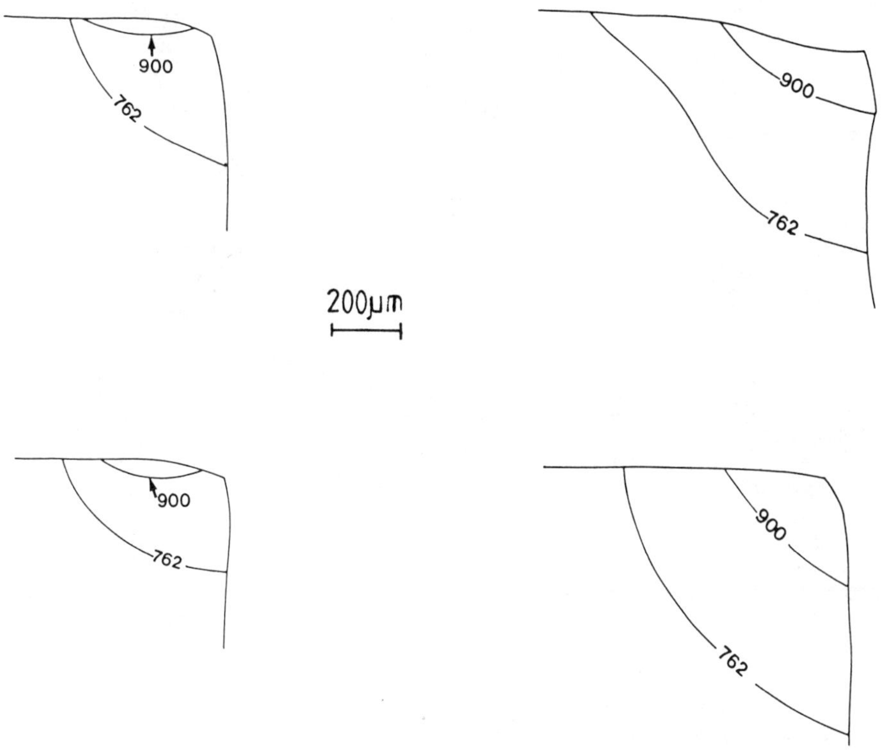

Fig. 16 Tool temperature maps derived after cutting:
(a) GA-iron with uncoated carbide at 200 m/min; (b) GA-iron with TiN coated carbide at 200 m/min; (c) SG-iron with uncoated carbide at 200 m/min; (d) SG-iron with TiN coated carbide at 200 m/min.
All temperatures in °C.

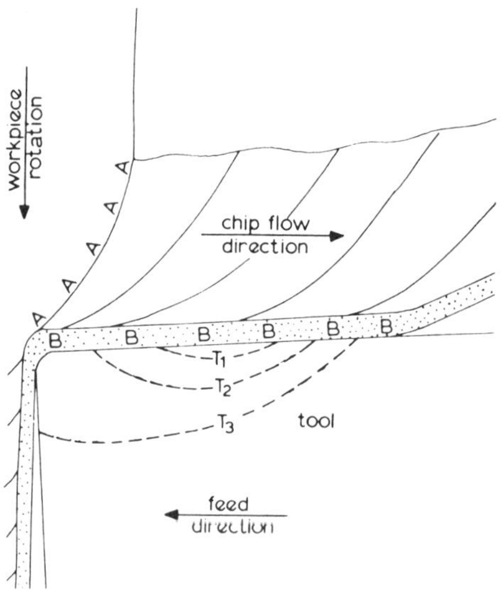

Fig. 17 Schematic metal cutting diagram, showing position of primary shear zone (A) and secondary shear zone (B). Note that isotherms, $T_1 > T_2 > T_3$.

original pearlite lamallae, were often resolvable, albeit distorted by plastic deformation (Fig. 13). This secondary shear zone is very different to that observed when cutting steel, which comprises submicron recrystallised ferrite and fine dispersions of cementite (Ref. 14), which are frequently only just, if at all, resolvable by optical microscopy. It is suggested that secondary shear, as well as primary shear, is severely restricted when cutting cast-iron types GA and SG, and that large plastic strains are accomodated by deformation of the graphite phase and by the formation of fractures in the vicinity of the graphite/pearlite interface. Hence, coherent contact between the chip and the tool rake face can be literally broken by fracture within the secondary shear zone. Therefore, in the case of cast-iron, the intrinsic strength of bonding between the chip and tool, across the main chip/tool interface, does not have the possibility to influence the contact length, as is the case when cutting steel.

THE FLANK/WORKPIECE INTERFACE - No evidence of a flank contact zone was visible in the new surface of the workpiece after quick-stopping coated carbides (Fig 3b), uncoated carbides and ceramics, even after a large wear land of 0.10 mm was accumalated on the tools, prior to quick-stopping. Instead, shallow grooves and voids, the latter often being elongated in the direction of workpiece rotation, were observed. The voids (arrowed in Fig. 3b), are indicitive of reduced interfacial constraint, suggestive of restricted interfacial movement. The constraint is sufficient, however, to gener-

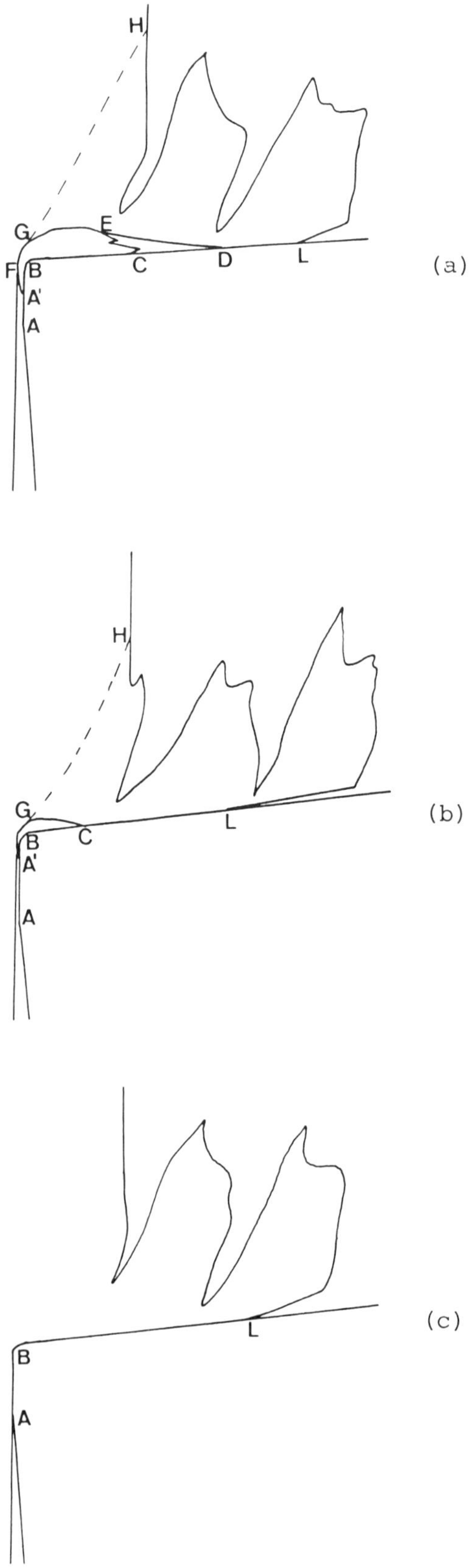

Fig. 18 Schematic depiction of chip formation when cutting:
(a) GA or SG-iron at 60 m/min;
(b) SG-iron at 200 m/min;
(c) GA-iron at 200 m/min.

Fig. 19 Metallographic section through outside edge of chip revealed after quick-stopping TiN/TiC coated carbide (type 2). Work-material:080M40; cutting speed: 180 m/min. Note absence of secondary shear zone.

ate a shear zone adjacent to the work-piece surface (Fig. 7b). The flank/work-piece interface is clearly different to that described for the peripheral contact made at the outer edges of the chip underside (zone E in Fig. 4), since very little shear takes place within this part of the chip (Fig. 19). Furthermore, atmospheric gases can penetrate the latter region of the chip/tool interface and cause the formation of notch wear, whereas atmospheric gases do not penetrate the flank/workpiece interface (under the cutting conditions evaluated here) as flank wear is not affected by atmospheric or externally applied gases.

CONCLUSIONS

(1) A continuous chip is produced when cutting steel, while a segmented chip is produced when cutting pearlitic cast-irons. Chip segmentation is caused by fracture in the vicinity of the graphite/matrix interface during primary shear.

(2) Large built-up-edge (BUE) is observed when cutting both types of cast-iron at 60 m/min. A small BUE was still observed at 200 m/min when cutting SG-iron, but was absent when cutting GA-iron.

(3) There are three distinct interfacial zones:
(a) the main chip/tool interface;
(b) peripheral chip/tool contact at the outer edges of the chip underside.
(c) the flank/workpiece interface.

(4) The main chip/tool interface is essentially of similar character for CVD coated carbides, uncoated carbides and ceramics, i.e., there is no evidence of interfacial movement. However, the strength of bonding between the chip and tool is greater when using uncoated cemented carbides, than when using coated carbides and ceramics.

(5) When cutting steel, the strength of bonding between the chip and tool, determines the size of the chip/tool contact length. Hence, chip/tool contact lengths are shorter for coated carbides and ceramics, than for uncoated carbides.

(6) When cutting cast-irons chip/tool contact lengths are determined in a different way; secondary shear adjacent to the chip underside, is curtailed by fracture in the vicinity of the graphite/matrix interface, effectively breaking the contact made between the chip underside and the tool rake face. In these circumstances the interfacial bond strength does not control the position where the chip leaves the tool. Hence, no difference in chip/tool contact length was observed in this case for coated carbides, uncoated carbides and ceramics.

(7) Chip/tool contact length determines the position and magnitude of the maximum rake face temperature, and the size of the heat affected zone (HAZ). Hence, coated carbides exhibit lower temperatures and a smaller HAZ, than uncoated carbides, when cutting steels, while when cutting cast-irons, no difference in the magnitude and distribution of temperature was observed between coated and uncoated carbides.

(8) The peripheral contact made by the outer 80 µm or so of the chip underside, with the tool rake face, involves sliding. Atmospheric gases often penetrate this portion of the chip/tool interface.

(9) Interfacial constraint is less across the flank/workpiece interface than across the main part of the chip/tool interface, and, there is evidence of restricted sliding. However, sufficient constraint exists to cause shearing of the workpiece ahead of the interface. Atmospheric gases do not pentrate this interface.

ACKNOWLEDGEMENTS

The authors wish to thank the Science and Engineering Research Council and Sandvik Hard Materials Ltd, for financing aspects of this work and the latter for permiss-to publish.

REFERENCES

1. Dearnley, P. A., and E. M. Trent, Met. Technol., 9, 60-75, 1982
2. Dearnley, P. A., Ph.D. Thesis, Univ. Birmingham, United Kingdom, 1980.
3. Williams, J. E., E. F. Smart and D. R. Milner, Metallurgia, 81, 3-10, 1970.
4. Dearnley, P. A., and A. N. Grearson, Unpublished Research, Sandvik Hard Materials Ltd, Coventry CV4 9UW, United Kingdom, May 1981.
5. Dearnley, P. A., Met. Technol., 10, 205-214, 1983.
6. T.I. Abar, Technical Literature, Hinxton, Cambs., United Kingdom.
7. Dearnley, P. A., To be published in Int. J. Mach. Tool. Des. Res., 1985/1986.
8 Williams, J. E., and E. C. Rollason, J. Inst. Metals, 98, 144-153, 1970.
9. Trent, E. M., "Metal Cutting", Butterworths, 2nd Edition, 1984.
10. Dearnley, P. A., Wear, 101, 33-68, 1985.
11. Kramer, B. M., and N. P. Suh, Trans. A.S.M.E., J. Eng. Ind., 102, 303-309, 1980.
12. Zorev, N. N., Conf. on "Int. Res. in Prod. Eng.", 42-49, A.S.M.E., New York, 1963.
13. Barrow, G., W. G. Graham, T. Kurimoto and Y. T. Leong, Int. J. Mach. Tool Des. Res., 22, 75-85, 1982.
14. Shelbourn, A. M., W. T. Roberts and E. M. Trent, Materials Science and Tech., 1, 220-226, 1985.

MACHINING FERROUS MATERIALS WITH CVD COATED CARBIDES. PART 2: WEAR MECHANISMS

P. A. Dearnley
W.I.S.E., Dept. Metallurgy & Materials
University of Birmingham
Birmingham, England

V. Thompson, A. N. Grearson
Sandvik Hard Materials Ltd.
Research Centre
Coventry, England

ABSTRACT

A broad variety of CVD coated carbides have been used to cut four types of steel and two types of cast-iron. In general, the application of hard CVD layers to cemented carbides is a very effective method for enhancing wear resistance, but the extent of the wear enhancement depends upon the coating material selected and the workpiece being machined. No single coating material is always superior to another. Hence, multi-phased coatings are a necessity. Examination of the worn tool surfaces, in conjunction with earlier observations on the interfacial conditions (Part 1), show that a variety of wear mechanisms can pertain when cutting ferrous materials, including: dissolution/diffusion, discrete plastic deformation/fracture, cracking and spalling, localised oxidation, and gross plastic deformation of the cutting edge.

IN PART 1 OF THIS WORK, close attention was paid to the interfacial conditions pertaining when cutting steels and cast-irons. In part 2, detailed observations of the wear rates and wear surfaces of CVD coated carbides, uncoated carbides and ceramics, used for cutting a variety of steels and cast-irons, are presented. By linking this information with that described in Part 1, a number of hypotheses are advanced for the principal wear mechanisms of all the tool materials examined.

It is shown that there is no single coating material, which resists all forms of wear better than another, and that the relative superiority of any particular coating material, is strongly dependant on the workpiece being machined. The work also confirms that the basic philosophy of applying multi-phased CVD coatings to cemented carbides has a sound basis, but the exact selection made depends upon the universality of application required.

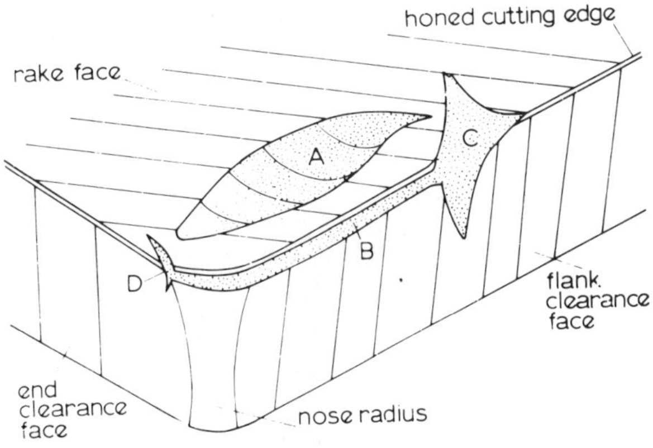

Fig. 1 Principal wear zones on a cutting tool: zone A = rake face "crater" wear, zone B = flank wear, zones C and D = depth of cut notch wear.

EXPERIMENTAL

TOOL MATERIALS - The tool materials used are given in Table 1 in Part 1.

WORKPIECE MATERIALS - The workpieces shown in Table 2 in Part 1 were used also in this investigation, together with the 070M55 and 316S16 steels shown in Table 1. The former steel was normalised, while the latter was annealed. All the workpieces were 150 mm dia. and 0.8 m long.

CUTTING TEST PROCEDURE - All cutting tests were made dry using VDF and DSG manual lathes and Swedturn 10 and 12 CNC lathes. The cutting conditions selected and the tool geometries used are cited in Table 2. When cutting 080M40 and 817M40 the positive rake angle geometry shown in Table 2, under column A, was used, while

Table 1 Composition and Hardness of Additional Work-Materials

Work-material	Mean hardness, HV 20	Composition, wt% C	Si	Mn	Ni	Cr	Mo	S	P
070M55	184	0.50	0.14	0.77	0.08	0.04	0.02	0.01	0.03
316S16	154	0.07 max	0.20-1.00	0.50-2.00	10.0-13.0	16.5-18.5	2.25-3.00	0.03 max	0.045 max

Table 2 Cutting Conditions and Tool Geometry.

Work-Material	Cutting speed, m/min	Feed rate, mm/rev	Depth of cut, mm
080M40	180	0.25	1.3
	240	0.25	1.3
	300	0.25	1.3
817M40	180	0.25	1.3
	240	0.25	1.3
	300	0.25	1.3
070M55	250	0.25	2.0
316S16	150	0.25	2.0
	170	0.35	2.0
	250	0.25	2.0
	350	0.25	2.0
GA-iron	60	0.25	2.0
	200	0.25	2.0
SG-iron	60	0.25	2.0
	200	0.25	2.0

Insert Geometry	A	B	C
Back rake angle	0°	-5°	-5°
Side rake angle	5°	-5°	-5°
Approach angle	15°	15°	15°
Trail angle	15°	15°	15°
End clearance angle	11°	5°	5°
Flank clearance angle	6°	5°	5°
Nose radius (mm)	0.8	0.8	1.2
Cutting edge radius (μm)	40	40	...
Edge chamfer (ceramics only)			1.2 mm x -20°

when cutting 070M55, 316S16 and both cast-irons, the negative rake angle geometry shown under column B was used. The ceramic tools were always used with an identical negative geometry but had an edge chamfer and a larger nose radius, as indicated under column C in Table 2. The type 10 uncoated carbides and all the coated carbides had a cutting edge radius ∿40 μm, whereas type 8 and 9 uncoated cemented carbides were sharp edged.

Old style tool holders were used, which incorporated a chip breaker, since all the tool "inserts" were of a plain top rake style,i.e., they did not have an integral chip control groove. The chip breaker of the tool holders was adjusted such that it was almost fully retracted, to prevent artificial shortening of the chip/tool contact length, while still maintaining a degree of chip control.

All the test edges were examined optically, prior to the commencement of tests, to check for soundness. All tool material types were then tested consecutively, for periods between 1 and 5 mins depending upon the rates of wear, so that tool wear could be gradually accumalated. This also allowed each tool type to be tested at different positions within the workpiece, so that any inconsistancies in wear due to workpiece inhomogeneity could be averaged out.

WEAR MEASUREMENT - Fig. 1 shows the principal wear zones on a cutting tool. Zone A represents rake face (crater) wear zone B flank wear, and zones C and D notch wear at the extremities of the

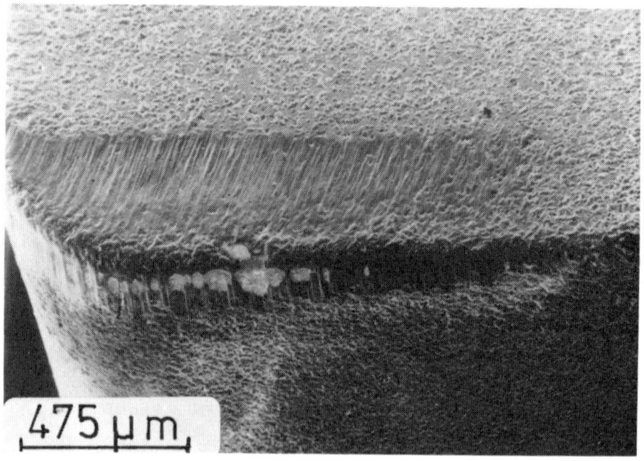

(a)

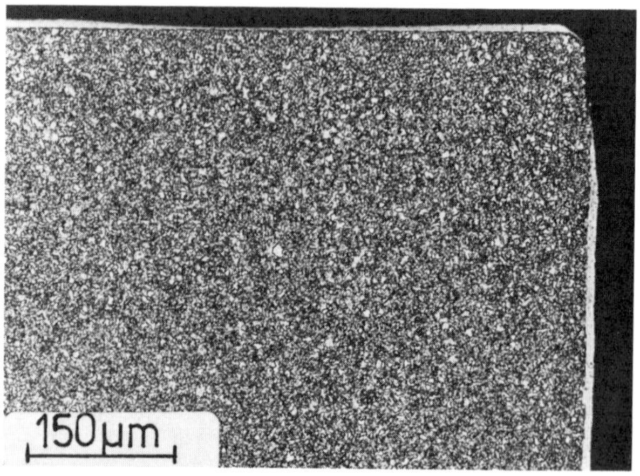

(b)

Fig. 2 Coated carbides after cutting 070M55 at 250 m/min showing greatest wear on flank face:
a) Al_2O_3 coated carbide (type 6)
b) TiN coated carbide (type 4).

depth of cut. All the worn surfaces were immersed in 50% HCl to remove adherent work material, prior to wear measurement. After cutting the 316S16 stainless steel it was necessarry to boil the HCl to facilitate rapid removal of the workpiece material.

Crater depth was measured using a light-section microscope, while flank wear was measured using a stereo binocular microscope. Wear of the coatings was more accurately determined by measuring the depth of wear on both flank (e.g., dimension "x" in Fig. 17) and rake faces, from metallographic sections prepared through the region of maximum wear.

GENERAL COMMENTS ON WEAR RATES

The rake and flank face wear rates of the coating materials, relative to the cemented carbides and Al_2O_3/ZrO_2 ceramics when cutting most of the ferrous work materials investigated at 200-250 m/min are summarised in Table 3. Wear rates of the coating and ceramic materials were often ~100 times less on the rake face than for uncoated cemented carbides. However, most coated carbides and ceramics were worn more rapidly on the flank face than on the rake face (Fig. 2), the exception being the TiC coated carbides, whereas the uncoated carbides were always worn more rapidly on the rake face. Hence, on the flank face, the coating and ceramic materials were frequently only worn ~10 times less rapidly than the uncoated cemented carbides.

The relative wear resistance of the principal coating materials, when used to cut all the ferrous materials, are summarised in Table 4. No single coating

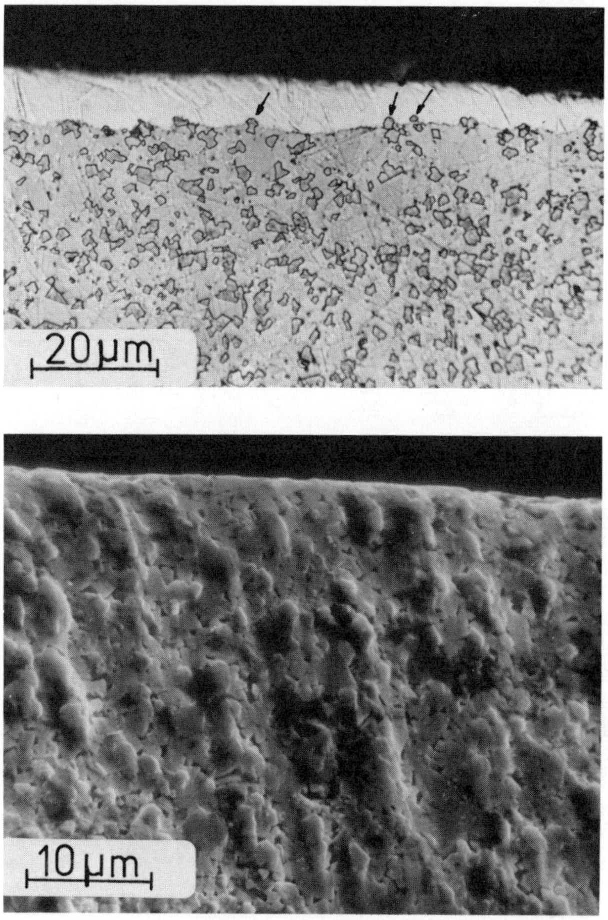

(a)

(b)

Fig. 3 Uncoated carbide (type 9) wear surfaces:
a) rake face after cutting 817M40 at 240 m/min. γ-phase arrowed.
b) flank face after cutting 070M55 at 250 m/min, SEM.

Table 3 Wear Rates of Various Coating and Tool Materials Used to Cut Ferrous Materials

WORN TOOL MATERIAL	WEAR RATES ON VARIOUS WORKPIECE MATERIALS (µm/min)									
	080M40 (240 m/min)		817M40 (240 m/min)		070M55 (250 m/min)		GA-IRON (200 m/min)		SG-IRON (200 m/min)	
	Rake Face	Flank Face	Rake Face	Flank Face	Rake Face	Flank Face	Rake Face	Flank Face	Rake Face	Flank Face
TiC coating (tool type 1)	0.48	0.48	0.20	<0.20	2.80	1.20	<1.0	<1.0	12.0	<8.0
TiN coating (tool type 3)	0.66	>1.80	1.00	~2.00	1.60	2.00	...	...	6.4	9.6
TiN coating (tool type 4)	....	....	....	....	2.20	>3.80	...	...	12.0	20.0
Al_2O_3 coating (tool type 6)	0.66	1.60	~1.20	1.80	1.80	2.20	1.4	2.0	3.0	5.0
Al_2O_3/ZrO_2 ceramic (tool type 11)	....	....	....	....	2.30	4.10	<0.5	1.7	1.3	4.4
Uncoated carbide (type 8)	318.0	38.0	....	....	....	....	72.0	3.8	80.0	3.5
Uncoated carbide (type 9)	42.0	38.0	300.0	40.0	168.0	40.0	42.0	2.8	56.0	2.4
Uncoated carbide (type 10)	....	....	....	....	....	....	8.0	2.4	360.0	188.0

Table 4. Realtive Wear Rankings of Principal Coating Materials.

Work-material	Rake Face Wear Rate	Flank Face Wear Rate
080M40 (0.4% C steel)	$TiN \simeq TiC \simeq Al_2O_3$	$TiN > Al_2O_3 > TiC$
070M55 (0.5% C steel)	$TiC > TiN \gtrsim Al_2O_3$	$TiN \gtrsim Al_2O_3 > TiC$
817M40 (0.4% C low alloy steel)	$Al_2O_3 > TiN > TiC$	$Al_2O_3 \gtrsim TiN > TiC$
316S16 (18-8 stainless steel)	$Al_2O_3 > TiC > TiN \simeq Ti(C,N)$	.. .. ..
GA-iron (flake graphite)	$Al_2O_3 > TiC$	$Al_2O_3 > TiC$
SG-iron (spheroidal graphite)	$TiC \gtrsim TiN > Al_2O_3$	$TiN > TiC > Al_2O_3$

Note: if wear is dissolution/diffusion controlled, expect wear rate of $TiC > TiN > Al_2O_3$

material was superior to the others for all the work materials cut, although on the whole TiN was most universal for rake face wear protection and TiC was most universal for flank wear protection, even though these coatings were often second best, in a specific application, e.g., when cutting SG-iron at 200 m/min, Al_2O_3 coatings were superior for combating both rake and flank face wear.

Once the coatings were worn through on either the rake or flank face, the substrate beneath was worn rapidly. However, both wear surfaces remained restricted in size and the overall wear rate of the substrate (in terms of volume loss of tool material) remained less than for the uncoated carbides. When cutting the plain carbon and low alloy steels, the coating at the cutting edge remained unworn (e.g., see Fig.16, the coating at the edge is arrowed "a"), whereas, the coating at the edge was often removed when cutting the cast-iron materials, especially SG-iron (e.g., see Fig. 29).

TOOL WEAR SURFACES WHEN CUTTING STEEL

UNCOATED CARBIDES - When cutting steels above the built-up-edge speed range ($\gtrsim$100 m/min), the rake and flank faces often had a similar appearance, in that the WC phase was preferentially worn while the solid solution (γ-phase) carbide remained proud (Fig. 3). However, when cutting 080M40 at 180 and 200 m/min the γ-phase was preferentially plucked out of the flank face by the adherent workpiece material. This is a form of attrition wear, first observed by Trent (Ref 1), while the preferential wear of the WC phase, relative to the γ-phase (Fig. 3) is generally accepted to be caused by the WC phase being preferentially dissolved in the adherent chip/workpiece (Refs 2,3,4), the γ-phase being undermined and removed en masse.

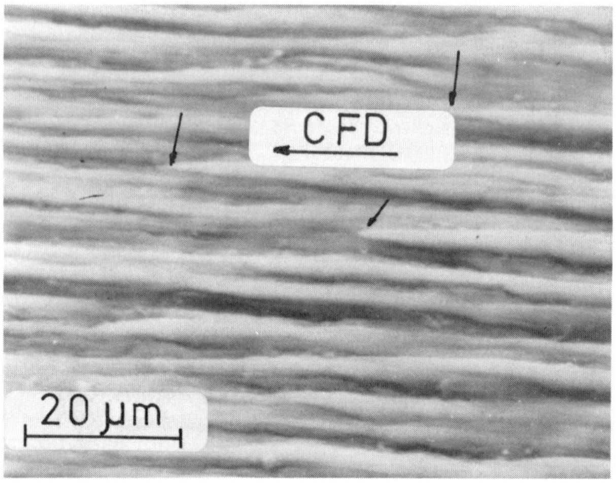

(a)

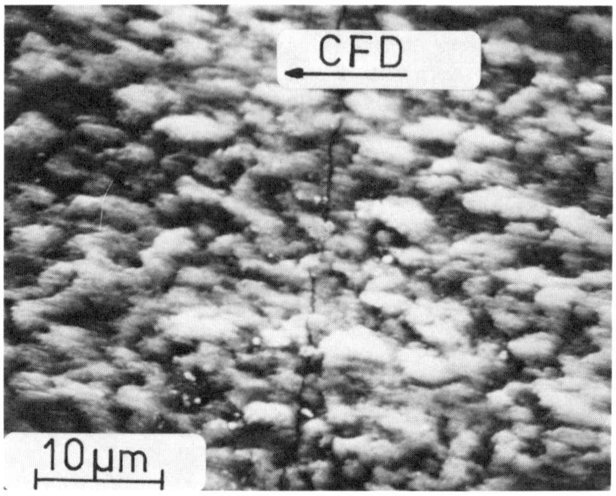

(b)

Fig. 4 Rake face wear surfaces:
a) Al_2O_3/ZrO_2 ceramic after cutting 070M55 at 250 m/min, ridge end points are arrowed.
b) TiC coating (type 1) after cutting 080M40 at 240 m/min showing original coating asperities deformed in direction of chip travel.

CFD = Chip Flow Direction

COATED CEMENTED CARBIDES AND CERAMICS - RAKE FACE WEAR - Ridge like features were commonly observed on the worn rake faces of all the coatings and Al_2O_3/ZrO_2 ceramics. They were always alligned parallel with the direction of chip flow (Fig. 4a). Examination of the early stages of this wear indicated that they often form via the plastic deformation of surface asperities (Fig. 4b). Many of the ridges tapered to a point (Fig. 4a) suggesting that ridge formation culminated in ductile fracture, as schematically depicted in Fig. 5. Ridge formation only

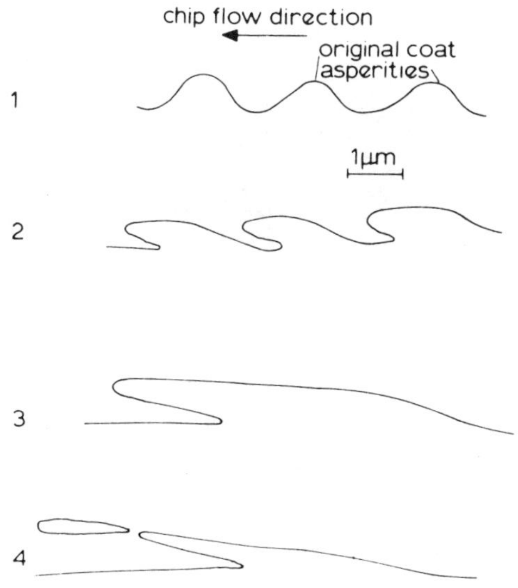

Fig. 5 Schematic depiction of ridge formation (sequence 1 to 3) culminating in ductile fracture (4)

appears to involve plastic deformation of the near surface, and, does not extend more than about 1 μm beneath the surface. Hence, this effect has been termed "discrete plastic deformation" (Ref. 5).

The size of the ridges on the Al_2O_3/ZrO_2 ceramic and Al_2O_3 coatings > TiC coatings > TiN coatings and were always most exaggerated at some distance from the cutting edge (Fig. 6a), corrolating with the position of highest rake face temperature and demonstrating that this process is temperature dependant.

The extent of ridge formation, was more marked when cutting the 316S16, 070M55 and 817M40 steels than the 080M40 steel. In the latter case, smoother wear surfaces were commonly observed (Fig. 7). When cutting 817M40 the initially ridge-like features, of a TiC coating (Fig. 6a) became smoothly worn after further cutting (Fig. 6b), indicating that the latter process competed with the former.

As the Al_2O_3 coating of tool type 5 became worn away, exposing the primary TiC coating beneath, the distribution of the remaining Al_2O_3 ridges (Fig. 8), suggested that the latter stages of wear of the Al_2O_3 coating was accomplished by decohesion at the Al_2O_3 coat/TiC coat interface. This 'peeling' type of mechanism is depicted schematically in Fig 9. This effect was never observed for the TiN or Ti(C,N) coatings of tool types 2 and 3 respectively, on becoming worn through to their adjacent TiC coating beneath. Instead the coatings were uniformly worn away.

When cutting the 070M55, 817M40 and 080M40 steels, all the CVD coatings were progressively worn through on the rake face, as schematically shown in Fig. 10. Once the substrate beneath the coating was exposed to the chip, it became worn in an identical manner to that observed for the uncoated carbides (Fig. 3).

After cutting the austenitic stainless steel (316S16) the coatings were less frequently uniformly worn. Instead large fragments of coating were often removed from the tool rake face (Fig. 11), especially for tool type 1. The early stages of this wear involved the development of numerous cracks in the coating, mainly adjacent to, and running parallel with the cutting edge (Fig 12), although a few cracks also formed perpendicular to the cutting edge.

It should be appreciated that all CVD coatings contain cracks, resulting from the relief of tensile stresses, during the cooling cycle of the CVD operation. During cutting, the distribution of cracks is modified (Fig. 13), Ref 6. When cutting plain carbon and low alloy steels, such cracks do not normally contribute to wear. However, when cutting 316S16 (and also Jethete M152, Ref 7) shear stresses are sufficiently great to cause the cracks lying parallel to the cutting edge (arrowed in Fig. 13) to enlarge (Fig. 12), eventually causing large pieces of coating to be torn from the tool surface (Fig. 11).

COATED CEMENTED CARBIDES AND CERAMICS - FLANK WEAR - After cutting 080M40, 817M40 and 070M55 steels the worn flank faces of TiN and TiC coatings appeared smooth at magnifications ~5K. This was also the case for the Al_2O_3/TiC coating of tool type 5, after cutting 080M40 at 180 and 240 m/min, Fig. 14. Upon cutting the other plain carbon, low alloy and stainless steels, however, ridge formation of the Al_2O_3 coatings (Fig 15) and Al_2O_3/ZrO_2 ceramics was observed, especially at cutting speeds 250-350 m/min. At these higher speeds, it was also possible to observe very small ridges on the worn flank faces of the TiC coatings.

For coated carbides flank wear was usually most rapid beneath the cutting edge radius, retaining the coating at

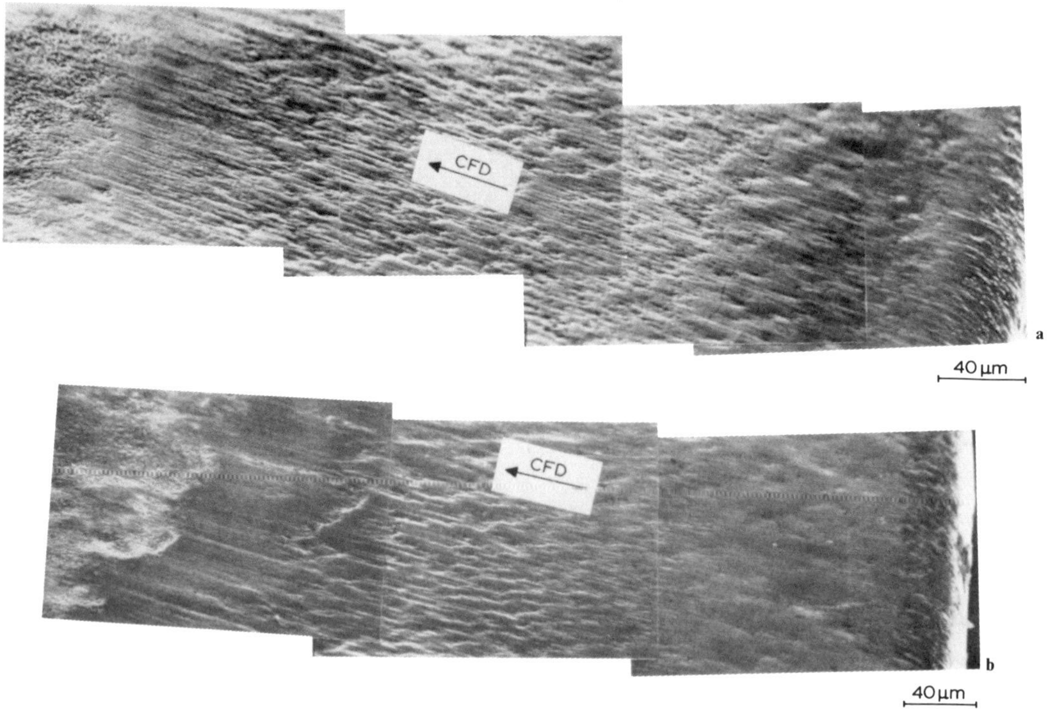

Fig. 6 Worn TiC coating (tool type 1) on rake face after cutting 817M40 at 180 m/min for: (a) 15 mins (note marked ridges in centre of rake face contact area) (b) 47 mins, showing smoothing off of ridges.
CFD = Chip Flow Direction

the cutting edge, which remains relatively unworn, arrowed A in Fig. 16, whereas sharp edged uncoated cemented carbides were worn most rapidly at the cutting edge (dimension 'x' in Fig. 17).

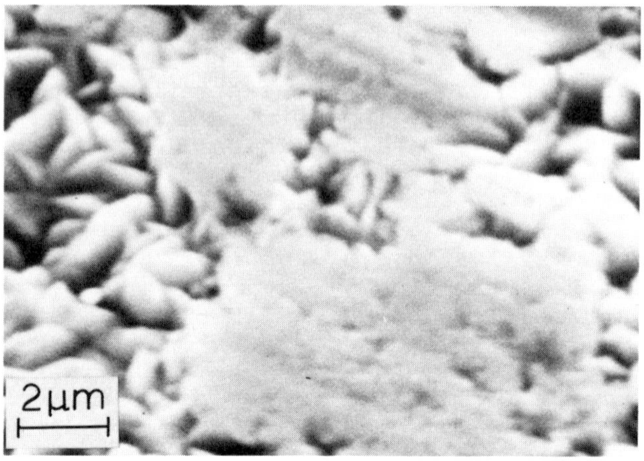

Fig. 7 Smoothly worn TiN coating of tool type 3 after cutting 080M40 at 300 m/min for 2 mins, SEM.

Once the coating becomes worn through on the flank face, the substrate beneath is worn in a fashion commensurate with the uncoated carbides, i.e., the WC grains are preferentially worn, while the γ-phase remains proud at the interface, Fig. 16.

COATED CEMENTED CARBIDES - NOTCH WEAR - Notch wear at the outside edge of the depth of cut was most rapid for the TiC coatings when cutting plain carbon (Table 5) and low alloy steels. Two distinct notch wear zones were visible (designated zones α and β in Fig. 18).

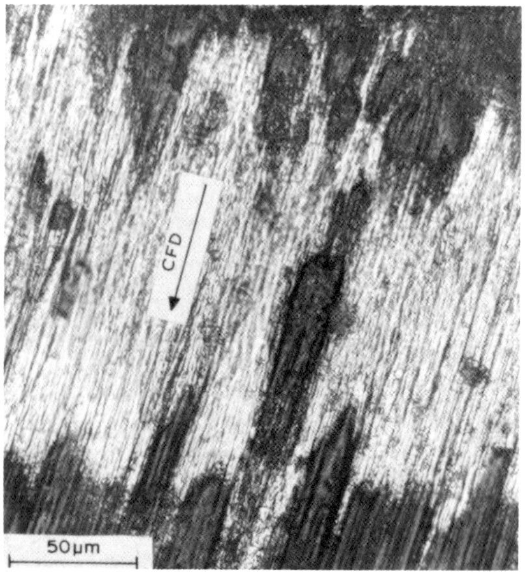

Fig. 8 Optical micrograph of tool type 5, showing ridges of Al_2O_3 (grey) lying across exposed TiC coating (white), after cutting 080M40 at 300 m/min for 2 mins.
CFD = Chip Flow Direction

Table 5 Notch Wear Rates of Coating Materials*

Tool Type	Coating Material	Wear Rate (μm/min)
1	TiC	6.0
3	TiN	0.5
5	Al_2O_3	0.3

*All tools used to cut 080M40 at 240 m/min

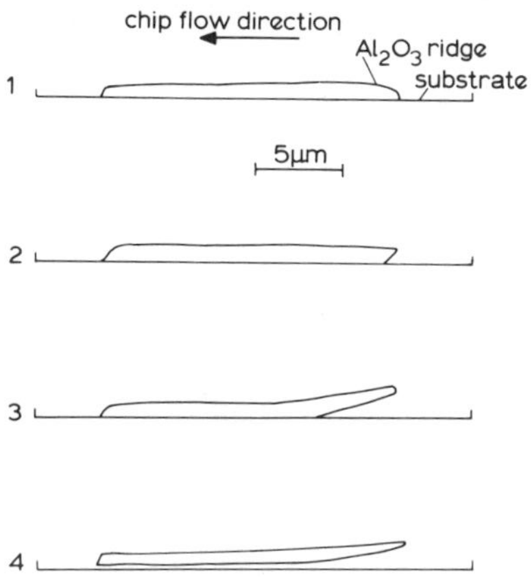

Fig. 9 Schematic depiction of decohesion mechanism of Al_2O_3 coatings, (sequence 1 to 4).

Zone α was smoothly worn , while zone β comprised numerous cracks, perpendicular to the direction of chip travel (Fig. 19) Such cracks ("P" in Fig 19) do not originate from the original cooling cracks ("O" in Fig 19) and are nucleated independantly. They are more finely spaced than the original cooling cracks. As they propogate, they merge, forming "islands" of coating, which are easily plucked from the tool surface, leaving behind small voids ("V" in Fig. 19).

When cutting plain carbon and low alloy steels, the TiN and Al_2O_3 coatings were much more resistant to notch wear development than TiC, Table 5. Zone α wear is very sensitive to atmospheric oxygen, since cutting the 080M40 steel

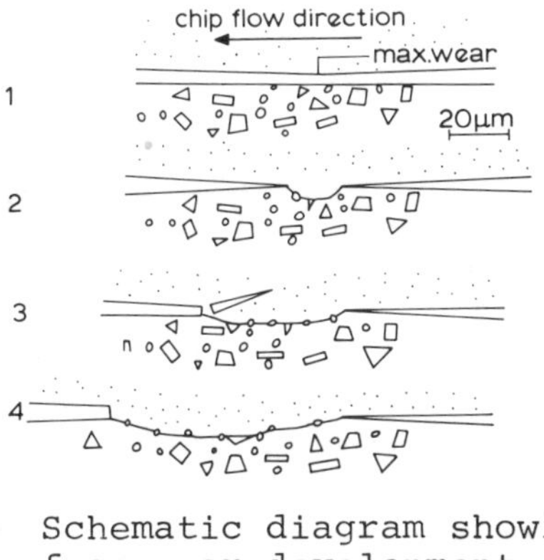

Fig. 10 Schematic diagram showing rake face wear development (sequence 1 to 4) of CVD coated carbides.

with an oxygen jet directed at the notch wear zone, promoted notch formation, Fig. 20 a,b, whereas, notch wear in zone α was practically eliminated when cutting in a jet of argon or nitrogen, Fig. 20 c,d.

Once the substrate beneath the coating was exposed to the chip, the WC phase was worn slightly less rapidly than the γ-phase.

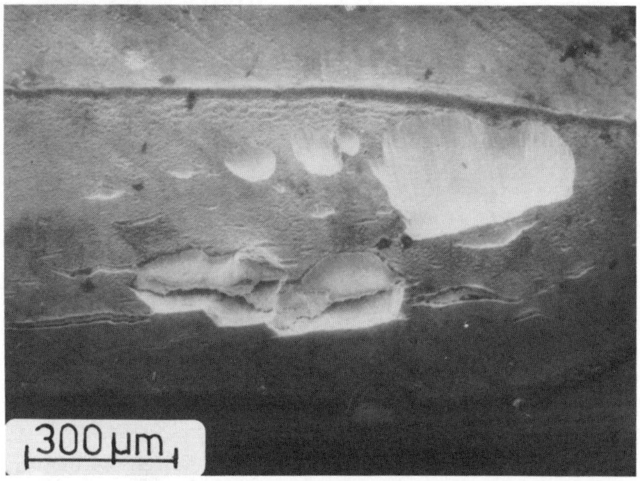

Fig. 11 Detail of rake face wear of TiC coated carbide (type 1) showing cracking and spalling of coating and adjacent substrate, after cutting 316S16 at 170 m/min, 0.35 mm/rev for 5 mins.

When cutting austenitic stainless steel, notch wear is predominantly caused by a fracture mechanism, analagous to zone β described above. However, the rate of notch wear formation, was far higher than that observed when cutting the plain carbon and low alloy steels. Furthermore, all the coating materials were badly affected, no single material being immune (Fig. 21).

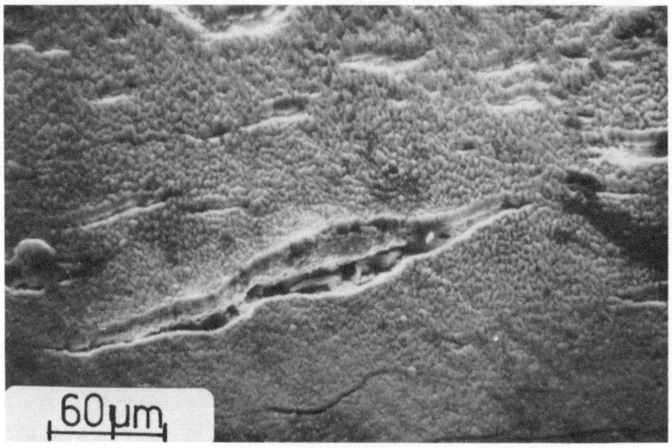

Fig. 12 Detail of Fig. 11 showing large cracks in TiC coating, oriented perpendicular to direction of chip travel.

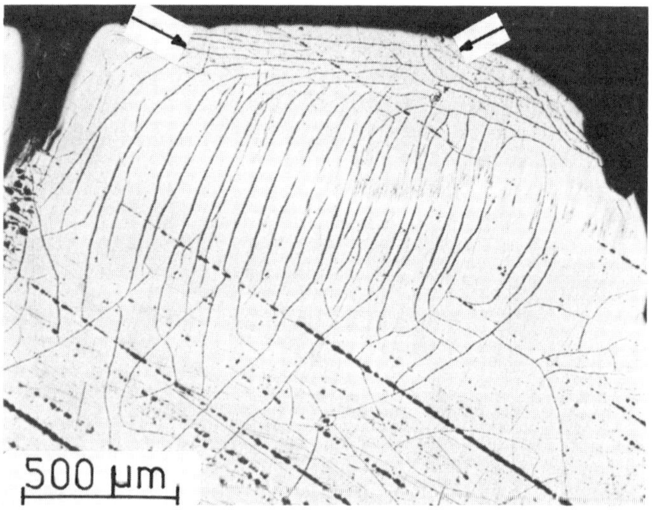

Fig. 13 Distribution of cracks on rake face contact area of TiC coated carbide (type 1), revealed by light polishing. Tool used to cut 080M40 at 180 m/min, for 15 mins.

COATED CEMENTED CARBIDES - PLASTIC DEFORMATION OF THE CUTTING EDGE - After cutting the low (817M40) and high (316 S16) alloy steels, plastic deformation of the cutting edge was always observed, especially in the vicinity of the nose radius. When cutting the low alloy steel, the coatings nearly always accomodated the plastic deformation of the substrate, as schematically depicted in Fig. 22. This effect was also observed after cutting the high alloy steel, but, when the deformation became severe, the coating began to spall (Ref. 7).

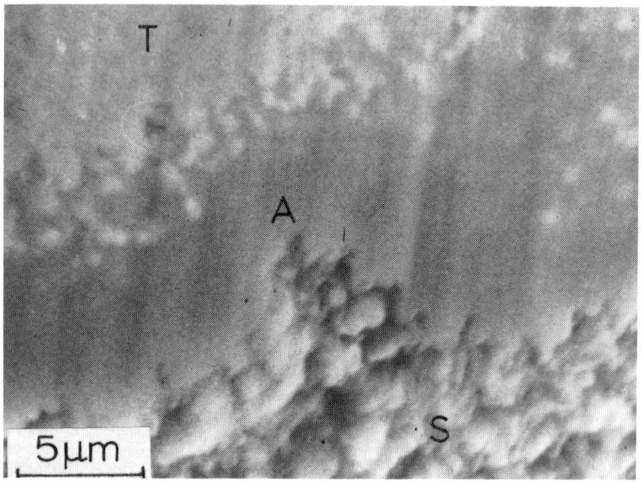

Fig. 14 Smoothly worn flank face of tool type 5 after cutting 817M40 at 180 m/min. T = TiC coat, A = Al_2O_3 coat and S = unworn flank.

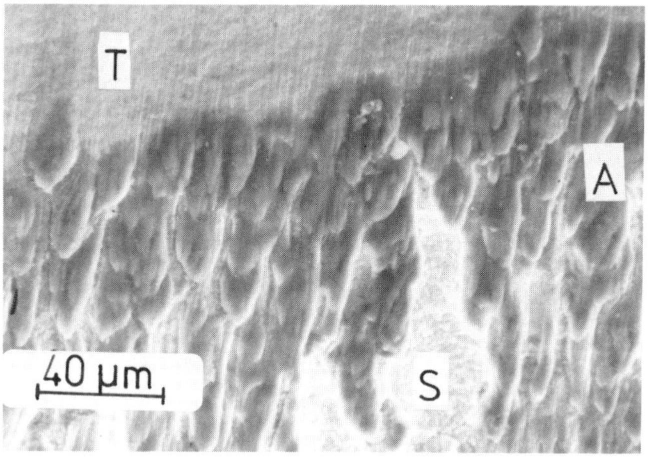

Fig. 15 Large ridges of Al_2O_3 (A) formed on flank of type 7 coated carbide after cutting 817M40 at 300 m/min, for 4 mins. T = TiC coat, S = exposed substrate.

TOOL WEAR SURFACES WHEN CUTTING CAST IRONS.

WEAR RATES - When cutting cast-irons wear rates are strongly dependant upon cutting speed. On the whole CVD coatings are unsuited for wear protection at 60 m/min, since built-up-edge (BUE) causes rapid coating loss, especially when cutting SG-iron. Hence, wear rates of the coated carbides often approach, if not exceed, those observed for uncoated carbides (especially types 6 & 7) as shown in Table 6. At higher speeds, however, e.g., at 200 m/min, CVD coated carbides are superior to uncoated carbides (Table 3)

TOOL WEAR SURFACES AT 60 m/min - Although BUE formation, could afford protection of the tool rake face beneath it (e.g., See Fig. 9 in Part 1), once the BUE became detached, fragments of coating or tool material, were also removed, Fig. 23. Coating loss was also observed on the rake face, in the zone lying be-

Table 6 Rake Face Wear Rates of Coated and Uncoated Carbides Used to Cut Cast-Irons at 60 m/min.

Tool Type	Coating Type	Wear Rate, GA-Iron (µm/min)	Wear Rate, SG-Iron (µm/min)
1	TiC	1.6	1.4
2	TiN/TiC	1.0	4.8
6	Al_2O_3	0.2	1.6
7	TiN/Al_2O_3/TiC	1.6	5.2
8	Uncoated	2.0	0.6
9	Uncoated	1.8	0.2
10	Uncoated	2.0	6.4

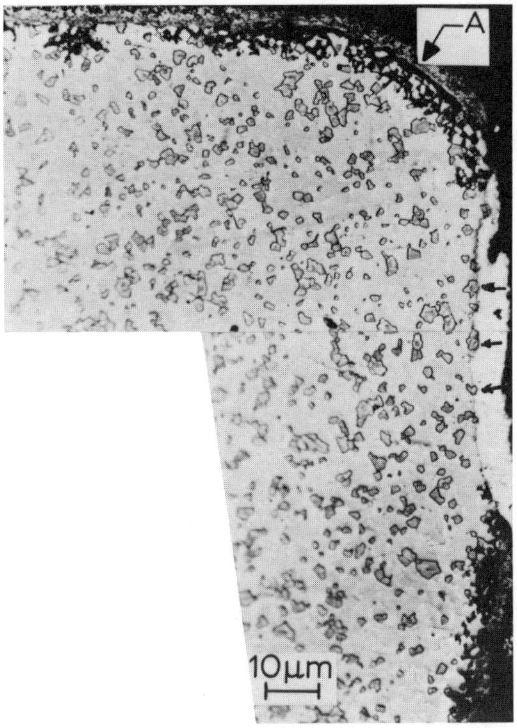

Fig. 16 Metallographic section of tool type 2, after cutting 080M40 at 240 m/min for 15 mins, showing unworn coating at cutting edge (arrowed A) and adjacent flank wear beneath. Note exposed substrate - γ-phase protrudes, arrowed.

hind the BUE, when cutting SG-iron (Fig. 24a), but this was much less widespread when cutting GA-iron (Fig. 24b). In the case of the Al_2O_3 coated carbide (type 6) coating loss was not observed at all when cutting GA-iron.

Irregular wear was also observed on the rake face behind the BUE, for the uncoated carbides containing the highest proportion of γ-phase (type 10). However the other uncoated carbides, were smoothly worn, the γ-phase being worn to the same extent as the WC phase, in the case of tool type 9. The flank faces of these tools appeared identical to the rake face behind the BUE.

The coating on the rake and flank faces which remained under the BUE, but was not removed en masse during BUE detachment, appeared smoothly worn at low magnification. At higher magnification, however, shallow scratch-like features were observed.

TOOL WEAR SURFACES AT 200 m/min

a) GA-IRON - At this speed no BUE was observed (see Part 1). The type 8 uncoated carbides were smoothly worn on both rake and flank faces, while, types 9 and 10 containing γ-phase, exhibited smooth wear surfaces in conjunction with uneven wear on both rake (Fig. 25) and flank

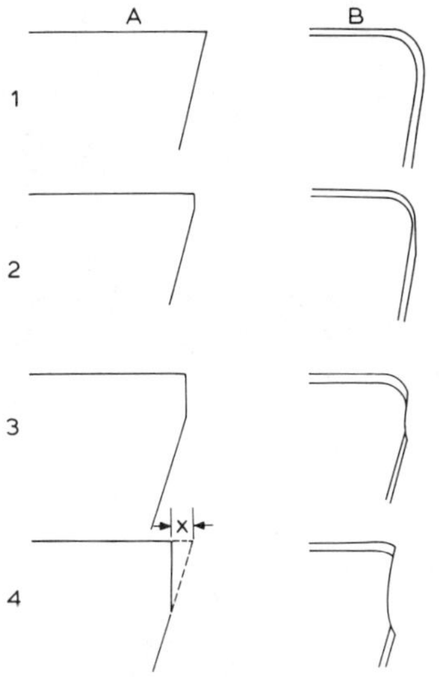

Fig. 17 Schematic depiction of essential differences between flank wear development of uncoated carbides (sharp edged), column A and coated carbides (edge radiused), column B, when cutting steel.

faces. Most coated carbides were smoothly worn on the rake and flank faces, and did not show any coating loss, Fig. 26a. However, some of the TiC coated carbides exhibited cracking and spalling of the coating, Fig. 26b. The cracks resembled those seen when cutting steel (arrowed in Fig. 13). When viewed in cross-section, the substrate was seen to be plastically deformed at the cutting edge, Fig. 27. It is possible that these cracks were enlarged during plastic deformation of the cutting edge. Certainly the deformation of the substrate caused these particular coatings to crack in a shear mode, on the flank face (arrowed in Fig. 27). However most of the other coatings accomodated the deformation of the substrate, as was the case when cutting most steels (Fig. 22).

When cutting GA-iron, the Al_2O_3/ZrO_2 ceramics were smoothly worn on the edge chamfer and flank face.

b) SG-IRON - All three uncoated carbide materials were unevenly worn on their rake faces, Fig. 28. Their flank faces, however, varied in appearance. Type 8 uncoated carbides were smoothly worn, while type 10 were unevenly worn. The type 9 uncoated carbides, however, exhibited both types of wear surface.

BUE was still present when cutting the SG-iron at 200 m/min (Part 1), which induced coating loss at the cutting edge, to varying degrees of severity, Fig. 29. This type of wear was also associated with coating loss, often sporadic in nature, on the adjacent flank face, Fig. 30. The Al_2O_3/ZrO_2 ceramics were smoothly

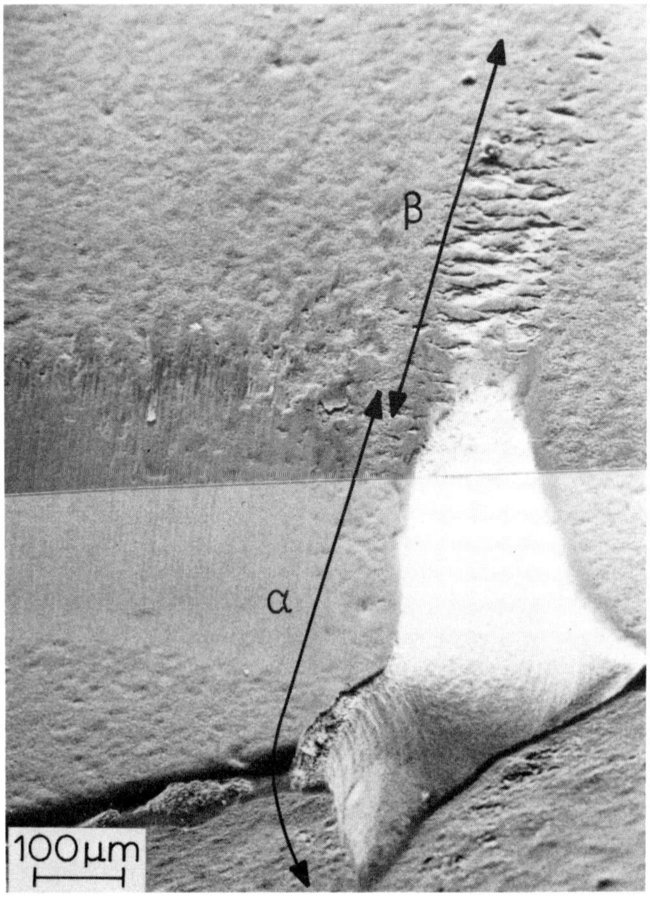

Fig. 18 Principal notch wear zones. TiC coated carbide (type 1) after cutting 080M40 at 180 m/min for 27 mins.

worn on the edge chamfer and flank face, but also exhibited micro-chipping at the cutting edge.

Rake face wear of all the coated carbides, in regions well away from the cutting edge, was smooth and rapid, especially for the TiC coatings, which were worn through after 30 secs, Fig. 31, and no coating remained after a further 30 secs cutting.

TiN coatings were more wear resistant than the TiC coatings, since the 1-2 µm layer on tool type 2, was worn through after 30 secs, compared with the 5-8 µm TiC coating of tool type 1, worn through after the same period. The rake and flank face wear resistance of the Al_2O_3 coatings was superior to TiN, hence the wear rate of TiC > TiN > Al_2O_3. This trend can be seen in Fig. 32. Note that the slightly thicker Al_2O_3 coating of tool type 6 (Fig. 32d) caused a reduction in rake face wear, compared with the slightly thinner Al_2O_3 coating of tool type 7 (Fig. 32c).

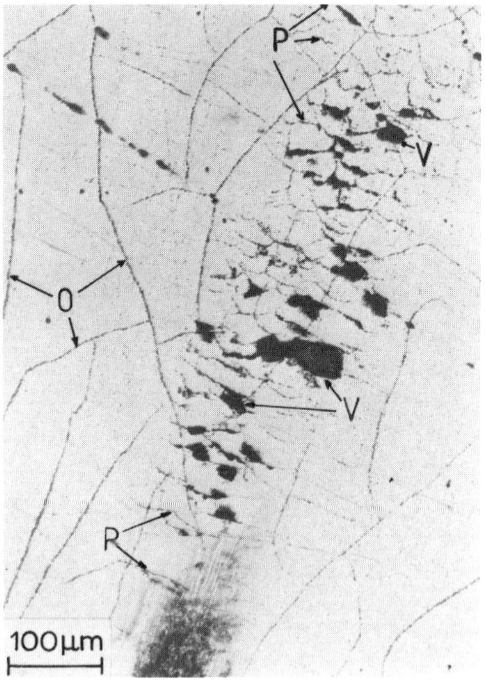

Fig. 19 Detail of notch wear (zone β) of TiC coated carbide (type 1) after cutting 080M40 at 180 m/min for 15 mins. Original cooling cracks (O), fretting cracks (P), and voids (V).

DISCUSSION

This work confirms that applying single or multiple CVD coatings of TiC, TiN and Al_2O_3 to cemented carbides, is in general, a very effective method for retarding tool wear during metal cutting. This is attributed to the very high wear resistant nature of these coatings, which are often worn as much as 100 times less than uncoated carbides, when cutting ferrous materials (Table 3). As shown in Tables 3, 4, 5 and 6 the relative wear resistance of a particular coating material depends upon the workpiece being machined, and upon the specific wear zone of the tool. The possible mechanisms of wear pertaining on the rake face, flank face and notch wear zones of the tools (zones A, B and C respectively in Fig.1) are discussed below, which provide explanations for some of the observed wear resistance rankings.

GRADUAL RAKE AND FLANK FACE WEAR - A major advantage of coating materials, and the reason for their selection, is that relative to WC (the principal constituent of uncoated cemented carbides) they are much less soluble in iron and steel (Table 7). Loladze (Ref. 2) and Trent (Ref. 3) were probably the first to appreciate that dissolution of cemented carbides in the chip and workpiece, during metal cutting, is a principal cause

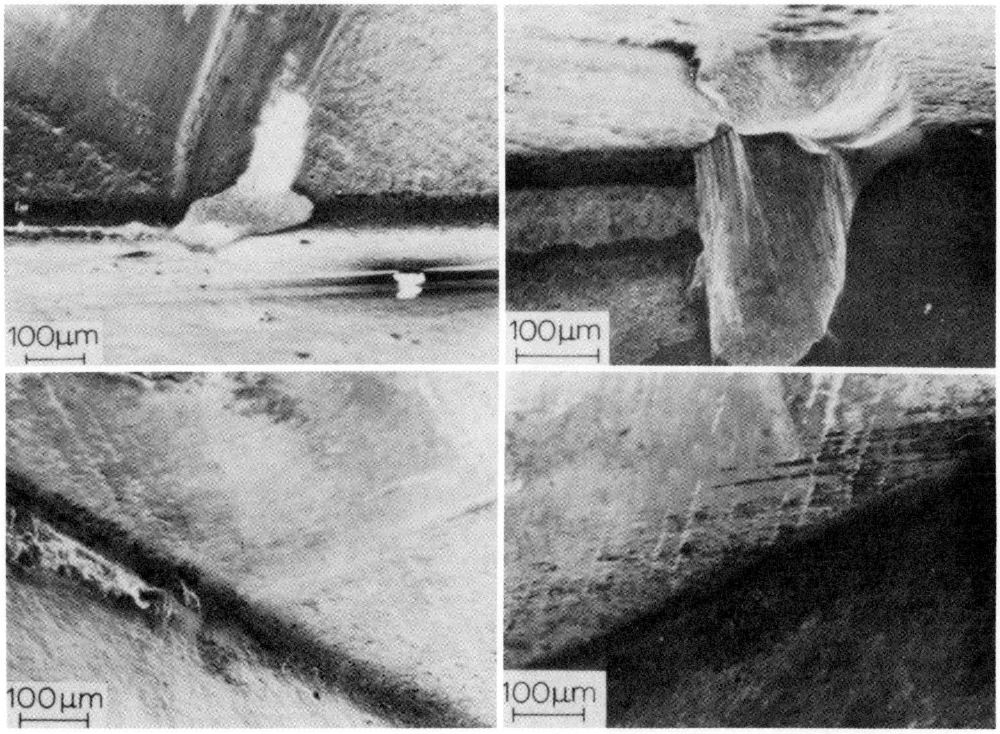

Fig. 20 Influence of Gaseous atmospheres on notch wear when cutting 080M40 at 300 m/min for 5 mins:
(a) TiN/TiC coated carbide (type 2), normal atmosphere; (b) as (a) but in oxygen jet; (c) as (a) but in nitrogen jet; (d) TiC coated carbide (type 1) in Ar jet.

of wear. Both workers attributed the reduced rake face crater wear (observed after adding γ-phase to WC-Co materials) to the fact that γ-phase is less soluble in iron than WC. The quick-stop experiments described in Part 1 showed that the largest part of the rake face contact area (of the coated and uncoated carbide and the Al_2O_3 ceramic tools) is in intimate contact with the chip underside. This type of contact is an essential pre-requisite for the diffusional transport of tool material atoms across the chip/tool interface, as too are the high temperatures (~900°C) generated when cutting steels and cast-irons at 200 m/min (Figs 14, 15 and 16 in Part 1).

It was recently proposed that the solubility product of a tool material in the workpiece material, provides the driving force for dissolution wear via diffusional transport (Ref 12). This hypothesis was developed further (Ref 5) suggesting that the solubility product in the chip-workpiece/tool interface (being proporional to the solubility product in the bulk work material) is the actual driving force for dissolution. This type of "dissolution/diffusion wear" is the most probable cause of the smoothly worn TiN and TiC coatings (Figs 7, 26a and 31), and, also the preferred wear of the WC phase relative to the γ-phase (Fig. 3).

The dissolution/diffusion wear model (Ref 5) predicts that, when rate determining, the coating materials will be worn in accordance with their solubility products in iron and steel (Table 7) i.e., the rate of wear of TiC > TiN > Al_2O_3.

However, the observed rake and flank face wear rates only followed this trend on two occasions (Table 4). Observations of the worn rake and flank faces often showed evidence of discrete plastic deformation of many of the coatings and also the Al_2O_3/ZrO_2 ceramics (Figs 4, 6 and 15), which probably culminated in ductile fracture in the manner depicted in Fig 5. This type of wear mechanism will be dependant upon both temperature and stress. Estimates of the rake face shear stress (Ref 6) indicate that this paramater is higher when cutting 817M40 than 080M40. High cutting stresses can also be inferred, by the fact that the cutting edges of the coated tools used to cut 817M40 were often deformed plastically, while the same tools were unaffected when cutting the 080M40 material. Note that in both cases the distribution and magnitude of temperature was verysimilar for the coated tools (Figs 14 and 15 in Part 1).

The higher rake face shear stress pertaining when cutting 817M40 promotes discrete plastic deformation of the coating near surface, and probably accounts for the relatively higher rate of wear of the Al_2O_3 when cutting this material. This trend was also evident when cutting 316S16 (Table 4). High cutting stresses were also in evidence when cutting this stainless steel, as all the coated carbides exhibited plastic deformation of the cutting edge.

For the TiC coatings the smoother wear, presumably caused by the dissolution/diffusion wear mechanism, competed with the discrete plastic de-

Table 7 Solubility Data for Tool Materials in Iron and Steel.

Tool Material		Solubility product, $(wt\ \%)^2$	Solubility limit wt %	Temp. ^oC	Phase
WC	(8)		7.0	1250	γ-Fe
TiC	(9)	1.4×10^{-2}	3.0×10^{-1}	1250	γ-Fe
TiN	(10)	7.1×10^{-4}	6.4×10^{-2}	1250	γ-Fe
TiN	(10)	1.0×10^{-5}	7.1×10^{-3}	1250	γ-Fe
Al_2O_3	(11)	1.0×10^{-14}	2.0×10^{-7}	1600	Liq Fe

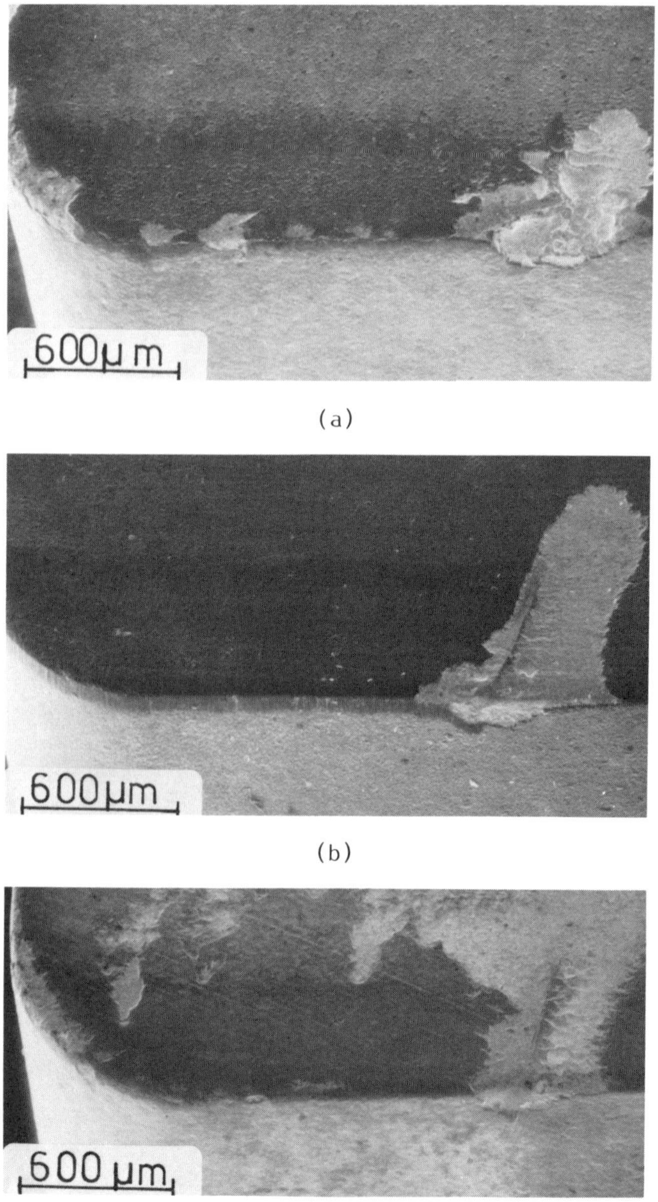

Fig. 21 General views of SPUN 120308 style coated tools after cutting 316S16 stainless steel at 150 M/min, 0.25 mm/rev for 1 min
(a) TiN coated carbide (type 4);
(b) Al_2O_3 coated carbide (type 6);
(c) TiC coated carbide (type 1)

formation mechanism, when cutting 817M40, since the initially ridged rake face (Fig 6a) became smoother after prolonged cutting (Fig 6b).

Since the dissolution/diffusion wear mechanism is strongly dependant upon temperature it is expected that wear caused by this mechanism will be greater on the rake face than on the flank face. This was the case for the uncoated carbides used to cut the steels at speeds ~250 m/min (Table 3), and both wear surfaces were consistant with wear by this mechanism (Fig 3). A similar trend was observed for the TiC coated carbides, even though discrete plastic deformation did contribute to some of the rake face wear. In the case of the Al_2O_3 and TiN coated carbides these coating materials were worn more severely on their flank faces than on their rake faces, Fig. 2, a trend which was also observed for the ceramic cutting tools. This strongly indicates that the flank wear was not rate controlled by dissolution/diffusion for these materials, although some wear by this mechanism is a possibility, especially in the case of the relatively more soluble TiN.

It is proposed that in the case of Al_2O_3 and TiN coatings, flank wear, as well as rake face wear is principally caused by discrete plastic deformation. Since temperatures are much less on the flank face smaller plastic strains are realised and hence such surfaces would superficially appear smooth, (zone A in Fig 14), at low cutting speeds. However, under more extreme cutting conditions plastic deformation is visible (Fig 15).

The high normal force which acts upon the tool in the vicinity of the cutting edge (Ref 13) will be experienced as a very high shear stress on the flank face, thereby promoting the observed high rates of flank wear.

Unfortunately there is not yet much reliable comparative high temperature shear strength data available for TiC, TiN and Al_2O_3. Hardness values suggest that the room temperature shear strength of TiC > Al_2O_3 > TiN (Ref 14). If this is also the case at 800^oC then it would explain why the flank wear rate of coatings of TiN> Al_2O_3 >TiC is very commonly observed (Table 4).

Neither dissolution/diffusion nor small strain discrete plastic deformation mechanisms satisfactorily explain the flank wear rate rankings of the coating materials observed when cutting SG-iron (Table 4). Many investigators have attributed flank wear, especially when cutting cast-iron, to abrasion. This does not seem to be a probable wear mechanism, since neither the steels nor the cast-irons contained particles of sufficient size or hardness to abrade the coating materials. It appears, therefore, that it is unlikely that a sound explanation can be advanced until a better understanding of the physico-chemical characteristics of coating materials at elevated temperatures is obtained.

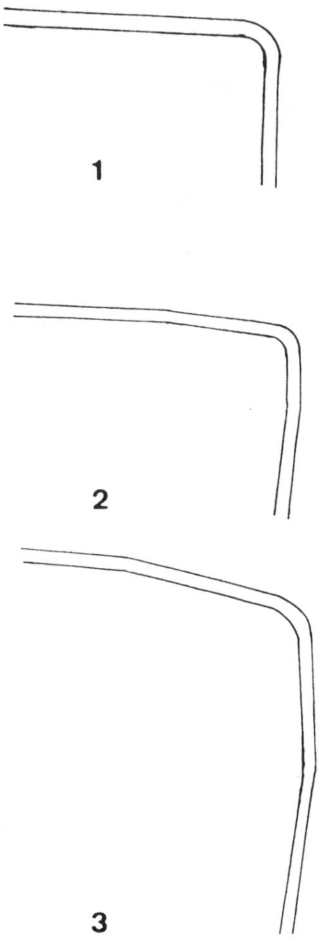

Fig. 22 Schematic depiction of plastic deformation of the cutting edge of coated cemented carbide cutting tools (sequence 1 to 3). Note that the coating deforms to accomodate the deformation of the substrate.

CATASTROPHIC WEAR PROCESSES - The cutting tests made on the cast-irons illustrate that built-up-edge (BUE) formation often severely limits the life of CVD coatings. The coating can be removed as large fragments, either, from beneath (Fig. 23) or behind (Fig. 24a) the BUE. Such effects can also happen when cutting steels at very low cutting speeds (where BUE is also created). However, it is generally possible to avoid BUE by selecting higher cutting speeds when cutting steel. This was not the case when cutting SG-iron, BUE still being observed at 200 m/min (see Part 1), resulting in coating loss at the cutting edge (Fig. 29).

Coating loss is also a significant wear problem when cutting austenitic stainless steel (Figs 11 and 12). Here, spalling of the coating appears to be initiated by fractures in the coating lying parallel to the tool edge (arrowed in Fig 13).

There is therefore an obvious need to improve coating adherence further. The majority of coatings were equally affected by the phenomena just described, although, when cutting GA-iron, the Al_2O_3 coated carbides (type 6) showed a marked resistance to spalling (Table 6).

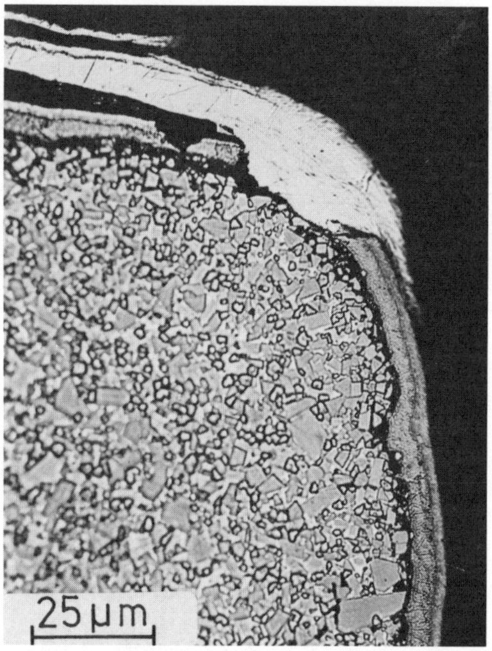

Fig. 23 Loss of coating at the tool cutting edge (tool type 2) after cutting GA-iron for 1 min at 60 m/min.

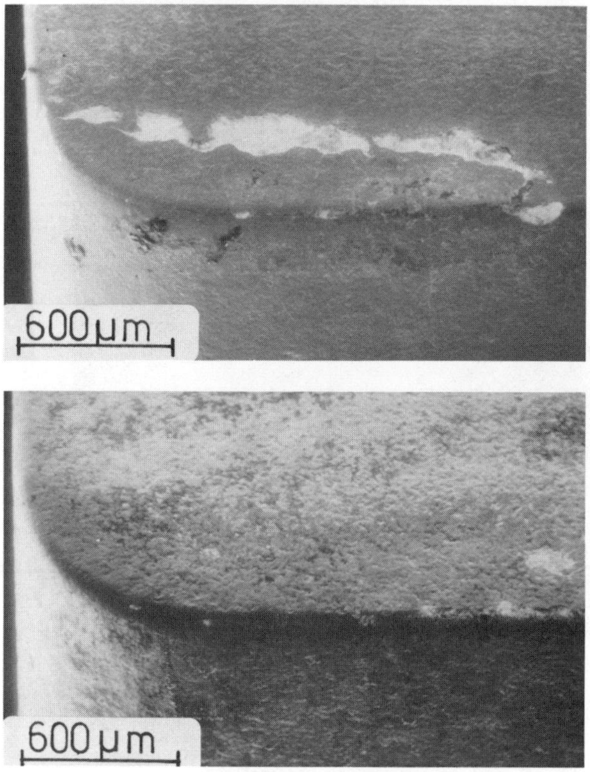

Fig. 24 Wear of coated carbide (type 2) tools used at 60 m/min to cut:

(a) SG-iron for 1 min;

(b) GA-iron for 1 min.

Fig. 25 Worn rake face of uncoated carbide (type 10) showing smooth and uneven wear. Tool used to cut GA-iron at 200 m/min for 15 secs.

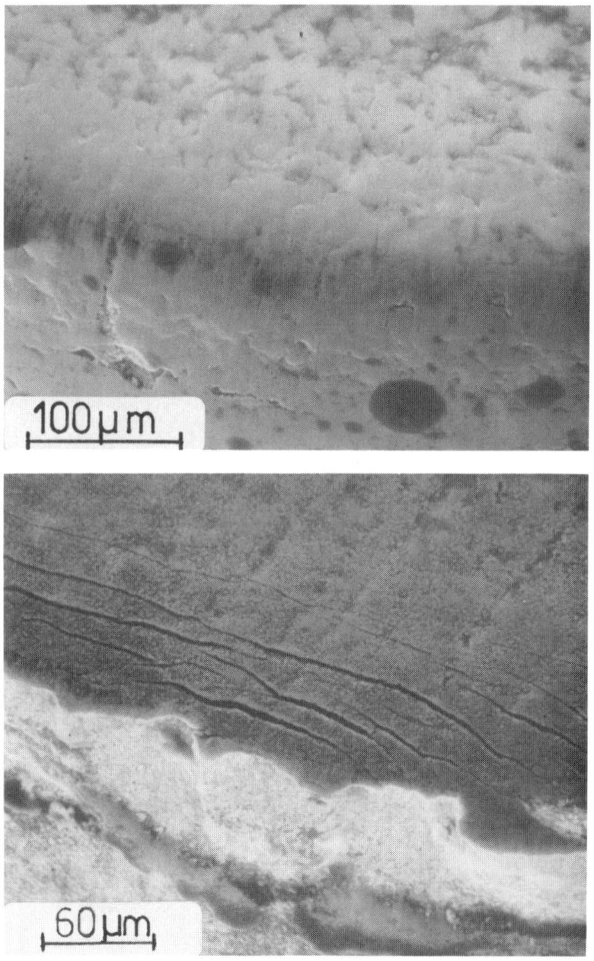

(a)

(b)

Fig. 26 Variability in coating adherence when cutting GA-iron at 200 m/min:

(a) TiN/TiC coated (type 2) after 5 mins

(b) TiC coated tool (type 1) after 1 min.

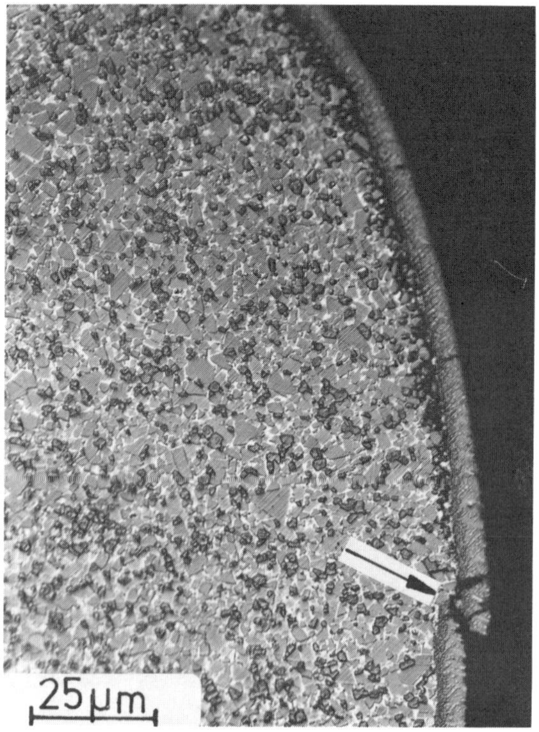

Fig. 27 Metallographic section showing shear fracture in TiC coating beneath cutting edge. Tool type 1 after cutting GA-iron at 200 m/min for 30 secs. Shear fracture is arrowed.

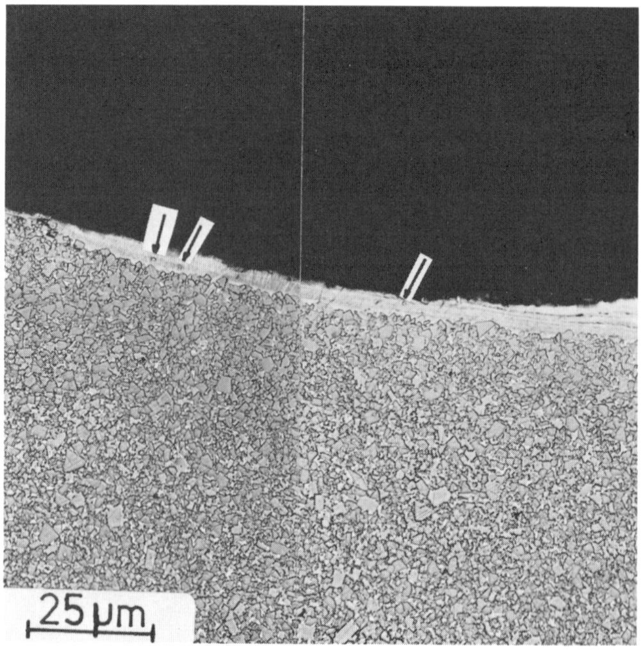

Fig. 28 Metallographic section showing detail of rake face "attrition" wear of WC grains. Tool type 8 (uncoated), after cutting SG-iron at 200 m/min for 30 secs. Adherent work material (white). Note detached WC grains within work material (arrowed).

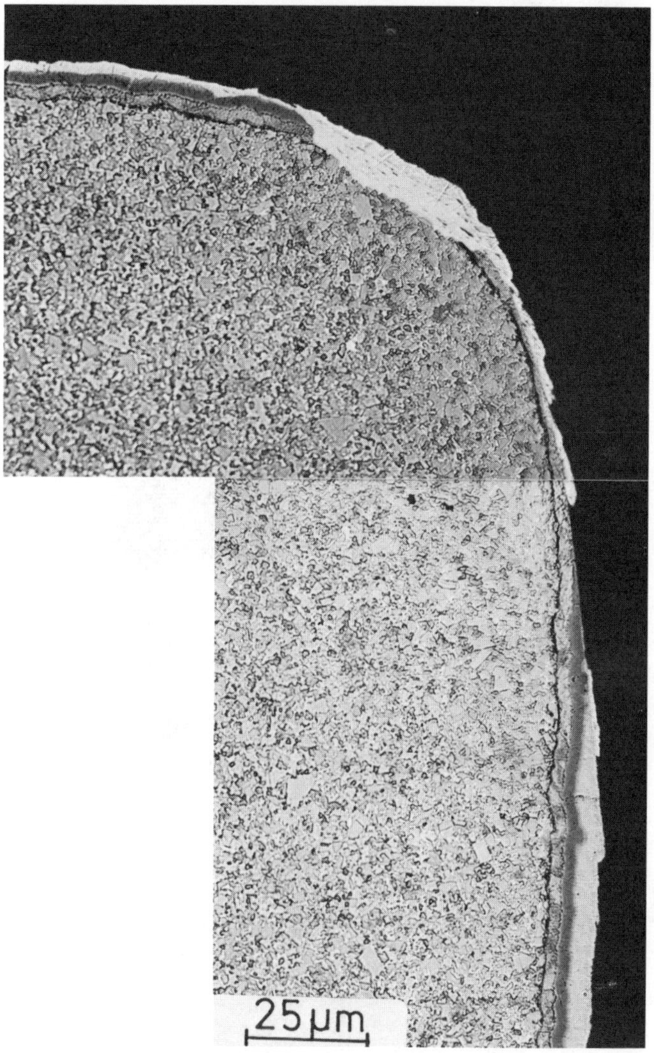

Fig. 29 Severe coating loss at cutting edge of TiN/Ti(C,N)/TiC coated carbide (type 3) after cutting SG-iron at 200 m/min for 1 min. Note rapid, but gradual wear of coating on flank face.

The factors which determine coating adherence are not properly understood, although, it is probably safe to state that residual stresses within the coating and in the vicinity of the coat/substrate interface play a significant, if not critical role. Certain workers believe that η-phase, at the coat/substrate interface retards coating adherence. However, it has been the experience of the authors that coating adherence is equally variable, whether η-phase is present or not.

NOTCH WEAR - The quick-stop experiments showed that the outer sides of the chip underside, make a sliding contact with the tool surface (zone E in Fig. 4 in Part 1). Atmospheric oxygen penetrates this portion of the chip/tool interface causing the smooth notch wear surfaces (zone α in Fig. 18), observed after cutting plain carbon and low alloy steels; proven by

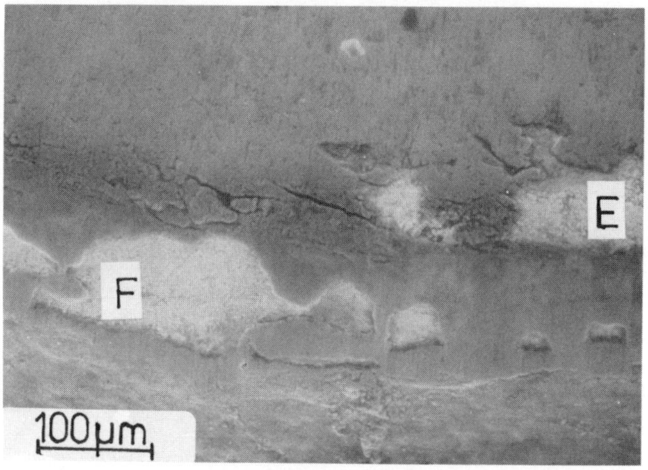

Fig. 30 Sporadic coating loss on cutting edge (E) and flank (F) associated with cracking of coating. TiN/TiC coated carbide (type 2) after cutting SG-iron at 200 m/min for 15 secs.

cutting in jets of oxygen (Fig. 20a,b), nitrogen (Fig. 20c) and argon (Fig. 20d). This type of wear obviously involves oxidation. It is easy to understand that Al_2O_3 coatings are very resistant to oxidation, but it is not so easy to appreciate why TiN should be more resistant than TiC (Table 5).

When TiC and TiN are exposed to oxygen at elevated temperature, oxygen ions diffuse into their respective lattices, displacing carbon and nitrogen atoms, respectively, which themselves diffuse out and are lost to the atmosphere (Refs 15 and 16). Such dissolution of oxygen by the lattice, appears to be more difficult for TiN than TiC (Ref 17), and it is therefore, probably for this reason that TiN coatings exhibit greater

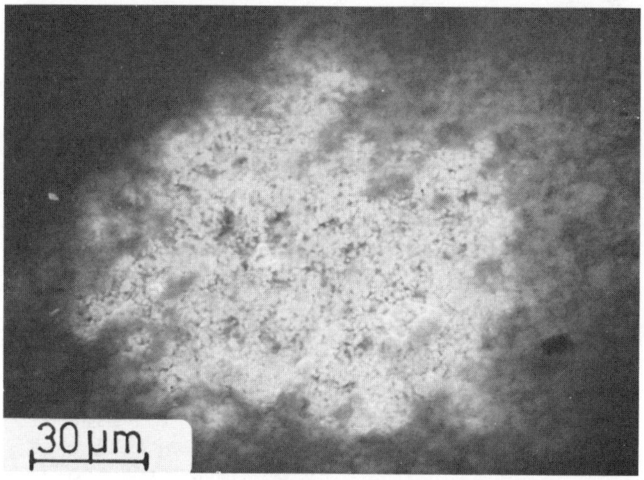

Fig. 31 Detailed view of rapidly and smoothly worn TiC coating on rake face and exposure of substrate beneath (centre). TiC coated carbide (type 1) after cutting SG-iron at 200 m/min for 30 secs.

(a) (b) (c) (d)

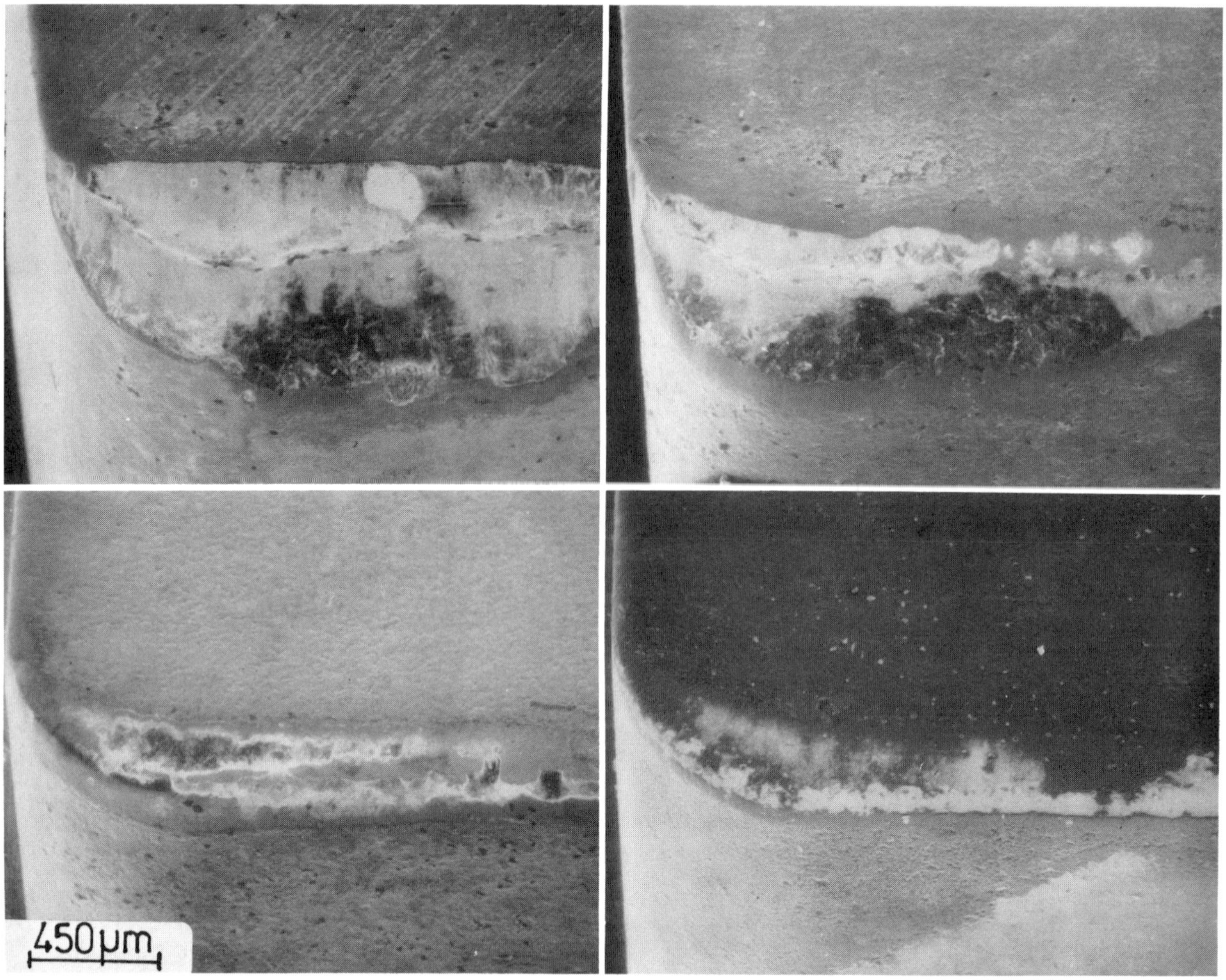

Fig. 32 Wear accumulated on coated cemented carbide tools after cutting SG-iron at 200 m/min for 2 mins:

(a) TiC coated carbide (type 1); (b) TiN/TiC coated carbide (type 2);
(c) TiN/Al_2O_3/TiC coated carbide (type 7); (d) Al_2O_3/γ coated carbide (type 6).

resistance to oxidation controlled zoneα notch wear than TiC coatings.

No visible traces of oxide, not even interference colours, could be seen on the zoneα wear zone. This implies that the oxidised coating is removed when the oxygen enriched region is perhaps only 100 to 200Å thick.

The cracking component of notch wear (zone β in Fig. 18)was probably caused by a fatigue mechanism. At cutting speeds ~180 to 300 m/min approximately 2000 to 4000 segments of chip (i.e., the segmented outside edge of the chip - see Fig. 19 in Part 1) per second pass over the tool rake face. Each segment will impart a small shear stress on the tool surface in the direction of chip travel, and hence the surface will experience ~2000 to 4000 cycles of shear stress per second. The pattern of crack formation in zone β (arrowed 'P' in Fig. 19) is very similar to that observed on materials subjected to fretting fatigue (Ref. 18).

When cutting austenitic stainless steel the entire notch wear surface appeared to be caused by fretting fatigue (Fig. 21). The rate of notch formation was also much more rapid than that observed when cutting plain carbon and low alloy steels, and Al_2O_3, TiC and TiN coatings were all badly affected (Fig. 21). It is suggested that the rapid notch formation is due to the fatigue shear stress being higher when cutting this material. There is insufficient time in this case for oxidation controlled zoneα type wear to show an observable effect.

SIGNIFICANCE OF NOTCH WEAR - In laboratory tests notch wear of CVD coated carbides does not normally limit tool life, since the depth of cut is usually larger than the nose radius of the tool. In many turning operations, coated carbides have to make many repeated cuts at several depth of cut settings. In such cases, accelerated wear of the cutting edge,due to oxidation controlled zoneα type wear, can be expected. Here, TiN and Al_2O_3 coatings would afford superior protection compared with TiC coatings. When cutting austenitic stainless steel, however, it is anticipated that all types of coating will be severely affected by fretting fatigue induced notch wear. Multiple depth of cut settings or a continuously varying depth of cut, as encountered when turning castings, would be especially detrimental to the life of the coatings. Hence, wherever possible, when cutting austenitic stainless steels, the depth of cut should be kept constant.

CONCLUSIONS

(1) In general, coating cemented carbides with single or multi-component layers of TiC, TiN and Al_2O_3 by CVD, is a very effective method for combating wear.

(2) The relative wear resistance of TiC, TiN and Al_2O_3 coatings is strongly dependant upon the workpiece being machined.

(3) Cutting tests performed on a broad range of ferrous workpieces indicate that TiN offers the most potential as a universally applicable coating, for rake face wear protection, despite it being second best in certain instances. Conversely, TiC is the most universally applicable coating for flank face wear protection. TiN also offers good wear protection from oxidation controlled notch wear. Hence, an optimal combination of TiN and TiC coatings, applied to cemented carbides, should confir the wear resistant virtues of both materials.

(4) Al_2O_3 was the most wear resistant coating when cutting SG-iron.

(5) The factors which determine coating adherence require further elucidation, since poor adherence limited the life of most coatings when cutting austenitic stainless steel. Built-up-edge formation, pertaining when cutting the GA- and SG-irons (especially at low cutting speed) also caused coating loss.

(6) All the coatings were uniformly worn when cutting plain carbon and low alloy steels above 150 m/min. In these cases, Al_2O_3/ZrO_2 ceramics, Al_2O_3 and TiN coatings were worn more rapidly on the flank face than on the rake face. However, TiC coatings and uncoated carbides were usually worn more rapidly on the rake face than on the flank face.

(7) Uniform wear of the CVD coatings when cutting the steels was probably caused by dissolution/diffusion and discrete plastic deformation/fracture mechanisms. Such mechanisms probably also contributed to wear when cutting the cast-irons but other wear processes as yet undetermined may also contribute to wear, particularly when cutting SG-iron.

(8) When cutting the steels, the TiN and Al_2O_3 coatings, and also the Al_2O_3/ZrO_2 ceramics are principally worn by the discrete plastic deformation/fracture mechanism (on both rake and flank faces). Even though evidence of this wear mechanism was visible on the TiC coatings, dissolution/diffusion wear was a competative,if not a rate determining wear mechanism, in this case.

(9) For the uncoated cemented carbides dissolution/diffusion wear usually prevailed on the rake and flank faces when cutting the steels, whereas, attrition contributed more significantly to the wear of both surfaces when cutting the cast-irons. The uncoated carbide grade with a high volume fraction of γ-phase was particularly susceptible to attrition.

(10) Notch wear of CVD coated carbides has two causes: (a) oxidation and (b) fretting fatigue cracking. Both types of wear process are observed when cutting plain carbon and low alloy steels. In these cases TiN and Al_2O_3 coatings offer most protection from oxidation controlled notch wear. When cutting austenitic stainless steel, fretting fatigue cracking is the predominant notch wear mechanism, and all three coating materials are poorly resistant to this type of wear.

(11) When cutting austenitic stainless steel with coated carbides, variable depth of cuts should be avoided, otherwise rapid coating loss will result.

ACKNOWLEDGEMENTS

The authors wish to thank the Science and Engineering Research Council and Sandvik Hard Materials Ltd, for financing aspects of this work and the latter for permission to publish.

REFERENCES

(1) Trent, E. M., J. Iron Steel Inst., 201, p.847, 1963.

(2) Loladze, T. N., J. Inst. Eng. (India), p.108, 1962.

(3) Trent, E. M., J. Iron Steel Inst., 201, p.1001, 1963.

(4) Naerheim, Y. and E. M. Trent, Met. Technol. 4, p.548, 1977.

(5) Dearnley, P. A., Surface Engineering, 1, (1), p. 43, 1985.

(6) Dearnley, P. A., Ph.D. Thesis, University of Birmingham, Birmingham, U.K.

(7) Dearnley, P. A. and V. Thompson, Proc. 11th Plansee Seminar, Vol 2, p.709, Reutte, Austria, May 1985.

(8) Edwards, R. and T. Raine, Proc. 1st Plansee Seminar, p.232, Reutte, Austria, 1953.

(9) Irvine, K. J., F. B. Pickering and T. Gladman, J. Iron Steel Inst., 205, p.161, 1967.

(10) Kunze, J. Met. Sci., 16, p.217, 1982

(11) Kiessling, R. and N. Lange, "Non-Metallic Inclusions in Steel", The Metals Soc., 1978

(12) Kramer, B. M. and N. P. Suh, Trans. A.S.M.E. J. Eng. Ind., 102, p.303, 1980.

(13) Barrow, G., W. G. Graham, T. Kurimoto and Y. T. Leong, Int. J. Mach. Tool Des. Res., 22, p.75, 1982.

(14) Dearnley, P. A., WEAR, 101, (1), p.33, 1985.

(15) Voitovich, R. F. and E. I. Golovko, Poroshk. Metall., 3, p.211, 1978.

(16) Pollard, F. H., and P. Woodward, Trans. Faraday Soc., 46, p.190, 1950.

(17) Dearnley, P. A., and E. M. Trent, Met. Technol., 9, p.60, 1982.

(18) Waterhouse, R. B., "Fretting Corrosion", Oxford, Pergamon Press, 1972.

WEAR BEHAVIOR OF TiB_2 AND BORON COATED CEMENTED CARBIDES

D. E. Graham
Carboloy Systems Department
General Electric Company
Detroit, Michigan, USA

Abstract

The effect of chemical vapor deposited TiB_2 and boron overlay coatings on the wear behavior of WC-Co alloys in Al-Si machining and rock abrasion applications was studied. During rock cutting, WC-Co was removed from uncoated samples by a complex mechanism involving plowing and fragmentation of WC grains. The TiB_2 and amorphous boron coatings, as well as a boron diffusion zone, had high mechanical abrasion resistance and coating removal occurred by mild abrasion. Once the boron containing coatings were breached, boride particles and the boron hardened cobalt binder at the coating-substrate interface were effective in holding the WC grains in place, preventing early pull-out.

During the cutting of abrasive Al-Si alloys, TiB_2 and boron coatings again provided a substantial increase in tool life, attributable to the hardness of these coatings.

Both coatings had a deleterious effect on strength; the degree of degradation depended on the presence or absence of a diffusion zone. Strength losses of 35 and 50% occurred in the absence and presence of a diffusion zone respectively.

THE WEAR OF CEMENTED CARBIDE tools is an important factor influencing the efficiency of metal cutting and rock drilling. In metal cutting, tool wear has a direct impact on tool consumption. In drilling, tool wear decreases the rate at which rock is penetrated and increases drilling forces which can lead to an increased tendency for tool fracture. It has been shown that tool wear can occur by a number of different processes, (1, 2) including surface spalling as a result of sudden impact, fatigue spalling as a result of multiple impacts, chemical dissolution, thermal fatigue, and abrasion. The actual mechanism that operates will depend on the cutting conditions and particularly on the properties of the metal or rock being cut. For example, if the rock is abrasive and has relatively low compressive strength (e.g., sandstone), abrasive wear will dominate.

Abrasion resistance can be increased by manipulating several structural and compositional parameters, including cobalt content, WC grain size, binder film thickness, and binder composition. By reducing cobalt content, reducing binder film thickness, decreasing WC grain size, and tungsten solution hardening the binder phase, the abrasion resistance can be increased significantly. Unfortunately, when the abrasion resistance is increased by making such adjustments, the toughness is reduced. Since toughness is indispensable in most applications, alternative means of increasing the abrasion resistance are desirable. This present study was conducted to determine the feasibility of increasing the abrasion resistance of relatively tough cemented carbides through the use of hard diffusion and/or overlay coatings.

There are several coatings and surface treatments incorporating boron which can be used to increase the hardness and/or lower the coefficient of friction of cemented carbides. Titanium diboride (TiB_2) is known to be an extremely hard refractory material and has high potential in applications where conditions of extreme wear are present. Additionally, its hardness is not as temperature dependent as most other refractories. Boron is a refractory element which is also known for its high hardness and strength and has been successfully deposited on various substrates. In addition to these overlay coatings, it has been reported (5,6) that the surface of a cemented carbide can be hardened by diffusing boron into the surface to form a diffusion coating, a treatment similar to other common processes, such as carburizing and nitriding. During the boronizing process, the boron is initially absorbed in solid solution by the cobalt-rich binder phase, but, as more boron is added and as the deposition time increases, several boride phases form. The composition and structure of the phases are dependent on the boronizing conditions and the composition of the substrate.

All of the coatings made for this study were produced using chemical vapor deposition (CVD) techniques. The CVD process is used commercially to produce wear- and corrosion-resistant diffusion and overlay coatings; the method is particularly useful when the thickness and/or homogeneity of the coating must be carefully controlled or when complicated shapes are to be coated.

II. EXPERIMENTAL

The CVD apparatus used to produce the coatings is shown schematically in Figure 1. The reactant gases were H_2, BCL_3, and $TiCl_4$. The H_2 was the carrier gas used to transport the $TiCl_4$ to the mixing chamber, which was heated slightly to prevent condensation of metal chlorides. The substrate cemented carbides were heated in a resistance furnace. The actual deposition conditions used for each of the coatings are given in Table I. Analytical procedures used to examine the coatings and substrates included optical metallography, X-ray diffraction, scanning electron microscopy (SEM), and microprobe analysis.

Three WC-Co alloys were used during the study and included two 6 wt. % Co alloys and a 16 wt. % Co alloy. The compositions and properties of these alloys are given in Table II. All of the microstructural parameters were determined by linear analysis of several SEM photographs, taken at random, of polished surfaces, using a method described elsewhere (8). The test specimens for the abrasion tests were domed compacts, 8 mm in diameter, and were made from the 16 % Co alloy and the fine grained 6 % Co alloy.

Transverse rupture bars 6.3 x 6.3 x 19 mm were made from the 16% Co and coarse grained 6% Co alloys for the determination of three-point bend strength in the coated and uncoated conditions.

The abrasion test, shown in Figure 2a, consisted of pushing a domed compact across a slab of sandstone (Shore hardness = 20.5) under cntrolled conditions so that the leading edge and one side of the dome were continuously crushing and rubbing the sandstone. A resultant wear scar is shown in Figure 2b. The cutting conditions employed

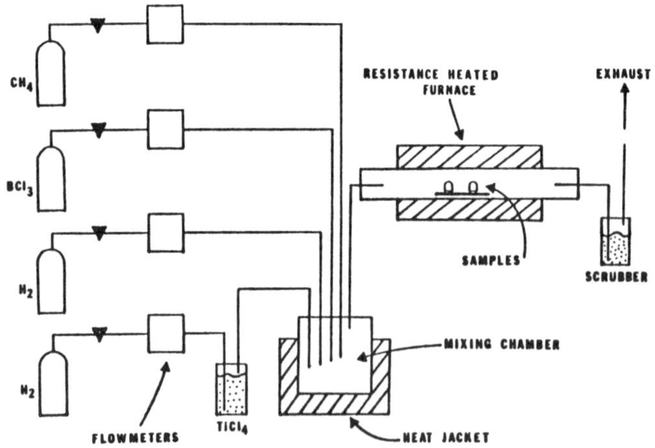

Fig. 1-Schematic of experimental CVD apparatus.

TABLE I

DEPOSITION CONDITIONS

Coating	Deposition Temperature (C)	Input Gas Partial Pressures H_2	BCl_3	$TiCl_4$
TiB_2	800	0.955	0.025	0.02
Boron	1000	0.975	0.025	-
CoWB	800	0.994	0.006	-
W_2B_5-CoB	900	0.99	0.01	-

TABLE II

ALLOY PROPERTIES

Composition	Hardness (Ra)	Grain Size (um)	Binder Film Thickness (um)
WC - 16% Co	86.7	2.5	0.8
WC - 6% Co	92.0	1.2	0.3
WC - 6% Co	92.7	0.8	0.2

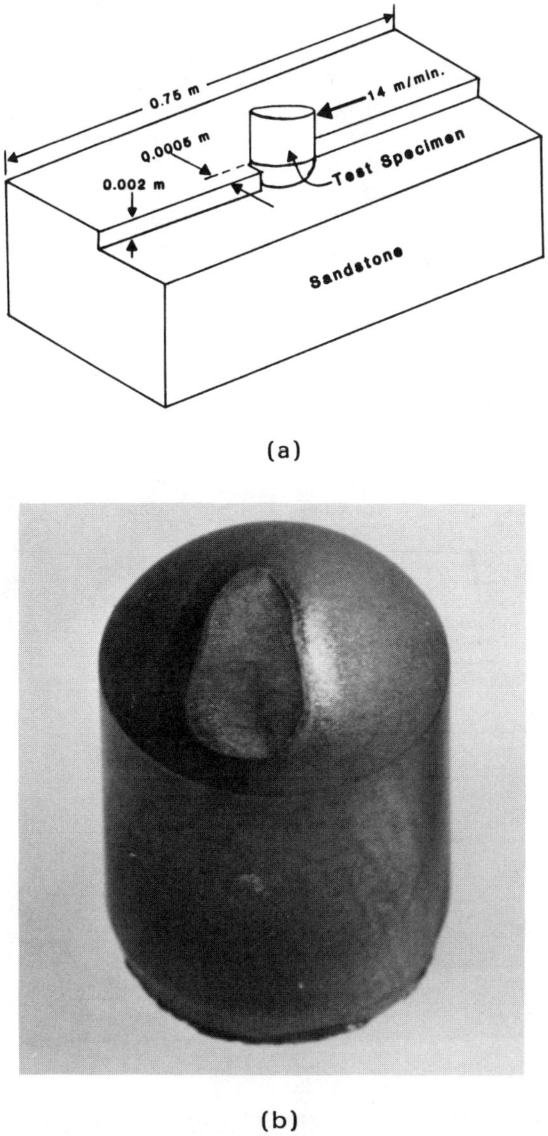

Fig. 2-a) Schematic of sandstone cutting test and b) resultant wear scar on domed compact.

throughout the testing were:

Cutting Speed......14 Meters/Min.
Depth of Cut.......2 mm
Cross-Feed.........0.5 mm
(Width of Cut)
Cutting Stroke.....0.75 Meters
(Length of Cut)

Each test consisted of 126 cutting strokes. The compacts were removed from the test apparatus for weighing every 20 strokes and were replaced so that the orientation was exactly the same every time. After each test a 1-mm layer of sandstone was removed from the surface of the slab using a hard, square cemented carbide insert. This helped to insure that a consistent surface was tested each time. No coolant was used during this study.

This test differs from the rotating disk test (modified Riley-Stoker) normally used to determine the abrasion resistance of cemented carbide. It was found that the combination of the rotating disk and Al_2O_3 grit wore through the coating too early in the test so that the abrasion resistance was only weakly dependent on the surface treatment. The sandstone cutting test is both a more sensitive test and more closely simulates wear mechanisms encountered during actual usage.

To assess the effectiveness of TiB_2 and boron coatings in preventing wear during cutting of abrasive metals, alloy 390, an aluminum silicon alloy containing 17 wt. % silicon was cut at surface speeds of 185 and 275 meters per minute (smpm) using a feed rate of 0.13 millimeters per revolution, and a depth of cut of 2.5 millimeters. A TPG 422 insert style was used in the unhoned condition.

COAT DEPOSITION

TiB_2

TiB_2 is most commonly produced by reducing halides -- most often the chlorides -- of titanium and boron with hydrogen. The synthesis reaction reported most often in the literature and the one used in this study is:

$$TiCl_4 + BCl_3 + 5\ H_2 \text{ -------} \\ TiB_2 + 10\ HCl$$

A thorough analysis of this reaction has been made by Besmann and Spear, (9) who calculated that, under the conditions used here, TiB_2 is easily the most stable phase (although species such as $TiCl_3$ and $HBCl_2$ may form temporarily). In the presence of titanium, elemental boron is not stable below about 1200° C.

X-ray diffraction analysis of the as deposited TiB2 indicated that the coating was crystalline and had a hexagonal structure with lattice parameter values of a=3.05 angstrons and c=3.19 angstroms. The surface morphology of the coating is shown in Figure 3a and a polished cross-section is shown in Figure 3b.

All of the TiB_2 coatings were deposited using the conditions specified in Table I. The effects of altering these conditions have been described by Besmann and Spear and by Blandenet et al. (10)

(a)

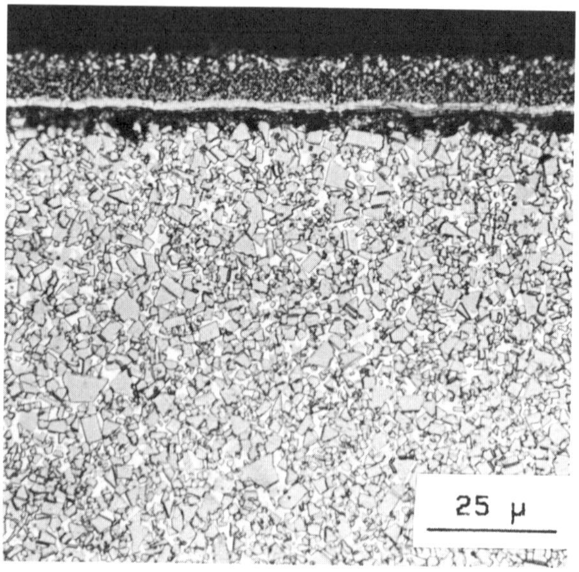

(b)

Fig. 3-a) SEM micrograph and b) microstructure of TiB_2 coating.

Boron

Elemental boron coatings were produced according to the reaction

$BCl_3 + 3/2\ H_2$

$B + 3\ HCl$

at temperatures from 900° C to 1200° C. This reaction has also been studied from thermodynamic and experimental viewpoints by several authors. (4, 5, 11) Consistent with most experimental studies using relatively low temperatures and high supersaturations, the coatings produced in this study were amorphous. The surface morphology and microstructure of the coatings are shown in Figure 4. As shown in Figure 4b, deposition of boron resulted in the formation of a boronized region under the coating. Attempts to limit boron diffusion into the substrate by initially depositing a TiC or TiN interlayer were only partially successful, and this technique was not used during this study. Thus, whenever the amorphous overlay coating is discussed it should be remembered that a W_2B_5-CoB type diffusion zone lies

(a)

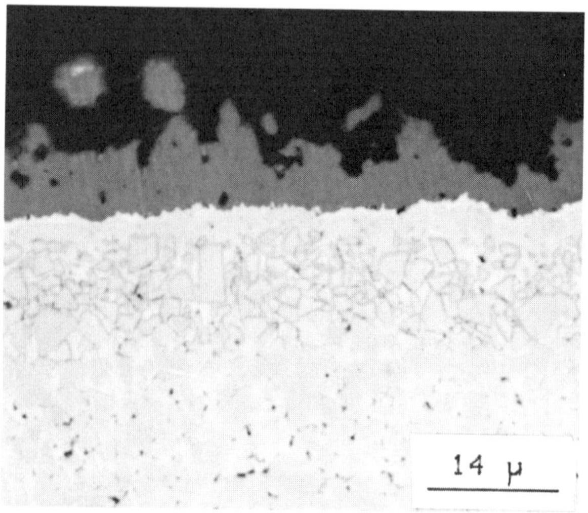

(b)

Fig. 4-a) SEM micrograph and b) microstructure of amorphous boron coating.

below the coating. The diffusion zone thickness was approximately the same thickness as the overlay coating. Whenever coating thicknesses are reported for the boron coating, the thickness of the diffusion zone is included.

Secondary reactions can take place inside the coating furnace if the substrate metal reacts with the gaseous halide. These reactions have been reported to result in severe corrosive attack of the substrate. However, indications of such reactions were not observed here.

Boronizing

Relatively few detailed studies have been published on the subject of boronizing WC-Co alloys. It is known, however, that when these alloys are boronized, three Co-B and four Co-W-B phases can form (6, 12) and that more than one boride phase can be present at one time. The composition, structure, and distribution of the phases depend on temperature, boronizing gas ratios, and substrate composition and grain size.

It is established that boronizing increases the hardness of cemented carbides, and the hardness increase is often reported to be directly related to the boron content. Two types of boron diffusion coatings were studied -- one with a low boron content and one with a high boron content. A CoWB surface layer was produced by using a low boron partial pressure and short boronizing times (see Table I), and a W_2B_5-CoB surface region was produced using a higher boron concentration and longer boronizing times. The structure of these coatings is shown in Figure 5.

Standard X-ray analysis and scanning electron microscopy have indicated that during the boronizing treatment the Co binder is changed to CoB. The border of the WC grains subsequently transformed to W_2B_5 (suggesting that C is at least partially replaced by B), given enough time.

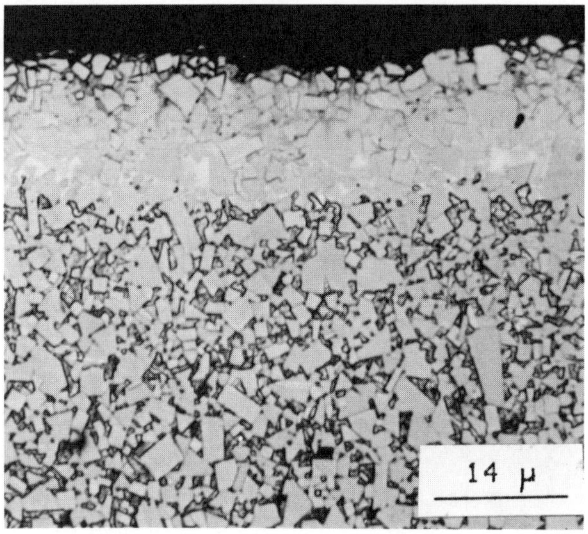

(a)

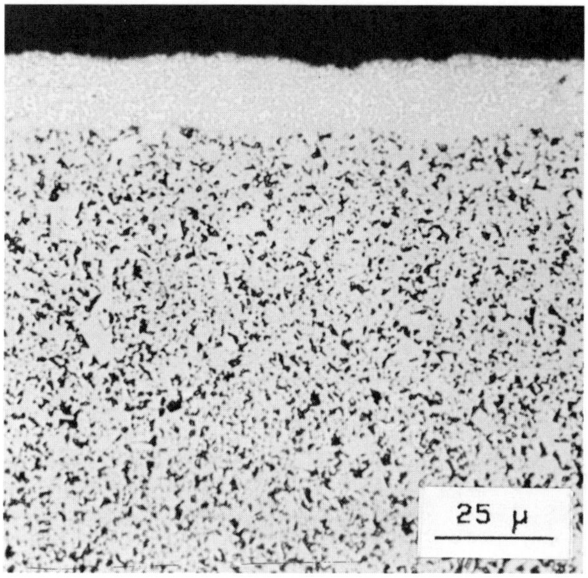

(b)

Fig. 5-Microstructure of a) W_2B_5-CoB and b) CoWB diffusion zones.

ABRASION TESTS

Sandstone Cutting Tests

The abrasion resistance of the uncoated samples (16 % Co and fine grained 6 % Co) was determined on the basis of the volume loss vs. cutting strokes data shown in Figure 6. Several of the 16 % Co samples were tested periodically throughout this study to establish reproducibility and consistency. The variation that was observed is shown. Most of the variation was attributed to the fact that the hardness of the sandstone varied throughout the slab. The scatter in the results was not so severe as to mask the effects of any of the coatings however.

As would be expected from the data in Table II, the fine grained 6 % Co alloy was much more resistant to volume loss than was the 16 wt. % Co alloy.

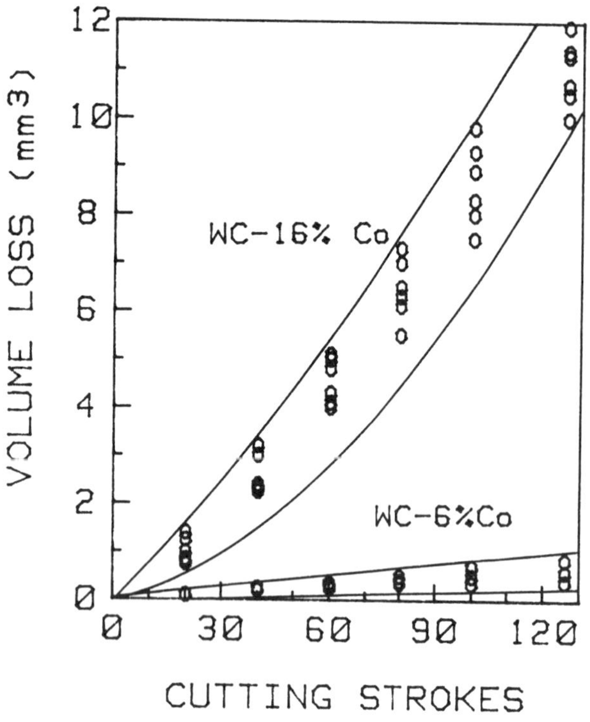

Fig. 6-Wear behavior of WC-16% Co and WC-6% Co in the sandstone cutting test.

For both of the alloys, an essentially linear relationship existed between the volume loss and the number of cutting strokes. An initial period of increasing slope (rate of volume loss) was observed for the 16% Co alloy and is believed to be a result of the domed geometry of the test specimen. Early in the test, the area of contact between the compact and the sandstone is limited but increases as the test progresses.

The effects of the various coatings are illustrated in Figures 7 and 8. The coating thickness for each test sample was 10 microns.

With the exception of the CoWB diffusion coatings, all of the coatings resulted in substantially increased abrasion resistance. The reason for the relatively high scatter in Figure 8b and poor abrasion resistance of the CoWB diffusion coating is not known, although it may be a result of the low and non-uniform boron content and a result of the porous nature of this coating.

A significant result of these tests was that the W_2B_5-CoB diffusion coating was essentially as effective as the boron overlay coating (with its associated diffusion zone) in preventing abrasion. This conclusion is

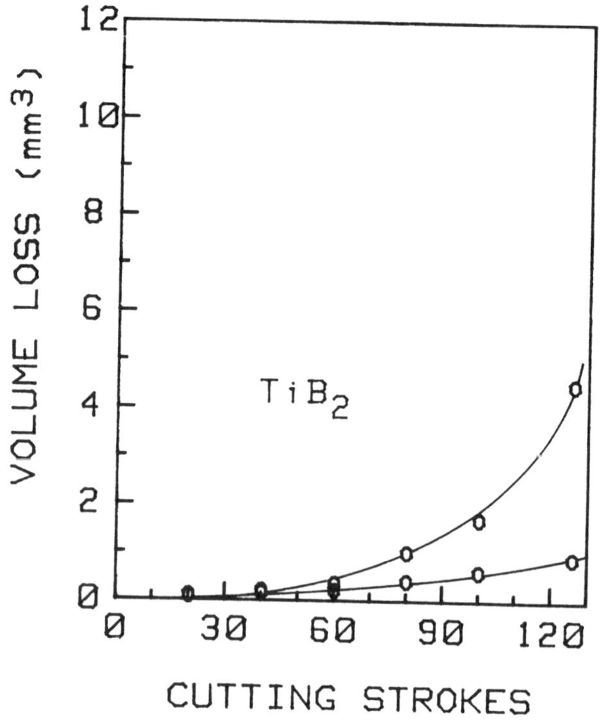

(a)

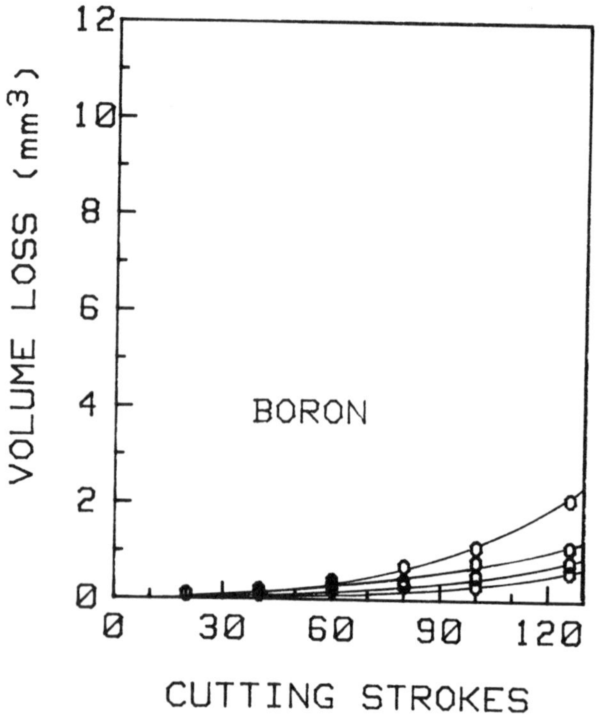

(b)

Fig. 7-Wear behavior of WC-16% Co coated with a) TiB_2 and b) amorphous boron.

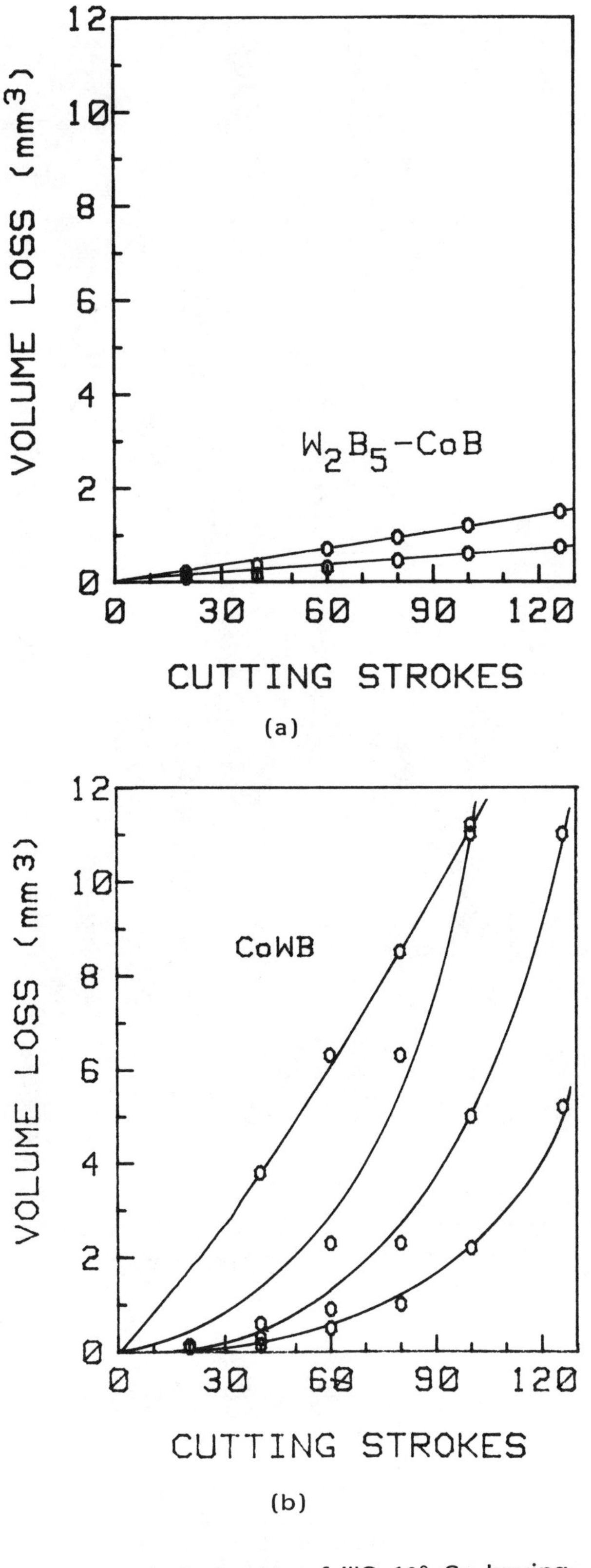

Fig. 8-Wear behavior of WC-16% Co having diffusion zones of composition a) W_2B_5-CoB and b) CoWB.

based on a comparison of Figures 7b and 8a. Reasons for this similarity will be discussed in the section <u>Observations on Wear Surfaces</u>.

The effect of coating thickness was studied for the TiB_2 and boron coatings. In each case it was found that, based on the sandstone cutting tests, abrasion resistance increased directly with increasing coat thickness, illustrated in Figure 9. In Figure 9 an "improvement ratio" is plotted against coating thickness. The "improvement ratio" is defined as the volume loss of the uncoated sample divided by the volume loss of the coated sample. An improvement ratio of 4 indicates that the uncoated sample lost 4 times the volume the coated sample lost.

Figure 9 indicates that amorphous boron provide almost twice the abrasion resistance that TiB_2 did. While this difference is greater than anticipated, the order is consistant with the expected hardness difference. Crystalline boron has a reported hardness of approximately 3200 kg/mm^2(13) compared to 3300 kg/mm^2 for TiB_2. Since it is known that amorphous materials are harder than their crystalline counterparts, amorphous boron should be harder than TiB_2.

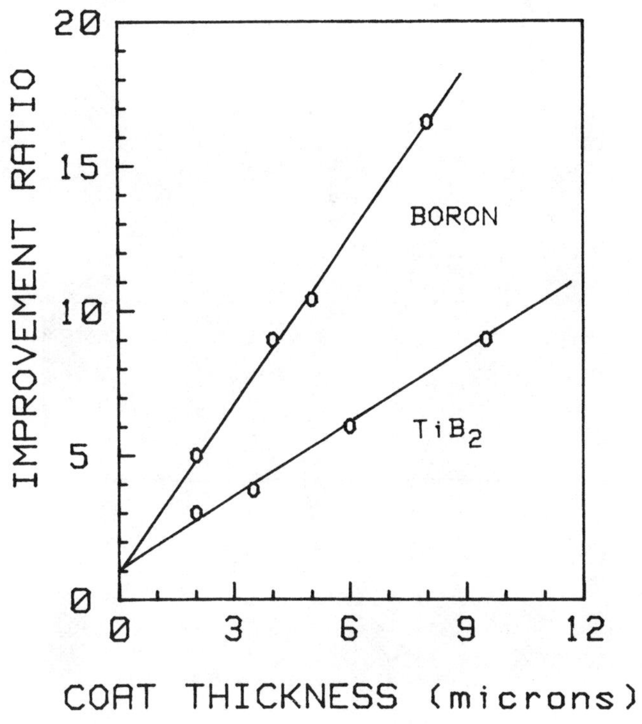

Fig. 9-Data illustrating the effect of TiB_2 and boron coating thickness on the wear behavior of WC-16% Co.

OBSERVATIONS ON WEAR SURFACES

Several uncoated 16 % Co samples were carefully polished to 1 micron diamond and rubbed across the sandstone surface for a distance of 10mm and subsequently examined in the SEM. As shown in Figure 10, the surface of the alloy was already grooved by quartz grain asperities in the sandstone via a "microcutting" mechanism. Cobalt was preferentially removed, probably by extrusion from between the tungsten carbide grains to accommodate deformation. (Co smearing along the sides of grooves was occasionally observed, which supports this supposition.) As abrasion continued, material was still being removed by the formation of grooves, but additional features such as pit formation (presumably a result of microspalling -- possibly due to fatigue), tungsten carbide fragmentation, tungsten carbide pullout, and wear debris were observed, as shown in Figure 11. This damage suggests that, on subsequent passes, material removal is probably facilitated by damage created during the first few passes during which the surface structure was disturbed and cobalt binder removed. These results are essentially consistent with those of other wear studies. (14, 15)

Fig. 10-Scanning electron micrograph illustrating the early wear behavior of WC-16% Co.

(a)

(b)

Fig. 11- a) and b) Scanning electron micrographs illustrating the intermediate stages of wear on the WC-16% Co alloy.

The CVD coatings wore away at a rate dictated by hardness which is consistent with SEM observations which showed that the coatings were worn away via a "polishing" type mechanism, where the grooves that were cut into the surface of the coating were very shallow. Once the coating was breeched, the mechanism discussed for uncoated carbide began to operate.

The wear mechanism which occurred on the compacts having a W_2B_5-CoB diffusion coating were more complex, as illustrated in Figure 12. This

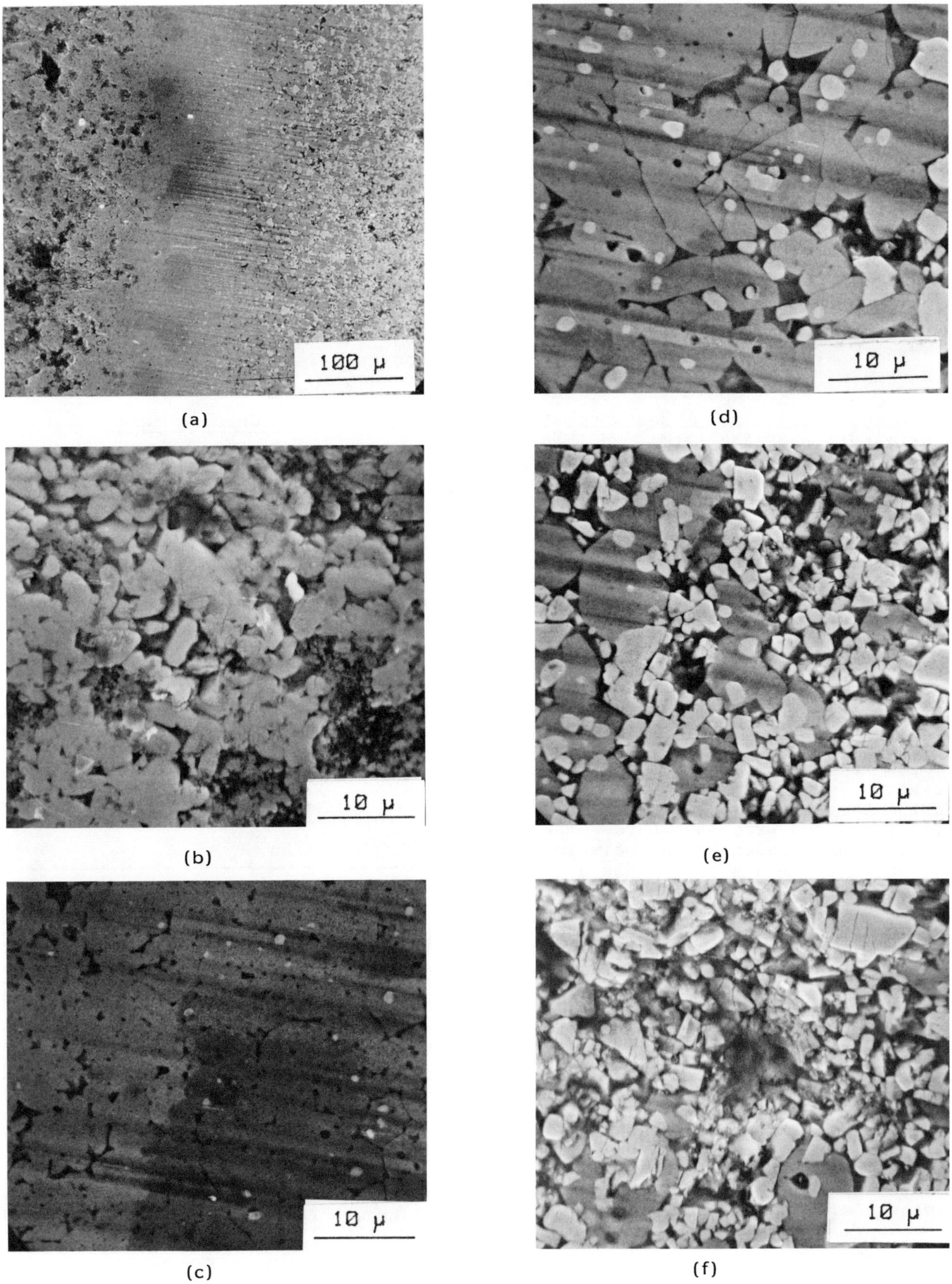

Fig. 12–a) to f) Scanning electron micrographs illustrating the progressive wear behavior of boron coated WC-16% Co. Micrographs b) to f) are taken moving from left to right on micrograph a).

figure is, in effect, a "taper polish" section of a compact boronized to W_2B_5-CoB and then coated with amorphous boron. In Figure 12a, a photograph of one side of the wear scar, the boron coating is at the far left, the boronized layer appears as the smooth region at the center of the figure, and the exposed substrate is at the right. Wear of the boron coating, which like TiB_2 occurs by a polishing mechanism, is shown more clearly in Figure 12b. This polishing mechanism continued to operate until the coating and the boron diffusion zone were completely worn away, the last stages of which are shown in Figures 12c and 12d. The gray area in Figures 12c and 12d is the W_2B_5-CoB structure, and the white areas are WC grains. It can be seen that these two phases wore at approximately the same rate. Additionally, it appears the boronized zone was effective in holding tungsten carbide grains firmly in their "settings", preventing premature fracturing and grain pullout. Most (although not all) of the cavities that appear formed during the boronizing process (see Figure 5) and were not the result of WC grain pullout. Eventually the boronized zone was worn through and the mechanism observed during the abrasion of untreated carbide began to operate. This is illustrated in Figures 12e and f.

On the basis of these observations, it seems likely that the boron diffused zone was effective in preventing abrasion for two reasons. First, the boride compounds were much harder than was the cobalt binder, thus resisting abrasion on the basis of increased hardness. Secondly, it seems reasonable to assume that the boride phases had increased strength compared to the original binder and therefore resisted material removal by preventing deformation and WC fracture. These two properties rendered this diffusion zone as resistant to this type of wear as was the amorphous boron overlay coating.

Rock Drill Simulator Tests

A rock drill simulator test was also used to assess the abrasion resistance of various surface treatments. During this test, a domed compact like that shown in Figure 2b was forced into the surface of a rock, pulled back out, indexed to a new area of the rock, and pressed into the surface again. Gabbro rock was used as the work material, and a 35,500 newton force was applied to the compact. Indexing occurred at a rate of 5 hertz. Abrasion resistance was determined by taking periodic profilameter traces throughout the test. Two such traces are shown in Figure 13 for uncoated and TiB_2 coated 16 % Co samples. These traces suggest that a slight increase in abrasion resistance resulted from the presence of the coating.

The boron-overlay-coated and boron-diffusion-coated samples broke up early in the test, which suggests that the surfaces of the compacts were embrittled by the presence of the boron. This is consistent with the transverse rupture data to be discussed in subsequent section.

Metal Cutting Tests

The results of the aluminum-silicon alloy machining tests are shown in Figures 14a and 14b. Only the fine grained 6 % Co alloy was used as a tool material during these tests. At both speeds, normal abrasive flank wear was the only operating wear mechanism. Both coatings provided substantial abrasion resistance, with the boron coating providing slightly more than the TiB_2 coating, especially at the lower cutting speed. At both speeds the time to 0.25 mm flank wear was at least doubled with the presence of a coating.

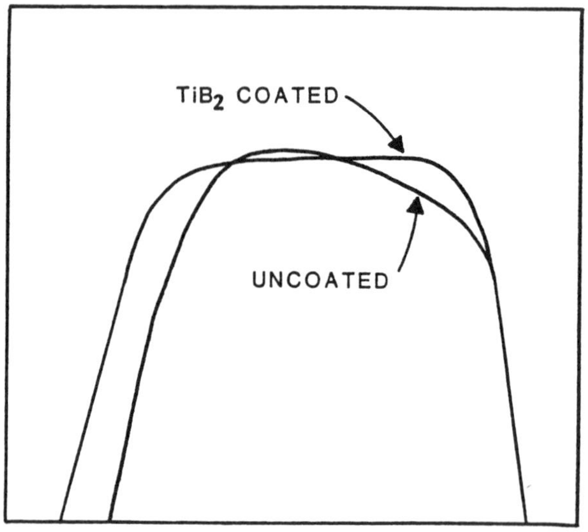

Fig. 13-Profilameter traces of uncoated and TiB_2 coated compacts after a rock drill simulator test.

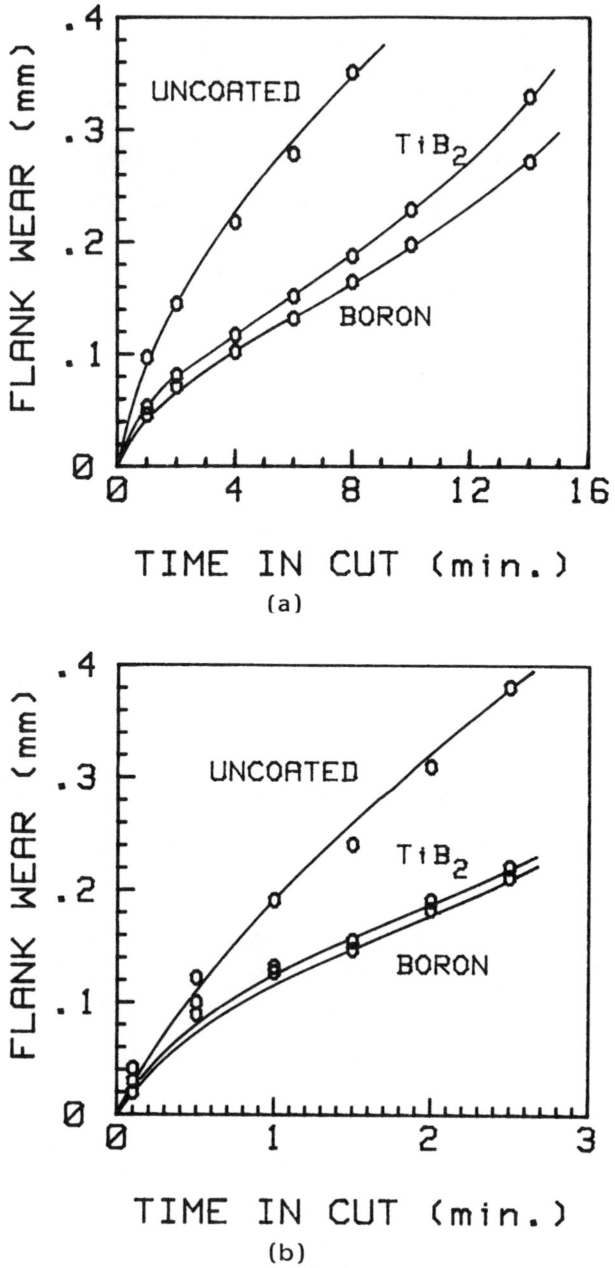

Fig. 14-Wear behavior of uncoated, TiB_2 coated, and boron coated SPG 422 inserts made from the WC-6% Co alloy. The work material was an aluminum-silicon alloy. Cutting speeds were a) 185 surface meters per minute and b) 275 surface meters per minute. For both tests the feed rate and depth of cut were 0.13 mm per revolution and 2.5 mm respectively.

BREAKAGE RESISTANCE

Transverse Rupture Tests

Transverse rupture test results are summarized in Table III. All of the test bars for a given grade were taken from a single powder batch and were given identical processing treatments. For all of the coated bars tested, the total coating thickness (which included coating plus diffusion zone for the boron coatings and total coating thickness for TiB_2-coated bars) was 25 um. All of the coatings significantly lowered the strength of both of the alloys tested. Amorphous boron and the W_2B_5-CoB diffusion coating both lowered the bend strength of the grades substantially more than did the TiB_2 coatings.

TABLE III

EFFECT OF SURFACE TREATMENTS ON BEND STRENGTH

Material	Transverse Rupture Strength (MPa)	Number of Samples
WC-16% Co	2750	10
Boron coated WC-16% Co	1240	5
WC-16% Co with W_2B_5 zone	1240	5
TiB_2 coated WC-16% Co	1655	5
WC-6% Co	2345	10
Boron coated WC-6% Co	1380	5
WC-6% Co with W_2B_5 zone	1240	5

The data in Table III show that the diffusion treatments had a proportionally greater effect on the 16% Co alloy even though the diffusion zone thicknesses were the same for both alloys. This phenomenon is probably related to the volume fraction of borides in the near surface region. Because of the greater cobalt content, there were more borides at the surface of the 16% Co alloy.

CONCLUSIONS

Several boron containing diffusion and overlay coatings were produced on WC-Co alloys and resulted in substantially increased abrasion resistance over uncoated material, as determined by sandstone and aluminum-silicon alloy cutting tests. TiB_2 and amorphous boron coatings increased the wear resistance of a 6% Co alloy two-fold in aluminum-silicon alloy tests. The same two coatings increased the wear resistance of a 16% Co alloy seven- and thirteen-fold respectively in a sandstone cutting application. A W_2B_5-CoB diffusion coating also increased abrasion ten- to eleven-fold in sandstone cutting tests.

All of these results were consistent with the estimated hardness of the coatings. Additionally, the abrasion resistance was found to be directly related to the coating thickness for all of the coating thicknesses tested.

It was also found that the breakage resistance of WC-16% Co and WC-6% Co alloys was substantially reduced by the presence of a coating, the reduction being greatest for boron and W_2B_5-CoB coatings and least for TiB_2. Thus, there is a trade-off between hardness and breakage resistance.

Because of the very low fracture strength of the amorphous boron coating and the boron diffusion coatings, it is felt that the range of usefulness for these coatings will be limited to non-impact applications such as rubbing or wear surfaces. The TiB_2 coating, however, appears a viable candidate material where a combination of impact and abrasive wear is encountered.

ACKNOWLEDGEMENTS

J. M. Conner provided experimental assistance by running the abrasion tests. Informative discussions with T. E. Hale, W. C. Yohe, and E. W. Goliber are gratefully acknowledged.

REFERENCES

1. H. J. Osburn, Powder Met., 12 p. 471.
2. J. Larsen-Basse, Powder Met., 16, (1973), p. 1.
3. K. Nakano, H. Matsubara, T. Imura, J. Less Common Metals, 47, (1976), p. 259.
4. A. A. Cochran and J. B. Stephenson, Met. Trans. 1, (1970), p. 2875.
5. J. Thebault, R. Naslain, and C. Bernard, "Chemical Vapor Deposition - Sixth International Conference", The Electrochemical Society, New York, 1977, p. 304.
6. R. Bonetti and H. E. Hinterman, "Chemical Vapor Deposition - Sixth International Conference", The Electrochemical Society, New York, 1977, p. 351.
7. A. V. linial and H. E. Hintermann, "Wear of Materials 1979", The International Cnference on Wear of Materials, ASME, New York, 1979, p. 403.
8. H. E. Exner, Powder Met., 13, (1970), p. 429.
9. T. M. Besmann and K. E. Spear, J. of Crystal Growth, 31 (1975), p. 60.
10. G. Blandenet, Y. Lagarde, J. P. Morlevat, and G. Uny, "Chemical Vapor Deposition - Sixth International Conference", The Electrochemical Society, New York, 1977, p. 330.
11. L. Vandenbuckle and G. Vuillard, "Chemical Vapor Deposition - Sixth International Conference", The Electrochemical Society, New York, 1977, p. 330.
12. K. Yamamura, M. Kasai, and H. Harigaya, Powder and Powder Metallurgy (Japan), 16, (1969), p. 230.
13. W. A. Powell, General Electric Report, 65-MPD-1, 1965.
14. R. I. Blombery, C. M. Perrot and P. M. Robinson, Mat. Sci. and Eng., 13, (1974), p. 93.
15. J. Larsen-Basse and E. T. Koyanagi, paper presented at the ASME-ASLE Joint Lubrication Conference, Minneapolis, MN, October 24-26, 1978, ASME Paper No. 78-Lub-19.

FRICTION AND SURFACE CHEMISTRY OF IMPLANTATION MODIFIED METAL/CARBIDE COMPOSITES

Sara A. Dillich
Worcester Polytechnic Institute
Worcester, Massachusetts USA

Irwin L. Singer
Naval Research Laboratory
Washington, D.C. USA

ABSTRACT

Dry sliding friction measurements were performed on model carbide/metal composite samples containing 24 to 42 volume % carbide. Wear debris, located preferentially on the carbides, were analysed using SEM/EDX and Auger Scanning Microscopy (SAM). The friction and wear of the composite samples were found to be influenced by the compatibility (degree of mutual solubility) of the carbides with the steel counterface, as well as by local inhomogeneities in the carbide distribution. Friction tests were also made on the samples following high fluence Ti implantation. Implantation resulted in increased friction and wear, possibly due to an implantation induced softening or degradation of the carbides.

CARBIDE STRENGTHENED ALLOYS, (e.g., Stellite alloys, tool steels, 440C steel) are valuable wear resistant materials. The carbides provide high hardness to the alloys, but little else is known about their influence on friction or on initial wear and transfer film formation. The intent of this study was to identify the role of the carbide phases in the friction and wear behavior of carbide-strengthened alloys. To this end, low speed, dry sliding friction measurements were made on model carbide/metal matrix composites tested against 52100 steel, followed by light and electron microscope examinations of the wear scars.

From previous analyses of commercially available composites (e.g. a Co-based superalloy, Stoody 3) it was realized that carbide/metal matrix microstructures are, in general, too fine to easily distinguish the tribological effects of different phases. For this reason, carbide/metal matrix composites formed by a new surface processing technique, the Laser Melt Particle Injection process (LMPI), were used in this study. The composite surfaces were prepared at the Naval Research Laboratory by injection of TiC and WC particles into Ti or Al alloy surfaces (1,2). These LMPI surfaces offer a unique opportunity for the investigatin of dry sliding friction and wear of transition metal/carbide composite surfaces. The large sizes of the carbides (on the order of 100 microns) allow easy identification of separate phases with both light and electron microscopes. Chemical profiling of debris and microwelds on friction tested surfaces are possible using high resolution scanning auger microscopy (SAM) in conjunction with ion sputtering. This cannot be easily performed on conventional metal-carbide composites (e.g. superalloys and cemented carbides) for which the carbides are typically an order of magnitude or more smaller. Thus, specific contributions of the carbide and matrix phases to the galling friction and wear of an alloy system can be investigated by taking advantage of the "magnified picture" provided by the LMPI samples.

Ion implantation is another surface modification technique which has been found to improve the tribology of engineering alloys (3-5). Because of the extreme shallowness of the implant depths (≃100-200 nanometers), implantation can also be used as a research tool, to assist in the determination of controlling wear modes operative at the surface during the initial stages of wear. High fluence Ti implantation was also successful in reducing friction and wear of a Co-based superalloy (6). However, the influence of implantation on the separate phases of metal/carbide alloy systems with conventionally sized carbides is not easily deduced. To aid in this determination, and in the characterization of the tribology of the steel - metal/carbide couples, friction tests were also made on the LMPI samples after high fluence Ti implantation.

EXPERIMENTAL PROCEDURE

A description of the laser processing technique and equipment can be found in References (1&2). Disks 1.27 cm. in diameter by 0.27 cm. thick were cut from carbide injected plates. Before testing, the surfaces were diamond polished to a 1 micron finish and cleaned with organic solvents. The matrix alloy compositions, as well as carbide types, sizes, and average volume fractions, of the test disks are listed in Table I.

Friction measurements were made in ambient air (40% - 60% humidity, room temperatures) using a ball on disk test geometry. The stationary balls (AISI 52100 steel, 1.27 cm. diameter) made contact with the moving LMPI disks. A sliding velocity of 0.1 mm/sec was used, with an applied normal load of 9.8 N. Except when noted, tests consisted of 20 unidirectional passes of a ball on a disk. The first pass of each test was 5 mm in length, while subsequent passes over the same track were 3 mm long. The applied normal load and the tangential frictional forces during testing were monitored continuously on a dual trace chart recorder. The coefficient of kinetic friction, μ_k, was calculated for each pass as the ratio of the tangential force to the normal force and may be considered constant ($\pm$ 25%) unless stickslip (i.e., large amplitude oscillations in the tangential force) is noted.

Polished LMPI disks were implanted in a modified Model 200-20A2F Varion/Extrion ion implanter with a hot cathode arc discharge type ion source. The samples were mounted onto a water cooled holder and the target chamber was cryogenically pumped to pressures of about 3×10^{-6} Torr or better. The disks were implanted with Ti ions at either a) 190 keV beam energy to a fluence of 5×10^{17} Ti/cm^2 or b) 100 keV beam energy to a fluence of 6×10^{17} Ti/cm^2.

Wear tracks on the disks were examined using Scanning Electron Microscopy (SEM) and light microscopy with Differential Interference Contrast (DIC). The elemental compositions of debris in the tracks and on the balls were examined by Energy Dispersive X-Ray analysis (EDX) using a JSM-U3 SEM equipped with a Kevex spectrometer. Composition versus depth profiles of worn surfaces were performed with two different Auger microprobes: one, a Perkin Elmer (PHI) model 600 microprobe, the other with a model 590 microprobe. Auger derivative spectra were taken with electron beam voltage at V = 2kV and modulation amplitude at V = 3 or 6 eV. Depth profiles were recorded during ion milling with a rastered 2 keV Ar ion beam.

Table I - Carbide Types, Sizes and Volume Fractions in Alloys Tested

Disk #	Alloy	Carbide Type	Size (μm)	Volume* %
80-5	Ti-6Al-4V	TiC	105-210	36 $\pm$ 2
80-20	"	WC	105-149	24 $\pm$ 4
80-21	"	WC	105-149	37 $\pm$ 5
80-31	6061 Al	TiC	74-88	31 $\pm$ 6
80-32	6061 Al	TiC	74-88	42 $\pm$ 5

* average values, from Ref. 2

RESULTS

A summary of the dry sliding friction coefficients, μ_k, and the EDX analysis of the wear debris on 1 and 20 pass tracks is shown in Tables II & III for nonimplanted and implanted surfaces, respectively. A range of values, instead of a single average value, listed for a friction coefficient in Table II & III denotes the occurrence of stickslip. EDX analysis of debris on ball scars from selected 20 pass tracks is summarized in Table IV.

FRICTION - In general, the nonimplanted disks showed lower μ_k values than did the implanted disks. The lowest friction coefficients ($\mu_k \cong 0.2$) after 20 passes were observed on the TiC/Ti sample 80-5). The WC/Ti disks on the other hand, exhibited

Table II - Coeff. of Friction and Wear Debris Analysis for AISI 52100 Balls Sliding Against Laser Processed Disks

Sample #	Test #	1st pass	Debris*	20th pass	Debris*
80-5 TiC/Ti	1	0.16	no	0.25	no
	2	0.18	no	0.18	no
	3	0.15	no	0.24	no
80-20 WC/Ti	1	0.23	no	.08-.42	Ti,Fe
	2	0.21	no	.20-.56	"
	3	0.20	no	.16-.44	"
80-21 WC-Ti	1	0.23	no	0.41	no
	2	0.16	no	0.46	no
	3	0.18	no	0.43	no
80-31 TiC/Al	1	0.17	no	0.25	no
	2	0.18	no	.20-.38	Al,Fe
80-32 TiC/Al	1	0.18	no	0.36	no
	2	0.18	no	0.35	no

* Identified by Energy Dispersive X-Ray analysis (EDX)

Table III - Coefficients of Friction and Wear Debris Analysis for AISI 52100 Balls Sliding Against Laser Processed Disks Implanted with Ti

Sample #	Implant+	Test #	1st pass	Debris*	20th pass	Debris*
80-5	a	1	.28-.56	Ti,Fe	0.35	Ti,Fe
TiC/Ti		2	.20-.52	"	0.34	"
	b	1	0.28	Ti,Fe	0.39	Ti,Fe
		2	0.24	"	0.38	"
		3	0.24	"	0.29[c]	"
80-20	b	1	.30-.60	Ti,Fe	0.30[d]	Ti,Fe
WC/Ti		2	0.42	"	.22-.48	"
		3	0.48	"	.32-.56	"
80-21	a	1	0.42	Ti,Fe	.12-.42	Ti,Fe
WC/Ti		2	0.48	"	.14-.46	"
80-31	b	1	0.32	Al,Fe,Ti	.20-.38	Al,Fe,Ti
TiC/Al		2	0.31	"	.20-.48	"
		3	0.33	"	0.36[c]	"
80-32	a	1	0.45	Al,Fe,Ti	.22-.50	Al,Fe,Ti
TiC/Al		2	0.42	"	.20-.50	"

+ a: $5x10^{17}$ Ti/cm^2 at 190 keV, b: $6x10^{17}$ Ti/cm^2 at 100 keV
* identified by Energy Dispersive X-Ray analysis (EDX)
c after 7 passes; d after 3 passes

much higher friction ($\mu_k \cong 0.45$), roughly twice as high as the TiC/Ti disk. The disk with the lowest carbide fraction (80-20, 24±4% WC) exhibited stickslip behavior as well. Intermediate values of friction coefficients ($\mu_k \cong 0.35$) were measured on the TiC/Al disks.

Ti implantation produced an increase in the μ_k values for each of the LMPI disks. Also, stickslip occurred earlier and more frequently during testing.

WEAR SCAR AND DEBRIS ANALYSIS - The majority of wear tracks on the nonimplanted surfaces could not be seen by either DIC or SEM microscopy. That is, no damage to the surfaces or debris particles could be detected. One exception was the low carbide WC/Ti disk, 80-20, which showed fine grooves in the Ti-6Al-4V matrix and wear debris along the tracks. Fig. 1 shows a portion of a 20 pass track covered by smeared wear debris. This debris, which is rich in Ti according to EDX analysis (Table II), must have been transferred from the matrix to the ball, then transferred to the carbides when they passed under the ball.

The effects of local variations in carbide concentrations on wear behavior were also seen during tests on the TiC/Al disk, 80-31. Although the average carbide volume for this sample was measured as 31 ± 6%, test #2 was made in a track containing only 24 ± 2% concentration (as revealed during DIC examination of the track), indicating that contact occurred primarily between the 52100 ball and the Al matrix phase. The wear track for test #2 contained debris which EDX analysis identified as rich in Al and Fe. In contrast, no wear scar or debris was observed in test #1 on the same disk, for which the local carbide concentration was higher (36 ± 2%) than the average value for the disk.

After implantation, all tracks made on the disks were clearly visible. In addition, wear debris containing Fe (from the steel ball) was found in the first pass track on each of the implanted disks (Table III). An illustration of debris on the single pass track of test #1 on WC/Ti disk 80-20 is seen in Fig. 2, which shows large debris fragments attached to carbides. Another example is the 20 pass track on disk

Fig. 1 - Wear track on nonimplanted WC/Ti disk, 80-20, 20 pass region

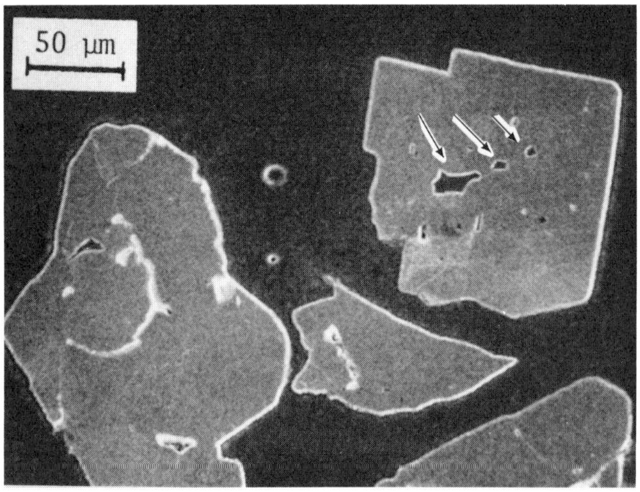

Fig. 2 - Debris on test track #1 on WC/Ti disk 80-20 after Ti implantation, one pass region.

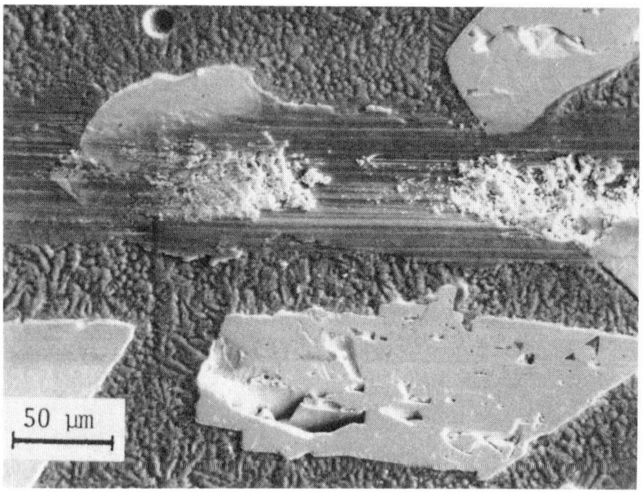

Fig. 3 - Wear track on WC/Ti disk 80-21 after Ti implantation, 20 pass region

80-21, shown in Fig. 3, illustrating the plowing damage deformation to the matrix phase. The Ti implanted TiC/Ti disk, 80-5, was the least degraded by wear. Although considerable debris was generated and deposited on the carbides, the matrix phase of this surface appeared undamaged (Fig. 4, left). The carbides in the implanted disk showed more damage after wear than in the nomimplanted disk. As seen in the DIC micrograph in Fig. 4 (right), sliding produced grooves in the TiC particles. Even more severe damage to the carbides was observed on the other implanted disks. Occasionally, carbide particles in the implanted tracks appeared to have been fractured and fragmented during testing, permitting fine carbide debris to be distributed in some of the tracks (Fig. 3).

Although no damage nor debris could be found on the majority of the nonimplanted tracks, SEM/EDX analysis of wear scars on the corresponding 52100 ball surfaces found evidence of some material transfer to the balls during testing (Table IV). For example, scars on the balls run against the nonimplanted TiC/Ti sample, 80-5 were detectable only by a slight discoloration. Nevertheless, the EDX spot analysis of the ball track indicated the presence of Ti. Similarly, inspection of the ball scar formed against the TiC/Al disk 80-32 showed both Ti and Al. In the case of the balls tested against the high carbide volume WC/Ti disk, 80-21, EDX showed a large W signal (See Fig. 5), indicating that particles from the WC phase adhered and transferred to the ball surfaces. In all cases much larger amounts of debris material were found on ball scars formed against implanted disks. The debris again consisted of an amalgam of disk and ball materials, as well as implanted Ti.

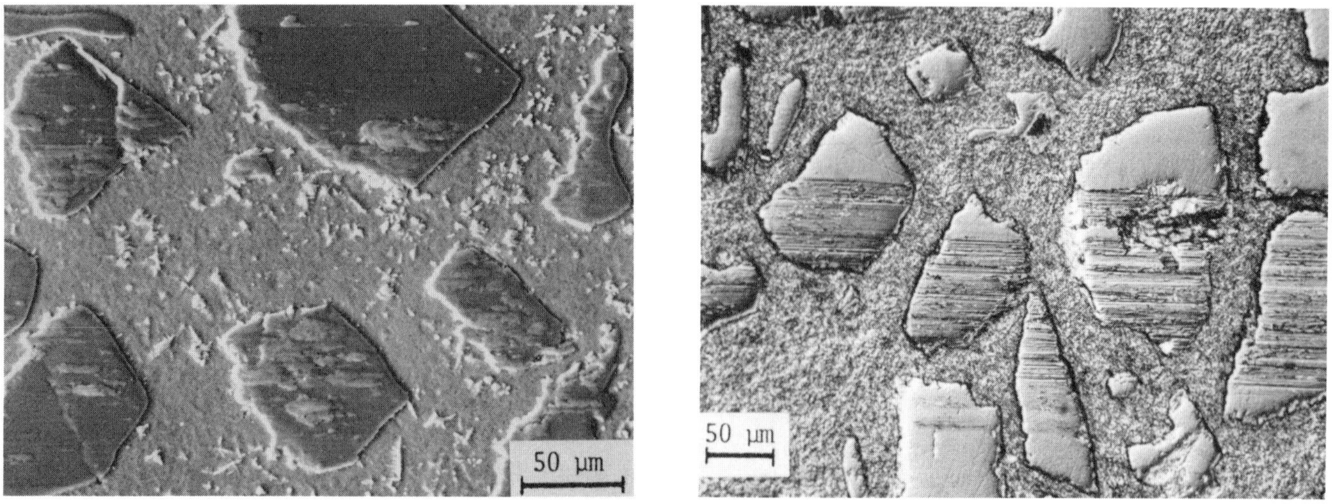

Fig. 4 - 20 pass regions of wear tracks on Ti implanted TiC/Ti disk, 80-5
Left: SEM micrograph Right: DIC micrograph

Table IV - EDX Analysis of Wear Debris on AISI 52100 Steel Balls

Counter Surface	Implanted	# of Scars Examined	Debris
TiC/Ti (80-5)	NO	2	Fe,Ti
TiC/Ti (80-5)	YES	3	Fe,Ti
WC/Ti (80-21)	NO	2	Fe,W,Ti
TiC/Al (80-31)	NO	1	Fe,Ti,Al
TiC/Al (80-32)	NO	2	Fe,Ti,Al

AUGER ANALYSIS - Auger sputter depth profiles of a portion of the 20 pass wear track on the Ti-implanted TiC/Ti disk are presented in Fig. 6. Three features of the track, shown in the circles in the SEM photo, were profiled concurrently during ion milling: a) non-worn Ti alloy matrix (away from the visible dendritic phase); b) a debris flake attached to a TiC particle; and c) a debris-free carbide surface. Both the Ti alloy matrix and the TiC particle showed changes in surface composition as a result of Ti implantation. The Ti alloy was diluted near the surface by O and C, presumably either by implantation mixing of a surface film TiO_xC_y or by implantation induced absorption of C and O from residual gases (7). The TiC acquired a shallow O layer, and as expected, a surface enriched in Ti. The debris particle, as illustrated in Fig. 6b, consisted of an outer layer rich in Fe oxide followed by an increasing proportion of Ti oxide. Both Fe and Ti were found only in an oxidized state, as deduced by the Ti and Fe line spectra; this portion of the d(NE)/dE spectra, obtained after 10 minutes of sputtering, is shown in the inset of Fig. 6b. As debris material was sputtered away, the underlying Ti-rich TiC surface appeared. Note that the sputter rate was increased after 20 and 40 min., in effect compressing the true depth scale (not shown) by factors of $\simeq 3$ then $\simeq 5$. A second debris flake was profiled at the same time and virtually identical compositions to that of Fig. 6b were obtained.

Auger sputter depth profiles were also obtained from four debris flakes stuck to the ball scar worn against the 20 pass track just analysed. The debris flakes were mixtures of Fe and Ti, and their oxides, but with only 1/3 to 1/2 the amount of O as the debris on the track. Two flakes had nearly equal amounts of Fe and Ti, and two had five times more Fe than Ti. The reduced O content of the debris flakes on the ball scar as compared to the disk track suggest that the debris oxidized after they had transferred to the track.

DISCUSSION

Debris formation on the nonimplanted disks was observed only on those low carbide tracks which also exhibited stickslip friction, i.e., tracks on the WC/Ti disk 80-20 and track #2 on the TiC/Al disk 80-31. This, and the presence of considerable disk material on balls run against the high friction, nonimplanted WC/Ti disks, indicates that the observed friction behavior can be correlated with adhesive interactions and material transfer between the contacts.

By the compatibility criteria, first introduced by Rabinowitz (8,9), friction couples with high compatibility (i.e., a high degree of mutual solid solubility) experience greater friction and wear than do those with a lower degree of compatibility. This may account then, at least partially, for the lower friction coefficients and wear of the TiC/Ti disk relative to the WC/Ti disks. That is, the solubility of WC in bcc Fe is much higher than that of TiC. For this reason, TiC tools used for the high temperature machining of steel show a much

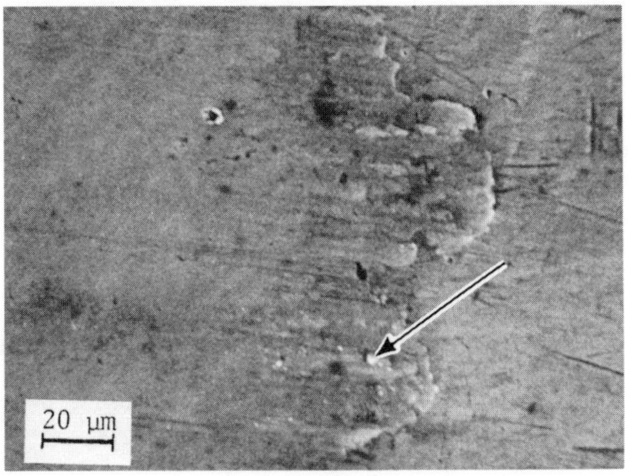

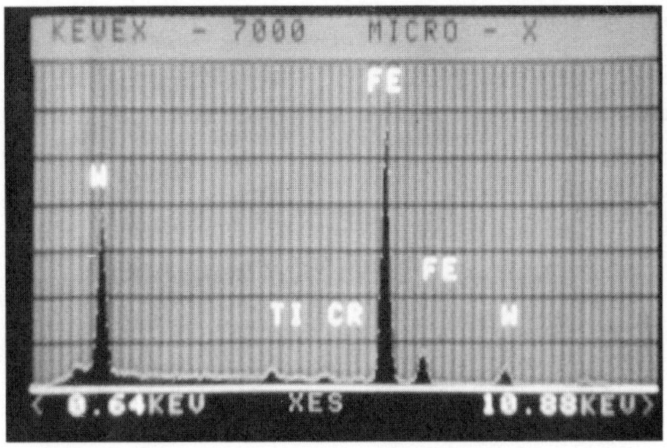

Fig. 5 - Left: Wear scar on steel ball tested against WC/Ti disk Right: EDX signal of a spot on the ball scar

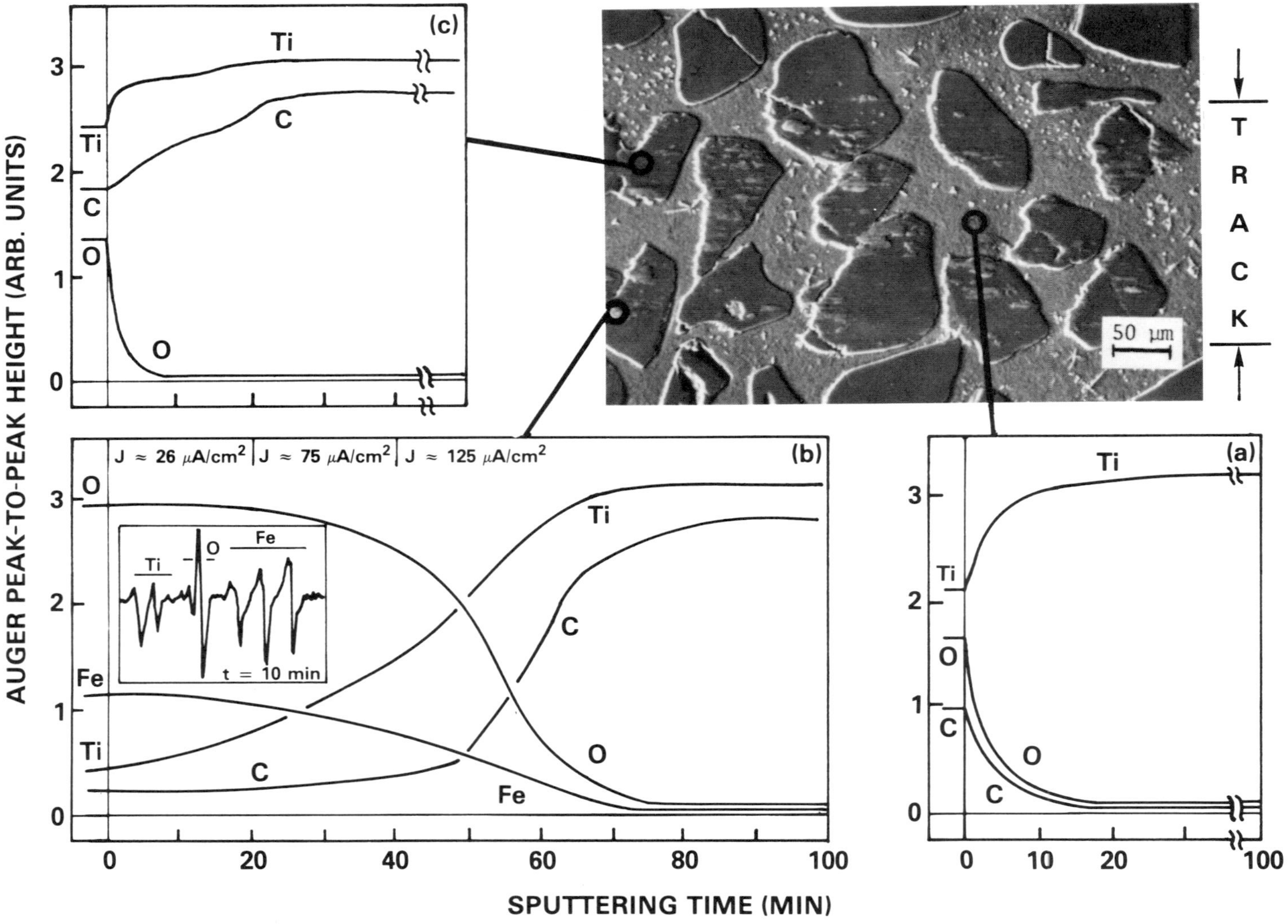

Fig. 6 - Auger sputter depth profiles of a portion of the 20 pass wear track on the Ti implanted TiC/Ti disk, 80-5. Three features of the track shown in the circles in the SEM photo, were profiled concurrently during ion milling: a) non-worn Ti alloy matrix (away from the visible dendritic phase); b) a debris flake attached to a TiC particle; and c) a debris-free carbide surface.

lower wear rate than do WC cutting tools (10). Although the contacts were, of course, at a much lower temperature during these tests, it is speculated that adhesive welding may have occurred as a result of the solution wear of WC in steel.

The difference in friction and wear between the TiC/Ti and WC/Ti samples may also be partially explained by the relative hardnesses of the TiC and WC particles (2600 kg/mm and 1500 kg/mm, respectively, Ref. 2), since, in general, harder materials exhibit superior friction and wear resistance. Also, a considerably lower degree of matrix hardening (28% maximum, Ref. 2) was achieved for the WC injected disks as a result of partial carbide dissolution during processing than for the TiC/Ti disk (122% maximum hardening, Ref. 2). Consequently, both the carbides and matrix of the WC/Ti disks were softer and more susceptible to wear than those of the TiC/Ti disk. The Al matrix of the TiC/Al disks was much softer than the Ti matrix samples (2), and did not harden appreciably as a result of laser processing (2). However the presence of the TiC particles allowed for lower friction and wear than for the WC/Ti disks.

The injected carbide particles provided the excellent wear resistance observed for the nonimplanted disks. The large carbide sizes and high carbide volume % of the disks insured that many, if not the majority, of contact events between ball and disk took place at the carbides. This explains, then, the lack of surface scratching, grooving or deformation on these samples during most of the tests made before implantation. On the low % carbide WC/Ti disk, 80-20, however, the matrix phase was more accessible to the ball, allowing for more debris accumulation on the ball to occur by the process of "prow formation", described by Antler and others (11-13). This process initiates by the adhesive transfer of material from the contact member with the larger surface area to the smaller contact, (e.g. from a disk to a ball). The transferred debris may then cause secondary wear by abrasion (12).

Apparently, in the case of tracks on disk 80-20, debris accumulated on the ball was able to produce abrasive damage and roughening on the matrix phase surface. After repeated passes over the test tracks, the debris was backtransferred to the disk so that, it is speculated, the onset of severe stickslip friction (at approximately the 10th pass) during testing corresponded to debris material on the ball making contact with debris smeared on the carbides, with resultant increases in area of contact and adhesive welding. The debris on the carbides appear to consist of stacked overlapping layers of material, a typical feature of wear debris produced by prow formation (11). The occurrence of stickslip and debris formation during test #2 on TiC/Al sample 80-31 (Table II) is further evidence that inhomogeneities in carbide concentration are the weakest link in the wear resistance of these carbide/metal composites, with the carbide poor regions providing the initiation points for adhesive wear.

Ti implantation of the LMPI disks promoted debris formation and increased friction during dry sliding as a result, it is suggested, of 1) softening of the carbide particles (discussed further below) and 2) enhanced reactivity of the disk surfaces due to the presence of excess Ti on both matrix and carbide phases. In contrast to the nonimplanted disks, the adhesive interactions between the balls and the disks were strong enough to produce high friction coefficients, occasional stickslip, and even some material transfer to the carbides, during the first passes on the implanted surfaces. The large amounts of debris generated on the implanted disks, and their chemical analyses, were further indications of an increased reactivity/affinity between the steel and the Ti implanted surfaces.

The extreme chemical activity of titanium insures strong adhesion between titanium alloys and contacting surfaces. Indeed it is the resulting high adhesive friction and galling wear during sliding contact which prohibits the use of these alloys in applications requiring wear resistance. Buckley attributes this high surface reactivity to the low d-band characteristic of the metallic bonding of titanium (14). That is, since relatively little d-band electron energy is involved in the bulk metallic bonding, very strong d-band interactions can take place between titanium atoms and contacting metal and insulator surfaces.

High fluence ion implantation has been found to cause softening in many ceramic and cermet materials (15,16). In particular, WC was found to soften after Ti implantation to a fluence of 5×10^{17} Ti/cm^2 at 400 keV (16). Ti implantation seems to have softened the carbides of the LMPI samples, as evidenced by the scratches on the carbides of the TiC/Ti disk after implantation (Fig. 5), and by the carbide debris found in several of the post-implantation tracks on other disks.

Apparently, the softer implanted layers wore off, contributing to the accumulation of debris on the ball surfaces during the initial passes of testing. Thus, decreased hardness of the carbide phase, and the increased adhesion, promoted sufficient debris accumulation on the ball to produce visible wear of the Ti and Al alloys. The large amounts of debris backtransferred to the disks (selectively to the carbides) is believed to have produced the severe stick-slip friction during the latter parts of the tests. The high coeffs. of friction of the implanted disks, then, reflect the increased forces required to shear the transfer film contacts at the ball/debris/disk interface.

The friction and wear properties of the TiC/Ti disk were degraded less severely than those of the other disks. The matrix phase appeared undamaged after testing, perhaps due to immediate adhesion and backtransfer of debris on the ball to the carbides. Analysis of debris on the TiC/Ti tracks and on the corresponding ball surfaces could not definitively determine the origen of the debris, since both matrix and carbide phases of the disk (as well as the implant species) consisted of Ti. However, the undamaged appearance of the matrix phase, as well as the presence of Ti on ball surfaces run against the nonimplanted TiC/Al disks, indicate that debris material was sheared initially from the implanted TiC particles.

Although there was considerable debris on the disk (Fig. 5), stickslip disappeared after the first few passes. Oxidation of the steel and disk debris after backtransfer to the disk (as indicated by the Auger data, Fig. 6), may have contributed to the prevention of the severe stickslip characteristic by diminishing the chemical reactivity at the contact interface.

CONCLUSION

1. High galling wear resistance during dry sliding against steel is provided by high carbide volume % LMPI carbide/metal composites. However, since carbide-poor regions serve as nucleation sites for adhesive wear, local inhomogeneities in carbide density are an important factor in the determination of galling wear resistance.

2. The lowest friction and wear behavior was exhibited by the TiC/Ti sample. This can be attributed to the superior hardness of this disk, and to the lower degree of compatibility (i.e. mutual solid solubility) of TiC/steel couples, as compared to WC/steel couples.

3. Stickslip friction, the result of adhesive microwelding at the contact interface, can occur when the reactivity of the materials is very great, as was the case on the first pass on the TiC/Ti disk following implantation . Severe stickslip, leading to high frictional losses, however, is most frequently indicative of a large amount of debris coverage on both counterfaces with resulting debris-debris contact and concommitant increases in contact area, adhesion and backtransfer.

4. High fluence Ti-implantation produced increased friction and wear of the LMPI disks as a result of the a) enhanced reactivity of the surfaces provided by the excess Ti and b) softening of the carbides.

ACKNOWLEDGMENT

We thank Jack Ayers and Bob Bolster for the use of the laser injected alloy samples and the Ion Implantation Branch at NRL for performing the ion implantation. One of the Authors (ILS) wishes to thank Ron Lee (Naval Surface Weapons Center) for the use of the model 590 analyzer and Nancy Finnigan at the Center for Microanalysis of Materials, U. of Illinois (DOE supported) for obtaining Auger data on the model 600 analyzer.

REFERENCES

1. Ayers, J. D., R. J. Schaefer and W. P. Robey, J. Met. 33(8), 19-23 (1981).

2. Ayers, J. D., and R. N. Bolster, Wear 93, 193-205 (1984).

3. Singer, I. L., and R. A. Jeffries, J. Vac. Sci. Technol. A1, 317 (1983).

4. Singer, I. L. and R. A. Jeffries, "Friction, Wear and Deformation of Soft Steels Implanted with Ti or N" in Mat. Res. Soc. Symp. Proc. vol 27, p. 667, Elsevier, N. Y. (1984).

5. Singer, I. L., "Tribomechanical Properties of Ion Implanted Metals" in Mat. Res. Soc. Symp. Proc. vol 27, p 585, Elsevier, N. Y. (1984).

6. Dillich, S. A., R. N. Bolster and I. L. Singer, "Friction and Wear Behavior of a Cobalt - Based Alloy Implanted with Ti or N" in Mat. Res. Soc. Symp. Proc. vol 27, p 637, Elsevier, N. Y. (1984).

7. Singer, I. L., and T. M. Barlak, Appl. Phys. Lett. 43, 457-459 (1983).

8. E. Rabinowicz, "Friction and Wear of Materials", 75, John Wiley and Sons Inc., N. Y. (1966).

9. Rabinowicz, E., ASLE Trans. 14, 198-205 (1971).

10. Kramer, B. M., ASME J. of Eng. Ind. 102, 303-309 (1980).

11. Antler, M. "Sliding Wear of Metallic Contacts" in Electrical Contacts 80, p 3, Illinois Institute of Technology, (1980).

12. Antler, M., ASLE Trans. 13, 2152-2161 (1962).

13. Cocks, M., J. of Appl. Phys. 33, 2152-2161 (1962).

14. Buckley, D. H., "Surface Effects of Adhesion, Friction, Wear and Lubrication" Tribology Series 5, 388-394, Elsevier, N. Y. (1981).

15. Dearnaley, G., B. James, D. J. Mazey and F. J. Minter, "The Micro Hardness of Nitrogen Ion Implanted Cobalt Cemented Tungsten Carbide", Proc. Plansee Conf., Reute, Austria (1985).

16. Burnett, P. J. and T. F. Page," The Effects of Ion Implantation on the Surface Mechanical Properties of some Engineering Hard Materials," in Science of Hard Materials II, Conf. Proc, Rhodes, Greece, 1984, in print.

LOW FRICTION AND ANTIFRETTING COATING INSIDE BORINGS BY MAGNETICALLY ENHANCED SPUTTERING

B. Zega
Battelle Geneva Research Centres
CAROUGE
Geneve, Switzerland

ABSTRACT

Many machine parts, such as receptor pinions in mechanical gear boxes, are exposed to severe fretting corrosion and wear between the bore and the shaft. Molybdenum-coated surfaces are known to behave better under such conditions. A new PVD-method is described allowing very uniform thick molybdenum coatings to be obtained on the internal cylindrical surfaces of diameters as small as 25 mm. The process is adaptable to long stacks of pinions or other ring-shaped parts. It can be applied in such a way that the bore alone is on the reduced pressure side during the deposition; the exterior of the pinion,including hardened teeth, remains accessible from the atmospheric side and can be cooled in a controlled maner.

MOLYBDENUM IS FREQUENTLY USED as wear resistant and antifretting coating. Its application by flame or plasma spraying has been a common practice for years. This well-mastered technique is very attractive, being applied in open air, deposits at high rates, and gives very satisfactory results in most cases. Nevertheless, the coatings obtained have a very rough surface and have to be reground to the final dimensional tolerances and suface finish. This additionnal operation increases considerably the cost of the coated part.

Physical vapour deposition (PVD), as electron beam evaporation, ion-plating and sputtering give dense and regular coatings with well-reproducible properties and thicknesses, but considered to be slow for metallurgical applications, where the lower useful thickness limit lies between 10 and 20 microns. Coating internal surfaces is a common problem to all mentioned methods, in particular when diameters are small. An additional problem appearing when parts are treated in vacuum or in a very rarefied atmosphere is the risk of overheating, and means for controlling the temperature in vacuum environement are very limited. Owing to the versatility of PVD-methods, solutions can be found where most other conventional systems fail.

In the present case, an attempt was made to reduce fretting and wear, appearing between receptor pinions and the shaft in gear boxes, by depositing a 10 to 20 microns thick molybdenum coating in bores of pinions, having been previously thermaly treated and ground to final dimensions.

Specific conditions of this case are:

- bore's diameter is between 25 and 40 mm, up to twice the diameter deep,
- pinions are made in hardenable steel and are hardened (generally carbonitrided) and then machined to final dimensional tolerances before the coating,
- dimensional tolerances are very narrow: $\pm$ 5 microns,
- quantities produced in the case of automotive industry are very large:thousands of pieces per day.

The proposed method fulfils these specifications in the following way:

- coating is obtained by magnetically enhanced sputtering from a small diameter, axial target,
- a long stack of superimposed pinions can be treated simultaneously, compensating in this way the relatively slow deposition rate,
- the magnetic field, needed for confining and intensifying the electrical discharge betwen the target (cathode) and the internal surface of substrates is obtained

by a high intensity current flowwing along the target,
- the stack of pinions is made vacuum-tight by inserted sealing rings; the temperature can in consequence be controlled by external cooling, e.g. by forced air.

PHYSICAL PRINCIPLES OF THE METHOD

When a negative bias is applied to a rod, installed in the center of a grounded metallic cylinder in presence of a reduced gas pressure (few Pa), a glow discharge appears provided the radial distance between the rod (cathode) and the wall (anode) is larger than the cathodic dark space. The rod is in this case bombarded by gas ions and the material sputterd away and deposited on surrounding walls. In normal sputtering conditions, cathodic dark space is 15 - 25 mm thick, limiting the use of the method to bores larger than 40 - 60 mm (taking into account the diameter of the target, which has to be water-cooled). By increasing the gas pressure, the discharge can be obtained in smaller diameters (1,2), but the quality of coatings obtained is inferior, unsuitable for mechanical applications.

When a magnetic field is superimposed at right angle to the (radial) electric field, electrons emitted from the cathode are deviated from their linear radial path towards the anode; by lengtening their path, the probability of collisions with gas atoms and production of ions is greatly increased. This principle is very successfully used in "post-magnetron" sputtering (3), where the magnetic field is obtained with large external coils, as well as in "planar" and "cylindrical" magnetron sputtering targets (4,5,), where internal permanent magnets produce intense field close to the cathode's surface. Owing to the strong gradient of the magnetic field intensity, the discharge is thus confined in a much narrower space around the target.

In our case, the magnetic field can neither be obtained by external coils, because of the shielding effect of steel pinions nor by internal magnets, the cathode's diameter being too small. In consequence, it can be obtained with the help of a high intensity current, flowing along the cathode. The concentric magnetic field lines force electrons to move not radially but along the cathode. The consequence of this longitudinal drift is an intensification of the discharge on one end of the cathode and a substantial thickness gradient of the obtained coating (fig 1/a). Sinultaneously, the discharge current intensity is increased by

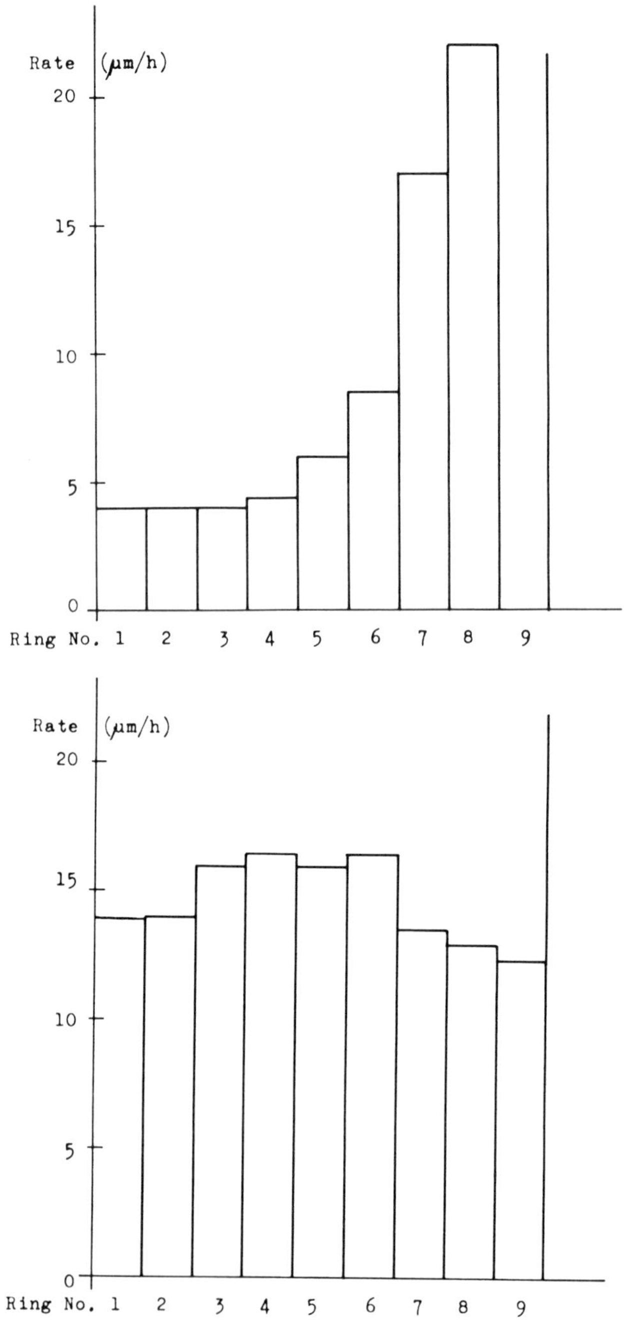

Fig.1 -Longitudinal distribution of coating thickness along the stack:
a - direct magnetizing current
b - alternating magnetizing current

- - -

30 -40 %. By using an alternating current (50 Hz), a satisfying compensation is obtained (fig 1/b) as well as a further increase of intensity (6).

Electrons drifted towards both ends have to be collected somewhere in order to close the current loop. With the stack grounded, pinions situated near both extremities would preferentially collect this electronic current and receive an additionnal heat input. For this reason, the stack is kept at floating potential during the deposition, and two ring anodes are provided at each end of the cathode, grounded or positively biased. The stack acquires a negative autobias of 20 - 40 volts, and its temperature remains fairly uniform during the cycle.

The presence of the high intensity alternating current has still another effect: it induces eddy currents in pinions and produces heating, in addition to that already caused by the heat of condensation of deposited metal and to plasma radiation.

A certain degree of heating is desirable, because it favours outgasing of surfaces before starting the deposition and improves the adhesion. But in a vacuum environment, where thermal exchanges are poor and cannot be controlled in a reproducible way, the temperature can rise rapidly over the allowable limit of 200° C.

In order to overcome this difficulty, the stack of pinions has been made vacuum-tight and a great freedom of temperature control from the atmospheric side has been obtained.

An additional way for improving the adhesion of the coating is to sputter-etch the substrate surface, by inversing the bias between the cathode and the stack for a few minutes.

EXPERIMENTAL SET-UP

BASIC ELECTRIC SCHEME of the system is shown in fig.2: the magnetizing current is supplied by a variac and a low voltage transformer with an isolated secondary winding, protected by a grounded shield. Intensities between 1000 and 1500 amperes are used. At higher intensities, the risk of plasma "pinch-off" exists.

High voltage is supplied by a d.c. generator, containing a simple doble-wave rectifier and no capacity elements in order to prevent arcs.In fact, high voltage bias is essential anly when the intensity of the magnetic field is near its maximum. As the magnetizing current is alternating, the double-wave rectified, unfiltered high voltage signal corresponds perfectly to needs. The only caution is to avoid substantial phase shift between both waves.

CATHODES (TARGETS) - Many possibilities exist for preparing targets. The simplest way is using molybdenum tubes, but this solution is expensive, and the final material yield relatively low, as the target's erosion has to be stopped a certain time before piercing, in order to avoid flushing with water of the vacuum system.

The most flexible way of preparing targets is depositing molybdenum by plasma spraying on a copper tubing. An effective protecting atmosphere is needed during spraying in order to avoid contamination with atmospheric gases and, in consequence, embrittlement of the coating. For the same reason, flame-spraying is excluded, the deposited molybdenum being to much contaminated.

VACUUM SYSTEM - The advantage of not neding a vacuum chamber is partly lost by the low conductance of the system. High capacity pumps are required, in order to eliminate desorbing gases before and during the deposition. Many stacks can be connected in parallel, but not in series on the same pumping group.

Gas (Ar) is preferentially supplied by electromagnetic valves, controlled by electronic flow-meters. Monitoring the pressure with vacuum-gauges appears very impractical in this low-conductance system.

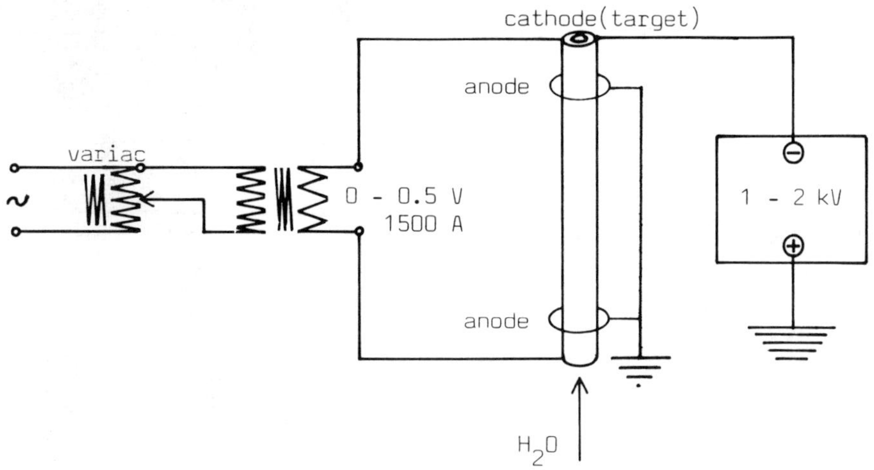

Fig. 2 - SCHEME OF THE ELECTRIC CIRCUIT (the stack of pinions surrounding the target is not represented)

SEALING RINGS - Special intermediate sealing rings have to be designed for each type of pinion. They contain grooves for Viton O-rings on both sides, centered on an appropriate pinion's surface, sufficiently smoth and regular to allow a good seal. By properly redesigning pinions with a view to coating them by this process, a groove for the O-ring can be provided, so no particular intermediate parts would be needed and one O-ring per pinion only.

Special adapting parts on both ends of the stack will remain necessary.

DEPOSITION CYCLE

A TYPICAL DEPOSITION CYCLE consists of the following operations:

1 - assembling the stack and pumping (durations on facilities available for simplifying assembling and on pumping capacities),
2 - preheating by induction: the magnetizing current of 1500 A is sent through the cathode for 5 - 10 minutes (depending on pinion size), in order to increase the temperature to about 190° C in the interior of pinions (the calibration is done with special pinions, having a deep radial bore for the thermocouple),
3 - ion-etching: argon pressure of about 5 Pa is established and a negative bias of 2 kV applied to the stack (with the target grounded) for 3 - 5 minutes. During this operation, the temperature can rise over 200°,so the forced air cooling is started,
4 - deposition: the negative bias is switched from the stack to the target and the discharge power level monitored (by adjusting the gas flow) to the value of 50 W/cm^2 of cathode surface.Under these conditions, the deposition rate is about 0.5 micron per minute when the bore-to-target diameter ration is as 4:1, or inversely proportional to other ratios.
5 - final cooling: with help of the forced air cooling, the temperature of the stack can dropped to 50° C in about 10 minutes.

PROPERTIES OF THE COATINGS OBTAINED

THICKNESS is very regular, reproducing truly the substrate finish, plus some microroughness, due to the crystallinity of the layer. This represents a few % of the total thickness and contributes rather favorably to the oil retention.
MICROHARDNESS MEASUREMENTS give different values, depending on the direction of the indentation. Knoop hardnesses of 600 have been observed by indenting the surface as coated, but less than 200 on metallurgical cuts of the same sample. This phenomenon seems to be due to the high anisotropy of the layer, showing a more or less columnar structure, but also to the compressive stress.
THE HARDNESS PROFILE of the coated pinion has been checked on coated samples and compared to uncoated ones (fig.3). Negligible differences have been observed when the cooling was correctly done.
SLIDING FRICTION MEASUREMENTS have been made on a pin-and-disk machine, with disks coated under very similar conditions, and steel balls. A standard machi-oil has been used as lubricant. Innitial friction coefficient of about 0.1 has rapidly decreased to 0.07 or even 0.06.

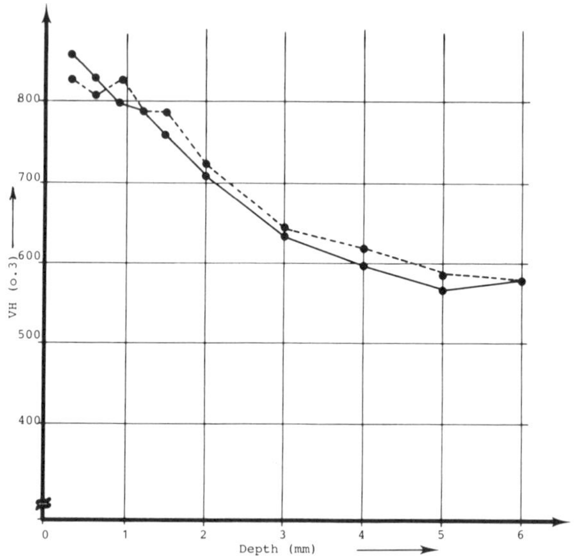

Fig.3 - HARDNESS PROFILE OF CARBONITRIDED PINION: full line -uncoated dotted line - after coating

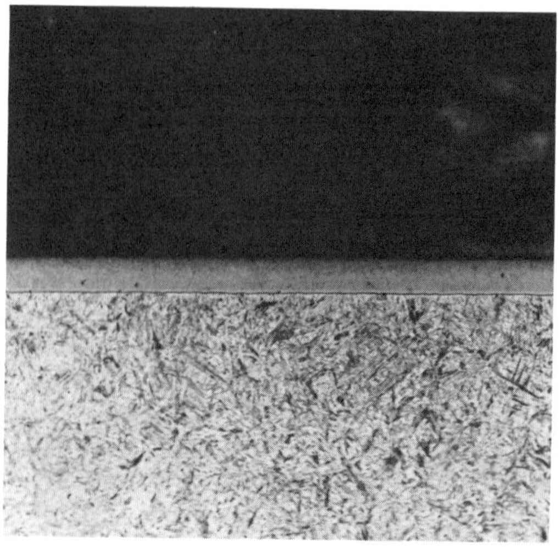

Fig.4 - METALLURGICAL CUT THROUGH THE COATED SURFACE
Thickness: 10 microns

BENCH TESTING of coated pinions has been performed with the corresponding gear engaged or disengaged. After 280 hours of mostly engaged position, when fretting is the most critical cause of corrosion, no sign of deterioration was present, even though a part of the thickness of the coating has benn worn out when run in disengaged position (turning freely on the shaft). These first provisional results are satisfactory and further development and testing are in due course.

CONCLUSIONS

The described coating system is simple and appears easily addaptable to high volume production. It is not necessarily limited to molybdenum. Other coatings, such as bearing metal, can be deposited inside relatively small diameters and at reasonably low temperature. Potential applications for corrosion protection in general have also to be considered.

REFERENCES

1. M.J.Witcomb, Thin Solid Films 74(1980) 223
2. I.P.Gibson, Thin Solid Films 83(1981) 27
3. J.A.Thornton, J.Vac.Sc.Technol.11/4(1974) 666
4. B.Zega, Proceed. 26th Ann.Tech.Conf.,Soc.of Vac.Coaters 1983,p.126
5. D.W.Hoffman Thin Solid Films 96(1982) 217
6. B.Zega, US Pat. 4,376,025

ANALYTICAL SIMULATION OF FRICTION RESISTANCE MODELED BY RIDGE MOBILITY

Yasunori Nakamura
Osaka Industrial University
Osaka, Japan
Visitng Scholar, Lehigh University
Bethlehem, Pennsylvania, USA

Yong-de Zhu
Xian Aero-Engine Factory
Xian, PRC
Visiting Scholar, Lehigh University
Bethlehem, Pennsylvania, USA

Betzalel Avitzur
Lehigh University
Bethlehem, Pennsylvania, USA

Abstract

A mathematical modeling for the simulation of friction resistance is developed by using the upper-bound approach. The friction between hard material (tool) and soft material is analyzed. The existence of wedge-shaped irregularities on the tool surface is assumed. Pressing these protrusions onto the workpiece and sliding the tool along the workpiece produces asperities on the surface of the workpiece. A surface deformation on the soft material is modeled by the interaction of asperities (ridge mobility) and the sublayer's plastic deformation. This ridge mobility and plastic deformation, together with local friction between tool and workpiece along the asperity interface, produces resistance to sliding.

The law of constant friction is used in local friction along the asperity interface. Three kinds of modeling, using two different velocity fields and one contact model between the tool and material, are used. Coulomb's coefficient of friction for moderate load and the constant friction factor for excessive load are determined as functions of the normal pressure, the interface friction factor, and the irregularities of interacting surfaces. The transition region from Coulomb's friction model to the constant friction model is also determined.

FRICTION IS ONE OF THE MOST IMPORTANT FACTORS in analyzing metal-forming processes. However, the knowledge of the frictional phenomenon during metal-forming processes is still very limited because friction between the tool and a workpiece is a complicated process involving mechanics, metallurgy, physics, chemistry, etc. Presently, Coulomb's law, or the law of constant friction stress, is used for the analysis of metal-forming processes.

Whenever two solid surfaces are in contact and slide against each other, a resistance to sliding is generated. Many patterns of relative motion and the plastic deformation of the irregularities on both surfaces are observed [1-3]. Wanheim [4] treated a ridge and its movement to explain the frictional phenomenon between the tool and a workpiece during metal forming. In a manner similar to Wanheim's ridge, we approach the friction phenomenon by using the interaction of a ridge asperity and a deformable asperity. An upper-bound approach is proposed. The movement of the ridge and the thin layer of plastic deformations beneath it leads to an energy consumption directly attributed to the frictional losses.

MODEL TYPE A [5]

To present the relative motion between the tool and the workpiece, we assume that the tool is stationary, and that the workpiece moves to the right at the velocity $\dot{U}$ (Fig. 1). In this

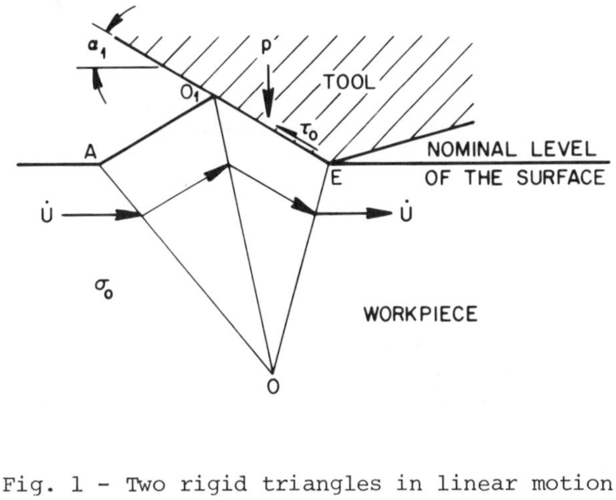

Fig. 1 - Two rigid triangles in linear motion

model, a two-dimensional plane strain flow is assumed. Because of the volume constancy, the incoming velocity is equal to the outgoing velocity.

Fig. 2 shows a two-dimensional plane strain model and its notations. The deformation region is divided into two triangular areas and five boundaries. Each triangle moves as a rigid body in linear motion.

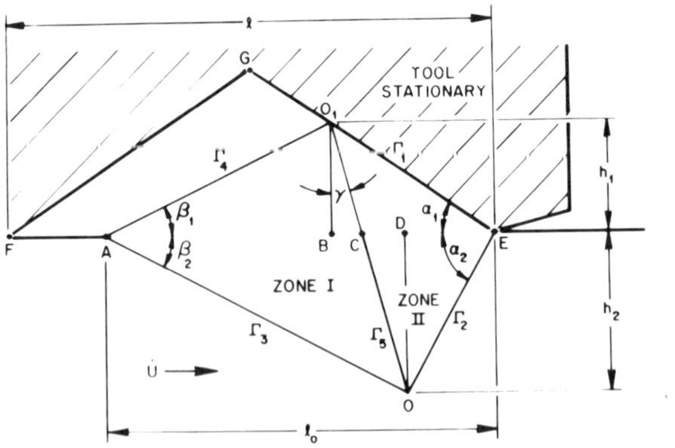

Fig. 2 - The friction model: the ridge and its separation into two rigid body zones

The bulge, forming the ridge AO_1E, is produced as a result of the pressure P acting on the interface. The higher the normal pressure, the larger is the volume of the bulge. In the first stage of increasing the normal pressure, the top of the ridge moves upward from E to G along the tool surface, and point A also moves to the left toward the point F (stage I). The top of the ridge reaches point G before point A reaches point F. After the top of the ridge reaches point G, point A continues to move toward point F (stage II).

To determine the relationship between the normal pressure and the frictional force, the equilibrium equation is used. Fig. 3 shows the forces on the interface between the ridge and the tool. The ridge shown is in plane strain, and it is of one unit width. Consider an individual wedge of length ℓ, and force P acting normal to the direction of sliding. The equilibrium equations in the x and y directions are given by:

$$F = N \sin\alpha_1 + T \cos\alpha_1 \qquad (1)$$
$$P = N \cos\alpha_1 - T \sin\alpha_1$$

where N is the normal force on the inclined surface of the ridge. T is the tangential force, which is obtained by using the constant friction factor m_o along the contact surface.

$$T = \frac{\sigma_o}{\sqrt{3}} m_o \frac{h_1}{\sin\alpha_1} \qquad (2)$$

F is the apparent frictional force, which is given by:

$$F = \frac{J^*}{\dot{U}} \qquad (3)$$

Substitution of Eq. (2) into Eq. (1) to solve for P gives:

$$\frac{P}{\sigma_o} = \left\{ \frac{F}{\sigma_o} - \frac{m_o}{\sqrt{3}} h_1 \frac{2}{\sin(2\alpha_1)} \right\} \cot\alpha_1 \qquad (4)$$

The relative power, the sum of the friction loss and all shear losses, is:

$$\frac{J^*}{\sigma_o \dot{U} \ell} = \frac{1}{\sqrt{3}} \left\{ \frac{1}{\sin(\alpha_1+\alpha_2)} \left(\frac{\sin\alpha_2}{\sin\alpha_1} m_o \frac{h_1}{\ell} + \frac{\sin\alpha_1}{\sin\alpha_2} \frac{h_2}{\ell} \right) + \frac{(h_2/\ell) C_2^2}{B} + \frac{(\ell_o/\ell)(h/\ell)(h_2/\ell)}{\cot\alpha_1 + \cot\alpha_2} \frac{A^2}{B} \right\} \qquad (5)$$

where:

$$A = \left\{ \left(1 + \frac{h_2}{h_1}\right)^2 + \left(\cot\alpha_1 - \frac{h_2}{h_1}\cot\alpha_2\right)^2 \right\}^{1/2}$$

$$B = \frac{h_o}{h_1} + \frac{\ell_o}{h_2} - (\cot\alpha_1 + \cot\alpha_2)$$

$$C_2 = \left\{ 1 + \left(\frac{\ell_o}{h_2} - \cot\alpha_2\right)^2 \right\}^{1/2} \qquad (6)$$

Symbolically this can be expressed as:

$$\frac{J^*}{\sigma_o \dot{U} \ell} = f\,(\alpha_1 h_1 m_o\ \alpha_2 \ell_o h_2) \qquad (7)$$

where α_o, m_o, and h_1 are input or independent parameters. The other parameters: h_2, ℓ_o, and α_2 are pseudo-independent parameters which are subject to optimization to minimize the relative power. The values of the pseudo-independent parameters are obtained by numerical optimization procedure.

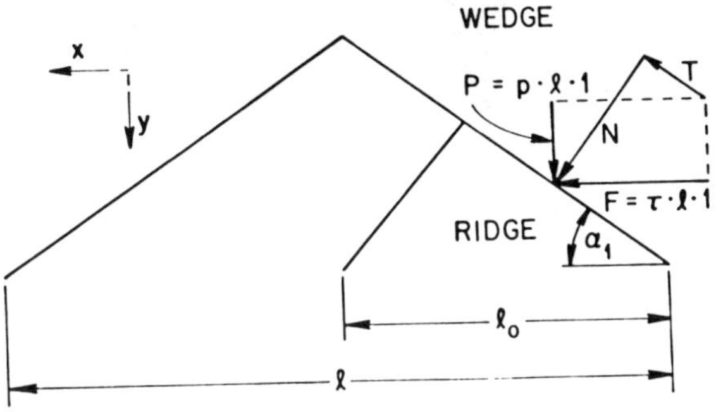

Fig. 3 - The force balance on the ridge

Fig. 4 shows graphically the effect of the interface friction factor m_o on the size and shape of the ridge. For a constant pressure, the ridge height h_1 increases and the depth of the plastic region h_2 decreases with increasing value of the friction factor m_o.

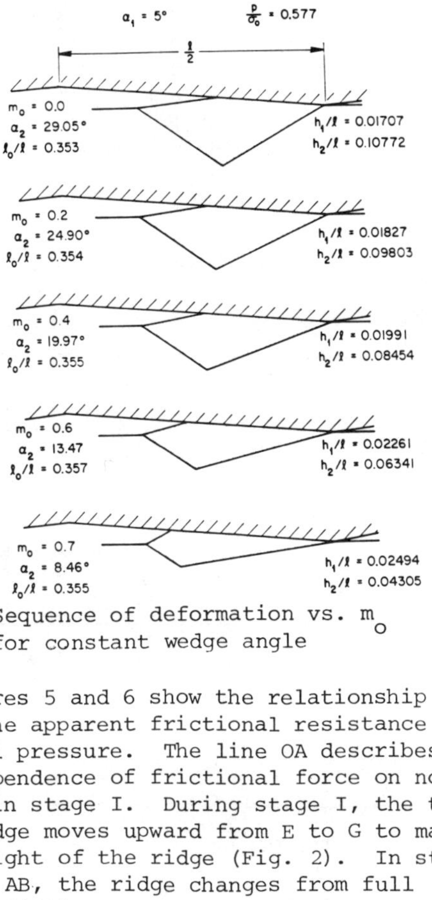

Fig. 4 - Sequence of deformation vs. m_o for constant wedge angle

Figures 5 and 6 show the relationship between the apparent frictional resistance and the normal pressure. The line OA describes the linear dependence of frictional force on normal pressure in stage I. During stage I, the top of the ridge moves upward from E to G to make a full height of the ridge (Fig. 2). In stage II, curve AB, the ridge changes from full height to filling up the wedge (point A comes to point F). The slope of the characteristic curve AB is smaller than that of OA. The horizontal lines were determined by considering the friction power loss along the surface Γ_4 (see Fig. 2). The frictional force will not exceed the value of the horizontal lines. This model can not apply beyond point B until the horizontal line is reached. Therefore, this region is shown by the broken curves.

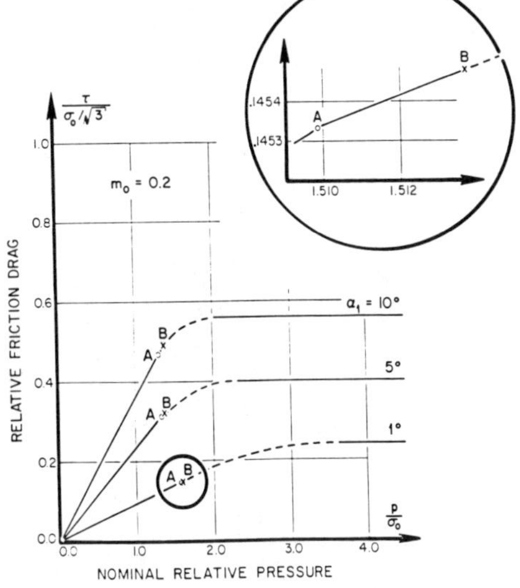

Fig. 5 - Relative friction drag vs. pressure and wedge angle

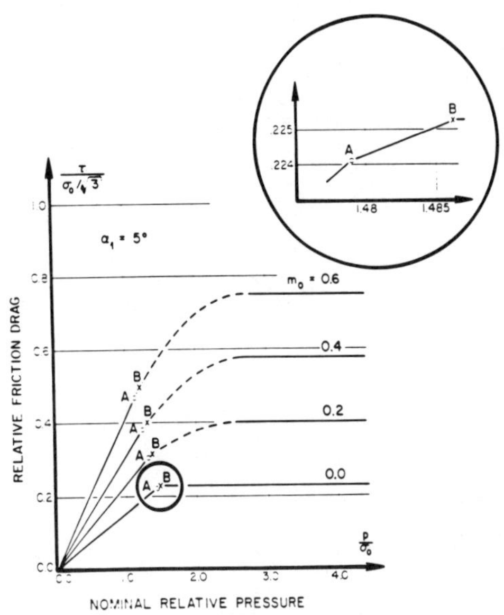

Fig. 6 - Relative friction drag vs. pressure and friction factor m

MODEL TYPE B [6]

In the type A modeling, the ridge and the plastic deformation region below the surface are described through the motion of two triangles, each acting as a rigid body in linear motion. In the type B modeling, the deformations are described through a more flexible velocity field which includes the velocity field of type A modeling as a special case. The two original triangles are separated by an intermediate triangle to form three rigid triangular bodies in linear motion.

Fig. 7 shows the three-triangular velocity field and notations. The power according to the three-triangular velocity field is derived as below.

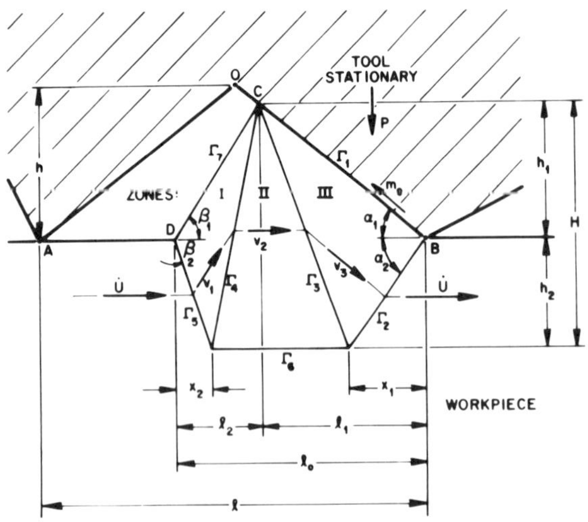

Fig. 7 - Tri-Triangular Velocity Field

The power dissipated in the boundaries Γ_1, Γ_2, and Γ_3 is:

$$\frac{J_1^*}{\sigma_o \dot{U} \ell} = \frac{1}{\sqrt{3}} \frac{1}{h_2 \cot\alpha_1 + X_1} \left\{ X_1^2 + h_2^2 + \frac{h_2}{H}[(\ell_1 - X_1)^2 + H^2] + m_o h_1 h_2 \frac{1}{\sin^2\alpha_1} \right\} \quad (8)$$

and that in Γ_4 and Γ_5 is:

$$\frac{J_2^*}{\sigma_o \dot{U} \ell} = \frac{1}{\sqrt{3}} \frac{1}{h_2 \cot\beta_1 + X_2} \left\{ X_2^2 + h_2^2 + \frac{h_2}{H}[(\ell_2 - X_2)^2 + H^2] \right\} \quad (9)$$

Because of the separation of triangles, the shear loss along the surface Γ_6 is also taken into account.

$$\frac{J_3^*}{\sigma_o \dot{U} \ell} = \frac{1}{\sqrt{3}} \left(1 - \frac{h_2}{H}\right)(\ell_o - X_1 - X_2) \quad (10)$$

The total power is:

$$J^* = J_1^* + J_2^* + J_3^* \quad (11)$$

Symbolically, this can be presented as:

$$\frac{J^*}{\sigma_o \dot{U} \ell} = f\,(\alpha_1\ m_o\ h_1\ \&\ X_1\ X_2\ \ell_o\ h_2) \quad (12)$$

where α_1, m_o, and h_1 are the independent parameters. The other four parameters: X_1, X_2, ℓ_o, and h_2 are called pseudo-independent parameters which are subject to optimization to minimize the total dispersing power. The parameter h_1 is treated as an independent parameter, but is directly related with the normal pressure. Using the next two equations:

$$\frac{\partial}{\partial X_1}(J^*) = 0; \qquad \frac{\partial}{\partial X_2}(J^*) = 0 \quad (13)$$

The optimal values of X_1 and X_2 that minimize the total power are derived and expressed as an explicit function of the independent process parameters. The optimal values of the other two pseudo-independent parameters are obtained by numerical procedure.

The type A model is just a special case of the type B model, and it is not the optimal one. As a result, the ridge of the type B model consumes less power than the ridge, with the equivalent size, of the type A model (Fig. 8). This behavior is attributed to the more gradual changes in velocities between the neighboring zones in the type B model. Because it has lower upper-bound solution, it is said that the three-triangular velocity field approaches the actual flow pattern more closely than the two-triangular velocity field.

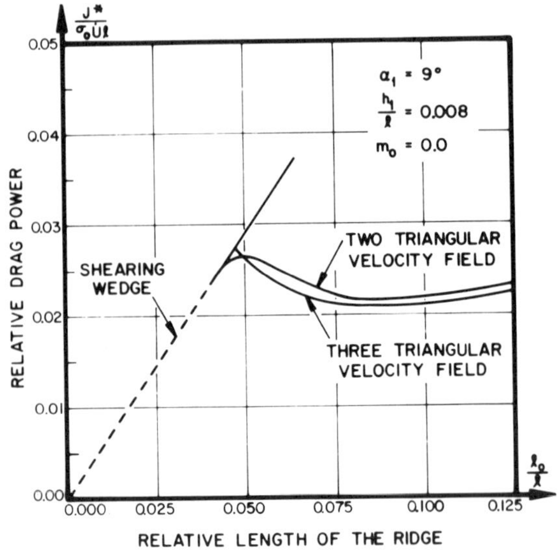

Fig. 8 - The comparison on power between two different models

The number of the pseudo-independent parameters, with respect to which the power expression of Eq. (12) is optimized by numerical procedure, is two, namely ℓ_o and h_2, while type A modeling has three such pseudo-independent parameters. Obviously, the smaller the number of the parameters for numerical optimization, the better and more precise is the calculation.

MODEL TYPE C [7]

In this type C modeling, stage III of the frictional state is introduced. Stage III commences after the wedge is fully occupied by the ridge (point B in Fig. 7). Stage III exists as the load increases and surface Γ_7 (OD) participates in carrying the load and contributes to the resistance to sliding. The characteristics of this stage III are not determined quantitatively in both type A and B modelings. They are accomplished under the following concepts:

(1) At the end of stage II (point B in Fig. 7), there is still no contact force acting between the tool and the workpiece along the surface Γ_7 (OD in Fig. 9).

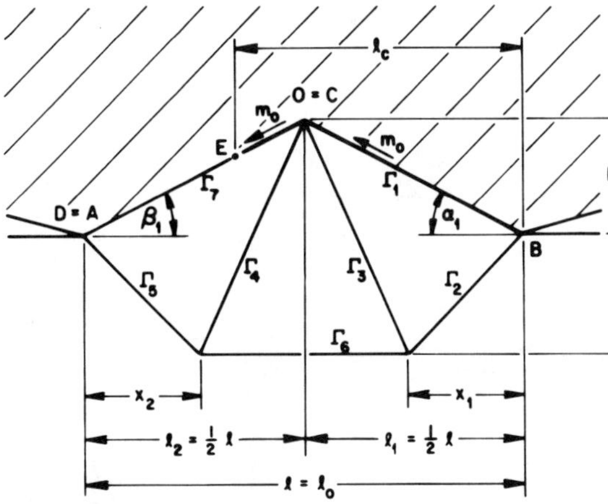

Fig. 9 - Contact model during stage III

(2) With a further increase in the vertical force P, the contact force along the surface Γ_7 can be denoted by the length OE on Fig. 9. Thus, the position of point E will determine the extent to which the load is transmitted through the surface Γ_7 and will determine the intensity of friction along this surface.

(3) The contact force acting on the portion OE has both a tangential component and a normal component as shown in Fig. 10. These force components are directly proportional to the length of OE. When point E reaches point D, the magnitude of these force components becomes equal to those acting along the other side of the wedge, namely along OB.

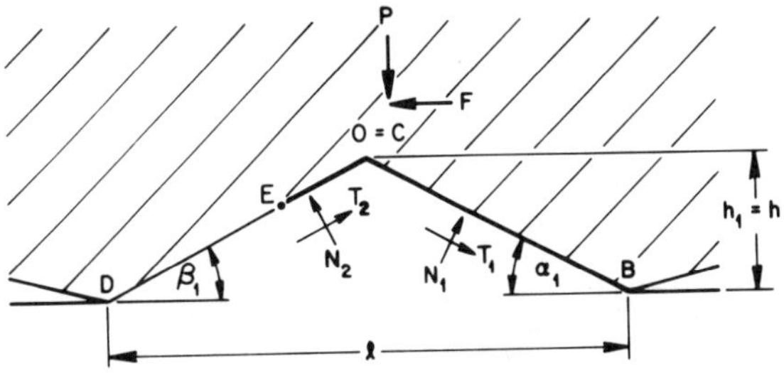

Fig. 10 - The force balance on the asperity of the tool during stage III

We defined the parameter r as the fraction of the tool area that is in contact with the workpiece. The parameter r is determined as the ratio between the length of BOE and BOD, and also by a ratio between ℓc and ℓ.

The equation of relative power becomes:

$$\frac{J^*}{\sigma \dot{U}\ell} = \frac{1}{\sqrt{3}} \frac{1}{h_2 \cot\alpha_1 + X_1}\{X_1^2 + h_2^2 + \frac{h_2}{H}[(\ell_1 - X_1)^2 + H^2] + m_o h_1 h_2 \frac{1}{\sin^2\alpha_1}\}$$

$$+ \frac{1}{\sqrt{3}} \frac{1}{h_2 \cot\beta_1 + X_2}\{X_2^2 + h_2^2 + \frac{h_2}{H}[(\ell_2 - X_2)^2 + H^2]$$

$$+ (2r-1)\, m_o h_1 h_2 \frac{1}{\sin^2\beta_1}$$

$$+ \frac{1}{\sqrt{3}}(1 - \frac{h_2}{H})(\ell_o - X_1 - X_2) \quad (r \geq 0.5) \qquad (14)$$

Symbolically, this can be expressed as:

$$\frac{J^*}{\sigma_o \dot{U}\ell} = f\,(\alpha_1\ m_o\ r\ \&\ X_1\ X_2\ h_2) \qquad (15)$$

where X_1, X_2, and h_2 are the pseudo-independent parameters subject to optimization.

Fig. 10 shows the forces acting on one ridge. To determine the relationship between the apparent normal pressure and the frictional force, the equilibrium equations are used. The equilibrium equations in the x and y direction are given by:

$$P = (N_1 + N_2) \cos\alpha_1 - (T_1 - T_2) \sin\alpha_1$$

$$F = (N_1 - N_2) \sin\alpha_1 + (T_1 + T_2) \cos\alpha_1 \qquad (16)$$

where N_1 and N_2 are the normal forces, and T_1 and T_2 are the tangential forces on the inclined surfaces of the ridge. T_1 and T_2 are obtained by using the constant friction factor m_o.

$$T_1 = \frac{\sigma_o}{\sqrt{3}} m_o \, h_1/\sin\alpha_1 \qquad (17)$$

We assume that the contact-pressure distribution is uniform along the contact surface. Thus, the relation between N_1 and N_2 is:

$$N_2 = (2r - 1) N_1 \qquad (r \geq 0.5) \qquad (18)$$

Eliminating N_1 and N_2 from Eq. (16), the normal force P is obtained as:

$$\frac{P}{\sigma_o} = \frac{r}{1-r} \cot\alpha_1 \frac{F}{\sigma_o} - 2\{(1-r) + \frac{r^2}{1-r} \cot^2\alpha_1\} \frac{m_o}{\sqrt{3}} h_1 \qquad (r \geq 0.5) \qquad (19)$$

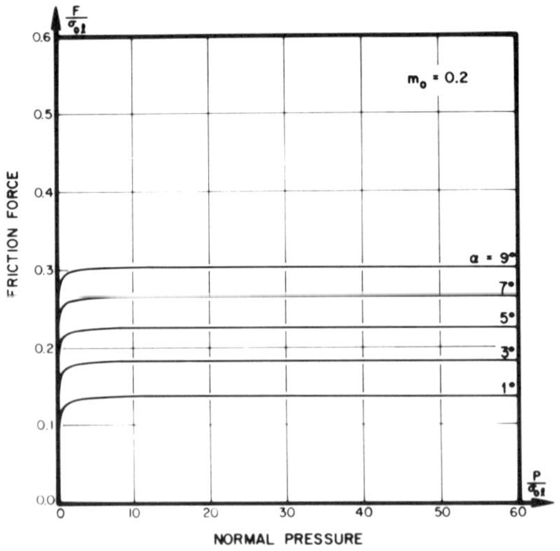

Fig. 11a - The relation between the friction force and normal pressure with wedge angle (α_1) as a parameter

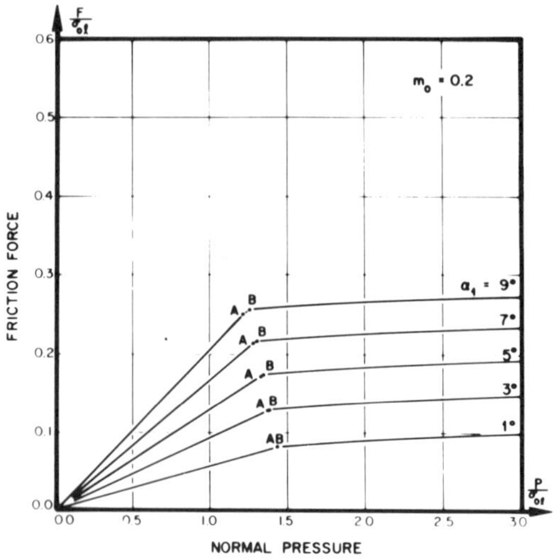

Fig. 11b - The relation between the friction force and normal pressure with wedge angle (α_1) as a parameter

Figs. 11a, 11b, 12a and 12b show the relationship between the apparent frictional resistance and the normal pressure. Point A indicates the end of stage I, and point B shows the end of stage II. After point B, during stage III, the gradient becomes smaller than in stage I. The greater the normal pressure, the smaller is the gradient of the curve. For very high pressure, the friction approaches asymptotically a constant value.

Figs. 13 and 14 show the relation between the depth of plastic region and the normal pressure. During stage I and II, the depth increases with an increase in normal pressure. But during stage III, the depth decreases. These negative slopes at stage III are explained by the increase in friction drag along the surface OA at higher pressure.

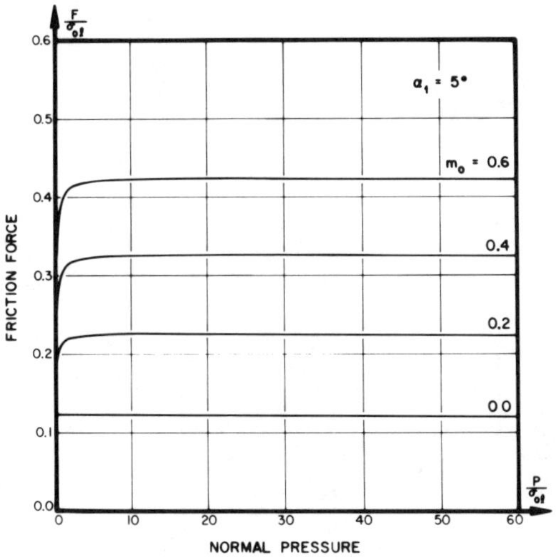

Fig. 12a - The relation between the friction force and normal pressure with friction factor (m_o) as a parameter

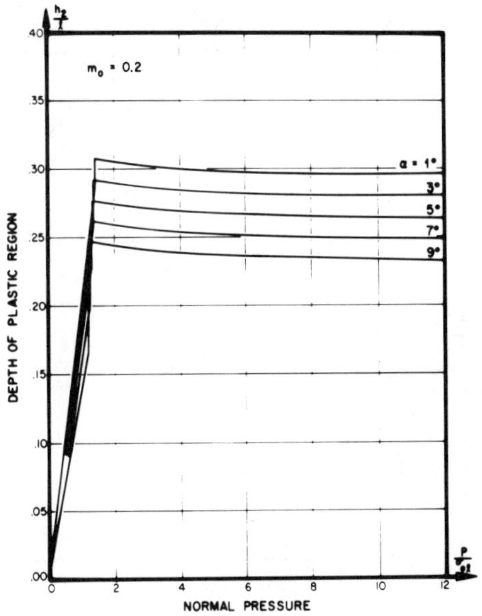

Fig. 13 - The relation between the depth of plastic region and the normal pressure with wedge angle (α_1) as a parameter

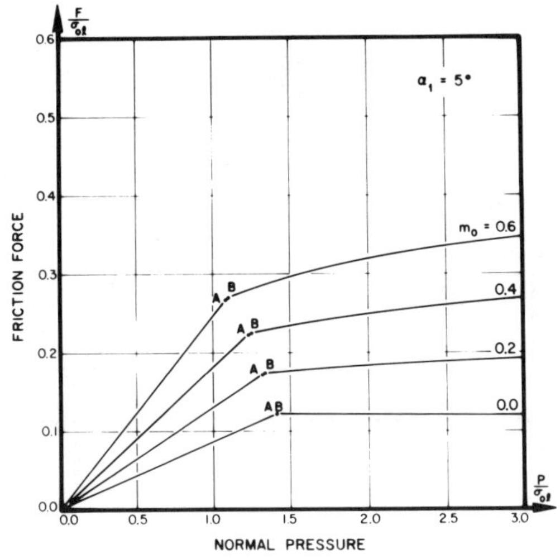

Fig. 12b - The relation between the friction force and normal pressure with friction factor (m_o) as a parameter

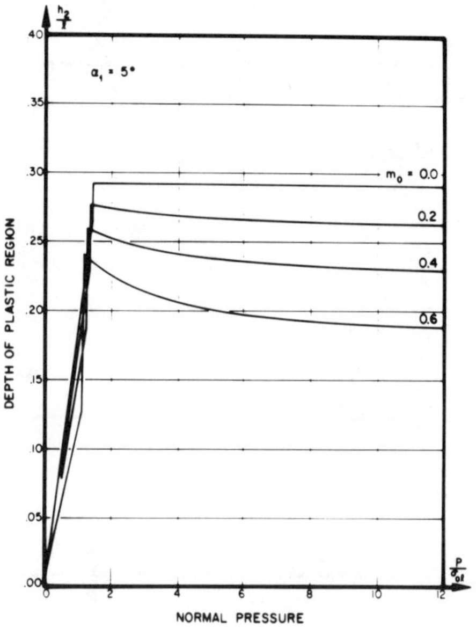

Fig. 14 - The relation between the depth of plastic region and the normal pressure with friction factor (m_o) as a parameter

The ratio between the frictional force and the normal pressure is traditionally defined as Coulomb's or Amonton's coefficient of friction μ. Figs. 15 and 16 show the relation between Coulomb's coefficient of friction and normal pressure.

From Figs. 11b and 12b, we can get the pressure limit, before which Coulomb's friction model can be adopted (point A). Figs. 17 and 18 show the pressure limit of Coulomb's friction model related with the local friction factor m_o and the wedge angle α_1.

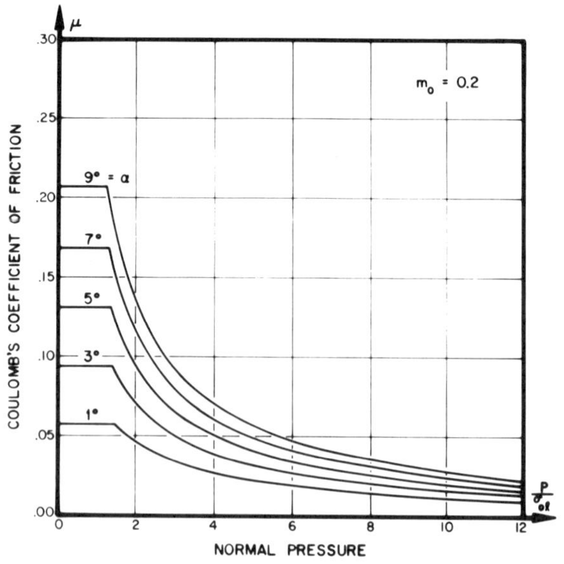

Fig. 15 - The relation between the Coulomb's co-efficient of friction and the normal pressure with wedge angle (α_1) as a parameter

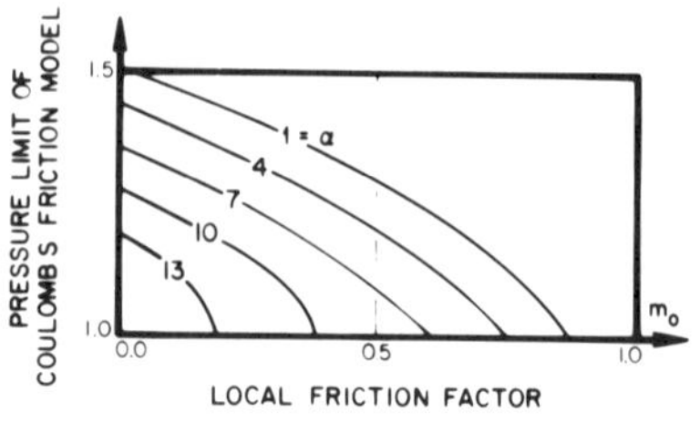

Fig. 17 - The relation between the pressure limit of Coulomb's friction model and local friction factor (m_o)

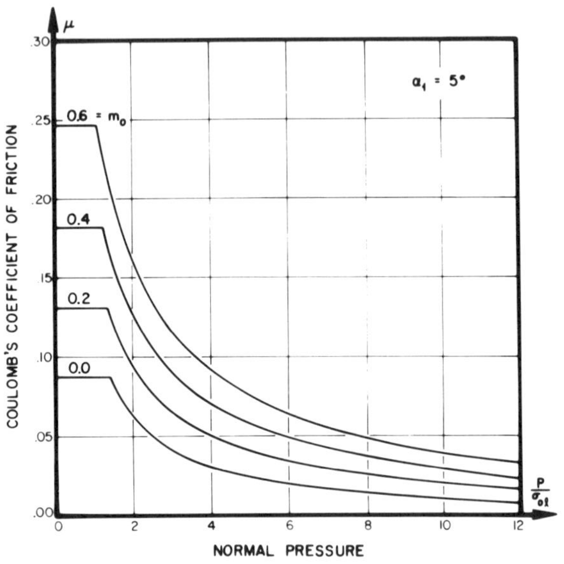

Fig. 16 - The relation between the Coulomb's coefficient of friction and the normal pressure with friction factor (m_o) as a parameter

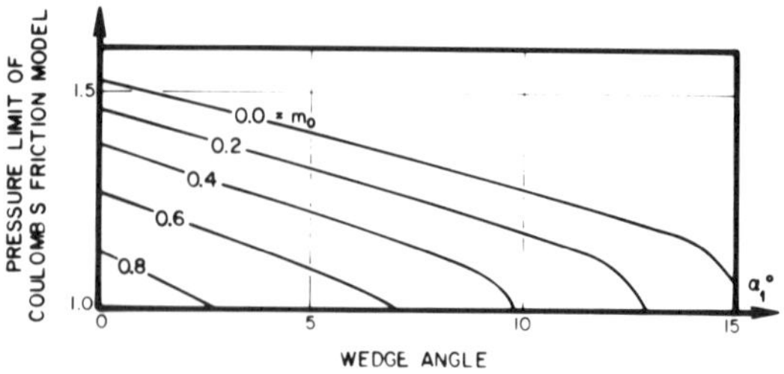

Fig. 18 - The relation between the pressure limit of Coulomb's friction model and the wedge angle (α_1)

CONCLUSIONS

The characteristics of global frictional resistance F as a function of load P are described in Figs. 11a, 11b, 12a, and 12b, where the wedge angle and the local resistance to sliding m_o are the parameters. The description of proportional Coulomb's coefficient of friction μ for moderate loads and constant friction factor m for excessive loads is determined quantitatively. It is strongly suggested from the result of this work that the frictional force depends mainly on the irregularities of interacting surfaces and the lubrication and surface chemistry.

REFERENCES

1. E. Rabinowicz, "Friction and Wear of Materials," John Wiley and Sons, New York (1966).

2. Suh, N. P., "An Overview of the Delamination Theory of Wear," WEAR 44 (1977), pp 1-16.

3. Tabor, D., "Friction - The Present State of our Understanding," Trans. ASME, J. Lub. Tech., 103, 169-179 (1981).

4. Wanheim, T., "Frikton ved høje Fladetyk" (in Danish), AMT-lincentiatarage, Technical University of Denmark (1969).

5. Avitzur, B., Huang, C. K. and Zhu, Y.D., "A Friction Model Based on the Upper-Bound Approach to the Ridge and Sublayer Deformations," WEAR, 95, 59-77 (1984).

6. Avitzur, B. and Zhu, Y. D., "A Friction Model Based on the Upper-Bound Approach to the Ridge and Sublayer Deformations - Update," Proc. of the NAMRI XIII Conf., May 19-22, Berkeley, CA, 103-109 (1985).

7. Avitzur, B. and Nakamura, Y., "Analytical Determination of Friction Resistance as a Function of Normal Load and Geometry of Surface Irregularities," submitted for consideration for publication (1985).

PLASMA-SPRAYED COATINGS FOR PISTON RINGS—STATE OF DEVELOPMENT AND APPLICATION POTENTIAL

Ulrich Buran, H.-Christian Mader, Dr. Martin Morsbach
Goetze AG
Burscheid, West Germany

Brian A. Newman
Goetze Corporation of America
Muskegon, Michigan, USA

HARD CHROMIUM-PLATED and flame-sprayed molybdenum-coated piston rings are standard equipment in today's production engines. The different behavior of these two coatings and the consequent limits of their application in engines are described in the Goetze technical paper "K 22" [1].

The demands made by engine manufacturers for piston ring coatings that are wear-resistant for 1 million kilometers in heavy duty truck engines or 20,000 operating hours in marine diesel engines cannot always be met by molybdenum or hard-chromium coatings. Moreover, flame-sprayed molybdenum coatings in diesel engines show partial signs of fatigue after lengthy periods of operation, which indicates that the limits of coating strength are reached under these loads. For all of the above applications, therefore, there is a requirement for piston ring coatings with considerably improved properties.

Of all the known coating methods, the plasma spraying technique is particularly suitable for solving this problem; because with this method, the range of coating materials is almost unlimited, and therefore selective coatings can be developed with specific properties.

Metallic, cermet and ceramic materials have been tested within the scope of our coating development program.

Plasma-sprayed coatings for piston rings and their metallurgical characteristics will be discussed below. Information will be given on the practical application of certain plasma-sprayed coatings on piston rings for various types of engines. This information is based on our experience gained in production engines and from test stand trials and field tests.

PLASMA COATINGS WITH HIGH RESISTANCE TO WEAR AND BREAK-OUT

The large number of coating combinations that were tested in engines were selected on the basis of metallurgical interrelationships and the engine load expected. Evaluation of the comprehensive test results from different engines showed that, even with the many coating materials tested, it is not possible according to the present level of knowledge to develop a "standard coating" that would be suitable for all engines.

Plasma-coated ring peripheries have to fulfill the following main requirements:

- High scuffing resistance
- High break-out resistance
- Low cylinder wear
- High wear resistance of coating
- High corrosion resistance
- Possibility of manufacturing sharp peripheral edges on full-face sprayed rings
- Low manufacturing costs

Conflicting properties arise in some cases if these requirements are to be met in full. It is therefore necessary to offer several plasma-sprayed coatings, the applicability of the coating being determined by the demand profile of the particular engine.

Figure 1 shows plasma-sprayed coatings used in production or about to be introduced in production. The coating designation is given under the first column. The abbreviations MP, MKP and KP indicate that metallic (M), cermet (MK) or ceramic (K) materials are applied by plasma spraying (P). The second column shows the coating family to which individual variants belong, and the last column shows the main constituents of the coatings.

Coating designation	Coating family	Main constituents
MP 43	Metallic mixed coating	Mo, NiCr B Si alloy
MP 96 A	Metallic mixed coating	Mo, Cr alloy NiCr alloy
MKP 81 A	Cermet mixed coating	Mo, Mo carbide NiCr alloy
MKP 107	Cermet mixed coating	Mo, Cr carbide NiCr alloy
KP 123	Ceramic mixed coating	Al oxide, Ti oxide, Alloying additions

Figure 1: Goetze plasma-sprayed coatings.

METALLURGICAL CHARACTERISTICS OF PLASMA-SPRAYED COATINGS AND HOW THEY ARE DETERMINED

PRINCIPLE OF THE PLASMA-SPRAYING PROCESS - The thermal energy required for melting the spray material supplies the "gas plasma" for plasma spraying. The "gas plasma" is formed by an electric arc between an electrode and a copper nozzle when a gas or gas mixture is introduced. Because of the high temperature of the arc, the "plasma" is formed by ionization and dissociation of the gas. This process causes the gas to heat up intensely and accelerate. On delivery from the nozzle the gas can reach several times the speed of sound, with the temperature at the centre of the plasma flame exceeding 15,000°C. During the spraying operation, the deposition material is introduced into the center of this flame by means of a transport gas. The spray material is melted inside the plasma flame and discharged at high speed onto the piston ring.

COATING CHARACTERISTICS - Because of the manufacturing processes used, the physical and mechanical properties of plasma-sprayed coatings are completely different from those of a solid material with the same chemical composition. A knowledge of these properties is important to the engine builder in determining whether the particular coating data will meet the demand profile of his engine.

The following coating characteristics are important:

- Chemical composition of the coating
- Coating structure
- Porosity
- Hardness
- Bond strength
- Torsional strength
- Bending strength
- Thermal shock resistance

These properties of plasma-sprayed coatings and how they are determined will be discussed below. Moreover, it will be shown how coating properties correlate to running behavior under particular loads.

COATING STRUCTURE - The structure of plasma-sprayed coatings has significant influence on the running behavior of piston rings in the engine. Particularly important here are the microstructure, the distribution of individual phases, the coating porosity, unmelted particles and oxide accumulations. These characteristics are best tested and evaluated on a metallographic cross-section of the coating.

Figures 2, 3 and 4 show typical macrographs of plasma coatings MP 96A, MKP 81A and KP 123. All microsections, except for coating KP 123, are etched with Murakami's solution in order to reveal the molybdenum phases. Molybdenum and molybdenum-rich phases change to a dark color after this etching.

In metallic and cermet plasma coatings the scuff resistance improves with increasing molybdenum content. However, plasma coatings consisting only of molybdenum have a lower wear resistance than flame-sprayed molybdenum coatings. This also applies to their break-out resistance under heavy loads.

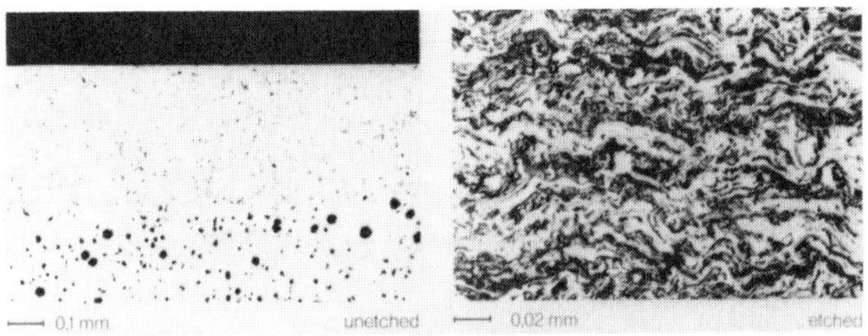

Constituents	Mo + Cr alloy + NiCr alloy		
Coating hardness (HV 2)	300 – 450		
Particle hardness (HV 0.05)	200 – 500	400 – 800	
Porosity (%)	≤ 10		
Application	Piston ring coating with very high wear and breakout resistance for high-speed engines.		

Figure 2: Plasma coating MP 96A.

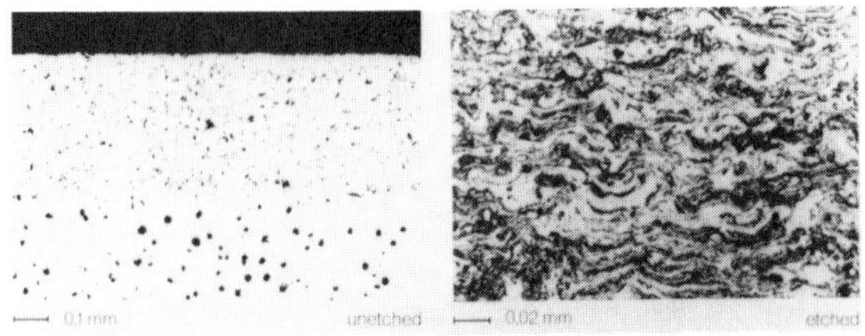

Constituents	Mo + Mo carbide + NiCr alloy		
Coating hardness (HV 2)	330 – 480		
Particle hardness (HV 0.05)	200 – 500	400 – 800	900 – 1400
Porosity (%)	≤ 10		
Application	Piston ring coating with high inherent wear resistance, low cylinder wear and elevated thermal loading capacity for high-speed diesel engines.		

Figure 3: Plasma coating MKP 81A.

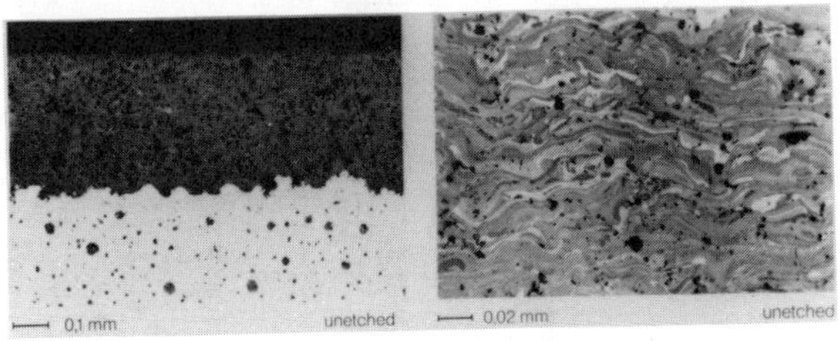

Constituents	Al_2O_3 + TiO_2 + Alloying additions
Coating hardnes (HV 1)	600 – 800
Particle hardness (HV 0.05)	900 – 1400 1100 – 1600
Porosity (%)	≤ 10
Application	Piston ring coating for extreme thermal loading for high-speed engines.

Figure 4: Plasma coating KP 123.

For these reasons, molybdenum mixed coatings are used for the surface treatment of piston rings. Phases with plastic or soldering properties are added to the molybdenum to improve break-out resistance. Wear resistance is increased by adding hard materials.

Compared with the MKP 107 coating, the MP 43, MP 96A and MKP 81A coatings have a higher proportion of molybdenum. This is illustrated by the area of the dark phases in Figures 2 and 3. These coatings have been developed mainly for engines in which the piston rings are subject to high thermal stresses.

Which of the coatings is the most suitable for any particular application depends on whether the deciding criterion is cylinder wear, coating wear or break-out resistance (see later section on evaluation of coatings).

The MKP 107 coating contains high proportions of hard material phases and relatively little molybdenum. This coating has been developed for medium speed diesel engines. In such engines it produces, under cyclic stresses, a coating wear with less spread and also an increased break-out resistance compared with the MKP 10 production coating.

Ceramic coatings, like coatings containing molybdenum, can also withstand high thermal stresses. Ceramic mixed coatings have a very high inherent wear resistance compared with coatings containing a high proportion of molybdenum. However, they are more brittle and more sensitive to thermal shock than cermet mixed coatings. Nevertheless, a sufficiently high break-out resistance can be obtained by designing the ring appropriately; and this will be discussed in more detail under the section "Design of piston rings with plasma-sprayed peripheries."

PHASE DISTRIBUTION - To evaluate the phase distribution in mixed coatings, i.e. the distribution of structural constituents of almost the same chemical composition, the microsections are suitably etched. In engine application, because of the inherent wear of the coating, a layered coating structure of different phases would lead to a sudden change in the material pairing of piston ring periphery/cylinder wall. The new friction pairings would have to re-adapt to one another each time, which could lead to defective running behavior of the rings.

The uniform phase distribution required, which would avoid the type of layer formation described, can only be achieved by selecting suitable spraying parameters and using a suitable type of spray material. These are key technological factors in coating development, especially with plasma-sprayed cermet coatings.

COATING POROSITY - Plasma-sprayed coatings generally have lower porosity than flame-sprayed molybdenum coatings, and this is one of the reasons for their higher inherent strength. On the other hand, there are reasons for designing piston rings with porous surfaces: oil reservoirs are formed which are beneficial under conditions of poor lubrication.

The porosity of plasma-sprayed coatings can be controlled by the spraying process and the type and form of the spray material. The degree of coating porosity is established in the relevant material specification. Goetze technical paper "K 33" [2] gives information on the testing of coating porosity.

UNMELTED PARTICLES AND OXIDE ACCUMULATIONS - The high temperature in the plasma jet is not, in isolation, sufficient to ensure that all sprayed particles are completely melted. Powder particles that do not remain long enough in the flame, are too big, or do not reach the center of the flame will be either unmelted or inadequately melted when they reach the coating. Unmelted particles generally have a looser bond with the adjacent coating constituents and consequently weaken the coating. Although it is not possible to completely avoid the inclusion of unmelted particles, their size and number can be controlled to the extent that their weakening effect is insignificant.

The total amount of unmelted particles in the piston ring coatings discussed here should be no more than 10%, and individual particles must be smaller than 100 μm.

Figure 5 shows inclusions of unmelted particles which have an acceptable bond with the adjacent coating constituents in the MP 43 coating.

Oxide accumulations have an unfavorable effect on the condition of metallic and cermet plasma-spray coatings. These accumulations are caused by the highly heated, metallic powder particles reacting with the oxygen in the atmosphere. Spraying in a protective gas atmosphere proves to be disadvantageous for piston ring coatings because of the associated reduction in hardness.

By selecting the correct spraying parameters, it is possible to control the process so that the required coating hardness is obtained and at the same time oxide accumulations are largely eliminated.

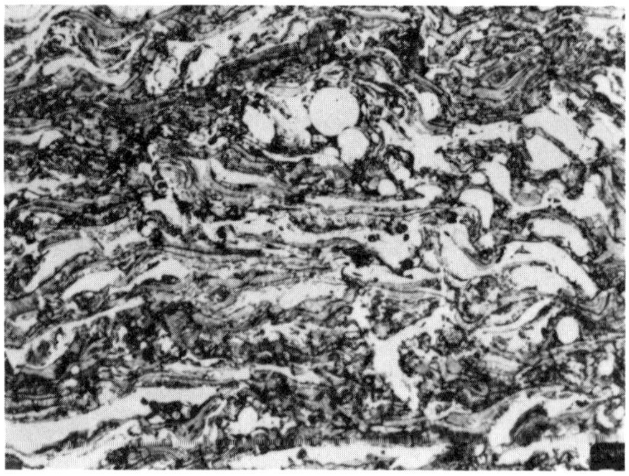

Figure 5: Unmelted particles with acceptable bond to the adjacent coating constituents in an MP 43 coating.

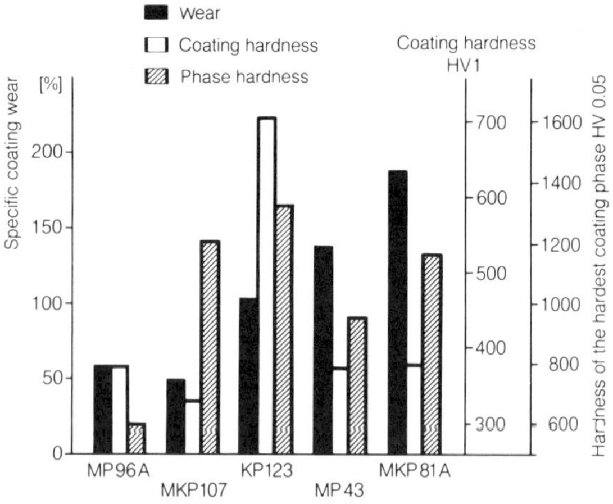

Figure 6: Coating wear and hardness of Goetze plasma-sprayed coatings.

COATING AND PHASE HARDNESS - Because of coating porosity and the inhomogeneous coating structure, special attention has to be paid to determining the hardness of plasma-sprayed coatings, since the measurements are very dependent on the applied indentor load.

Hardness measurement serves as a check on the finished coating quality. The Vickers HV2 or HV1 method is used.

Individual particles or phases are identified by microhardness measurement (HV). The test load is normally 0.05 kp; but smaller loads can also be used depending on the size of the phase.

Further details on determining the hardness can be obtained from our "Guidelines for Testing Sprayed Coatings on Piston Rings". Coating and phase hardness values are defined in the relevant material specifications.

Occasionally the hardness of materials is given as a measure of their wear resistance. Figure 6 shows the coating and phase hardness, and the engine wear characteristics of our plasma coatings, arranged according to increasing coating wear values. As expected, because of the inhomogeneity of sprayed coatings, a dependent relationship between wear and hardness cannot be established.

COATING ADHESION - With the plasma spraying process, the bond strength between coating and base metal is a result of mechanical interlock, physical adhesion, chemisorption and partial diffusion. Therefore for good adhesion of the sprayed particles, the following factors should be taken into account:

- The impact speed of the particles should be as high as possible.
- The spray particles should be molten on impact.
- The spray particles should not have an oxide film.
- The surface of the base metal should be roughened and activated.
- The coefficients of thermal expansion of the coating and base metal should be as close to one another as possible.

The last-mentioned requirement is particularly difficult to realize with coatings containing molybdenum.

The conditions listed depend partly on the spraying parameters and partly on the surface preparation before the spraying process and also on the type and quality of the spray material.

Big differences between the thermal expansion coefficients of the sprayed coating material and the base metal will cause shearing stresses at the transitional area; this reduces the bond strength. In such cases, a ductile bond coating can compensate for the development of thermal stresses. This problem is treated in Goetze technical paper "K 20" [3].

The spraying data for good bond strength is part of the special expertise involved in the development of piston ring peripheral coatings.

The bond strength test to DIN 50 160 provides values for establishing orders of priority. It is not used as a test method with the ring.

The following tests are made on the piston ring:

- Gap spread test to simulate assembly of the piston ring onto the piston.
- Twist test to explain behavior during torsional stressing.
- Thermal shock test to determine thermal fatigue resistance.
- Thermal crack test to determine the crack propagation temperature.

As these stresses do not arise individually during engine operation, but are superimposed, the results of these individual tests do not provide any definite conclusion on the break-out behavior of the rings in the engine.

Figure 7 shows a table of the values determined for Goetze plasma coatings.

Coating designation	Adhesive strength to DIN 50 160 [N/mm²]	Value for gap spead test 90,9 x 3 [mm]	Twist value for ring 90,9 x 3 [<°]	Crack propagation temp. [°C]
MP43	≥ 25	115	55	180
MP96A *	≥ 30	140	85	210
MKP81A*	≥ 35	120	75	200
MKP107	≥ 30	105	40	200
KP123 *	≥ 30	90	65	185

* with intermediate layer

Figure 7: Technical bond strength values according to different test methods.

DESIGN OF PISTON RINGS WITH PLASMA-SPRAYED PERIPHERIES

Depending on the properties of the coating, thermally-sprayed piston rings require special design features if their full potential functional advantages are to be realized. Plasma-sprayed rings are closely based on the design of flame-sprayed rings.

DESIGNS - The basic design types of rings with spray-coated peripheries are shown in Figures 8 and 9.

These periphery designs are not dependent on the ring type or cross-section; i.e. in addition to plain compression rings, they are used for keystone rings, tapered rings, etc.

DESIGN DETAILS

Inlaid Periphery - The inlaid design (Figure 8) enables perfectly sharp peripheral edges to be formed at the base metal of the ring, irrespective of the peripheral coating.

The shape of the inlay is determined by the required coating thickness, the run-out angle of the inlay α and the run-out radius of the inlay R. By selecting an inlay run-out angle that is as flat as possible, i.e. by selecting the largest possible angle α and designing the radius R as large as possible, the break-out resistance of the coatings of inlaid rings is improved (see Goetze technical paper "K 20") [3].

The angle α and the radius R are limited by the absolute ring width h_1, the coating thickness required, and the permissible size of the axial width (h_x) of the uncoated part of the periphery. The following values of α and R are nowadays standard for compression rings fitted in the top groove:

α = (15°) / 30° / 45° / (60°)

R = 0.1 mm / 0.3 mm / 0.5 mm / 1 mm.

The values for h_x are normally 0.2 mm min. or are dimensioned at 0.2 ± 0.1 mm or 0.3 ± 0.1 mm.

For rings of greater axial width h_1, an optimized inlay shape is used in which the higher values for α and R are preferred. Such rings with, for example, α = 45° and R = 1 mm are in use from $h_1 \geq 4$ mm. Peripheral chamfers prevent the values for h_x from becoming too large.

Half-Inlaid Periphery (Figure 8) - The half inlaid designs are similar to the inlaid designs. By means of an optimized production procedure, an extremely flat inlay run-out is obtained, with α = 70° and R = 1 mm, to achieve the highest possible break-out resistance. Because of the large angle α, the size of the base metal land at the lower peripheral edge has to be at least 0.4 mm. Moreover, it should be taken into account that, because of the flat run-out of the inlay, the value of h_x increases in relation to peripheral wear at the ratio of 1:3.

Keystone rings with sprayed peripheries and this optimized inlay shape are successfully used in large numbers in diesel engines for heavy duty trucks.

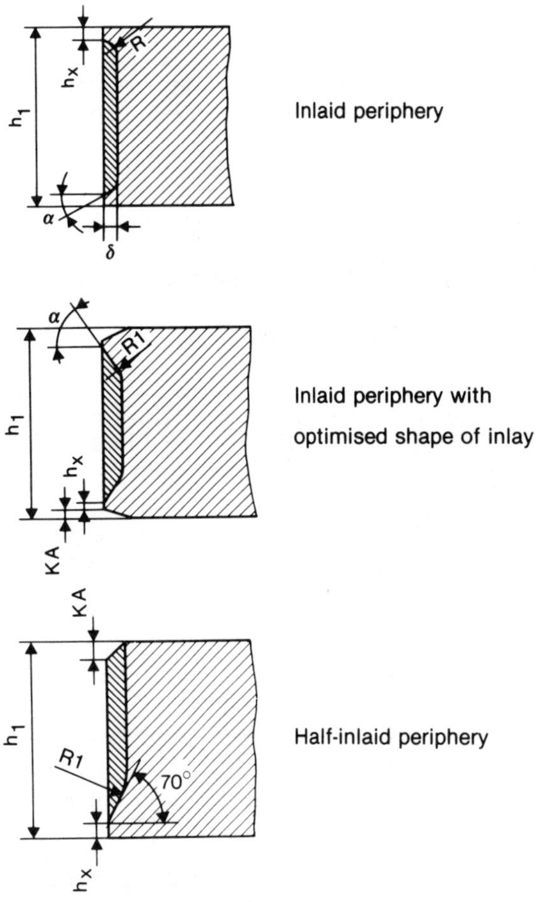

Figure 8: Designs for inlaid coating.

Full-Face Sprayed Periphery (Figure 9) - Of the three design types presented here, the full-face sprayed periphery has the highest break-out resistance in engine application. This can be explained on the one hand by the fact that maximum adhesion to ring base metal can be obtained by spraying the coating material onto the ring at right angles to periphery; on the other hand, due to the absence of inlay wall area, it is not possible for tensile or shearing stresses to build up between the coating and the base metal of the ring. With inlaid rings, as a result of the different coefficients of thermal expansion of the coating and base metal, the temperature increase that takes place in the piston ring is alone sufficient to cause such a large build-up of stress between the material pairing that the coating resistance can be exceeded and cracks can develop. (For detailed information see Goetze technical paper "K 20" [3].)

Exclusive application of the full face design is limited by the need in manufacturing to produce chamfered peripheral edges. It is not generally possible to produce full face rings with perfectly sharp peripheral edges. The advantageous effects of this feature on oil control and blow-by are well known.

The standard design for a full-face sprayed ring as shown in Figure 9 has peripheral edge chamfers of 0.3 ± 0.1 mm x 45°. In the interim period, the development of appropriate production methods has made it possible to limit the chamfer on the lower edge to a max. of 0.2 mm x 35°. The upper chamfer is normally larger, for example, 0.35 ± 0.15 mm x 45°.

Piston rings with an optimized lower peripheral edge of almost-sharp design, as shown in Figure 9, with a chamfer of 0.2 mm max. x 35°, have meanwhile proved outstandingly successful in quantity production. The almost-sharp peripheral edge controls oil consumption and blow-by to almost the same extent as the perfectly sharp design of a half-inlaid ring and provides the additional important advantage of high break-out resistance.

INTERMEDIATE COATING BETWEEN BASE METAL AND PLASMA PERIPHERAL COATING - To increase the adhesion and break-out resistance of various plasma coatings, intermediate coatings are applied to the base metal. The thickness of these intermediate coatings is about 10-30 µm. Pure ceramic plasma coatings particularly require intermediate ductile bond coatings.

COATING THICKNESSES AND BASE METAL MATERIALS - For plasma-sprayed piston rings of nominal diameter up to 200 mm the following coating thicknesses are commonly used:

0.1 mm min. / 0.15 mm min. / 0.2 mm min.

A coating thickness tolerance of + 0.2 mm is considered necessary for quantity production.

Minimum coating thicknesses greater than 0.2 mm are only possible to a limited extent. The inherent stresses in the coating increase with the thickness of the coating and consequently reduce coating strength and the potential loading capacity

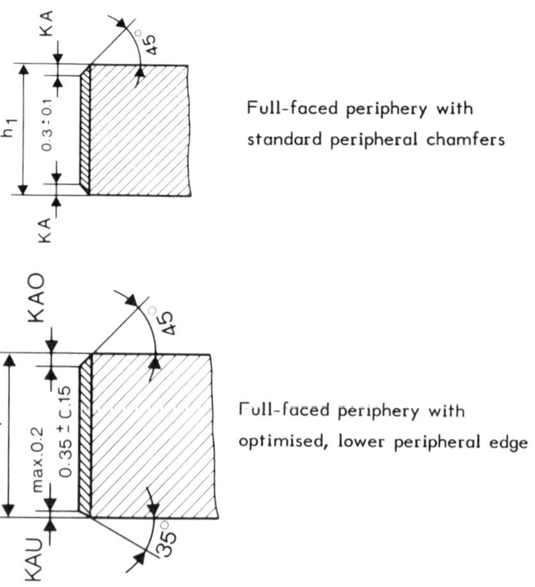

Figure 9: Designs for full-face coating.

during engine operation. For piston rings in high-speed engines, a minimum coating thickness of 0.3 mm nominal value should not be exceeded.

For piston rings of larger nominal diameter for medium-speed and slow-speed diesel engines, coating thicknesses of up to 0.4 mm min. can be specified. However, the coating thickness possible is also determined by the type of spray coating, i.e. composition, structure, and hardness.

Goetze plasma-spray coatings can be applied to all piston ring materials. This includes all gray cast irons with lamellar graphite, spheroidal graphite cast irons, and steel to Goetze material specification KS 3. Plasma-coated piston rings are mainly used in the top piston groove. In high-speed engines, top compression ring base material is usually either spheroidal graphite cast iron or steel. In medium-speed and slow-speed engines, top ring materials are unalloyed or alloyed gray cast iron or, occasionally, spheroidal graphite cast iron.

TESTING THE COATING IN THE ENGINE

TEST PROGRAM - The engine performance of the coatings was determined by measurement of ring wear, cylinder wear, and coating break-out resistance.

The wear tests were carried out in a four-cylinder passenger car diesel engine, naturally aspirated.

Engine characteristics:

N_{max} = 53 KW at n = 4400 rev/min

Swept volume = 2.4 l

Test data:

Brake load: full load at n_{max}

T_{wa} = 120°C ≙ T_{cyl} = 180-200°C at TDC top ring reversal position.

Total running time = 83.5 h ≙ 11,000 vehicle kilometers.

See Figure 10 for further details of the program.

The cylinder wall temperature T_{cyl} was measured continuously in all four cylinders by means of thermocouples.

This so-called hot test produced well-defined, measurable wear levels of ring peripheries and cylinder bores at the TDC area of the top piston ring. It is therefore assumed that, under the above conditions, starved lubrication and mixed friction set in especially at the TDC area as a result of localized and temporary penetrations of the oil film. This leads to an increased adhesive wear.

No.	Operating programme	Number of cycles	Time
1	Production run-in	1	6 min.
2	Quality back-up run	1	140 min.
3	Power curve	1	1 hour
4	Hot test	960	80 hours

One cycle of no. 4 is made up of:

Point	Speed rev./min.	Brake load	Time min.	T_{WA} °C	T_{oil} °C
1	4400	Full load	4.5	120±5	140±5
2	1000	Zero load	0.5	120±5	

Oil: Rotella Z 30

Figure 10: Hot test in 4-cylinder car diesel.

The peripheral wear was determined at measuring points 10°, 90°, 180°, 270° and 350° from the ring joint. The average value of the wear levels at these individual measuring points was defined as total wear. Similarly, the total cylinder wear was determined from the wear levels at the measuring points 0°, 45°, 90°, 135°, 180°, 225°, 270°, and 315°, with 0° and 180° being aligned with the pin axis and 90° and 270° at right angles to pin axis.

The coating break-out tests were run in a turbocharged 5-cylinder passenger car diesel engine.

Engine characteristics:

N_{max} = 81 KW at n = 4200 rev/min

Swept volume = 3.01

Test data:

Total running time = 4 x 150 cycles = 240 h

The test was interrupted every 150 cycles for wear measurement. See Figure 11 for further details.

No.	Operating programme	Number of cycles	Time (h)
1	Initial run-in	1	0.20
2	Torque measurement	1	0.50
3	Break-out test	150	58

One cycle of no. 3 is made up of:

Point	Speed rev/min.	Brake load	Time min.	T_{WA} °C	T_{oil} °C	Remarks
1	–	–	10	25±5	25±5	Oil cooler with separate pump
2	–	–	30 sec.	25±5	25±5	Ignition
3	1000	–	30 sec.	25±5	25±5	Starting
4	4350	full load	10	90±5	150-10	
5	3000	partial load	1	90±5	150-10	Supercharger cooling phase
6	1200	zero load	2	90±5	150-10	Supercharger cooling phase

Oil: SX-30

Figure 11: Break-out test in 5-cylinder car diesel.

The break-out of sprayed coatings is evidently caused by fatigue processes resulting from frequent mechanical and thermal cyclic stresses. A break-out test was therefore developed during which the mechanical and thermal loading of the coating varied rapidly between extreme values.

The break-out behavior of the coatings was determined by measuring the individual areas of break-out with a microscope and calculating the total area over the entire periphery of the ring. To better evaluate the break-out mechanism, this analysis was made after every 150 cycles (see Figure 11). The break-out behavior of the individual test coatings was finally classified after a total running period of 600 cycles.

The piston ring pack used in the engines was comparable with the production design except for the peripheral coatings of the top and second rings.

ANALYSIS AND INTERPRETATION OF THE RESULTS - Within the past few years, a total of more than 100 different types of coatings have been tested in engines over a total of about 40,000 hours of operation. Most of these tests were used for preselecting promising coatings. Those coatings which were successful in the screening tests were then subjected to additional engine testing to provide statistical confirmation of the results.

Wear Behavior - Figure 12 shows the wear behavior of piston ring peripheries and cylinder bores, arranged according to the amount of the cylinder wear. This shows that, in relation to cylinder wear, the peripheral wear of the piston ring acts in an opposite direction. In other words:

> Low peripheral wear of the piston ring causes increased cylinder wear, and vice versa. At the point of intersection of this "scissors" pattern (see Figure 12), the cylinder wear and peripheral wear of the piston rings are the same.

It therefore appears to be impossible because of the tribological interrelationships between the paired components to achieve, with the materials known at present, minimal cylinder wear in conjunction with low peripheral wear of the piston rings.

The interrelationship between ring peripheral wear and cylinder bore wear shown in Figure 12 is specific to the program and the engine. It can therefore be assumed that, for example, the starved lubrication conditions in the hot test become less pronounced as cylinder wall temperature decreases and more pronounced as temperature increases. Accordingly, the "scissors" pattern, which in the hot test is mainly determined by adhesive wear, would probably shift up or down in the diagram with changes in cylinder wall temperature and at the same time close or open out further.

A substantial change in the wear relationships shown to exist during the hot test can especially be expected with very low cylinder wall temperatures during the so-called cold operation. The initial test results for cylinder wall temperatures of $T_{cyl} < 50°$ show that the pattern of wear resistance of individual coating materials found during the hot test can shift during cold operation. This can probably be attributed to abrasive and corrosive wear constituents, in addition to adhesive wear, becoming increasingly active during low temperatures. With high cylinder wall temperatures, this places different demands on the wear resistance of the materials than adhesive wear. These relationships will be discussed separately.

Other important factors influencing the pattern of wear levels of different material pairings of piston ring peripheries and cylinder bores are the oil and fuel quality, cylinder material, the microsurface and macrosurface of cylinder bore and piston ring periphery, etc.

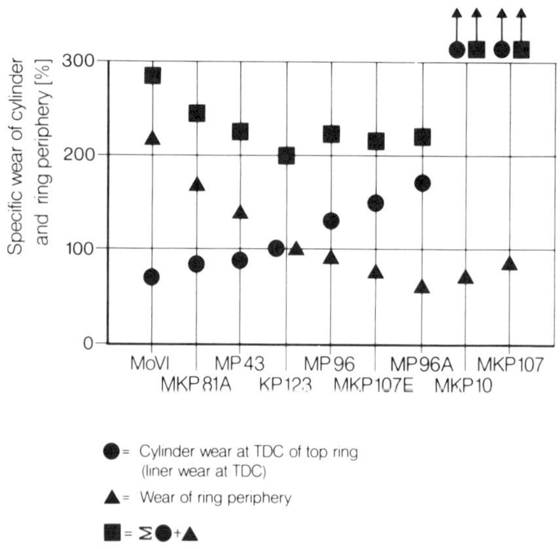

Figure 12: Cylinder and peripheral wear during the hot test.

In this context, the question arises of whether wear investigations on different coating materials are at all meaningful with any type of model equipment. For if the wear pattern shown in Figure 12 could be applied even partially to all tribological systems, which according to all our previous experience, however, can be considered out of the question, then all further wear investigations would simply be superfluous. But if tribological systems that are different from a design and loading aspect have correspondingly different, i.e. specific, wear patterns, then results from model equipment can likewise only be evaluated on this equipment-specific basis. These results cannot, therefore, be applied to other tribological systems or can only be applied if there is a chance affinity in the wear processes.

Break-Out Behavior - Although a satisfactory wear behavior of the pairing of piston ring and cylinder is a necessary condition, it is frequently not an adequate condition for the suitability of a peripheral coating for engine application. The resistance of the peripheral coating against fatigue, i.e. its break-out resistance, during frequent mechanical and thermal stress reversals also plays an important role. Figure 13 gives an overview of this problem that exists between wear behavior and break-out behavior. Comparison of Figure 12 with Figure 13 will show that materials with advantageous wear characteristics of cylinder bore or piston ring periphery can only be used to a limited extent, because they do not have a sufficiently high resistance to mechanical and thermal cyclic stresses. An important development aim is therefore to make such coatings more resistant to break-out while at the same time maintaining their wear characteristics. This especially applies to ceramic coatings.

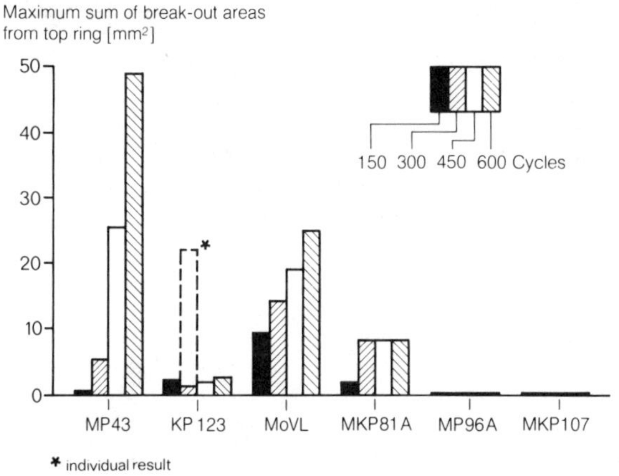

Figure 13: Maximum areas of coating break-out from the top ring (inlaid design).

The break-out tests were interrupted and evaluated after every 150 cycles. By using this procedure, we could recognize significant differences in the break-out mechanisms of different peripheral materials. For example, flame-sprayed molybdenum began to show incipient fatigue after 150 cycles. As running time increased, this led first to small areas of irregularly distributed break-out and then to constantly increasing large areas of break-out (see Figure 14). Although in the long run such a break-out behavior can detach larger particles of the molybdenum coating from the periphery of the piston ring, under no circumstances will this lead to damage of the cylinder liner; because during such a break-out process, small particles are detached successively and not in one piece over a large area of coating.

By comparison, very ductile materials generally showed no signs of fatigue up to 600 cycles. However, break-out did take place with some materials between 450 and 600 cycles. Because of the high ductility of these materials, such break-out inevitably took place over a large area and consequently resulted in the scoring of the cylinder. Comparable production coatings of other piston ring manufacturers showed similar break-out characteristics.

The break-out mechanism described here was previously found only in applied engine tests under heavy thermal and mechanical cyclic stresses. Without thermal cyclic stresses, no significant break-out occurred, even during long periods of full-load operation. In other words: the break-out resistance of a peripheral profile will face more exacting requirements in an engine that is subject to frequent cold starting than in an engine that is mainly used for long-distance operation.

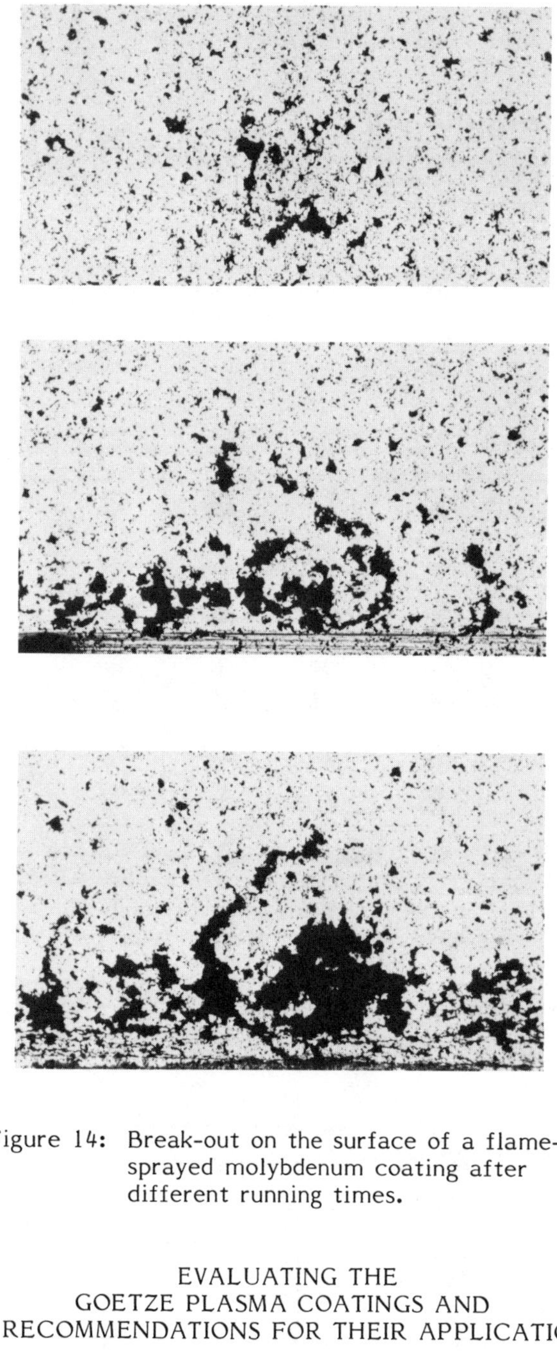

Figure 14: Break-out on the surface of a flame-sprayed molybdenum coating after different running times.

EVALUATING THE GOETZE PLASMA COATINGS AND RECOMMENDATIONS FOR THEIR APPLICATION

The important properties of plasma coatings were treated from a metallurgical viewpoint in in the early sections of this paper. We discussed the test methods developed for determining the coating behavior during engine operation and also the relationship between the wear of the piston ring periphery and the cylinder bore and the effect of this relationship on the break-out behavior of the coatings. The design and design limits of piston rings with plasma-coated peripheries were also analyzed. The individual plasma coatings for the various applications will now be evaluated and recommendations given for their selection.

As before, we emphasize that the full-face design should be preferred, irrespective of the type of coating, to achieve the greatest possible coating break-out resistance.

Figure 15 shows an evaluation of the three coating properties that are particularly important in practical application and also the cylinder wear induced. For a better comparison of the properties of plasma coatings, we have also included in the table galvanically applied hard-chromium and flame-sprayed molybdenum coatings. The symbols used for this evaluation refer to the chromium coating as basis.

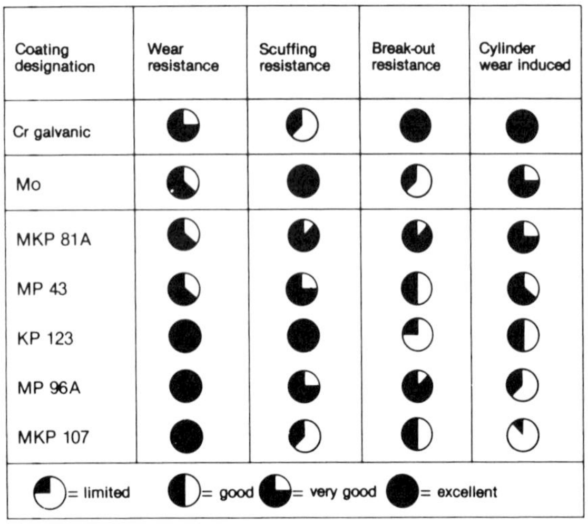

Figure 15: Goetze peripheral coatings for piston rings - evaluation of coating properties.

Coating designation	Special coating properties	Recommended applications
MKP 81A	- Higher inherent wear resistance - Causes less cylinder wear - Increased thermal loading capacity - Increased breakout resistance	High-speed diesel engines Also in cases of high thermal stress reversal
MP 43	- High scuffing resistance - Good relationship between inherent coating wear and induced cylinder wear	High-speed diesel engines Especially for engines whose life expectancy is determined by the inherent wear of the ring
KP 123	- Increased inherent wear resistance - Highest thermal loading capacity	High-speed engines, but also medium and low-speed engines For engines working chiefly in continuous service
MP 96A	- Very high inherent wear resistance - Very high breakout resistance	Spark ignition engines Can also be used for high-speed diesel engines with wear-resistant cylinder liner
MKP 107	- Very high inherent wear resistance	Medium and low-speed engines Not suitable for use in high-speed engines

Figure 16: Application recommendations for plasma coatings.

Figure 16 gives recommendations for application in accordance with the overall evaluation shown in Figure 15.

This table only contains recommendations or limits for the application of Goetze plasma coatings. The coatings should then be selected for any particular application by carrying out tests in the engine. The application requirements have to be matched with the properties of the plasma coating.

SUMMARY

Plasma-coated piston rings will find increasing application because the hard-chromium coatings and flame-sprayed molybdenum coatings used almost exclusively up to now can no longer keep pace, for the reasons stated, with the demands for longer life and increased loading capacity.

The information and data presented on the Goetze plasma coatings is intended to help the user in the engine industry to gain a better understanding of the application potential, and also the limits of application, of plasma-coated piston rings.

The five plasma coatings presented clearly show that it is not possible according to the present level of knowledge to have a universal plasma coating that equally meets all requirements. Therefore, the different demand profiles of the engine on the piston rings will require in future the availability of a range of plasma coatings with different properties.

REFERENCES:

[1] H. Beyer and U. Buran
"New Plasma Coatings for Long-Life Piston Rings" Goetze Technical Paper "K 22"

[2] H.J. Neuhaeuser
"Zur Porositatsbestimmung von thermischen Spritzschichten durch quantitative Bildanalyse" Goetze Technical Paper "K 33"

[3] H. Beyer and H.J. Neuhaeuser
"SEM Studies of Cracking in Thermally Sprayed Piston Rings"
Goetze Technical Paper "K 20"

METALLURGICAL ASPECTS OF THE WHEEL/RAIL SURFACE CONTACT

M. N. Bassim, David Shum, S. Yannacopoulos, M. R. Bayoumi, M. C. Chaturvedi
Metallurgical Sciences Laboratories
Department of Mechanical Engineering
Univeristy of Manitoba
Winnipeg, Manitoba, Canada

Abstract

Study of the surface contact problem between a railroad wheel and rail is of significant importance in improvement of the load-carrying capabilities of freight and passenger trains as well as their speed. Such a complex problem is generally analysed from a mechanical and tribiological aspect by determination of methods to improve the adhesion between wheel and rail as well as by a metallurgical analysis of the wear and fatigue problem incurred in these railroad components. In this investigation, a review of these two aspects of analysis of the surface contact problem is presented. More specifically, the effort in improving the wear resistance of rails and wheels by using low-alloy steels, different microstructures as well as most recent developments in the use of powder metals and shape memory alloys in the construction of the wheel and rail is outlined.
The adhesion between rail and wheel is generally improved either by track oriented solutions such as friction particles and debris, chemical decontaminants and plasma discharge or by train oriented solutions such as monomotor truck or coupled drives, independent motors for each axle and weight transfer between axles. The present study addresses these various aspects.

RAILROAD TRANSPORTATION has been an effective means of transportation of bulk merchandise for many years. The last increase in energy cost in recent years, however, has necessitated accelerated research toward the solution of the problems associated with this means of transportation. In addition, desire for higher speeds and heavier loads has complicated the picture even more. It has been realized that two of the main factors that influence the performance of a locomotive are the wear resistance of both wheel and rail and the adhesion of the wheel to the rail.
to the rail. Apparently improvement of one of these two factors results in deterioration of the other. The major attention of research has been focussed on the parameters which influence the wheel/rail interface with respect to wear and adhesion such as stresses developed, natural or artificial lubrication, slippage and creepage when starting from rest and during braking and materials for wheels and rails.

The study of the contact problem between rail and wheel has been only recent and presents a high degree of complexity. Several phenomena take place simultaneously at this interface including wear, friction and adhesion. Thus, beside metallurgical means for improving these properties, mechanical and electrical approaches including weight transfer systems, microslip control devices and electrical control systems are used to improve the adhesion between the wheel and rail.

One of the factors that can be manipulated to influence the wear and friction behavior of the wheel/rail interface is the material used for the construction of these two components. In order to combat the wear problem due to heavier loads and higher speeds, the train wheel has evolved from the one made of the wooden spokes and rims with wrought iron tread to that made of cast iron and carbon steel(1). Advances in metallurgy have resulted in the development of new compositions and heat treatments for steels used for wheels and rails.

FACTORS AFFECTING WEAR AND ADHESION

STRESSES - The classical theory of elasticity treats the contacting bodies, wheel and rail, as ideal media exhibiting perfect elastic behavior and having boundaries which are perfectly straight, circular, etc[2]. Friction contact has also been studied extensively as a problem in mathematics and as an example of numerical analysis applied to finite element methods[3]. Plastic flow plays a significant role in the working life of rails. On "heavy haul" railways, bulk plastic flow of the head of the rail, particularly in curves, limits the load carrying capacity of the axle and developments are in progress to use head hardened and alloy steel rails to raise their yield stress. Corrugations which occur under heavy axle loads have been found to arise through unsteady plastic deformation. Even with relatively light axle loads, there is ample evidence that severe plastic flow takes place in a thin layer close to the running surface of the rail induced by the tangential contact forces which arise from traction, braking, guidance and cornering. It is generally agreed that plastic flow plays a role in the wear process and in rolling contact fatigue, but the detailed mechanisms are far from clear. Plastic deformation gives rise to residual stresses and there is growing conviction that residual stresses are important in contact fracture and fatigue, where the applied stresses are largely compressive[4-7]. Since plastic deformation leads to residual stresses which tend to inhibit further plastic flow during subsequent cycles of load, the possibility exists for the residual stresses to lead to steady state in which the cyclic deformation is entirely elastic. This process is known as shakedown and the maximum load under which this can occur is the shakedown limit [7-9].

LUBRICATION - The purpose of injecting lubricant layer between wheel and rail surfaces is to utilize its viscous nature to reduce the maximum pressure between the individual asperities and to minimize the sliding friction throughout the entire zone of contact. Its beneficial effect in reducing rail curve adhesive wear is recognized by the fact that a track-side mounted lubricator is a permanent element in the track structure of heavy haul railway lines in North America. The rail lubricant is usually calcium or lithium soap based grease containing up to 12% graphite or 3% MoS_2. The lubricant composition, thickness and location on the rail head are dependent on lubricator location, type of traffic, climatic conditions and other operational parameters. The variation in the thickness and the rheological properties of the lubricant contaminated layer determines the type of wheel/rail lubrication regime. High pressures and temperatures reached at wheel/rail asperities in the contact zone could affect the rheological properties of the contaminant lubricant mixtures and the surface-lubricant physical and chemical interactions. They also degrade the chemical characteristics of oily fluid contaminants and affect their reaction with the basic lubricants.

Lubrication of the gauge face only of the outside rail of a curve can yield benefits in rail performance on both the inside and outside rails. On the outside rail, gauge face wear and height loss rates are reduced. On the inside rail, head height loss rates and corrugation occurence are reduced. Gage face wear and corrugation occurence exhibit a metallurgical/lubrication interaction such that premium rails benefit far less from lubrication, both in an absolute and a relative sense, than does the standard carbon steel rail. However, much of the potential benefit of lubrication is wasted because of the intercession of rail head fatigue as the primary cause for rail replacement. Thus, efforts to improve the overall level in wheel/rail lubrication as a means of energy savings may lead to unexpected rail fatigue problems. The use of metallurgical cleaner and mechanically stronger rail steels can compensate to some extent for this shift in behavior. Rail profile grinding on a programmed basis offers great potential benefit both by decoupling of tread and flange force effect[10-12].

SLIPPAGE AND CREEPAGE - The relative sliding at the wheel/rail interface is very important throughout the journey of a locomotive and it becomes more pronounced at starting from rest and during braking. The term creepage is used to identify the deviation of actual sliding behavior from the ideal pure rolling between wheel and rail. Lateral creepage takes place mainly at the curved sections of the railway. Longitudinal creepage takes place at starting from rest and during braking and it has been defined as the relative velocity between the two bodies if the two bodies were not quite in contact in order to compensate for distortion in the contact zone[13]. Even though the effect of low slippage and creepage as reflected by low adhesion would result in failure to control locomotives during braking, the most serious problems arise from rail burns when starting from rest and wheel flats during braking[14].

CAUSES OF LOW ADHESION

There are in general two major groups of factors that contribute to low adhesion: natural and man-made causes.

NATURAL CAUSES

1. Moisture - Moisture is probably the most significant factor in low adhesion situations. As the temperature drops below the freezing point, any moisture present on rail tops turns to ice that predictably reduces the coefficient of adhesion (COA) available. Moisture enhances the corrosion of steel rails to produce various forms of oxides

on the rail top. Although the oxides themselves have the potential to increase adhesion at the expense of wear of the rail, oxides combined with the right amount of moisture turn into a paste that drastically reduces the COA available. On one occasion, spray from a weed killing train reduced adhesion from 0.32 to 0.20. With tap water, a drop of 0.23 was recorded with the average drop in the range of 0.05 to 0.15. Overall, the average adhesion on dry rails was about 0.30 and it dropped to 0.22 for wet rails. Furthermore it was found that the lowest value of COA was obtained when there was high humidity present rather than when bulk water was present on the rail. In one study, the transition from dry rails through moist to wet was accompanied by a change of COA from 0.40 to 0.15 and back to 0.25(16). The results serve to confirm that the lowest friction will be encountered on running rails in mist or when dew forms rather than when the rails are thoroughly wetted by rain.

2. Fallen Leaves - Few leaves fall on the rails themselves to cause low adhesion, and it has been shown that the majority of leaves are sucked in by the slip stream of a train set subsequently caught in the wheels. Although the leaves as bulk material lower the COA available, it is the black pulp left behind as a result of the crushing action of wheels and moisture that proves to be the most dangerous situation. In the presence of moisture, chemical and bacterial reactions transform them into surface active oily deposits. In damp conditions adhesion coefficients, as low as 0.05 have been reported (15, 17) when the leaf affected portion of the track includes a 25% gradient and curve radius of 300-400 meters. Infrared absorption spectra of the oily deposits on the rails were analyzed to trace them back to the presence of these leaves(17).

MAN MADE CAUSES

1. Contaminants - For liquid contaminants, the bulk properties of the fluids, such as viscosity, are not considered important and, instead, chemical properties largely determine how strongly the compounds are bonded to the steel to provide protective, lubricating surface layers that may inhibit corrosion, but unfortunately also lower the COA. The majority of oily deposits have been shown to be fuel oil in origin, with fallen leaves and other lubricating oils used in trains contributing to the problem. The oils must be surface active and adhere to the rail surface to significantly lower the COA. Industrial pollutions from surrounding industries also contribute to lower COA through reactions with the moisture in the air and subsequently settling on the rail tops. Some types of contaminants are specific to certain types, eg. coal slurry from mines or dust from cement works. Terminal areas are also badly affected with the presence of stationary cars that tend to concentrate their contaminants, such as lubricating and fuel oils, in one spot. The constant starting and stopping of trains in these areas serve to harden the contaminant layers on the rail tops.

2. Design Considerations - Available adhesion is known to decrease as the train speed goes up, and it is becoming generally accepted that this is due to the interaction between track irregularities and vehicle dynamics(14,15). Adhesion has been found to be less affected by train speed on continuously welded rails under dry rail conditions adhesion seems to be independent of speed(19) as demonstrated in Figures 1,2.

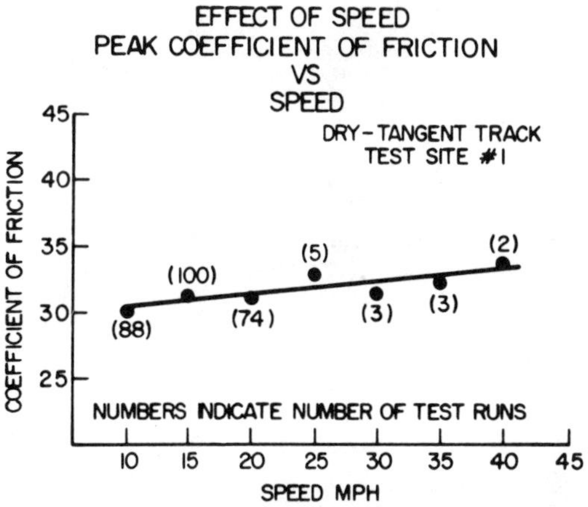

Figure 1 - Effect of speed - percent creep at the peak, [1].

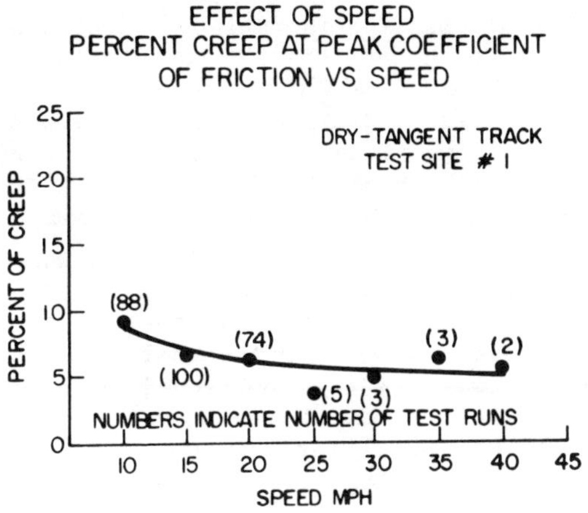

Figure 2 - Effect of speed - peak coefficient of friction, [1].

On laboratory testing machines where low frequency dynamic effects can be minimized, friction was found to be unaffected by changing the rolling speed(15). In addition, adhesion levels decrease for increasing track curvature, with the peak creep value increasing with increasing curvature(19) as shown in Fig. 3. Studies in England have shown that the level of contaminant on curved tracks are similar to that over straight tracks(15). In conventional train designs torque reaction can lead to a loss of adhesion of about 7%(14). Reactive forces at the wheel rail interface tend to lift the leading rail car up and load the trailing rail cars down due to the couple developed between the coupler of a train car and the wheel forces. Unequal distribution of loads within a train car also tend to produce varying COA over the traction surfaces, and introduces additional instability into the system.

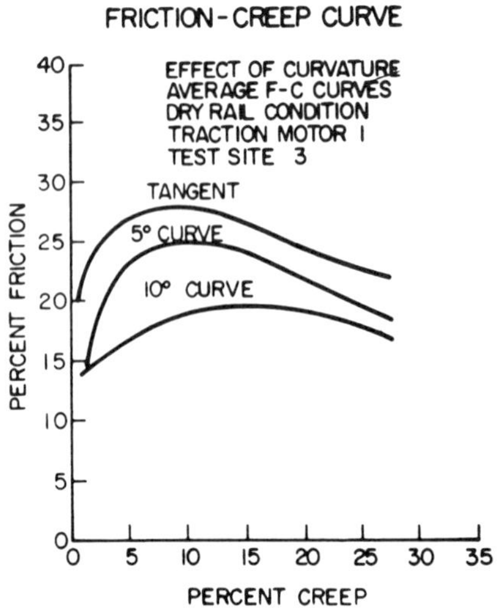

Figure 3 - Effect of curvature-dry rail, [1].

3. Operational Considerations - Roughness of the wheels were once thought to be a factor in the level of available adhesion. New wheels tend to exhibit higher adhesion values. However the decrease in adhesion due to wear of wheels have been shown to be probably due to the geometry change of the wheel from an ellipse to that of a circle(15). The wheel and rail profiles tend to change with time. Experience has shown that rail/wheel contact is not a broad line across the whole head of the rail but more nearly approached to point contact. These changed profiles tend to aggrevate the situation by lowering the already diminished COA.

ROLE OF MATERIALS ON RAILWAY WHEEL/RAIL TRIBOLOGY

One of the factors that can be manipulated to influence the wear and friction behavior of the wheel/rail interface is the material used for the construction of these two components. Advances in metallurgy have resulted in the development of new compositions and heat treatments for steels used for wheels and rails. However, the limitations of plain carbon steel have necessitated the exploration of alloy steels for this application.

In order to achieve hardness requirements for increased wear resistance new heat treatments resulting in pearlitic, bainitic or martensitic microstructures are being developed (20). Also, the influence of various alloying elements, such as Cr, Mo, V etc, on the mechanical properties of rail/wheel steels is being explored by various investigators (21-23).

STANDARD STEEL WHEELS AND RAILS - In order to deal with various situations encountered in the railroad industry the American Society for Testing and Materials (ASTM) has standardized a number of steels and provided the industry with standard manufacturing, inspection and testing techniques as well as compositional and hardness limits(24-25). The various steels as found in the ASTM A 504-83 and ASTM A 583-81 for wrought carbon steel wheels and cast carbon steel wheels respectively are:

Class U - general service where an untreated wheel is satisfactory.

Class L - high-speed service with more severe braking conditions than other classes and light wheel loads.

Class A - high-speed service with severe braking conditions, but with moderate wheel loads.

Class B - high-speed service with severe braking conditions and heavier wheel loads.

Class C - (1) service with light braking conditions and high wheel loads.
(2) service with heavier braking conditions where off-tread brakes are employed.

Class U wheels are untreated and Class L, A, B and C designate wheels with heat treated rims. The compositional and rim hardness limits are shown in Tables 1 and 2.

TABLE 1 Chemical Requrements (6,7)

	Composition. %
Carbon:	
Class U	0.65 to 0.77
Class L. Max	0.47
Class A	0.47 to 0.57
Class B	0.57 to 0.67
Class C	0.67 to 0.77
Manganese	0.60 to 0.85
Phosphorus, max	0.05
Sulfur, max	0,05
Silicon, min	0.15

TABLE 2 Rim Hardness Values (6,7)

Class	Brinell Hardness, H.B. min	max
L	197	277
A	255	321
B	277	341
C	321	363

EFFECT OF MATERIAL HARDNESS ON WEAR

Each form of rail wear is a result of a process in which the wheel surface applies a set of normal and tangential forces on the rail surface layer, thereby deforming the surface layer material elastically or plastically and generating wear particles. Metal loss can be caused by a number of wear mechanisms which may occur independently or simultaneously. They all, however, depend on the surface structure, the time-dependent triaxial state of stress and plastic flow in highly lubricated regions.

In the analysis of the large amount of data obtained from a great variety of wear tests, many attemps have been made to establish quantitative relations between wear and material properties and process parameters. The wear data obtained have been plotted against independent variables such as load, time, speed, temperature, etc, and the trends observed have been studied, together with observations of worn surfaces and wear particles generated.

With additional data and observations constantly being made, it is now readily recognized that there are a number of mechanisms and processes, depending on the particular conditions, that may be responsible for wear. In contrast to earlier models of wear based on adhesion of contacting asperities, attention is now being focussed on the changes taking place below the surface of the contacting bodies. This indicates the significant role of factors such as subsurface plastic flow, subsurface generation of voids and cracks, fracture behavior of surface layers and the role of cyclic loading, state of stress, environment, surface energy, etc.

In view of the many factors involved it is obvious that generalizations concerning wear are very difficult to make. This is especially true when one considers the fact that there are different sliding systems leading to wear, as indicated by the great variety of friction and wear devices available both commercially and in various laboratories.

Most investigations have been devoted to dry or lubricated sliding wear and lubricated rolling wear. Relatively few investigations appear to have been carried out concerning dry rolling wear, particularly as it pertains to steel wheel and rail systems in railroad applications. Some of the significant work has been reported by Dearden [26], Halling and Brothers[27], Ishii et all[28] Krause and Scholten[29,30] and Aronov and Kalpakjian[31].

Some of the wheel and rail wear results from the tests at the Facility for Accelerated Service Testing (FAST)[32] are analyzed in light of both adhesive and fatigue wear theories. This analysis suggests that the wheel/rail wear processes occurring in the unlubricated state contain one or more components other than adhesive wear alone. Whether or not increases in the hardness of one component will cause an increase or decrease in the wear rate of the mating component appear possibly to be related to the mechanism of wear, the magnitude in the change in hardness and the relative change in the wear rate of the component which shows a change in hardness. A metallurgical or metallic-lubrication interaction which has been observed to occur[32] when lubrication is applied has been found to be consistent with a concept of friction-dependent alteration of surface contact stresses when modified by the inclusion of a material sensitive parameter. Utilizing the FAST wear data in an analysis based upon observations of Rabinowicz[33] one can show that both improvement or deterioration of wear performance of one component might be possible by increasing the hardness of the mating component. The strong metallurgical-lubrication interaction that has been observed to occur in rails can be understood in terms of transition from full or partial elasto-hydrodynamic (EHD) lubrication to mild wear with the possibility of severe wear occurring above a limiting value at least for the standard rail.

A strong effect of equivalent carbon content and rail hardness upon rail wear rate has been observed in a large number of papers in the open literature, for example ref. (12). A metallurgical-lubrication interaction influences both gage face wear and corrugation behavior such that standard carbon rail benefits to a much greater extent from the presence of lubrication than does the premium rail. A strong dependence of rail wear rate on hardness is observed in the dry regime which seems to be consistent with an observed shear band cracking mechanism of fatigue nature [12]. Typical wear rate as a function of hardness under dry and lubricated conditions is shown in Fig. 4.

ROLE OF MATERIAL ON FRICTION, SLIPPAGE AND CREEPAGE.

Steel wheel and rail friction is a problem that has created a great deal of interest both in the past and in the current technical literature. Its importance is increasing in the measure that more effective locomotive design requires increasing load transfer between wheel and rail.

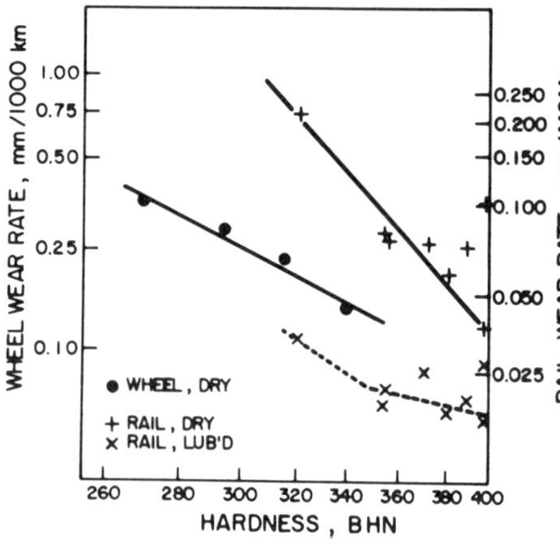

Figure 4 - Wear Rates as a function of hardness, [12].

The stresses and relative displacements at the wheel/rail interface are of primordial importance in most basic technical problems facing railway technology. The wheel and rail wear, braking, acceleration, headway limits, draw bar pull, hunting and other oscillation and stability problems depend on the interaction phenomena taking place at the wheel/rail interface. In spite of the intensive efforts, both theoretical and experimental, a complete understanding is still lacking.

Evidence has been cited[34] that rust is a major railhead debris, being found to possess similar surface characteristics. Particles are observed to spread on the wear band in wet weather and to be worn off in the dry laboratory experiments showed that debris has little effect on friction except when mixed with an oil. Friction is then reduced depending on the quantity of oil and the surface area of the particles. A considerable proportion of oil is needed to reduce friction to its lowest value, and railhead debris normally bears so little it helps maintain high adhesion in dry conditions.

Water from the atmosphere affects all rails. Laboratory experiments were described[35]which showed that water reduces friction between rail and tire steel surfaces. The lowest friction occurs when the ratio of water to debris is small, and a viscous (non-Newtonian) paste is formed.

The friction coefficient plays an important role in the transition from mild wear to severe wear which was reported earlier[32]. Conceptual view of change in wear mechanism as friction coefficient changes is shown in Fig. 5. Similar transition in wear mechanisms in the braking mode was reported[36]. The typical wear rate vs coefficient of friction in the braking mode is shown in Fig. 6. The discontinuity of the straight line at about 0.14 coefficient of friction indicates the transition from one wear mechanism to another. In the range of mild wear the friction coefficient is very small. That indicates its dependence on the material properties and that wear is governed by another mechanism. The x-ray examination[36] of the wear particles (Fig. 7) showed that intensities ratio of the Fe_2O lines to the Fe lines are sharply decreased with increase in the friction coefficient. Therefore the discontinuity of the line in Fig. 6 is associated with the transition from oxidative wear to adhesive wear mechanisms.

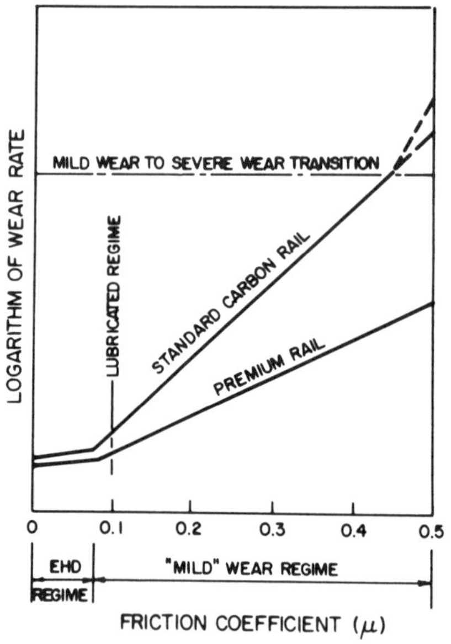

Figure 5 - Conceptual view of effect of change in wear mechanism as friction coefficient changes, [32].

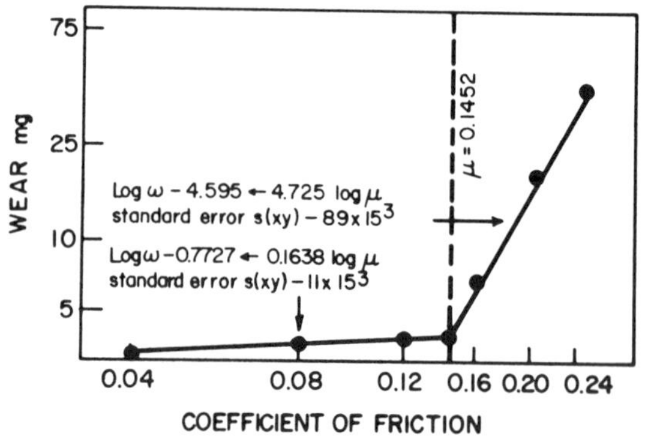

Figure 6 - Wear rate vs. coefficient of friction in the braking mode, [36].

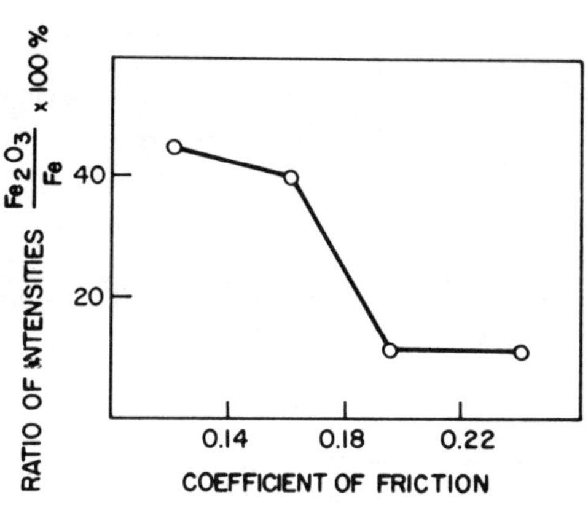

Figure 7 - Ratio of intensities of Fe_2O_3 lines to the Fe lines vs. coefficient of friction, [36].

The effect of slip ratio on wear behavior of wheel steels has been discussed by Hirokawa et al[5]. The ratio of the weight loss under 3% slip to 10% slip vs hardness is shown in Fig. 8. It can be seen that little effect of slip ratio is recognized where hardness is less than 220Hb and that it becomes remarkable with the increase of hardness. This behavior may depend on the experimental condition, i.e., whether the Hertzian stress is higher than the yield strength of wheel steel or not. This means that when an applied axle load is too high, the expected wear resistance of the high hardness wheel may not be obtained under large slippage such as on a small curvature track line.

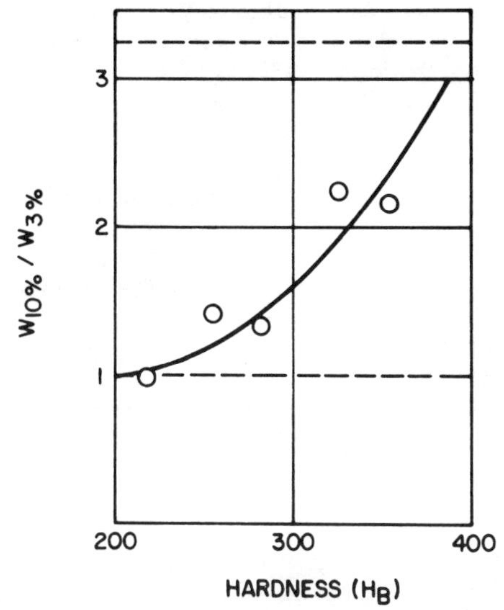

Figure 8 - Effect of slip ratio on wear behavior of wheel steels, [23].

In the open literature, it has been demonstrated extensively the extent to which the traction/creep relationships of wheel/rail contact are influenced by surface contamination, oxide layers, roughness, plastic deformations and the preferred crystal orientations (textures) they cause and the dynamic variations of vertical load, while studies relating the material properties and friction, creepage and slippage are still lacking.

NEW DEVELOPMENTS

The desire for increased cost efficiencies in railway transportation in recent years has made the exploration of materials with superior properties unavoidable. The complicated and at times unpredictable profile of the wheel/rail interface has made research very difficult since various factors influencing this interface must be considered. In general, it may be said that the achievement of increased wear resistance and adhesion would be a very beneficial one. However, great complications arise since many factors resulting in wear resistance have detrimental effects on adhesion. For the above mentioned reason, research in recent years has been focused on either wear rate reductions or adhesion improvement.

Since the early age of railway transportation it has been realized that materials with improved wear resistance must be developed to deal with requirements for higher speeds, heavier loads, and lower maintenance costs. In recent years many investigators have been actively involved in the development of new materials or heat-treatments for railway wheels and rails using as a general rule the fact that increased hardness results in superior wear resistance.

Suggestions have been made for use of guided rubber tires which would provide improved traction and braking, and magnetically levitated systems but they seem to have limitations due to susceptibility to snow and ice[1].

It has been demonstrated that traction/creep relationships are influenced by the elastic properties as well as plastic deformation characteristics of the material[37]. The influence of composition as reflected by carbon content reduction[37] and microalloying additions such as chromium[20-22] molybdenum[22], silicone[20], and aluminum[21] on mechanical properties of railway steels has been studied rather extensively. The development of a new wheel (Fig. 9) based on the principles of powder metallurgy by ASEA[38] is worthy of special attention. Based on the fact that different parts of the wheel are subjected to different loads and are utilized for different purposes a wheel having different properties along the tire and flange profile was fabricated. powder used for the tire is mixed with particles of hard material which provide good adhesion. The inside of the wheel flange is heat-treatable and becomes harder than the rest of the wheel. Also for wear reduction powder giving a lubricating effect may be used. The outside of the wheel flange which is subjected to severe impact is made of powder providing desirable strength and ductility. Although testing is still going on, the results obtained so far are encouraging.

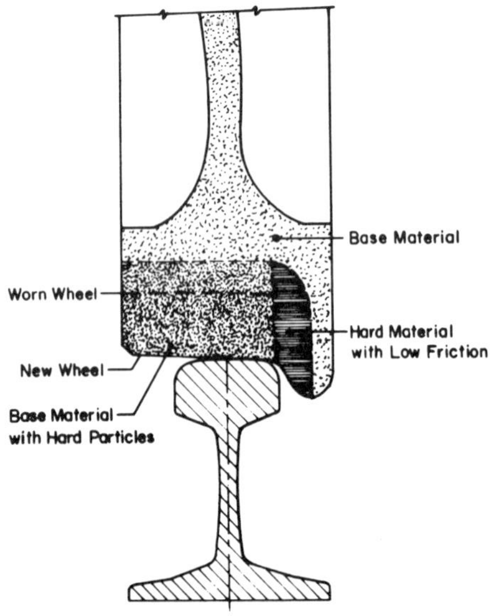

Figure 9 - ASEA wheel, [38].

EFFECT OF CHEMICAL COMPOSITION AND MICROSTRUCTURE ON HARDNESS AND WEAR BEHAVIOR

The fact that the wear behavior of the wheel/rail interface depends on the hardness of the two components has been demonstrated in the past. However, the belief that the wear resistance increases with increasing hardness needs further clarification. The above statement is true only within the same microstructural group as it has been supported by K. Hirakawa et al[23] and others[20]. It has been generally agreed upon that pearlitic microstructures provide higher wear resistance than matensitic microstructure at similar hardness values[23,39] and that bainitic microstructures are inferior to both[20,23,39]. It has also been shown that fine pearlite is more desirable than coarse pearlite from a wear resistance point of view[20,23]. Fine pearlitic microstructures may be achieved with additions of silicon, chrominum[20,23] and vanadium[21]. It has also been shown that the process by which fine pearlite is obtained, i.e., rolling or heat-treatment, does not affect the wear behavior since this behavior is a result of the homogeneity of the microstructure only[20].

In addition to the above, other factors such as material anisotrophy [37], population and orientation of non-metallic inclusions[22,40,41] etc. may affect the wear and/or adhesion behavior appreciably.

CONCLUSIONS

It is concluded that most of the research effort in studying the effect of metallurgical variables on the rail/wheel interface deals with improvement of wear of the wheel and rail. Adhesion improvement is achieved through mechanical and/or electrical means rather than by improving the quality of the rail and wheel material. Finally, powder wheels show some promise in new wheel and rail design.

ACKNOWLEDGMENTS

The authors would like to acknowledge the Canadian Institute of Guided Ground Transport, Queen's University, Kingston, Ontario for support of this investigation.

REFERENCES

1. May, C.C., CIM, 23rd Annual Conference of Metallurgists, 1984, K4-10934.
2. Gladwell, G.M.L., "Development of the classical theory of contact stress", Contact Mechanics and Wear of Rail/wheel Systems, J. Kalousek, R.V. Dukkipati and G.M.L. Gladwell, eds. University of Waterloo Press, 1983.
3. Kikuchi, N. "Friction Contact Problems by Using Penalty Regularization Methods", ibid 2.
4. Hertz, H., "On Hardness", Miscellaneous Papers, MacMillan, London, 1896.
5. Eason, G. and Shield, R.T., "Plastic Indentation of a Semi-Infinite Solid by a Perfectly Rough Circular Punch", ZAMP vol. 11, 1960.
6. Skalski, K., "Contact Problem Analysis of an Elastic-Plastic Body", Warsaw Technical University Publications, No. 67, 1979.
7. Johnson, K.L., "Inelastic Contact: Plastic Flow and Shakedown", ibid 2.
8. Johnson, K.L., "A Shakedown Limit in Rolling Contact", Proc. 4th U.S. National Congress of Applied Mechanicas, Berkeley, California, 1962.
9. Johnson, K. L. and Jefferis, J.A., "Plastic Flow and Residual Stresses in Rolling and Sliding Contact", Proc. Mech. Engineer Symposium on Rolling Contact Fatigue, London, 1963.

10. Report of subcommittee 8, "Effect of Rail Lubrication on Train Operation", Appendix G, AEEA Economics of Railway Location and Operation, 1940.
11. Fujinawa, I., "Flange and Rail Lubrication", Railway Gazette, Dec. 1, 1967.
12. Steele, R.K. and Devine, T.J., "Wear of Rail/wheel Systems", ibid 2.
13. Pearce, T.G. and Rose, K.A., "Tangential Force-Creepage Relationships in Theory and Practice", ibid 2.
14. Barwell, F.T., "Wheel to Rail Adhesion", ibid 2.
15. Collins, A.H. and Pritchard, C., "Recent Research on Adhesion", REJ, Sept. 1972, pp. 19-34.
16. Pearce, T.G. and Rose, K.A., "Tangential Force-Creepage Relationships in Theory and Practice, ibid 2.
17. Ohno, K., Tsuchiya, J., Ogawa, Y. and Shinkawa, M., Quarterly Reports, Vol. 24, No. 3, 1983, pp. 135-136.
18. Marta, H.A. and Mels, K.D., Trans. ASME, J. Engineering for Industry, Aug. 1969, p. 839.
19. Longston, C.F. and Itami, G.S., Journal of Engineering for Industry, Vol. 102, Aug. 1980, pp. 275-281.
20. Heller, W. and Schweitzer, R., Heavy Haul Conference, Colorado Springs, 1984, (82-HH-26).
21. Parsons, D.E., Malis, T.F., Boyd, J.D., "Microalloying and Precipitation in Cr-V Rail Sheets", CANMET, Report MRP/PMRL 83-53 (OP-J), August 1983.
22. Fergredo, D.M., Kalousek, J., Connors, E.F., "The Effect of Wear on Crack Formation and Work Hardening in Bainitic Cr-Mo Rail Steel Wear Disks, With and Without Martensite", CANMET, Report: MRP/PMRL 82-60 (TR), December 1982.
23. Hirakawa, K., Toyama, K., Susuki, S., Hamazaki, A., Heavy Haul Conference, Colorado Springs 1984, (82-HH-59).
24. Standard Specification for Wrought Carbon Steel Wheels, ASTM Designation:" A 504-83.
25. Standard Specification for Cast Steel Wheels for Railway Service, ASTM Designation A 583-81.
26. Dearden, J., "The Wear of Steel Rails and Tires in Railway Service", Wear, 3(1960), pp. 43-59.
27. Halling, J., Brothers, B.G., "Wear Due to Microslip Between a Rolling Body and Its Track", ASME paper 64-Lub-30, 1964.
28. Ishii, K., Oda, N. and Nishioka, N., "Evaluation of Wear Testing", ASME STP 446, 1968, pp. 115-132.
29. Krause, H. and Scholten, J., "Factors Infuencing the Frictional and Wear Behavior of the Wheel/Rail System", Int. Wheelset Congress, Tokyo, October 1975, paper no. 7.
30. Krause, H., and Scholten, J., "Wear of Titanium and Titanium Alloys Under Conditions of Rolling Stress:, Int. Conf. on Wear of Materials, St. Louis, Mo., April 1977.
31. Aronov, V. and Kalpakjian, S. "Wear Kinetics of Rail and Wheel Steels in the Dry Friction Condition", Wear, 61 (1980), pp. 101-110.
32. Steele, R.K., ASLE Trans. vol. 25, 3(1981) pp. 400-409.
33. Rabinowicz, E., "Wear Coefficients - Metals", in Wear Control Handbook, ASME, 1980, pp. 501-502.
33..Beagley, T.M., McEwen, I.J. and Pritchard, C., "Wheel/Rail Adhesion - The Influence of Railhead Debris",. Wear, 33(1975), pp. 141-152.
35. Beagley, T.M. and Pritchard, C., "Wheel/Rail Adhesion - the Overriding Influence of Water", Wear, 35 (1975) pp. 299-313.
36. Aronov, V. and Pons, M., "The Wear Mechanism of the Railroad Wheel Steel in Free Rolling Braking Conditions", Proceeedings of the Second International Heavy Haul Railway Conference, Colorado Springs, Colorado, U.S.A., Sept. 25-29, 1982.
37. Krause, H. Poll, G., Contact Mechanics and Wear of Rail/Wheel Systems, University of Waterloo Press, 1982, pp. 353-372.
38. P/M Railway Wheels, ASEA, private communication
39. Mutton, P.J. Epp, C.J., Railway Engineering Symposium, Melbourne, 30 August - 1 September 1983, pp. 216-221.
40. Jamison, W.E., Lubrication Engineering, vol. 36, 7, pp. 401-411.
41. Clayton, P., Steele, R.K., Processes of Wear at the Wheel/Rail Interface, private communication.

CHARACTERISTICS OF COATINGS PRODUCED USING A NEW HIGH VELOCITY THERMAL SPRAY TECHNIQUE

K. V. Rao
Cabot Corporation
Kokomo, Indiana, USA

Abstract
JET KOTE™* is a new thermal spray technique which relies on continuous internal combustion of oxygen and a fuel gas to produce high velocity exhaust jet stream. Powder is injected into this pressurized gas stream where it is heated and accelerated toward the part to be surfaced. Particle velocities of 1800 to 2600 feet/second (548 to 792 meters/second), depending on material, have been estimated. Coatings of several varieties of materials like tungsten carbide + cobalt, TRIBALOY® alloy T-800 and HASTELLOY® alloy C have been made using this high velocity spray system.

In this paper the salient features of the JET KOTE spraying system and the characteristic properties of coatings of different materials made using this new technique are discussed.

THERE ARE VARIOUS TECHNIQUES available to produce coatings for tribological applications. Some of the popular techniques include wire spraying, electroless plating, plasma spraying, the detonation gun, and ion plating methods such as chemical vapor deposition (CVD) and physical vapor deposition (PVD). Cost effectiveness combined with high quality coatings have become key requirements of many commercial thermal spray operations. The JET KOTE™ system is one of the recently developed thermal spray systems that meets these requirements. JET KOTE technique is a high velocity continuous combustion process which offers the advantages of a simpler operation and reduced operating costs while producing coatings comparable to the detonation gun process. This paper deals with the description of the JET KOTE process and some of the characteristic properties of coatings produced using this system.

*JET KOTE, TRIBALOY, and HASTELLOY are trademarks of Cabot Corporation

JET KOTE PROCESS DESCRIPTION

A schematic drawing of a JET KOTE II system is shown in Figure 1. The major components of the system are:

a) a control console to accurately control the combustion gases,
b) JET KOTE II torch which is a means to transfer the energy of combustion into a jet and then into powder particles, and
c) a powder feeder.

The system relies on continuous internal combustion of gases producing a high velocity exhaust jet stream. Combustion takes place (see Figure 2) by burning a fuel gas (can be propylene, propane, or MAPP), and oxygen injected continuously under pressure into a water cooled combustion chamber. Hydrogen can also be used when a non-carbon main fuel is desired. After combustion, the hot gases are discharged through a flow expansion nozzle and the powder is fed axially and centrally into this exhaust jet stream. It is confined in a water-cooled nozzle bore for a distance of six inches (15.2cm) or more. The nozzle design ensures adequate heating of powders while it is accelerated toward the part to be surfaced. Argon or nitrogen can be used as carrier gas and hydrogen is used for a pilot flame prior to main ignition. With unique JET KOTE II gun design, a high velocity flame is obtained as seen by shock diamonds shown in Figure 3.

Depending upon the feed pressures [60psig (413KPa) or 90 psig (620KPa)] of the combustion gases, gas speeds upon exiting from the nozzle can be in the range of 4000-5000 ft/sec (1220-1524 m/sec). Particle velocities of 1800 to 2600 ft/sec (548 to 792 m/sec) depending on material, have been estimated. Flame temperature measurements have not been made; however, calculated temperatures range from 5000 to 5800°F (2760-3200°C). The JET KOTE-II system has been found to be "operator friendly," producing good quality coatings over a range of operating parameters. Typical spray rates for tungsten carbide-cobalt are 8-12 lb/hr (3.6 - 5.4 kg/hr) and the spray distance varies 6-9 inches (15 to 23cm). The noise level during this operation is about 120 dB.

PROPERTIES AND STRUCTURE OF JET KOTE COATINGS

The JET KOTE II system has been successfully used to produce high quality coatings of tungsten carbide-cobalt of various compositions, TRIBALOY alloys, STELLITE alloys, HASTELLOY alloys, and a wide range of other consumables. A series of tests on several types of coatings (produced using propylene fuel at 60psi pressure) have been conducted. The results are presented in Table 1. Microhardness measurements were made using a 300gm load. Bond strengths of the coatings were determined using the ASTM procedure C633-79. For each coating, four identical test specimens and one epoxy check sample were pulled. The maximum epoxy strength was found to be about 11,000psi. Density values were determined using an oil impregnation test as per ASTM procedure B328-80. The dry sand abrasive wear tests were performed according to ASTM procedure G65-81. The tests were conducted at 30-pound loads using AFS 50/70 sand abrasive. Five hundred revolutions were used for most of the coatings except WC-Co coatings for which 2000 revolutions were required to obtain any appreciable volume loss.

The extremely high particle velocities developed in the JET KOTE II system result in coatings with high bond strengths and densities. For example, JET KOTE coatings of WC-12% Co typically exhibit densities of 13 to 13.7gms/cm^3. These values are higher than the density of 12.5 gms/cm^3 reported for plasma spray coatings and comparable to the 13.5gm/cm^3 value quoted for WC-13% Co coatings produced by the detonation gun technique. JET KOTE coatings of WC-Co also exhibit high hardness numbers. At a propylene fuel pressure of 60 psig (413 KPa), the hardness values for WC-12% Co (1160 DPH_{300}) and WC-17% Co (1060 DPH_{300}) are comparable to the detonation gun coatings (1150 and 1050 DPH 300, respectively) of WC-Co with similar compositions. WC-Co coatings with much higher hardness numbers can be achieved, if so desired, by using 90 psig (620 KPa) propylene fuel pressures.

The microstructures of typical JET KOTE coatings of WC-12% Co are shown in Figures 4a, 4b, 5a and 5b. Figures 4a and 4b show coatings made using propylene fuel while Figures 5a and 5b were made with hydrogen fuel. These micrographs illustrate the denseness of the coatings and the significant amount of retained carbides. The higher hardness numbers obtained with JET KOTE seem to be related to the high degree of retained carbides in the coatings. JET KOTE system is not limited to the application of tungsten carbide. Some impressive coatings have been demonstrated for TRIBALOY alloys, STELLITE alloys, HASTELLOY alloys and other carbides. Figures 6a and 6b show the microstructure of TRIBALOY alloy T-800 coatings produced using the JET KOTE system. These JET KOTE coatings have a higher hardness, 650 DPH_{300}, compared to that typical of a plasma sprayed coating, 550 DPH_{300}. The microstructure of HASTELLOY alloy C corrosion resistant coatings applied, using the JET KOTE system, is shown in Figures 7a and 7b and the properties of these coatings are listed in Table 1.

Tungsten carbide-cobalt and TRIBALOY alloy T-800 coatings are widely used for tribological applications. The coatings produced by the JET KOTE process are of high quality and can be made at a reasonable cost. The JET KOTE system can be purchased outright and is suitable for use in a thermal spray job shop environment. Several applications of JET KOTE have reached commercial scale. Components where JET KOTE coatings are used include ball valves, turbine rotors, wire drawing capstans, boiler tubes, Banbury mixer journals, injection molding screws, knife blades, etc.

1. CONTROL CONSOLE
2. PENDANT
3. JET KOTE TORCH
4. HOSE BUNDLE
5. POWDER FEEDER
6. INLET HOSES
7. TRIGGER LINES (OPTIONAL USE)
8. PUMP

JET KOTE II SURFACING SYSTEM

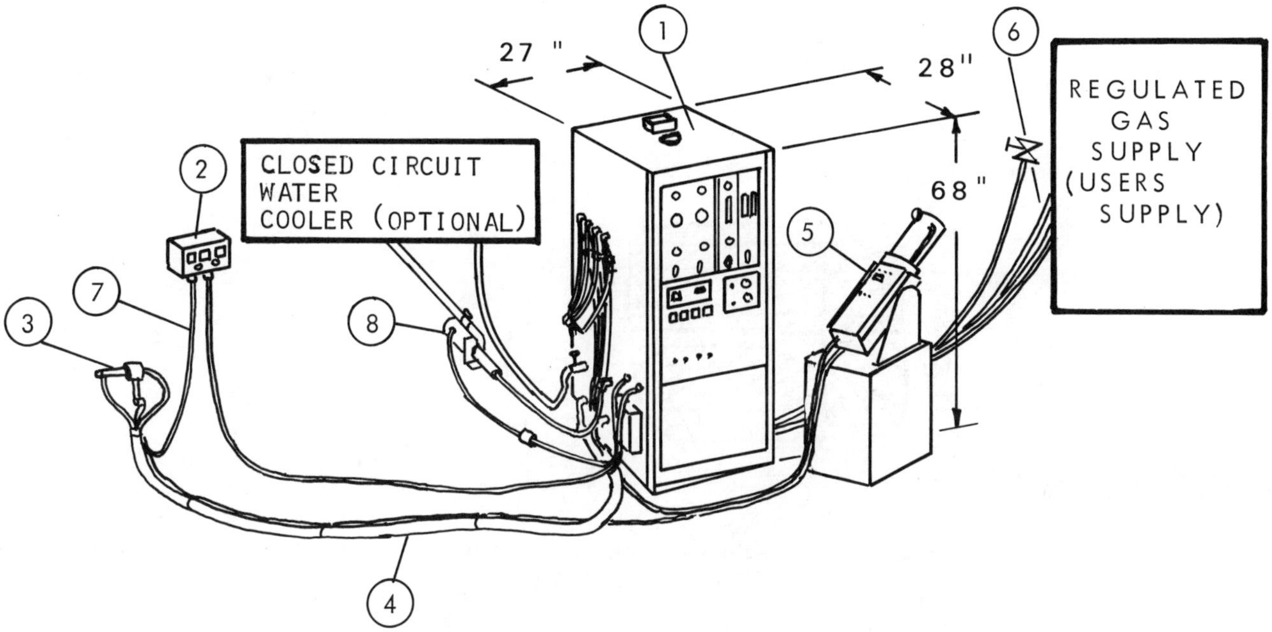

Figure 1. A schematic drawing showing the components of the JET KOTE II system

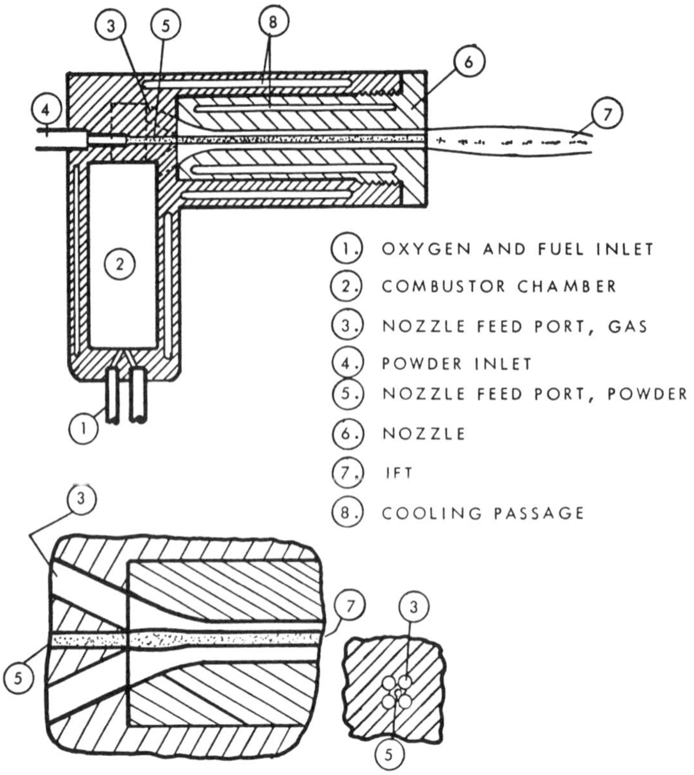

Figure 2. A schematic drawing showing the JET KOTE II torch

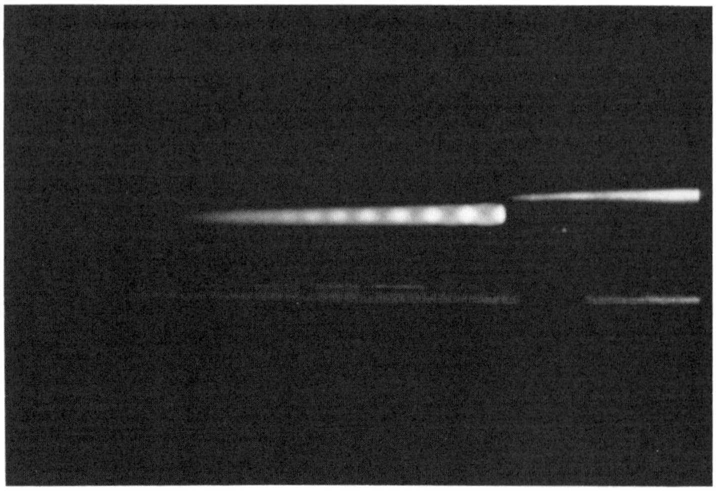

Figure 3. JET KOTE flame with its characteristic shock diamonds

TABLE 1

Properties of JET KOTE Coatings

Alloy Composition	WC-12% Co	WC-17% Co	TRIBALOY® alloy T-800	HASTELLOY® alloy C
Hardness (DPH 300)	1160 (60 psi C_3H_6) 1429 (90 psi C_3H_6)	1056 (60 psi) 1283 (90 psi)	648	385
Bond Strength (psi)	> 11,400	> 11,250	> 8,300	> 7,600
Density (Gms/cm^3)	13.01 (60 psi) 13.68 (90 psi)	12.63	7.65	7.85
Surface finish (microinch RMS) As-coated	160 - 240	140-200	200-300	200-300
As-ground	6-10	10-20	6-12	3-5
Abrasive Wear (Adj. volume loss cm^3)	4.4 2000 rev.	4.7 2000 rev.	12.2 500 rev.	34.3 500 rev.

(no measurable loss at 500 rev.)

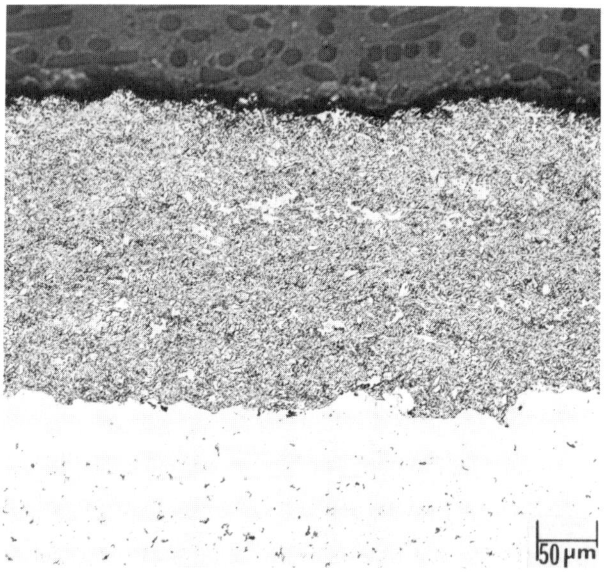

Figure 4a: Optical micrograph of WC-12% Co coating produced using propylene fuel.

Figure 5a: Optical micrograph of WC-12% Co JET KOTE coating produced using hydrogen fuel.

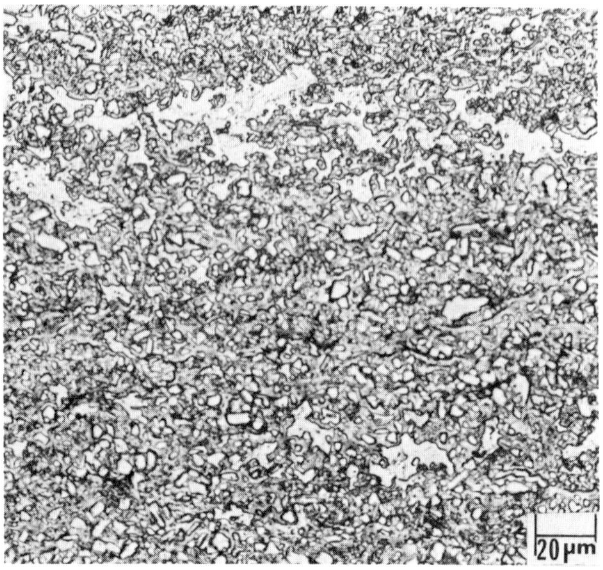

Figure 4b: Optical micrograph of WC-12% Co coating produced using propylene fuel.

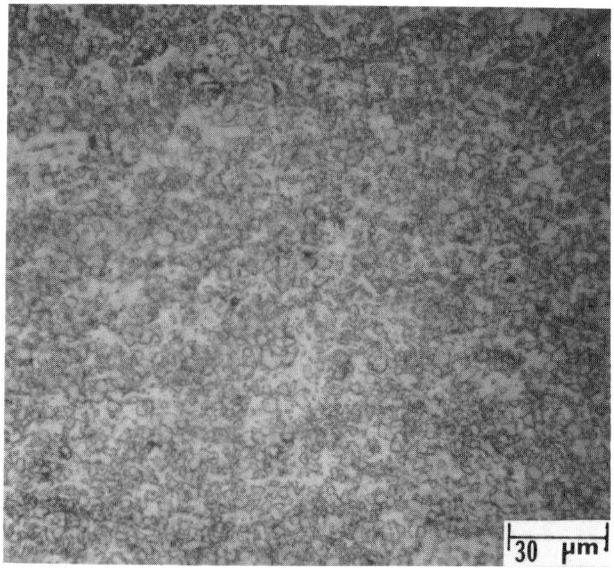

Figure 5b: Optical micrograph of WC-12% Co JET KOTE coating produced using hydrogen fuel.

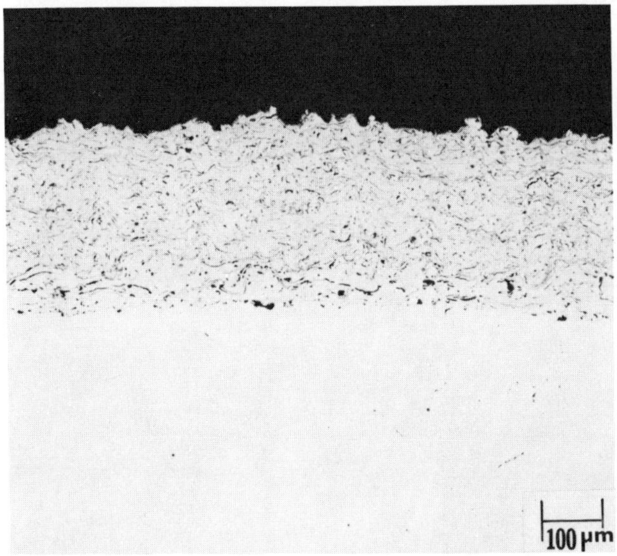

Figure 6a: Optical micrograph showing the microstructure of JET KOTE coating of TRIBALOY alloy T-800.

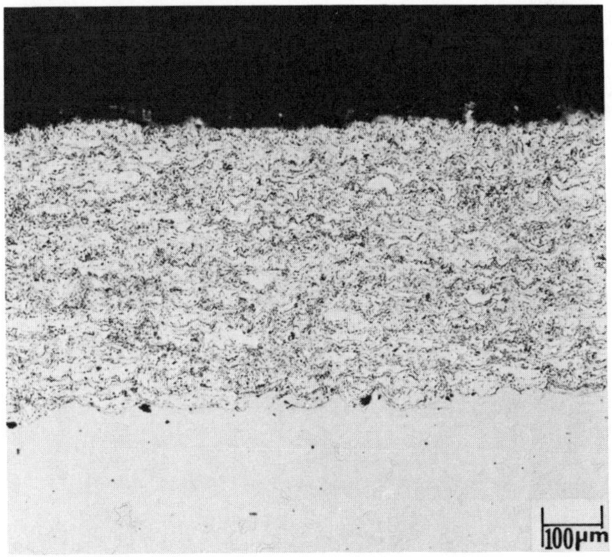

Figure 7a: Photomicrograph of JET KOTE coating of HASTELLOY alloy C.

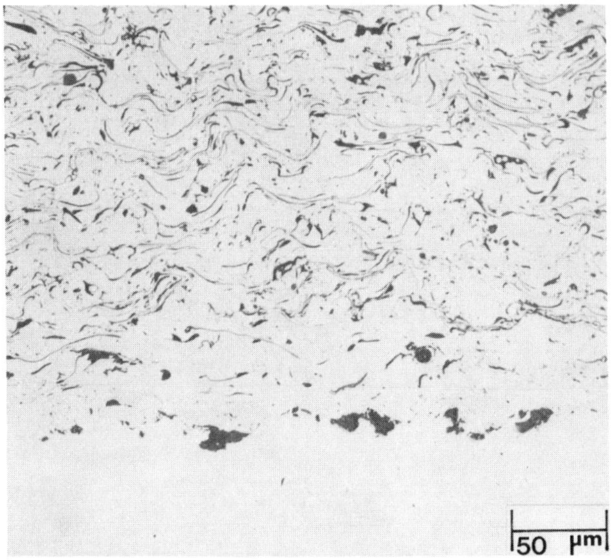

Figure 6b: Optical micrograph showing the microstructure of JET KOTE coating of TRIBALOY alloy T-800.

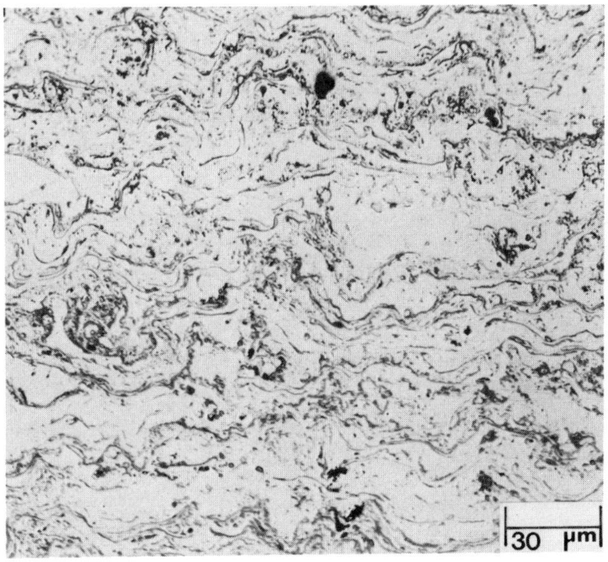

Figure 7b: Photomicrograph of JET KOTE coating of HASTELLOY alloy C.

FLAME SPRAYED CERAMIC COATINGS AS AN ECONOMIC ALTERNATIVE TO SPRAY AND FUSE COATINGS FOR FRICTIONAL AND CORROSIVE USE

F. T. Jaeger
Eutectic + Castolin
Pointe Claire, Quebec, Canada

ABSTRACT

Spray and fuse coatings have been used for many years in various industries on parts to resist mechanical and corrosive wear. Because of the nature of this coating process, there are applications not feasible for reasons of cost, distortion, or change in physical and mechanical properties of the base metal. There is also a limited service temperature. Refractory, oxide type, ceramic coatings offer an economic alternative with improved service life, even at very high temperatures. New improved technology and systems allow applications to be applied automatically or manually on-site.

WITH THE ADVENT of high energy and versatile flame spray systems (up to 130 MJ/hr - equal to that of a 36 KW plasma unit), a variety of oxide ceramic powders may be easily applied to metallic surfaces previously considered difficult, or of such poor quality, it had limited usage. These new systems obtain their energy by combustion from manifold oxygen and acetylene supply. The powder is quantity fed into the combustion flame cone where the intense temperature (3125°C to 3275°C) and concentrate heat melt the ceramics. Dry compressed air assists by adding it's kinetic energy to propel the molten particles at high speeds (up to 250 M/S) to impinge the prepared substrate producing a dense and uniform coating. (FIGURE I)

Surfaces need only be clean, roughened and preheated (≃100°C). This prevents contamination from interfering with the impact and adhesion of the molten particles to the substrate. In most cases a metallic coat of an exothermic alloy is applied first to produce a stronger bond, give the surface a suitable texture, and lower residual stresses.

Ideal applications for ceramic coatings are in seals or packing areas of motors, pumps, and compressors that are subject to wear by friction and compression. Shafts and sleeves used in these areas may also be subjected to corrosion, abrasion/erosion and heat; particularly when there is, or could be, a lack of lubrication. Excellent service life and unlimited part size, shape, base metal, internal or external applications, make flame sprayed ceramic coatings advantageous over spray and fuse coatings. The costs are easily offset by large savings in production, maintenance and inventory.

Spray and fuse coatings which are used for these applications are difficult to apply, limited to size, shape, base metal and service conditions. This paper outlines the advantages, cost savings, and wear resistance mechanisms that make flame sprayed ceramic coatings an attractive alternative.

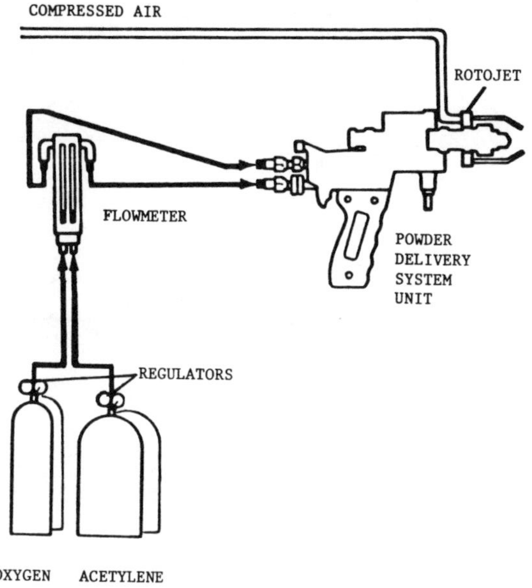

FIGURE I - Typical set-up for a high-energy, flame spray system.

TABLE I - Common Oxide Ceramics Used In Flame Spraying

Type	Chemical Formula	Melting Temperature	Macro Hardness Rockwell C *	Max. Service Temperature
Alumina	Al_2O_3	2085 °C	55	1500 °C
Titania	TiO_2	1820 °C	60	550 °C
Chromium Oxide	CrO_3	2260 °C	65	550 °C
Zirconium Oxide	ZrO_2	2600 °C	25	1700 °C
Magnesium Zirconate	$MgZrO_3$	2200 °C	30	1400 °C

* Note: RC converted from R15N

MATERIALS

There are many ceramic oxides used with the high energy, flame spray systems. Some of the common ceramics are listed in TABLE I along with their physical properties. Fused combinations are often used for certain characteristics, for example: the combination of Alumina/Titania takes advantage of thermal insulative, high service temperature and abrasion/erosion resistance properties of Alumina, along with the hardness, glazing and low coefficient of friction of Titania. Sometimes a metallic (usually a bonding alloy) is blended with the ceramic and sprayed to produce a deposit known as ''cermet''. This combination improves coatings resistance to thermal shock.

Metal alloys used for spray and fuse applications are either nickel or cobalt base having boron and silicon for fluxing agents; chromium and sometimes molybdenium for corrosion resistance; carbon, tungsten and iron for hardness and toughness. Tungsten carbide particles may be blended into the matrix to give deposits increased abrasion resistance. TABLES II and III present some of the common grades used in sealing and packing areas.

It should be noted at this point that the terminology ''spray and fuse'' is misleading. Actually, the flame spraying system is used only to attach the metal particles to the surface (usually performed at about 300°C). Fusing is in fact a ''Torch Brazing'' process in which the deposit is melted (about 1125°C +) on to the surface forming a shallow alloy bond with the substrate (usually less than 0.3 mm deep) and a 100% metallurgical bond. The fluxing oxides (borax and silica) form small nodular inclusions just above the fusion zone. (See FIGURE II.)

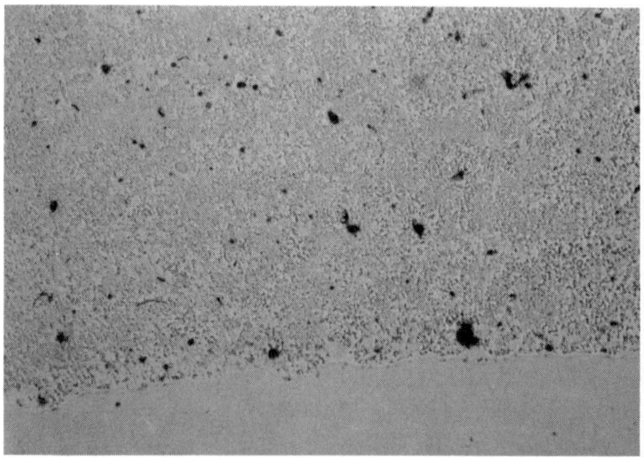

FIGURE II - 80 X enhanced micrograph of nickel base spray and fuse alloy revealing small nodules of borax and silica, 0.3 to 0.6 mm above mild steel surface in a coating 1.2 mm thick.

TABLE III - Nickel Base Alloy Group

Nominal Chemistry					Macro Hardness	Melting
Carbon	Chrome	Silicon	Boron	Iron	Rockwell C	Temperature
0.3%	9%	3.0%	1.5%	2%	30 - 45	1050-1200°C
0.45%	12%	2.5%	2.5%	3%	35 - 40	1000-1190°C
0.6%	13%	4.0%	3.0%	4%	45 - 50	980-1170°C
0.7%	15%	4.5%	3.5%	4%	55 - 62	1020-1075°C

TABLE III - Cobalt Base Alloy Group

Carbon	Chrome	Nickel	Silicon	Boron	Iron	Tungsten	Macro Hardness Rockwell C	Melting Temperature
0.4%	19%	17%	0.8%	0.8%	1.0%	4%	34 - 40	1150-1230°C
1.5%	25%	30%	1.2%	1.5%	0.5%	4%	38 - 45	1100-1200°C

SPRAY AND FUSE COATINGS

Fusing of nickel and cobalt alloy coatings require the surface to exceed 1125°C and slow cooled from above 900°C to prevent cracking. This is accomplished by emersion of the part in vermiculite or by wrapping it in a suitable insulated blanket for 12 hours plus, until it reaches handling temperatures (below 60°C). This procedure creates several problems.

DISTORTION - Since the part must be handled above 900°C (red hot), long and thin walled tubing or structurally weak objects sag, bend, knick, etc., during the holding periods.

SIZE AND DIMENSIONAL STABILITY - Hollow structures will shrink considerably during the cooling cycle due to the annealing of the base metal and the compressive load applied by the contraction of the coating. Allowances must be made on the outside diameter and for finishing in the inside. It is difficult and impractical to spray and fuse diameters larger than 15 cm solid and 30 cm tubing. (See FIGURE III.)

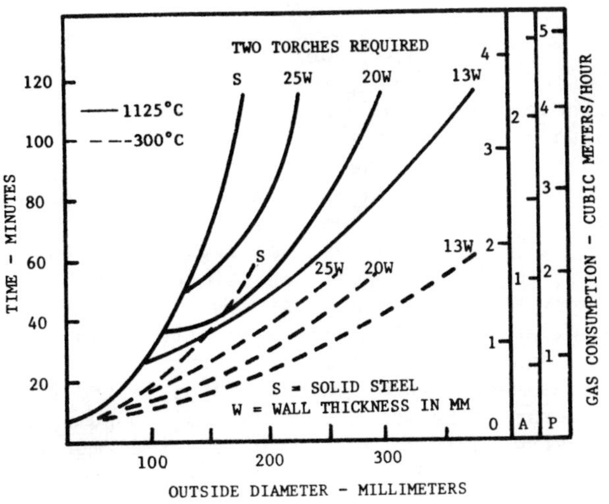

FIGURE III - Gas consumption and time for preheating and fusing, spray and fuse coatings; mild steel 180 mm long. Oxyacetylene torch at 60 MJ/hr for preheating at 3000°C and oxypropane torch at 100 MJ/hr for fusion at 1125°C. O = oxygen, A = acetylene, P = propane.

METALLURGY - With ferritic alloys, annealing of the base metal can be expected. Heat treating is not practical. However, some strengthening is obtained by the compressive load of the coating (if circumferential).

With austenitic alloys, the result will be distortion and annealing. In the case of austenitic stainless steel, some ferrite may convert to austenite (1100 - 1200°C) or convert to brittle sigma phase (550 - 900°C) causing hot cracking. Also carbide precipitation (sensitizing) is quite common (450 - 900°C). See FIGURE IV. Low carbon or stabilizers such as niobium may solve this problem.

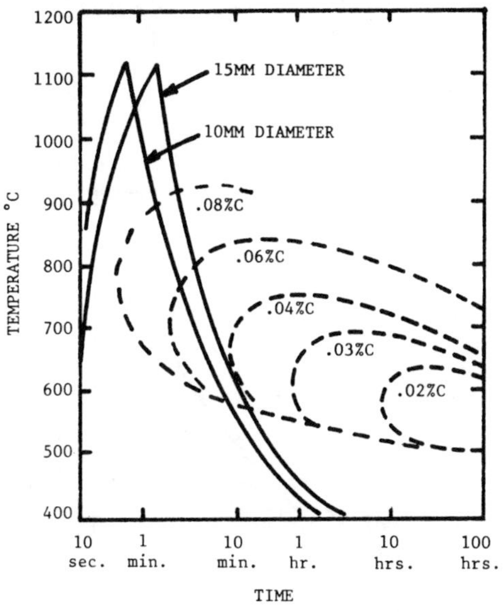

FIGURE IV - Sensitized range of 18-8 stainless steel depending on carbon content. Solid lines represent the thermal cycle of fusion and slow cooling of solid and heavy walled cylindrical parts.

COSTS - Preparing and estimating spray and fuse application(s) requires several steps in the procedure for consideration. A few of them (for single coatings) are: size, area and coating thickness, preparation, preheating, fusing, handling, finishing outside and inside, etc. With the high costs of labour, gasses, materials, equipment, etc., one must be very careful, particularly because considerable time and gas is spent on heating and fusing. FIGURE III gives some idea of the amount of time and gasses required. By applying a dollar rate, such as the example given below, the total cost of a job can be determined.

Examples:

- solid steel - 25 to 150 mm diameter,
- hollow steel tubing - 100 to 300 mm outside diameter, 60 to 25 mm wall thickness,
- coating - 150 mm in length, 0.75 mm in finish thickness. Finishing by grinding, machining inside,
- allowances - 13 mm on each side of length for overspray and 0.13 mm oversize in thickness for grinding.

Factors used:

- labour - \$25/hr (0.42/min) including direct overhead,
- gasses - oxyacetylene for preheating and spraying (\$0.62/min); oxypropane for fusing (\$0.16/min); spraying rate at 15.4 cm^3/min oxyacetylene,
- alloy powder - cost \$0.20/cm^3.

FIGURE V summarizes the total time and costs of above examples using the allowances and cost factors mentioned. Analysis of results are:
- labour: 65% to 85% of total costs,
- gasses: up to 32% of total costs,
- alloy: only 8% to 12%

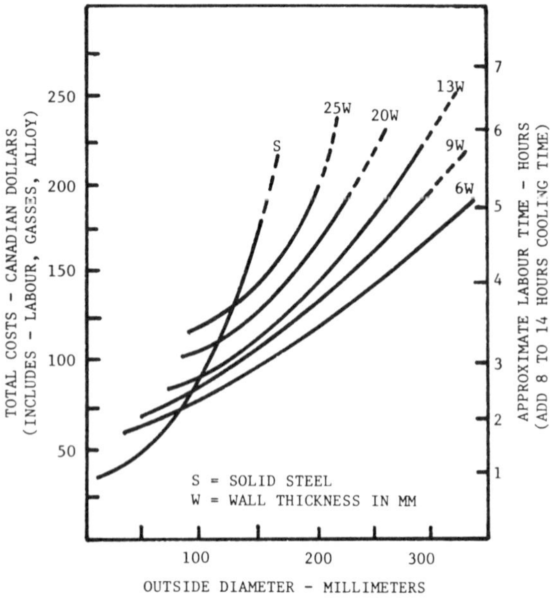

FIGURE V - Relative costs and labour time of spray and fuse coatings on solid and hollow mild steel involving preparation, preheating, spraying, fusing, handling, grinding O.D., and machining I.D.

PERFORMANCE - Hard nickel-chrome and cobalt-chrome alloy coatings improve the service life of sealing and packing areas when compared to stainless steel and other hard facing metal alloys. The formation of a thin layer of chrome oxide (less than 0.005 mm thick) on the surface produces a low co-efficient of friction and corrosion resistance even at high temperatures. However, Ni-Cr and Co-Cr alloys, even though they are a metal, have relatively poor thermal conductivity (about 15 and 12 W/mK° and lower at higher temperatures). Without sufficient lubrication or coolant, surface temperatures will increase very rapidly, and the alloys become soft and subject to plastic flow. See FIGURE VI. They can no longer support the hard (> Rc 60) chrome-oxide layer and failure occurs by brittle facture and galling. (It is for this reason that with water pumps, a small amount of leakage is acceptable in the packing areas. Clean solution is injected continuously to flush out abrasive particles.)

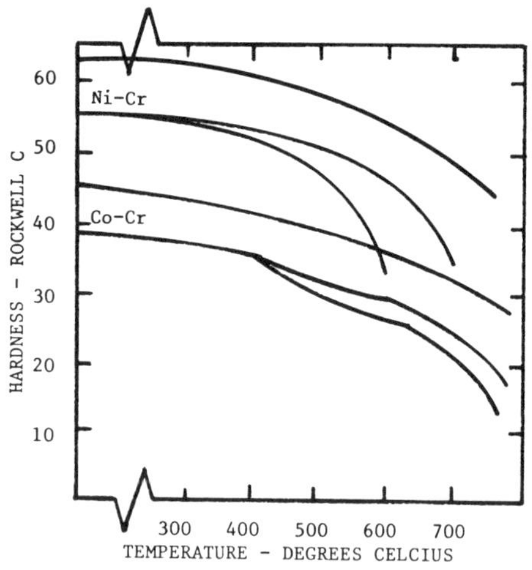

FIGURE VI - Two examples of nickel and cobalt, spray and fuse alloys' hardnesses at elevated temperatures. Spread on the lower end is the result of holding time 1 to 4 minutes (creep).

CERAMIC COATINGS

These coatings of less than 0.6 mm of refractory have excellent performance for seals and packing areas. There does not appear to be any real limitation to application in size or shape except that of economics.

Unlike spray and fuse coatings, deposits are used in the ''as-is'' condition. No secondary steps, such as fusing and refinishing insides, are required. The base metal is not affected and outside final size is obtained by grinding. A roughened surface is preheated to 100°C before spraying. Time and gas consumption are substantially lower. Much larger diameters can be accomodated without auxiliary equipment. (See FIGURE VII.)

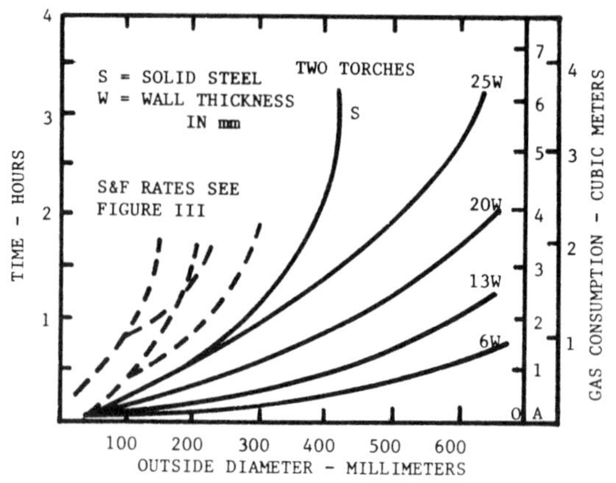

FIGURE VII - Gas consumption and time for pre-heating solid and tubing 80 mm long to 100°C using oxyacetylene 60 MJ/hr torch. O = oxygen, A = acetylene.

COSTS - Even with the higher costs of exothermic bonding alloy and ceramic powder, the total costs are much lower than those coatings made with the spray and fuse process. FIGURE VIII illustrates total costs as compared to spray and fuse coatings using the same size applications and larger. The only differences are:

- exothermic bond coat: \$0.75/cm³, 0.13 mm thick, spray rates at 3.9 cm³/min.
- ceramic top coat: \$0.33/cm³, 0.53 mm total thickness leaving 0.08 mm allowance for grinding, spray rate at 5.4 cm³/min.
- gasses: oxyacetylene for spraying - \$0.70/min. (preheat same as S&F)

Analysis of results are:

- labour: 42 to 70% of total costs,
- gasses: 10 to 35% of total costs,
- powders: 25 to 35% of total costs.

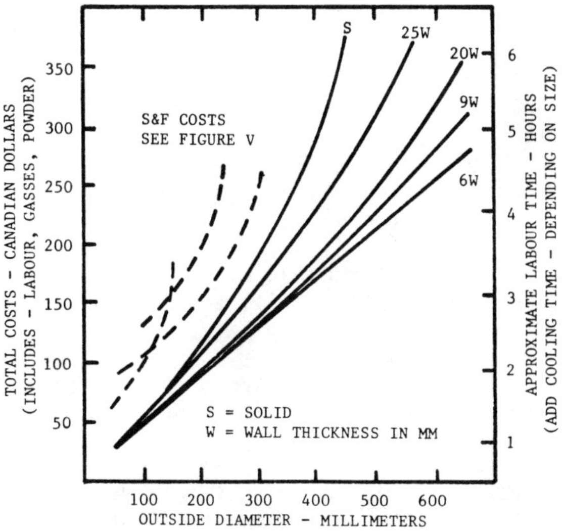

FIGURE VIII - Relative costs and labour time of ceramic coating on solid and hollow mild steel involving preparation, preheating, spraying, handling and grinding.

WEAR/STRENGTH MECHANISMS - Previous papers by a number of authors report that resistance to frictional and erosive wear improves considerably at high temperatures - in the 1000°C range. Also the surface hardness and density increase with temperatures (1000°C plus) and pressure. A report by Ritter, Rosenfeld and Jakus - ''Erosion and Strength Degradation of Alumina'' - offers the following results and explanation.

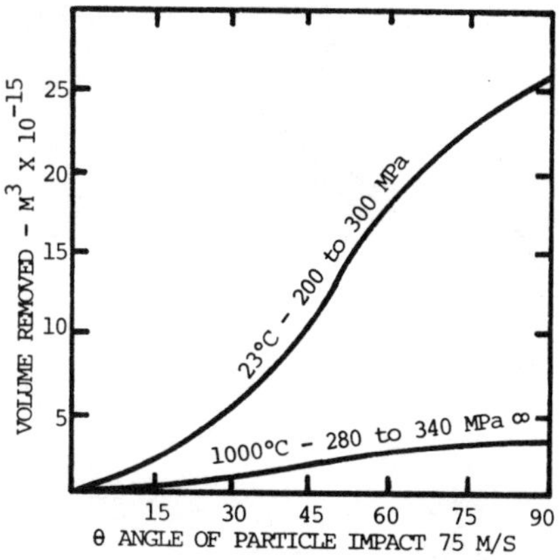

FIGURE IX - Erosion (abrasion/impact) damage is characterized by pits created through intergranular chipping. The results are based on kinetic energy of impact particles on grain boundaries. At higher temperatures (1000°C), residual stresses around the impact are tempered somewhat.

Scott of Advance Materials Laboratory, Australia, investigated the frictional properties of Alumina and Zirconium Oxides (among other ceramics) and concludes that wear life improved when contact was dry. Frictional heating at contact regions is undoubtably important in achieving improved wear resistance.

Because finish by grinding is performed with lubricant, no significant change in coating structure is expected (except for smoothness). It is probable that any stress relieving or glazing developed on the surface during spraying is ground away to a softer structure underneath. For this reason, there must be care in obtaining final size and suitable surface structure at the same time.

Thermal conductivity tests made during deposition indicate that ZrO_2 and Al_2O_3 improve their insulative quality with coating thicknesses more than 0.2 to 0.4 mm and temperatures above 250 to 275°C. TiO_2 (and oxides containing TiO_2) do not show any substantial improvement until above 300°C and 0.5 mm thick; while CrO_3 does not appear to change at all!

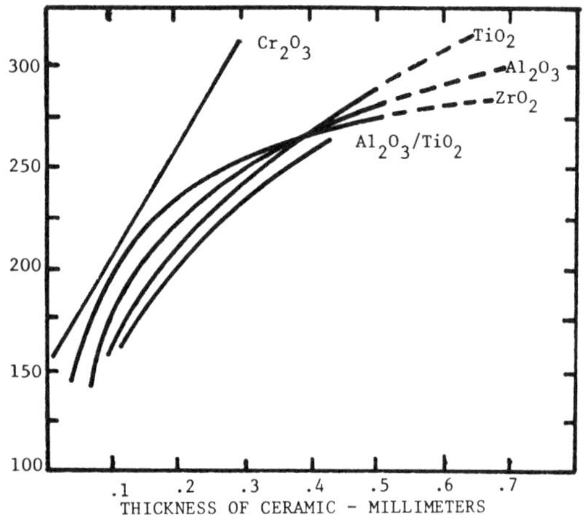

FIGURE X - Coating surface temperatures increase as thickness increases. Deposits made on 38 mm diameter X 250 mm long mild steel shafting using a 60 MJ/hr oxyacetylene torch.

Inconsistency in spray rates, deposition efficiencies, delay in temperature readings, as well as ceramic powder conductivity in the flame, resulted in considerable variation. Additional tests, using more accurate control and temperature response, are required to verify the above assumptions. However, there appears to be sufficient evidence that at least some of the coating performance quality is the result of deposition.

FIELD PERFORMANCE - Field tests with ceramic coated packing sleeves used in pumps for mining, processing, and pulp mill industries, proved to out-wear spray and fuse coatings 3 to 1 and more. Sometimes the coating out-wore the packing itself. Investigation revealed that the packing was tightened regularly under routine maintenance (as normal for spray and fused coatings). This problem was solved by using fixed or cast packing not requiring any adjustments. Alumina-Titania coatings have become popular because of the insulative qualities of alumina and the polish developed in service by Titania.

CONCLUSION

Refractory oxide coatings offer a practical alternative to spray and fuse coatings for the following reasons:

- a larger variety of ceramic coatings can now be flame sprayed by high energy systems easily in the shop or field,
- coatings can be produced without affecting part shape, metallurgical, mechanical, or corrosion resistant properties,
- costs of coatings are substantially lower; 25% to 50%, depending on size or shape.
- there does not appear to be any practical limit to size or shape of part to be sprayed,
- ceramic coatings last 3 times (or more) longer in packing areas. This improved performance is the result of surface conditioning (glazing and stress relieving) at elevated temperatures. It is achieved somewhat during spraying but more so from contact frictional heat. Further tests are required to verify the above statements.

ACKNOWLEDGEMENT

The author wishes to thank the cooperative efforts of the Research and Development Department; particularly Mr. Mike Smolak, who contributed to the investigation of flame sprayed thermal conductivity characteristics.

REFERENCES

1. AWS Specification, C2.13-70, ''Flame Spraying of Ceramics''

2. AWS ''Brazing Manual'', Third Edition (1976)

3. Eutectic+Castolin Flame Spraying Operation Manuals ''TeroDyn 2000 and TeroDyn 3000'' (1985)

4. AWS Specification A5.13-80, ''Surface Welding Rods and Electrodes''

5. Althouse A.D., Turnquist C.H., Bowditch W.A., ''Modern Welding'' (1976), section 28-15, 16, 17

6. Materials Engineering/Selector Guide (1982), section C, ''Flame Spray Coatings - Refractory Oxides''

7. Lundgvist B., ''Welding Stainless Steels'', Part I, Canadian Welder and Fabricator, (Jan. 1983)

8. ASM Metals Handbook, Volume 5, Ninth Edition, ''Ceramic Coating - Flame Spraying'', page 538 to 542

9. ASM ''International Conference on Wear Materials'', (Vancouver 1985), edited by K.C. Ludema, papers:

   - Ritler S.E., Rosenfeld L., Jackus K., ''Erosion and Strength Degradation in Alumina'', page 1
   - Scott H.G., ''Friction and Wear of Zirconia at Very Low Sliding Speeds'', page 8
   - Larsen-Brasse J., ''Binder Extrusion of Sliding Wear of WC-Co Alloys'', page 39
   - BhaHacharyya S., Agarwala V.S., Kim K.Y., ''Wear-Corrosion of Steels Under Cyclic Operation in a Corrosive Liquid Environment'', page 162

- Miyoshi K., Buckley D.H., Rengrtoft G.W.P., Ishigski H., ''Surface Effects of Corrosive Media on Hardness, Friction Wear of Materials'', page 302
- Soemantri S., McGee A.C., Finnie I., ''Some Aspects of Abrasive Wear at Elevated Temperatures'', page 338

PROCESS AND ENGINEERING BENEFITS OF SPUTTER ION PLATED TITANIUM NITRIDE COATINGS

Michael H. Jacobs
Abar Ipsen International
Feasterville, Pennsylvania, USA
(UK address:
Hinxton Hall, Hinxton
Essex, England)

Abstract

Titanium nitride (TiN) coatings are established in the metal cutting and metal forming industries where they enhance the operating lives of tools and offer substantial gains in productivity. Several technologies for physical vapour deposition (PVD) of TiN have been developed. Coatings, which are usually 1 to 3 microns thick, are hard, chemically inert and create low friction against mating materials.

This paper concentrates on a particular PVD technology called sputter ion plating (SIP). An outline is given of the equipment and the salient features of the process; emphasis is placed on the capability to coat large, mixed workloads on stationary fixtures with uniform, high quality layers of TiN.

The main theme of the paper is to illustrate the engineering benefits of SIP TiN coatings by reference to a number of case histories. Applications include tooling used for metal cutting, metal forming, slitting and punching; also applications are included where the performances of bearings and fasteners are improved.

SURFACE ENGINEERING OF TOOLING used in the metal working industry by the application of very thin, hard coatings is growing in application and importance. The lives of metal working tools are frequently improved many fold with associated, important gains in productivity during metal working operations.

Titanium nitride (TiN) is currently the most popular hard coating. There are two families of processes for depositing TiN: chemical vapour deposition (CVD) and physical vapour deposition (PVD). CVD is a high temperature process, ~1000°C (1850°F), in which a gaseous metal chloride (usually titanium tetrachloride) is reacted with hydrogen and nitrogen (ammonia) to form TiN. The high temperature of the reaction and the subsequent need to heat treat CVD coated steel tooling, to obtain the necessary substrate hardness, causes some distortion. This normally limits CVD coatings to steel tooling toleranced greater than approximately +/- 0.05 mm. In marked contrast, PVD is a low temperature coating process carried out at less than 500°C (930°F). This temperature is below the tempering temperature of high speed steels; consequently PVD coating normally produces insignificant distortion and can be applied to high toleranced tooling. PVD coating is normally the last step in tool manufacture.

The family of PVD technologies for depositing TiN include sputtering, electron beam evaporation and cathodic arc deposition. Sputter ion plating (SIP) is a patented(1) technique which employs a glow discharge to sputter titanium, from an extended source, whereupon it reacts with nitrogen to form TiN. SIP has a number of advantages related both to the execution and the application of the process, as further elucidated below.

In this paper an outline is given of the salient features of sputter ion plating (SIP) together with a brief description of the equipment. Virtues of SIP which gives it a competitive edge include:

- good throwing power
- stationary fixturing
- mechanically 'simple' construction
- ability to coat uniformly components of different sizes and shapes as one workload
- ability to coat large components
- excellent replication of the substrate surface finish
- excellent gold colour
- high equipment 'up' time
- minimum maintenance

In addition, the coatings have good adherence, high integrity, very high hardness, low coefficient of friction and a uniform, micro-crystalline structure.

The main theme of this paper is to illustrate

the engineering benefits of SIP TiN coatings. This is partly achieved by reference to the physical and mechanical properties of the coatings, as they influence metal cutting and metal forming, and partly by reference to selected applications. Examples include tooling for cutting, slitting, punching and forming metal; also illustrated are bearing and anti-seizure applications.

PROCESS OF SPUTTER ION PLATING

Sputter ion plating takes place in a vacuum chamber which contains a soft vacuum of nitrogen and argon (pressure ~0.1 mbar). Figure 1 is a schematic of the SIP process. The workload to be coated is jigged in a stationary fixture which is top-loaded into the vacuum chamber. In principle the chamber can be of any size, to accommodate components of any diameter or length, provided adequate titanium source electrodes are distributed within the work chamber.

As the name suggests, sputter ion plating depends on a 'sputtering' mechanism for its operation. A glow discharge is excited within the chamber so that energetic positive ions strike titanium source electrodes which line the internal surfaces of the chamber; also additional titanium rods may be incorporated within the chamber. The titanium electrodes form an extended cathode source from which titanium atoms are displaced, or sputtered, and some of these will land on the anodic workload. The nitrogen in the process gas then reacts with the deposited titanium to form titanium nitride.

The process has excellent throwing power, even with a stationary workload, because of:

- the extended source of titanium
- the use of a soft vacuum which creates multiple gas scattering of the sputtered titanium atoms; hence, unlike all competitive PVD processes, SIP is not a line-of-sight process.

As shown in Figure 1, the SIP process involves a three-electrode system with the workload biased below the potential (H.T.) of the small anodes. The consequence of this arrangement is illustrated in Figure 2. In addition to the major sputtering which occurs at the earthed titanium cathodes, low energy sputtering occurs at the surfaces of samples during simultaneous deposition of TiN. This low energy 'bias sputtering' has two important functions:

- impurities are removed from the coating
- the microstructure of the coating is refined.

Figure 3 shows LIMA (Laser induced Ion Mass Analysis) peaks for biassed and unbiassed coatings, the latter showing carbon, oxygen, sodium, potassium and calcium as impurities; these are all elements associated with greases and dust. Unbiassed coatings are dark brown, or even black, whereas correctly biassed coatings are of excellent gold colour. Unbiassed coatings tend to be columnar in structure and tend to be of inferior toughness and lower hardness compared with the equiaxed, microcrystalline structure of biassed coatings (see Figures 4 and 5).

The SIP process can be used to coat complex shapes without the need to rotate components during the coating cycle (a difficult and expensive operation inside a hot vacuum chamber). The process is amenable to the use of large chambers, in which components of widely different geometry and size may be coated simultaneously, see Figure 6. This is a very important virtue of SIP which differentiates it from other PVD technologies.

SIP EQUIPMENT AND PROCEDURES

To ensure good adhesion of the coating the components to be coated must first undergo a rigorous inspection and pre cleaning procedure. The presence of grinding burns, burrs, oxide, grease, accidental contamination or residual salts from a salt bath heat treatment will have a very detrimental effect on the coating performance. Adhesion will be seriously reduced, and burrs break off when a tool is first used and expose regions of uncoated substrate. Chemical and/or mechanical cleaning may be required.

Cleaned samples are loaded into the SIP chamber, which is then evacuated by a mechanical rotary pump and booster pump combination. The workload is heated (via external heaters) to 300°C in a stream of gettered, high purity argon; this helps to further clean the samples. The final stage of cleaning is an in situ 'ion clean' in an argon glow discharge, with the workload cathodic with respect to the chamber.

Next, a thin layer of titanium is deposited onto the workload (to improve adhesion) prior to sputter ion plating with TiN. Coatings of TiN are usually about 2 microns thick. Internal thermocouples ensure that unacceptable temperatures (i.e. greater than about 500°C) are never achieved.

APPLICATION OF SIP COATINGS

Titanium nitride coatings are being used in a wide variety of applications in order to enhance the performance of components. A good 'rule of thumb' is that TiN coatings will normally appreciably improve the performance of a component that already (uncoated) performs acceptably and finally fails through wear. Badly designed tools, tools that are misused, or tools that fail by fracture rather than wear, are unlikely to benefit from TiN coating.

The performance improvement of a coated component depends very greatly on the nature of its application. The characteristics of SIP TiN coatings which can contribute to enhanced performance are:

- very high hardness; ultramicrohardness in excess of 4000 Hv has been measured(2)
- low coefficient of friction; against mild steel, for example, the coefficient of friction is ~0.2

- excellent surface finish; the coating retains the smoothness of the substrate (see Figure 7), which is extremely important in bearing applications.
- coating is pure TiN; it is impossible because of the nature of the sputtering process for undesirable droplets of titanium to be incorporated in the coating, also biassing of the workload within the plasma ensures ultraclean TiN.

CUTTING TOOLS - Table 1 lists some "typical" increases in tool life achieved with SIP TiN coatings; the word "typical" is put in parentheses because the enhancement depends very much on the precise application. In some machining operations, such as gear cutting, the more important criterion is for the tool to survive one complete shift. This is because tool changing is time consuming and costly through lost production. Also, expensive tools must be capable of being resharpened without a dramatic loss of the benefits of coating. Figure 8 illustrates two types of gear shapers; test results showed that resharpening only reduced the life improvement from x2.5 to x2.0.

SIP TiN coated drills have demonstrated up to 10 times improvement in life compared with conventional steam tempered drills, with approximately 80% of this improvement in life retained after resharpening (Figure 9). Alternatively, a coated drill may be used at a 50% higher cutting speed than a steam tempered drill with about the same life. The real benefit on inexpensive tools like drills lies not in the improved economics from longer tool life, but rather higher productivity and less downtime. Furthermore, the quality of holes drilled (in terms of surface finish) is significantly improved, see Figure 10; hence an additional engineering benefit is that a subsequent reaming operation, in some instances, may be omitted.

Figure 11 compares wear of coated and uncoated thread chasers applied to thread cutting of AISI 321 stainless steel. Figure 12 illustrates how the morphology of the chip is effected by the coating. Balling-up of the chip indicates more efficient cutting and less heat generated at the cutting tip; however, a chip breaker must be ground in the cutting edge to prevent the formation of potentially dangerous chip balls. This is an example of a more general principle that "tools may have to be re-designed to take full advantage of the engineering benefits of TiN coatings".

Figure 13 illustrates the action and tooling employed for in-line tube shearing. The important result here was not only was the life of the shears increased by a factor of 3 but the quality of the cuts was much improved.

The above applications all apply to steel tooling. Figure 14 shows TiN coated opthalmic scissors, where the base material is high purity titanium (for medical, hygiene reasons). Tests, which are still in progress, are showing that coated scissors remain sharp for very much longer. For this application it is also important to note that TiN is chemically inert.

METAL FORMING - Tools used for metal punching are enhanced in performance by TiN coatings. Again, the degree of benefit depends very much on the application. Figure 15 shows punches used for forming slots in spanners, where a three-fold improvement in life was experienced.

Figure 16 shows TiN coated tungsten carbide drawing plugs used in tube making. Coated and uncoated plugs demonstrated similar lives, but the very important engineering benefit obtained with the TiN coated plug was a much improved surface finish on the bore of drawn tubes.

BEARINGS - The friction coefficient of TiN running against TiN is extremely low, ~0.14. This low figure(2) implies a high quality of surface finish, readily obtained by SIP. Very dramatic improvements in bearing life may be obtained, particularly in abrasive wear situations where stress levels are low. Figure 17 illustrates a tool steel punch which reciprocates in a tool steel guide. Life improvements better than x100 have been obtained.

The bearing properties of a coating are strongly influenced by pick up of the mating material. Figure 18 shows stainless steel gears used in an aerospace application. TiN coating of one gear out of a pair produced no life enhancement, but coating of both mating gears gave a very appreciable improvement in life, because pick-up was eliminated.

ANTI-SEIZURE COATINGS - The chemical inertness and low friction properties of TiN have proved to be very useful in preventing seizure of fasteners used in hot vacuum systems. The excellent throwing power of SIP readily allows the threads of small nuts to be coated, Figure 19.

SUMMARY AND CONCLUSIONS

The equipment required for sputter ion plating is relatively simple - a mechanically pumped vacuum chamber, a process gas control unit, external heating and cooling facilities, and a power supply for the glow discharge; hence it is relatively inexpensive compared with other PVD processes. Costs are important because engineering benefits of coatings inevitably will be judged in terms of economic effectiveness.

Other factors which have a direct bearing on low process costs for SIP are stationary fixturing and the ability to accept large and mixed workloads. The inherent high hardness and excellent surface quality of SIP TiN coatings also impact on the extent of the engineering benefits.

This paper provides examples of benefits of SIP TiN coatings for metal cutting, metal forming, bearing and anti-seizure applications.

The many virtues of SIP TiN differentiates it from other PVD processes and makes it very attractive from cost and quality points of view.

REFERENCES

1. Sputter ion plating was pioneered at the Atomic Energy Research Establishment, Harwell, England, and is licenced to Abar Ipsen International; see Coad, J.P., P. Warrington, R.B. Newbery and M.H. Jacobs, Materials and Design, 6, No.4
2. Data measured and supplied by The National Centre of Tribology, England.

ACKNOWLEDGMENTS

The author is indebted to many colleagues within Abar Ipsen International and TI Research Laboratories, Cambridge, England, for assistance in the preparation of this paper. He wishes especially to thank Mark Podob of Abar Ipsen Industries for agreeing to present the paper at the ASM meeting in Toronto, October 1985 on behalf of the author.

Table 1 - Typical increases of life for SIP TiN coated steel cutting tools

Tool	Increase in Life
End mills	x5
Shear blades	x3
Parting-off tools	x3.5
Gear shapers	x2.5
Slitting saws	x10
Segmented hobs	x3
Monobloc hobs	x3
Twist drills	x10
Thread cutting dies	x3

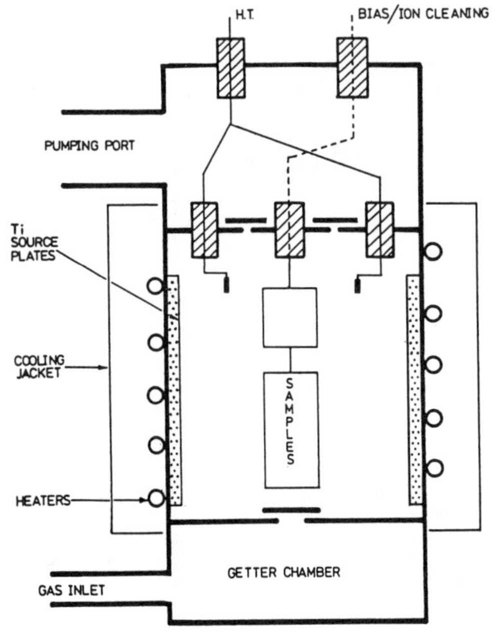

Fig. 1 - Schematic view of the sputter ion plating (SIP) system

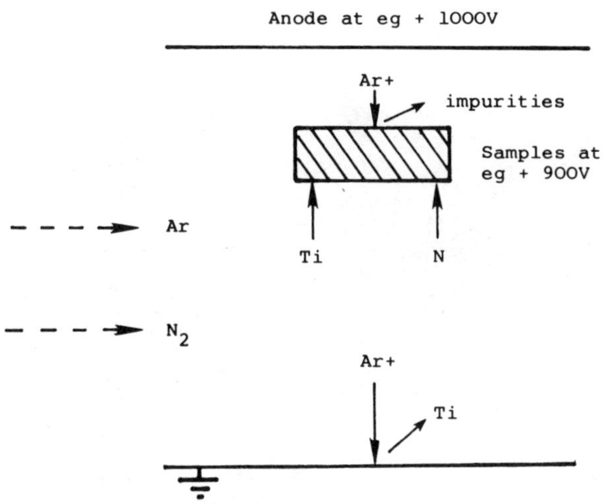

Fig. 2 - Diagrammatic representation of the SIP coating process

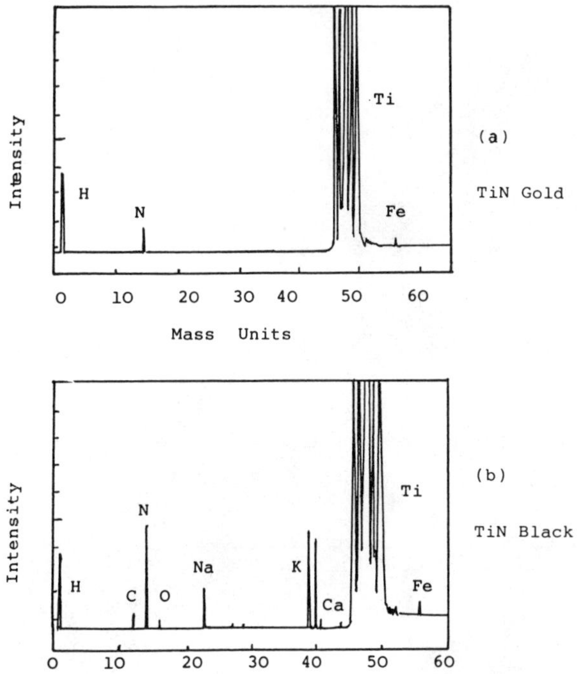

Fig. 3 - LIMA traces from coatings with (a) an applied bias voltage and (b) no bias

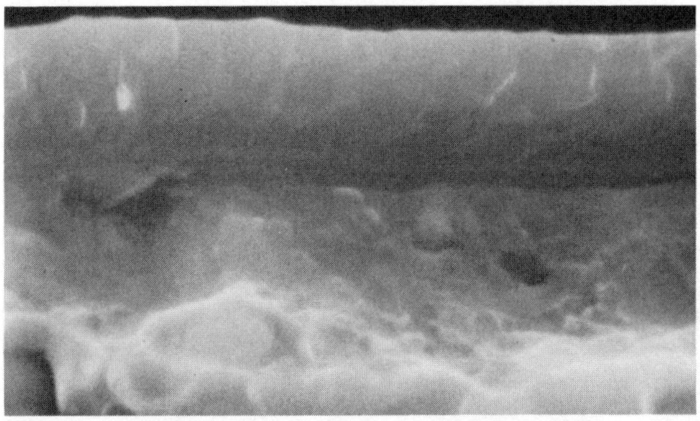

(a)

(b)

Fig. 4 - SEM micrographs of (a) biassed and (b) unbiassed fractured coatings showing the columnar structure of the latter

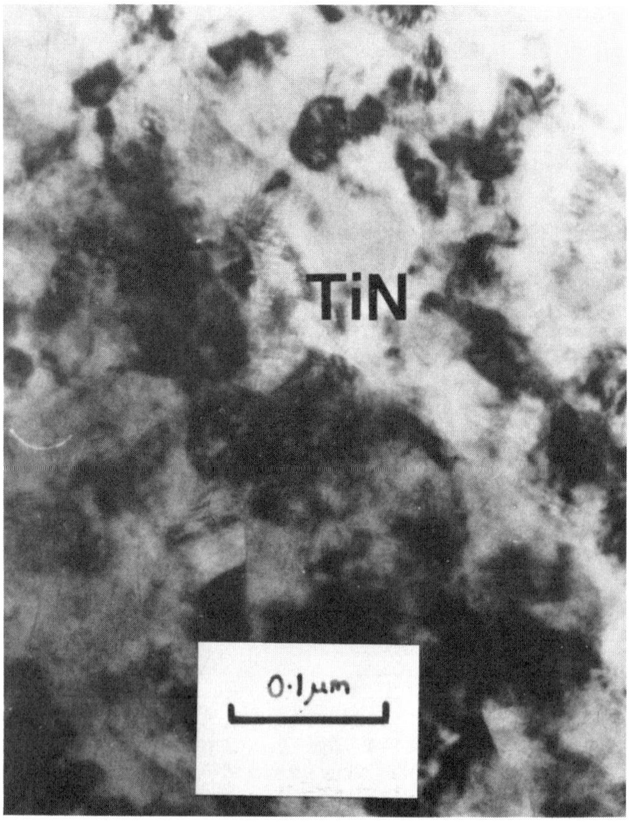

Fig. 5 - High resolution TEM micrograph of a thin foil showing equi-axed microcrystalline structure

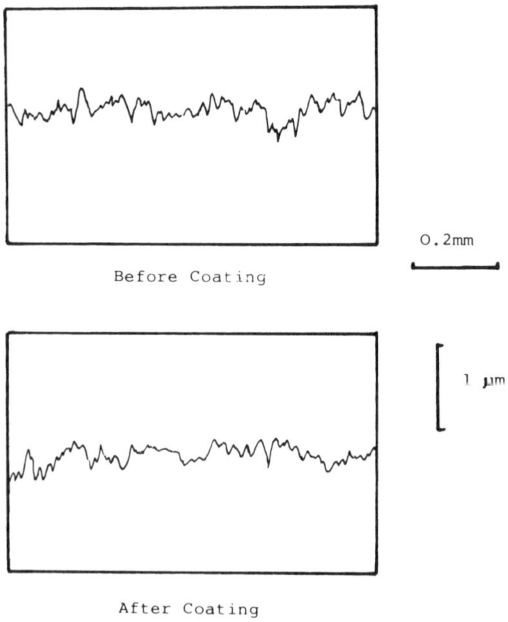

Fig. 7 - Roughness profiles from a tool surface before and after coating

Fig. 6 - A mixed workload processed as one batch

Fig. 8 - Two types of gear shapers

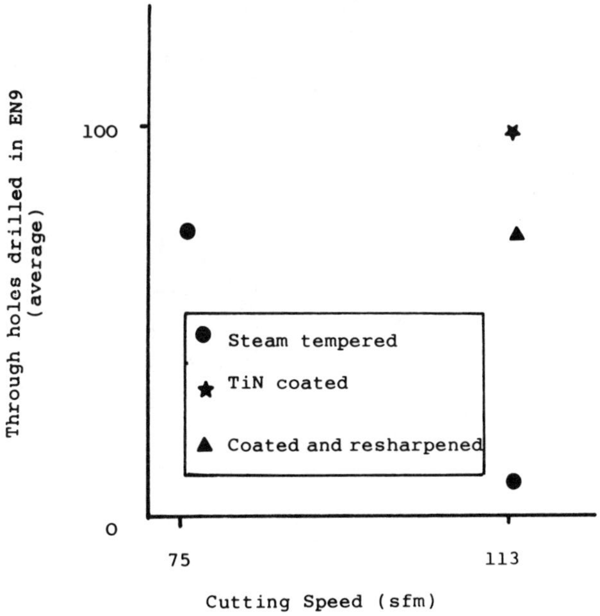

Fig. 9 - 6 mm drill performance comparisons

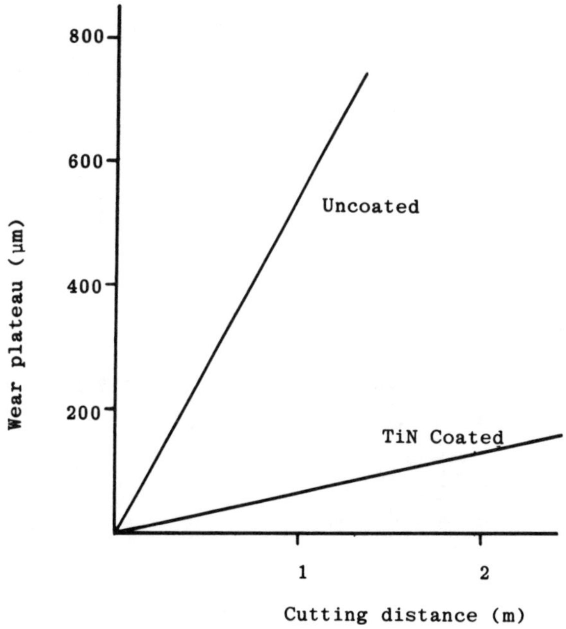

Fig. 11 - Wear of thread chasers

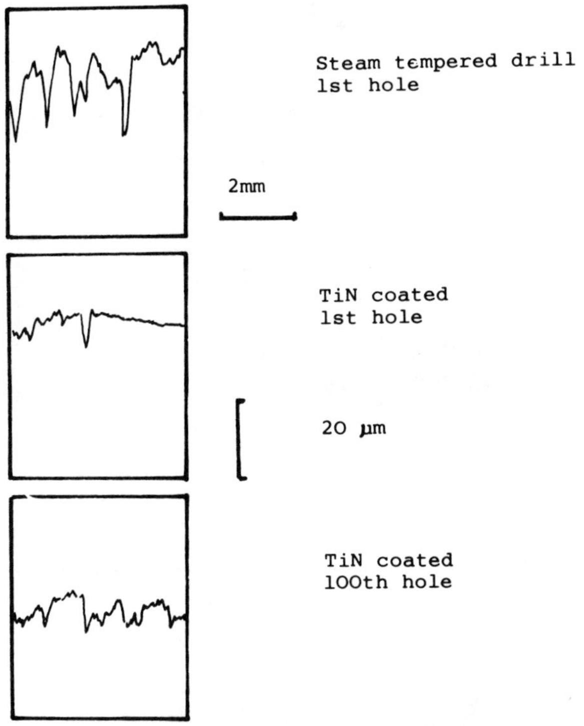

Fig. 10 - Comparison of roughness of drilled holes

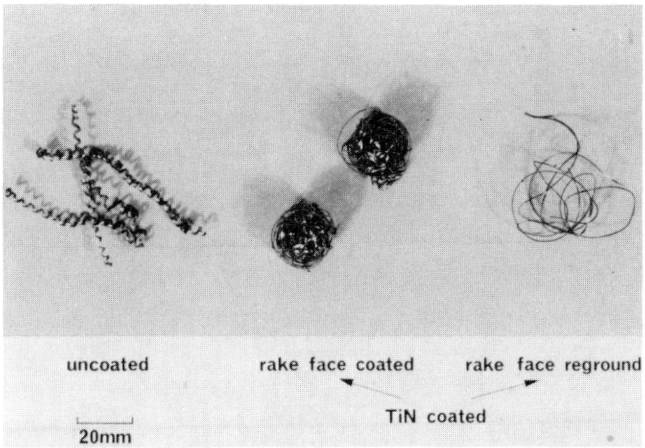

Fig. 12 - Chips from uncoated and coated thread chasers

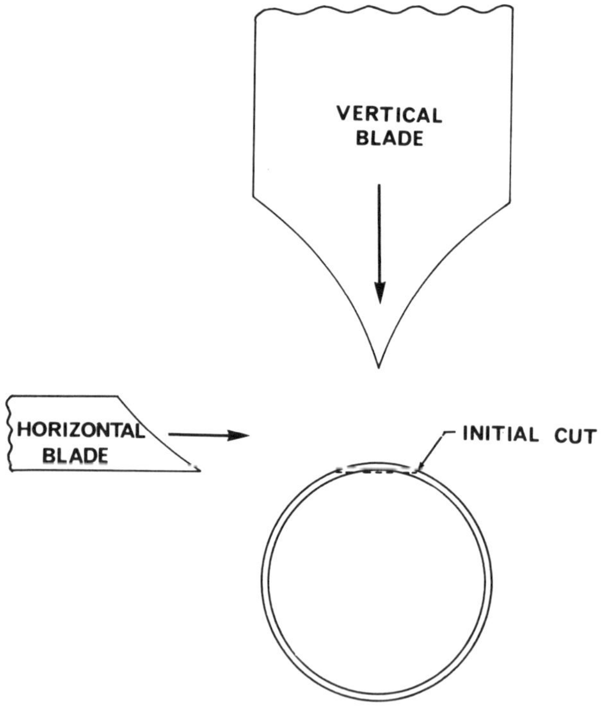

(a)

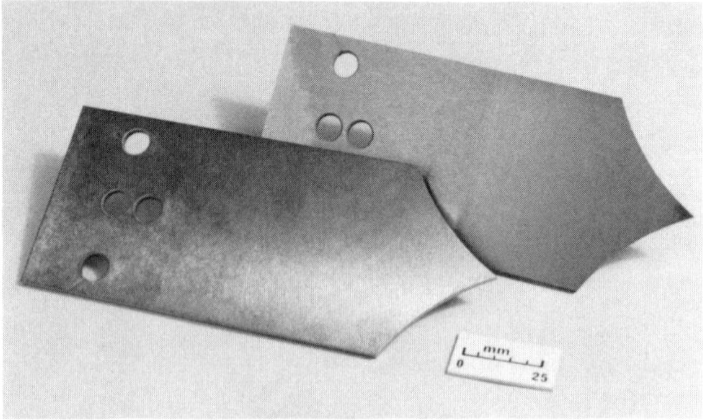

(b)

(c)

Fig. 13 - In-line tube shearing: (a) diagram showing action of vertical (b) and horizontal (c) shears

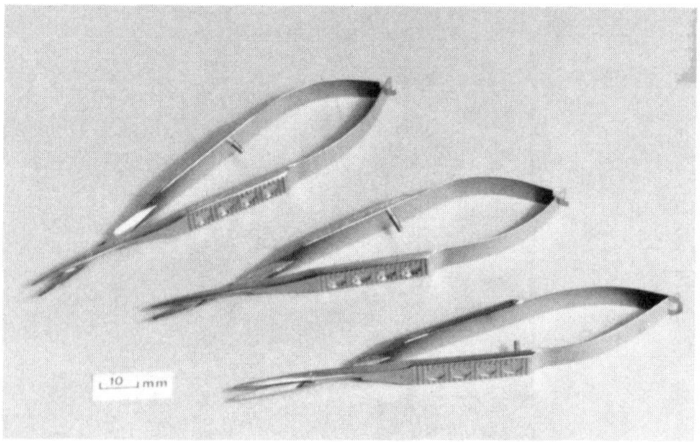

Fig. 14 - Opthalmic scissors - TiN coated titanium

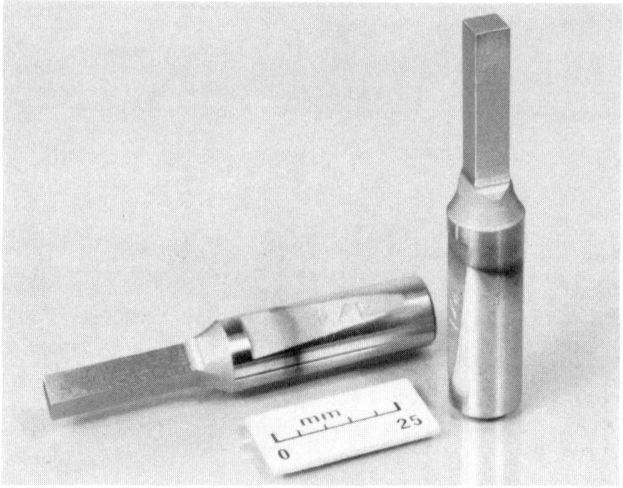

Fig. 15 - Tool steel punches

Fig. 16 - Tungsten carbide tube drawing plugs

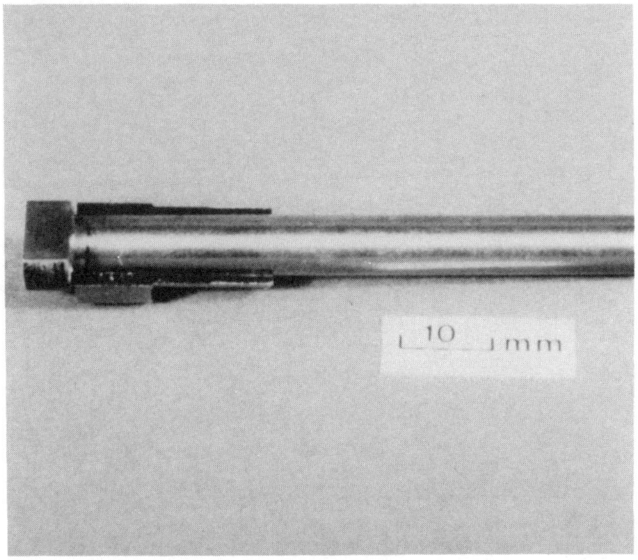

(a)

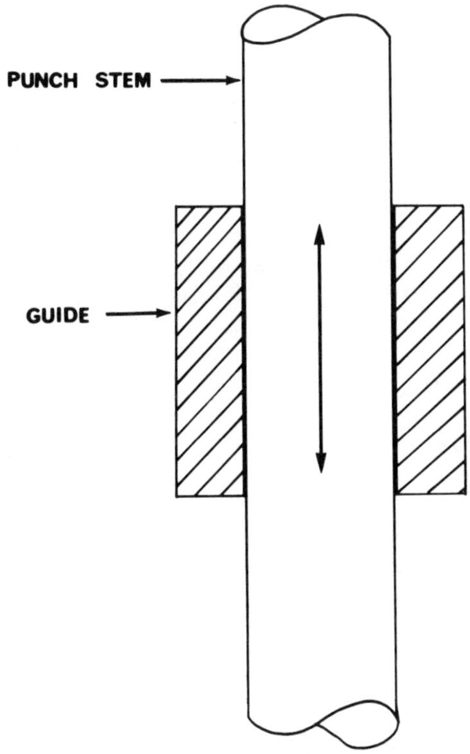

(b)

Fig. 17 - Reciprocating punch in guide:
(a) guide sectioned to illustrate action (b)

Fig. 18 - TiN coated stainless steel gears

Fig. 19 - TiN coated small nuts

TRIBOCOR 532N–A NEW MATERIAL FOR ENVIRONMENTS INVOLVING WEAR AND CORROSION

P. F. Ziegler and J. J. Rausch
Fansteel Inc.
North Chicago, Illinois USA

Abstract

Tribocor 532 is an alloy having the composition Nb-30Ti-20W which is produced by the conventional melting techniques used for other refractory metal alloys. The cast alloy is highly fabricable and can be readily formed into various and complex part shapes by cold working, machining and welding methods.

Surface nitriding of the finished shape produces a structure of coherent, hard nitrided phases dispersed in a refractory metal alloy matrix (Tribocor 532N). During nitriding, titanium nitride forms preferentially and grows as platelets within the grains of the base metal alloy. As the metal matrix adjacent to the platelets becomes depleted in titanium, niobium enters into the reaction, resulting in a final structure of complex nitrides in a niobium-tungsten rich matrix. The reaction produces a surface hardness in excess of low binder content hard metals (e.g., WC-4.5%Co). The depth of hardness and the gradient from the surface to the core metal can be varied over considerable limits with different nitriding treatments.

The corrosion resistance of Tribocor 532N has been found to be outstanding in a number of different corrosive media, including hydrochloric, nitric, phosphoric and sulfuric acids, and chloride solutions. The corrosion rates over a wide range of concentrations and temperatures are far less than for common wear grades of cemented tungsten carbide. Excellent resistance to stress corrosion cracking has also been found in tensile and bend test samples evaluated in chloride and alkaline environments.

TUNGSTEN CARBIDE HARD METALS with various binder systems have long been used for parts operating in environments involving both wear and corrosion. However, there is a growing need for materials combining greatly improved resistance to corrosion in severe environments with wear resistance at least equivalent to cemented carbides. For example, although low binder content hard metals are still frequently employed for wear resistance, corrosion is a common failure mode for many parts such as seals, valve seats, flow-meter components, and other hard trim used in sour gas, oil well, and chemical industry environments. The nitrided ternary alloy, Nb-30Ti-20W, offers considerable promise for such applications(1). The base alloy is cast by the conventional techniques used for other refractory metal alloys. Prior to nitriding the material is easily formed and readily machined, allowing the manufacture of complex shapes not practical for some P/M products. However, when a manufactured part is then nitrided at temperatures above 1250°C, the alloy develops surface hardness exceeding cemented carbides combined with demonstrably superior corrosion resistance. This paper defines the useful alloy composition range, the nitriding kinetics of a selected ternary alloy, and the consequent microstructural changes which occur. Selected physical and mechanical properties and corrosion test data are presented. Some potential application are also described.

BASIS FOR ALLOY SELECTION

Alloys in the Nb-Ti-W system form series of continuous solid solutions over broad composition ranges.[2] The composition map (Figure 1) shows characteristics of nitrided materials in various regions of the system. Nitrides in the titanium and niobium-rich corners form continuous discrete nitride layers which crack and spall under relatively low thermal or mechanical stress. In the tungsten-rich corner, the concentration of nitride formers is insufficient to develop hard, wear-resistant materials. The polygon-bounded area in the figure defines the primary composition ranges where nitriding produces materials with coherent surface layers having graded hardness extending from the surface inward. Within this region, nitrided materials combine excellent wear and corrosion resistance.

Compositions outside the polygon, when nitrided, show limited performance for various reasons. Those compositions

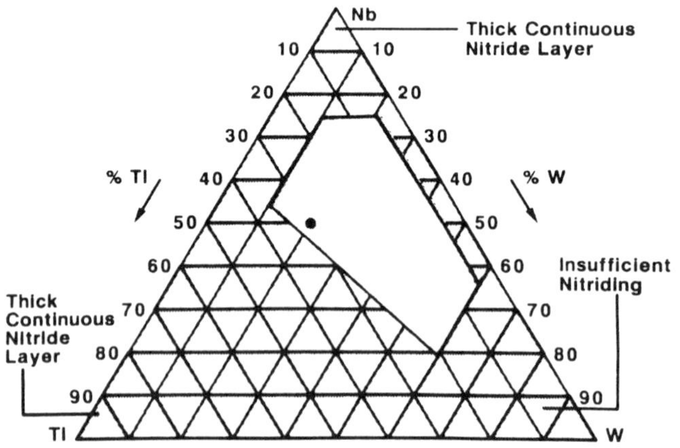

Fig. 1- **TRIBOCOR COMPOSITION MAP**

having less than 10 percent tungsten, for example, are lower strength since there is less refractory metal matrix to support the hard nitride phase. Compositions richer in titanium than those in the preferred region develop a coarser nitride structure resulting in diminished mechanical performance.

Within the constraints of the preferred composition region, the specific alloy, Nb-30Ti-20W ("●" on Figure 1), was selected for the following reasons:

(1) Titanium is the lowest cost alloy constituent; it decreases density and aids fabricability. Therefore, the titanium content should be as high as possible.

(2) Tungsten strengthens the alloy, but higher levels increase density, elevate melting point, and diminish fabricability.

For certain uses the performance of the nitrided alloy may be improved by elemental substitutions. For example, zirconium[3] or hafnium[4], substituted for part of the titanium improves the high temperature properties and oxidation resistance of the nitride phase. Likewise tantalum substituted for niobium, or molybdenum for tungsten, increases corrosion resistance in certain environments.

Compositions within the preferred polygon boundary need not be restricted to use as homogeneous alloys. For example, favorable results have been achieved by surface diffusing both Nb-W and Ta-W binary alloys with titanium[5]. These materials, when nitrided, yield the desired composite microstructures.

NITRIDING REACTION KINETICS

The reaction of the Tribocor 532 alloy and nitrogen follows a classical parabolic rate relationship between weight gain and time over the temperature range 1350°C to at least 1900°C. This relationship can be expressed as:

$$W^2 = k_p t$$

Where W = nitrogen weight gain, mg/cm^2 of alloy surface

t = exposure time, min.

k_p = rate constant, mg^2/cm^4-min.

The variation of k_p with the reciprocal of temperature is plotted in Figure 2. In obtaining these data, pure nitrogen gas at atmospheric pressure was used. The deviation from linearity in the low temperature region is probably explained by a change in rate-controlling mechanisms, for example, from grain boundary diffusion to volume diffusion. The calculated activation energy for the linear portion of the plot is 64,900 cal/mol N_2 absorbed. These rate constants can be used to calculate, with considerable accuracy, the time required for various weight gains as a function of temperature. Figure 3 shows the calculated time-temperature relationship for 1700°, 1800°, and 1900°C.

The kinetic relationships apply to situations where the base metal is nitrided to less than 50% of saturation, where saturation is defined as the nitrogen required to convert all titanium and niobium to their respective mononitrides. When nitriding exceeds 50% saturation, the reaction rate decreases rapidly.

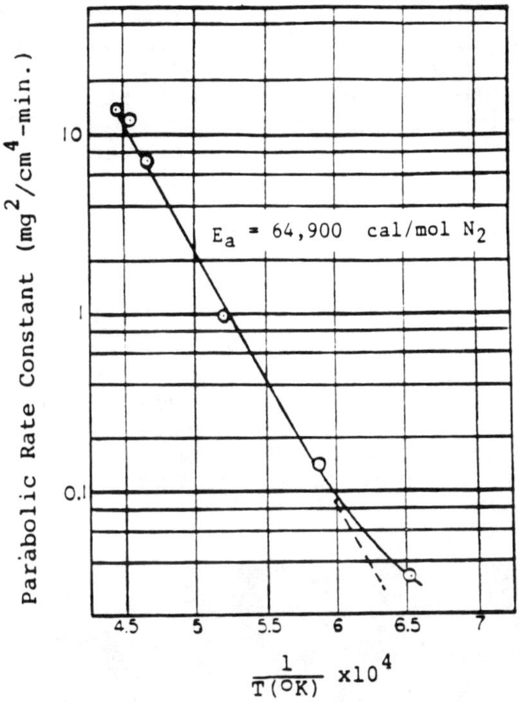

Fig. 2 - Parabolic Rate Constant vs. Reciprocal Temperature

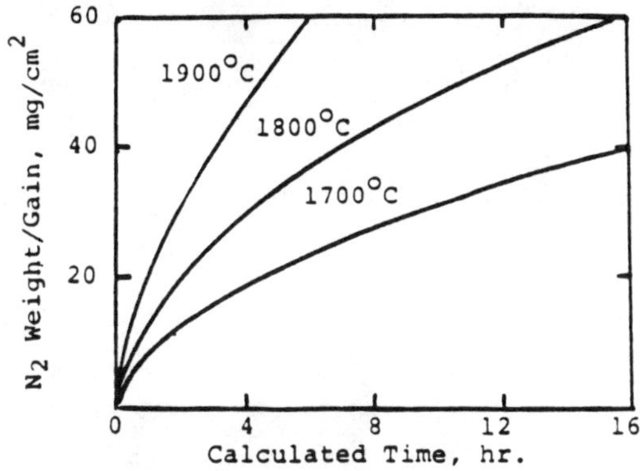

Fig. 3 - Effect of Temperature on Rate of N_2 Weight Gain

When Tribocor 532 is nitrided, a relatively small and predictable expansion occurs which is directly related to the weight gain. For example, a rod 6.5 mm in diameter increases in both diameter and length by approximately one percent at 30-35 mg/cm2. When nitrided to 50-55 mg/cm^2, these dimensional changes approach two percent. However, certain geometries will not grow uniformly due to constraint or plastic deformation of the base alloy induced by the hard nitride surface layer during growth. As an example, a very small diameter rod, when nitrided, will grow in length but not in diameter. In identical parts, nitrided under the same conditions, dimensional changes are always constant. Therefore, it is possible to compensate for dimensional changes when the part is fabricated. During nitriding, absorption is higher at corners and edges, resulting in greater expansion in these regions than in the adjacent flats. This growth can be reduced by radiusing prior to nitriding or removed by subsequent honing. Neither the grain size nor the amount of cold work in the base metal alloy significantly affects growth during nitriding.

Nitriding conditions affect surface finish to a limited degree. Tribocor 532 parts with finishes between 0.05 and 0.40 µm show surface roughening of approximately 0.10 µm after nitriding. However, lapping or honing readily re-establishes the finish.

MICROSTRUCTURE AND PHASE SEPARATION

The annealed microstructure of Tribocor 532 is shown in Figure 4.

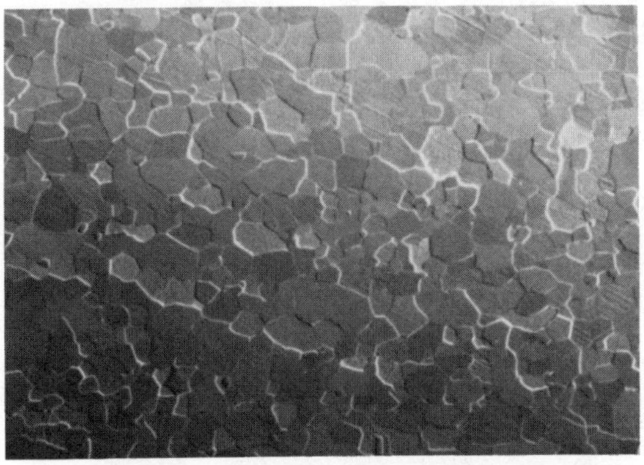

Fig. 4 - Annealed Tribocor 532 100 µm

Among the constituents, titanium is the most mobile and has the greatest affinity for nitrogen. These elements determine the surface structure of the nitrided alloy. By analogy with the known reaction sequence for titanium oxidation in similar alloys(6), we surmise nitrogen reacts with the base alloy in the following manner. Nitrogen stabilizes the hexagonal, α-titanium solid solution(7). The latter segregates by first forming on the surface and then growing along grain boundaries as intergranular platelets. Eventually transgranular segregation also occurs. As nitrogen absorption proceeds, the depth to which α-titanium is formed increases as does the amount of α-phase which is converted to cubic TiN. The resulting structure is, therefore, one of phases rich in titanium and nitrogen that have grown in a b.c.c. β-phase matrix now depleted in titanium and, therefore, enriched in niobium and tungsten. Figure 5a illustrates the microstructure obtained by nitriding at 1900°C to a weight gain of 30 mg/cm^2. The adjacent x-ray maps, Figure 5b, 5c, and 5d, show the distribution of titanium, niobium and tungsten in the same area. Segregation of niobium and tungsten into the matrix is apparent. Nitrogen is partitioned preferentially into the titanium-rich areas.

The outer portion of the diffusion zone that contains the highest nitrogen content is shown at higher magnification in Figure 6. From this back-scattered electron image (BEI), a difference in the darkness of the nitride platelets is detected in the region from the outer surface to approximately 20 µm deep, indicating the presence of a higher atomic number element in the platelets. Energy dispersive spectrometric analyses (EDS) of these platelets confirm a higher niobium concentration in the more heavily nitrided outer zones. Further evidence of the interaction between niobium and previously-formed TiN platelets may be seen in Figures 7a and 7b. Figure 7a is the microstructure formed in <u>Tribocor 532N</u> at 50 mg/cm^2 and shows obvious <u>coring</u> based on the color differences in the nitrided layer that appear in the back-scattered electron image. The differences in composition between the dark and lighter areas are shown in Figure 7b.

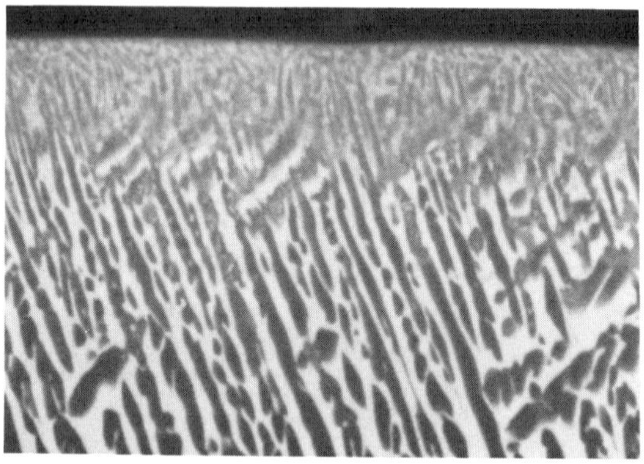

(a.) Back-Scattered Electron Image

(b). Titanium X-ray Map

(c.) Niobium X-ray Map

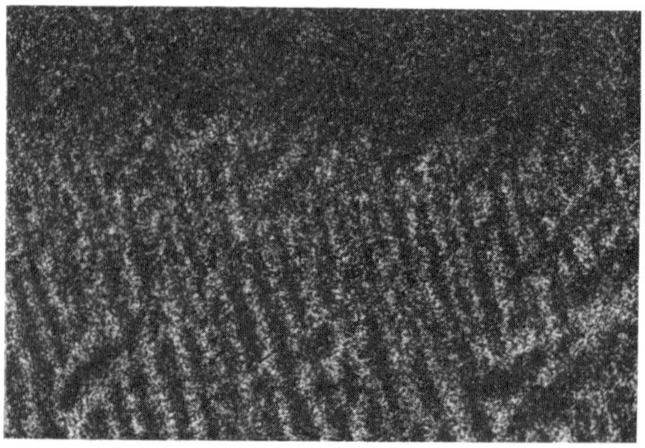

(d.) Tungsten X-ray Map

Fig. 5. BEI and X-ray Maps of <u>Tribocor 532N</u> at 30 mg/cm^2 N_2 10 µm

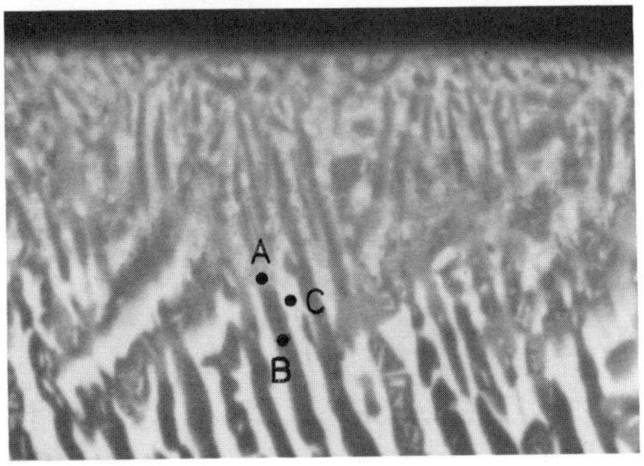

Fig. 6 BEI of Tribocor 532N at 30 mg/cm^2 — 10 μm

Composition	A	B	C
Niobium	56.0	32.0	53.6
Titanium	38.2	61.4	16.5
Tungsten	5.8	6.6	29.9

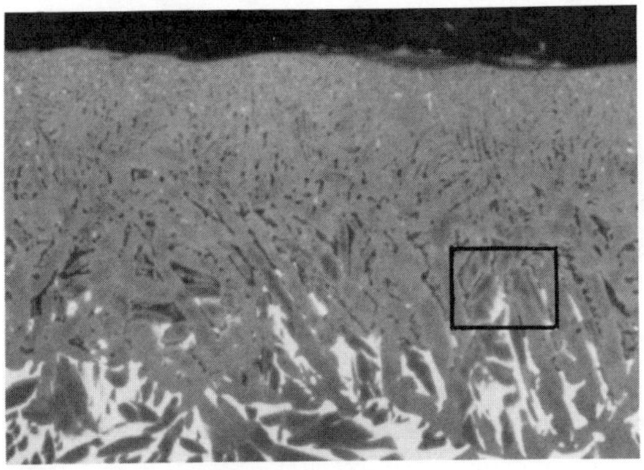

Fig. 7a. BEI of Tribocor 532N at 50 mg/cm^2 N_2 — 10 μm

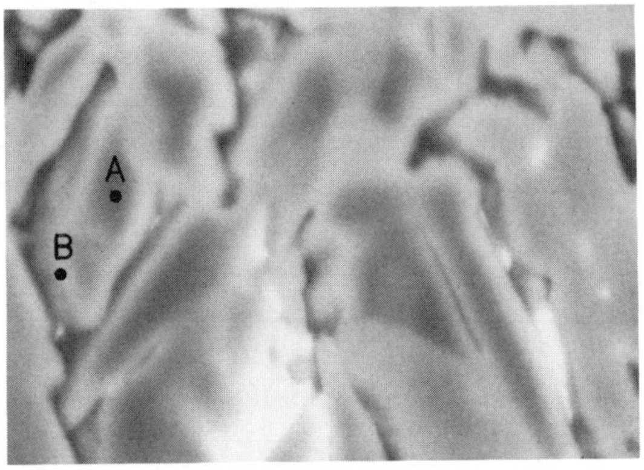

Fig. 7b. Inset Region — 10 μm

Composition	A	B
Niobium	31.5	56.5
Titanium	65.9	38.4
Tungsten	2.6	5.1

As the platelets are scanned more distant from the surface, a rather sharp transition to higher titanium content is found. At about 70 μm depth (Figure 8a), the high titanium content of the platelets and sharp compositional variation between the platelets and matrix are distinctly apparent.

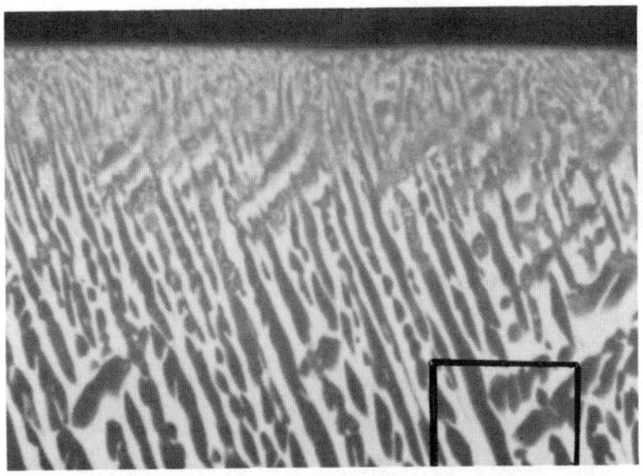

Fig. 8a. BEI of Tribocor 532N at 30 mg/cm^2 N_2 — 10 μm

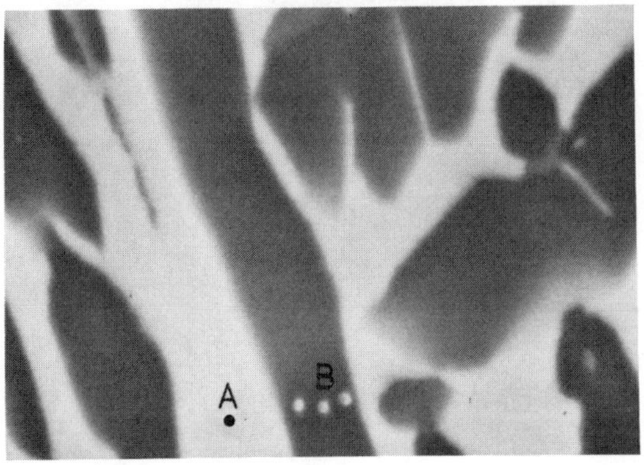

Fig. 8b. Inset Region — 1 μm

Composition	A	B
Niobium	65.1	18.2
Titanium	3.1	78.0
Tungsten	31.9	3.8

Note: Probe dots on Figs. 6, 7b and 8b represent focal points of analysis.

Continued nitriding results in a marked increase in the thickness of the high nitride content surface zone and coarsening of the lower platelet structure. (Compare Figures 9a and 9b.)

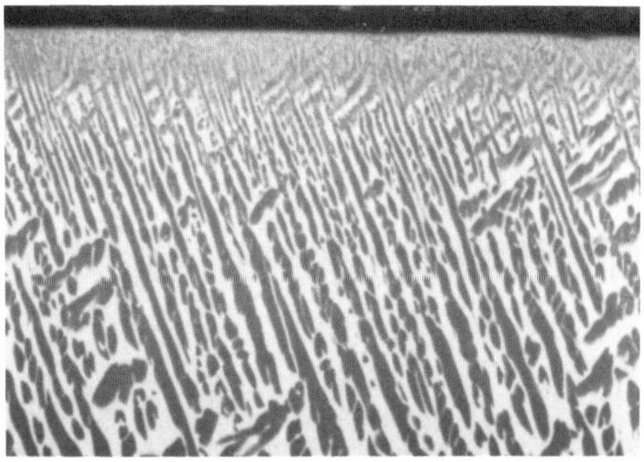

Fig. 9a. 30 mg/cm^2 N_2 10 µm

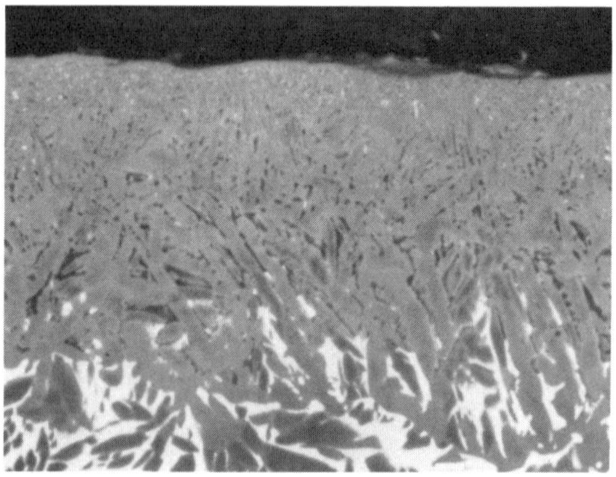

Fig. 9b. 50 mg/cm^2 N_2 10 µm

Fig. 9 Structure of Tribocor 532N

CORROSION RESISTANCE

The corrosion resistance of Tribocor 532N to aqueous solutions is excellent in all of the environments in which it has been evaluated to date. The corrosion rates are compared versus cemented tungsten carbide (4.5 and 10% cobalt) in acid media in Figures 10 and 11. At room temperature, the nitrided alloy shows no detectable corrosion.

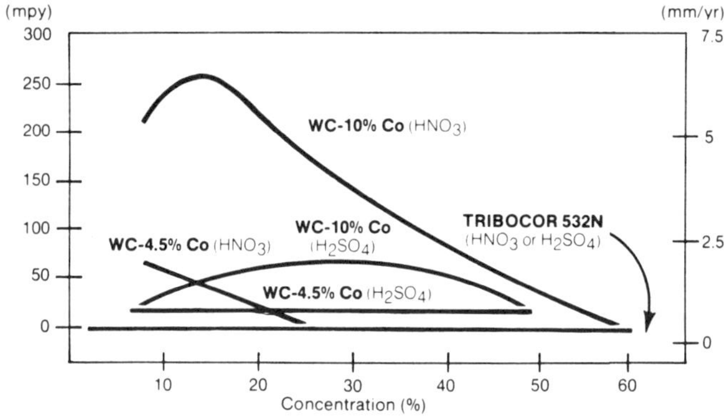

Fig. 10. Corrosion Rate Vs. Acid Concentration (Room Temperature)

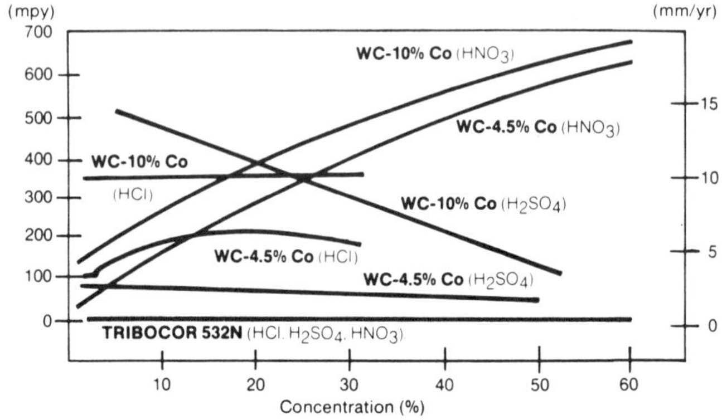

Fig. 11 Corrosion Rate Vs. Acid Concentration (Boiling)

In the boiling acids, under the most severe conditions, penetration rates are less than 0.5 mils per year whereas rates of attack are unacceptably high at all concentrations for the carbides. In 1 to 10 percent solutions of ferric and cupric chloride at ambient temperature (Figure 12) Tribocor 532N is unattacked. Both cemented carbide compositions are severely corroded.

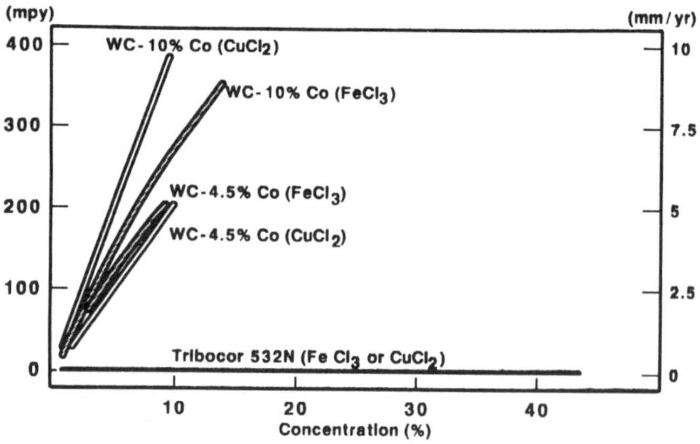

Fig. 12 Corrosion Rate Vs. Chloride Concentration (Room Temperature)

For a number of other media including a mixed HF-HCl and NACE* sour solution, the material is also remarkably resistant, e.g.:

penetration rate - mils per year

Environment	Tribocor 532 N	Cemented Tungsten Carbide 4.5%Co	10% Co	16% Co
85% H_3PO_4(25°C)	0.06	0.2	0.6	
10% NaCN (25°C)	0.0	0.18	0.10	
Sea Water (25°C)	0.0	0.49	0.38	
3%HF-13% HCl (25°C)	0.16	.32	3.3	
*5% Na Cl + 0.5 CH_3COOH, Sat'd with H_2S (25°C)	0.0	0.4	0.67	3.7
(Boiling)	0.03	13.6	13.4	
10% NaOH (Boiling)	0.38			0.26

Stress-corrosion tests were conducted in a number of media using C-ring and tensile specimens. In all the tests, there was no evidence of susceptibility to stress corrosion cracking induced by the environments.

TRIBOCOR 532 N
STRESS CORROSION TEST DATA
C-RING AND TENSILE SAMPLES

Environment	Stress MPa	Time hr	Results
3.5% NaCl Boiling	138	240	No Failures
42% Mg Cl_2 Boiling	138	240	No Failures
10% NaOH Boiling	138	240	No Failures
*5% NaCl + 0.5% CH_3COOH + Sat'd H_2S - Boiling	138	720	No Failures

The oxidation behavior of Tribocor 532N in air at elevated temperatures is influenced by the degree of nitriding. The higher the nitrogen content, the greater the stability. Curves of weight gain versus time at 427 and 532°C for the base metal alloy and for two levels of nitrogen absorption are shown in Figure 13. The base alloy oxidizes parabolically at both temperatures and develops a dark blue scale that turns grey with increasing exposure. The nitrided materials that oxidize most rapidly undergo a series of color changes from blue through grey, dark green, green and ultimately yellow. The yellow scale is presumed to be rich in tungsten. Its presence is generally associated with a sharp increase in the rate of oxidation.

The data in Figure 13 define the temperature limitations imposed by thermal oxidation for long service life exposure. However, for short-term applications, such as cutting tools, oxidation will not result in failure for times of 60 minutes or less at 760°C and 15 minutes at 871°C. This is based on an oxidation weight gain of less than 0.65 mg/cm^2 under these test conditions.

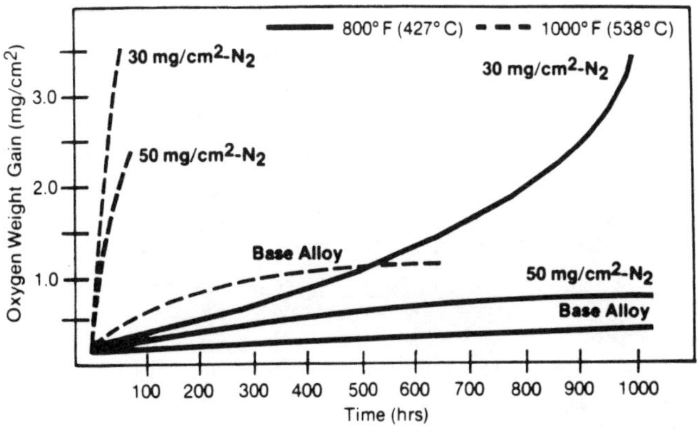

Fig. 13 Tribocor Oxidation Kinetics

HARDNESS AND WEAR RESISTANCE

The surface hardness of Tribocor 532N is in excess of 2200 DPN (Diamond Pyramid Numerals) and is essentially independent of nitrogen weight gain. However, the depth to which this hardness is maintained is strongly dependent on this parameter. High hardness is achieved only when there is a high volume fraction of nitride phase relative to the alloy matrix. Figure 14 illustrates DPN hardness versus distance from the surface (planar specimens) for three different nitriding conditions. Hardness above 1800 can be maintained to a depth of over 75 μm when the alloy is nitrided to 50 mg/cm^2 nitrogen absorption. The volume fraction of nitrides decreases uniformly with distance from the surface resulting in a smooth hardness transition progressing inward. The matrix phase is relatively soft, generally in the range of 250-500 DPN. The matrix hardness is increased by the absorption of some nitrogen, an increase in tungsten content associated with titanium depletion, and precipitation of fine nitride particles at high nitrogen saturation. Projections based on machining tests and high temperature tensile data indicate Tribocor 532N maintains high hardness at temperatures beyond those where cemented carbides have utility.

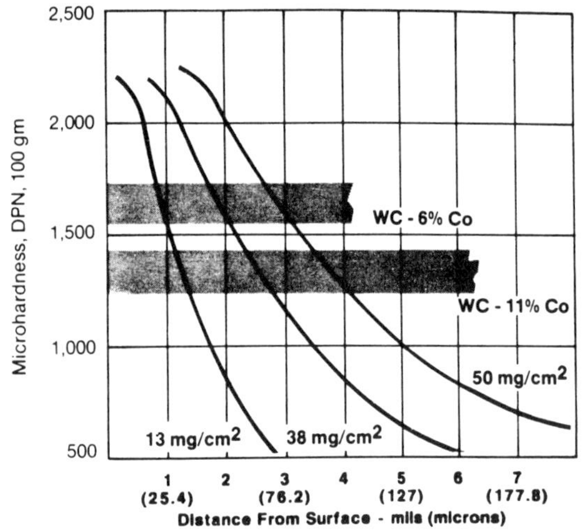

Fig. 14 - Microhardness Vs. Distance from the Surface

The abrasive wear resistance of the nitrided material has been found to be excellent when tested in accordance with ASTM G-65 (silica sand/rubber wheel). Figure 15 compares the volume loss for Tribocor 532N and other materials after 6000 revolution tests.

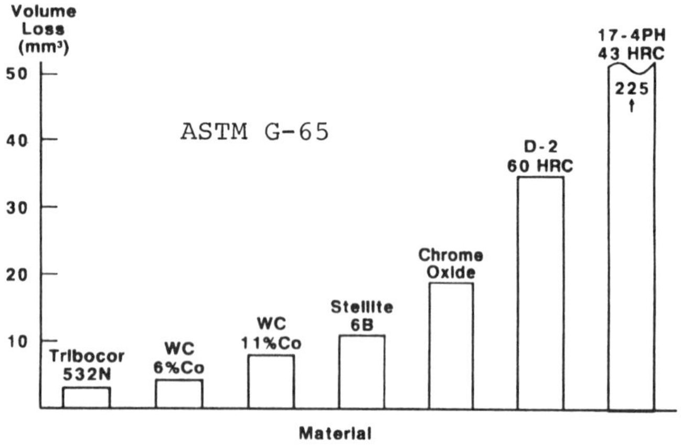

Fig. 15 - Abrasive Wear Rates

The unlubricated sliding wear resistance has also been determined by ring-on-block tests(8) at temperatures to 350°C under a Hertzian stress of 118 MPa. After 50,000 cycles wear rates were very low with no evidence of galling on Tribocor 532N blocks in contact with rings of tungsten carbide, Stellite Alloys 3 and 6 and type 316 stainless steel.

The dynamic coefficient of friction for Tribocor 532N against itself has been measured using a hemispherical pin rotating against a block. A constant value of 0.2 was determined during a 5000 cycle test with loads varying up to 1000g.

MECHANICAL AND PHYSICAL PROPERTIES

Table I lists some mechanical and physical properties of Tribocor 532 both as annealed and after nitriding to 50 mg/cm^2. The base alloy as annealed has high ductility and can be fabricated to complex shapes at ambient temperatures using conventional techniques, such as forging, extrusion, rolling, drawing, heading, drilling, turning, grinding, and welding. Reductions in area of 70-75 percent between anneals are normal with 100 percent recrystallization accomplished by annealing at 1250°C for one hour in vacuum.

As nitrided, the alloy lacks ductility and is non-homogeneous. Therefore, tensile values are not a meaningful property measurement. Youngs modulus and compressive strength are both higher for the nitrided alloy; these properties increase over the base alloy with increasing nitrogen contents.

Table I

ROOM TEMPERATURE MECHANICAL PROPERTIES

	Tensile Strength MPa	Yield Strength MPa	Elongation %	Hardness DPN	Youngs Modulus MPa	Compressive Strength MPa
Base Alloy As Annealed	671	638	21	220	92,400	734
Tribocor 532N 50 mg/cm^2 - N_2					267,000	1585

PHYSICAL PROPERTIES OF TRIBOCOR 532N AND CEMENTED TUNGSTEN CARBIDE

	Specific Gravity g/cc	Magnetic Permeability (VAC = 1.0)	Thermal Conductivity W/m - K		Thermal Expansion x10^{-6}/°C		Dynamic Coefficient of Friction
			25°C	500°C	25°C	500°C	532N/532N
Tribocor 532N	7.5	~ 1.0	42	50	7.4	9.0	0.2
WC - 6% Co	14.9	2.6	100	88	3.8	4.9	
WC - 11% Co	14.3	4.4	92	84	4.9	6.1	
WC - 20% Co	13.5	8.7	117	92	5.7	7.0	

Transverse rupture strength values for Tribocor 532N range from 550 to 830 MPa compared to approximately 1550 MPa for WC-4.5%Cobalt.

As shown in the Table, the nitrided alloy is approximately one-half as dense and has a much lower thermal conductivity than cemented tungsten carbide. It is also virtually non-magnetic. The thermal shock resistance of the material is exceptional, being able to withstand repeated rapid heating to at least 650°C and water quenching with no adverse effects.

SUMMARY

There are a number of characteristics which make Tribocor 532N an attractive candidate for wear and corrosion resistant applications. These include:

- High surface hardness
- Deformation resistance at elevated temperatures (up to about 850°C)
- Ease of fabricability of the base metal sufficent to permit the use of conventional metalworking practices to form complex shapes
- Excellent corrosion resistance
- No magnetic permeability
- Low thermal conductivity
- Excellent thermal shock resistance
- Density about one-half that of cemented carbides
- More economical wear part manufacturing costs

Several potential applications are indicated in Figure 16 as related to the requirements for both wear and corrosion resistance. Evaluations of Tribocor 532N have been undertaken in a number of these uses. The expected utility of the alloy is not limited to being a substitution material for cemented carbides. Because of its unique combination of properties, wear part users should have greater design freedom, thereby improving component performance and reliability.

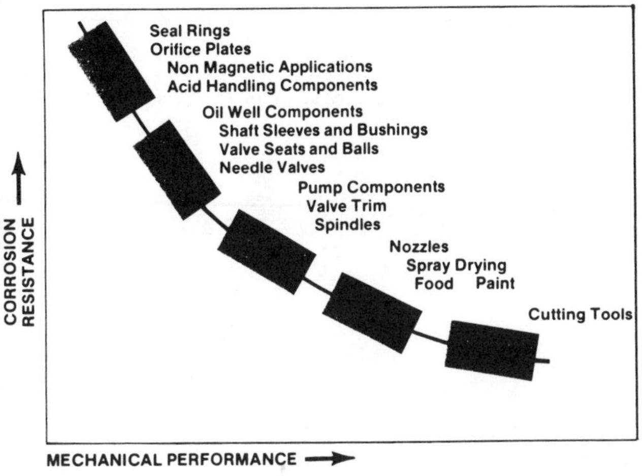

Fig. 16 - Application Requirements

ACKNOWLEDGMENTS

Certain property data for Tribocor 532 and Tribocor 532N reported herein were obtained by Fisher Controls, Marshalltown, Iowa. The coefficient of friction data were obtained from Sandia National Laboratories. Use of these data is appreciated.

REFERENCES

1. J. B. Lambert, P. F. Ziegler, and J. J. Rausch, RM-22, 11th Plansee Seminar, Vol. 1, Metallwerk Plansee, 1985.
2. L. Kaufman and H. Nesor, "Calculation of Superalloy Phase Diagrams: Part 3", Met. Trans, 6A, 2123-2131 (1975)
3. U.S. Patent No. 3,549,429
4. U.S. Patent No. 4,026,730
5. U.S. Patent No. 3,801,381
6. W. D. Klopp, D. J. Maykuth, and H. R. Ogden, "High Temperature Oxidation Behavior of Complex Tantalum Alloys", High Temperature Materials II, Interscience Publishers, 1963
7. M. Hansen, Constitution of Binary Alloys, 2nd Edition, McGraw-Hill, 1958
8. R. C. Tucker, Jr. and A. E. Miller, "Low Stress Abrasive and Adhesive Wear Testing", Selection and Use of Wear Tests for Metals, ASTM, STP-615, 1976

DIFFUSION CARBIDE COATINGS FORMED IN MOLTEN BORAX SYSTEMS (REACTION IN BORAX BATH AND PROPERTIES OF CARBIDE COATED STEEL)

Tohru Arai, Hironori Fujita, Yoshihiko Sugimoto, Yukio Ohta
Nagakute-cho
Aichi, Japan

Abstract

By use of the molten borax bath ordinarily employed for boriding, it is possible to form carbide coatings on carbon containing materials such as steels, nickel alloy, cobalt alloy, cemented carbide and carbon. Coatings, thus formed, consist of VC, NbC, TaC, TiC, Cr_7C_3, etc. and are far superior to those by conventional surface hardening techniques in resistance to wear, seizure, oxidation, corrosion, etc.
Some additives in the borax bath can produce carbide coating or borided layer (boride of major element in the substrate) and some of them produce none. This fact can be understood by taking account of the free energies for carbide formation and oxide formation of element added into the borax baths.
Carbide coated materials by the process are quite excellent in tribological, mechanical and chemical properties and not at all inferior to TiC coating by CVD in these properties and performances in industrial applications. Unlike some of conventional hardening techniques, these highly improved properties are not at the expense of lower mechanical strength. No significant difference is found between ordinarily hardened steel and that coated by this process.

IN RECENT YEARS, three new technologies have been developed to apply thin coatings of very hard material, such as carbides, nitrides and oxides, to dies and tools. Three techniques are: (1) Chemical vapor deposition (CVD); (2) Physical vapor deposition (PVD) (ion plating, reactive sputtering, etc.); and the salt bath immersion process developed by us.(1) The salt bath process involves immersion of materials to be coated in a molten borax bath (usually 850-1050°C) that contains the relevant carbide forming element (usually vanadium, niobium or chromium), under ambient atmosphere. The temperature of the molten borax bath is generally adjusted to be around the hardening temperature of the substrate steel; so that coating and substrate hardening are performed in the one operation. Thus, the salt bath process can be said to be a kind of the salt bath heat treatment for steels. Of course, slow cooling can be done easily.

Unlike CVD and PVD, which are highly confined to die and tool application, the salt bath process involves relatively simple equipment (e.g. no requirements for vacuum or controlled atmospheres) and low in coating cost. Thus, the process has been used to small machine parts such as fastness, roller chains and pumps as well as to dies and toolings. The carbide forming element (CFE), which dissolves in the molten borax, reacts with carbon in the material thus building up gradually a layer of carbide. Typical coating thickness is 2-15 μm. The borax bath features good bath stability and life.

REACTION IN MOLTEN BORAX BATHS

As carbide coatings are formed through the reaction between CFE in the borax bath and carbon atoms in substrates, the salt used in the process comprises a mixture mainly of borax plus compounds containing CFE.

The boriding process by use of the molten borax bath containing some additives had been reported before the development of the carbide coating process. Then, to make clear the relationship between a kind of additives and coatings formed, we examined the surface layers on metals immersed in various kinds of borax bath added with powders of alloys and pure metals containing relevant elements.(1)(2)

As a result, we conclude, the relationship between kinds of formed layers and additives can be clarified, by taking the free energy of oxide formation and carbide formation into account. The boride layers formed in the bath contain elements which have a smaller free energy of oxide formation than B_2O_3 as shown in Fig. 1, through the diffusion of boron atoms, thus produced by the reduction of B_2O_3. The carbide layers can easily be formed on carbon containing materials in the bath contain elements have relatively small free energy for carbide formation, namely CFE, as shown in Fig. 2. If the elements with small free energy for carbide formation also have a small energy of oxide formation than B_2O_3, there may be an opportunity for both carbide formation and boride formation. Titanium, tantalum and manganese belong to this category.(2)(3) As illustrated in Fig. 3, carbon content in steel and amount of such an element added in the borax bath are controlling factors to determine the coatings formed.

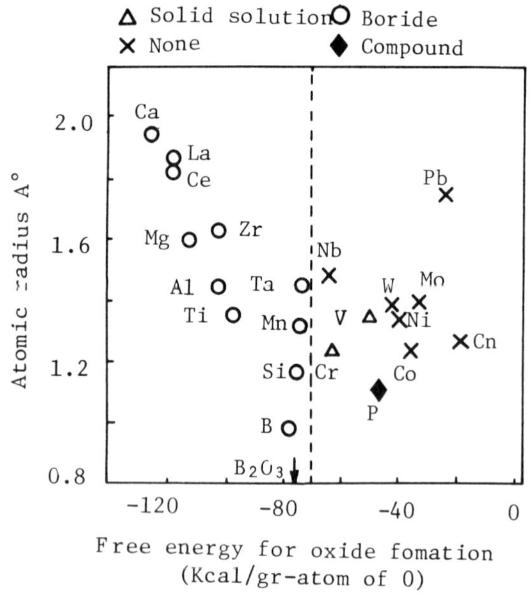

Fig. 1 Relationship between free energy for oxide formation and atomic radius of elements added in borax bath and layers formed on pure iron at 1273 K

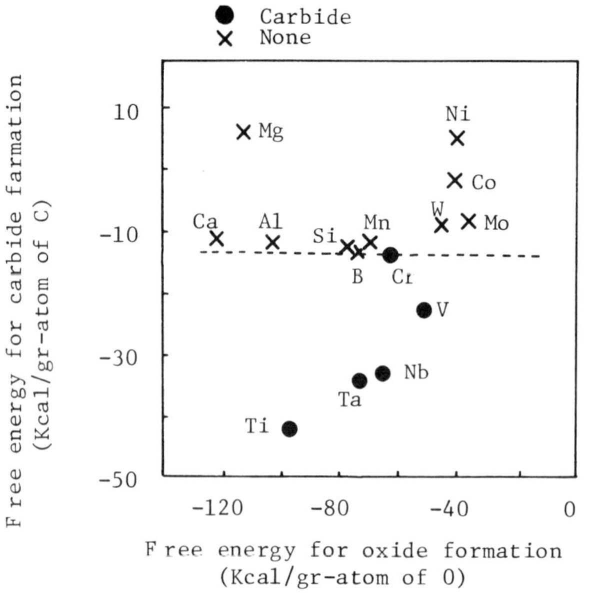

Fig. 2 Relationship between formation free energies of elements added in borax bath and carbide formation on graphite at 1273 K

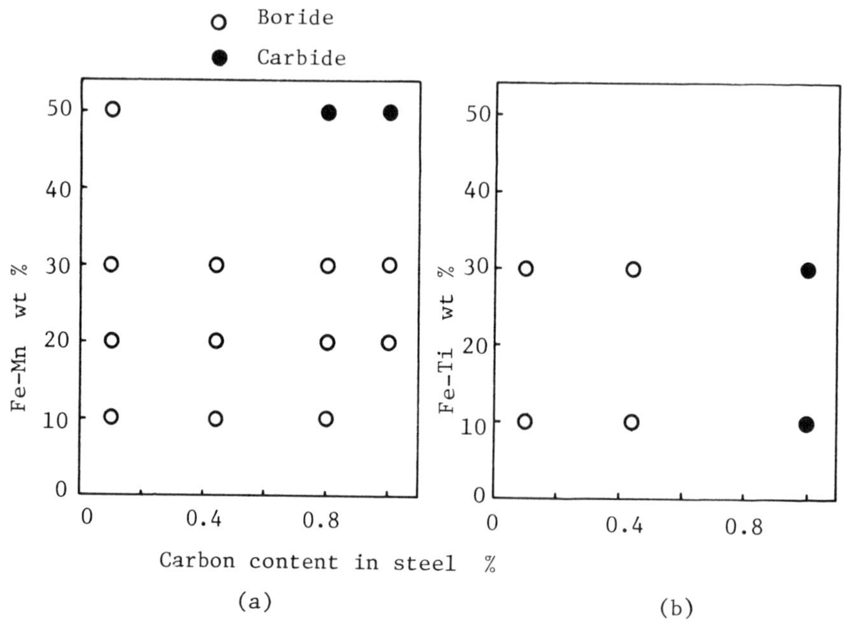

Fig. 3 Relationship between amount of Fe-Mn (a) and Fe-Ti (b) added in borax bath and layers formed on plain car steels containing different carbon content at 1273 K

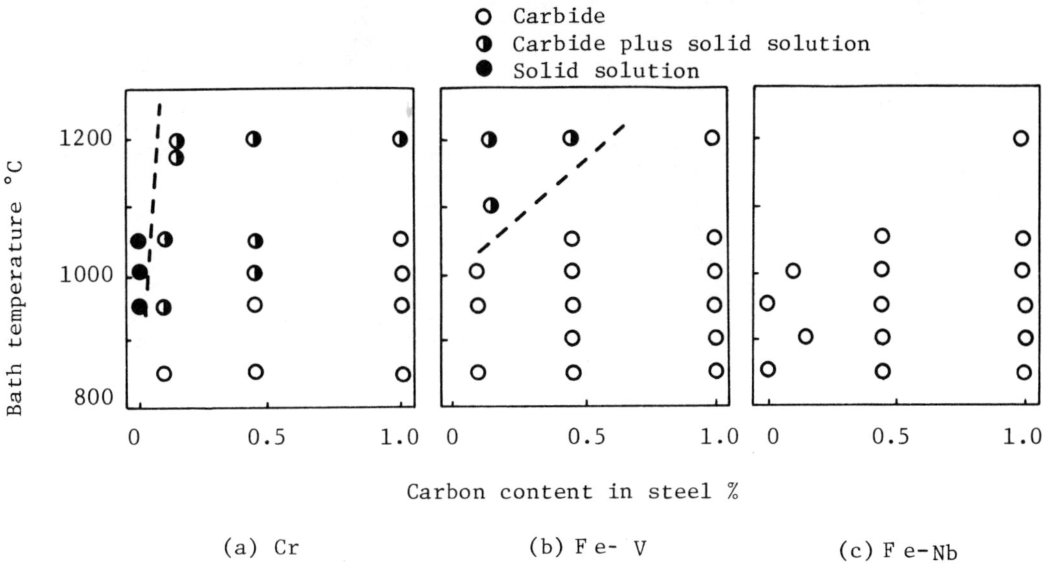

Fig. 4 Relationship between temperature of borax baths added with appreciable amount of Cr, Fe-V and Fe-Nb powders and layers carbon content, at 1273 K

In the bath containing elements which have larger free energies both for carbide and oxide formation, neither carbide nor boride layers can formed. However, if the said elements should have relatively small atomic size, solid solution layers can be formed. Even in the case of CFE, solid solution layers can be formed on metals. Unless the metals have the critical amount of carbon enough to produce carbide layers, which is dependent on the carbide, bath temperature,(4) etc. as exemplified in Fig. 4.

Fig. 5 graphically illustrates the relationship between the kinds of layers formed on high carbon steel in the borax bath and the free energies both for carbide formation and oxide formation.

Figures similar to Fig. 2 and Fig. 5 have been produced with nickel alloys and cobalt alloys containing no carbon and appreciable carbon, respectively.

To dissolve CFE into the borax baths, not only metal powders but also CFE oxide can be introduced. However, the additive containing the elements which have smaller free energy for oxide formation than said oxide, namely the reducing agent, must be added together with CFE oxide and too much reducing agent results in forming a boride layer due to excess boron generated in the bath through the reducing of B_2O_3, as shown in Fig. 6.(5)

We found, moreover, the fact that complex carbide layers are formed on metals in the baths added which both CFE oxide and metal powders containing other CFE, irrespective of comparative free energy for oxide formation, as exemplified in Fig. 7. Complex carbide layers are formed also in the bath added with metal powders containing two or three different CFE, to obtain a combination of advantageous properties.(6)

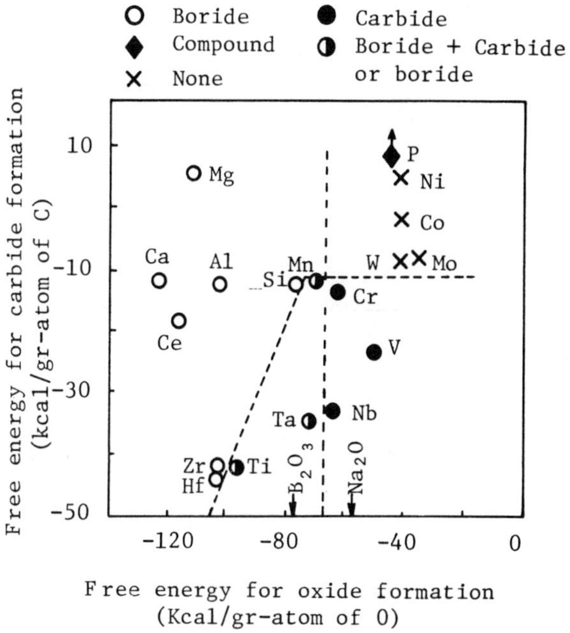

Fig. 5 Relationship between formation free energies of elements added in borax bath and kinds of layers formed on high carbon steel at 1273 K

CARBIDE LAYER FORMATION

While immersed in the bath, carbide layers are formed on the surface of materials through a reaction between CFE dissolved in the fused borax and carbon in the materials. The carbide layers grow as carbide forming reaction is maintained by the continuous supply of carbon atoms to the surface of the coating by the base metal.

The thickness of the carbide layer can be precisely controlled by the length of immersion time, bath temperature and base metal composition. Immersion time ranges from 30 minutes to 10 hours to obtain the optimum thickness 5-15 μm for most application.

Carbide, thus formed, are VC, NbC, TaC, TiC, Mn_5C_2, Cr_7C_3, $Cr_{23}C_6$, etc. Detailed explanation on this items have been already reported by us[1] and results of further researches will be introduced in other opportunities.

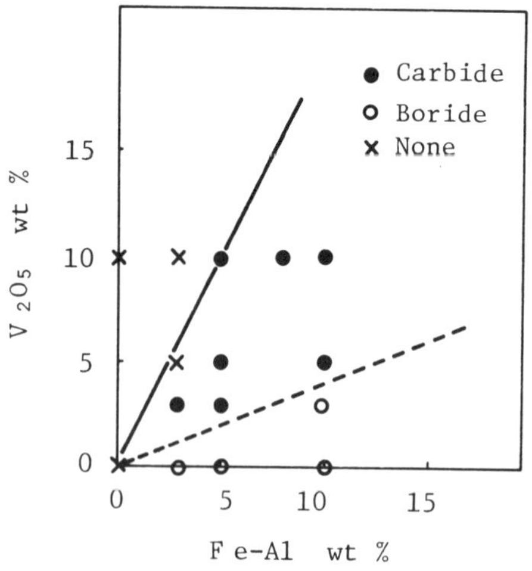

Fig. 6 Relationship between amount of vanadium oxide and ferro-aluminium added in borax bath and layers formed on high carbon steel (W1) at 1223 K

PROPERTIES OF CARBIDE COATED MATERIALS

Materials, coated with vanadium carbide and niobium carbide, are far superior to cemented carbides and materials surface hardened with conventional processes such as plating, nitriding, thermal spraying, etc. in surface hardness, resistance to wear, seizure, corrosion, etc. They are also not at all inferior to materials with TiC coating produced by CVD method. Coatings by this process are also superior to TiN coatings by PVD and CVD in adhesion to substrate. Fig. 8-13 and Table 1 will give evidence of this description.

Chromium carbide coatings are inferior to

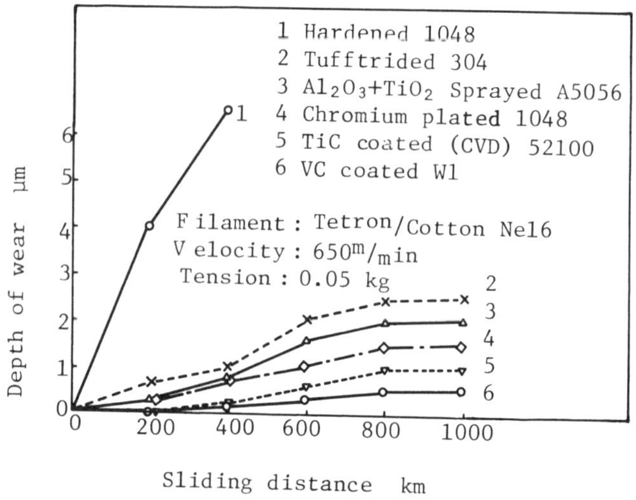

Fig. 8 Comparative depth of wear scar on cylindrical specimens on which textile fibre is running contineously (N. Miyamoto et al)

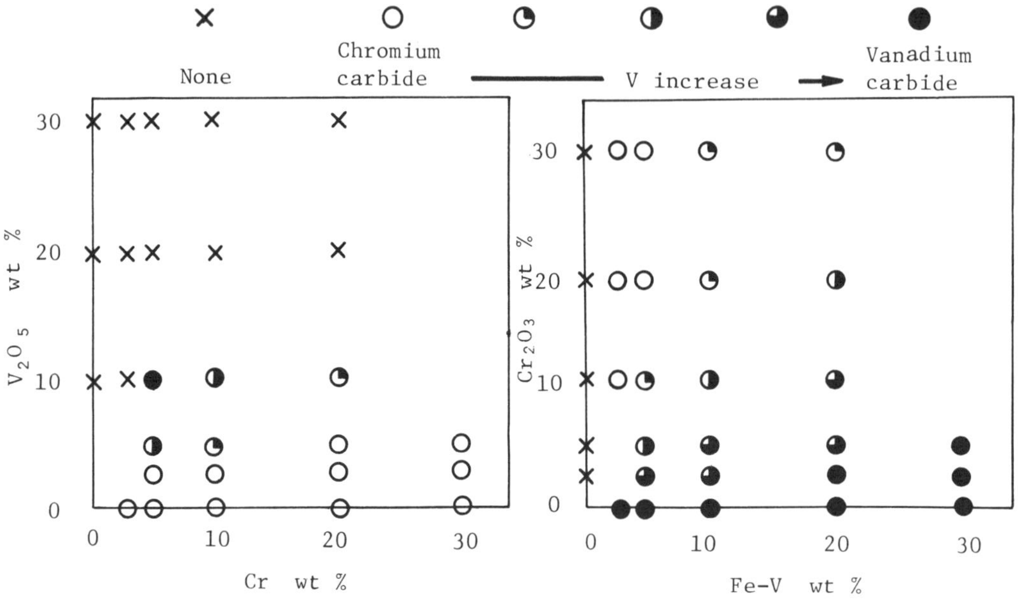

Fig. 7 Relationship between amount of oxides and metal powders added in borax bath and layers formed on high carbon steel (W1) at 1223 K

vanadium carbide and niobium carbides coatings in surface hardness and resistance to wear, while they have excellent resistance to high temperature oxidation as shown in Fig. 14.

It should be emphasized that excellent properties, as mentioned above, are produced without being detrimental to the bulk strength such as toughness, fatigue strength(7), etc. as shown in Fig. 15-17. Thus, carbide coated materials are characterized by a proper combination

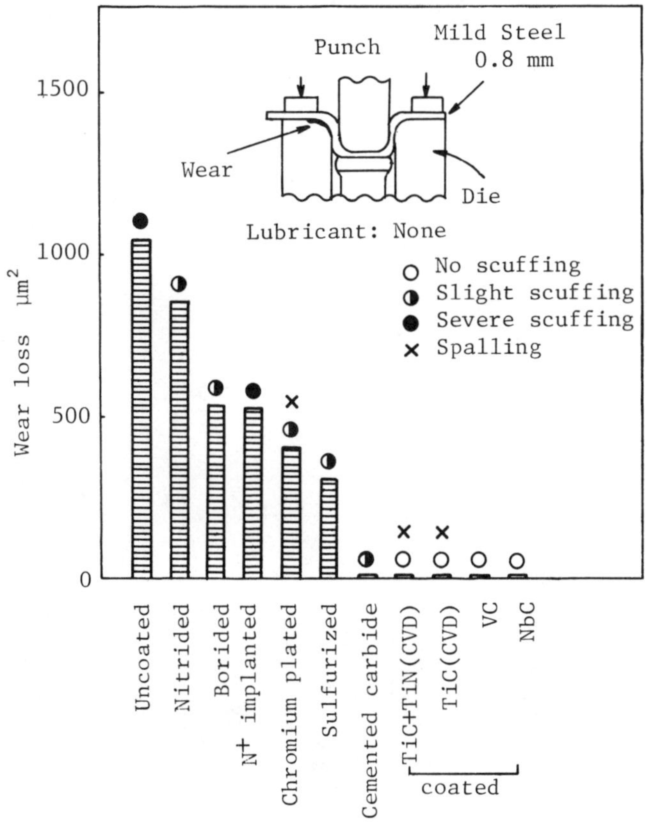

Fig. 9 Comparative cross-sectioned area worn, scuffing and spalling on die radius in the sheet steel bending test

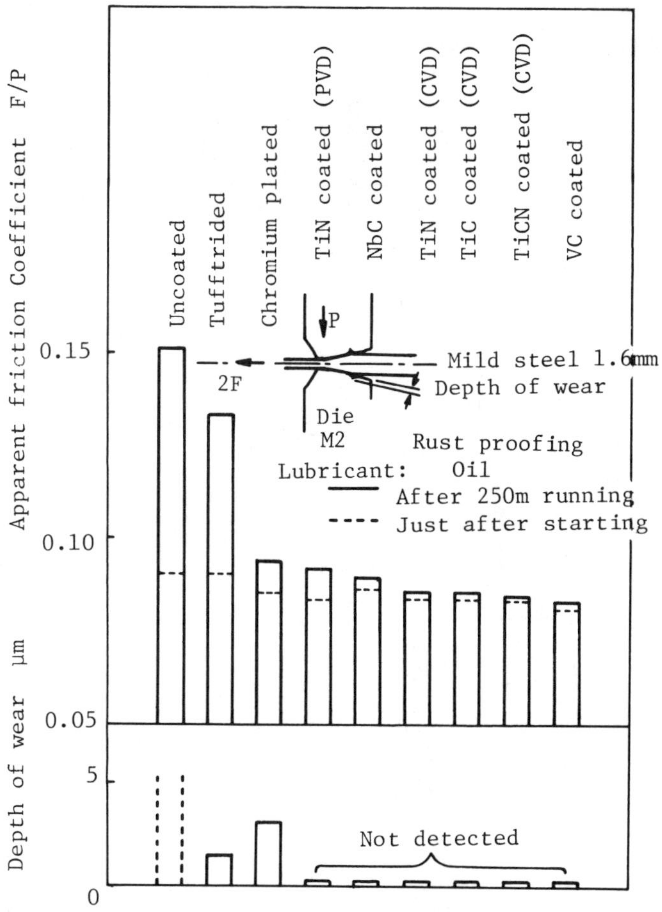

Fig. 10 Comparative friction coefficient and depth of wear on dies in the sheet steel ironing test

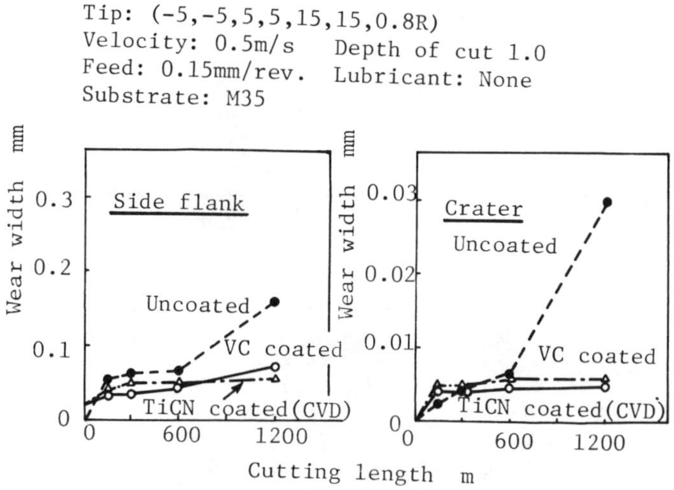

Fig. 11 Comparative width of wear on throw-away tips in the turning test

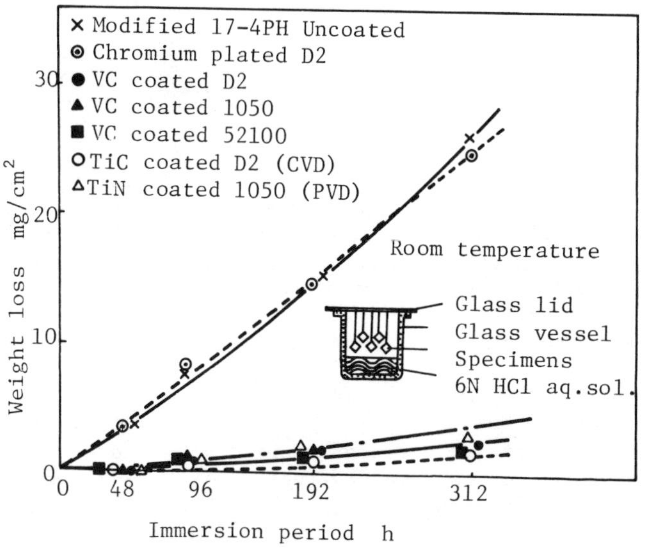

Fig. 12 Comparative loss of weight by corrosion in hydrochloric acid vapour

Table 1 - Comparative loads (kg) at which scoring was observed when successively increased load was applied without lubricant (Sliding velocity: 2.6 m/s)

Countering 50 mm dia × 10	Block (10 × 10 × 10 mm) D2 Hardened	Cemented Carbide -	1045 Chromium plated	1045 Nitrided	W1 Borided	D2 TiC coated (CVD)	W1 VC coated	W1 NbC coated	W1 Chromium carbide coated
304 -	2.7	7.7	3.6	4.2	10.6	7.5	7.3	7.7	4.2
-	4.1	4.0	-	-	3.4	-	11.8	11.2	-
Gray cast iron									
VC coated	-	-	-		>14	-	>14	>1.4	-
NbC coated	-	-	-	-	>14	-	-	>14	10.3
Chromium carbide coated	-	-	-	-	-	-	11.1	13.0	>10.5
Al-20% Si	2.5	2.1	1.8	1.7	3.1	-	3.0	2.4	3.1

> : over the maximum load can be supplied by the test machine

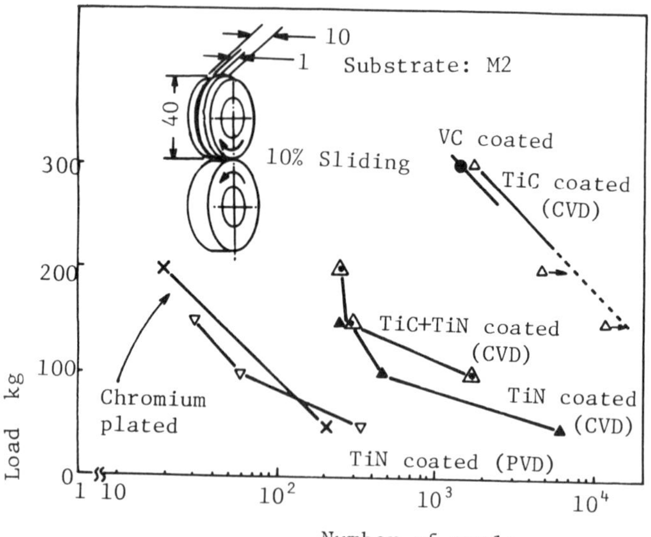

Fig. 13 Comparative number of cycle at which spalling occurred in the rolling test with sliding

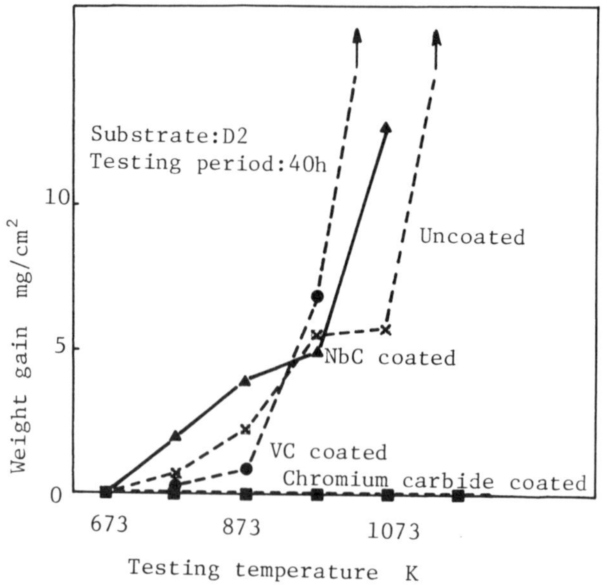

Fig. 14 Comparative weight gain in the high temperature oxidation test

of excellent surface properties inherent in carbides and high strength inherent in substrate materials. The excellent properties of carbide coated materials has led to the widespread application of the process to wide variety of industrial articles requiring good resistance to wear, seizure, oxidation, corrosion, etc.

SUMMARY

The molten borax added with some additives can produce carbide coatings and borided layers on materials immersed therein. The relationship between kinds of formed layers and additives can be clarified by taking the free energy of oxide formation and carbide formation into account. Thin layers of VC, NbC, Cr_7C_3 or other carbides are formed through combination of carbide forming element in the bath and carbon in substrate.

The carbide coating technique based on this phenomenon is highly competitive with CVD and PVD techniques in properties of coated materials and practical applications.

REFERENCES

(1) Arai, T., J. Heat Treating, 1, 15-22 (1979)
(2) Arai, T., M. Mizutani and N. Komatsu, J. Japan Inst. Metals, 39, 247-255 (1975)
(3) Arai, T., Y. Sugimoto and N. Komatsu, J. Japan Inst. Metals, 41, 68-72 (1977)
(4) Arai, T., H. Fujita, M. Mizutani and N. Komatsu, J. Japan Inst. Metals, 40, 925-932 (1976)

(5) Arai, T., Y. Sugimoto, M. Mizutani and N. Komatsu, J. Japan Inst. Metals, 41, 332-338 (1977)
(6) Arai, T., H. Fujita and N. Komatsu, J. Japan Inst. Metals, 41, 438-444 (1977)
(7) Child H. C., S. A. Plumb and G. Reeves, 2nd Int. Conf. Heat Treatment, 21 Sept. 1982

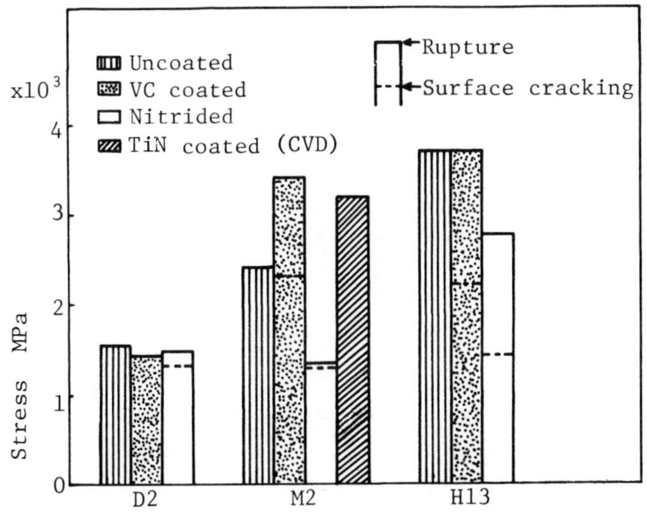

Fig. 15 Comparative rupture and surface cracking stress in the static bending test (H.C.Child et al)

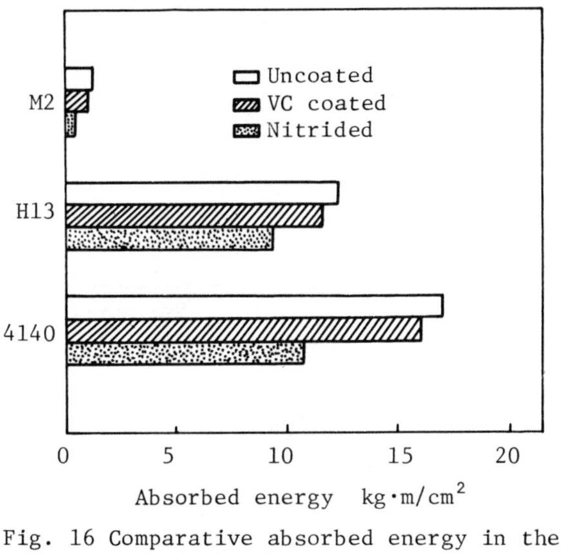

Fig. 16 Comparative absorbed energy in the dynamic bending test

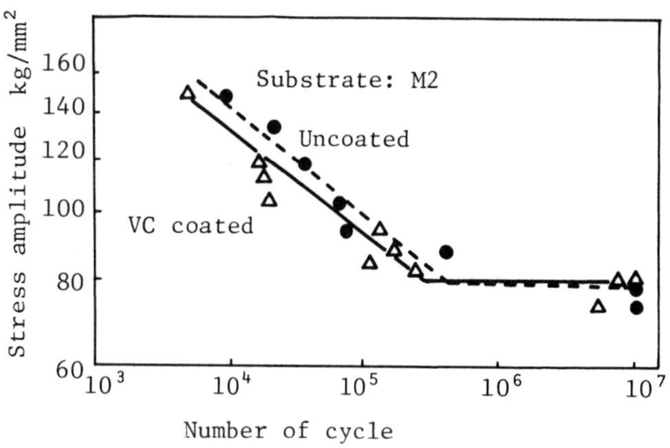

Fig. 17 Comparison of fatigue strength in the rotating bending test

LOW TEMPERATURE DEPOSITION OF TRIBOLOIGICAL COATINGS

J. T. Prater
Pacific Northwest Laboratory
Richland, Washington, USA

Abstract

The results of microstructural and tribological studies performed on sputter-deposited coatings of $ZrO_2 \cdot 20Y_2O_3$, TiB_2 and B-18 wt.% Si are reported. In each case, deposition parameters were identified that offered good tribological properties while maintaining low deposition temperatures (< 400°C) to avoid distortion and softening of prior heat-treated substrates. Zirconia deposits were prepared with a pronounced columnar microstructure that provided excellent thermal fatigue and wear resistance making them candidates for heat engine applications. The TiB_2 coatings were the hardest coatings prepared. These provided good wear resistance on steel substrates. The B-Si coating displayed excellent wear resistance and good adherence to Cu and Al alloys.

MICROSTRUCTURAL AND TRIBOLOGICAL PROPERTIES of $ZrO_2 \cdot 20Y_2O_3$, TiB_2 and B-Si coatings are reported. Each was sputter deposited at low deposition temperatures to minimize the potential for distorting or softening the substrates. Zirconia was studied because of its demonstrated success in gas turbine applications. The boron-based TiB_2 and B-Si were selected because of their high reported hardnesses(1). In each case the effects of deposition conditions on coating hardness, crack toughness and adherence were examined to evaluate the resistance of these deposits to surface attrition and spallation.

Tribological coatings are usually used in combination with hardened alloy substrates. This can impose certain restrictions on the coating process. Certainly the least disruptive and frequently the only feasible processing approach is to perform the coating step following the final machining and heat treatment of the part. However, this requires that the deposition process not exceed temperatures where softening or distortion of the substrate occurs, typically below 300-400°C. Identifying coating techniques that meet this criteria and still produce deposits with good tribological properties was the major objective of this study.

Ceramic compounds, e.g. oxides, nitrides, carbides and borides represent the hardest and thus the most promising compounds for development as tribological coatings (1). Based on the reported hardnesses of TiB_2 (3350 kg/mm^2) and B_3Si (5000 kg/mm^2), it was expected that excellent wear properties could be realized. Physical vapor deposited zirconia coatings, similar to those under development for gas turbines, were also evaluated.

The development of segmented zirconia coatings is particularly noteworthy. Coatings prepared by electron beam physical vapor deposition (EB-PVD) processes, that have a columnar (segmented) microstructure oriented normal to the surface, have exhibited outstanding thermal-cycle durability and good erosion resistance (2). In EB-PVD coatings this excellent combination of properties was only achieved by depositing at temperatures near 1000°C. These high deposition temperatures are acceptable for the superalloys used in gas turbines, which retain their mechanical strength to these temperatures. However, for coating automobile and diesel engine components it would be desirable to apply columnar ZrO_2 at substantially lower deposition temperatures, e.g. 500°C or less.

EXPERIMENTAL PROCEDURE

COATING PREPARATION - A list of the coatings prepared for this study is presented in Table 1. All of the wear test specimens were prepared in the sputter-deposition system shown in Figure 1. This equipment was capable of sputtering simultaneously from two independent, opposing planar

Table 1 Sputter Deposited Coatings

Deposition Experiment	Coating	Substrates	Deposition Temperature	Substrate Bias:Fixturing	Coating Microstructure
T-1	15 μm TiB_2	Cu or C-steel disks	30°C	floating(unbiased): nonrotating	dense (noncolumnar)
T-2	12 μm TiB_2	Cu or C-steel disks	250°C	floating:nonrotating	dense (noncolumnar)
T-3	15 μm TiB_2	A-6 steel domes	30 or 250°C	floating:rotating	columnar
B-1	15 μm B-12 Si	Cu disks	240	floating:nonrotating	dense (noncolumnar)
B-2	15 μm B-18 Si	Cu-Be disks	240	floating:nonrotating	dense (noncolumnar)
B-3	9 μm B-12 Si	Cu disks	240	-35 V:nonrotating	dense (noncolumnar)
B-4	13 μm B-18 Si	Cu-Be domes	240	floating:rotating	columnar
Z-1	15 μm $ZrO_2{\cdot}20Y_2O_3$	X-40 wedges	300°C	12 μm floating/3 μm at -35V:rotating	columnar ZrO_2 w. dense outer layer
Z-2	24 μm $ZrO_2{\cdot}20Y_2O_3$	X-40 wedges	300°C	floating:rotating	columnar ZrO_2

targets of up to 23 cm in diameter. In most cases only the lower target position was utilized. Up to seven wear test specimen could be positioned in the center of the chamber and rotated at 20 rpm. The bias voltage and temperature of each sample could be controlled independently. Alternatively, when only one surface of a sample was to be coated it was positioned in a fixture located in the upper target position.

The TiB_2 and Si-B coatings were prepared using either a 15 cm (6") diameter TiB_2 or 23 cm (9") diameter B-Si pressed powder targets. All targets were hot isostatically pressed from the respective powders. The substrate to target distance was 5.5 cm and the sputtering gas was research grade Krypton maintained at a pressure of 0.4 Pa (3 mTorr). Prior to the start of coating deposition, the substrates

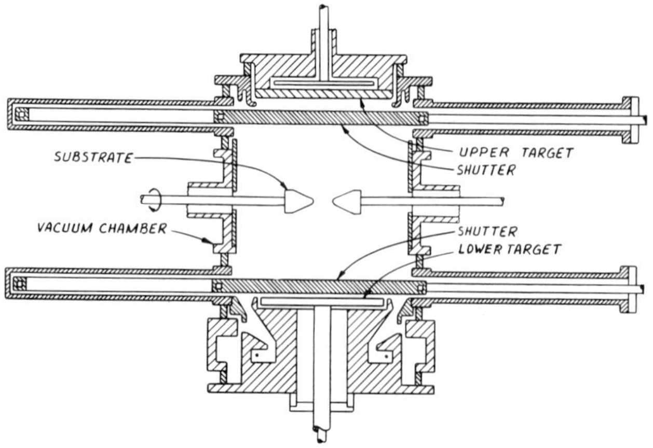

Figure 1. Schematic of the Sputtering System.

were etched for 10 minutes. This step cleaned the surface and improved coating adherence. The deposits were made using enhanced thermionically-supported plasma-discharge d.c. sputtering. Target power densities of 1.3-1.8 watts/cm^2 were used for all coatings. While higher power densities (factor of 4) were possible, target heating would eventually create excessive thermal stresses and cause target cracking. The selection of 1.5 watts/cm^2 was made to provide reasonable deposition rates (0.3 to 0.6 μm/hr on rotating substrates; 2 μm/hr on nonrotating surfaces) while maintaining low thermal stress levels in the target.

The zirconia coatings were produced by conventional r.f. diode sputtering techniques using the same system as described above. The sputtering targets were 23 cm (9") diameter steel base plates with .125 mm (.05") thick surface layers of plasma-sprayed $ZrO_2{\cdot}20Y_2O_3$. The sputtering plasma was composed of either Kr or a 50:50 mixture of Kr-O_2 maintained at a pressure of 5.5-8.5 Pa (40-60 mTorr). Coatings were deposited at a rate of about 0.4 μm/hr onto rotating substrates. A maximum target power density of 1.3 W/cm^2 was used. The substrates could be biased (typically -35 V r.f.) during deposition. Prior to deposition of the zirconia a 50 μm thick NiCrAlY bond layer was deposited onto the substrates.

COATING CHARACTERIZATION - Coating compositions and microstructures were determined using a scanning electron microscope (SEM) equipped with an energy dispersive X-ray analyzer (EDX). Hardness measurements were made using a Leitz Indent Hardness tester with a 300 gm load. Wear tests were performed in a ring-on-block Falex-1 wear tester. Flat test blocks were secured in the test fixture and then a rotating ring of hardened S-10 steel (HR_c=60) was brought into contact with the sample surface. The 3.5-cm ring rotated at 72 rpm and was loaded to provide a contact weight of 68 Kg (150 lbs). During the test, the samples were immersed in an oil bath that remained at about 45°C. The test was terminated after 5000 ring rotations. The extent of wear that occurred during the test was determined by measuring the width (or area) of the wear scar that was generated on the test block. A narrow wear scar indicated that little coating had been abraded.

RESULTS AND DISCUSSION

The deposits were evaluated in terms of their microstructural appearance and the results of some general hardness and wear experiments. The results of these studies are reported below.

COATING MICROSTRUCTURE - The deposit microstructure varied significantly depending upon the positioning of the sample during deposition. Deposition onto flat stationary substrates resulted in a dense equiaxed-grain structure. In contrast, when the substrates were rotated during deposition a geometric shadowing mechanism (3) resulted in the development of a columnar microstructure, as shown in Figure 2, oriented perpendicular to the substrate surface. This was precisely the microstructure that has proven to be so desirable in the EB-PVD zirconia thermal-barrier coatings. In zirconia deposition experiment Z-1, a substrate bias was applied during the latter portion of the deposition to provide a limited amount of ion bombardment of the surface. This disrupted the growth of the columns, and produced a dense equiaxed-grain structure, as shown in Figure 3.

In the coatings that were rotated during deposition wart-like defects developed, particularly in the boron-containing coatings. These defects had an inverted conical morphology, initiating at a point near the base of the deposit and expanding through the deposit thickness. Deposits made onto flat, nonrotating, substrates did not develop these growth defects. Usually, application of a small substrate bias (-35V) to the rotating substrates inhibits growth of such defects by providing a limited amount of ion bombardment that disrupts their growth. In the case of TiB_2, application of a negative bias greater than -20V produced black powdery deposits with reduced adherence. Deposits made at lower biases were metallic in appearance and adherent, but contained the conical defects. For the B-Si coatings biases up to -35V had no effect on the coating microstructure.

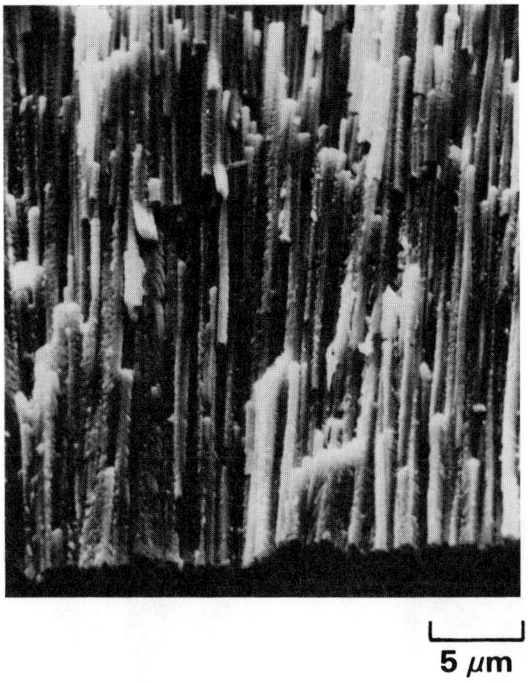

Figure 2. SEM Micrograph of a Fracture Surface of a Sputter-Deposited $ZrO_2 \cdot 20Y_2O_3$ Layer Showing the Segmented (Columnar) Microstructure.

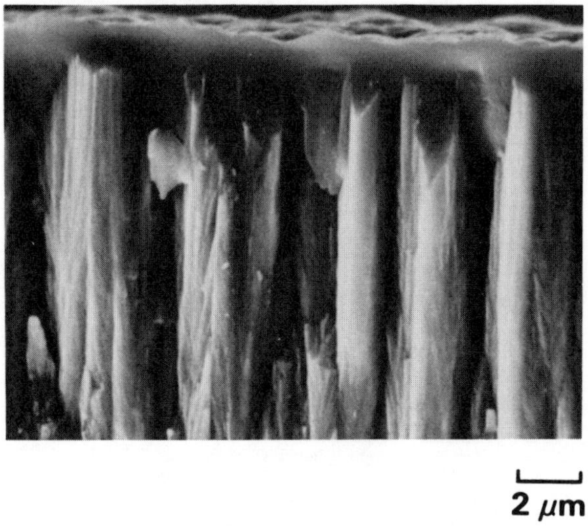

Figure 3. SEM Micrograph of the Dense Layer Deposited Over the Columnar Ceramic Structure in Deposition Experiment Z-1.

TITANIUM DIBORIDE - The TiB_2 coatings were prepared at two deposition temperatures: 30°C for water-cooled substrates and 250°C for uncooled substrates. Deposits prepared on cooled substrates were slightly harder than those on the uncooled substrates. Average hardnesses were about 3000 and 2800 DPH, respectively. No significant difference in hardness was detected for coatings prepared on flat vs rotated substrates despite the differences in microstructure.

Examination of metallographic cross sections of the TiB_2 coatings showed that the uncooled samples were much less prone to cracking and had slightly fewer growth defects, Figure 4. In preparing metallographic cross sections of the TiB_2 coatings some separation along the coating-substrate interface was observed. The compositions of TiB_2 coatings were all 70 wt.% Ti, which is slightly richer (1%) in Ti than stoichiometric TiB_2. X-ray diffraction analysis identified TiB_2 as the only phase present.

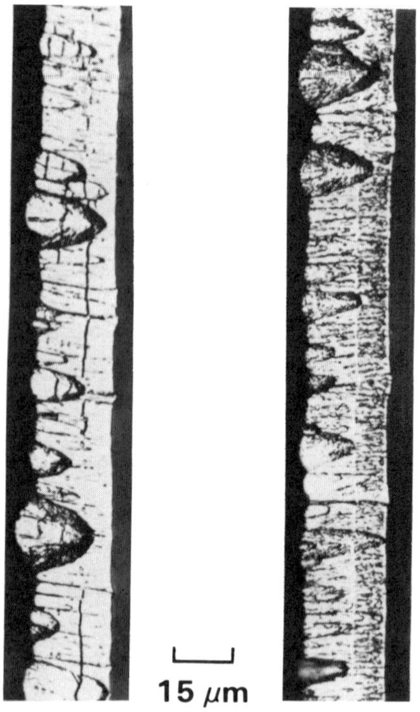

(a) (b)

Figure 4. Etched Cross Section of TiB_2 Coatings (T-3) Prepared with a Substrate Temperature of a) 30°C and b) 250°C. Note the presence of cracks in the coating prepared at the lower temperature.

BORON-SILICON - Earlier work by Busch (4) indicated that the hardness of B-Si deposits varied with Si content from a high of 9500 DPH at 9 wt.% Si down to 4000 DPH at 45 wt.% Si. These were flat deposits prepared on cool (80°C) hardened steel (HR_c=60) substrates. Hardness dropped significantly to about 2500 DPH when the deposits were annealed at 500°C for 1 h, suggesting that the exceptionally hard coatings were associated with a metastable phase. These hard deposits were brittle and prone to cracking.

Deposition at higher temperatures improved coating adherence. Still, adherent B-12 wt.% Si coatings deposited on A-6 steel, H-13 steel and Ti substrates appeared to be limited to approximately 3 μm. Thicker deposits were prone to peeling. Thicker 15 μm deposits were easily prepared on Cu and Al. The Si content of the coating did not appear to affect coating adherence.

A harder Cu-1.8 Be alloy (C17300) was later included as a substrate material. This alloy could be hardened to HR_c=45 by heat treating at 320°C for 3 hours. To prevent overaging, deposition temperatures were maintained below 300°C. Adherent 15 μm thick, B-Si coatings were successfully deposited on the Cu-Be (HR_c=45), as shown in Figure 5.

Figure 5. Etched Cross Section of B-18 Si Deposited onto a Cu-Be Substrate in Deposition Experiment B-4.

Coatings prepared at higher deposition temperatures showed no strong dependence of hardness on Si contents for compositions of 12, 18 and 25 wt.% Si. The hardness values averaged about 1300 DPH. The higher Si contents did provide the targets with a low melting point component that improved the compaction of the powder and afforded the target greater durability. Thus, a target composition of B-22% Si was used in the wear studies. Coatings prepared from this target were analyzed at B-18% Si, indicating a slight shift in composition.

X-ray diffraction patterns of the B-Si coatings were of poor quality because of the low scattering cross sections of B and Si which, in turn, produced very low intensities. However, there were several distinct peaks which did coincide with the patterns for B_6 Si and B.

STABILIZED ZIRCONIA - The intent of the zirconia study was to produce a fine-grained columnar microstructure at low deposition temperatures. Previous work demonstrated that this structure provided excellent thermal cycle and erosion resistance (2). The low temperature sputter-deposited coatings appeared to have the desired microstructure, as previously shown in Figure 1.

During deposition experiment Z-1 the outermost 3 μm of zirconia was deposited with a -35V r.f.-induced d.c. bias applied to the substrate. This produced a dense closeout layer over the columnar microstructure. This denser structure was expected to improve the wear resistance of the coating and restrict infiltration of unwanted gaseous and liquid species down the open intercolumnar gaps.

The thermal-cycle tolerance of the zirconia coatings was tested by subjecting both as-sputtered and vacuum-annealed (1080° for 8 h) samples to a series of ten 10-min thermal cycles to 1000°C. In the first five cycles, the samples were air cooled to 150°C; in later cycles, the samples were quenched into water. No coating spallation was observed.

COATING WEAR - The results of the wear test are presented in Table 2. All of the coatings studied exhibited substantial improvements in wear resistance compared with uncoated steel substrates. The TiB_2 coatings were the hardest coatings prepared and provided the C-steel substrates good wear resistance. The factor of 4 difference in wear scar width between T-1 and the C-steel samples actually represented a 10-fold reduction in the volume of material removed. The TiB_2 deposits made onto the cooled substrates performed slightly better, perhaps because of its slightly greater hardness, but the difference was small. While the wear mechanism was not clearly identified, some debris was found which indicated that coating delamination occurred. An example of the debris is shown in Figure 6. If delamination rather than surface abrasion was the principal mode of degradation then gains afforded by higher values of coating hardness could well be offset by increases in their brittleness.

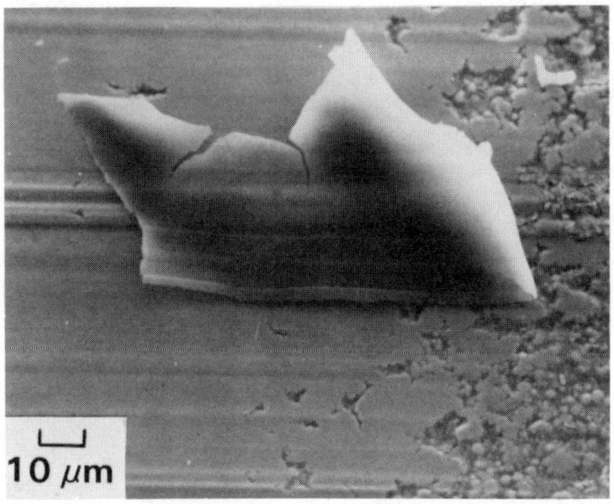

Figure 6. West Debris Observed on TiB_2 Coatings Following the Wear Test.

The B-Si coatings demonstrated excellent wear resistance. The unbiased coatings exhibited better wear resistance than those deposited with a -35V bias. All the B-Si samples that were wear tested were prepared on unrotated flat disks and therefore were composed of a dense equiaxed (noncolumnar) microstructure. It is expected from previous work on the effect of sputtering parameters that the bias introduced additional compressive stresses into the deposit and these may have detrimentally affected the wear resistance of the biased coating.

The wear resistance of ZrO_2 coatings was outstanding. As expected, the application of the dense ZrO_2 outer layer significantly improved wear resistance. The application of a bias near

Table 2. Wear Test Results

Sample		Description	Hardness (DPH)	Scar Area (width x length, mm^2)
TiB_2				
	T-1	cold, unbiased, noncolumnar	3000	6.6
	T-2	hot, unbiased, noncolumnar	2800	8.6
B-Si				
	B-1 and -2	hot, unbiased, noncolumnar	1300	4.0
	B-3	hot, -35 V bias, noncolumnar	1300	7.8
ZrO_2				
	Z-1	columnar microstructure with dense outer layer	1200	2.5
	Z-2	columnar microstructure	1000	4.3
Standards				
	C-steel	HR_c = 16	200	26
	H-30 steel	HR_c = 31	300	11

the end of deposition experiment Z-1 produced a dense ZrO_2 layer over the columnar ZrO_2 understructure that greatly improved the overall wear resistance of the coating.

CONCLUSIONS

1. Zirconia and boron-based coatings with excellent wear resistance can be prepared at low deposition temperatures using sputtering.

2. Zirconia coatings with a columnar microstructure exhibit an excellent combination of thermal cycle durability and wear resistance that make them good candidates for heat engine applications. Deposition of a dense closeout layer above the columnar microstructure further enhances the wear resistance of the coating.

3. B-Si coatings have excellent promise as wear resistant coatings for Cu and Al alloys.

4. TiB_2 can provide steel substrates with significant improvements in wear resistance.

ACKNOWLEDGMENTS

This work was supported by the U.S. Department of Energy, Office of Coal Utilization, under Contract DE-AC06-76RLO-1830.

The author would like to thank H.E. Kjarmo for performing the scanning electron microscopy; R.H. Beauchamp for preparing the metallographic specimens; A.G. Graybeal for conducting the wear and hardness tests and R.F. Stratton for preparing the sputtered coatings.

REFERENCES

1) A.A. Ivan'ko, Handbook of Hardness Data (1971), Keter Press, Jerusalem, Israel.

2) J.T. Prater and E.L. Courtright, "Combustion Zone Durability Program - Material and Coating Development for Alternative Fuel Applications", PNL-5259, (1985).

3) J.W. Patten, Thin Solid Films, 63 (1979), p. 121-129.

4. R. Busch, U.S. Patent 3,856,537.

SURFACE SEGREGATION
AND
SELF-RENEWING COATINGS

EQUILIBRIUM SURFACE SEGREGATION —RECENT ADVANCES IN THEORY AND EXPERIMENT

Paul Wynblatt, Daniel A. Steigerwald
Department of Metallurgical Engineering
and Materials Science
Carnegie-Mellon University
Pittsburgh, Pennsylvania, USA

Abstract

This paper is divided into two principal sections. The first section reviews early theoretical approaches used to model the equilibrium surface composition of alloys, then proceeds to summarize recent theoretical advances which have been made in the description of this phenomenon. The second section reviews briefly early experimental work and follows with a discussion of recent experimental studies which have probed deeper into interesting details of surface compositional behavior.

THE EQUILIBRIUM SURFACE COMPOSITION of materials generally differs from their average or bulk composition, a phenomenon that is commonly referred to as surface segregation. Over the past decade, significant advances have been made in understanding the factors which control the equilibrium composition of metallic alloys, and a start has been made at addressing this phenomenon in the context of other materials, such as metal oxides. In the case of alloys, numerous experimental studies have been reported in the literature and a reasonable theoretical basis has been developed for the interpretation of experimental results. However, there still remain important areas of both experimental and theoretical study which are, as yet, relatively unexplored.

An understanding of the factors which control surface composition is clearly important in phenomena such as oxidation, corrosion, thermionic emission, tribology, catalysis, etc., which are dominated by surface characteristics. These phenomena relate in turn to many of the technological settings in which self-renewing coatings are deemed to play an important role.

This paper will focus on the equilibrium surface composition of alloys. The emphasis on alloys stems principally from the fact that most work to date has been performed on that class of materials, and that many of the fundamental concepts developed for metallic systems can be extended to other classes of materials. However, it is important to recognize that differences in the nature of chemical bonding in different material classes will certainly modify the detailed behavior with regard to surface composition. For example, electrostatic effects in ionic solids which may affect surface composition have no parallel in metallic systems. Some of these issues will be covered in detail in other papers presented at this conference.

Among the phenomena of interest in the context of self-renewing coatings are some which take place under conditions so far removed from equilibrium that an understanding of equilibrium surface composition may be essentially irrelevant. An example in point is the case of the surface composition which develops during high flux, high energy bombardment by heavy charged particles. In this instance, the removal of atoms from the surface by sputtering phenomena, coupled with the solute drag which results from annealing out of the large excess of point defects generated near the surface, may result in steady state surface compositions that bear little resemblance to the equilibrium values. In other relevant phenomena, such as erosion and wear, surfaces that are freshly exposed by these processes will tend to revert to a segregated state under a driving force which should be virtually identical to that for equilibrium segregation. Thus, in these instances, an understanding of the factors which control equilibrium surface composition and its driving force should be of prime importance.

1. THEORY

1.1 EARLY APPROACHES - Before discussing recent approaches used to model surface segregation phenomena, it is useful to review briefly some earlier approaches in order to place more recent developments in their proper perspective. While this paper is not intended to provide a comprehensive review of theory, it touches upon the major concepts which have been proposed for the description of the equilibrium surface composition of alloys.

1.1.1 Gibbs' Treatment - Gibbs [1] was the first to point out the existence of an "excess concentration" associated with interfaces. In his rigorous thermodynamic treatment of interfaces, he developed relationships between the surface tension of a solution, the surface excess concentrations and the chemical potentials of the various components. However, this treatment is of limited usefulness for the interpretation of the surface composition of solid alloys, because of the experimental difficulty of measuring the surface tension of solids, and its dependence on compostion and/or temperature.

1.1.2. Early Treatments Based on the Regular Solution Model - The inherent difficulties in the application of the Gibbsian formalism has led, in more recent times, to the development of a family of approximate models based on microscopic statistical thermodynamic treatments of surface composition. The earliest of these treatments [2,3] assumed the existence of a system consisting of two phases, a "bulk phase" and a "surface (or interface) phase", and adopted the framework of the regular solution approximation. In this approximation, the free energy of a solution is considered to consist of two contributions, the first arising from an ideal entropy of mixing, and the second from a finite (i.e. non-ideal) internal energy or enthalpy of solution. With this framework, the atom fraction of solute in the surface phase of a binary solid solution, X^S, may be expressed as:

$$\frac{X^S}{1 - X^S} = \frac{X^b}{1 - X^b} exp\left[\frac{-\Delta H_{seg}}{kT}\right] \quad (1)$$

where X^b is the atom fraction of the solute in the bulk phase, ΔH_{seg} is the enthalpy of segregation of the solute, and k and T are the Boltzmann constant and the absolute temperature, respectively. Physically, the enthalpy of segregation represents the enthalpy change which results when a solute atom located in the bulk phase is exchanged with a solvent atom in the surface phase, and may be regarded as the driving force for the segregation of the solute species to the surface*. Whereas all models of this type lead to expressions of the form of Eq. (1), each model produces different expressions for ΔH_{seg}.

Defay et al.[2] applied this approach to the equilibrium composition of a liquid surface. Their model employed a nearest neighbor bond model to describe the internal energy of the solution, and assumed that the atoms of the liquid were located on the points of a lattice, i.e. they used a solid-like model for the liquid. In addition, they took the surface phase to consist of a single plane of atoms in equilibrium with a semi-infinite bulk phase. Their results may be stated as [4]:

$$\Delta H_{seg} = (\gamma_B - \gamma_A)\sigma + \frac{2\Delta H_m}{Z X^b(1-X^b)} \times\left[Z_\ell(X^b - X^S) + Z_v(X^b - 1/2)\right] \quad (2)$$

where γ_A and γ_B are the surface tensions of the pure A and B components in the binary solution (component B taken to be the solute), σ is the area per surface atom in the solution, ΔH_m is the enthalpy of mixing of the solution, and Z_ℓ, Z_v and Z are coordination numbers. Z_ℓ is the number of bonds made by an atom to its nearest neighbors which lie within the plane of the surface, Z_v is the number of neighbors which lie in the adjacent atom plane, and Z is the total coordination number of an atom in the bulk.

McLean[3] applied a similar approach to the equilibrium composition of grain boundaries. He also used a monolayer representation of the interface, but assumed that the driving force for segregation was the elimination of the elastic strain energy associated with a solute atom in the solvent matrix. In addition, McLean postulated that the elastic strain energy of a solute atom disappears at the interface, i.e.:

---

*In general, ΔH_{seg} may be either positive or negative, corresponding to either depletion or enrichment of the solute in the surface phase, respectively.

$$\Delta H_{seg} = -E_{el}, \quad (3)$$

where E_{el} is the elastic strain energy associated with a solute atom in the bulk, and can be estimated from elasticity theory descriptions of a misfitting sphere in an elastic continuum [5].

As was pointed out by Burton and Machlin [6], neither of the above two models is able to yield qualitatively correct predictions of the segregating component in a solid binary alloy. This is because the model of Defay et al. [2] considers only "chemical" contributions to the heat of segregation whereas McLean [3] ignores all but "elastic" contributions. While the neglect of elastic effects by Defay et al. is proper for the case of liquid solutions, application of the model to solid solutions requires modification to include the elastic term considered by McLean. These considerations were first put forward by Wynblatt and Ku [4,7] in a unified model which combined both elastic and chemical effects to yield a heat of segregation of the form:

$$\Delta H_{seg} = (\gamma_B - \gamma_A)\sigma + \frac{2\Delta H_m}{ZX^b(1-X^b)}\left[Z_\ell(X^b-X^s)+Z_v(X^b-1/2)\right] - E_{el}, \quad (4)$$

where; $E_{el} = 24\pi KGr_Ar_B(r_A-r_B)^2/(4Gr_A + 3Kr_B)$,

K is the bulk modulus of the solute, G is the shear modulus of the solvent, and r_A and r_B are the atomic radii of the solvent and solute, respectively. This model was quite successful in predicting which element will segregate to the surface of a given binary solid solution alloy, and even yielded reasonable estimates of the heat of segregation for the limited data available at that time [4].

An approach similar to that of Wynblatt and Ku [4,7] was developed later by Miedema [8] in which it was also postulated that the heat of segregation should consist of three parts, in the same spirit as Eq. (4), i.e. a surface tension term, a heat of mixing term, and a strain energy term.

1.1.3 Other Nearest Neighbor Bond Model Approaches - The regular solution model ignores all entropy contributions in excess of the ideal entropy of random mixing. However, experimental studies have shown that the excess entropy of segregation in real solutions can be significant[7]. For real solutions, Eq. (1) should be rewritten as:

$$\frac{X^s}{1-X^s} = \frac{X^b}{1-X^b} exp\left[\frac{-\Delta F_{seg}}{kT}\right], \quad (5)$$

where: $\Delta F_{seg} = \Delta H_{seg} - T\Delta S_{seg}$, and ΔS_{seg} is the excess entropy of segregation. While formal expressions of the excess entropy of segregation can be derived to the same level of detail as the heat of segregation of Eq. (4)[4,7], this entropy can rarely be evaluated fully because of lack of data on the temperature dependence of the surface tension of solids.

Another approach for dealing with the entropy of segregation more completely than in the regular solution model involves the use of Monte Carlo simulation. Sundaram and Wynblatt [9] have used this approach, in conjunction with a nearest neighbor bond model, to show that it is possible to account for deviations from random mixing which occur in solutions with non-zero enthalpies of mixing, while simultaneously avoiding the simplifying assumptions inherent in any analytical form of the excess entropy. In addition, the Monte Carlo method is not limited to the calculation of the composition of a single layer surface, and is able to provide information about the near-surface composition profile of segregated surfaces.

Analytical models which include deviations from random mixing in both the heat and entropy of segregation, through the inclusion of short range order parameters, have been developed by Kumar et al. [10] and by Moran-Lopez and Falicov [11]. The above treatments [10,11], however, have dealt exclusively with chemical effects and have ignored the elastic contribution to the free energy of segregation. While elastic effects can be included in these approaches, in principle, the task is difficult and has not been attempted.

1.2 RECENT APPROACHES

1.2.1 Extensions of the Regular Solution Model - As described in section 1.1.2, the regular solution approach was used within a framework which assumed the existence of a surface phase consisting of a single atom layer. While regular solution multilayer surface treatments of equilibrium composition were developed earlier, for example by Williams and Nason [12], their treatments could only readily be applied to simple surfaces, such as the {111} and {100} surfaces of fcc solutions. A more general treatment, which is applicable to any {hkl} surface, has been developed by Lee and Aaronson [13]. In this formalism, the atom fraction of solute in the ith atomic layer, X^i (where i = 1 is the topmost surface layer) is given by:

$$\frac{X^i}{1-X^i} = \frac{X^b}{1-X^b} exp\left[\frac{-\Delta H^i_{seg}}{kT}\right], \qquad (6)$$

where ΔH^i_{seg} is the enthalpy change accompanying the exchange of a solute atom from the bulk (i.e. a region where X^i is virtually identical with X^b) with a solvent atom in the ith layer.

In order to write expressions for ΔH^i_{seg}, it is necessary to define an index j which is referenced to the ith atom layer, in both directions. The maximum value of j is denoted by J and represents the farthest layer which still contains nearest neighbors of atoms lying in the ith atom layer. For the case of fcc solutions, ΔH^i_{seg} may be expressed for i ≤ J as:

$$\Delta H^i_{seg} = \frac{1}{8}[\gamma_B a_B^2 - \gamma_A a_A^2] \sum_{j=1}^{J} Z^j$$
$$+ \frac{2\Delta H_m}{ZX^b(1-X^b)}\left[X^b Z - X^i Z^i - \sum_{j=1}^{J} Z^j X^{i+j}\right.$$
$$\left. - \sum_{j=1}^{i-1} Z^j X^{i-j} - 1/2 \sum_{j=i}^{J} Z^j\right] - E^i_{el}, \qquad (7a)$$

and for i > J:

$$\Delta H^i_{seg} = \frac{2\Delta H_m}{ZX^b(1-X^b)}\left[X^b Z - X^i Z^i\right.$$
$$\left. - \sum_{j=1}^{J} Z^j(X^{i+j} + X^{i-j})\right] - E^i_{el} \qquad (7b)$$

Here a_A and a_B are the lattice constants of the pure A and B components, Z^i is the number of nearest neighbors of an atom in the ith layer which also lie in the ith layer (i.e. Z^i is equivalent to Z_ℓ of Eq. (2)) Z^j is the number of neighbors of an atom in the ith layer which lie in the jth layer, and E^i_{el} is the elastic energy associated with a solute atom lying in the ith layer, and may be estimated by procedures described elsewhere [13,14]. This approach, which allows the calculation of layer-by-layer compositions of any {hkl} surface, provides a means of evaluating the anisotropy of surface composition, as will be described in a later section on experimental results.

1.2.2. Other Nearest Neighbor Bond Model Approaches - Extensions of the Lee and Aaronson treatment [13] to account for excess entropy effects has recently been reported by Steigerwald et al. [14]. They employed an earlier formalism [7] to write expressions for the entropy of segregation which parallel the form of Eqs. (7) for the enthalpy. However, in order to evaluate the anisotropy of the entropy of segregation it was necessary to resort to an empirical scheme which detracts somewhat from the predictive capabilities of the approach. An example of the results obtained by this method is shown in Table 1.

Table 1

Calculated Composition Profile for a (211) Surface of a Ni-0.1 a/o Au Alloy at 850°C (after Ref. 14).

Layer No. (i)	X^i (a/o)	ΔH^i_{seg} (eV/atom)	$\Delta S^i_{seg}/k$ (e.u.)
1*	10.98	-0.74	-2.79
2*	1.95	-0.46	-1.72
3*	0.47	-0.24	-0.96
4	0.14	-0.04	-0.13
5	0.11	-0.01	-0.03
6	0.10	0.00	0.00

Improvements have also been reported [15,16] of earlier treatments incorporating short range order parameters [10]. Kumar [15] has shown that the heat of segregation to the topmost surface layer can be written (in a form compatible with Eq. (2)) as:

$$\Delta H_{seg} = (\gamma_B - \gamma_A)\sigma + \frac{\Delta H_m}{ZX^b(1-X^b)}\left[Z_\ell(1-2X^S)\,\alpha_o\right.$$
$$\left. + 2Z_v((1-X^b)\,\beta_o - 1/2) - Z(1-2X^b)\,\alpha\right], \quad (8)$$

where α, α_o and β_o are short range order parameters. An expression for the excess entropy of segregation can also be written in terms of these parameters. For a disordered solution, these short range order parameters are defined to equal unity. Under these conditions, Eq. (8) reduces to Eq. (2) and the excess entropy of segregation vanishes. Kumar et al. [16] have applied a similar approach to a multilayer model of the {120} stepped bcc surface. However, this treatment does not provide relations for surfaces of general orientation, although it can, in principle, be applied to other surface orientations. Also, as in the case of single-layer surfaces, elastic contributions to the enthalpy of segregation have been neglected.

1.2.3. Electron Theory Approaches - There

*Layers lacking complete coordination.

have been a few attempts to predict the surface composition of alloys on the basis of electronic theory. For example, Mukherjee et al. [17] have minimized the free energy of an alloy with respect to surface compositions in a calculation where the internal energy is expressed as a tight-binding Hartree Hamiltonian. The entropy was assumed to have the usual ideal form. Calculations for Cu-Ni alloys give good agreement with experiment. This approach is worthy of further investigation, as its potential for broad applicability to alloys less ideal than Cu-Ni is not known at this time.

Another approach taken by Rosengren and Johansson [18] has demonstrated that for certain alloys, the enthalpy of segregation to a surface may be equated to the surface core-level binding energy shift (SCS) of the solvent atom. The SCS is the difference in core level binding energy of a surface atom and that of a bulk atom. This approach is restricted to binary alloys whose components are adjacent in the periodic table, such that the solvent has atomic number Z and the solute has atomic number Z + 1. Although this is a very stringent restriction, it provides a possible alternative method for measuring surface segregation enthalpies, which may be useful in certain alloy systems.

1.2.4. Approaches Based on Interatomic Potentials - The description of metallic crystals by interatomic potentials can provide a powerful tool for improving nearest neighbor bond model treatments of solid state phenomena. The advantages of the interatomic potential approach over the nearest neighbor bond model include the ability to account for interactions beyond the first neighbor shell, the possibility of allowing for local distortions of the crystal lattice in the vicinity of "defects" (such as impurity atoms, surfaces, etc.) and a capability for calculating the total energy of a solid solution, without the need for explicit separation of the energy into chemical and elastic effects.

Until recently, interatomic potentials had not been widely applied to the estimation of the surface composition of alloys, because of the failure of most pair potentials to produce correct estimates of the surface energies which enter explicitly into the surface heat of segregation. However, this problem seems to have been resolved by the recent development of two interatomic potential schemes with improved descriptions of surface properties. These are a set of empirical interatomic potentials for fcc metals proposed by Baskes and Melius [19], which include a volume dependent energy term in addition to the usual pairwise interactions, and a group of functions based on the so called "embedded atom method" developed recently by Baskes and Daw [20]. The former have been used by Wynblatt and Steigerwald [21] to model the anisotropy of surface segregation, and the latter are being used at this time for similar calculations [22]. By combining the approach of interatomic potentials with either Monte Carlo or molecular dynamics modeling schemes, it is possible to account for both entropic as well as enthalpic contributions to the free energy of segregation.

1.3. CONCLUDING REMARKS ON THEORY - In summary, there has been continued and sustained effort, over the past decade or so, to improve previous models and develop new approaches for the description of surface segregation phenomena. This body of work has advanced the state of understanding considerably, but as will be seen in the following section, significant challenges remain in the detailed interpretation of experimental results.

2. EXPERIMENTS

2.1 EARLY EXPERIMENTS - Reliable measurements of surface composition have only been possible since the 1960's, when ultra-high vacuum (UHV) capabilities (i.e. access to pressures of the order of 10^{-10} torr) became commercially available. Prior to that time, the means of preparing and maintaining clean surfaces were limited to a handful of laboratories, using home-built experimental systems. With the emergence of routine UHV capabilities, a large number of spectroscopies for the chemical characterization of surfaces were rapidly developed, and many of these were soon applied to the measurement of surface composition in alloys as well as other materials. The spectroscopies which have been used most widely in this context include Auger electron spectroscopy (AES), low energy ion scattering (LEIS), x-ray photoelectron spectroscopy (XPS) and atom-probe field ion microscopy (AP-FIM).

Most of the early measurements on alloys were confined to qualitative determinations of the identity of the segregating component in binary alloys, and insufficient attention was paid to the achievement of near-equilibrium surface compositions. Even in experiments where the importance of equilibration was recognized, experimenters frequently chose to confine their measurements to single temperatures.

The previous section on theory has shown that, at the present time, the enthalpy of segregation is the only quantity that can be reliably predicted from models of surface segregation. Without a knowledge of the associated entropy of segregation, the equilibrium surface composition of an alloy at a given temperature cannot be estimated, and only the temperature dependence of the equilibrium surface composition is given to prediction with any confidence (see Eq. 5). Thus, much of the early data on the surface composition of alloys did not provide an adequate base on which to test the validity of different modeling approaches, except for crude qualitative comparisons.

While it is not the intent here to provide a comprehensive compilation of experimental results, some examples of both early and recent measurements of the enthalpy of segregation in binary alloys are listed in Table 2.

2.2 RECENT EXPERIMENTAL ADVANCES - More interesting recent experiments have addressed a number of issues that go beyond the mere measurement of average equilibrium surface composition. These include assessments of the effects of surface orientation (i.e. anisotropy) on surface composition, the measurement of variations in surface composition between the edges of a terrace (i.e. steps) and the center of terraces, the determination of the near-surface composition profile, the estimation of the heat of segregation by means of measurements of so called surface core-level binding energy shifts, and the evaluation of the effects of particle size on the surface composition of small alloy particles. The following sections provide additional details on these approaches.

2.2.1. Anisotropy of Surface Composition - Most models of the surface segregation phenomenon indicate that the orientation of the surface will affect its segregation behavior. This may be seen clearly from expressions for the heat of segregation, such as Eq. (7), where the parameters Z^i, Z^j and J are all orientation dependent. A number of experiments have addressed the anisotropy of equilibrium surface composition [14,32,33], but only Steigerwald et al.[14] have measured the heat of segregation over a broad range of surface orientations. Their experiments on a Ni-0.1 a/o Au alloy have indicated the existence of a large degree of anisotropy, with heats of segregation ranging over a factor of three.

These measurements were performed on a polycrystalline foil. The foil was equilibrated in a scanning Auger microprobe, and the surface compositions of 38 different grains were then determined in a scanning electron microscope by means of selected area electron channeling patterns. The variations of heat of segregation and surface composition at 850 °C, over the standard stereographic triangle, are illustrated in Figs. 1 and 2. Attempts to model these results by three different methods, including a nearest neighbor bond model[14] and two approaches based on interatomic potentials[34], all failed to produce the observed pattern of anisotropy shown in Figs. 1 and 2. One may conclude from these results that the effects of anisotropy on surface composition may be too subtle to be reflected by currently available models of surface composition.

2.2.2. Variation of Surface Composition across Surface Structural Features - While the above mentioned studies of surface segregation anisotropy represent attempts to relate surface composition to surface structure, they do so only in an average way, in the sense that the composition of all the structural features of a high index orientation, such as terraces, steps and kinks, are ultimately represented by a single value of surface composition. In contrast, the technique of atom-probe field ion microscopy allows the determination of the composition associated with each specific surface feature. The feasibility of such studies was demonstrated by Tsong et al.[35].

In these studies the variation in composition from the edge of a terrace (i.e. a step) to its center have been measured in Ni-3.4 a/o Cu alloy. The results have shown, for example, that annealing in nitrogen leads to a copper depletion at the very edge of a {111} plane. Near the plane edge is a highly copper enriched region, whereas the center of the plane is moderately enriched in copper. While a number of models[13,21] are capable of predicting the variation of the heat of segregation from the edge to the center of a terrace, a test of these models would require measurements of the temperature dependence of the composition variation, something that has not yet been accomplished.

2.2.3. Near-Surface Equilibrium Composition Profiles - Two principal approaches have been used to probe near-surface equilibrium composition profiles. The first approach has taken advantage of the variation in depth resolution associated with Auger transitions of different energies. The most recent application of this method is due to King and Donnelly[33], who extracted the compositions of the two topmost layers in (100) (110) and (111) surfaces of several Ag-Au alloys. Reasonable agreement was found between the experimental results and a Monte

Table 2
Selected Values of the Enthalpy of Surface Segregation in Binary Alloys

Alloy Solvent(Solute)	Experimental Method	Sample Type	ΔH_{seg} for solute (kJ/mole)	Reference
Au(Ni)	AES	Polycrystal	-4	7
Fe(C)	AES	(100)crystal	-84	23
Fe(Sn)	AES	Polycrystal	-46	24
Ni(Au)	AES	(111)crystal	-50	25
	AES	Polycrystal	-63	7
	AES	Polycrystal	-94	14
Ni(C)	AES	(100)crystal	-46	26
Ni(Fe)	AES	Polycrystal	0	27
Ni(Pd)	AES	Polycrystal	-29	28
Pt(Rh)	AES	Polycrystal	0	29
Rh(Pt)	LEIS	Polycrystal	-21	30
Sn(Pb)	LEIS	Polycrystal	-21	32

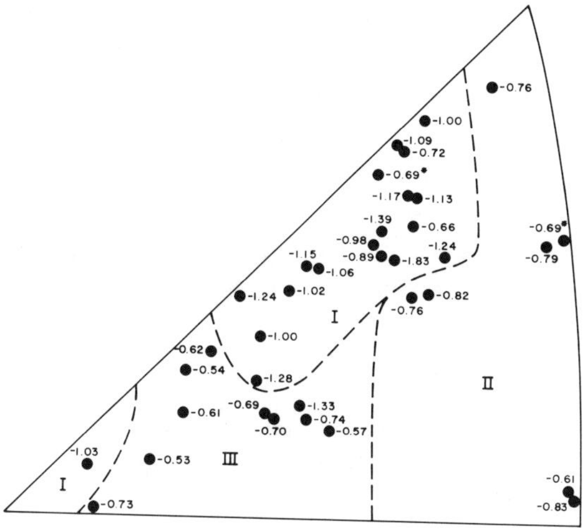

Figure 1. Stereographic triangle showing experimentally determined heats of segregation (eV). Regions marked I, II and III indicate orientations having high, moderate and low (absolute values) of the heat of segregation, respectively. (After Ref. 14)

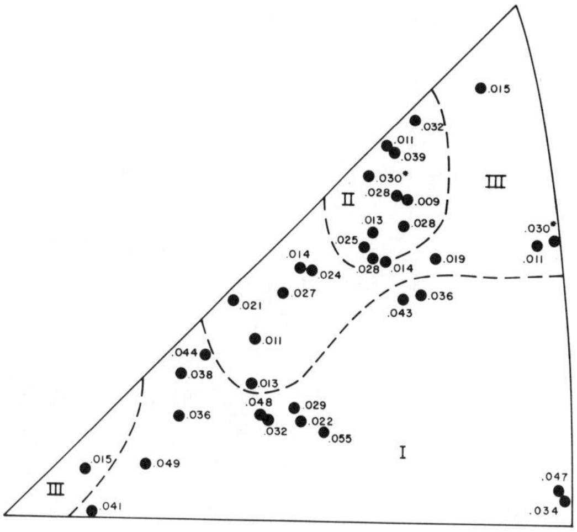

Figure 2. Stereographic triangle showing experimentally determined surface atom fractions of Au at 850°C. Regions marked I, II and III indicate orientations having high, moderate and low surface compositions, respectively. (After Ref. 14)

Carlo model. However, the deconvolution of the Auger spectra used to obtain composition profiles introduces a fair amount of inaccuracy into the layer-by-layer compositions.

The other approach is based on atom-probe field ion microscopy, and is potentially more precise. In addition, it can yield the composition of any number of near-surface atom layers. This method has been used quite effectively by Tsong and co-workers to study Cu-Ni[36] and Pt-Au[37] alloys. In the latter, significant gold enrichment over the bulk composition was found in the four topmost surface layers of a (100) surface. Unfortunately, no determination was made of the temperature dependence of the near-surface composition profile, thus limiting possible comparisons with models.

2.2.4. Estimations of Enthalpies of Segregation from Surface Core-level Binding Energy Shifts - The prediction that enthalpies of surface segregation can be approximated by surface core-level binding energy shifts [18] has been discussed in section 1.2.3. This approach has been used recently by Egelhoff[38] in measurements of the SCS of Ni, which should correspond according to theory to the enthalpy of segregation of Cu to a Ni surface.

The measurement was performed on a Ni(100) surface by monitoring the binding energy of the Ni $2p_{3/2}$ XPS peak. The binding energy of this peak was first determined on the clean surface to yield contributions from both surface and bulk Ni atoms. Next, the binding energy was redetermined after depositing a four-monolayer overlayer of various metals (Cr, Mn, Fe, Co) on the nickel surface; this is expected to eliminate contributions from nickel surface atoms, and yields the bulk binding energy. The surface binding energy is then extracted from the difference between these two measurements. While the binding energy shift was affected by the nature of the overlayer, the discrepancy only amounted to 10% of the average measured shift of -0.46 eV.

This experimental approach is no easier to execute than the conventional approach of measuring the temperature dependence of surface composition, and in view of the limited data available at this time, it is too early to assess how reliable it might be when tested in other alloys. However, this approach provides an alternative to the conventional experimental method in certain alloys, and may yield other useful thermochemical data for alloy surfaces[38].

2.2.5. Effect of Particle Size on the Surface Composition of Small Particles - Most of the supported metal catalysts used for chemical processing and automotive emission control systems consist of small metal particles supported on metal-oxide substrates. In view of the increasing use of alloys in the metallic phase of these catalysts, and of the importance of surface composition for catalytic performance, it is surprising that little experimental work has been performed on the surface composition of small particles. Groomes and Wynblatt[39] have recently addressed the issue of the particle size dependence of surface composition in a Ni-1 a/o Pd alloy. They found a five-fold change in the Pd surface atom fraction over a range of particle diameters from 250 to 1000Å, as illustrated in Fig. 3. The effects of particle size on surface composition were modeled within a mass conservation framework. This modeling approach predicts correct qualitative trends, but does not account in detail for the observed experimental behavior as shown in Fig. 4. This lack of detailed agreement has been postulated to arise from neglect of capillarity effects in the modeling approach.

2.3 CONCLUDING REMARKS - The experimental work reviewed in this section indicates interesting recent directions in research aimed at developing a better fundamental understanding of the equilibrium surface composition of alloys. It also points to several areas where improvements in theory are required for both the rationalization of existing experimental results and the development of improved predictive capabilities. Thus, while significant progress has been made both theoretically and experimentally in the understanding of equilibrium surface segregation phenomena, important challenges still remain to be addressed.

Acknowledgement

The authors acknowledge with thanks support for this work by the National Science Foundation under grant DMR 8420616.

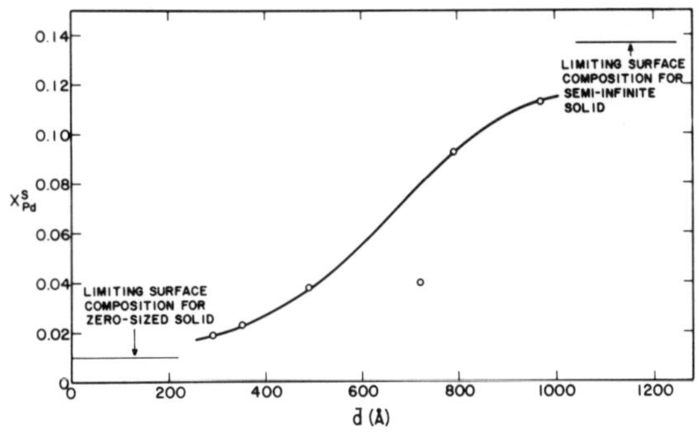

Figure 3. Summary of experimental results showing surface atom fraction of palladium, X^S_{Pd}, plotted versus average particle diameter, d̄. The figure also gives the average palladium atom fraction, labelled: "limiting surface composition for zero-sized solid" and the surface composition of a Thick foil, labelled: "limiting surface composition for semi-infinite solid". (After Ref. 39)

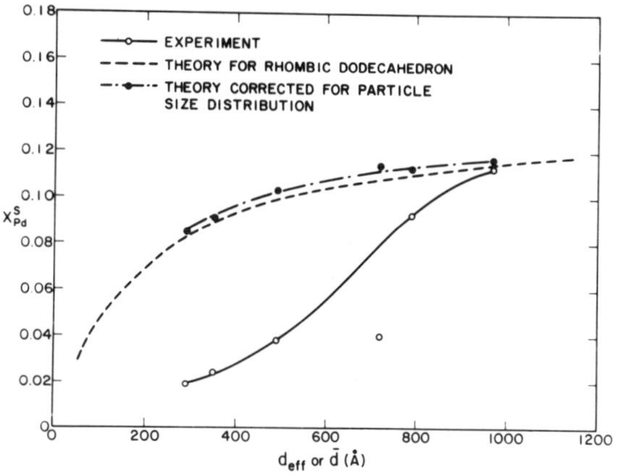

Figure 4. Comparison of experimental results (solid line) with calculated results (dashed line), and with calculated results corrected for the experimental particle size distributions (dot-dash line). (After Ref. 39)

References

1. J.W. Gibbs, The Scientific Papers of J. Williard Gibbs, vol. 1 (Dover, NY 1961).
2. R. Defay, I. Prigogine, A. Bellmans and D.H. Everett, Surface Tension and Adsorption (Wiley, NY 1966).
3. D. McLean, Grain Boundaries in Metals (Oxford Press, London, 1957).
4. P. Wynblatt and R.C. Ku, Interfacial Segregation, Eds. W.C. Johnson and J.M. Blakely (ASM, Metals Park, OH, 1979) p. 115.
5. J. Friedel, Advan. Phys. 3, 446 (1954).
6. J.J. Burton and E.S. Machlin, Phys. Rev. Lett., 37, 1433 (1976).
7. P. Wynblatt and R.C. Ku, Surf. Sci. 65, 511 (1977).
8. A.R. Miedema, Z. Metallk. 69, 459 (1978).
9. V.S. Sundaram and P. Wynblatt, Surf. Sci. 52, 569 (1975).
10. V. Kumar, D. Kumar and S.K. Joshi, Phys. Rev. B 19, 1954 (1979).
11. J.L. Moran-Lopez and L.M. Falicov, Phys. Rev. B 18, 2542 and 2549 (1978).
12. F.L. Williams and D. Nason, Surf. Sci. 45, 377 (1974).
13. Y.W. Lee and H.I. Aaronson, Surf. Sci. 95, 277 (1980) and Acta Met. 28 539 (1980).
14. D.A. Steigerwald, S.J. Miller and P. Wynblatt, Surf. Sci. 155, 79 (1985).
15. V. Kumar, Phys. Rev. B 23, 3756 (1981).
16. V. Kumar, C.E.T. Goncalves da Silva and J.L. Moran-Lopez, Phys. Rev. B 23, 2752 (1981).
17. S. Mukherjee, J.L. Moran-Lopez, V. Kumar and K.H. Bennemann, Phys. Rev. B 25, 730 (1982).
18. A. Rosengren and B. Johansson, Phys. Rev. B 23, 3852 (1981).
19. M.I. Baskes and C.F. Melius, Phys. Rev. B 20, 3197 (1979).
20. M.S. Daw and M.I. Baskes, Phys. Rev. Lett. 50, 1285 (1983); Phys. Rev. B 29, 6443 (1984).
21. P. Wynblatt and D.A. Steigerwald, Surf. Sci. 150, 289 (1985).
22. S.M. Foiles, private communication (to be published: Phys. Rev. B).
23. H.J. Grabbke, G. Tauber and H. Viefhaus, Scripta Met. 9, 1181 (1975).
24. M.P. Seah and C. Lea, Phil. Mag. 31, 627 (1975).
25. J.J. Burton, C.R. Helms and R.S. Polizzotti, J. Vac. Sci. Technol. 13, 209 (1976).
26. L.C. Isett and J.M. Blakely, Surf. Sci. 58, 397 (1976).
27. J.J. Burton and R.S. Polizzotti, Surf. Sci. 66, 1 (1977).
28. D.A. Mervyn, R.J. Baird and P. Wynblatt, Surf. Sci. 82, 79 (1979).
29. W.B. Williamson, H.S. Gandhi, P. Wynblatt, T.J. Truex and R.C. Ku, AIChE Symposium Series No. 201, 76, 212 (1980).

30. F.L. Williams and G.C. Nelson, Appl. Surf. Sci. 3, 409 (1979).
31. G.C. Nelson and G.A. Borders, J. Vac. Sci. Technol. 20, 939 (1982).
32. K. Wandelt and C.R. Brundle, Phys. Rev. Lett. 46, 1529 (1981).
33. T.S. King and R.G. Donnelly, Surf. Sci. 151, 374 (1985).
34. D.A. Steigerwald, Ph.D. Thesis, Carnegie-Mellon University (1985).
35. T.T. Tsong, Y.S. Ng and S.B. McLane, J.A.P. 51, 6189 (1980).
36. Y.S. Ng, T.T. Tsong and S.B. McLane, Surf. Sci. 84, 31 (1979).
37. T.T. Tsong, Y.S. Ng and S.B. McLane, J. Chem. Phys. 73, 1464 (1980).
38. W.F. Egelhoff, Phys. Rev. Lett. 50, 587 (1983).
39. D.O. Groomes and P. Wynblatt, Surf. Sci. 160, 475 (1985).

SEGREGATION AND ORDERING ON GOLD ALLOY SURFACES STUDIED BY LOW ENERGY ION SCATTERING AND LOW ENERGY ELECTRON DIFFRACTION

T. M. Buck, E. G. McRae
AT&T Bell Laboratories
Murray Hill, New Jersey, USA

ABSTRACT

This review describes surface segregation of gold in two alloy systems which exhibit very different surface, and also bulk, behavior. In Cu-Au alloys the differences in bond energies, a major driving force for segregation, is small. There is also a negative enthalpy of mixing which favors mixing and ordering and suppresses segregation. On the other hand, the Ni-Au system provides a rather extreme case for surface segregation having a positive enthalpy of mixing and a large difference in bond energies. The methods used to study these surfaces were low energy ion scattering (LEIS) and low energy electron diffraction (LEED). LEIS has outstanding surface selectivity and is sensitive to short-range structure, while LEED is sensitive to longer-range structure. On $Cu_3Au(001)$ there is a c(2x2) ordered structure with a 50-50 Au-Cu first layer below T_c = 390°C. The surface composition (of layers 1 and 2) and ordering were studied over temperature ranges above and below T_c. On Ni(110)-0.8% Au-segregated 1x1 surfaces with Au in substitutional sites were observed at low coverage while close-packed overlayers with 7x7 periodicity formed at the saturation coverage of ~1 ML.

THE TERM SURFACE SEGREGATION appearing in the title of this session has been applied in recent years to any enrichment of the surface of a solid by one of its component elements. The detection and study of such effects has become possible only in fairly recent times with the advent of modern surface analytical techniques and UHV systems. The process by which the surface enrichment occurs is, in reported cases, usually the migration of that element of a metal alloy which reduces the surface free-energy in ultrahigh vacuum (UHV). However, in the presence of a reactive gas such as oxygen another element with a greater affinity for oxygen may segregate (1,2). Preferential sputtering by ion bombardment is another type of segregation which has been studied extensively (3). Kelly (4) has recently proposed that it involves thermodynamic or Gibbsian segregation. That is, in many if not all cases, preferential sputtering is just bombardment-induced Gibbsian segregation.

In the most numerous published examples of surface segregation a metal segregates from an alloy, but non-metallic impurities such as carbon and sulfur are also known to segregate from metals (5) upon heating, and segregation in semiconductor materials has been reported recently (6,7). In its various manifestations surface segregation has important implications in corrosion (2), catalysis (8), microelectronics, (6,7) and, as suggested by the session title, in self-renewing coatings.

In the present paper we will confine the discussion to equilibrium surface segregation and the related phenomena of ordering and other phase transitions on single crystal surfaces of binary metal alloys at elevated temperature in UHV. We will further concentrate on two particular alloys of gold, Cu_3Au and Ni −0.8% Au which illustrate rather strikingly the major thermodynamic influences on both segregation and ordering. Particular attention will be paid to the composition and atomic arrangement or structure of the (001) and (110) surfaces of these alloys as studied by low energy ion scattering (LEIS) and low energy electron diffraction (LEED). Heretofore there has been little study of the structure of segregated surfaces, although its importance has been recognized (5). In the absence of surface structure information, theories of surface segregation (9-12) are largely limited to models in which the segregating atoms simply replace host atoms at substitutional sites. However, as will be shown, segregation is not always substitutional.

DRIVING FORCES FOR SURFACE SEGREGATION AND ORDERING

The present understanding of equilibrium surface segregation theory is reviewed by Wynblatt and Steigerwald in these proceedings (13). Therefore, in this section the discussion of theory will be limited to those aspects which appear to be especially relevant to segregation in the gold alloys under consideration.

The Regular Solution Theory is a convenient starting point in thinking about surface segregation. Although it is inadequate in a number of cases, either qualitatively or quantitatively, it is based on concepts which are important and which appear in more complete formulations. The segregation phenomenon occurs in order to reduce surface free-energy, when the temperature is high enough to permit the migration of the segregating element to the surface. The free-energy difference between the two states, one in which $X_A^s = X_A^b$ before segregation, and the other in which $X_A^s > X_A^b$ after segregation, where X_A^s and X_A^b are atom fractions of element A at the surface and in the bulk, is given by

$$\Delta F_s = \Delta H_s - T\Delta S_s \qquad (1)$$

The enthalpy term, containing bond energies, favors placing atoms with weaker bonds at the surface while the entropy term $-T\Delta S_s$ opposes the ordering effect of segregation so that, in general, the coverage is not complete, although it is very nearly so in some cases. The pure element bond energies thus give a strong clue as to which element will segregate, while the enthalpy or heat of mixing, positive or negative, is an indicator of the degree of segregation. The so-called regular solution parameter Ω, as defined in

$$H_{AB} = \Omega + \frac{1}{2}(H_{AA}+H_{BB}) \qquad (2)$$

is proportional to the enthalpy of mixing. The H's are the enthalpies, negative for an attractive interaction, of the A-B, A-A, and B-B bonds, and are usually derived from experimental enthalpies of mixing and vaporization (9,14). For an ideal solution $\Omega = 0$ by definition, so that H_{AB} is the arithmetic mean of the elemental bond enthalpies. However, for a regular solution $\Omega \neq 0$ and may be positive or negative. If $\Omega > 0$ demixing is favored, a miscibility gap gap appears in the bulk phase diagram, and surface segregation of the weaker bonding element is greater than for the ideal solution. Conversely, for $\Omega < 0$ mixing is favored; A-B bonds are preferred over A-A or B-B, and long-range ordering may occur in the bulk for compositions with small integral ratios of the components. Segregation is expected to be weaker than for an ideal solution since it reduces the number of A-B interactions. This preference for A-B bonds should also lead to depletion of the segregating element in the second layer (9).

In a regular solution the entropy of mixing, or configurational entropy, is that of an ideal solution:

$$\Delta S_m = -R(X_A \ln X_A + X_B \ln X_B) \qquad (3)$$

in which R is the gas constant. This simplification, which assumes random mixing, is called the "zeroth approximation." A more realistic approach which recognizes that $\Omega \neq 0$ implies preferential association of either like or unlike atoms is called the "first-order" (15) or "quasi-chemical" (16,17) approximation. This refinement made little difference in the Ni-Cu case (18) but might be significant in other cases.

The surface composition is calculated by minimizing free-energy with respect to surface-layer composition taking into account the difference in coordination number between an atom in the bulk and one at the surface. In the zeroth-order approximation, the equilibrium surface composition is given (18,9,19) by

$$\ln\left(\frac{1-X_s}{X_s}\right) = \frac{|\Delta H_{v1}-\Delta H_{v2}|m}{RT} - \qquad (4)$$

$$-\frac{2\Omega}{RT}\left[\frac{1}{2}mZ-(m+\ell)ZX_b + \ell ZX_s\right] + \ln\frac{1-X_b}{X_b}$$

in which X_s and X_b are the surface and bulk concentrations of the element with the higher heat of vaporization, $|\Delta H_{v1} - \Delta H_{v2}|$ is the absolute difference between the heats of vaporization of the pure elements, m is the fraction of nearest neighbors of an atom lying in a plane above or below it, ℓ is the fraction in the plane, and Z the coordination number in the bulk.

Certain failures of regular solution theory in qualitative and quantitative predictions for particular alloys pointed to a deficiency in this approach: it did not take account of atom size differences which might contribute a strain energy driving force for segregation to the surface such as McLean (20) had used to explain grain boundary segregation. Several authors have combined the driving forces of lattice strain and bond energies in various ways finding better agreement with experiment. Wynblatt and Ku (11) used bond energies derived from bulk thermodynamic properties and solute strain energies based on an expression of Friedel (21). Abraham et al. (22,23) also combined the bond energy and lattice strain driving forces, to map combinations of bond strength ratio and atom size ratio where solvent or solute segregation should occur. However, they did not use bulk thermodynamic properties but rather employed a Monte Carlo method and a Lennard-Jones potential. The model does not assume that elastic strain energy is eliminated for a solute atom at the surface and it finds a larger driving force for segregation of large solute atoms than for small atoms, since the lattice distortion is less pronounced for undersize atoms. Comparing with experimental results, predictions of this model were 85% successful. Miedema (10) has used another approach to the problem of combining bond energy and atom size effects which predicts which element will segregate quite successfully and also predicts ratios of surface to bulk concentration. The surface segregation predictions are derived from semi-empirical calculations of surface energies of the pure metals and bulk heats of mixing which in turn are based on considerations of electron density and electronegativity, or chemical potential of electrons, at the boundaries of Wigner-Seitz atomic cells. Surface energies are adjusted for atomic area $V_{at.}^{2/3}$, and size mismatch strain energy is taken into account but is not given much weight. Miedema provided a convenient table of predictions which Hamilton (24) and Chelikowsky (25) have extended considerably, making additional comparisons with experiment and with predictions based on surface enthalpy alone (25). These tables and those of the other authors mentioned provide reasonably reliable predictions which are useful in practical situations, especially in the absence of experimental results. However, it must be admitted that a few recent results are very surprising and puzzling, in particular those of Sakurai et al. (26) who observed surface segregation of Ni at the Cu-rich end of the Ni-Cu phase diagram, i.e., $X_b^{Cu} \geqslant 0.85$. This is definitely not in accord with expectations. As is evident from Table I the difference in bond energies, assumed proportional to ΔH_v, clearly favors Cu segregation and the difference in atomic radii is too small to account for the reversal. This and a few other conceivable explanations were considered and discarded by Sakurai et al. (26).

Frankenthal and Siconolfi (27) have reported complicated behavior in the Pb-Sn system with Pb segregating in a normal manner from an Sn-rich alloy, but with the Pb-rich alloy exhibiting surface enrichment of Sn below 170°C and depletion above. To explain this they treated this alloy as a non-ideal (and non-regular) solution and, noting that Sn and Pb have different crystal structures, considered crystallographic transformation energy as well as bond-energy and strain energy.

An interesting contrast with the more common segregation behavior is observed in the case of an ordering alloy, i.e. one in which the A and B atoms occupy preferred lattice sites in order to maximize the number of nearest neighbors of the opposite species. Long-range order-disorder behavior of atoms in the bulk has been known and studied by x-ray diffraction, thermochemistry, and electrical conductivity measurements for many years (17,29,30), but relatively little has been done until recently on long-range order of alloy surfaces. Ordering and surface segregation have been investigated theoretically by Moran-Lopez et al. (31,32) and others more recently (see below). In general, there is expected to be competition between the ordering and surface segregation effects. The same A-B bonding preference manifested by the negative enthalpy of mixing which drives bulk ordering also acts to oppose surface segregation which would reduce the number of unlike nearest neighbors.

The understanding of surface segregation is not yet complete, as illustrated by the exceptions cited above to the generally accepted rules. However, the thermodynamic data and atomic radii for Au, Ni, Cu

Table I - Thermodynamic Values and Atomic Radii for Some Gold Alloys

Element	ΔH_v(14) [Kcal/Mol] 298°K	$r_{at.}$ [Å] (28)	Integral Heats of Mixing (14) $0.5A + 0.5B = A_{0.5}B_{0.5}$		
			Alloy	Temp (K)	ΔH[Cal/g.at.]
Cu	80.5	1.28	Au Cu	800	−1221
Ni	102.8	1.25	Au Ni	1150	+1807
Au	88	1.44	Cu Ni	973	+425
Ag	67.9	1.44	Ag Au	800	−1111

and their alloys given in Table 1 permit some fairly confident qualitative predictions regarding the alloys of particular interest here. For the Ni-Au system, Au is clearly favored to segregate, assuming that elemental bond energies are proportional to heats of vaporization. The atom size difference will also favor Au segregation, at least in a Ni-rich alloy. Furthermore, the large positive enthalpy of mixing should favor de-mixing and surface segregation, and indeed several studies have shown very strong segregation of Au from Au-Ni alloys (11,33-38). A disagreement between AES and LEIS results have appeared with regard to the Au-rich alloys: Wynblatt and Ku (11) reported Ni segregation from dilute Ni in Au alloys while three groups using LEIS (34,37) or LEIS (TOF) (38) found Au enrichment.

The size difference between Ni and Au raises an interesting question: can the Au atoms segregating from a dilute solution in Ni occupy substitutional Ni sites at the surface, especially at high coverage? Burton et al. (36) obtained (6x6) and (13x13) hexagonal LEED patterns from a Ni (111)-Au segregated surface, indicating overlayer structures with the same symmetry and orientation as the Ni (111) surface but with larger unit cell size, which they explained by a simple atom size mismatch argument. LEIS (TOF) and LEED results on Ni (110) Au and Ni (001) Au (33) segregated surfaces to be described below show that the atom size influence is felt on these surfaces also. At close to one monolayer the Au atoms cannot fit into a 1x1 substitutional arrangement.

The data for the Au-Cu system in Table 1 shows only a small difference in ΔH_v values, which slightly favors Cu segregation. However, the atom size difference would favor the larger Au atom (22,23) while the negative enthalpy of mixing should weaken the segregation drive for either element. In fact, relatively weak Au segregation was found by AES on polycrystalline films (39) and on a single-crystal Cu_3Au (111) surface (40) and attributed (10) to surface energy difference adjusted for atom size. LEIS studies (41) also showed Au segregation from $Au_{0.1}Cu_{0.9}$ and $Au_{0.9}Cu_{0.1}$ single crystal surfaces (111), (110) and (001). There was no significant dependence on surface orientation. Studies of individual layer composition and ordering on Cu_3Au surfaces by LEIS and LEED will be described below.

In the Au-Ag system, segregation of Ag is favored on the basis of the heat of vaporization. Atom sizes are equal, and some degree of ordering is favored by the negative heat of mixing. In fact, Ag segregation has been found by AES (42) and LEIS (44), and surmised from the rounding of Au particles caused by 0.25% Ag impurity addition (43).

SURFACE ANALYSIS TECHNIQUES

The phenomena of surface segregation, ordering, and other surface phase transitions of interest here occur in the first or first few layers of the surface, as is well documented in the literature cited above. Analytical techniques for their study therefore must possess unusually high surface sensitivity. The three techniques which seem to have been used most in segregation studies have been AES, LEIS and the Atom-Probe Field-Ion microscope, with AES the most popular for composition analysis of segregated surfaces. An extensive survey of methods for surface and grain boundary analysis was published by Joshi (45). Comparisons of results of AES and LEIS with Miedema's predictions are included in ref. 46. AES is convenient, widely available, and has good elemental specificity and lateral resolution. Damage by the electron beam should not be significant in alloy segregation studies. On the other hand, quantitative interpretation of AES is relatively complicated and depth resolution not easily defined, although advances in depth profiling have been made recently (47,48). AES gives no information about the lateral interatomic arrangement of the surface. The atom-probe field ion microscope (49,50) has single atom-layer depth resolution over as many as 13 layers (49) and provides quantitative depth profiling, but with some difficulty in accumulation of adequate statistics which, however, has recently been alleviated by an improvement in detection efficiency (50).

LEED The most common application of LEED (51) is in monitoring the long-range order of crystal surfaces. LEED may be used to determine the dimensions of the unit mesh (the surface unit cell) in the case of ordered surface phases and to characterize surface phase transitions.

These applications depend on a very simple, approximate form of diffraction theory called kinematical (single-scattering) theory, similar to that used in bulk-crystal structure determination by x-ray diffraction. LEED may also be used to determine the geometrical arrangement of surface atoms within the unit mesh of ordered surfaces, but this calls for extensive measurements of LEED intensities and their analysis by large-scale computations based on dynamical (multiple-scattering) diffraction theory. The analysis can be prohibitively expensive if the unit mesh contains more than a few (≈ 10) surface atoms. For larger unit-mesh dimensions, an attractive alternative is to use LEED observations interpreted by kinematical theory, in conjunction with another technique sensitive primarily to the short-range correlations of atom positions. The applications of LEED to the surface-segregation structures described in this review were made in conjunction with LEIS.

The following hypothetical two-dimensional examples will serve to introduce the essential ideas underlying these applications of LEED.

Order-Disorder Transitions – Figure 1a shows an ordered alloy AB made up of atoms A and B lying on

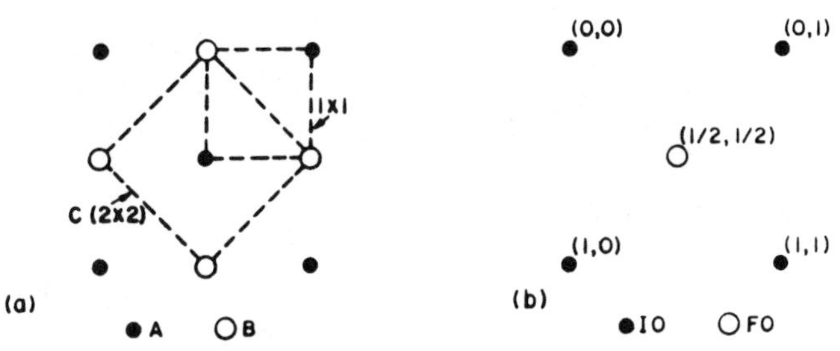

Fig. 1 (a) Hypothetical two-dimensional alloy (b) Schematic LEED pattern showing integer-order (IO) and fractional-order (FO) beams.

sites whose periodicity is represented by a square net.

Following established conventions, the unit meshes of the sites alone and of the entire structure are designated 1x1 and c(2x2), respectively. The LEED pattern (Fig. 1b) consists of integer-order (IO) beams such as (1, 0) and fractional-order (FO) ones such as (1/2, 1/2). Suppose now that the degree of long-range order is reduced by interchanging A and B atoms. A measure of the long-range order is provided by the parameter $M_1 \equiv \langle p \rangle$ where $\langle\ \rangle$ denotes a value averaged over sites, and for each site p can take the value +1 or −1; p = + 1 if an A(B) atom is on an A(B) site, otherwise p = − 1. According to kinematical theory, the intensities of fractional-order LEED beams are proportional to M_1^2. Therefore, within the kinematical approximation, the variation of the long-range order parameter with respect to temperature T may be determined by measuring fractional-order beam intensities. In particular, a continuous order → disorder transition is observed in LEED as a continuous decrease of intensity and eventual disappearance of all fractional-order beams as T increases.

LEED may also be used to study the short-range order. A measure of short-range order is provided by the correlation function

$$P(\mathbf{r}) \equiv \langle p_o p_r \rangle - \langle p_o \rangle^2 \qquad (5)$$

where p_o and p_r denote values of p for a site at the origin and a site at position **r**, respectively. According to kinematical theory, the LEED intensity near a FO beam is proportional to

$$M_1^2\, \delta(\mathbf{q}) + \hat{P}(\mathbf{q})$$

where **q** denotes the surface-parallel momentum transfer relative to the FO beam center, and $\hat{P}(q)$ is the Fourier transform of the correlation function. In the special case where $P(\mathbf{r})$ has the exponential form $\exp(-kr)$, $\hat{P}(\mathbf{q})$ has the Lorentzian form,

$$\hat{P}(\mathbf{q}) = \hat{P}(0)\ k^2/(k^2+q^2)\ . \qquad (6)$$

In this case the width (HWHM) of the Lorentzian component of the FO beam profile is the inverse correlation length k characterizing the short range order.

Geometrical Structure Determination – Figure 2a shows an overlayer structure where adsorbed atoms are arranged on sites forming a square net to give a 1x7 unit mesh. In the LEED pattern (Fig. 2b) the positions of the beams are determined by the 1x7 periodicity, but the distribution of intensity among the fractional order beams depends on the details of the geometrical arrangement of adsorbed atoms.

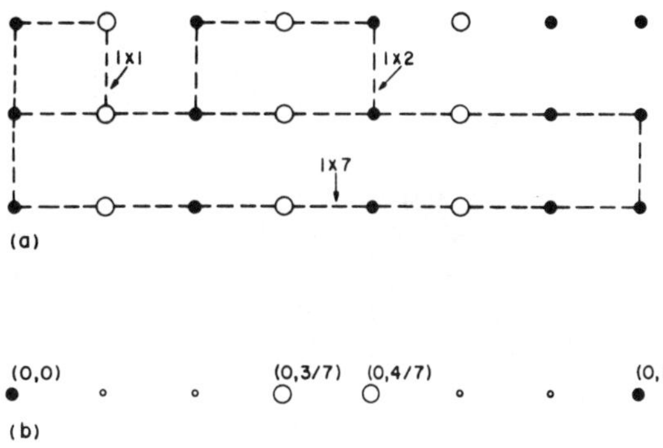

Fig. 2 (a) Hypothetical 1x7 surface structure in which the unit mesh contains 1x2 subunits. (b) Schematic LEED pattern. The sizes of the open circles indicate the relative intensities of fractional order beams.

In general the observed intensities must be analyzed in detail to extract structural information, but there is one fairly common situation where useful conclusions can be drawn merely from inspection of the pattern – namely where the unit mesh contains several identical or nearly identical subunits. In this case the kinematical expression for the distribution of intensity among fractional-order beams contains a modulation factor which peaks at or near the positions where there would be beams if the surface were entirely made up of these subunits. For example, in Fig. 2a the 1x7 unit mesh contains three 1x2 subunits. Therefore, as shown in Fig. 2b, the fractional-order intensity is largest for beams near positions such as (0, 1/2) – e.g. (0, 3/7) and (0, 4/7) beams. The identification of a unique structural subunit reduces the structure-determination problem to one amenable to a short-range probe more convenient than LEED intensity analysis.

LOW-ENERGY ION SCATTERING Low energy ion scattering (LEIS) or ISS is one such short range probe which is complementary to the long-range structure sensitivity of LEED.

The earliest version of LEIS, using noble-gas ions and an electrostatic analyzer, was used early in the study of surface segregation (44,54). The basic concepts and measurements of LEIS are quite simple (Fig. 3).

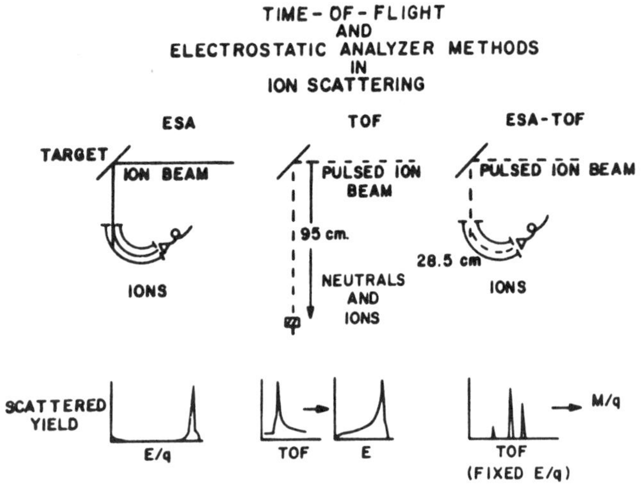

Fig. 3 Schematic diagrams of electrostatic analyzer (ESA) and time-of-flight (TOF) techniques for low-energy ion scattering. Typical spectra below. The ESA-TOF method alloys m/q identification of secondary ions removed by the beam.

A monoenergetic, well-collimated beam of ions with energy of ~0.2-10 keV strikes the target surface; most of the ions penetrate into the bulk, but a small fraction are reflected back, being repelled by the screened Coulomb forces of individual atoms in close encounters near the surface. The energy distribution of particles scattered off at some particular angle provides information on the mass, or chemical identity, and the number of surface atoms through the energy and magnitude, respectively, of peaks in the spectrum. In the case of single-crystal surfaces, structure information can also be derived, from the angular dependence of scattered yield. Detailed information about LEIS or ISS generally, or about the several versions of the technique may be found in reviews (55-57) and other sources: LEIS (TOF) (46,58), ICISS (59), ALICISS (60) and blocking-cone display (61).

The time-of-flight version, LEIS (TOF), differs from noble-gas electrostatic energy analysis in that it avoids uncertainties due to neutralization in scattering since both the neutrals and ions are collected. Differences in neutralization due to different target atoms or to penetration depths or multiple scattering should not interfere with quantitative interpretation as does occur in electrostatic energy analysis of noble-gas ions. However, other factors which might cause problems are: a) the accuracy of scattering cross-sections employed b) enhancement of scattering yield from sub-surface layers by "focusing" effects c) dependence of detector efficiency on scattered energy or charge state and d) background subtraction in the analysis of single scattering (SS) peaks; i.e., does the peak area or height include only SS events for which scattering cross-sections are presumably known, or does it also include multiple scattering for which cross-sections are less well defined? These questions have been investigated and dealt with in a previous paper (62) and the analysis presented here makes use of those results.

The technique for composition analysis of individual layers depends essentially on orienting the sample so that atoms in upper layers shadow those in lower layers from incoming ions, or block the ions (or neutrals) scattered toward the detector from deeper atoms. Thus, one scatters ions from atoms in the first layer only, while they shadow all the others. For an FCC (001) surface this would require orienting the scattering plane parallel to the [001] azimuth and using polar incidence angle (measured from the surface) $\psi = 45°$, which is equivalent to channeling along a <110> axis going in, and also blocking along a <110> axis on the way out if $\theta_L = 90°$. One then turns the crystal to an angle which exposes both the first and second layers to the ion beam for a second spectrum at equal ion dose. On the FCC (001) surface this would be the [110] azimuth at $\psi = 35°$, again a channeling direction, the <111> axis. Atoms of the first two layers shadow the deeper layers and the exit direction along <112> is a blocking direction. The first measurement gives the first layer composition and the difference between the two measurements gives the second layer composition after correction for any focusing effect i.e., concentration of ion flux on second layer atoms between shadow cones of first layer atoms. Correction should also be made for different incidence angles, i.e. for different surface area and therefore different number of atoms exposed to the beam at different ψ.

A few expressions used in layer composition analysis follow. For a single-crystal binary alloy $A_xB_{(1-x)}$ with an unknown surface composition, the first layer A/B ratio is given by

$$N_1^A/N_1^B = (Y_1^A/Y_1^B)\ (\sigma_B/\sigma_A)/d \qquad (7)$$

in which $N_1^{A(B)}$ is the number of first layer A or B atoms exposed to the beam, $Y_1^{A(B)}$ is the yield of singly scattered ions and neutrals (sum of counts in the TOF peak), σ is the differential scattering cross-section derived from the Molière approximation to the Thomas-Fermi potential, and d corrects for the fall-off in detector sensitivity with energy between the A and B peaks. If no foreign atoms or vacancies are present, the concentration of A in the first layer is then

$$X_1^A = (N_1^A/N_1^B)/(1 + N_1^A/N_1^B) \qquad (8)$$

The second layer composition is given by

$$X_2^A = F\left\{\frac{(Y_{1+2}^B - Y_1^B)}{(Y_{1+2}^A - Y_1^A)}\ \frac{Y_1^A X_1^B}{Y_1^B X_1^A} + F\right\}^{-1} \qquad (9)$$

in which $F \equiv f_B/f_A$ and

$$f_B = \left(\frac{Y^B_{l+1} - Y^B_l}{Y^B_l} \; \frac{X^B_l}{X^B_2} \right)$$

is the focusing factor for B atoms in the 2nd layer, i.e., the factor by which the scattering yield from them is enhanced by focusing. A similar relation holds for element A. F = 1 was assumed in the Cu_3Au (001) analysis (see below) without serious error, since unequal focusing on Au and Cu in the second layer, i.e., $f_{Cu} \neq f_{Au}$, is most likely to occur when the surface is ordered, but that is just the condition in which there is very little Au in the second layer anyway. In a case for which this simplification is not valid, computer simulation could be used to calculate unequal f's. The incidence angle normalization is made by multiplying the [110]35° yield by sin35/sin45 = 0.81.

Cu_3Au

Cu_3Au is a classic ordering alloy (29). It undergoes a bulk order-disorder transition with critical temperature T_c equal to 663K (52a). In its ordered state ($T << T_c$) the crystal has the atoms on fcc sites to give the $L1_2$ structure in which alternate (001) layers have the compositions 50-50% CuAu and 100% Cu as shown in Fig. 4.

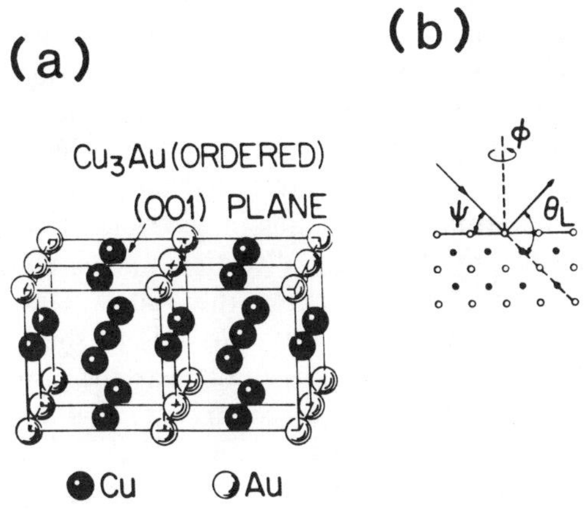

Fig. 4 Arrangement of Au and Cu in ordered bulk Cu_3Au crystal.

For $T >> T_c$ the Cu and Au atoms are distributed randomly on fcc sites. The bulk order-disorder transition was early reported to be of the first-order type in which the long-range order parameter jumps discontinuously at $T = T_c$, but the persistence of short-range order for $T > T_c$ and the observation of other critical phenomena characteristic of a continuous (higher-order) transition indicate that there is a continuous component as well (53).

$Cu_3Au(001)$ SURFACE The TOF and corresponding energy spectra for the ordered $Cu_3Au(001)$ first layer, measured in the [001]45° direction, and shown in Fig. (5a,b) immediately demonstrate that the top layer contains both copper and gold and the peak sums inserted into Eq. (7) and (8) give the composition $Cu_{0.495}Au_{0.505}$.

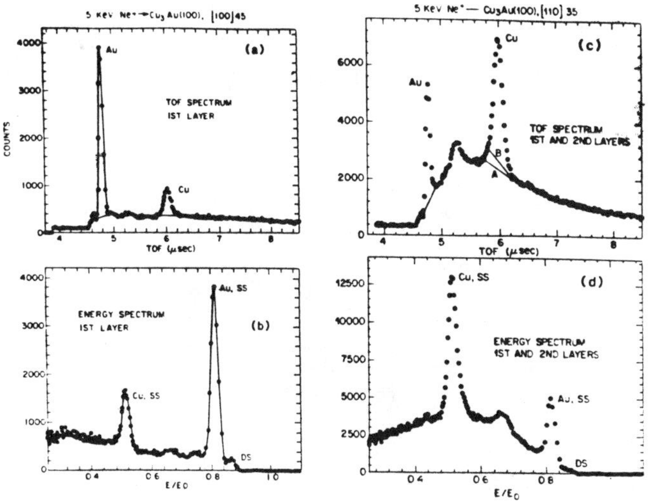

Fig. 5 (a) Time-of-flight spectrum for Ne scattered from $Cu_3Au(001)$. Single scattering occurs only on first layer. E_0 = 5 keV. (b) Energy spectrum derived from TOF spectrum. (c) TOF spectrum scattering occurs only on first and second layers. E_0 = 5 keV. (d) from TOF spectrum (62).

To expose the first and second layer the sample was turned 45° to the [110] azimuth and tilted to ψ = 35°. In the resulting TOF and energy spectra (Fig. 5c,d), the Cu peak is much larger than for first layer scattering while the Au peak has changed very little, owing primarily to the fact that the second layer is nearly pure Cu. Additional factors are an increase in co-planar double scattering and zig-zag scattering between first and second layers which contribute to the shoulder between the Cu(SS) and Au(SS) peaks, and the focusing effect which increases the single scattering yield from layer 2. More on these effects as studied by computer simulations of D. P. Jackson may be found in reference 62.

First and second layer compositions at the $Cu_3Au(001)$ surface are shown as functions of temperature in Fig. 6. LEIS(TOF) measurements were made at temperature after annealing to reach equilibrium for times ranging from 60-70 hr. at 25°C to 2-3 hr. above 450°C. The first layer gold concentration is constant at ~0.52 up to 400°C and then decreases gradually, presumably heading for 0.25 at sufficiently high temperature. Interplay between ordering and segregation is evident in the shape of this curve.

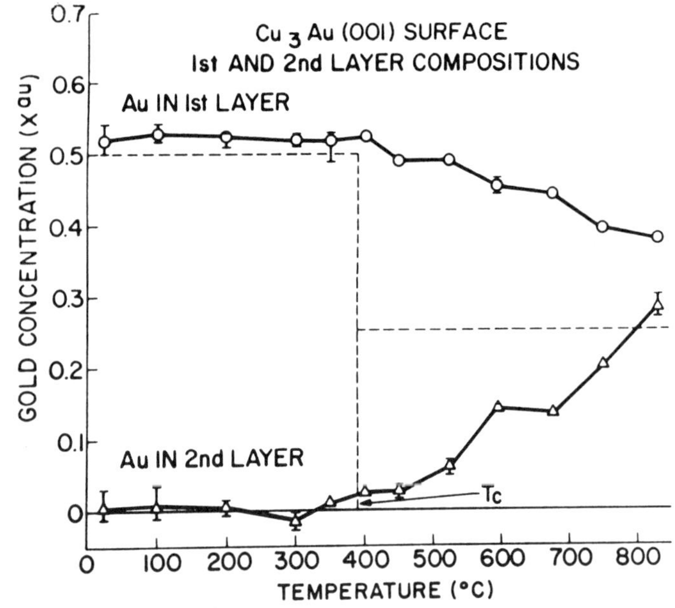

Fig. 6 Compositions (X^{Au}) of first and second layers of $Cu_3Au(001)$ as function of annealing temperatures. Dashed lines depict hypothetical abrupt transition to average bulk compositions in both layers at T_c (63).

If there were no ordering tendency the Au concentration as f(T) would normally be expected to exhibit a monotonic downward trend with increasing temperature, in accord with the entropy term in the free-energy of segregation $\Delta F_s = \Delta H_s - T\Delta S_s$, since segregation itself from a random mixture, is an ordering process. However, the Cu-Au ordering evidently dominates below $T_c = 390°C$, while Au surface segregation prevents an abrupt drop in X_1^{Au} to 0.25 at T_c, and the second layer Au concentration does not rise abruptly beyond T_c owing to the Cu-Au bonding preference.

The gold segregation tendency must be responsible for the fact that the 50-50 Au-Cu layer was invariably on top after numerous sputter-cleaning and annealing sequences.

Additional ion-scattering evidence of the ordered arrangement of Au and Cu atoms at the surface is provided by the angular dependence of Ne single scattering yield from Cu and Au, shown in Figs. 7a and 7b. $\Sigma_{ss}(Cu)$ and $\Sigma_{ss}(Au)$ are plotted separately against azimuthal angle ϕ for incidence angle $\psi = 45°$. Considering Fig. 7a for $E_o = 5$ keV, the broad minima centered around [001] represent first layer scattering and the Au/Cu yield ratio = 2.93 corresponds to $Au_{0.5}Cu_{0.5}$. At [110] where both the first and second layers are exposed the Cu curve peaks while $\Sigma_{ss}Au$ is at a minimum between two peaks. The yields at [110] correspond to a second-layer composition of $Au_{0.06}Cu_{0.94}$.

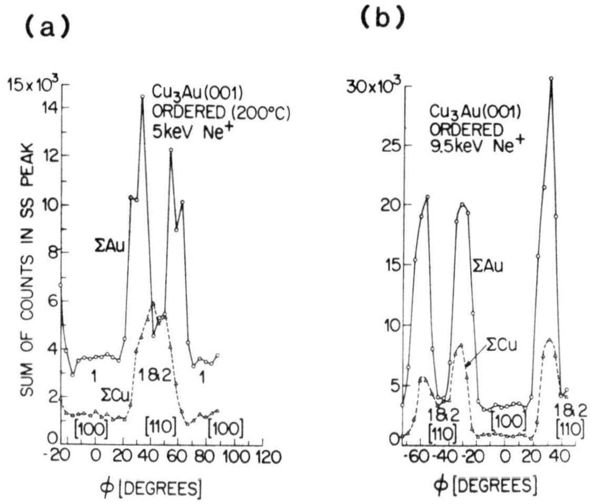

Fig. 7 Single scattering yield of Ne from Au and Cu as function of azimuthal angle ϕ. Incidence angle $\psi = 45°$. Ordered surface. (a) $E_0 = 5$ keV. (b) 9.5 keV. The crystal was remounted for scan (b) shifting the scan limits by about 50° (63).

The two peaks in Σ(Au) centered at 30° from the nearest [001] azimuths are not duplicated in the Σ(Cu) curve for $E_o = 5$ keV but maxima do appear in Σ(Cu) at 30° from [001] for 9.5 keV incident energy (Fig. 7b). An explanation is found in terms of third layer scattering by shadow-cone analysis as illustrated in Fig. 8 which shows maps of ψ–σ combinations for which scattering of Ne through 90° by Cu or Au atoms in the third layer is prohibited by shadowing and/or blocking. The maps are based on a calculation (64) which determines the radius of the shadow cone in a plane normal to the beam direction and containing the target atom, and compares the radius with the distance in this plane from the cone axis to the target atom. Blocking cones of atoms in or near the outgoing path are also taken into account. The difference between Au and Cu scattering at 5 keV evidently results from the fact that in the ordered arrangement a Cu atom in the third layer is shadowed by two Au atoms in the first layer (one row over) whereas third-layer Au atoms are shadowed by Cu atoms. The Au shadow cones are larger, with a radius of 1.4Å vs 1.0Å for Cu at a distance of 4.0Å and there is "hole" in Fig. 8a for Au at $\psi = 45°$, $\phi = 30°$ but none for Cu (8b). However, for 9.5 keV Ne^+ (Fig. 8d) the shadow cones are smaller and the Cu atoms are exposed as well as the Au.

For $T < T_c$ the $Cu_3Au(001)$ surface gives a c(2x2) LEED pattern consistent with a simple termination of the bulk structure shown in Fig. 4 (65,66). This result together with that of LEIS indicates that the outermost layer has the same c(2x2) structure as shown in Fig. 1a.

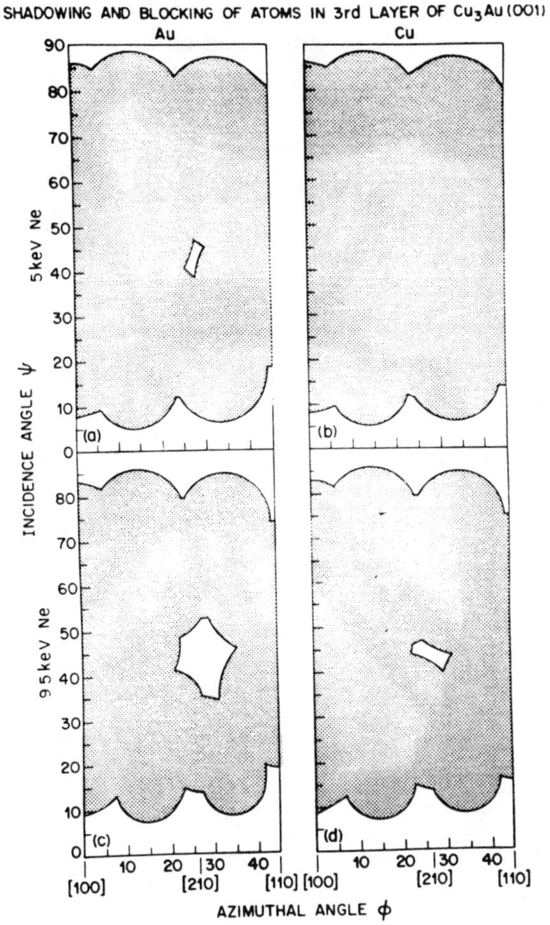

Fig. 8 Shaded areas map calculated combinations of ψ and ϕ for which single scattering of Ne through 90°, by Cu or Au atoms in the third layer, is prohibited by shadowing and/or blocking. $E_0 = 5$ keV in (a) and (b), 9.5 keV in (c) and (d) (63).

Observations of LEED intensities as a function of T indicate that the Cu_3Au(001) surface undergoes an order-disorder transition with a critical temperature equal within error (±2K) to the bulk critical temperature T_c (65,67).

With regard to the nature of the transition, the LEED results can be interpreted in different ways depending on the method of allowing for the Debye-Waller factor in the dependence of fractional-order beam intensities on T (67). When the Debye-Waller factor was assumed to be the same for fractional-order as for integer-order beams, the phase transition was found to be continuous with the surface long-range order parameter M_1 varying like $(T_c-T)^{\beta_1}$ for $T < T_c$ where $\beta_1 = 0.23 \pm 0.02$. A second interpretation is that the surface undergoes a transition to a relatively "soft" state (lower Debye temperature) near 473K. In this picture the transition is discontinuous. The resolution of this question calls for intensity measurements at narrower temperature intervals all the way from room temperature to beyond T_c.

LEED observations of the short-range order characteristics of Cu_3Au(001) were also reported (67). The correlation length k^{-1} and the Lorentzian weight $\hat{P}(0)$ were determined as a function of T by analyzing the (1/2, 1/2) beam profiles. It was observed that k^{-1} increased significantly when T_c was approached from either above or below, while $\hat{P}(0)$ varied only slowly with T in the same range (Fig. 9).

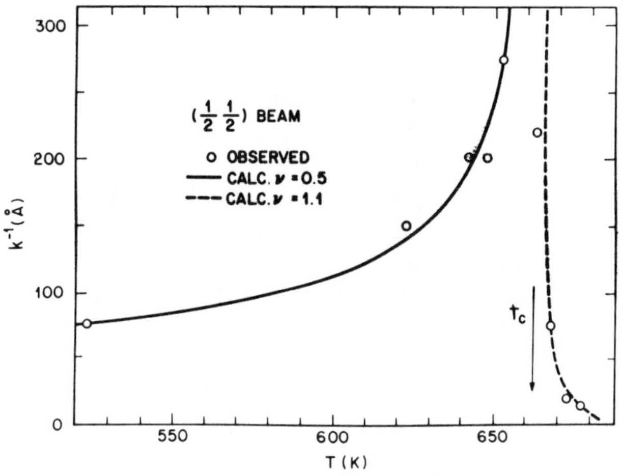

Fig. 9 Temperature dependence of the correlation length k^{-1} determined from (1/2, 1/2) beam profiles (circles) compared to that calculated by $k^{-1} \alpha |T_c/(T-T_c)|^{\nu}$ (67).

The nature of the (001) surface phase transitions for fcc ordering alloys A_3B has been studied theoretically (68-70).

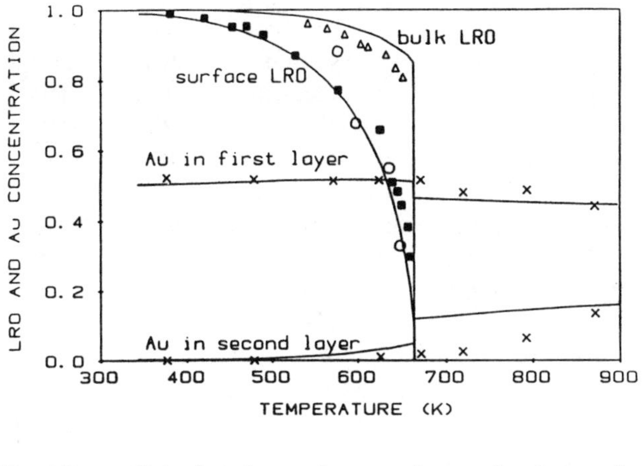

Fig. 10 Calculated and experimental data for Cu_3Au. Full lines show the results of cluster calculations (69). Experimental points refer to: the bulk long-range order (LRO) parameter (Δ), (52(b)); the (001) surface LRO parameter (○) [65] (■) (67); the Au atom fraction in the first two layers (×) (63).

In recent calculations by a cluster method, (69) the form of temperature dependence of the long-range order parameter and of the surface composition was studied for various assumed values of the ratios μ/V, $\Delta E/V$ and V^{00}/V where μ is the bulk chemical potential, ΔE is the effective surface segregation energy and V^{00}, V are the effective nearest neighbor pair interaction energies between surface atoms and between bulk atoms, respectively. Figure 10 summarizes specific results of cluster calculations in comparison with experimental LEIS results and results derived from LEED for Cu_3Au(001). In the calculation, the value of V was chosen to fit the observed bulk critical temperature T_c, and the values of V^{00}/V and $\Delta E/V$ were chosen to reproduce the respective observed values of the surface critical temperature and surface composition for $T << T_c$. As shown in Fig. 10, the main trends in the surface Au atom fraction are well reproduced, although for the second layer the atom fraction is significantly lower than calculated and there is no experimental evidence for the small discontinuities predicted. The calculation also reproduces the variation of surface order parameter M_1 according to one possible interpretation of LEED, namely $M_1 \propto (T_c-T)^{0.23}$ for $T < T_c$ (67).

While the calculation substantially succeeds in explaining the observed variations of surface composition and surface long-range order from a common viewpoint, it fails to reveal any particular significance in two salient experimental facts. These are that within respective experimental errors: (a) for $T << T_c$ the (001) surface atom layer has the same composition as a bulk (001) CuAu layer, and (b) the surface critical temperature is equal to the bulk critical temperature. However, an analysis of order-disorder transitions of semi-infinite systems with first-order bulk transitions (70a), indicates that the surface should undergo a continuous transition with critical temperature exactly equal to the bulk critical temperature as observed. The same analysis provides predictions of the behavior of the correlation length k^{-1} and the Lorentzian weight $\hat{P}(0)$ as a function of T for $T \approx T_c$, namely $k^{-1} \propto |T-T_c|^{-0.5}$ and $\hat{P}(0) \approx$ constant (70b,c). This is in qualitative accord with the observation of singular behavior of k^{-1} but not of $\hat{P}(0)$, and contrasts with analogous bulk-crystal transitions for which both k^{-1} and $\hat{P}(0)$ exhibit singularities (71).

Cu_3Au(110) SURFACE In contrast with the (001) surface of Cu_3Au, the (110) surface does not exhibit long-range order, according to LEED studies (66). If the surface were an ordered extension of the bulk it also would have alternating 50-50 Au-Cu and pure Cu layers at the surface when $T \leqslant 390°C$. Instead, according to LEIS(TOF) (72) $X_1^{Au} \sim 0.40 - 0.42$, and $X_2^{Au} \sim 0.25 - 0.27$ at temperatures between 100°C to 600°C, i.e. there was weak Au enrichment of the first layer while the second layer had the average bulk composition even though, below $T_c = 390°C$, the bulk must have been ordered, with alternating 50-50 and pure Cu layers. The absence of long-range order at the surface is presumably related to the fact that a gold atom in an ordered first layer of the (110) surface would have only two Cu atoms as nearest neighbors in the first layer whereas it has four on the (001).

Ni-0.8% Au

Ni-0.8% Au ALLOY - (110) SURFACE When this crystal is sputtered and then annealed at 700°C, the surface coverage of Au increases to a maximum value of 1.00 ± 0.05 ML after 1h (1ML = 1 monolayer = the number of surface Ni atoms per unit area or 1.141×10^{15} cm^{-2} in the case of Ni(110)) (73). Strong segregation is of course expected in the Ni-Au system if the segregation principles outlined above are valid.

The nature of the Au-covered surface is indicated by the LEED pattern in Fig. 11a. One can say immediately that it is not a 1x1 surface structure, i.e., the Au atoms are not in substitutional Ni sites.

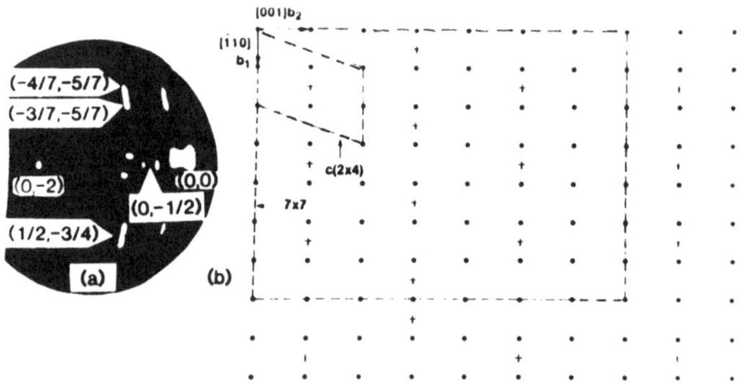

Fig. 11 (a) LEED pattern for a Ni(001) surface covered with Au, electron energy 50 eV. The crystal is oriented so that the line joining the (00) and (0,−2) spots is parallel to [001]. Streaks located near (0, −1/2), (±1/2, −1/4) and (±1/2, −3/4) positions are visible. The other prominent extra spots in this pattern are due to Au(321)−3x1 facets of Au crystallites. (b) Unit mesh and structural subunits indicated by LEED patterns. Dots denote Ni(110)−1x1 net points. b_1 and b_2 denote that basic net vectors. The 7x7 and c(2x4) unit meshes are outlined by broken lines. The relative positions of the centers of identical, approximately c(2x4) subunits in different 7x7 unit meshes are indicated by crosses (73).

This is not surprising in view of the atom size difference between Ni and Au. The main features are streaks centered on positions corresponding to a 7x7 unit mesh outlined in Fig. 11b. The streaks centered at the (−4/7, −5/7) and (−3/7, −5/7) positions overlap to give the appearance of a single streak. However, when the crystal is annealed for a long time at low temperature (≈ 200°C) they revert to discrete spots. The directions of the streaks are consistent with disorder in the spacing of 7x7 domain boundaries that are parallel to [001]. Other structures coexisting with 7x7 domains include Au-induced Ni(112)−1x1 facets and Au crystallites (74). The pair of spots to the left of the (0, −4/7) streak are due to Au(321)−3x1 facets of crystallites.

The gross nature of the 7x7 surface may be inferred from the LEED pattern using kinematical theory as illustrated by the 1x7 example above (Fig. 2). The 7x7 unit mesh was found to be made up of c(2x4) subunits as sketched in Fig. 11b.

The geometrical arrangement of atoms within these subunits was determined by LEIS (TOF) in a mode which differed in two ways from that used to study the $Cu_3Au(001)$ surface. A scattering angle θ_L of 135°, rather than 90°, was employed, together with lower incidence angles for azimuthal scans. The approach is similar to those of Bronckers and de Wit (75) and Aono (59) who used noble-gas ions with an electrostatic analyzer and to that of Niehus and Comsa (60) who used alkali ions with an ESA. The latter two groups have used mainly polar scans of scattered intensity rather than azimuthal scans. The large scattering angle reduces multiple scattering and blocking effects. Aono has called the technique ICISS, for impact collision ion scattering spectrometry.

An azimuthal scan at a very low incidence angle may be thought of as demonstrating two-dimensional channeling, in contrast to the three-dimensional channeling used in the $Cu_3Au(001)$ experiments. The scans of gold peak height in Fig. 12 show minima at angles where the incident beam and the scattered beam to the detector are aligned with major rows of atoms, with each atom shadowing the next in line. As the sample is rotated about its axis normal to the surface, a dip in the scattered yield is produced as atoms of a major row enter the shadows of their preceding neighbors traverse, an arc subtended by a chord of the shadow cone cross-section, and finally emerge on the opposite side of the shadow-cone. An atom entering or emerging receives a concentrated flux of ions due to the focusing effect at the edges of the shadow cone. In Fig. 12 the minima occur at azimuthal angles 0° (the [001] azimuth), 30°, 60° and 90° (the [$\bar{1}$10] azimuth). These 30° separations suggest a hexagonal arrangement of the gold atoms.

The LEIS and LEED results, and the 1 ML coverage determined by Rutherford backscattering (RBS), suggest a nearly close-packed structure like that shown in Fig. 13a, where each of the c(2x4) subunits indicated by LEED is associated with a set of six Au atoms forming a hexagon with a seventh Au atom at

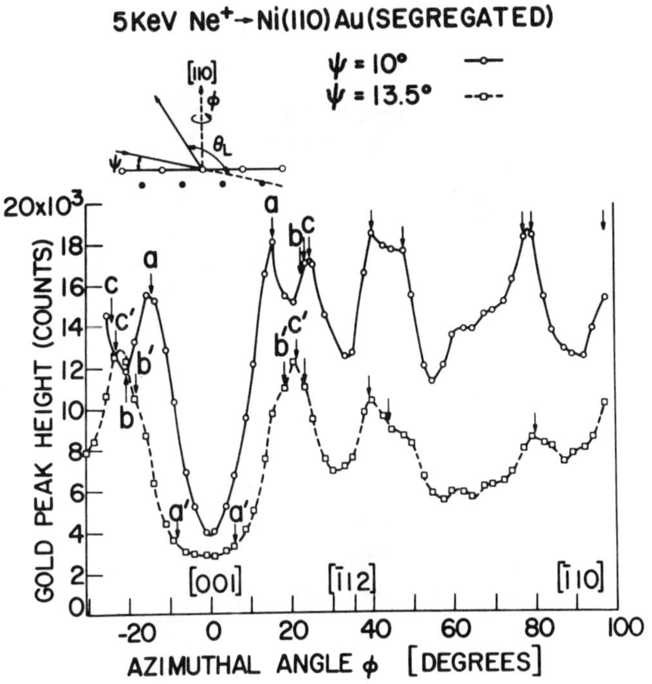

Fig. 12 LEIS single-scattering yield (peak height) from segregated gold atoms on Ni(110) as function of azimuthal angle ϕ for two incidence angles ψ = 10° and 13.5°. Ne^+ ion energy E_0 = 5 keV, scattering angle θ = 135°. The arrows indicate calculated angles of maxima in the yield. Insert: scattering geometry (73).

the center (73).

The azimuthal scans were compared with the model by means of a calculation (77) using the impulse approximation (78) and the Molière approximation to the Thomas-Fermi potential. The angular dependence of the scattering intensity was computed for the case of scattering by two atoms, with the trajectory of the scattered particle not necessarily confined to one plane. The most prominent feature of the azimuthal scans in Fig. 12 is the dip centered at ϕ = 0, the [001] direction of the Ni(11)) surface. This dip is wider for polar incidence angle ψ = 13.5° than for ψ = 10°, contrary to expectation if Au atoms in the row were all at the same height, since the arc traversed by the shadowed atom across the shadow cone would be shorter at higher ψ (Fig. 13b). The observed behavior is attributed to differences in height of Au atoms along Ni[001] with the Au atoms forming a "stile" peaked at the center of a hexagon. Thus, a narrow dip for the upward steps, with maxima indicated by arrows a and a′ in Fig. 12 appears for ψ = 10° but is washed out by the finite angular resolution of the experimental system for ψ = 13.5° because the second atom is too close to the edge of the shadow cone.

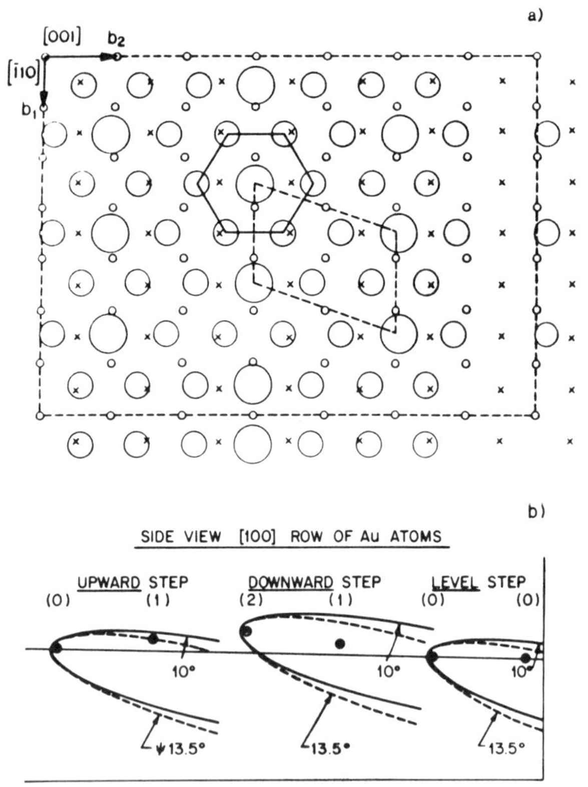

Fig. 13 (a) Arrangement of Au overlayer atoms on Ni(110) as determined by LEED and LEIS, plan view. Small circles denote top-layer Ni atoms, crosses denote hollows between Ni atoms, larger open circles denote Au atoms. The 7x7 and c2x4 unit meshes, defined relative to the Ni(110) unit mesh vectors b_1, b_2 are outlined by broken lines. A representative hexagon formation of six Au atoms with a seventh Au atom (large open circle) at the center is outlined. The Au atoms are at increasing heights h_0, h_1, h_2 in order of increasing distance from the nearest hollow site. The dimensions assumed in drawing the atom positions were the Ni(110) dimensions $b_1 = 2.49$Å, $b_2 = 3.52$Å, and the nearest-neighbor distances: Au-Au 2.85Å and Au-Ni 2.69Å (73).

(b) Side view of Au atoms along [001], and cross-sections of shadow-cones. The atoms at heights h_0, h_1, h_2 (see Fig. 13a, caption) are numbered 0, 1, 2 (73).

The shadowing of (0) by (0) then dominates, producing a broader dip with maxima indicated by arrows b(b′). Maxima corresponding to the shadowing at downward steps are marked by arrows c(c′). The relative prominence of the [001] dip as compared with the [1̄10] dip is consistent with the same picture; at [001] only the up-steps of the stiles are exposed to the beam while at [1̄10] both up- and down-steps are exposed. Calculations for the dips at [1̄10] and also those centered near 60° and 30° gives the positions of maxima for the dominant dips indicated by unlettered arrows for the atom spacings of the model in Fig. 13a.

In the overlayer model (Fig. 13a) most of the Au atoms form zig-zag chains running along the [1̄10] channels provided by the Ni(110) surface. The surface-projected Au-Au distances are close to the bulk distance (2.88Å). The resulting misfit with the Ni(110) surface in the [001] direction is accommodated by leaving every seventh [1̄10] channel vacant, giving the unit-mesh dimension $7b_2$. The misfit in the [1̄10] direction is accommodated by antiphase boundaries parallel to [001] at average interval $7b_1$. The antiphase boundaries correspond to the interruption of the zig-zag chains along the [1̄10] channels.

At 340°C there is a transition from the 7x7 to a 2x7 surface. It has been suggested (74) that the transition might be related to a lateral expansion of the Au overlayer relative to the substrate. This could stabilize uninterrupted zig-zag chains relative to zig-zag chains that are periodically interrupted as in the proposed room-temperature structure.

The 7x7 LEED pattern is observed only for Au coverages in excess of 0.8ML. For lower coverages a variety of other LEED patterns are relatively prominent, indicating that the close-packed structure does not form until the coverage approaches 1 ML.

LEIS (TOF) results for weak segregation (10-15% ML) caused by annealing 1-2 hr. at 300°C differed markedly from those for 1 ML Au coverage, in that azimuthal scans for Au and Ni were nearly identical in shape, and the widths of Au dips at [001] and [1̄10] corresponded to Ni spacings along those directions. LEED patterns were 1x1. Thus, the Au at lower coverage appears to be substitutional in the first layer.

Ni-0.8% Au ALLOY - (001) SURFACE In earlier LEIS (TOF) experiments (33) the Ni-0.8% Au(001) surface also exhibited very strong Au segregation when heated, e.g., to 700°C, for several hours. In those experiments with 5.0 and 9.5 keV Ne^+ beams, a scattering angle of 90°, and incidence angles of 20°, 35°, and 45°, the Au atoms were also found in a single layer, with none detected in deeper layers, beyond the bulk concentration. Azimuthal scans of Ne scattering from Ni beneath the Au layer at the three different ψ's were very similar to those for scattering from the clean Ni(001) surface. In other words, shadowing of Ni atoms by the Au layer was very much like that by a Ni(001) 1x1 first layer. Furthermore, computer simulations of spectra for scattering from the first layer and the first and second layers, assuming gold in substitutional sites, reproduced the experimental spectra fairly closely. Ion scattering behavior therefore seemed consistent with a substitutional segregated Au layer. However, LEED patterns from the segregated surface were not 1x1 but were indexed as c(2x6), and have more recently been indexed as (2x12). The correct modelling of this surface evidently requires further LEED analysis plus low incidence angle ion scattering ($\psi = 5 - 15^\circ$) to detect the subtle displacements of Au atoms from substitutional Ni sites which escape detection at higher incidence angles.

SUMMARY AND CONCLUSIONS

The low energy ion scattering and low energy electron diffraction results together with other published information demonstrate distinct differences in surface behavior between Au-Ni and Au-Cu alloys which are consistent with expectations based on thermodynamic properties and radii of the elements. Segregation from the Ni-0.8% Au alloy is very strong, reflecting the weaker bond energy of Au compared with Ni, the larger atomic radius of Au, and the positive enthalpy of mixing in the Ni-Au system. At high coverage, close to 1 ML, the segregated Au layer is non-substitutional on both Ni(001) and Ni(110). On Ni(110), LEIS and LEED analysis lead to a model of a close-packed Au overlayer in which Au atoms form zig-zag chains along the $[\bar{1}10]$ semi-channels of Ni(110). Au atoms in a adjacent chains combine in hexagons, each of which is capped by an extra Au atom to give a nearly close-packed arrangement. At low coverage the Au is present in substitutional sites. Ni-0.8% Au(001) also produced ~1 ML saturated Au layers of which the structure is not yet determined.

In Cu-Au alloys the small bond-energy difference favors Cu segregation but Au segregates, presumably because of the larger size of the Au atom. Segregation is suppressed and ordering is promoted by the negative enthalpy of mixing in this system. $Cu_3Au(001)$ exhibits long-range order, with the 50-50 Cu-Au layer on top and a second layer of pure Cu below $T_c = 390^\circ C$. First and second layer Au concentrations X_1^{Au} and X_2^{Au} gradually approach each other as T is increased beyond T_c. No confident classification of the surface order-disorder transition as first-order or continuous can be made with the present LEED evidence; additional measurments are needed. $Cu_3Au(110)$ does not exhibit long range order; $X_1^{Au} \approx 0.40$ and $X_2^{Au} \approx 0.25$ from 100°C to 600°C.

References

1). E. E. Latta and H. P. Bonzel, "Interfacial Segregation" Ed. W. C. Johnson J. M. Blakely, 381-404, Am. Soc. Met., Metals Park, Ohio (1979).

2). G. W. Graham, Surface Sci. *137*, L79-L83 (1984).

3). H. Wiedersich and P. R. Okamoto, "Interfacial Segregation" Ed. W. C. Johnson and J. M. Blakely, 405-432, Am. Soc. Met., Metals Park, Ohio (1979).

4). R. Kelly, Mat. Sci. and Eng. *69*, 449-455 (1985).

5). J. M. Blakely and H. V. Thapliyal, "Interfacial Segregation" Ed. W. C. Johnson and J. M. Blakely, 137-174 Am. Soc. Met., Metals Park, Ohio (1979).

6). A. E. T. Kuiper, G. C. J. van der Ligt, W. M. van de Wijgert, M. F. C. Willensen, and F. H. P. M. Habraken, J. Vac. Sci. Technol. B 3(3), 830-835 (1985).

7). E. J. van Loenen, L. W. M. Frenken, and J. F. van der Veen, Appl. Phys. Lett. *45*, 41-43 (1984).

8). J. H. Sinfelt, J. L. Carter, D. J. C. Yates, J. Catal. *24*, 283-296 (1972).

9). F. L. Williams and D. Nason, Surface Sci. *45*, 377-408 (1974).

10). A. R. Miedema, Z. Metallkd. *69*, 455-461 (1978).

11). P. Wynblatt and R. C. Ku, Surface Sci. *65*, 511-531 (1977); see also P. Wynblatt and R. C. Ku, "Interfacial Segregation," Ed. W. C. Johnson and J. M. Blakely, Am. Soc. Met., Metals Park, Ohio (1979) 115-136.

12). R. A. van Santen and M. A. M. Boersma J. Catal. *34*, 13 (1974).

13). P. Wynblatt and D. A. Steigerwald, ASM Materials Week '85 Proceedings.

14). R. Hultgren, P. D. Desai, D. T. Hawkins, M. Gleiser and K. K. Kelley, "Selected Values of Thermodynamic Properties of Binary Alloys," Am. Soc. Met. New York (1973); R. Hultgren, P. D. Desai, D. T. Hawkins, M. Gleiser, K. K. Kelley, and D. G. Wagman, "Selected Values of Thermodynamic Properties of Binary Alloys and the Elements," Am. Soc. Met. New York (1973).

15). E. A. Guggenheim, "Mixtures," Clarendon Press, Oxford (1952).

16). R. A. Swalin, "Thermodynamics of Solids," Chapt. 7, John Wiley and Sons, New York (1972).

17). J. W. Christian, "The Theory of Transformations in Metals and Alloys" Part 1, Chapt. 6, Pergamon press, New York (1975).

18). H. H. Brongersma, M. J. Sparnaay, and T. M. Buck, Surface Sci. *71*, 657-678 (1978).

19). W. M. H. Sachtler, R. A. van Santen, Appl. Surface Sci. *3*, 121-144 (1979).

20). D. McLean, "Grain Boundaries in Solids," Clarendon Press, Oxford (1957).

21). J. Friedel, Adv. Physics *3*, Appendix (1954).

22). F. F. Abraham, N. H. Tsai, G. M. Pound, Surface Sci. *83*, 406-422 (1979).

23). F. F. Abraham, Phys. Rev. Lett. *46*, 546-549 (1981).

24). J. C. Hamilton, Phys. Rev. Lett. *42*, 989-992 (1979)

25). J. R. Chelikowsky, Surface Sci. *139*, L197-L203 (1984).

26). T. Sakurai, T. Hashizume, A. Simbo, and A. Sakai, Phys. Rev. Lett. *55*, 514-517 (1985).

27). R. P. Frankenthal and D. J. Siconolfi, Surface Sci. *119*, 331-348 (1982).

28). A. F. Wells, "Structural Inorganic Chemistry," 3rd ed., p. 984 Clarendon Press, Oxford (1967).

29). F. Smolukowsky, "Handbook of Physics," pp. 8-97, Ed. E. U. Condon and H. Odishow, McGraw-Hill, New York, (1967).

30). B. E. Warren, "X-ray Diffraction," Chapt. 12 Addison-Wesley, Reading (1969).

31). J. L. Moran-Lopez and K. H. Bennemann, Phys. Rev. B. *15*, 4769-4780 (1977).

32). J. L. Moran-Lopez and L. M. Falicov, Phys. Rev. B *18*, 2542-2548 (1978).

33). T. M. Buck, I. Stensgaard, G. H. Wheatley, and L. Marchut, Nucl. Inst. and Meth. *170*, 519-525 (1980).

34). M. J. Kelley, P. W. Gilmour, D. G. Swartzfager, J. Vac. Sci. Technol. *17*, 634-637 (1980).

35). F. L. Williams and M. Boudart, J. Catal. *30*, 438-443 (1973).

36). J. J. Burton, C. R. Helms and R. S. Polizzotti, Surface Sci. *57*, 425-430 (1967).

37). P. Biloen, R. Bouwman, R. A. van Santen, H. H. Brongersma, Appl. Surface Sci. 532-542 (1989).

38). T. M. Buck, G. H. Wheatley, R. Seals and Y.-S. Chen, unpublished.

39). J. M. McDavid and S. C. Fain, Jr., Surface Sci. *52* 161-173 (1975).

40). C. G. Shaw and S. C. Fain, Jr., "Proceedings of the Seventh International Conference and the Third International Conference on Solid Surfaces," p. 2315, Ed. R. Dobrozemsky, F. Berg & Son, Vienna, 1977.

41). M. J. Sparnaay and G. E. Thomas, Surface Sci. *135*, 184-198 (1983).

42). S. H. Overbury and G. A. Somorjai, Surface Sci. *55*, 209-226 (1976).

43). B. E. Sundquist, Acta Met. *12*, 585-592 (1964).

44). G. C. Nelson, J. Vac. Sci. Technol. *13*, 512-513 (1976).

45). A. Joshi, "Interfacial Segregation," 39-109, Ed. W. C. Johnson and J. M. Blackely, Am. Soc. Met., Metals Park, Ohio (1979).

46). T. M. Buck, "Chemistry and Physics of Solid Surfaces IV," 435-464, Ed., R. Vanselow and R. Howe, Springer-Verlag, New York (1982).

47). R. P. Frankenthal and D. J. Siconolfi, Surface and Interface Anal., in press.

48). M. Alonso, F. Soria and J. C. Sacedon, J. Vac. Sci. Technol, 1598-1602 (1985).

49). Y. S. Ng, T. T. Tsang, S. B. McLane, Jr., Phys. Rev. Lett. *42*, 588-591 (1979).

50). T. Sakurai, T. Hashizume, and A. Jimbo, Appl. Phys. Lett. *44*, 38-40 (1984).

51). L. J. Clarke, "Surface Crystallography," John Wiley and Sons, New York (1985).

52). (a) J. M. Cowley, J. Appl. Phys. *21*, 24-30 (1950).
(b) R. Feder, M. Mooney and A. S. Nowick, Acta Met. *6*, 266-277 (1958).

53). M. F. Collins and H. C. Teh, Phys. Rev. Lett. *30*, 781-784 (1973).

54). H. H. Brongersma and T. M. Buck, Surface Sci. *53*, 649-658 (1975).

55). T. M. Buck, "Methods of Surface Analysis," 76-102, Ed. A. W. Czanderna, Elsevier Sci. Pub. Co. New York (1975).

56). E. Taglauer and W. Heiland, Appl. Phys. *9*, 261-275 (1976).

57). E. P. Th. M. Suurmeijer and A. L. Boers Surface Sci. *43*, 309-352 (1974).

58). S. B. Luitjens, A. J. Algra, E. P. Th. M. Swermeijer, and A. L. Boers, Appl. Phys. *21*, 205-214 (1980).

59). M. Aono, Nuc. Instrum. Meth. Physics Res. *B2*, 374-383 (1984).

60). H. Niehus and G. Comsa, Surface Sci. *140*, 18-30 (1984).

61). R. S. Williams and J. A. Yarmoff, Nucl. Intstrum. Meth. Physics Res. *218*, 257-265 (1983).

62). T. M. Buck, G. H. Wheatley and D. P. Jackson, Nucl. Instrum. Meth. Physics Res. *218*, 257-265 (1983).

63). T. M. Buck, G. H. Wheatley and L. Marchut Phys. Rev. Lett. *51*, 43-46 (1983).

64). L. Marchut, T. M. Buck, G. H. Wheatley and C. J. McMahon, Jr., Surface Sci. *141*, 549-566 (1984).

65). V. S. Sundaram, B. Farrell, R. S. Alben and W. D. Robertson, Phys. Rev. Lett. *31*, 1136-1139 (1975); V. S. Sundaram, R. S. Alben and W. D. Robertson, Surface Sci. *46*, 653-671 (1974).

66). H. C. Potter and J. M. Blakely, J. Vac. Sci. Technol, *12*, 635-642 (1975).

67). E. G. McRae and R. A. Malic, Surface Sci. *148*, 551-580 (1984).

68). V. Kumar and K. H. Bennemann, Phys. Rev. Lett. *53*, 278-281 (1984).

69). J. M. Sanchez and J. L. Moran-Lopez, Surface Sci. in press.

70). (a) R. Lipowsky and W. Speth Phys. Rev. B28, 3983-3993 (1983);
(b) R. Lipowsky, J. Appl. Phys. *55*, 2485-2490 (1984);
(c) R. Lipowsky, private communication;

71). J. Als-Nielsen and O. W. Dietrich, Phys. Rev. *53*, 706-710 (1967); O. W. Dietrich and J. Als-Nielsen, Phys. Rev. *153*, 711-717 (1967); J. Als-Nielsen and O. W. Dietrich, Phys. Rev. *153*, 717-721 (1967).

72). T. M. Buck, G. H. Wheatley and L. Marchut, unpublished.

73). D. O. Boerma, T. M. Buck, E. G. McRae, R. A. Malic and G. H. Wheatley, unpublished.

74). E. G. McRae and R. A. Malic, unpublished.

75). R. P. N. Bronckers and A. G. J. de Wit, Surface Sci. *104*, 384-404 (1981).

76). M. Aono, Y. Hou, C. Oshima and Y. Ishizawa, Phys. Rev. Lett. *49*, 567-570 (1982).

77). D. O. Boerma, T. M. Buck and G. H. Wheatley, unpublished.

78). M. T. Robinson and I. M. Torrens, Phys. Rev. B9, 5008-5024 (1974).

SURFACE PRECIPITATION AND ITS APPLICATION TO COATINGS

K. Nii and K. Yoshihara
National Research Institute for Metals
Meguro-ku, Tokyo, Japan

Abstract

The precipitation phenomena on the surfaces of two types of stainless steels were observed when they were heated at high temperatures in vacuum. One was the precipitation of titanium carbide on the surface of commercial 321 stainless steel and the other was the precipitation of boron nitride on the surface of laboratory-melted 304 stainless steel doped with nitrogen and boron(304-NB) and that doped with nitrogen, boron and cerium(304-NBCe). In the case of 304-NBCe stainless steel, a boron nitride layer almost uniformly covered the surface at the eraly stage of heating. Such surface precipitation phenomena will be applied to control the structure and properties of metal surfaces and will be a promising process for materials design. The titanium carbide film deposited by ion plating on the surface of 321 stainless steel was stable against the thermal cycle. This means that the precipitated titanium carbide could be a bonding agent between ceramics and metals. The surface covered with boron nitride was inert to the adsorption of gas. Therfore, stainless steel doped with nitrogen, boron and cerium will be a candidate material for vacuum vessels.

IN THE BULK OF stainless steels, there are usually precipitates such as carbides and nitrides. If these compounds precipitate on the surface of stainless steels from the bulk by heating in vacuum, this process is applicable to surface coating of stainless steels, and if these compounds exfoliate from the surface, they can precipitate again from the bulk of stainless steel only by annealing in vacuum. Consequently, the surface covered with these compounds has a self-healing ability.

In this experiment, surface precipitation behaviors were investigated of commercial 321 stainless steel, laboratory-melted 304 stainless steel doped with nitrogen and boron(304-NB) and that doped with nitrogen, boron and cerium(304-NBCe) using AES and XPS. Titanium carbide and boron nitride were expected to precipitate on these alloy surfaces, respectively.

Such surface precipitation phenomenon will be applied to control the structure and properties of metal surfaces and will be a promising process for materials design.

EXPERIMENTALS

SPECIMENS - Specimens of 10mm × 10mm were cut out from a 0.8mm commercial 321 and 304 stainless steel sheets. Specimens were also cut out from the laboratory-melted 304-NB and 304-NBCe stainless steels. The average grain size of laboratory-melted stainless steels was 80μm. The chemical compositions of stainless steels are shown in Table 1.

SURFACE ANALYSIS - The specimens were polished with emery paper and diamond paste (0.25μm). After being degreased in acetone, specimens were attached to a hot stage; the details of the hot stage were given elsewhere

Table 1 - Chemical Compositions of Stainless Steels(mass%)

	C	Si	Mn	P	S	Ni	Cr	Ti	N	B	Ce
321	0.037	0.81	0.93	0.033	0.006	9.81	17.29	0.42			
304-NB	0.054	0.30	1.05	0.02	0.007	8.47	19.47		0.160	0.010	
304-NBCe	0.044	0.49	1.64	0.008	0.0024	9.49	18.67		0.170	0.008	0.004
304	0.06	0.46	0.81	0.027	0.007	8.67	18.04		0.033		

(1). The specimens were heated on the back side by electron bombardment in a vacuum of 10^{-7} Pa. The change of surface composition of a specimen was continuously observed at high temperatures using AES(PHI 10-155) which has primary electron beam diameter of 100μm. The shape and heights of Auger peaks above 100eV were not influenced by electron bombardment heating. The energy of the primary electron was 3keV and the modulation amplitude was 6V.

The concentration of elements on the surface was calculated using the published relative sensitivity factors for the pure elements(2). This method neglects the effect of chemical states on Auger peak shape and of escape depth of Auger electrons. Therefore, the surface compositions obtained with this method are only the approximated values, but are accurate enough for the semi-quantitative analysis.

The distribution of elements on the surface were determined using SAM(PHI 15-110A; primary electron beam diameter was 3μm and PHI 15-255G; 0.25μm). The chemical states of elements were determined using XPS(PHI 15-255G; X-ray source was MgKα).

COATING OF TITANIUM CARBIDE - Titanium carbide was deposited by using an ion plating process: Titanium was vaporized and ionized in an atmosphere of acetylene, partial pressure of which was 10^{-1}Pa. Titanium carbide was formed and deposited on a specimen. The thickness of deposited titanium carbide film was about 1.5μm. The specimens coated with titanium carbide were heated in vacuum at 1100K for 1.8ks(30min), then cooled to room temperature and kept for 0.6ks(10min). This thermal cycle was repeated 4 times. The surface of the specimen was observed by using SEM.

GAS DESORPTION BEHAVIOR - A specimen, size of which was 5mm × 80mm × 0.8mm, was installed in a vacuum chamber with a sample heating system and a quadrupole mass filter. A regulated DC power was supplied through a specimen for heating, and temperatures were measured by means of a chromel-alumel thermocouple which was spot welded to the center of the sample. The heating rate was 50K/s and the base pressure was 10^{-3}Pa. The change of the peak height of mass number of 18(H_2O) was observed during heating process.

PRECIPITATION OF TITANIUM CARBIDE

Metallic materials have been coated with ceramics for use under severe conditons. However, ceramic films on metals often to exfoliate and this is a big problem to be overcome. In the bulk of 321 stainless steel, precipitates of titanium carbide exist. If the annealing condition is suitable, titanium carbide can also precipitate onto the surface. Precipitation of titanium carbide on surface will alter the structure and properties of interface of ceramic coated materials, and it will possibly increase the adherence between metals and ceramic films.

The changes of surface compositions of a commercial 321 and 304 stainless steels were observed using AES at high temperatures, and the adherence of deposited titanium carbide film on these steels was evaluated by thermal cycle test.

PRECIPITATION BEHAVIOR ON 321 STAINLESS STEEL - After the surface was cleaned by argon ion sputtering, the temperature was rapidly raised up to 1100K and kept at that degree. The change of the surface composition is shown in Fig.1. The ordinate is the surface concentration calculated by using relative sensitivity factors and abscissa is heating time measured after arriving at a given temperature. The concentrations of titanium and carbon increased as heating time was prolonged. These elements were not replaced by the most surface active element, sulfur. At the initial stage of heating, phosphorus and oxygen also segregated on the surface. As heating time was prolonged, however, phosphorus and oxygen were replaced by sulfur due to the difference in surface activity. At the final stage of heating, titanium, carbon and sulfur remained on the surface.

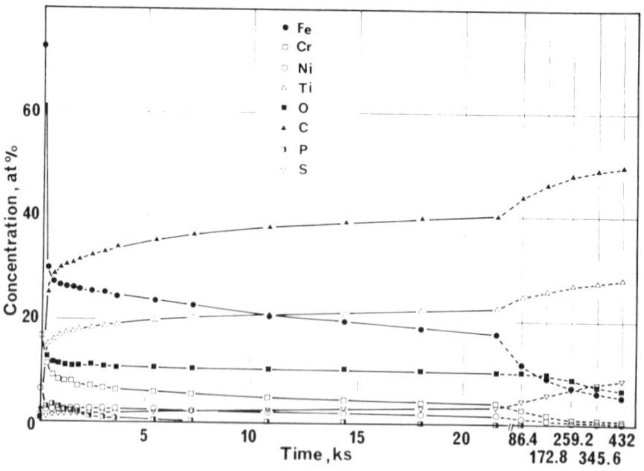

Fig.1 - Change of the surface composition of 321 stainless steel at 1100K

The change of surface composition of commercial 304 stainless steel at 1000K was measured with the same method as 321 stainless steel. The result is shown in Fig.2. At the initial stage of heating, phosphorus segregated. As heating time was prolonged, sulfur segregated, and replaced phosphorus since sulfur is more surface active than phosphorus (1).

After 321 stainless steel had been heated at 1100K for 432ks(120hr), it was transferred to the apparatus for SAM through exposure to the air. The SAM images are shown in Fig.3. In the SEM image of Fig.3, an unevenly precipitated layer is found on the surface, which has many holes of about 1μm in diameter. It is shown that SAM image of titanium and that of carbon almost completely coincide. SAM images of iron, sulfur and oxygen were the

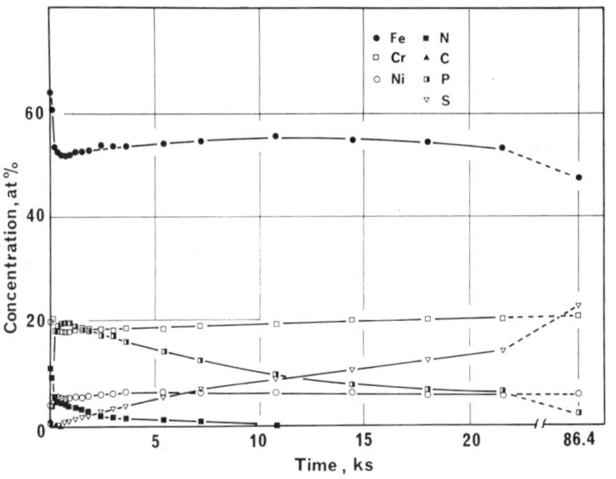

Fig.2 - Change of the surface composition of 304 stainless steel at 1000K

negative images of those of titanium and carbon and the distributions of iron, sulfur and oxygen coincided with that of holes of the precipitated layer. Sulfur, however, spread more widely than iron and oxygen. The results of point Auger analyses at point A (precipitated layer) and at point B (hole) in Fig.3 are shown in Fig.4. At point A (precipitated layer), titanium and carbon were observed and the shape of Auger peak of carbon coincided with that of titanium carbide(3). At point B (hole), sulfur and oxygen in addition to the alloy elements were observed. Consequently, it was shown that the precipitated layer was titanium cabide and the bottom of the hole was 321 stainless steel, on which phosphorus, oxygen and sulfur segregated. As heating time was prolonged, these segregants were replaced one by one in the order of surface activity, and finally the most surface active element, sulfur, remained at the bottom of the hole.

Moreover, sulfur diffused over the outside of the hole.

The thickness of the precipitated layer was measured by the argon ion sputtering at room temperature (acceleration voltage; 3kV, current density; $0.6A/m^2$). The sputtering rate for titanium carbide layer was estimated from that for the chromium deposition layer, thickness of which had been known. In this experiment, the sputtering rate was about 0.1nm/s. The changes of Auger peak heights of titanium and carbon of the specimens, which were heated at 1100K for 432ks(120hr) and 86.4ks(24hr) respectively, with sputtering time are shown in Fig.5. The changes of Auger peak heights of titanium and that of carbon were similar. The thickness of the precipitated layer was estimated from the sputtering time when the Auger peak height started to decrease abruptly. Figure 6 shows the relation between the thickness and the heating time at 1100K. The thickness of the precipitated layer increased in proportion to the square root of heating time and became about 0.05μm after being heated at 1100K for 432ks. Therefore, the schematic illustration of the surface of 321 stainless steel after being heated at 1100K for 432ks (120hr) is shown in Fig.7.

The surface of the specimen which had been heated at 1100K for 432ks(120hr) was sputtered by argon ion at 1100K for 540s(9min). At 1100K the concentrations of titanium and carbon did not decrease during the sputtering (acceleration voltage; 3kV, current density; $0.6A/m^2$). This is because titanium and carbon are supplied to the surface at a sufficient velocity by the diffusion during the sputtering at high temperature.

Titanium carbide also precipitated at 900K, but the rate was smaller than that at 1100K.

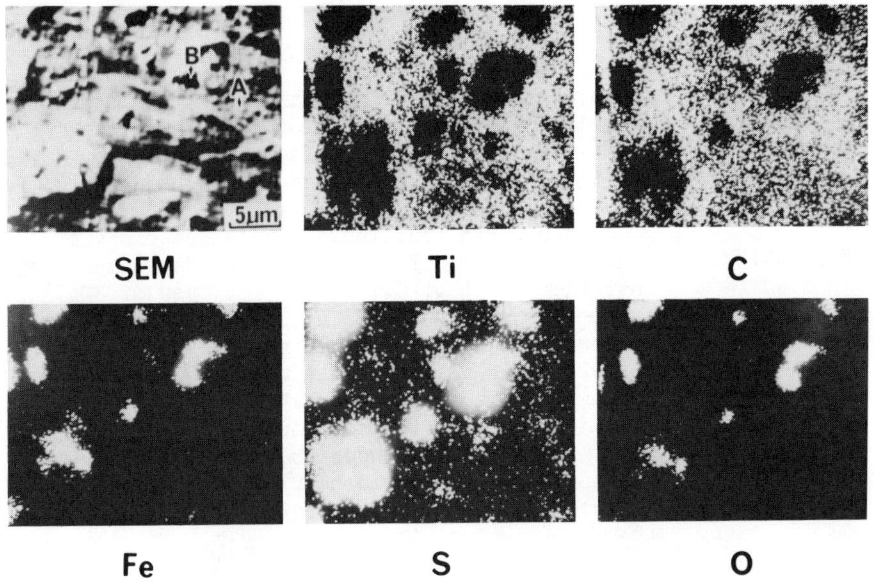

Fig.3 - The SAM images of 321 stainless steel after annealing at 1100K for 432ks

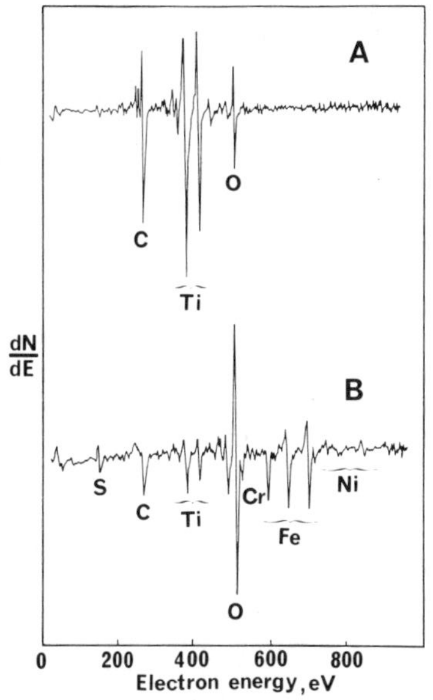

Fig.4 - Auger spectra from points A and B indicated in SEM image in Fig.3

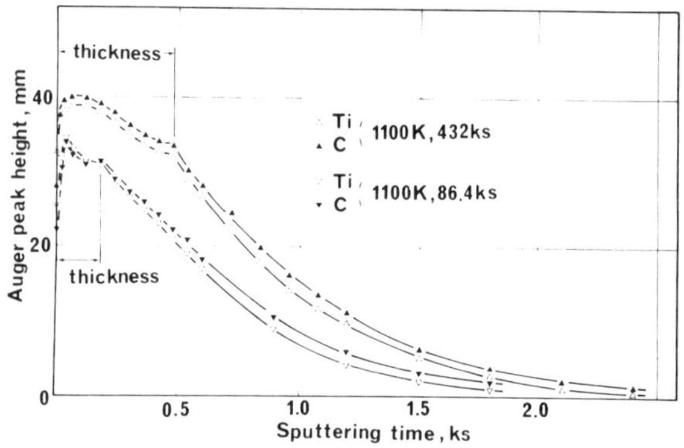

Fig.5 - The change of Auger peak heights of titanium and carbon with sputtering time

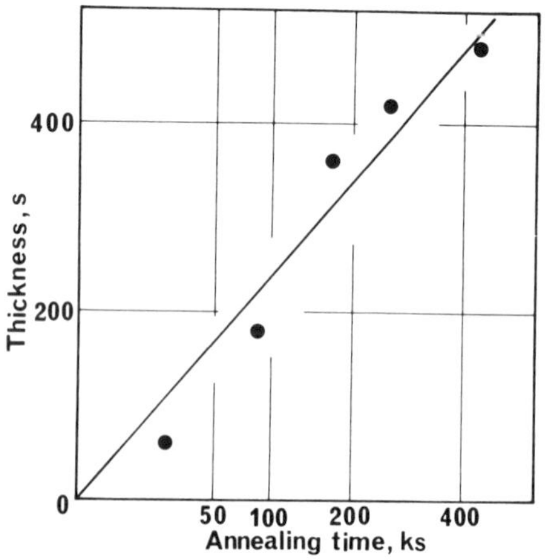

Fig.6 - The relation between the thickness of titanium carbide layer and the square root of annealing time at 1100K

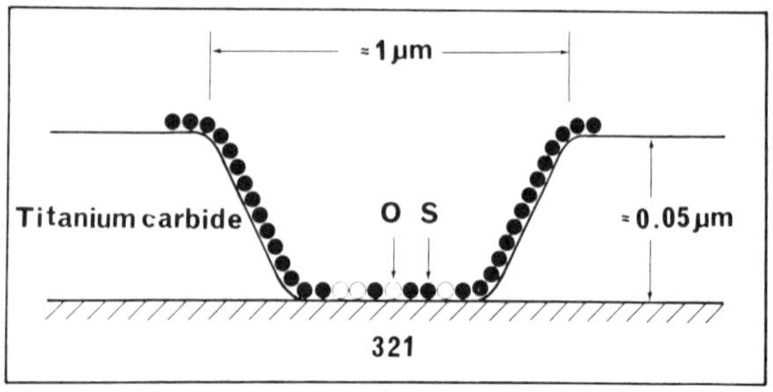

Fig.7 - The schematic illustration of titanium carbide layer on 321 stainless steel

MECHANISM OF THE PRECIPITATION OF TITANIUM CARBIDE - Thermodynamic consideration showed that compound which was precipitated in the bulk comes onto the surface in order to reduce the strain energy.

The precipitation of titanium carbide on the surface occurs in the following processes: 1. The titanium and carbon dissolved in the bulk diffuse onto the surface. 2. The titanium and carbon combine into titanium carbide and precipitate on the surface, because the precipitation on the surface is thermodynamically more stable than that in the bulk. 3. The titanium carbide in the bulk dissolves into matrix because the concentrations of titanium and carbon became under their solubility limits in the bulk. 4. The titanium and carbon diffuse onto the surface. The parabolic character of the growth rate of the thickness of the precipitated layer suggests that the rate will be controlled by the diffusion of titanium and carbon in the precipitated titanium carbide layer.

ADHERENCE OF DEPOSITED TITANIUM CARBIDE ON 304 AND 321 STANLESS STEELS- 321 stainless steel had been heated at 1100K for 86.4ks(24hr) before ion plating of titanium carbide. After deposition of titanium carbide on 304 and preheated 321 stainless steels by ion plating, the thermal cycle tests were given(1100K-room temperature) in vacuum. The SEM images of the surfaces after the test were shown in Fig.8. The cracks generated because of the thermal stress between deposited titanium carbide and steels. The average width of cracks of

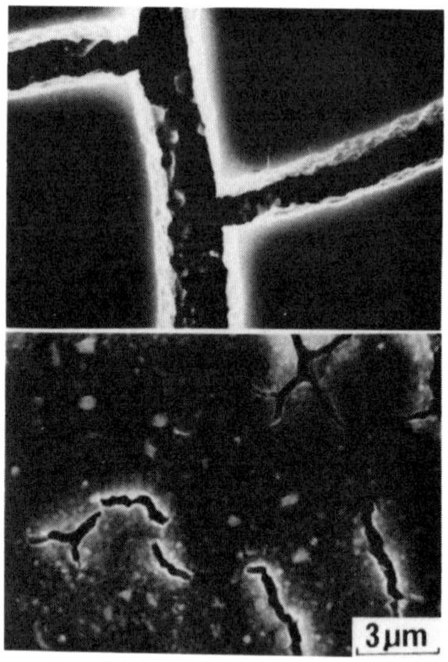

TiC on 304

TiC on preheated 321

Fig.8 - The SEM images of titanium carbide coated stainless steels after thermal cycle test

titanium carbide deposited on 304 stainless steel was much wider than that on 321 stainless steel, and also the crack on 304 stainless steel was continuous. This means that the thermal stress were reduced in the case of 321 stainless steel.

The depth profiles of coated steels are shown in Fig.9. The thickness of deposited carbide was about 0.1μm. In the case of 304 stainless steel, the change of the composition occurred sharply at the interface, while that of 321 stainless steel did not. The gradual change of the composition at the interface could reduce the thermal stress.

These results shows that precipitated titanium carbide could be a binding agent between ceramics and metals.

PRECIPITATION OF BORON NITRIDE

Stainless steels are widely used as vacuum

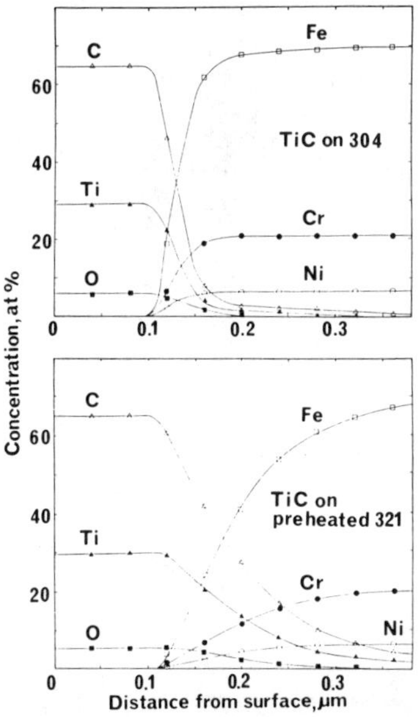

Fig.9 - The depth profiles of titanium carbide coated stainless steels

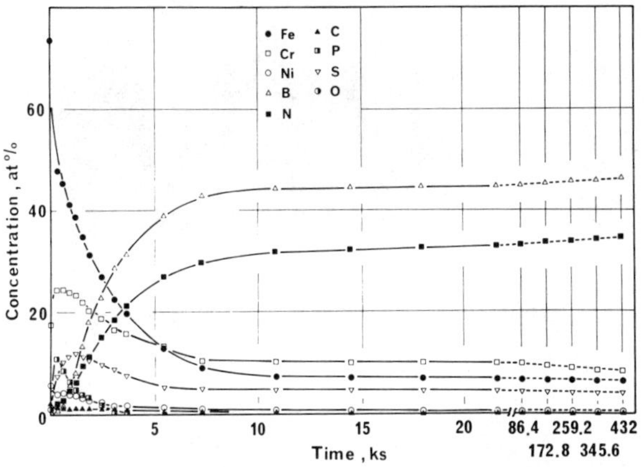

Fig.10 - Change of the surface composition of 304-NB stainless steel at 1100K

vessel materials. The most important properties for vacuum vessel materials are the adsorption and desorption rates of gases. These properties are strongly affected by the surface compositions of materials.

Boron nitride is considered to be inert to the adsorption of gases(4). Therefore, if the surface of stainless steel is covered with boron nitride, this surface will be a suitable vacuum vessel wall. Stulen et al. reported that boron nitride spottily precipitated on the surface of austenitic stainless steel doped with 0.272wt% of nitrogen(5). The most important requirement of developing a inert vacuum vessel wall is complete coverage of the surface of the wall with boron nitride.

Sulfur is known to be a very surface active element, so that it may interfere with the precipitation of a boron nitride layer on the surface. In this experiment, therefore, two types of stainless steels were laboratory melted. One was 304 stainless steel doped with nitrogen and boron (304-NB) and the other was 304 stainless steel doped with nitrogen, boron and cerium (304-NBCe). Cerium was doped because it reacts with sulfur to form a stable compound in the bulk.

The changes of surface compositions of these stainless steels were observed using AES and XPS at high temperatures in vacuum, and the quantity of adsorbed water vapor on these was measured by flash desorption method.

PRECIPITATION BEHAVIOR ON 304-NB STAINLESS STEEL - After the surface was cleaned by argon ion sputtering (acceleration voltage; 3kV, current density; 0.6A/m^2), the temperature of the specimen was rapidly raised and maintained at 1100K. The change of the surface composition with time is shown in Fig.10. On heating, the surface concentration of phosphorus first increased. As heating continued, the surface concentration of sulfur increased and that of phosphorus decreased. Then, the concentrations of boron and nitrogen increased, while that of sulfur started to decrease and those of iron, chromium and nickel also decreased. On the surface of 304-NB stainless steel, boron and nitrogen were not replaced by sulfur even after prolonged heating.

The shapes of Auger peakes of boron and nitrogen on the surface of the specimen heated at 1100K for 432ks(120hr) were similar to those of boron nitride which were reported by Stulen et al(5). The values of K levels of boron and nitrogen from this specimen observed with XPS were 191.2eV and 398.7eV, respectively. These values were well coincided with those of boron nitride which were reported by Stulen et al. and Hamlin et al.(boron; 190.4eV(5), nitrogen; 398.2eV(6)). The K level of elemental boron was 186.7eV(7). From these results, it is concluded that enriched boron and nitrogen on the surface existe as boron nitride.

The specimen was also heated in vacuum at 1100K and the changes of distributions of elements on the surface were observed with SAM. The Auger images of boron(169eV), nitrogen (380eV), iron(703eV) and sulfur(152eV) on the surface of the specimen heated at 1100K for various times up to 21.6ks(6hr) are shown in Figs.11, 12 and 13. Auger images of boron and nitrogen completely coincide, and those of iron and sulfur are the negative images of those of boron and nitrogen(Fig.11). The area covered with nitrogen was increased as heating time was prolonged(Fig.12) and eventually, boron nitride replaced segregated sulfur (Fig.13).

The thickness of the boron nitride layer was estimated by argon ion sputtering. In this experiment, the sputtering rate was about 0.1nm/s. The change of Auger peak heights for boron and nitrogen with sputtering time are similar. The thickness of the boron nitride layer was estimated from the sputtering time required for Auger peak height to decrease abruptly. From this measurements, the thickness of boron nitride was estimated to be about 3.5nm.

Boron nitride also precipitated at 1000K, but the rate was smaller than that at 1100K.

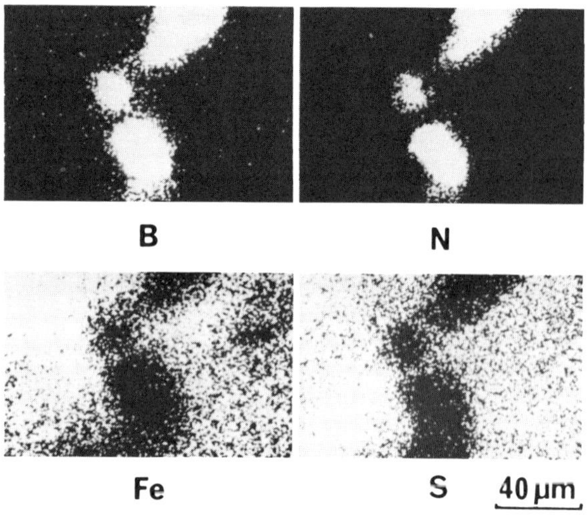

Fig.11 - Auger images of boron, nitrogen, iron and sulfur on the surface of 304-NB stainless steel heated at 1100K for 0.6ks

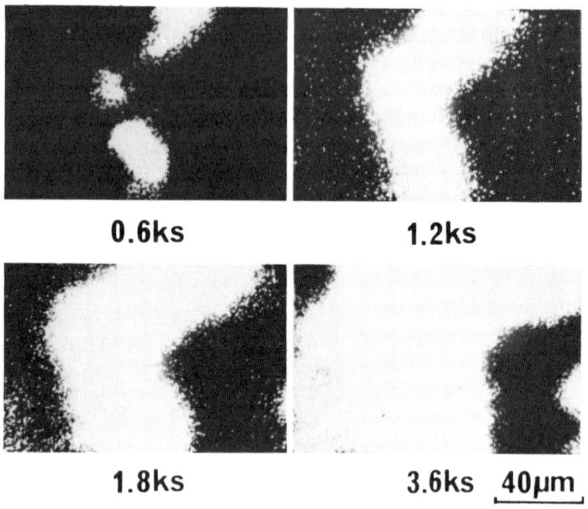

Fig.12 - Change of Auger image of nitrogen on the surface of 304-NB stainless steel with heating time at 1100K

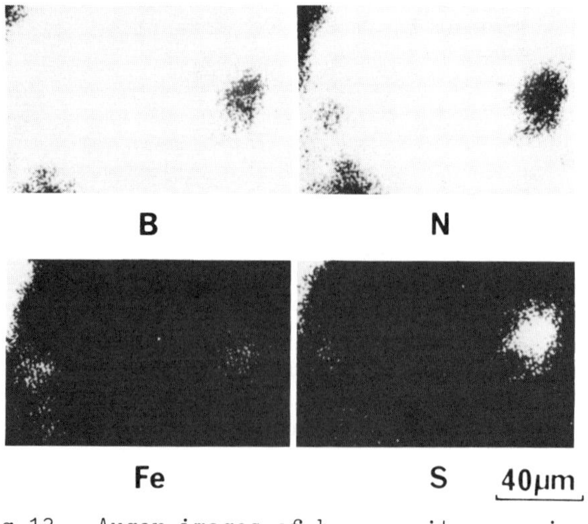

Fig.13 - Auger images of boron, nitrogen, iron and sulfur on the surface of 304-NB stainless steel heated at 1100K for 21.6ks

PRECIPITATION BEHAVIOR ON 304-NBCe STAINLESS STEEL - After the surface was cleaned by argon ion sputtering, the temperature of the specimen was rapidly raised to 1100K and kept at that temperature. The change in surface composition with time is shown in Fig.14. Contrary to the case of 304-NB stainless steel,

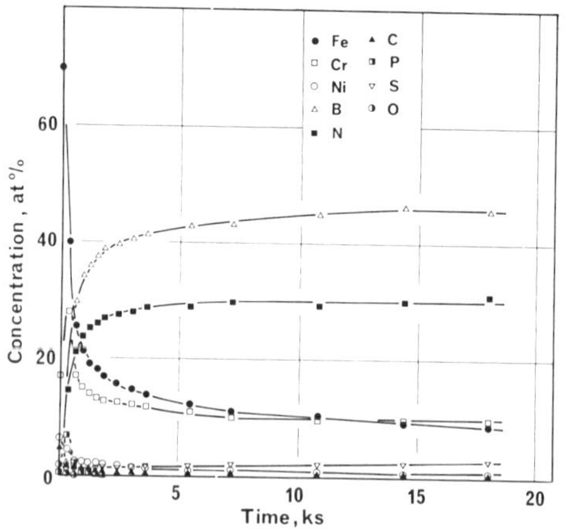

Fig.14 - Change of the surface composition of 304-NBCe stainless steel at 1100K

the segregation of sulfur to the surface was suppressed, and boron and nitrogen were enriched rapidly. The shape of Auger peaks of boron and nitrogen were similar to those of boron nitride, suggesting that precipitation of boron nitride was enhanced on the surface of 304-NBCe stainless steel.

The distributions of elements on the surface were observed with SAM. The Auger images of boron and nitrogen on the surface of the specimen heated at 1100K for 0.6ks (10min) and for 1.2ks(20min) are shown in Fig.15, where boron nitride is shown to have precipitated more uniformly on the surface of 304-NBCe stainless steel than on that of 304-NB stainless steel.

THE SPREADING MECHANISM OF BORON NITRIDE ON 304-NB STAINLESS STEEL - When 304-NB

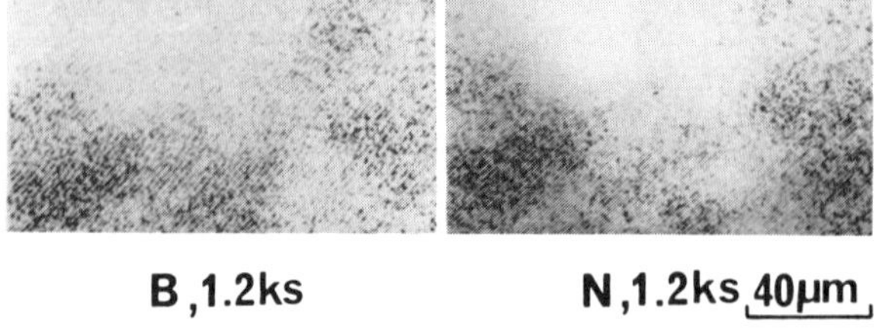

Fig.15 - Auger images of boron and nitrogen on the surface of 304-NBCe stainless steel heated at 1100K for 0.6ks and 1.2ks

stainless steel was heated in vacuum, a thin boron nitride layer spread on the surface. Boron nitride replaced segregated sulfur.

From Fig.12, it was possible to draw lines of the periphery of boron nitride layer on the surface at various heating time. The lines are shown in Fig.16. In the bulk of stainless

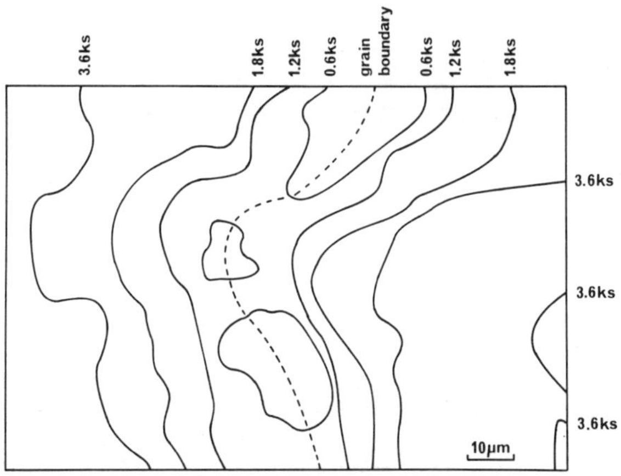

Fig.16 - Contour lines of boron nitride layer on the surface of 304-NB stainless steel at various heating time

steel doped with boron and nitrogen, boron nitride precipitates near grain boundaries(8). Therefore, the precipitation of boron nitride on the surface must originate from the grain boundary. However, it was impossible to get direct evidence of the grain boundary site in this experiment. Therefore, a grain boundary is assumed to exist near the center of the area covered with boron nitride shown in Fig 16. Figure 16 shows that the distances between the periphery of the boron nitride layer and the grain boundary increase almost linearly with time for 1.8ks(30min). However, the spreading velocity becomes small after 1.8ks(30min). This means that at the initial stage of heating, boron nitride spreads at a constant velocity. Dissolved boron and nitrogen in the bulk should be supplied to the surface by grain boundary diffusion and form boron nitride on the surface. Surface precipitated boron nitride is thermodynamically more stable than that in the bulk because of an absence of stress, and thus the precipitation on the surface continues. Then the concentration of dissolved boron and nitrogen in the bulk decreased near the surface. The supplying velocity from the bulk was large at the initial stage of heating, so the spreading velocity of boron nitride layer was controlled by a surface wetting mechanism. However, after prolonged heating, the diffusion flux of boron and nitrogen became smaller, decreasing approximately with the square root of time. Therefore, the spreading velocity was assumed to be controlled by diffusion.

THE SPREADING MECHANISM OF BORON NITRIDE ON 304-NBCe STAINLESS STEEL - When 304-NBCe stainless steel was heated in vacuum, almost all of the surface was quickly covered with boron nitride. In this stailess steel, boron nitride precipitates on the very stable cerium oxi-sulfide particles which precipitates inside of grains(9). In this alloy, precipitates of boron nitride exist uniformly inside of grains and not on grain boundaries. Since the sources of boron and nitrogen are dispersed in the bulk of steel, boron and nitrogen were supplied to the surface uniformly from the bulk, when this steel was heated in vacuum.

Sulfur may interfere with the spreading of boron nitride layer on the surface because it is a very surface active element. Sulfur was, however, strongly trapped in the bulk by cerium in 304-NBCe stainless steel. Therefore, the segregation of sulfur was suppressed. This effect also assists the spreading of boron nitride layer. Manganese sulfide is also reported to precipitate inside of grains with boron nitride precipitated around it(10). However, manganese sulfide is not so stable to hinder the segregation of sulfur on the surface of the alloy doped with manganese (11). Therefore, in the case of 304-NB stainless steel, boron nitride was not expected to be bound to manganese sulfide.

GAS DESORPTION BEHAVIOR OF 304-NBCe STAINLESS STEEL -304-NB stainless steel was heated at 1100K for 21.6ks(6hr) in vacuum, and then was exposed to the air for 3.6ks(1hr) (temperature; 298K, relative humidity; 50%). The surface of 304-NB stainless steel was only partially covered with boron nitride. The Auger images of boron, nitrogen and oxygen (510eV) for the specimen after it was exposed to the air are shown in Fig.17. From Fig.13 and Fig.17, it is inferred that oxygen adsorbs hardly on the area covered with boron nitride, but it adsorbs on sulfur segregating area.

304-NBCe stainless steel was heated at 1100K for 0.6ks (10min) in vacuum, and then was exposed to the air for 3.6ks(1hr). This specimen was installed in an apparatus for flash desorption and the change of the peak height of mass number of 18 was observed during

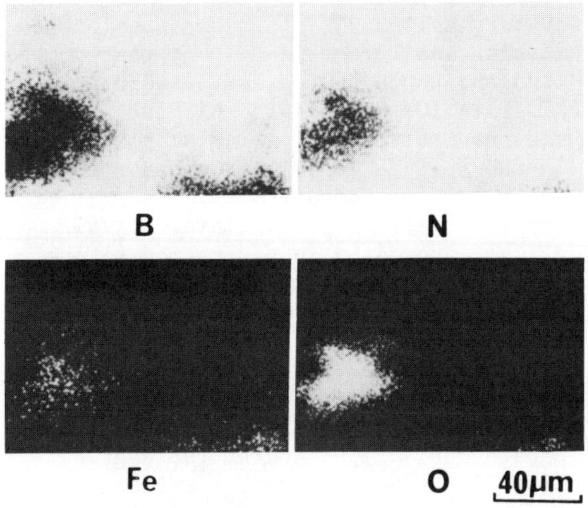

Fig.17 - Auger images of boron, nitrogen, iron and oxygen on the surface of 304-NB stainless steel after exposed to the air

heating process. For comparison, commercial 304 stainless steel was heated at 1100K in vacuum, and the desorption behavior was also observed under the same conditions.

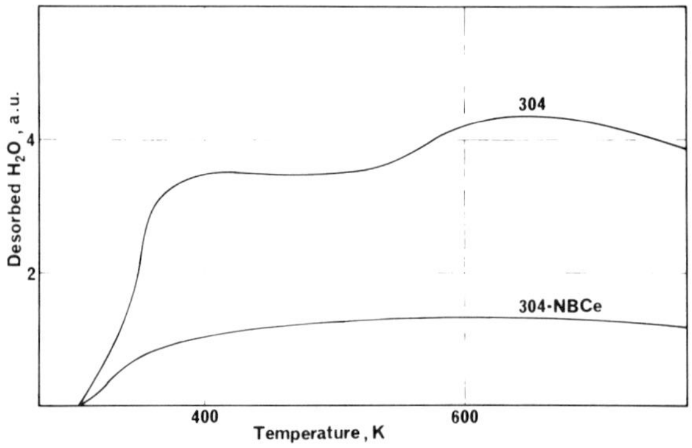

Fig.18 - Gas(water vapor) desorption behavior of 304 and 304 NBCe stainless steels

Fig.18 shows the flash desorption behavior of 304-NBCe and 304 stainless steels. The ordinate is the observed peak height of mass number of 18(arbitrary unit) and the abscissa is temperature. As for 304 stainless steel, two peaks were observed. One is at about 400K, which means physisorption, and the other is about 800K, which means chemisorption. In the case of 304-NBCe stainless steel, no peak was observed and the quantity of desorbed water vapor was much less than that of 304 stainless steel. This means that 304-NBCe stainless steel is a superior candidate material for vacuum vessels.

CONCLUSIONS

The precipitation behaviors on the surface of two types of stainless steels were investigated at high temperatures in vacuum using AES and XPS. One was the precipitation of titanium carbide on the surface of commercial 321 stainless steel and the other was the precipitation behavior of boron nitride on the surface of laboratory-melted 304 stainless steel doped with nitrogen and boron(304-NB) and that doped with nitrogen, boron and cerium(304-NBCe).

PRECIPITATION OF TITANIUM CARBIDE -

(1) The precipitation of titanium carbide was found on the surface of 321 stainless steel above about 900K in vacuum.

(2) The thickness of precipitated titanium carbide layer increased in proportion to the square root of heating time and became about 0.05μm after being heated at 1100K for 432ks (120hr).

(3) The precipitated layer was not unifom and had many holes of about 1μm in diameter. The bottom of a hole was 321 stainless steel, on which phosphorus, oxygen and sulfur segregated. As heating time was prolonged, these segregants were replaced one by one in the order of the surface activity, and finally the most surface active element, sulfur remained on the bottom of the hole.

(4) The titanium carbide precipitated on the surface was not removed by the argon ion sputtering at 1100K. This was because titanium and carbon were supplied to the surface at a sufficient velocity by the diffusion during the sputtering at high temperatures.

(5) The titanium carbide film ion plated on the surface of 321 stainless steel was stable agaist the thermal stress. This means that the precipitated titanium carbide could be a bonding agent between ceramics and metals.

PRECIPITATION OF BORON NITRIDE -

(1) Boron nitride precipitated on the surfaces of 304-NB and 304-NBCe stainless steels above about 1000K.

(2) For 304-NB stainless steel, a non-uniform, thin layer of boron nitride formed, originated from the grain boundary and spread over the surface.

(3) In the case of 304-NBCe stainless steel, the born nitride layer almost uniformly covered the surface at the initial stage of heating. This is because the source of boron and nitrogen is dispersed uniformly in the bulk steel and in this steel cerium formed a stable compound with sulfur to inhibit its surface segregation.

(4) Gas desorption behavior of 304-NBCe stainless steel was compared with that of commercial 304 stainless steel. The surface covered with boron nitride was inert to the adsorption of gas. Therefore, 304-NBCe stainless steel will be a candidate material for vacuum vessels.

REFERENCES

(1) K.Yoshihara, M.Kurahashi and K.Nii, Trans. Japan Inst. Metals 21,425(1980).

(2) Physical Electronics Industries Inc., Data Sheet No 7315,(1973).

(3) C.C.Chang, "Characterization of Solid Surfaces", p.509, Ed. by P.F.Kane and G.B. Larrabee, Plenum, New York (1974).

(4) H.E.Thomas, R.N.Ramsey and R.A.Pierotti, J. Chem. Phys. 59,6163(1973).

(5) R.H.Stulen and R.N.Bastasz, J. Vac. Sci. Technol. 59,6163(1973).

(6) K.Hamlin, G.Johansson, V.Gerius, C.Nording and K.Siegbahn, Phys. Scr. 1,277 (1970).

(7) G.Mavel, J.Escard, P.Costa and J. Castaing, Surface Sci. 35,109(1973).

(8) Y.Kawabe and R.Nakagawa, Tetsu to Hagane 53,467(1967).

(9) T.Funakoshi, T.Tanaka, S.Ueda, M. Ishikawa, N.Koshizuka and K.Kobayashi, Tetsu to Hagane 63,105(1977)

(10) M.Tanino, S.Funaki, H.Komatsu and E.Chow, Tetsu to Hagane 65,577(1980)

(11) K.Yoshihara and K.Nii, J. Japan Inst. Metals 45,283(1981).

APPLICATIONS OF SELF-RENEWING COATINGS TO IMPROVED VACUUM MATERIALS, HYDROGEN PERMEATION BARRIERS AND SPUTTER-RESISTANT MATERIALS

A. R. Krauss
Argonne National Laboratory
Argonne, Illinois, USA

ABSTRACT

The phenomena of Gibbsian segregation, radiation-induced segregation and radiation-induced precipitation modify the surface composition and properties of alloys and compounds. In some cases, the change in properties is both substantial and useful, the most notable example being that of stainless steel. When surface-modifying phenomena are investigated as a class, a number of additional materials emerge as candidates for study, having potential applications in a number of technologically important areas. These materials are predicted to produce self-sustaining coatings which provide hydrogen permeation barriers, low-sticking and stimulated desorption coefficients for vacuum applications, and low-Z, sputtering-resistant surfaces for fusion applications. Several examples of each type of material are presented, along with a discussion of the experimental verification of their properties and the status of the corresponding applications development program.

IN A GREAT MANY AREAS of technology the properties desired of the surface or near-surface region of a material do not correspond to the properties of the bulk material. In such cases the surface must be suitably modified. Relevant surface and near-surface properties include corrosion behavior, gas permeability, gas adsorption and desorption properties, ion-impact sputtering, catalytic behavior, and properties which depend on sub-surface impurity concentration such as semiconductor surface conductivity.

A number of techniques are available for modifying surface and near-surface properties, producing films and modified layers of varying thicknesses. Coating, plating, and cladding techniques produce coatings ranging in thickness from a few Angstroms to arbitrarily thick layers. The coating composition is often totally unrelated to the substrate composition. In some cases, the coating and substrate are sufficiently dissimilar that the coating peels as the result of, e.g., thermal shock. Although there are techniques to promote bonding between dissimilar materials, the integrity of the interface is a major concern for these processes.

Chemical reaction techniques are used in processes such as anodizing to produce a film with high bond strength to the substrate. Such techniques are normally used in systems in which the thickness tends to be inherently self-limiting, usually not exceeding a few hundred Angstroms.

Ion implantation produces an altered region in which the properties and composition merge in a continuous fashion with those of the substrate material. The thickness of the altered layer depends on the mass and energy of the implanted species but seldom exceeds a few microns. The technique may be used in an obvious manner for, e.g., semiconductor doping but can also be used for such purposes as surface hardening.

Willard Gibbs predicted in 1892 [1] that at thermodynamic equilibrium, the surface of an alloy tends toward the elemental composition corresponding to the minimum surface free energy. We have applied relatively simple models to predict the surface composition of a number of alloy systems. Several of these calculations predict the existence of very strongly segregating alloy systems in which self-sustaining solute coatings form for very dilute solute concentrations. We will discuss here some of the systems which appear to have technological significance and, where possible, mention experimental results related to the accuracy of the calculations and the potential uses of the alloy.

The two principal factors affecting the surface free energy are the bond enthalpy and the lattice strain energy.[2] The first factor tends to produce a surface which is enriched in the alloy component having the lowest heat of sublimation. If there is significant mismatch in atomic radii of the alloy components, the lattice strain tends to drive the minority component to the surface. To a first approximation, the modified surface composition occurs only in the uppermost atomic layer [3] although more careful analysis shows deviations from bulk composition to a depth of 2-3 atomic layers.[4] The solute concentration profile at equilibrium depends on the properties of the solid solution.

Calculated solute concentration profiles for Al-Li and Cu-Ni alloys are shown in Figs. 1 and 2. For Au-Ni, the gold concentration of the

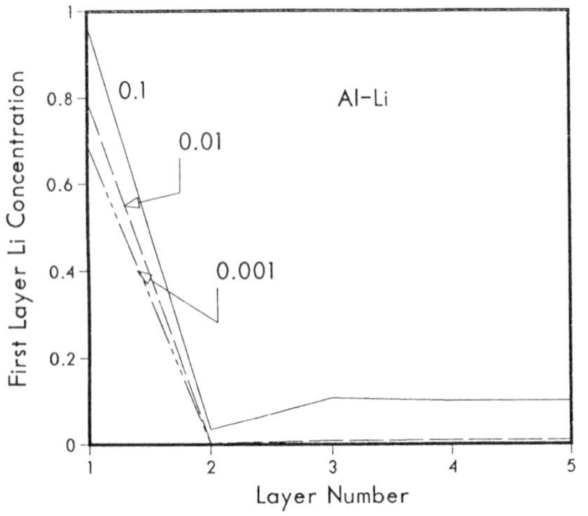

Fig. 1 - Thermal equilibrium depth profile of lithium in Al-Li alloys consisting of 0.1, 1, and 10 atm.% Li vs. temperature. Calculated according to the model of Williams and Nason.[4]

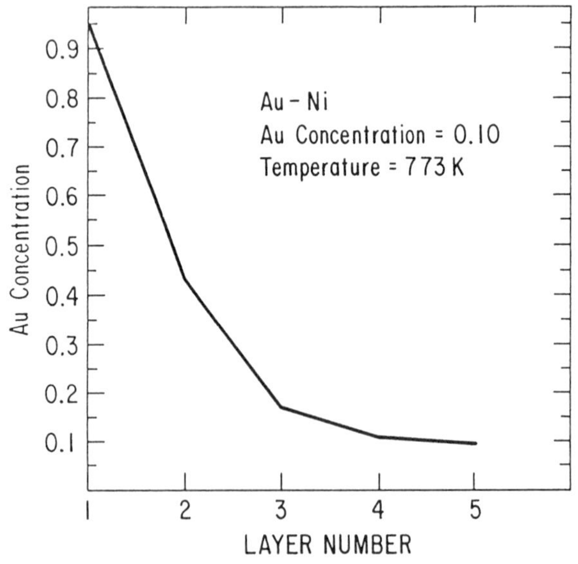

Fig. 2 - Thermal equilibrium depth profile of nickel in Cu-Ni and lithium in Al-Li alloy as a function of temperature. Calculated according to the model of Williams and Nason.[4]

first atomic layer is predicted to be nearly 100%, decreasing monotonically to the bulk concentration by the third atomic layer. For Al-Li, the second layer lithium concentration is below that of the bulk alloy, rising to the bulk value by the third or fourth layer.

The most common example of Gibbsian segregation as a means of surface modification is found in stainless steel. Although stainless steels normally consist of a number of elements, they can be thought of as being largely a Fe-Cr binary alloy. Figure 3 shows the chromium fraction of Fe-Cr calculated according to a model

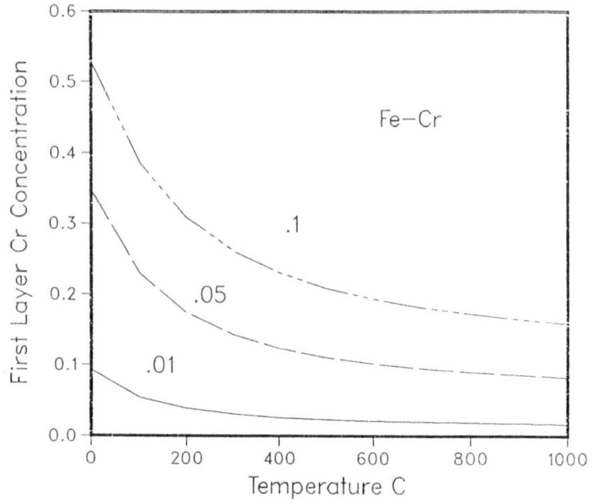

Fig. 3 - Concentration of Cr in the first atomic layer of Fe-Cr alloys vs. temperature. Calculated according to the model of Miedema.[5,6]

based on Miedema's alloying rules [5,6], which incorporate both bond enthalpy and size mismatch effects. It is predicted that a binary alloy containing 10 at.% Cr will have a surface consisting of 53% Cr at room temperature, decreasing at higher temperatures. Experimental measurements on stainless 410 samples containing 11.9 at.% Cr in the near-surface region [7] find that the surface contains 63±10 at.% Cr at 500C.

In another well-known application, Cu-Au alloys are used for electrical contacts. Gold segregates to the surface, providing an oxidation-resistant layer. If the gold layer is scratched by, e.g., contact insertion, the gold layer will spontaneously reform. The gold layer is therefore self-forming and self-sustaining. If however, the surface is scratched while exposed to oxygen at high temperature, copper oxide will form, inhibiting the formation of a new gold layer.

A number of other strongly segregating alloy systems present potentially significant technological advantages but are relatively unknown. We present here a discussion of the properties of some of these materials, along with a discussion of the experimental and developmental work to date on these systems.

RESULTS

ALKALI METAL COATINGS - We have examined the segregation properties of several alkali metal-bearing binary alloy systems. For those systems which form a solid solution, the alkali metal component segregates very strongly, forming a complete overlayer for even very dilute alloys. Several of these systems have interesting technological applications.

One such alloy is copper-lithium. Because of its high thermal conductivity, copper is an excellent material for use in high heat flux areas such as the limiters or divertor plates of magnetic confinement fusion devices. These areas are however subject to severe erosion by ion impact sputtering. Some of the sputtered atoms are ionized in the plasma and accelerated by the unipolar sheath potential between the plasma and the limiter, returning to the surface with a kinetic energy up to ten times that of the effective edge temperature of the plasma. A runaway self-sputtering condition can therefore occur for moderate plasma edge temperatures.[8] Tungsten, although less than ideal in some respects, is much more resistant to self-sputtering than copper. The self-sputtering yields for copper and tungsten as a function of impact energy, as calculated by the computer code TRIM [9], are shown in Fig. 4. Self-sputtering yields in excess of unity occur at 340 eV for copper and at 1000 eV for tungsten. However, because of the high charge state acquired by tungsten impurities in a fusion plasma, the maintenance of a plasma edge temperature low enough to avoid runaway self-sputtering of tungsten poses an extremely difficult design problem.

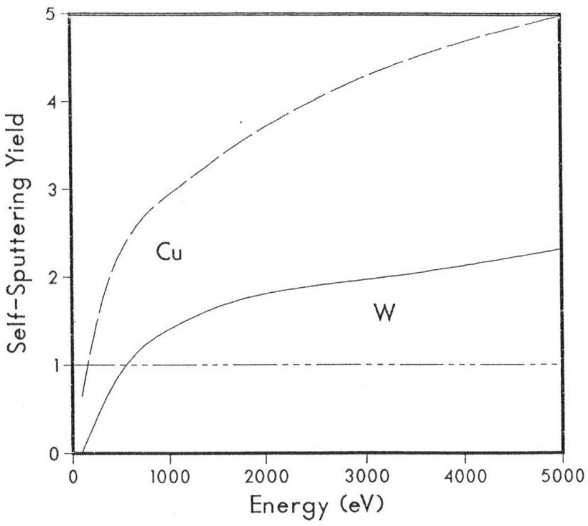

Fig. 4 - Cu and W self-sputtering yield vs. impact energy.[25]

The use of self-sustaining coatings in copper-lithium alloys has been investigated [10,11] as a means of reducing the self-sputtering of a limiter or divertor structure. As shown in Fig. 5, thermodynamic equilibrium corresponds to a surface consisting almost entirely of lithium. It is theoretically expected that lithium resides on the surface as an ionic impurity, [12] and it has been experimentally demonstrated [13] that the lithium sputters largely as secondary ions. The unipolar sheath potential acts to trap the secondary ions at the limiter surface, thereby lowering the erosion rate of the lithium coating in the presence of the electric fields associated with a fusion environment.

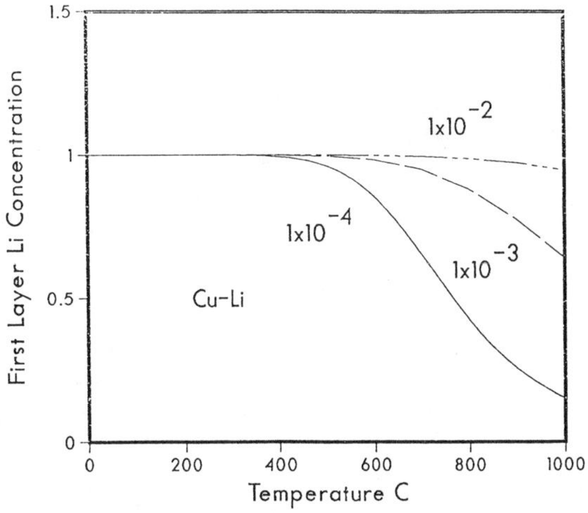

Fig. 5 - Concentration of Li in the first atomic layer of Cu-Li alloys vs. temperature. Calculated according to the model of Miedema.

TRIM calculations predict, and experiments [14] confirm, that sputtered atoms originate almost entirely in the first 1-2 atomic layers. It is therefore expected that even the very thin layers produced by Gibbsian segregation would substantially reduce the erosion rate of the major alloy component. In practice, radiation-related effects such as radiation-enhanced diffusion, radiation-induced segregation, preferential sputtering and recoil implantation result in subsurface solute enrichment and surface depletion relative to the value associated with pure Gibbsian segregation.[15]

The TRIM calculations of the sputtering yield for copper incident on a surface consisting of pure copper and copper covered with 1 and 1.5 monolayers of lithium are shown in Fig. 6. Depending on the thickness of the lithium overlayer, the energy corresponding to runaway copper self-sputtering may exceed 3 KeV, thereby eliminating the self-sputtering problem in fusion reactors.

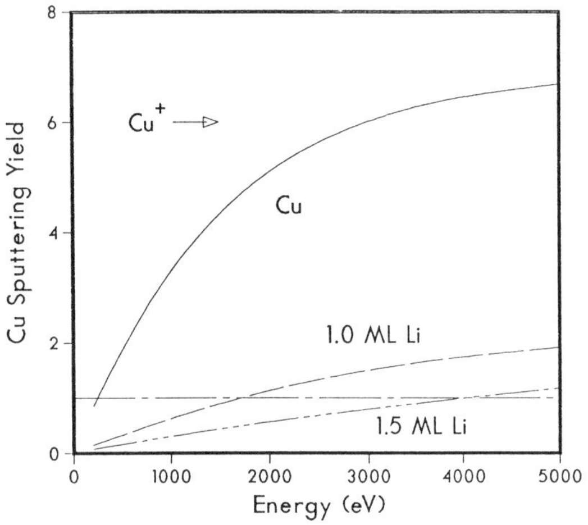

Fig. 6 - Calculated sputtering yield for copper ions incident on pure copper and on materials consisting of 1 and 1.5 monolayers of lithium on a copper substrate.

Preliminary experiments [16] using weight loss and spectroscopic techniques as a means of measuring the effective erosion rate of Cu-Li in the presence of a plasma sheath potential have shown a 2-6x reduction in the rate of copper erosion.

Alternative strategies for reducing the erosion of a limiter based on tungsten or tantalum are shown in Fig. 7. The structure consists of porous tungsten or tantalum, prepared by powder metallurgical or plasma spray methods, and deposited on an actively cooled substrate of Cu, W, or Ta. The porous layer is impregnated with either an alkali metal alumino-silicate similar to the materials used to produce alkali ion sources [17] or with an alkali metal-bearing alloy such as Cu-Li or Al-Li.

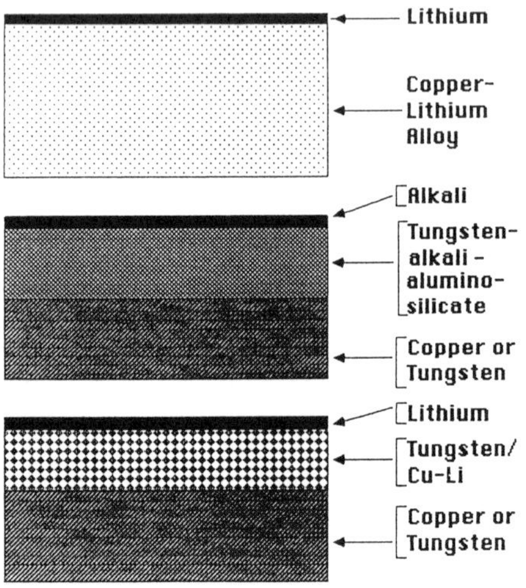

Fig. 7 - Schematic representation of the near-surface composition of Cu-Li binary alloy and composite structures consisting of W (Ta) and a binary alkali metal-bearing alloy or an alkali metal alumino-silicate on an actively cooled substrate.

When heated by, e.g., a high flux of plasma particles, the alkali metal diffuses along the pores and intergranular surfaces forming an alkali metal coating. The degree to which the alkali metal overlayer is expected to reduce the substrate erosion depends on the actual density of the alkali metal at the surface. Experiments are in progress at our laboratory to determine the alkali metal atom density at the surface. However, preliminary TRIM calculations indicate a possibility of producing materials with extreme sputtering resistance in a fusion environment by this technique.[18]

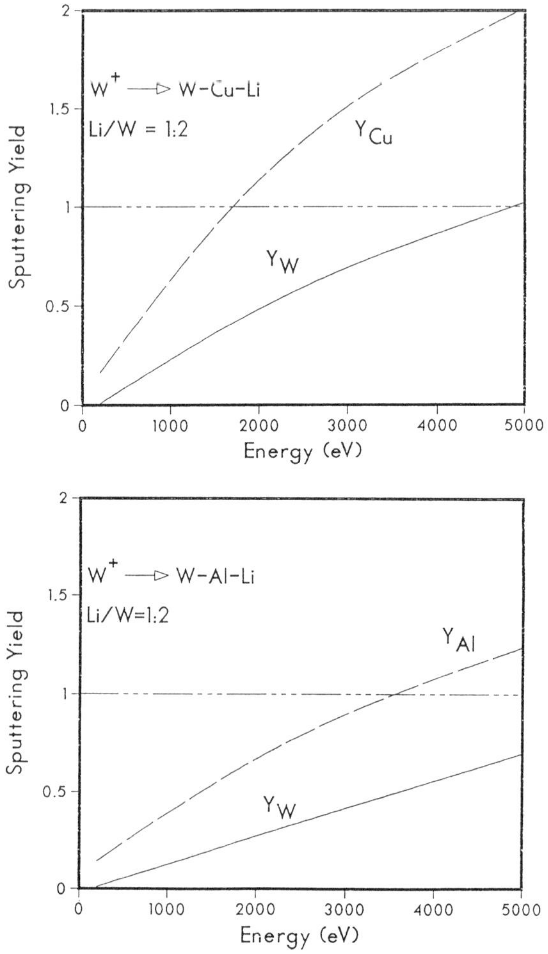

Fig. 8 - Calculated sputtering yields for W-Cu-Li and W-Al-Li composites. The lithium layer is arbitrarily assumed to contain one-half as many atoms as the W-Al or W-Cu substrate.

Alkali ion sources based on the alumino-silicate filler are commercially available [19] and are capable of supplying alkali metal atoms to the surface at rates in excess of $10^{16} cm^{-2} sec^{-1}$ [17] with lifetime limited only by the total available alkali metal inventory. For the alloy-filled matrix, the refractory metal structure allows high temperature operation with attendant high alkali metal diffusion rates, and insures excellent thermal conductivity while maintaining the stability of the filler metal melt layer in the event of an unplanned temperature excursion such as that caused by a plasma disruption. The filler metal used and the ratio of tungsten to filler metal also determines the sputtering properties of the composite material. For example, Fig. 8 shows the predicted self-sputtering yields of the two components of two binary metal-lithium composites. Because of the mass mismatch between tungsten and aluminum, the W-Al-Li composite has a much lower self-sputtering yield than does the W-Cu-Li composite.

In a totally different vein, it has been observed [27] that the strong segregation of lithium in silicon can be used as a means of creating low work function surfaces for use as photocathodes. Current techniques used for the activation of photo-emitting surfaces rely on the cesiation of the surface as a separate step in production. Self-forming layers of lithium in silicon have demonstrated quantum efficiencies of about 10%.

ALUMINUM COATINGS - The calculated first atomic layer equilibrium composition of several aluminum-bearing binary alloys is shown in Fig. 9. These alloys represent some of the materials currently under investigation in our laboratory. For the alloys shown, very low bulk aluminum concentrations are sufficient to insure the formation of an aluminum coating. As can be seen, by comparison with Fig. 3, aluminum appears to segregate to the surface of iron much more strongly than does the chromium which provides the protective coating of stainless steel. The formation of self-sustaining aluminum coatings has interesting technological consequences.

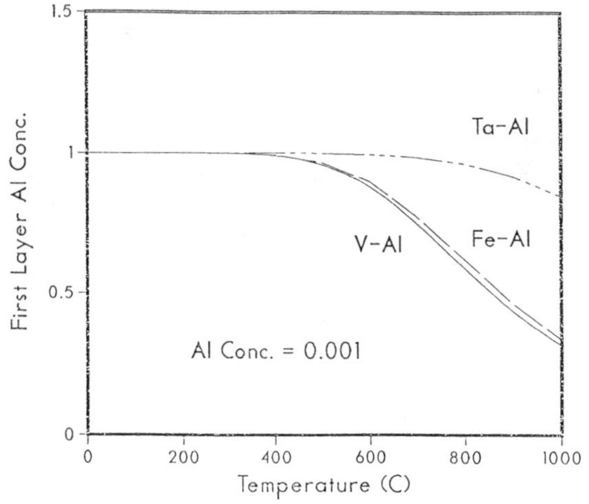

Fig. 9 - Calculated first monolayer concentration of aluminum in Fe - 0.1 at.% Al, V - 0.1% at.% Al, and Ta - 0.1 at.% Al alloys.

The ultimate pressure attainable in stainless steel ultrahigh vacuum systems is limited by the bulk diffusion of hydrogen from the atmosphere. VanDeventer and Maroni [20,21] have noted that the addition of 2 at.% aluminum to 400 series stainless steels reduces the hydrogen permeation rate by about 2 orders of magnitude, and that oxidation of the surface reduces the hydrogen permeability by an additional factor of 10. The reduced hydrogen solubility associated with such low bulk aluminum concentration is not nearly adequate to explain the size of this effect, and it is to be assumed that the reduced permeability is a surface effect resulting from the formation of an aluminum overlayer in accord with the calculations shown in Fig. 9. If an aluminum oxide-like layer is formed, an even greater reduction in hydrogen permeation is expected. The hydrogen permeability through aluminum oxide is so low that the permeation rate is limited by the completeness of the layer, rather than its thickness. Self-healing properties are clearly desirable, and thick Al_2O_3 layers are in fact undesirable as hydrogen permeation barriers because of their tendency to crack and separate from the substrate. The use of self-sustaining aluminum layer formation in aluminum-bearing steels, therefore, represents a possible avenue to improved vacuum materials.

A similar problem is encountered for high-pressure applications involving the use of steel in hydrogen atmospheres. Hydrogen permeation represents a stumbling block to the operation of high-temperature Stirling engines which use hydrogen as a working medium. However, a number of aluminum- modified iron-based alloys have demonstrated superior performance [22] in this regard.

Fusion technology provides several areas in which hydrogen permeation barriers are needed. Because of the extensive experience in the fabrication of large stainless steel vacuum vessels, stainless steel is the preferred material for the vacuum envelope of large near-term fusion devices. However, the leakage of tritium through such a structure is substantially greater than the maximum permissible emission to the environment. The use of aluminum-modified stainless steel vacuum vessels appear to be a means of circumventing this problem.

As another example, vanadium has excellent thermal-mechanical properties and extremely low neutron activation. If it were not for the fact that it is subject to oxidative corrosion and hydrogen embrittlement, vanadium would make an excellent first wall material for a fusion reactor. Tantalum would make an excellent high heat flux material for fusion applications except that it is plagued by hydrogen-related problems similar to those of vanadium. As shown in Fig. 9, aluminum is expected to segregate very strongly in both of these metals. Auger

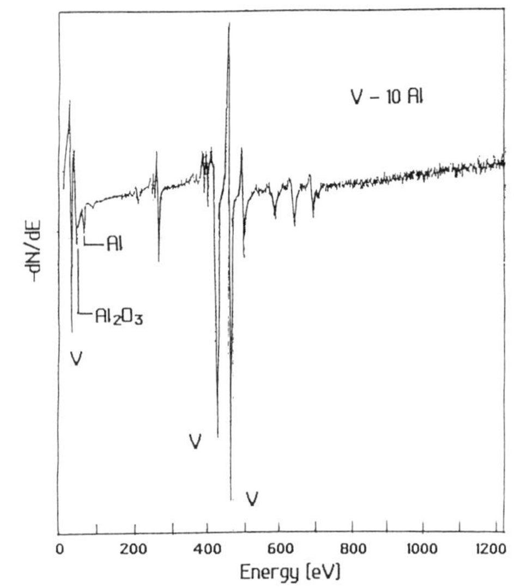

Fig. 10 - Auger spectrum of V - 10 at.% Al alloy at room temperature. Calibration against standard reference tables [26] indicates that the aluminum is equivalent to 1.05 monolayers on a pure V substrate.

spectroscopy has been used to observe the room temperature formation of an aluminum oxide layer one monolayer thick on V-10 Al (Fig. 10). This alloy has in preliminary tests demonstrated considerably better oxidation resistance than pure vanadium, and the formation of an aluminum or an aluminum oxide layer even one monolayer thick is expected to provide substantial resistance to hydrogen permeation.

GOLD COATINGS - For some vacuum applications, the critical surface property is the electron- or photon-induced desorption rate, rather than the thermal outgassing rate. Synchrotron light sources represent one such example. Several devices have recently experienced considerable difficulty reaching design goals as a result of beam-induced gas desorption. One way of circumventing such problems is to provide a surface which is as close to atomically clean as possible. Such a procedure is diametrically opposite to standard vacuum practice which relies on the low-sticking coefficient of oxidized stainless steel or aluminum surfaces. Unfortunately, oxide surfaces have extremely large electron- and photon-induced desorption cross sections.

Of all the metals, gold and silver come closest to nonadsorbing surfaces. For both economic and structural reasons, it is desirable to use the noble metal as a coating on a material such as stainless steel or aluminum, which can be welded and formed by conventional techniques. The coating must however be dense, uniform, and devoid of trapped gases or voids which may result from standard coating or plating techniques. One possible approach to producing such a layer on a structural steel vacuum wall is by making use of the segregation phenomenon to produce the gold layer. The solubility for gold is very low but, as shown in Fig. 11, it is predicted that extremely dilute alloys of Au in Fe, Ni, and Cr will form self-sustaining gold surfaces. Recent experimental results by Morita [23] and by Burton [24] have confirmed the formation of gold overlayers by segregation in Fe and Ni. It should be noted that the operating temperature is very important for some of these systems. If the temperature is too low, the formation rate of the gold layer is diffusion limited. If the temperature is too high, the surface gold concentration is limited by evaporative loss.

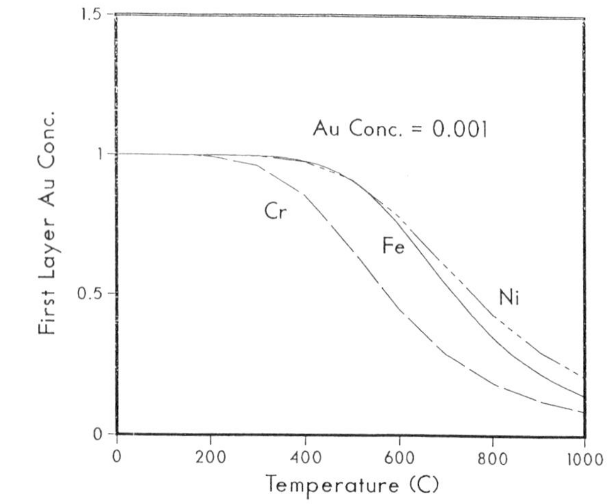

Fig. 11 - First monolayer gold concentration in alloys consisting of 0.1 at.% Au in Fe, Cr, and Ni.

SUMMARY

Although isolated instances of self-sustaining coatings in technological applications are well known, there does not appear to be general recognition of the possibility of systematically using segregation phenomena to produce "monolayer engineered" materials. We have identified several areas of application for such materials and, in some cases, have been able to experimentally verify the predicted properties. Several materials are being developed with the intent of producing engineering materials with specific properties based upon their segregation behavior.

REFERENCES

1. J. W. Gibbs, Trans. Connecticut Acad. Sci. 3, 108 (1875-76)

2. McLean, David, "Grain Boundaries in Metals," Clarendon Press, Oxford (1957)

3. P. Wynblatt and R. C. Ku, Surf. Sci. 65, 511 (1977)

4. F. L. Williams and D. Nason, Surf. Sci. 45, 377 (1974)

5. A. R. Miedema, Z. Metallkund 69, 455 (1978)

6. A. R. Miedema and J. W. F. Dorleijn, Surf. Sci. 95, 447 (1980)

7. Y. S. Ng and T. T. Tsong, Surf. Sci. 78, 419 (1978)

8. C. Boley, J. N. Brooks and Y. Kim, Argonne National Laboratory, Fusion Power Report, ANL-FPP-TM-171 (1983)

9. J. P. Biersack and L. G. Haggmark, Nucl. Instrum. Meths. 174, 257 (1980)

10. A. R. Krauss and D. M. Gruen, J. Nucl. Mater. 85/86, 1179 (1979)

11. A. R. Krauss and D. M. Gruen, J. Nucl. Mater. 93/94, 686 (1980)

12. J. W. Gadzuk, J. K. Hartman, and T. N. Rhodin, Phys. Rev. B4, 241 (1971)

13. A. R. Krauss, D. M. Gruen and M. Venugopalan, J. Nucl. Mater. 122/123, 1425 (1984)

14. M. F. Dumke, T. A. Tombrello, R. A. Weller, R. M. Honsley and E. H. Cirlin, Surf. Sci. 124, 407 (1983)

15. R. A. Johnson and N. Q. Lam, Phys. Rev. B13 4364 (1976), N. Q. Lam, G. K. Leaf and H. Wiedersich, J. Nucl. Mater. 88, 289 (1980)

16. R. Conn, D. Goebel, Y. Hirooka, University of California at Los Angeles, Los Angeles, CA; D. M. Gruen, and A. R. Krauss, Argonne National Laboratory, 9700 S. Cass Ave., Argonne, IL, to be published.

17. J. P. Blewett and E. J. Jones, Phys. Rev. 50, 464 (1936)

18. A. R. Krauss, D. M. Gruen, A. B. DeWald, and M. Valentine, to be published in the Proceedings of the 32nd National Symposium of the American Vacuum Society, Houston TX, Nov. 18-22, 1985

19. O. Heinz and R. T. Reaves, Rev. Sci. Inst. 39, 1229 (1968)

20. E. H. VanDeventer and V. A. Maroni, J. Nucl. Mater. 113, 65 (1983)

21. E. H. VanDeventer, V. A. MacLaren, and V. A. Maroni, J. Nucl. Mater. 88, 168 (1980)

22. S. Bhattacharyya and W. Peterman, DOE/NASA/0006-2 (May 1983)

23. K. Morita, Nagoya Institute for Plasma Physics, Nagoya 464, Japan, private communication.

24. J. J. Burton, C. R. Helms and R. S. Polizzotti, J. Vac. Sci. Technol. 13, 204 (1976)

25. Energy Dependence of the Yields of Ion-Induced Sputtering of Monatomic Solids, N. Matsunami, Y. Yamamura, Y. Itikawa, N. Itoh, Y. Kazumata, S. Miyagawa, K. Morita, R. Shimizu, and H. Tawara, IPPJ-AM-32, Institute of Plasma Physics, Nagoya University, Nagoya, Japan (1983)

26. Handbook of Auger Electron Spectroscopy, edited by L. Davis, N. MacDonald, P. Palmberg, G. Riach and R. Weber, Perkin-Elmer, Minnesota, (1978)

27. M. Lagues, J. L. Domange, and J. P. Hurault, Sol. St. Comm. 12, 203 (1973)

PREPARATION AND SOME PROPERTIES OF Cu-Li ALLOYS CONTAINING UP TO 20 AT .% Li

The submitted manuscript has been authored by a contractor of the U. S. Government under contract No. W-31-109-ENG-38. Accordingly, the U. S. Government retains a nonexclusive, royalty-free license to publish or reproduce the published form of this contribution, or allow others to do so, for U. S. Government purposes.

M. H. Mendelsohn, A. R. Krauss, D M. Gruen
Argonne National Laboratory
Argonne, Illinois, USA

ABSTRACT

Lithium strongly segregates to the surface of Cu-Li alloys, thus substantially lowering the Cu sputtering yield relative to pure Cu. Use of Cu-Li limiters or divertors in Tokamaks can therefore be expected to be beneficial in limiting high-Z plasma impurity influx.

A large scale (100-200g) method for the preparation of Cu-Li alloys is described. Analysis reveals that on solidification from the melt stratification occurs which leads to compositional inhomogeneity. The results are discussed in the light of the Cu-Li binary phase diagram and rationalized on the basis of large density differences between Cu and Cu-Li solid solutions. It is concluded that obtaining homogeneous Cu-Li solid solutions is a nontrivial task.

IT HAS BEEN PROPOSED that erosion of fusion materials by sputtering and the generation of plasma impurities can be significantly reduced by using materials selected to maximize the secondary-ion fraction.[1] In the presence of a plasma sheath potential at the tokamak limiter or divertor surfaces, the escape probability into the plasma should be small for secondary ions attempting to overcome the retarding potential. Furthermore, secondary ions emanating from the first wall, the limiter, or the divertor are in all likelihood redeposited without entering the plasma as a result of deflection by the toroidal magnetic field. Calculations incorporating the effects of a sheath potential at a tokamak limiter or divertor [2,3], of the tangential toroidal magnetic field at the wall [3,4], and of the time dependent erosion and redeposition processes at the limiter and divertor surfaces [4] have predicted low erosion rates and, therefore, strongly reduced plasma impurity effects due to sputtered ion species.

One approach suggested for maximizing the secondary-ion fraction is to maintain an alkali-metal overlayer on selected materials. In this way, secondary-ion fractions exceeding 90% might be achieved.[1] In order to avoid the thermal, mechanical, and replenishment problems associated with coatings or claddings, alloy systems have been proposed that would produce self-sustaining secondary-ion emitting overlayers in a fusion irradiation environment via surface segregation.[3,5]

Initial studies, focused on lithium-copper alloys dilute in lithium as candidate materials, have shown that self-sustaining secondary-ion emitting lithium overlayers form on predominantly copper substrates which in turn significantly reduce the erosion rate in the presence of a negative potential.[6,7,8] Preliminary studies on the PISCES facility at UCLA, which simulate the edge plasma conditions in a fusion reactor, are in accord with the earlier experimental and theoretical results.[9]

The detailed behavior of lithium overlayer formation and its maintenance by surface segregation under ion bombardment conditions can be expected to be strongly correlated with the basic physico-chemical and metallurgical properties of the Cu-Li binary system. Unfortunately, there exists only a very sparse data base regarding this system. Up to the present, there have been only three papers published relating to the Cu-Li phase diagram.[10,11,12] In Ref. 1 only the liquidus curve in the Cu-Li system was determined. Klemm and Volavsek [11] describe the "Mischkristalle" or solid solution of Li in copper. They determined the solid solution range to extend to a composition limit of about 20 at.% Li (or Cu_4Li) by reacting lithium with copper powder. Old and Travena [12], on the other hand, reacted liquid Li with solid copper and found only a "face-centered cubic intermediate phase" whose composition is near Cu_4Li.

This discrepancy will be discussed more fully in the light of our results where a preparative method was used which involved heating above the melting point of copper. It will become apparent in the course of the discussion that the preparation and characterization of homogeneous Cu-Li alloy samples is a nontrivial task. The preparation of Cu-Li alloys for possible applications in a fusion reactor environment was done on the 100-200g scale and was based on solidification from the liquid state. The results and current status of the work are described below.

EXPERIMENTAL - An appropriate amount of Cu OFHC rod is machined as a straight-walled, flat-bottomed crucible whose inside diameter is 5/8". A Cu OHFC press fit cover of about 1/8" thickness is machined to seal the Cu crucible. After cleaning in 20-25% HNO_3 solution and washing with distilled water and acetone, the Cu crucible and cover are taken into a He-filled glovebox. An appropriate amount of Li, which includes a 10-20% excess for allowance of Li losses, is added to the crucible. We have found it convenient to add the Li in the form of Li-filled copper tubing (0.60" OD) obtained from Foote Mineral Company. In this case, an appropriate length of Li-filled copper tubing is cut in the dry box and slipped into the pure copper crucible. The crucible top is then pressed into place in the dry box. The crucible may now be safely removed from the dry box. Ambient air appears not to react with the enclosed Li metal. The Li-containing copper crucible is placed inside a BN crucible, used to contain the Cu-Li melt. The BN crucible is itself placed in a quartz tube whose ID is about 1-3/4". The quartz tube is evacuated with a mechanical pump and backfilled with UHP argon. The entire assembly is placed in a vertical position in a resistance-heated furnace at about 350°C for about 16 hours. This preheat treatment has been found substantially to reduce the amount of Li lost during the subsequent higher temperature treatments. The quartz tube is removed from the furnace and the furnace temperature is then raised to about 1100°C. The quartz tube assembly is then lowered into the furnace and the temperature monitored until the melting point of Cu is again reached at the heating coils. The sample is maintained in the liquid state for about 20 mins and subsequently cooled rapidly by removing it from the furnace to the ambient air. After removal from the crucible, the sample is inverted and remelted. Disks ~1/6" thick are cut from the 1" diameter boule with a low-speed cutting wheel. These are polished using various grades of diamond powder. Lithium analyses have been performed by atomic absorption measurements after dissolving the samples in nitric acid solutions. Optical microscopic observations have been made with a Bausch & Lomb microzoom microscope.

RESULTS

Several rods of 1" diameter have been prepared and analyzed. These results are given in Table I. Three samples were sectioned and analyzed with results displayed graphically in Fig. 1.

DISCUSSION

Before proceeding with a discussion of the results obtained in the present work, it is worthwhile to review current knowledge about the Cu-Li phase diagram. Recent thermodynamic calculations of this system by A. D. Pelton serve as a focal point for this review.[13] The equilibrium phase diagram (Fig. 2), reproduced from the work of Pelton, is based on that of Klemm and Volavsek.[11] Thermodynamic calculations give a solubility of Cu in liquid Li at the eutectic temperature of about 0.01 mol%. The very flat shape of the liquidus suggests that there may be a region of liquid-liquid immiscibility. A metastable liquid-liquid immiscibility gap has in fact been calculated by Pelton with a consolute point very close to the liquidus.

An appreciable solubility of Li in solid Cu has been observed by X-ray analysis [11] following quenching in Fe crucibles. Neither Pastorello [10] nor Klemm and Volavsek [11] found any evidence of intermediate phases, nor did we obtain evidence in our X-ray work for an intermetallic phase. On the other hand, Old and Travena [12] identify Cu_4Li as a distinct phase. The discrepancy may be due to the way in which Old and Travena carried out their preparative procedure which appeared to involve what effectively amounted to reacting massive solid copper with an excess of liquid lithium. In some of their preparations, lithium contents of less than 20 at.% should have resulted if the reaction had been carried out to equilibrium. The results of Old and Travena can be understood, it seems to us, if one assumes that the reaction of liquid lithium with solid copper is rapid up to the composition ~20 at.% Li. However, diffusion into the underlying Cu substrate appears to be slow so that attainment of equilibrium solid solution compositions by temperature annealing would take a very long time.

Below about 500°C, the Li-saturated solid Cu phase is in equilibrium with virtually pure Li so that the Li activity is essentially unity. However, the solubility of Li in solid Cu continues to increase as the temperature is decreased below 500°C.[11] This behavior requires a rather large excess entropy. As a result, the thermodynamically calculated solvus (dashed line on Fig. 2) also shows retrograde solubility. The lattice constants of quenched solutions were measured at compositions up to 19 mol% Li.[11] Results were presented only in the form of a graph which is reproduced in Fig. 3. The measured variation with composition is nearly, but not quite, linear. Molar volume measurements were in good agreement with molar volumes calculated from the lattice parameters, thus indicating a substitutional solution. The increase in lattice parameter with Li content in Fig. 3 is much less than that calculated from an additivity of molar volumes. A partial molar volume of Li in Cu of 8.2 ± 0.1 ml/mol was obtained compared with the molar volume of pure Li of 13.0 ml/mol. The large observed solubility of up to 22 at.% Li is perhaps surprising in view of the large difference in the atomic volumes of Li and Cu. The inertness of Li in the substitutional solid solution with respect to its reactivity with water is perhaps indicative of the presence of Li in a partially ionized form which would lead to a reduced atomic volume.

We return now to a consideration of the analytical results shown graphically in Fig. 1. Lithium analyses were performed on 1" diameter boules approximately 6 cm long as a function of length along the boule. Top and bottom refer to the location of the analytical sample with respect to the positioning of the boule in the BN crucible used in the melting and solidification of the alloy.

Three boules (A,B,C) with nominal Li compositions of 6, 12, and 15 at.%, respectively, were prepared and analyzed. It can be seen that in each case the lithium content decreases sharply going from the top to the bottom of the boule. It is very likely that during the melting procedure some Li was lost due to thermal evaporation accounting for the fact that the average composition of the boules is less than the nominal composition based on the amount of Li added initially. The inhomogeneity in boule C is particularly striking with the top composition exceeding the nominal Li content by a few percent being very close to the limiting solubility of Li in Cu.

To understand these results, one must examine the density of the Cu-Li solid solutions. The densities of the solutions are easily calculated using the lattice constant information shown in Fig. 3. It turns out that they vary essentially linearly with Li content. Compared to the density of Cu, which is 8.92 g/cm^3, the density of 20 at.% Li-Cu is 7.09 or 20.5% less than that of pure Cu. Consider the likely situation occurring during the solidification of boule C. Beginning with a melt containing 15 at.% Li, the temperature falls to the liquidus and solid solution begins to crystallize with a starting composition close to 5 at.%. There is then an approximately 10% difference in density between liquid and solid with the expectation that the solid will rapidly sink to the bottom of the crucible. The solid solution composition follows the solidus line shown in Fig. 2 while the supercooled liquid becomes enriched in Li. Both solid and liquid decrease in density while preserving a large density difference thus giving rise to the observed inhomogeneity in composition. The Li content of the solid continues to rise to approach the limiting solubility of Li in Cu. Although the density gradient mechanism just presented goes a long way to rationalize the results shown in Fig. 1, certain characteristics of the samples indicate a more complex behavior.

In no case were we able to prepare samples which appeared homogeneous under low resolution microscopic examination. Using coaxial illumination, all samples observed contained numerous small dark areas whose volume fraction appeared to increase with increasing Li concentration. The nature of this "dark" phase has not yet been elucidated, but it appears to be strongly enriched in Li and may in fact be due to liquid Li trapped in the sample during the solidification process.

The procedure employed in our work allows one to prepare large amounts of Cu-Li alloys in a convenient way but ways must be found to overcome the extensive segregation which occurs due to large density gradients during solidification so that compositionally homogeneous material can be obtained.

REFERENCES

1. A. R. Krauss and D. M. Gruen, J. Nucl. Mater. 85/86, 1179 (1979)

2. A. B. DeWald, J. N. Davidson, A. R. Krauss, and D. M. Gruen, J. Nucl. Mater. 111/112, 448 (1982)

3. A. R. Krauss and D. M. Gruen, J. Nucl. Mater. 103/104, 239 (1981)

4. Argonne National Laboratory Report, ANL-FPP/TM-194 (October 1984)

5. A. R. Krauss and D. M. Gruen, J. Nucl. Mater. 93/94, 686 (1980)

6. A. R. Krauss, D. M. Gruen, N. Q. Lam, and A. B. DeWald, J. Nucl. Mater. 128/129, 570 (1984)

7. A. R. Krauss, D. M. Gruen, and A. B. DeWald, J. Nucl. Mater. 121, 398 (1984)

8. A. R. Krauss, D. M. Gruen, and M. Venugopalan, J. Nucl. Mater. 122/123, 1425 (1984)

9. A. R. Krauss and D. M. Gruen, Argonne National Laboratory, Argonne, IL; D. Goebel and R. Conn, University of California at Los Angeles, Los Angeles, CA, unpublished work.

10. S. Pastorello, Gazz. Chim. Ital. 60, 988 (1930)

11. W. Klemm and B. Volavsek, Z. Anorg. Allgem. Chem. 296, 184 (1958)

12. C. F. Old and P. Travena, Metal. Science 15, 281 (1981)

13. A. D. Pelton, University of Montreal, Montreal, Quebec, Bulletin of Alloy Phase Diagrams, published by NBS and ASM, in press.

Table I. Summary of Furnace Melted Cu-Li Samples

Initial Li(at.%)	Melting Temp.	Melting Time(min.)		Analysis At.%	(Li)		Comments
10	1100°C	15	BC:6.2	BO:6.2	MC:7.5	MO:7.7	preheated pre arc-melted
18	1100°C	~20	M:12.5				pre arc-melted
5	1110°C	30					preheated
5	1110°C	20	B:4.1	M:4.2	T.4.1		~130g preheated
10	1110°C	30					
10	1100°C	20	B:10.3		T.9.1		~130g
10	1100°C	22	B:2.3		T:17.3		~136g preheated
5	1110°C	33					preheated
5	1110°C	20	B:5.3		T:5.2		~130g
5	1090°C	~30/29	B:3.0		T:5.3		~245g
10	1060°C	30/45	B:8.9		T:12.0		~250g
15	1050°C	30/35	B:5.7		T:19.5		~244g
Graphite crucible result:							
1	1120°C	≤8	B:0.1	MO:0.7 MC:0.7	T:1.6-1.8		preheat for only 30 min.

B = bottom piece
T = top piece
M = middle piece
R = remelted sample

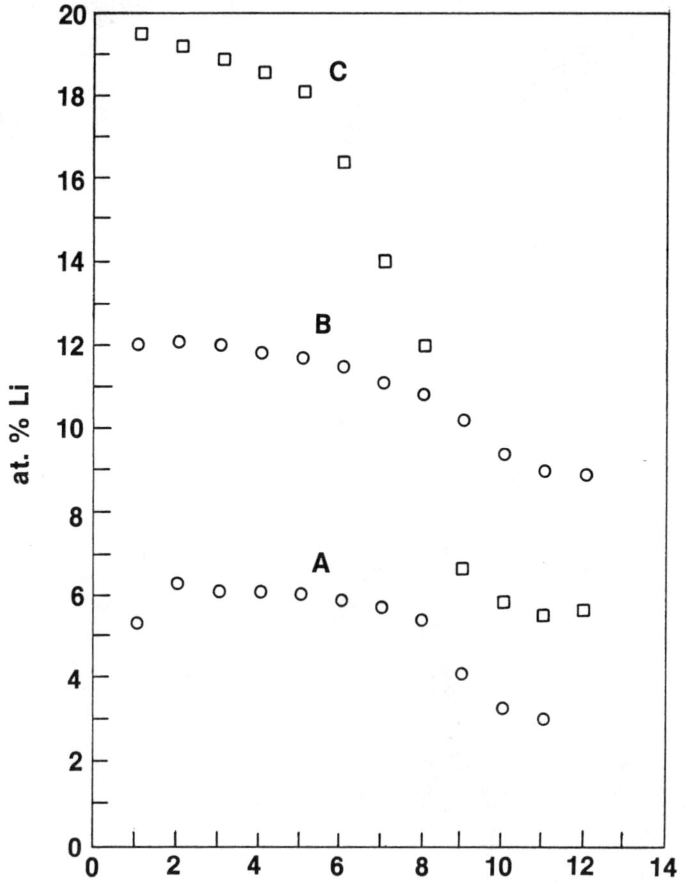

Fig. 1 - Lithium Concentration Profiles for Three Cu-Li Boules (nominal Li concentrations = 6%, 12%, 15%)

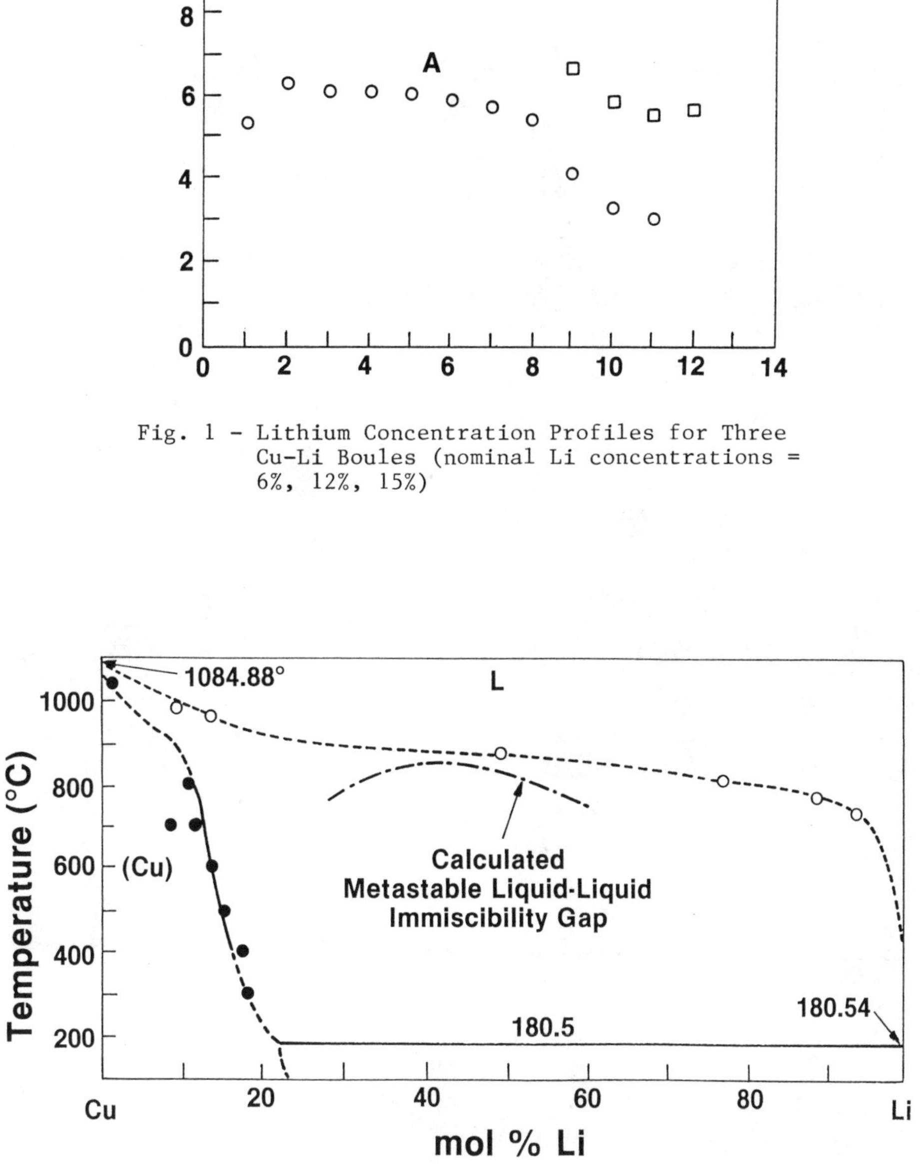

Fig. 2 - Assessed Cu-Li Phase Diagram (after Ref. 13)

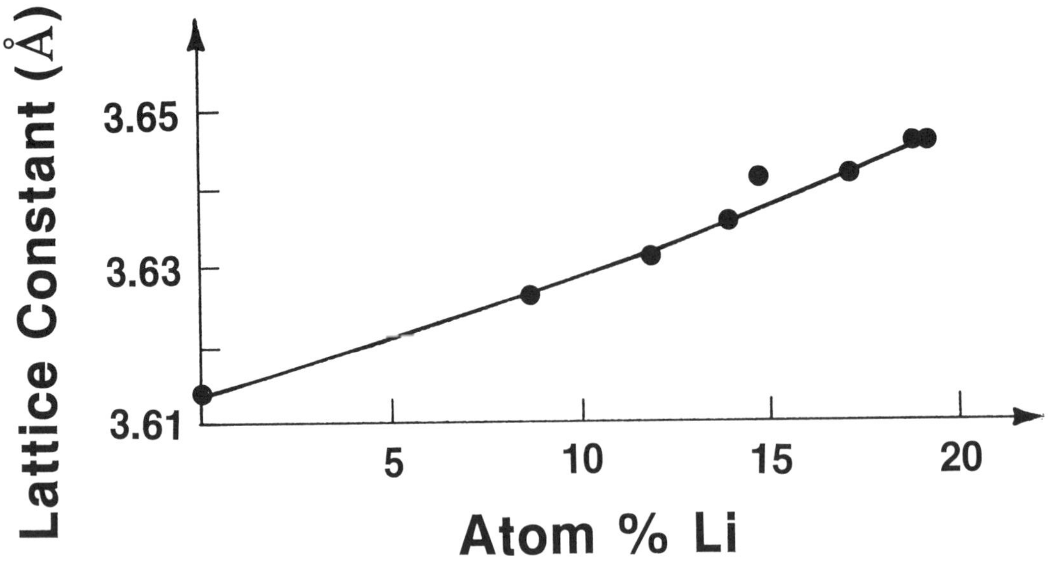

Fig. 3 - Lattice constant of Li-Cu solid solution vs. Li content from Ref. 11.

EFFECT OF THE SURFACE SEGREGATION ON THE ION DAMAGES OF ALLOYS DURING SPUTTERING AT ELEVATED TEMPERATURES

H. Shimizu, S. Ichumura, K. Morii, H. Murakami
Electrotechnical Laboratory
Umezono, Sakura, Ibaraki, Japan

Abstract

Surface ion damage was studied for Cu/Ni, Co/Ni, Fe/Ni and Fe/Cr alloys during noble gas ion bombardment at elevated temperatures using the techniques of Auger electron spectroscopy (AES) and X-ray microanalysis (XMA). Results related to the effect of surface segregation on (a) the formation of altered layer, (b) the angular distribution of the sputtered atoms, and (c) grain boundary erosion are presented.

For Cu/Ni alloy, a long range depletion of copper on the order of 100 nm in depth was observed. From the depth analysis of the copper depleted zone, ion enhanced diffusion coefficients were determined. The angular distributions of atoms sputtered from the alloys were studied as a function of temperature and average grain diameter. It was found that the segregated atoms have a tendency to be sputtered preferentially in the direction of the target surface, except for the case of Fe/Cr alloy. After sputtering of the Cu/Ni alloys at 873 K, the grain boundaries were no longer straight and some had developed zigzag shapes 2 - 3 µm in width. The cross section of the specimen showed the existence of large voids of 20 - 30 µm in depth just beneath the eroded grain boundaries.

I. Introduction

Many experimental and theoretical studies have been developed to understand sputtering phenomena of monoatomic materials. Yields for many ion-target combinations have been calculated as a function of energy, species and incident angle using theoretical model and agreement is reasonably good with experimental work. Recently, a number of studies on alloys have been carried out. However, the results have shown that their behaviour, not unexpectedly, is considerably more complex especially at elevated temperatures.

The high temperature performance of alloys during sputtering are considered to be very important from the view points of plasma-surface interactions and also fabrication of tailored surface materials. The characteristics of the ion irradiated surfaces will be affected to some extent by the thermodynamical properties of the surface and also by radiation damage, resulting in large concentration changes in the surface layer due to ion enhanced diffusion and surface segregation.

Major parameters governing the damage and erosion of the ion bombarded near surface layers of alloys include the type of projectile, its energy, total ion dose, beam density, target temperature and micro structure of target materials. In order to understand the mechanism of the ion induced surface erosion, the present authors have used AES during sputtering at elevated temperatures to investigate surface and near-surface compositional changes of various alloys, including copper-nickel which was used as a model system . In this paper experiments were performed using mainly Cu/Ni alloys[1]. The ion enhanced diffusion coefficients and apparent activation energy were calculated in order to determine the diffusion mechanism. The role of surface segregation on the angular distribution of the sputtered components was investigated[2], and finally the compositional and morphological changes along grain boundaries were investigated[3]. In addition to the results for Cu/Ni alloys the results for other kinds of binary alloys (Fe/Ni, Fe/Cr, and Co/Ni) are mentioned.

II. Experimental

Specimens of various kinds of alloys (Cu/Ni, Fe/Ni, Co/Ni and Fe/Cr) were made by arc melting. The ingots were cold rolled into foil 0.3 - 0.2 mm thick. The foils were annealed at various temperatures up to near the melting point in a quartz ampule filled with argon gas to depress the evaporation of volatile element in order to remove the strain caused by rolling and to enlarge the average grain diameters . In order to produce the copper-nickel alloy specimens, the material was first cold-rolled, resulting in foils with very fine grains. This material was then vacuum annealed at 800 - 900°C to produce a sample (a) with 50 - 100 μm grain size. The second sample (b) was produced by cold-rolling to 99% of the original thickness of the sample (a) and annealing at a temperature just below the melting point. This sample had 2 - 3 mm grain diameter. The average grain diameters were calculated from the total number of grains in the optical micrograph area. The SEM observation of the surface revealed the existence of many annealing twin boundaries, i.e., one grain consisting of a few smaller grains.

Sputtering experiments were carried out in a UHV chamber which was evacuated with a turbo molecular pump. Its back ground pressure was $2 - 5 \times 10^{-10}$ torr after baking out. Argon gas was introduced up to 1×10^{-4} torr during sputtering. The surface composition was analyzed by using a scanning type AES with a lateral resolution of 5 μm. In this report the average compositional changes over the area of 50×35 μm^2 were measured by scanning the electron beam. The area covered the surface of a small grain of the specimen. Ion bombardment was performed using a conventional ion gun. In order to measure the net ion beam current accurately a tantalum aperture of 5×5 mm^2 was placed in front of the sample surface. The noble gas ion beam density was 0.4 - 80 $\mu A/cm^2$ (Ne, Ar, Kr and Xe) and the sample was heated by passing ac current directly during sputtering. The sample temperature was measured with a Pt/Pt•Rh thermocomple.

The angular distribution of sputtered atoms was measured with a semi cylindrical shape Ta collector which was 10 mm away from the center of the target sample surface. The ion beam impinged on the sample at normal incidence through a rectangular window (2×2 mm^2). Angular distributions of the sputtered atoms were investigated by an energy dispersive type X-ray microanalyzer (EDX). To get good S/N ratio in the X-ray spectra for deposited thin films it was necessary to sputter the specimen at least 12 hours using the argon ions (3 keV, 40 $\mu A/cm^2$).

To understand the morphological changes along the eroded grain boundaries beneath the sputtered surface the specimen was cut with a diamod saw. Then the cross-section was polished with diamond paste (quater μm in average diameter) to reduce scratches. Furthermore, ion bombardment procedure was carried out at room temperature to make the grain boundaries clear.

III. Results and Discussion

A) Surface composition changes during sputtering at elevated temperatures

When the copper-nickel alloys of various composition were bombarded with 3 keV noble gas ions at room temperature the nickel concentration in the surface layer determined by AES increased due to the selective sputtering of copper and reached equilibrium values which depended on the ion species (Ne, Ar, Kr, and Xe). Segregation of copper to the sputtered surface was observed with good reproducibility whenever the sputtered surface was annealed at temperature above 423 K. The equilibrium surface concentration of copper is independent of both the ion species and beam densities, being dependent only on the annealing temperature. High temperature sputtering is of great interest because the surface composition is mainly determined by the competition between the segregation of copper to the surface and the preferential removal of copper from the surface leaving the layers depleted in copper.

It is well known that atoms segregate to the grain boundaries and that in the temperature range between 0.3 and 0.5 of the absolute melting point of the materials, the diffusion coefficient along the boundaries is much more larger than that of the normal volume diffusion. This suggests that the diffusion phenomena along the grain boundaries play an important role on the compositional changes of the sputtered surfaces. For the analysis of the role of diffusion, the effects of ion beam densities and the average grain diameter on the surface composition changes were studied at 573 K and 873 K using 40 at%Ni-Cu type (a) and (b) specimens under the same sputtering conditions. Sputtering was started after the specimen temperature reached a steady state. Both the sputtering and annealing were stopped simultaneously after 30 minutes of bombardment. The AES measurement was carried out at room temperature. Before each bombardment experiment the surface was sputtered at room temperature with an ion beam density of 40 $\mu A/cm^2$ to remove the altered layers produced during the previous high temperature sputtering.

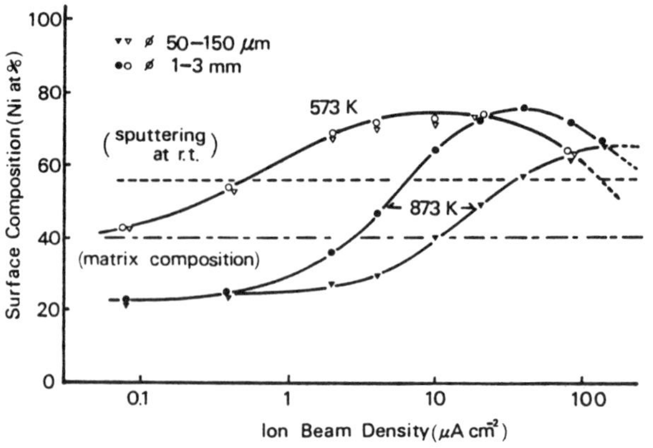

Fig.1 The effect of argon ion beam densities on the surface composition of 40 at%Ni-Cu alloy specimens with varied grain diameters. Ion bombardment was carried out for 30 min at 573 and 873 K.

In Fig.1 the surface compositional changes determined by AES are shown as a function of ion beam density. At 873 K the copper concentration on the fine grained type (a) specimen surface was higher than that of a coarse garined type (b) specimen at low ion beam densities. Above an ion beam density of 100 μA/cm² both specimens showed almost the same composition. Sputtering at 573 K indicated no significant difference between two specimens regardless of ion beam current density. It is noted that the copper flux to the surface of a fine grained specimen is larger at 873 K, probably as the result of fast diffusion along grain boundaries.

B) Analysis of the ion-enhanced diffusion coefficients

When the copper-nickel alloys of various compositions were bombarded with a high flux of 3 keV noble gas ions at 873 K the nickel concentration in the surface increased. In order to understand the mechanism of ion induced effects on the volume diffusion coefficient (D^*_{vol}), a 40 at%Ni-Cu alloy of a coarse grained type (b) specimen was sputtered at 40 μA/cm² for 1 h at different temperature. The depth analysis of the altered layer was carried out by AES and ion bombardment. The results are shown in Fig.2.

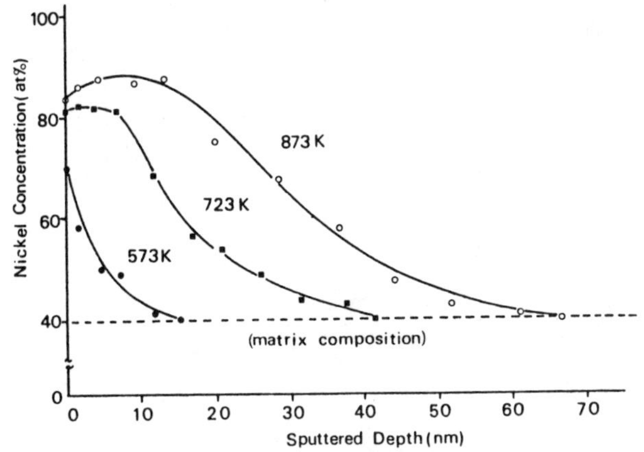

Fig.2 Effect of sputtering temperature on the depth profile of copper depleted region in a 40 at%Ni-Cu alloy coarse grained type (b) specimen, which was bombarded by a 40 μA/cm² beam of 3-keV argon ion for 1 h.

It was found that the copper depleted layers were 15,42 and 65 nm for sputtering at 573, 723 and 873 K, respectively. At steady state, the nickel profile was described by using the following equation

$$\theta(X)=C+[\theta(0)-C]\exp(-X\cdot V/D^*_{vol}), \qquad (1)$$

where $\theta(0)$ and $\theta(X)$ are nickel concentration at the surface and distance X from the surface, respectively. C is the nickel concentration in the matrix and V is the sputtering recession velocity. In this work V is 0.16 nm/s for a 40 μA/cm² beam of 3 keV argon ion.

In Fig.3 ion-enhanced volume diffusion coefficients (D^*_{vol}) are shown. A large enhancement of the copper diffusion was evident especially at low temperature. The apparent activation energy for ion induced volume diffusion coefficient was only 5.4 ± 0.5 kcal/mol, while the activation energy of the normal volume diffusion was 60 - 65 kcal/mol. The extrapolated diffusion constant at room temperature is 2×10^{-16} cm²/s. This value suggests that even at room temperature copper atoms can move to the topmost layer from the second or third layer within a few seconds. Namely, the outer surface layer seems to be affected by the copper segregation phenomenon during sputtering at low temperature so long as the recession velocity is not too large.

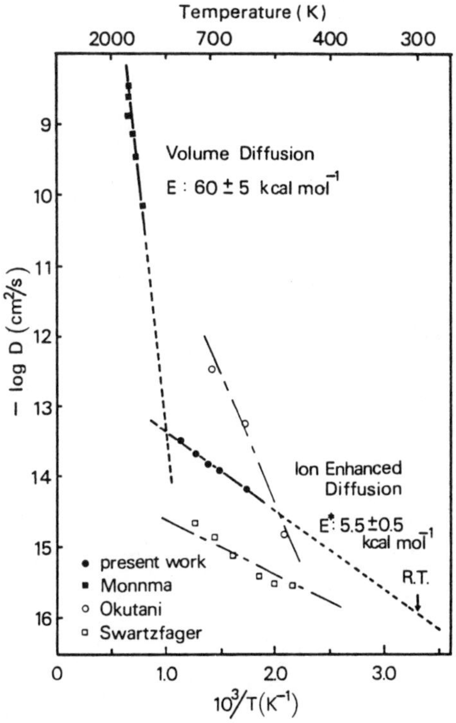

Fig.3 The effect of temperature on the enhanced diffusion coefficients for copper of a coarse grained 40 at%Ni-Cu alloy specimen. The normal volume diffusion coefficients were reported by Monma et al.[4,5].

C) Effect of segregation on the preferential ejection of sputtered atoms

The role of surface segregation on the angular distribution of sputtered components from binary alloys were studied using a 50 at%Ni-Cu alloy as a function of average grain diameter and temperature. Three types of the alloys (cold rolled, fine grained, and coarse grained specimens) were chosen to investigate the role of the copper diffused from the eroded boundaries. As mentioned for the results in Fig.1 amount of effused copper atoms seemed to be significantly effected by the average grain diameter.

Fig.4 shows the angular dependence for a coarse grained type (b) specimen of the quantity $X_{Ni}/(X_{Ni}+X_{Cu})$ where X_{Ni}, X_{Cu} are the relative amount of Ni and Cu collected on the Ta collector plate at the ejection angle θ. The ejection angle of the sputtered atoms is measured from the direction to the target normal. It is noted that nickel atoms are liable to be ejected in the direction to the target normal both at room temperature and 873 K and that the tendency of the preferred normal ejection of nickel atoms is enhanced at 873 K.

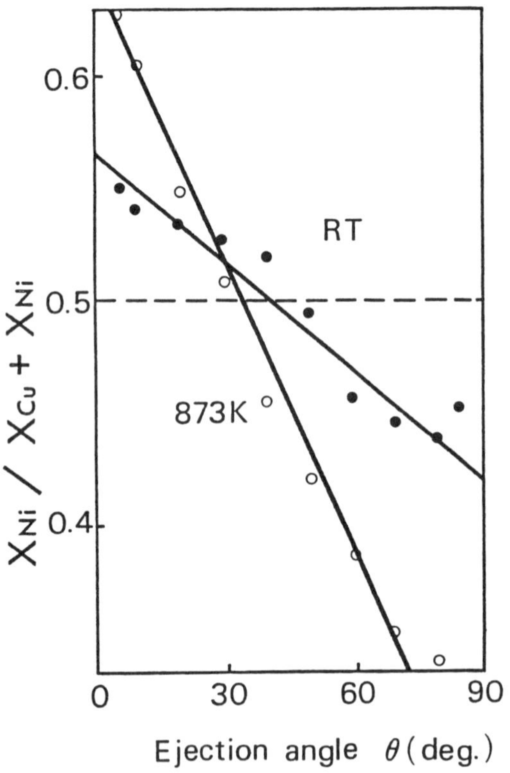

Fig.4 Dependence of Ni concentration in the deposited film on the ejection angle of atoms sputtered from a coarse grained 50 at%Ni-Cu alloy. Ejection angle is measured from the normal direction of the sample. Ar ions (energy 3 keV, density 40 μA/cm²) impinge on the sample from normal directions.

The same experiments are also carried out using fine grained type (a) and cold rolled specimens. The results are shown in Fig.5(a) and (b). For the fine grained spcimen, Fig.5a shows the same room temperature angular profile as Fig.4. However, at 873 K, the coarse-grained sample (Fig.4) exhibits a steeper angular profile than the fine-grained sample.

Observation of the sputtered surface by scanning electron microscopy (SEM) gave additional information about the effect of the specimen surface structure as shown in Fig.6(a), (b) and (c). As the average grain sizes of specimens become smaller the roughness of the eroded surface increased during sputtering at 873 K. It seems for the cold rolled specimen

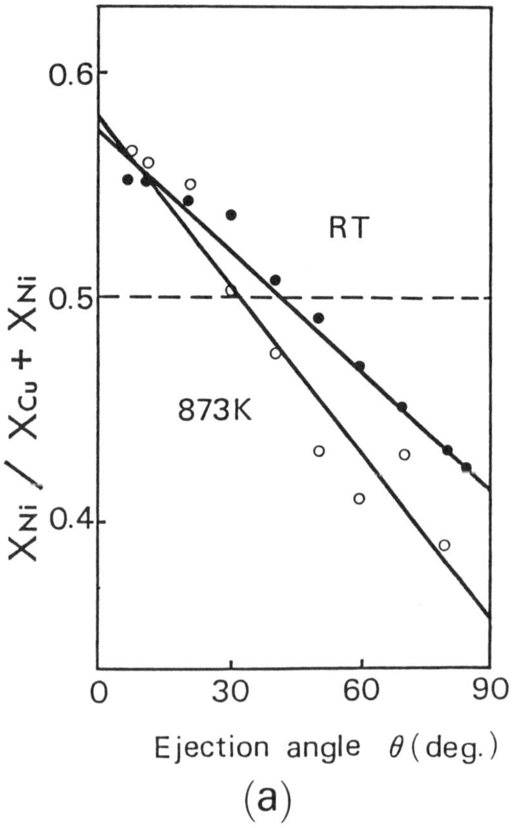

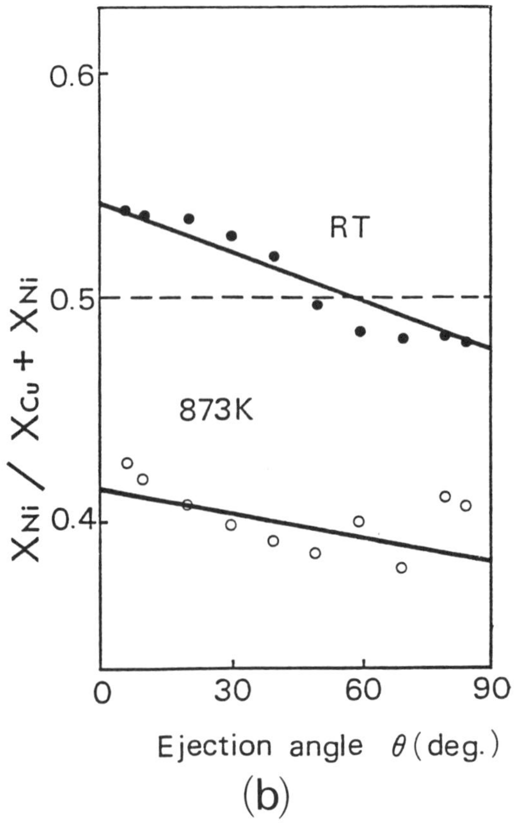

Fig.5 Dependence of Ni concentration in the deposited film on the ejection angle of atoms sputtered from fine grained (a) and cold rolled (b) specimens after sputtering with 3 keV of Ar ions (40 μA/cm²).

that ion irradiation and annealing promote the recrystalization of grains in the size of a few μm. Instability of each grain surface and boundary structure promoted the rapid diffusion of copper resulting in a high surface copper concentration. Comparing the cold-rolled material with the fine-grained sample (a) and the coarse grained sample (b), we find that the cold-rolled sample has a much higher rate of copper ejection. Because of the increase in surface roughness for the cold-rolled sample, there is a much smaller angular variation of the composition of the sputtered atoms as shown in Fig.5(b).

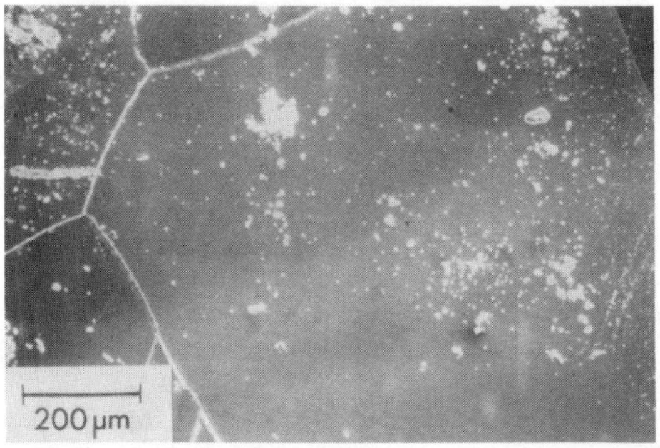

(a)

(b)

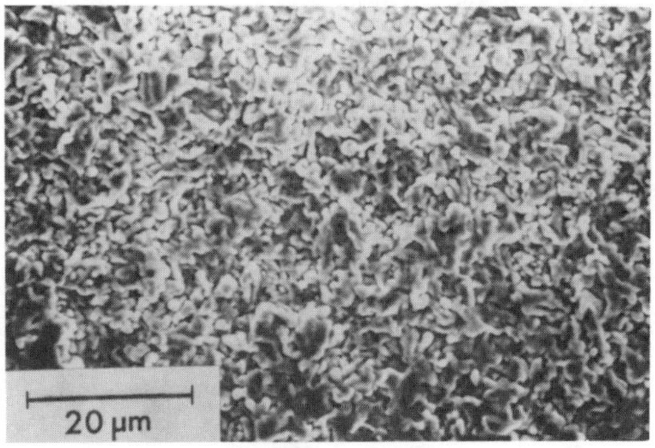

(c)

Fig.6 SEM images of the 50 at%Ni-Cu alloy surface sputtered by 3 keV argon ion (40 μA/cm²) at 873 K for 4 h. (a) a coarse grained specimen, (b) fine grained specimen, and (c) cold rolled specimen.

Similar experiments were performed for other binary alloys of Co/Ni and Fe/Ni. For the Co/Ni alloy the angular dependence of the Co concentration of the sputtered materials depended on the sample temperature during sputtering. Although it was reported that no segregation occurs for Co/Ni alloy, in fact, AES measurement revealed slight enrichment of Ni on the surface heated at 573 K after room temperature sputtering. Consequently, the Co concentration of the sputtered atom flux is peaked along the sample normal in a manner simlar to Cu-Ni alloys. For Fe/Ni alloys, segregation is not thermodynamically expected and was not detected by AES. Consequently, there was no observed angular variations of the concentration of the sputtered atom flux.

The role of the segregated copper atoms was studied by Monte Carlo calculation and it was found that preferred ejection of Ni was in the direction of surface normal[6]. The experimental results for Ni/Cu alloys are explained by the copper segregation phenomenon and surface roughness.

D) Grain boundary erosion and observation of its cross section

Figure 7 shows an SEM micrograph of a fine grained 40 at% Ni-Cu specimen after sputtering at 873 K by 40 μA/cm² of Ar ions for 4 h. It is clearly shown that the grain boundaries moved in

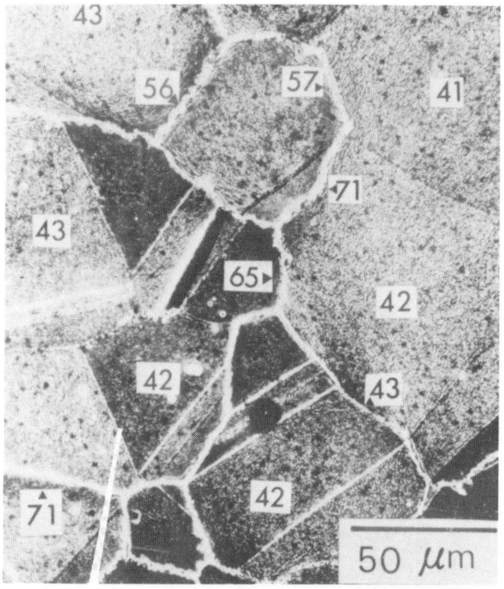

Fig.7 SEM images of the 40 at%Ni alloy surface sputtered by 3 keV argon ion (40 μA/cm²) at 873 K for 4 h. Point analysis of nickel concentration (at%) determined by XMA method.

a zigzag shape and small cavities a few μm in diameter were formed. The analysis of these voids and grain boundaries by XMA indicated an enrichment of nickel by about 15 - 30 at% as shown in the figure. No enrichment of nickel was found on the twin boundaries, while the surface of the grain was enriched with nickel by only a few at%. Since the depth resolution of XMA is the order of 1 μm, the copper depletion zone along the eroded grain boundaries is much more extensive than that of the grain itself. Namely, the diffusion of copper along the grain boundaries is enhanced during high temperature sputtering. Such eroded boundaries and compositional alteration were easily discerned by SEM and XMA after slight sputtering (40 $\mu A/cm^2$, Ar^+) for 10 min at 873 K, although the ion fluence amounted to the removal depth of no more than 100 nm. Nickel enrichment along grain boundaries was found at all sputtering temperatures above 573 K. However, as the temperature fell, the eroded boundary width and nickel concentration decreased.

To understand the morphological changes of the eroded grain boundary beneath the sputtered suface, the fine grained type (a) specimen was cut with a diamond saw after sputtering at 873 K with a 3 keV argon ion beam of 40 $\mu A/cm^2$ for 24 h. In Fig.8 an SEM micrograph of the cross section is shown. The figure shows significant structual changes as follows: (1) the width of the eroded grain boundaries are more than a few μm, (2) the depth of the eroded boundaries are 20 - 30 μm, and (3) the grain boundaries migrate and the behaviour of individual boundaries is apparently random. (4) recrystalization of small grains (a few μm) were observed near the eroded boundary.

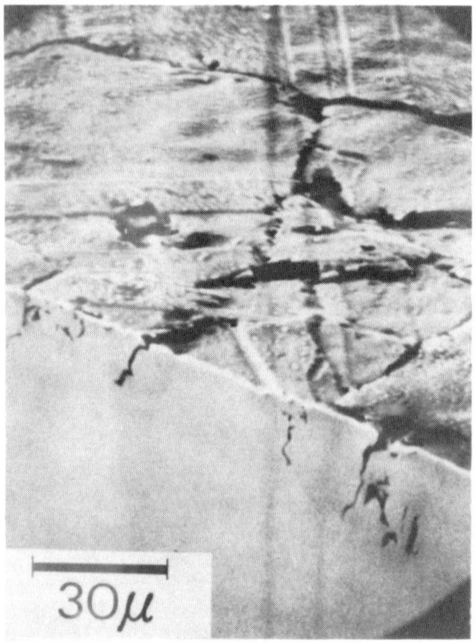

Fig.8 An SEM micrograph for the cross section of an eroded fine grained 50 at%Ni-Cu specimen.

There is as yet no clear understanding of what causes these variations in bounary structures. An important question, for the understanding of the ion induced grain boundary erosion is the diffusion profile of the alloy elements behind an altered boundary. The compositional changes determined by XMA along the eroded grain boundaries showed a large enrichment of nickel. The concentration gradient abruptly reached the bulk value at the end point of the eroded region. Considering the copper depleted region in the grain surface to be 100 nm, the diffusion coefficient along the eroded grain boundaries appears to be 10^3 - 10^4 times larger than the enhanced volume diffusion coefficient. Of course the length of the eroded region is far beyond the primary ion range and the knock on atom trajectories.

The reason for copper depletion in the deep region of the boundary is believed to be as follows: The copper atoms, arriving at the surface, are sputtered off directly at the exit site of the grain boundary and are replenished by rapid grain boundary diffusion. At first the width of the boundary is only a few atomic layers, but widens as the copper atoms spread out on the surface of the grain. The eroded boundaries were also found for Ne^+ sputtering. However, in the case of Xe sputtering drastic changes were not recognized. Such erosion phenomena were also observed in the Fe/Cr alloy system. For this material extensive chromium enrichment was observed by AES after room temperature sputtering followed by anneating. However, for high temperature sputtering, XMA failed to indicate chromium depletion, unlike the Cu-Ni alloy. For Co-Ni and Fe-Ni alloys, segregation did not occur and consequently there were no enrichment/depletion phenomena associated with sputtering.

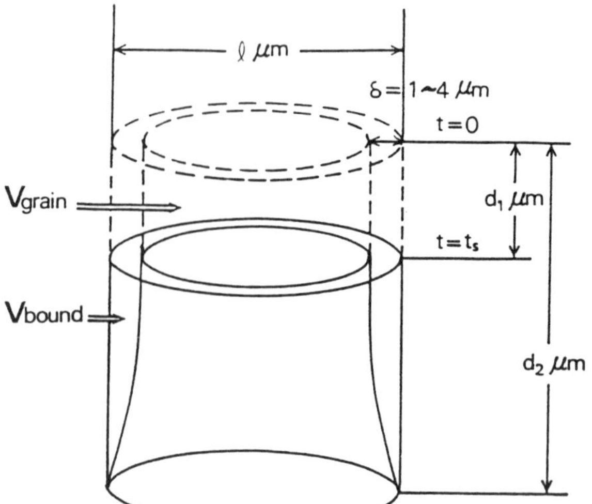

Fig.9 A schematic model to evaluate the role of effused atoms from eroded grain boundary.

Here, the effect of the eroded boundaries on the copper flux to the surface is considered. Assuming the grain diameter (ℓ) and the width of the eroded boundary (δ), the ratio of the exposed area of the eroded boundary to the total surface area of the grain is $4\delta/\ell$. For the fine grained type (a) specimen the value is less than 10%, however, the copper flux $[(4\delta/\ell)\ D^*_{vol}(\frac{\partial c}{\partial x})]$ is no more negligible due to the rapid diffusion from eroded grain boundaries. Using a very simple model as shown in Fig.9, the ratio of the contribution of the volume of diffusing atoms leaving voids ($V_{boundary}$) against the removed total volume (V_{grain}), is calculated as follows;

$$R = \frac{V_{boundary}}{V_{grain}} = \frac{4\delta}{\ell} \quad \frac{d_1}{d_2} \qquad (2)$$

d_1: thickness of surface removal layers
d_2: thickness of eroded grain boundary

During sputtering for 24 h with 3 keV argon ions (40 $\mu A/cm^2$), the thickness of the removal layer (d_1) was about 14 μm, while the depth of eroded boundary (d_2) was more than 30 μm. Assuming that the width of eroded boundary (δ) was 1.5 μm, the contribution of the diffused atoms become 18 and 36% for the specimens having 100 and 50 μm average grain diameter, respectively.

Concluding remarks

The paper revealed that the copper diffusion is much enhanced beyond the trajectory range of impinging ions[7]. Characteristic copper segregation affected the ejection of sputtered atoms and formation of eroded boundaries. To some extent, such a segregation property during high temperature sputtering has some effects on the surface erosion of other alloy systems. This phonomenon is very important not only from the view point of ion erosion during high temperature sputtering but also for the purpose of tailoring surface properties. The ability to control the depth and concentration of an alloyed zone is potentially useful for the surface engineering purposes[8].

Acknowledgement

The authors wish to thank Drs. K. Nakayama and M. Ono of the Electrotechnical Laboratory and Prof. Y. Ishida of the University of Tokyo. They are also extremely grateful to Mr. Agari of Tsukuba University for help with experimental work.

References

1) H.Shimizu et al.: J.Appl.Phys. 53 (1982) 3044.
2) S.Ichimura: J.Nucl.Mat. 128 & 129 (1984) 601.
3) H.Shimizu et al.: J.J.Appl.Phys. 19 (1980) L671.
4) K.Monma et al.: Nippon Kinzoku Gakkaishi 28 (1964) 192 (in Japanese).
5) D.B.Butyrymowicz et al.: J.Phys.Chem.Ref. Data Vol.5, No.1 (1976) 103.
6) R.Shimizu and T.Okutani: Proc. 5th Symp. on Ion Sources and Ion Appl.Technol. (1980, Tokyo) p.273.
7) N.Q.Lam et al.: Surf.Sci. 149 (1985) 5-17.
8) R.Kossowsky: ed "Surface Engineering" (Martinus Nijhoff Publishers, Pordrecht 1984).

CRITERION FOR SELF-SUSTAINING COATING IN METAL- OR CARBIDE-COATED GRAPHITE

K. Morita, H. Ohno
Department of Crystalline Materials Science
Faculty of Engineering
Nagoya University
Furo-cho, Chikusa-ku, Nagoya, Japan

Abstract

We have studied the condition for self-sustaining coating of nickel and titanium carbide layer deposited on graphite with segregated carbon atoms. Both the concentration of carbon atoms at the surface and the sputtered flux of carbon atoms from the nickel and titanium carbide deposited on the graphite substrate are calculated as a function of ion flux. It is shown that the critical ion flux for self-sustaining coating depends on the rate constant of segregation to the bombarded surface, the diffusion constant, and the rate constant of dissolution from graphite into the deposited layer. The experimental results on the ion flux dependence of the sputtered flux of carbon and on the critical ion flux are compared with the calculated results and the rate of segregation of carbon on the Ni surface is determined to be 57 s^{-1} at 650°C and that on the TiC surface to be 0.1 s^{-1} 900°C. It is pointed out that segregation from the grain boundary to the surface of the deposited layer, grain boundary diffusion and dissolution from graphite into the grain boundary are important factors for self-sustaining coating.

1. Introduction

Graphite is a candidate material for limiters in tokamak-type fusion devices and has been tested under various conditions. The enhanced erosions observed at high temperatures, namely chemical sputtering[1-5] and radiation-enhanced sublimation[6,7] are its drawback. Some examinations to reduce the chemical sputtering have been done by means of coating the graphite surface with a metal layer[8] or metal-carbide layer.[9] It has been shown that a thin nickel layer deposited on graphite reduces the reaction probability of CH_4 formation responsible for chemical sputtering by a factor of 4-5.[8] Moreover, it has been found that the yield of CH_4 formation for the TiC-layer deposited on graphite is much less than that for graphite.[9] In those cases the suppression of sputtering of coated high-Z elements is a main problem. On the other hand, recent probe experiments with large fusion devices shows that the graphite limiters are covered by deposition of metal atoms sputtered off from the first walls and the neutral injection liners.[10] Such depositon of metal layer under irradiation at elevated temperatures varies erosion behavior of graphite because of modification of the surface properties as well as reduces the chemical sputtering, as described above. Therefore, it is required to know how the erosion of graphite is modified by deposition of metal layer.

When metal layer is accumulated on graphite under irradiations with the high flux of multi-particles and energy, carbon atoms are dissolved into the metal layer from graphite. In some cases metal-carbide phases[11] are produced due to thermally activated and/or, radiation-enhanced processes. Furthermore the carbon atoms are segregated on the surface of the metal or metal carbide layer as well. In recent papers,[12,13] it has been shown that when nickel-layer deposited graphite is bombarded with Ar^+ ions below certain current density at elevated temperatures, continuous coverage of the surface with the carbon atoms segregated in a self-sustaining way protects the nickel layer from being sputtered. The protection of the deposited metal layer, which reduces the chemical sputtering of graphite, from resputtering by the self-sustaining coating indicates that the graphite limiter may show a favorable performance. Although erosion and deposition of graphite limiter in fusion devices are much more complicated, it is important to study, using simple systems, in which metal or metal-carbide layer the self-sustaining coating takes place and what is the limitation to the self-sustaining coating: the thickness of the layer to be protected, temperature, ion flux and so on.

The coverage of the material surface with segregated layer is essential for the self-sustaining coating. The surface coverage seems to be determined apparently by a balance between the sputtering flux and the fluxes of segregation from inside to the surface and of dissolution in The reverse direction. Extending the theoretical treatment of the sputtering of binary alloys,[14] we have shown that the surface concentration of segregated atoms depends on the rate constants of segregation and dissolution at the surface, the diffusion constant and the sputtering (or desorption) cross-section,[13] based on an assumption that the rate constants of segregation and dissotion at the interface between the graphite and deposited layer are sufficiently large compared with those at the surface. The critical ion flux and critical thickness for the self-sustaining coating have been expressed as a function of the relevant rate constants.

In this paper, we present in the first place an improved treatment including the rate constants of segregation and dissolution at the interface to evaluate the surface concentration of segregated atoms and to describe the criterion for the self-sustaining coating in detail. Nextly, we show briefly the experimental results on the surface segregation of carbon atoms on nickel-deposited graphite obtained by means of the AES technique. Furthermore we present the experimental results, obtained by means of the RBS technique, on the critical ion flux, for a TiC-deposited graphite[15] bombarded with 1 keV hydrogen ion beam at 900°C and for a nickel-deposited graphite bombarded with 5 keV argon ion beam at 650°C. Comparing the experimental values with the calculated ones of the critical ion flux, the rate constants of segregation and dissolution, key factors are estimated.

2. Criterion for Self-Sustaing Coating

We suppose a carbon-metal composite layer on graphite, of which carbon atoms tend to be segregated at the surface. The surface is usually covered with an amount of carbon atoms at given elevated temperature because of Gibbsian segregation. When such a surface is bombarded with energetic ions at the elevated temperature, a balance between the sputtering and segregation of carbon atoms at the surface leads to a steady state. The segregation may be enhanced by ion bombardment.[16] Such a problem has been previously treated by the present authors.[13] We extend the treatment more generally to discuss the critical ion fluxes for the self-sustaining coating obtained experimentally.

We consider a composite system composed of carbon atom and metal atom, where the fractional concentrations of the carbon atoms in the bulk and at the top-surface are represented by $C_b(x,t)$ and $C_s(t)$, respectively, where x is the depth from the surface and t is the time after the start of ion bombardment. Referring to Fig.1, the surface concentration $C_s(t)$ of main interest in this study is obtained by solving the following equations:

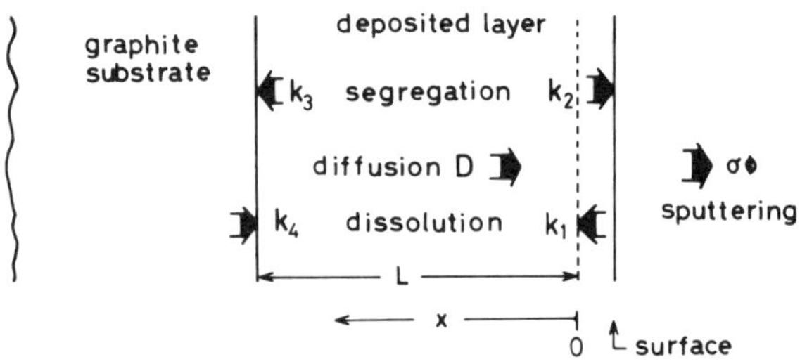

Fig. 1 Processes inducing fluxes of carbon atoms migrating in a metal-carbon composite layer, with a thickness L, deposited on graphite under ion bombardment.

$$\frac{dC_s(t)}{dt} = k_2 C_b(0,t) - k_1 C_s(t) - \sigma\phi C_s(t) \tag{1}$$

$$\frac{dC_b(x,t)}{dt} = D\,\frac{d^2 C_b(x,t)}{dx^2} \tag{2}$$

where k_2 and k_1 are the rate constants of segregation and dissolution given in the units of s^{-1}, D is the diffusion constant of carbon atoms, σ is the sputtering cross-section for carbon atoms and ϕ is the ion flux. The boundary condition at x=0 for eq.(2) is given by the relation

$$D\left.\frac{dC_b(x,t)}{dx}\right|_{x=0} = k_2 a\, C_b(0,t) - k_1 a\, C_s(t) \tag{3}$$

Furthermore, the boundary condition at the back-surface of the composite layer is similarly given by the relation

$$D\left.\frac{dC_b(x,t)}{dx}\right|_{x=L} = k_4 a\, C^{o}_{subst} - k_3 a\, C_b(L,t) \tag{4}$$

where C^{o}_{subst} is the fractional concentration of carbon atoms in graphite, namely unity, k_4 and k_3 are the rate constants of dissolution from the graphite into the deposited layer and its reverse process, respectively, L is the thickness of the deposited layer and a is the planar spacing. Eq.(4) is different from the boundary condition used in the previous treatment based on the assumption that the dissolution rate is enough large compared with the diffusion rate. In practice, when deposited layer is polycrystalline, the short circuit grain boundary diffusion is a dominant process for carbon atoms to be supplied to the surface. The radiation-enhanced process is less important, when thickness of the deposited layer is sufficiently large compared with the range of bombarding ions. Thus, $C_b(x,t)$ should be replaced by $C_g(x,t)$, D by D_g, k_2 by $k_2^{!}(2a/r)$, where C_g, D_g and k_2' and k_3' are the fractional concentration at the grain surface, the grain boundary diffusion constant, and the rate constants of segregation from the grain boundary to the surface and to the interface, respectively and r and a are the spherical grain radius and the average thickness of the grain boundaries.

Eqs.(1) and (2) can be solved easily at a steady state, where $dC_s(t)/dt=0$ and $dC_g(x,t)/dt=0$, as observed experimentally in the previous work. The steady state solution for the fractional concentration $\xi(\phi)$ of the carbon atoms is given by

$$\xi(\phi) = \frac{C_s^o}{1 + (\sigma\phi/k_1)(1 + Q_1 + Q_2)} \qquad (5)$$

where $C_s^o=(k_2'k_4/k_1k_3')C_{subst}^o$, $Q_1=k_2'/k_3'$ and $Q_2=k_2'(2a/r)/(D_g/La)$. C_s^o is the fractional surface concentration at Gibbsian equilibrium segregation and Q_2 represents the ratio of the segregation rate from the grain boundary at x=0 to the surface to the diffusion rate from the graphite to the surface through the grain boundary. The removal flux for carbon atoms from the specimen is expressed by multiplication of $\xi(\phi)$ by $N_s\sigma\phi$, as follows;

$$R(\phi) = \frac{N_s\sigma\phi C_s^o}{1 + (\sigma\phi/k_1)(1 + Q_1 + Q_2)} \qquad (6)$$

where N_s is the surface density of graphite of monolayer thickness. Since the relevant rate constants are mainly contributed by the thermally activated process, eq.(6) indicates that if the relation between R vs ϕ is obtained experimentally, σ can be determined from the slope of $R(\phi)$ at low ϕ and moreover $(1+Q_1+Q_2)/k_1$ can be determined from the saturation level at high ϕ.

We consider the conditions for the self-sustaining coating. The condition of $Q_1<1$ and $Q_2<1$, namely $k_2'<k_3'$ and $k_2'(2a/r) < (D_g/La)$ is the segregation-limited, the condition of $Q_1<1$ and $Q_2>1$, namely $(D_g/La) < (2a/r)k_2' < (2a/r)k_3'$ is the diffusion-limited case and the condition of $Q_1>1$ and $Q_1<1$, namely $(D_g/La) > (2a/r)k_2' > (2a/r)k_3'$ is the dissolution-limited case. The critical thickness for the segregation- and dissolution-limited cases is geven by the relation $Q_2=1$, namely $L_c=D_g/(2a^2/r)\cdot k_2'$.

Using eq.(5), the condition for the self-sustaing coating is derived as follows: 1) $C_s^o=1$ and 2) $(\sigma\phi_c/k_1)(1+Q_1+Q_2) \le 0.1$. The condition 2) represents the competition of enrichment due to segregation with the removal due to sputtering. The condition 2) indicates that the critical ion flux for the self-sustaining exists. When $Q_2<<1$, namely $L<<L_c$, ϕ_c depends on k_2' or k_4 and not on L. When Q >1, ϕ_c depends on L. Therefore, the experiments to determine the rate constants of k_2' and k_1 from the critical ion flux had better be done under a condition of $L<L_c$. When the carbon layer segregated at the surface of deposited layer has the same structure as the graphite, one can regard that $k_2'=k_3'$, thus $Q_1=1$. In such a case the critical ion flux for the self-sustaining coating is by a factor of 2 smaller than that for the segregation-limited case.

3. Experimental

The specimens used were both nickel-deposited and TiC-deposited graphites. The former specimen was prepared by vacuum deposition of nickel layer (~5000 Å) on mechanically polished Poco graphite with a size of 10x15x2 mm^3. The nickel layer deposited was tested by means of a 1 MeV transmission electron microscope. It was found that the nickel film is polycrystalline and the average size of crystalline grains is 200-300 Å in diameter. The latter TiC layer deposited was prepared by chemical vapor deposition and the thickness was about 3000 Å. The chemical composition of the TiC layer was analysed by both EPMA and RBS techniques. The EPMA analysis showed that the composition ratio of carbon to titanium is about 1.0. On the other hand, the RBS analysis showed that the ratio is also 0.94. The SEM observation of the TiC surface showed that the average grain size is 1000-2000 Å in diameter.

Experiments on nickel-deposited graphite were carried out using a conventional UHV chamber at a base pressure of 5×10^{-10} torr equipped with a manipulater and a sputter ion-gun. The specimen, in contact with a ceramic heater placed on the manipulater, was bombarded with 5 keV Ar^+ beam in perpendicular to the surface. Ar^+ ion current density was uniform over a spot of 5-6 mm in diameter and varied in the range between 1 and 40 $\mu A/cm^2$. The surface composition of the specimen was determined by means of the AES technique. The depth-distribution of carbon atoms in the specimens was determined using the RBS-measurement with 1.3 MeV H^+ beam of 1 mm in diameter. The sputtering yield of nickel was obtained from the difference between the depth-distribtuions before and after the ion bombardment, as a function of the ion flux.

Irradiation of TiC-deposited graphite with 1 keV hydrogen ion beam was done in Toshiba Corp. using the experimental apparatus which consists of a low-energy ion beam system and a UHV chamber. The detailed description of the apparatus was given elsewhere.[17)] The hydrogen ion beam was generated by a duo-plasmatron and the current density at the target was varied in the range between 20 and 700 $\mu A/cm^2$. The angle of the beam to the target normal was 50° and the irradiation area was elliptically in shape, 3x4.5 mm. The pressure in the chamber during hydrogen ion bombardments was about 3×10^{-8} torr. The TiC-deposited graphite specimen, mounted on the tantalium strip heater, could be heated up to 1000°C. The temperature was measured with a Pt-PtRh thermocouple which was fixed in a hole prepared in the specimen. The specimen irradiated was analysed using the RBS-measurement with

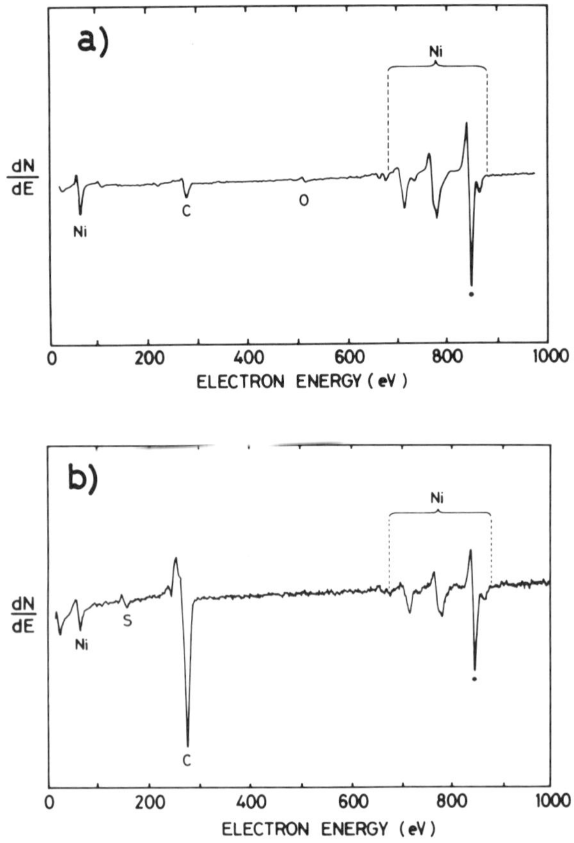

Fig. 2 Auger electron energy spectra from a nickel layer deposited on graphite: a) after bombardments with 5 keV Ar^+ ions at room temperature, and b) after heating for 2.5 hrs at 650°C.

2 MeV He^+ beam of 0.4 mm in diameter. The RBS-energy spectra were obtained at several points inside and outside of the irradiated area on the specimen. The sputtering yield of titanium atoms was determined from the difference between the thicknesses of of the TiC layer in the irradiated and non-irradiated parts of the specimen.

4. Experimental Results

4.1. Ni-deposited graphite

The Ni-deposited graphite specimen as-prepared was bombarded with 5 keV Ar^+ beam at room temperature in order to remove the oxide layer at the surface. Immediately after the ion bombardment, the specimen was heated for 2.5 hours at 650°C. Before and after the heating, the specimen was tested by means of the AES and RBS techniques. In Fig.2, are shown typical AES spectra from a Ni-deposited graphite specimen. In the spectrum a) obtained before the heating, one can find a main signal from nickel and small signals from residual impurities of carbon and

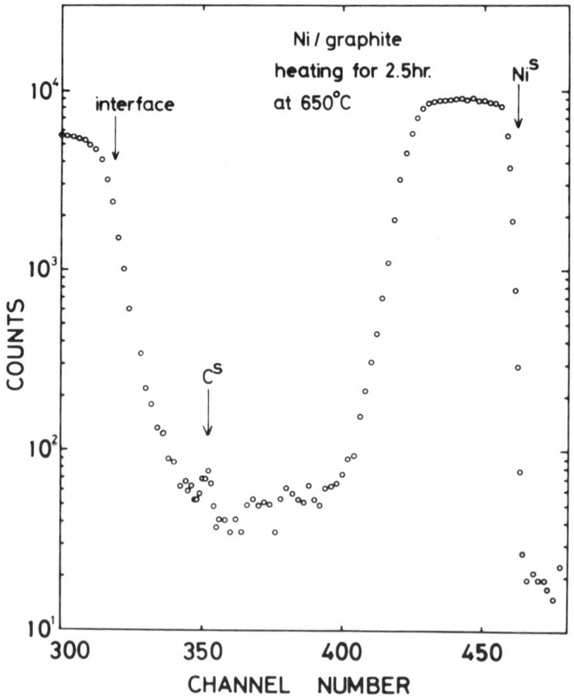

Fig. 3 RBS-energy spectrum of 1.3 MeV H^+ ion beam from a nickel layer deposited on graphite heated for 2.5 hrs at 650°C.

oxygen at the surface. In the spectrum b) after the heating, one can see clearly that the carbon signal is grown by the heating. The growth of carbon signal is ascribed to the carbon atoms which are dissolved into the nickel layer from graphite and are segregated to the surface via grain boundary diffusion. The shape of the carbon signal observed was identified to be the same as that for graphite. This result corresponds to the experimental obsevation by Shelton et al.[18] that at temperatures below 1065 K the surface of nickel bearing carbon is covered with segregated (or precipitated multilayer) graphite. Thus, one can conclude that the segregated carbon layer in this case is a graphitic layer.

The segregation of carbon atoms is also found in a typical RBS energy spectrum of 1.3 MeV H^+ beam from the specimen, shown in Fig.3. In this spectrum, H^+ ions backscattered from the nickel layer form a broad peak and those from a graphite substrate form a flat continuum below the channel marked by the interface. A clear small peak, which is produced by the heating, at the channel marked by C^S indicates the carbon atoms segregated to the nickel surface. Small counts between the marks of the interface and C^S indicate that the concentration of carbon atoms solute in nickel layer is very small. From the area of the small surface peak, the effective thickness of segregated carbon layer is estimated to be about 2 monolayers.

Since the width of the nickel peak is proportional to the thickness of the nickel layer, the sputtering yield of nickel was determined

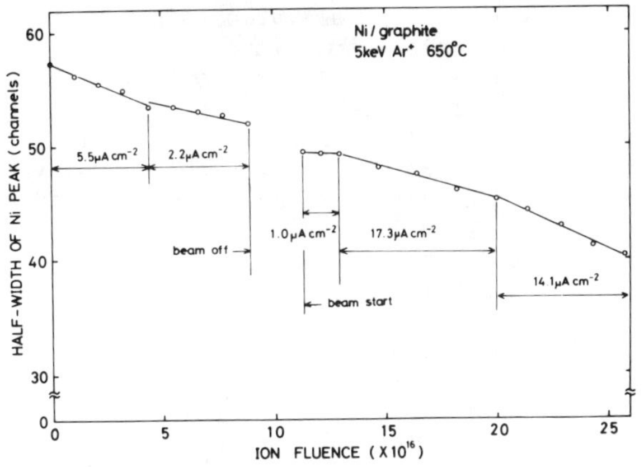

Fig. 4 Variation of the width of Ni peak with the ion fluence for a Ni-deposited graphite bombarded with 5 keV Ar^+ ion beams of different current densities at 650°C.

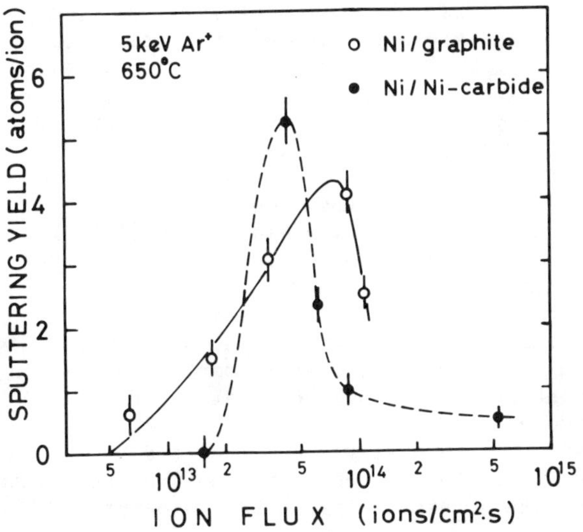

Fig. 5 Ar^+-ion flux dependence of the sputtering yield for nickel atoms from a Ni-deposited graphite (o) and from a Ni/nickel-carbide bi-layer (o), bombarded with 5 keV Ar^+ ions at 650°C.

from reduction in the width caused by the 5 keV Ar^+ ion bombardments. The reduction in the width by the bombardment at different current densities at 650°C is shown in Fig.4. It is seen from Fig.4 that the width of nickel peak at each current density decreases linearly with increasing the ion fluence. It was observed also that the surface peak of segregated carbon layer keeps almost a constant level at each case. These facts indicate that the Ni-deposited graphite attains at a steady state during each Ar^+ bombardment. The sputtering yield of nickel was calculated from the slope of linearly decreasing

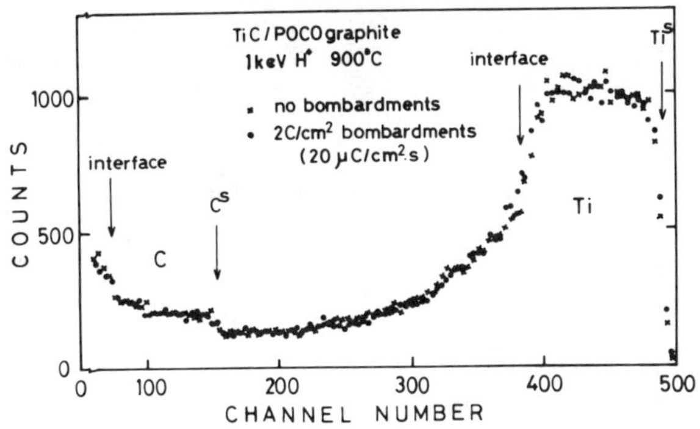

Fig. 6 RBS-energy spectra of 2 MeV He^+ ion beam from a TiC-deposited graphite bombarded with 1 keV H^+ ion beam of 20 $\mu A/cm^2$ at 900°C: non-bombarded area (x) and bombarded area (o).

width. The experimental results were presented in the previous paper.[13] Therefore, only the experimental results for the Ni-depostited graphite and self-supporting Ni/nickel-carbide bi-layer at 650°C are shown in Fig.5. One can see that the sputtering yield of nickel atoms for both specimens are almost zero at ion fluxes below $5x10^{12}$ ions/$cm^2 \cdot s$ and $1.7x10^{13}$ ions/$cm^2 \cdot s$, respectively.

4.2. TiC-deposited graphite

Since it is already known that the carbon layer segregated at the surface of TiC-deposited graphite heated above 700°C and it has a graphitic strucutre,[17] we have investigated, by means of the RBS technique, only the partial sputtering yield of titanium atoms. In Fig.6, typical RBS energy spectra of 2 MeV He^+ beam from a TiC-deposited graphite are shown which is bombarded with 1 keV H^+ beam of 20 $\mu A/cm^2$ by a fluence of 2.0 C/cm^2 at 900°C. In Fig.6, the counts between the 200 and 500 channels represent scattering from Ti atoms in the specimen and the counts below the 160 channels that from carbon atoms. The closed points indicate the spectrum obtained at the outside of the irradiation area and the open circles that obtained at the center. The counts for Ti and C between marks of the interface and Ti^S (or C^S) represent scattering from the deposited TiC layer and the counts for Ti tailed to the lower channels from the interface means that Ti atoms are distributed into graphite substrate. Comparing both spectra, it is found that the width of the Ti peak is not reduced by the irradiation, although an appreciable reduction in the peak width was produced by the same irradiation at room temper ature. Moreover the counts for Ti atoms at the channels around the mark of the interface in the irradiated spectrum are found to be increased. These results indicate that the hydrogen ion bombardment of the specimen at 900°C and at 20 $\mu A/cm^2$ does not reduce the thickness of the TiC layer, but it increases the composition ratio of

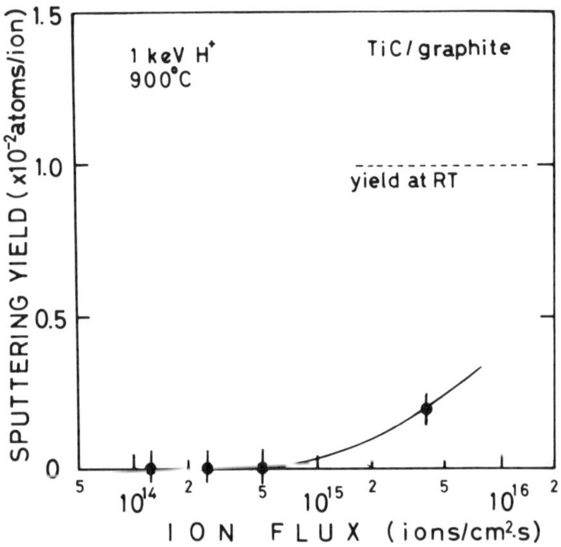

Fig. 7 H^+-ion flux dependence of the sputtering yield for Ti atoms from a TiC-deposited graphite bombarded iwth 1 keV H^+ ions at 900°C. The dotted horizontal line represents the sputtering yield for the H^+ bombardment at room temperature.

Ti/C at the interface, namely removes the carbon atoms from the interface of the specimen. It is also seen that the ratio of the average counts of Ti and C for the TiC layer, namely the composition ratio of Ti/C is not changed by the hydrogen ion bombardment. From these facts, one can conclude that the sputtering of Ti atoms from the TiC-deposited graphite is prohibited by the self-sustaining coating under irradiation at a current density of 20 μA/cm^2 at 900°C. Similar results[15] are observed for the specimens irradiated at the ion current densities of 40 and 80 μA/cm^2.

On the other hand, it is found that the width of Ti peak for the specimen irradiated at the higher ion current density is clearly reduced. Moreover, the composition ratio of Ti/C at the interface is also found to be increased. The sputtering yields of Ti atoms obtained at different ion current densities at 900°C and at room temperature are shown as a function of the ion flux in Fig.7. The sputtering yield,~10^{-2} atoms/ion, of Ti atoms at room temperature is found to agree well with the result at normal incidence obtained by Brossa et al.,[18] when taking into accout the factor caused by the difference between the incident angles in both experiments. It is seen from Fig.6 that the sputtering yield of Ti atoms at 900°C is almost zero at low ion fluxes and seems to increase with increasing the ion flux beyond a critical ion flux of $5x10^{14}$ ions/cm^2·s.

5. Discussion

It has been experimentally shown that when Ni-deposited graphite, Ni/nickel-carbide bi-layer and TiC-deposited graphite are bombarded with energetic ions at certain elevated temperature, the sputtering of metal atoms is substantially suppressed by the self-sustaining coating with segregated carbon layer below certain critical ion flux. The citical ion fluxes for the Ni-deposited graphite and Ni/nickel-carbide bi-layer bombarded with 5 keV Ar^+ ions at 650°C are $5x10^{12}$ ions/cm^2·s and $1.7x10^{13}$ ions/cm^2·s, respectively. On the other hand, the critical ion flux for the TiC-deposited graphite bombarded with 1 keV H^+ ions at 900°C is $5x10^{14}$ ions/cm^2·s. Here, using the criterion for the self-sustaining coating derived, we discuss which parameter makes a main contribution to the critical ion flux.

When the substanial suppression of metal sputtering takes place, one can regard that the condition 2) of $(\sigma\phi_c/k\ )(1+Q_1+Q_2) \lesssim 0.1$ is satisfied. As described above, the values of and $(1+Q_1+Q_2)/k_1$ are determined from the curve of the removal flux $R(\phi)$ obtained experimentally, since the relevant rate constants k_2' is mainly caused by the thermally activated process in this case. In a previous paper,[13] we have obtained the removal flux of carbon atoms as a function of the ion flux of 5 keV Ar^+ beam using the self-supporting Ni/nickel-carbide bi-layer film, of which the thickness is sufficiently small compared with L_c. The values of σ and $k_1/(1+Q_1+Q_2)$ are determined to be $0.43x10^{-15}$cm^2·and 0.17 s^{-1}, respectively. Using these values, the critical ion flux is computed to be $4x10^{13}$ ions/cm^2·s, which corresponds well to the value of $1.7x10^{13}$ ions/cm^2·s determined from the experimental result on the sputtering yield of nickel atoms. Therefore, one can conclude from this fact that the phenomenological treatment developed in the present study is applicable to evaluate the critical ion flux for the self-sustaining coating.

Nextly, we consider which parameter makes a main contribution to the critical ion flux. If the rate constants of k_1, k_2', k_3' and k_4 are available, it is easily known from the condition 2). Unfortunately, there exist no available data on the rate constants. When the thickness of the deposited layer, L, is much smaller than L_c, as shown later, the condition 2) is rewritten into the relation $\phi_c=0.1(k_1/\sigma)/(1+k_2'/k_3')$. The relation indicates that the difference between the critical ion fluxes observed for the two nickel-carbon composite systems is due to the difference in the values of k_2'/k_3', since the values of σ and k_1 relevant to the surface of the deposited layer are the same for the two systems. As shown in Fig.2, the carbon layer segregated at the nickel surface is a graphitic phase. This result indicates that for the Ni-deposited graphite the dissolution rate constant k_1 of carbon atoms from the graphitic layer at the surface into the nickel layer may be regarded to be the same as the dissolution rate constant k_4 from the graphite into the nickel layer, namely it holds that $k_2'=k_3'$ at eqilibrium segragation. On the other hand, for the Ni/nickel-carbide bi-layer, the value of k_3' is expected to be much larger than the value of k_2', because the nickel-carbide phases are meta-stable, thus the dissolution rate constant k_4 from the carbide into the nickel layer being much larger than the dissolution rate constant k_1 from the stable graphitic phase. In such a case, the ratio k_2'/k_3' is sufficiently small compared with unity. Based on these considerations, the critical ion flux for the Ni/nickel-carbide bi-layer is expected to be by a factor of 2 larger than that for the Ni-deposited graphite, thus the factor of 2 corresponding well to the difference in the critical ion fluxes observed experimentally for the two systems.

Using that $Q_1(=k_2'/k_3') \sim 0$ for the Ni/nickel-carbide bi-layer, the dissolution rate constant k_1 is determined to be 0.17 s^{-1} at 650°C. For Gibbsian equilibrium segregation, $C_s^0/s=(2a/r)k_2'/k_1$ is suggested to hold,[19] where s is the solubility of carbon in nickel in the units of the fractional concentration. The solubility of carbon in nickel at 650°C is experimantally determined by Coad and Riviere[20] to be 0.30 at%. Thus, the value of $(2a/r)k_2'$ is computed to be 57 s^{-1}. So, we can confirm that the condition of $Q_2 << 1$ is satisfied in the present study. The nickel layer deposited is polycrystalline, thus the grain boundary diffusion constant having to be used. Since it is not available, we use the bulk diffusion constant of carbon in nickel,[21] $5x10^{-10}$ cm^2/s at 650°C. For the thickness of the deposited nickel layer, ~5000 Å, Q_2 is computed to be 0.11. Since the grain boundary diffusion is, in general, larger than the bulk diffusion constant, the value of Q_2 indicates that the values of k_1 adn k_2' determined are very reasonable.

Here, we discuss, briefly, the critical ion flux for the TiC-deposited graphite. It has been also observed that the carbon layer segregated at the TiC surface is a graphitic phase. Thus, in this case, we may put also that $k_2'/k_3'=1$. However, we have no available experimental data on k_1 and σ. Therefore we estimate the value of k_1 from the experimental critical ion flux. The sputtering cross-section σ for the carbon layer at the TiC surface is approximated by the physical sputtering cross-section for graphite,[22] $\sim 10^{-17} cm^2$. When $Q_2 << 1$, the value of k_1 is computed to be 0.05 s^{-1} for $\phi_c=5x10^{14}$ $ions/cm^2 \cdot s$. For Gibbsian equilibrium segregation, the relation $C_s^0/C_g(0)=(2a/r)k_2'/k_1$ is regarded to hold. Using that $C_s^0 \sim 1$ and $C_g(0) \sim 1/2$, the value of $(2a/r)k_2'$ is obtained to be 0.1 s^{-1}. The grain boundary diffusion constants of C in TiC have been experimentally determined by several authors, $5x10^{-13}$ cm^2/s at 900°C.[2,3] Using these values, Q_2 is computed to be 0.1 for the thickness of the TiC layer, ~3000 Å, used in the present study. This indicates that the value of k_1 estimated is also very reasonable.

Finally we emphasize that the critical ion flux and the critical thickness for the deposited layer to be protected from resputtering can be evaluated using the phenomenological treatment developed in the present study and that the rate constant of segregation from the grain boundary to the surface and the grain boundary diffusion constant included are important parameters Especially available data on the rate constants of segregation and dissolution are required.

Acknowlegements

The authors are grateful to Professor N. Itoh for helpful discussions.

References

1) J. Roth, J. Bohdansky, N. Poschenrieder and H. K. Sinha, J. Nucl. Mater. 63, 222 (1976)
2) N. P. Busharow, E. A. Gorbatov, V. M. Gusev, M. I. Guseva and Yu. V. Martynenko, J. Nucl. Mater. 63, 230 (1976)
3) C. M. Braganza, S. K. Erents and G. M. McCraken, J. Nucl. Mater. 75, 220 (1978)
4) J. N. Smith, Jr and C. H. Meyer, Jr, J. Nucl. Mater. 76/77, 193 (1978)
5) R. Yamada, K. Nakamura, K. Sone and M. Saidoh, J. Nucl. Mater. 95, 278 (1980)
6) J. Roth, J. Bohdansky and K. L. Wilson, J. Nucl. Mater. 111/112, 775 (1982)
7) V. Philipps, K. Flaskamp and E. Vietzke, J. Nucl. Mater. 111/112, 781 (1982)
8) V. Philipps, K. Flaskamp and E. Vietzke, J. Nucl. Mater. 122/123, 1440 (1984)
9) S. Sukenobu and Y. Gomay, J. Vac. Sci. Technol. A1, 920 (1983)
10) E. Taglauer, J. Nucl. Mater. 128/129, 141 (1984)
11) H. J. Goldshmidt, Interstitial Alloys, Butterworth Pub.Co., London (1967)
12) K. Morita, T. Tsuchiya, M. Hayashibara and N. Itoh, J. Nucl. Mater. 116, 63 (1983)
13) K. Morita, H. Ohno, M. Takami and N. Itoh, J. Nucl. Mater. 128/129, 903 (1984)
14) N. Itoh and K. Motira, Rad. Effects 80, 163 (1984)
15) S. Sukenobu, Y. Gomay, H. Ohno and K. Morita, to be published.
16) A. R. Krauss, A. B. Dewald, D. M. Gruen and N. Q. Lam, Rad. Effects 89, 129 (1985).
17) S. Sukenobu and Y. Gomay, J. Nucl. Sci. Technol. 21, 366 (1984)
18) F. Brossa, J. Bohdansky, J. Roth and A. P. Martinelli, J. Nucl. Mater. 93/94, 474 (1980)
19) J. C. Shelton, H. R. Patil and J. M. Blankely, Surf. Sci. 43, 493 (1974)
20) J. P. Coad and J. C. Riviere, Surf. Sci. 25, 609 (1971)
21) T. A. Mussaro and E. E. Peterson, J. Appl. Phys. 42, 5534 (1971)
22) J. Roth, J. Bohdansky and W. Ottenberger, J. Nucl. Mater. 85/86, 1077 (1979)
23) S. Sarian, J. Appl. Phys. 39, 3305 (1968)

DEPENDENCE OF SPUTTERED ATOM ANGULAR DISTRIBUTION ON THE DEPTH OF ORIGIN

T. D. Andreadis
Chemical and Nuclear Engineering Department
Univeristy of Maryland
College Park, Maryland, USA
and
Naval Research Laboratory
Washington D.C., USA

M. L. Roush, G. Chambers
Chemical and Nuclear Engineering Department
University of Maryland
College Park, Maryland, USA

O.F. Goktepe
White Oak Laboratory
Naval Surface Weapons Center
Silver Spring, Maryland, USA

ABSTRACT

Sputtering yield angular distributions were calculated using the computer simulation code EVOLVE. Simulations are reported for Ga and In targets as well as for a composite target consisting of one surface-monolayer of In on a Ga substrate. This study indicates that atoms sputtered from subsurface layers of a target have angular distributions which are more oriented towards the surface normal than the distribution due to surface sputtered atoms.

In the present study, the EVOLVE model uses a depth dependent displacement energy model. Sputtering yields were calculated as a function of ion-beam energy for Ni and Ge targets and are compared with experiment.

INTRODUCTION

The theorectical description of sputtering by Sigmund [1] indicates that the sputtered atoms originate from the uppermost 5-10 Å of sputtered samples. The detailed features of the sputtering are affected by both the composition in this surface layer and the gradient of composition there. It was predicted by Sigmund et al. [2] that sputtered atoms of one kind which originate beneath a surface-enriched layer of another kind of atom will exhibit a narrow angular distribution peaked at the surface normal. This feature of the angular distribution arises due to the surface layer reducing the fraction of the atoms originating deeper in the sample which exit at oblique angles.

Measurements of sputtered atom composition as a function of ejection angle have been used to indicate the presence of a segregated layer of indium on a liquid-phase Ga-In eutectic [3] and of copper on a CuPt alloy [4] during Ar^+ ion bombardment.

Angular distributions of sputtered particles have been measured for a variety of combinations of target and bombardment species [3-6]; the resulting distributions characterized by $\cos^n\theta$ with values of n commonly between 1 and 2 but extending to as large as 5 [6].

The process of sputtering can be characterized in terms of models which involve an attractive force at the surface parameterized in terms of a surface binding energy, E_s, and a local potential well which must be overcome to set an atom in motion within the solid, a feature parameterized in terms of a displacement energy, E_d.

Computer simulation codes (e.g. see references 7-9) are frequently used to study the effects which arise from particular assumptions or from particular facets of the model of the process. These calculations have shown [10-12] the sputtering yield to be quite sensitive to assumed values for E_s and E_d.

The current work was undertaken to examine the sensitivity of predicted angular distributions of sputtered atoms to the depth of origin of the sputtered atoms and to the assumptions made concerning E_s and E_d.

CODE DESCRIPTION

The Monte Carlo code, EVOLVE [9], is used here to simulate atomic collision cascade effects in the ion bombardment of a solid. EVOLVE treats the transport and scattering of ions (and recoil atoms) in a solid as in a random walk problem. The trajectories of individual particles are followed in the form of straight-line segments joining points of interaction. Atoms in the target are treated as randomly located.

Atom-atom interactions are treated in the binary approximation. If the struck atom receives kinetic energy in excess of a displacement energy, E_d, the atom leaves its site with a kinetic energy $T-E_d$ where T is the energy loss of the incident atom. Beam atoms and recoiling target atoms are all tracked through the target until they leave the surface or lose sufficient energy to fall below 1.2 eV. A moving atom may escape from the target on reaching the surface by overcoming the surface binding energy, E_s, which was represented by using a planar model [13].

The scattering angle for individual interactions is dependent upon a randomly selected impact parameter and is found by integration of the classical central-potential scattering integral. The Thomas-Fermi potential with the screening function of Moliere [14] was used as the interaction potential.

A calculation of electronic energy loss was carried out for each segment of path length using the formulation of Lindhard and Scharff [15].

SURFACE BINDING AND DISPLACEMENT ENERGIES

The displacement energy is often separated into two parameters, the threshold energy to produce a recoil, E_v, and the energy lost in removing the atom from its position, E_b. Many calculations are done with $E_b = 0$. EVOLVE calculations combine these two parameters into E_d, the displacement energy.

The parameters E_d and E_s are a simplified representation of the summation of the potentials in a material. They take into account an averaging of effects from very complex force fields. Since they arise from the same source, the interatomic potentials, it is difficult to deconvolute the two effects.

It is plausible to assume that a depth dependence exists for the displacement energy. Since there are fewer neighboring atoms at the surface, atoms deeper into the target are more tightly bound in their locations than atoms at the surface. A depth dependent displacement energy model is used in this study.

In addition to the depth dependence, the displacement energy is related to the surface binding energy by adapting a formulation recently discussed by Biersack and Eckstein [16]. Using the computer program TRIM.SP those authors showed that the sputtering yield did not change when E_b and E_s were varied, as long as their sum was held constant; this sum being normally set equal to the heat of sublimation, H_s.

The model used in this study is shown in Figure 1. The displacement energy has a value E_{dt} over an interval bounded by the surface and a depth Z_t, a depth equal to the nearest neighbor distance. From this point the displacement energy increases linearly to the bulk displacement energy E_d at the depth Z_s. Equation (1) indicates how the quantities E_s and E_{dt} are specified.

$$E_s = E_{dt} = \tfrac{1}{2} H_s \quad (1)$$

Values of bulk displacement energy and heat of sublimation were taken from the compilation of Andersen [17].

SPUTTERING YIELD CALCULATIONS ON Ni AND Ge

One necessary criterion for an acceptable displacement energy model is that it result in reasonable values for the sputtering yield. Sputtering yield calculations were made for Ni and Ge targets using the proposed displacement energy model. Incident ion energies were in the 1 - 10 keV range. In addition, calculations were made at different values of Z_s for the Ni target in order to see the sensitivity of the sputtering yield to the model parameters.

Nickel and Germanium targets were selected because of their different properties under bombardment. Nickel maintains its crystal structure during ion bombardment whereas LEED studies indicate that the lattice structure of Ge deteriorates rapidly [18], becoming amorphous during on bombardment.

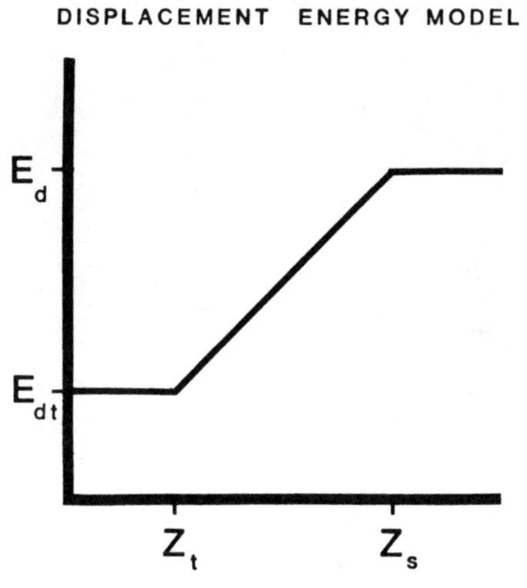

Fig. 1 - The depth dependent displacement energy model used in the EVOLVE calculations

Figure 2 shows a comparison of calculations of sputtering yield with experimental results [19] for Ar bombardment of Ge. The distance Z_t was 0.244 nm while Z_s was three times as large. The surface binding energy was 1.9 eV while the displacement energy of bulk atoms was 7.6 eV.

Figure 3 displays a similar comparison of calculations of sputtering yield with experimental results [19] for Ne bombardment of Ni. The distance Z_t was 0.249 nm while Z_s was chosen as twice as large and as three times as large for the two calculations. The surface binding energy used was 2.23 eV and the displacement energy for bulk atoms was 20 eV.

This depth variation for the displacement energy can be seen to produce sputtering yields which are in reasonable agreement with experiment when Z_s is chosen 2.5 - 3.0 times the value of Z_t, the nearest neighbor distance.

ANGULAR DISTRIBUTION OF THE SPUTTERING YIELD

Simulation calculations were utilized to examine the variation in angular distribution of sputtered particles with depth of origin. Conditions were chosen to represent those of the experiments of Dumke et al. [3], where variations in angular distribution were interpreted as arising from a monolayer of In which was found to form on the surface of a Ga-In alloy.

Three different samples were

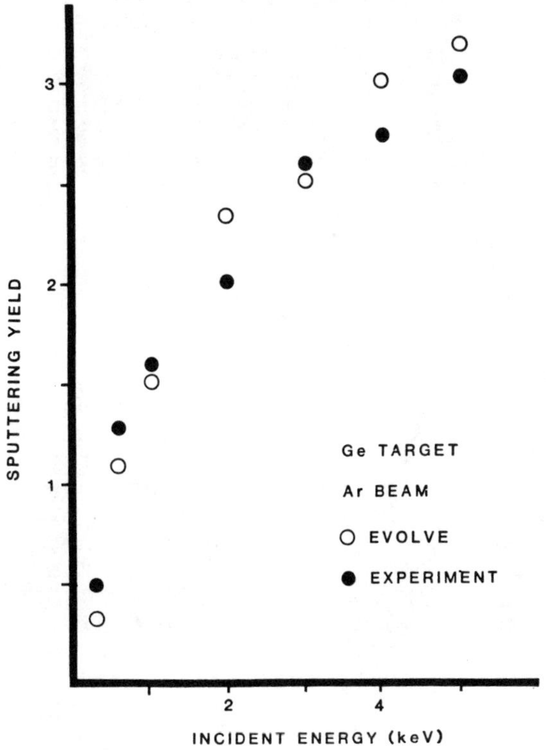

Fig. 2 - EVOLVE calculations of the sputtering yield for Ge bombarded at normal incidence by Ar ions. Experimental data is from ref. 19

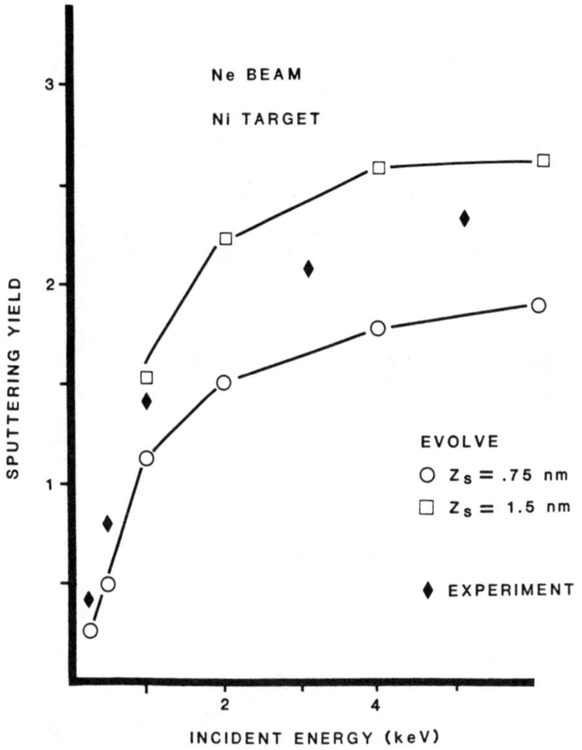

Fig. 3 - EVOLVE calculations of the sputtering yield for Ni bombarded by Ne atoms. Calculations for Z_s values of 0.75 nm an 1.5 nm simulated for these calculations. Gallium, indium, and gallium covered with a monolayer of indium were bombarded by 15-keV Ar ions. Five thousand incident particle histories were calculated for each target sample. Details concerning each sputtered particle were recorded and angular distributions compiled for sputtered particles originating within a selected depth interval. Depth intervals of 0.30nm and 0.24nm were selected for In and Ga respectively, representing the distance to the nearest neighbor atoms. Surface binding energies of 1.25 and 1.3 eV were used for In and Ga respectively. Displacement energies were taken to be 15 eV for In and 12 eV for Ga obtained from the compilation of Anderson [17].

Figure 4 shows the sputtered atom angular distributions of the three top layers of the In/Ga sample. Each individual angular distribution was fit with a function of the form $A\cos^n\theta$. The angular distributions grow more strongly peaked at the surface normal as the depth of origin is increased. The fits range from n=0.88 for the top layer to a value of approximately 2.4 for the fifth layer and all deeper layers. Although the uncertainties increase with depth because of the decreasing yield of sputtered atoms, there appears to be a region for which associated angular distributions are the same.

The angular distributions and their variation with depth of origin were found to be very much the same for the simulations of sputtering of pure In and Ga.

While the simulation calculations have shown the qualitative depth dependence anticipated, the calculated angular distributions are not as sharply peaked as was observed by Dumke et al. [3]. For pure samples the simulation results gave n equal to about 1 rather than the experimental observation of about 2. Similarly, the EVOLVE results for Ga sputtered from under a monolayer of In gave an n of 1.2 as compared with experimental observation of 3.6.

The reason for these discrepancies is not known at this time. An area of possible explanation may involve the treatment of surface binding energy. It has been noted [20] that reducing the surface binding to zero can change the angular distribution of sputtered atoms from a $\cos\theta$ to a $\cos^2\theta$ distribution.

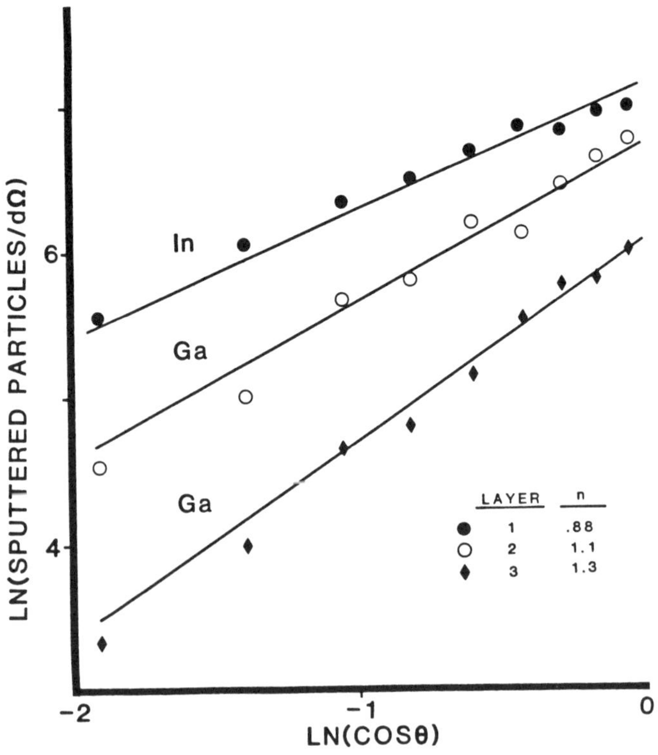

Fig. 4 - EVOLVE calculations of the angular distribution of atoms sputtered from a Ga target with a surface layer of In. Solid curves are line fits of the form $A\cos^n\theta$. Bombardment was with 15-keV Ar at normal incidence

It is well known that sputtered particles originate predominantly from the first few atom layers of the target. The displacement energy model used produced this basic feature.
The fractional contribution to the total yield for the first four layers calculated by EVOLVE was .44, .27, .12, and .06 for the In/Ga target. This gives a drop to 1/e in the first three atom layers.

CONCLUSION

A depth dependent displacement energy model was introduced in this study. It was demonstrated that variation in the dependence of the displacement energy affects the sputtering yield; the modeling used provided agreement with experimental observations.

The depth dependent displacement energy model also gave a reasonable variation of sputtering yeild with depth. In addition, this modeling demonstrated increasing peaked angular distributions as the depth of origin was increased.

ACKNOWLEDGMENT

Computer time for this study was made available by the Computer Science Center of the University of Maryland. We also wish to acknowledge, with pleasure, informative discussions with M. Rosen of the Naval Research Lab.

REFERENCES

[1] P. Sigmund, Phys. Rev. 184 (1969) 383.
[2] P. Sigmund, A. Oliva and G. Falcone, Nucl. Instr. Methods 194 (1982) 541.
[3] M.F. Dumke, T.A. Tombrello, R. A. Weller, R.M. Housley and E. Cirlin, Surf. Sci. 124 (1983) 407.
[4] H.H. Anderson, V. Chernysh, B. Stenum, T. Sorensen and H.J. Whitlow, Surf. Sci. 123 (1982) 39.
[5] K. Besocke, S. Berger, W.O. Hofer and V. Littmark, Rad. Effects 66 (1982) 35.
[6] J. M. Lambert et.al., IEEE Transactions on Nuclear Science NS-130 (1983) 1285.
[7] M.T. Robinson and I. M. Torrens, Phys. Rev. B9 (1974) 5008.
[8] W. Eckstein and J. Biersack, Nucl. Instr. Meth. B2 (1984) 550.
[9] M. L. Roush, T.D. Andreadis and O.F. Goktepe, Rad. Effects 55 (1981) 119.
[10] M.L. Roush, T.D. Andreadis, F. Davarya and O.F. Goktepe, Nucl. Instr. Meth. 191 (1981) 135.
[11] H.H. Andersen, in: Ion Implantation and Beam Processing, eds., J.S. Williams and J. M. Poate (Academic Press, Sydney, Australia, 1984).
[12] O.F. Goktepe and M. L. Roush, Nucl. Instr. Meth. B7/8 (1985) 803.
[13] M. Hou and M. T. Robinson, Appl. Phys. 18 (1979) 381.
[14] J. M. Torrens, Interatomic potentials (Academic Press, New York, 1972)
[15] J. Lindhard and M. Scharff, Phys. Rev. 124 (1961) 128.
[16] J. Biersack, W. Eckstein, Appl. Phys. A (1984) 73.
[17] H. H. Andersen, Appl. Phys. 18 (1979) 131.
[18] M. T. Robinson in: Topics in Applied Phys, Sputtering by Particle Bombardment I, ed., R. Behrich (New York, 1981) p. 78.
[19] H. H. Andersen, H. L. Bay in: Topics in Applied Physics, Sputtering by Particle Bombardment I, ed., R. Behrich (New York, 1981) 145.
[20] M. Rosen, R. Bassel, Nucl. Instr. Meth. Phys Res B2 (1984) 592.

SURFACE MODIFICATIONS AND COATINGS —GENERAL SESSIONS

FUNDAMENTALS OF ION IMPLANTATION

G. Dearnaley
Harwell Laboratory
Didcot, England

ABSTRACT

Ion implantation and related surface treatments using energetic ion beams are being employed increasingly either as a means of preparing materials for research or for industrial applications. Most of the applications so far have been for protection against wear or corrosion.
Ion implantation is a highly non-equilibrium process in which many point defects are created which can influence the location and distribution of implanted atoms. There may be removal of surface atoms by sputtering, and heating effects may accelerate radiation-enhanced diffusion and segregation. Coatings can be mixed wholly or partially into the substrate by ion bombardment and the properties of coatings can be modified by ion bombardment during or after deposition.
The basic mechanism of strengthening of metals, and of achieving corrosion resistance are summarized. The non-equilibrium nature of the process allows insoluble species to be introduced, and surfaces can be rendered amorphous: both these factors lead to unusual mechanical or corrosion behavior.

THE PROCESS OF ION IMPLANTATION has now become a widely-used technique for the modification of near-surface properties of all classes of solid materials, including metals, semiconductors, polymers ceramics and ceramic composites. Suitably chosen ions are accelerated in vacuum to energies that are generally between 10 and 200 keV and are directed towards the material surface, into which they become embedded. The depth to which modification can be achieved in this way is very limited, being from 10 nm to 1 μm in most cases, but it has been shown that this surface layer plays a critical part in establishing the course of long term degradation during wear, fatigue or corrosion. Moreover, substantial changes in material properties can be achieved with the introduction of only minor concentrations of certain atoms. This was the case when ion implantation was introduced, during the 1960s for the doping of semiconductor devices, and the precision and versatility of the method gave substantial improvements in device yield.

In this paper we shall concentrate on the modification of the properties of metals, with relevance to the important processes of wear, friction and corrosion. The development of industrial-scale plant for the treatment of metal tools and components is the subject of another paper in this group of symposia (1).

It is important to recognize the non-equilibrium nature of ion implantation, by contrast with thermochemical methods of modifying surfaces or the use of bulk alloying. Ions of the energies mentioned will displace many atoms in their path (the displacement energy being typically ∿ 20 eV); the introduction of foreign atoms is not constrained by the normal considerations of solubility or diffusivity; intense ionization phenomena occur in insulating materials and in the oxide films on metal substrates. Some of the more interesting applications of ion implantation exploit this non-equilibrium aspect, for example to introduce insoluble species or to create a highly disordered or amorphous surface layer.

In this short review, it is possible only to outline the main features of ion implantation and the resulting alteration of some important properties of metals and alloys. Because of the close relationship in technique, a description will be given of the use of energetic ion beams for the mixing of overlay coatings or for the modification of coatings which may, for instance, be deposited during ion bombardment.

These treatments may seem to resemble ion plating or other forms of plasma-assisted deposition, but there is a significant distinction. The relatively high vacuum conditions during ion implantation enable the achievement of ion energies that are orders of magnitude greater than those in plasmas, and the result is a far greater departure from thermal equilibrium accompanied by greater ion penetration.

ION RANGE

The mean depth of penetration of ions into solids, measured normal to the surface, is termed the mean projected range, R_p, and it is determined by the processes of energy loss which are (i) excitation and ionization of electrons surrounding target atoms and (ii) the elastic process of energy transfer to the atom as a whole, often causing it to recoil with a substantial energy of its own, sufficient to cause secondary displacements. Most of the statistical fluctuation in ion range comes from the elastic energy transfers which control the 'straggling' in range or mean square deviation of what is, to a first approximation, a Gaussian distribution about R_p (Fig. 1). Theoretical calculations of ion ranges were carried out by Lindhard et al. (2) on the basis of the Thomas-Fermi model of the atom, and the consequent interatomic potential during the elastic collision of two atoms. More recently, with progress in computer speed and availability, it has been possible to develop a program for a Monte Carlo simulation of ion penetration (3) which agrees well with experiment and is more reliable in terms of the distribution of ion range that determines the concentration profile of implanted atoms. This TRIM code (3) is therefore being more and more widely used as a means of gaining both quantitative and immediately visual information regarding ranges, ion collisions and lateral spreading of a collimated incident beam.

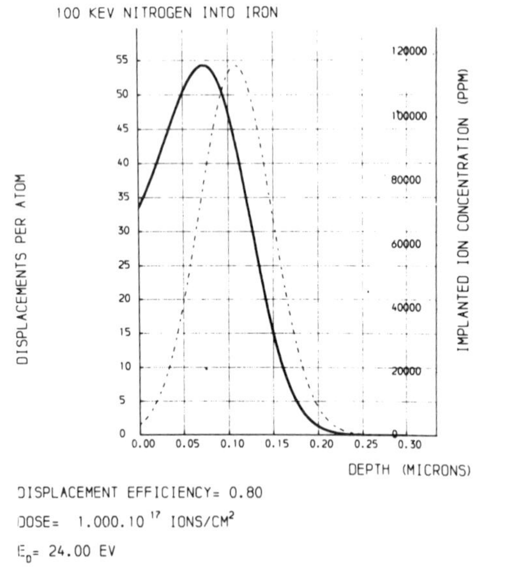

Fig. 1 - The distribution of ions (dotted curve) and damage (solid curve) induced by 100 keV N^+ implantation into iron. The range distribution is approximately Gaussian.

RADIATION DAMAGE

The violent collisions caused by energetic ion bombardment displace many atoms by primary and secondary encounters. The long-range and non-directional nature of interatomic bonding in metals (and ionic materials) aids the restoration of crystallinity: vacancies and interstitials (the primary defects) recombine and leave only a relatively small amount of residual disorder in most cases. This is more evident above a temperature approximately half the melting temperature T_m on the absolute scale, when vacancy migration becomes significant. Because interstitials move more easily there is a vacancy-rich core (Fig. 2) around the ion trajectory, and many forms of defect or defect-impurity interaction occur in the surrounding region. Residual vacancies and interstitials tend to aggregate and such clusters collapse to form platelets which, because of their appearance in the electron microscope, are called 'loops'. At high levels of irradiation such loops, which are bounded by edge dislocations, entangle with each other to form a microstructure reminiscent of highly cold-worked metal. Given appropriate 'sinks' e.g. for interstitials, and in the presence of helium impurities, vacancy clusters may be stabilized and can grow to form 'voids'. Conditions for void growth can exist in fast reactors in which the helium is created by (n,α) reactions, and so a good deal of research on ion-induced damage in metals and alloys has been carried out in order to simulate much longer periods of service in reactors. This body of knowledge is now most useful in understanding ion bombardment effects and material modification. As we shall discuss in more detail later, there is an important distinction between a fast neutron irradiation and that due to monoenergetic ions in that, in the latter case, point defects are created within a narrow lamina rather than homogeneously. As a consequence of this, there is a flux of mobile defects away from the peak of the damage production profile, situated at a depth around 0.7 R_p (Fig. 1).

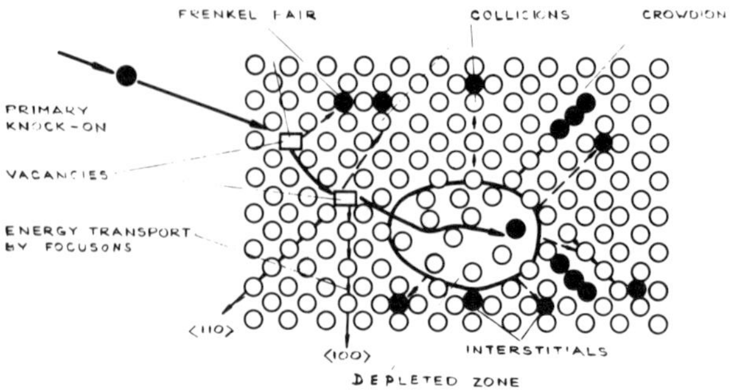

Fig. 2 - Various forms of damage created by ion bombardment of a metal lattice. Notice the vacancy-rich core near the end of the trajectory.

SPUTTERING

Collisional energy transfer to atoms lying within the first one or two atomic planes at the surface may cause them to have sufficient velocity to escape the cohesional binding potential. The emission of these 'sputtered' atoms is centred around the surface normal and falls off more or less as the cosine of the angle. Likewise, ions incident at angles other than normal induce sputtering which increases as the inverse cosine of the direction (with respect to normal): the reason for this lies in their ability to transfer energy more effectively to the near-surface region from which escape is possible.

Sputtering coefficients i.e. the number of atoms released per ion increase with the atomic number of the projectile (because more energy is deposited) and increase also with the atomic number of the target material, but in this case the increase is not monotonic but mirrors the cohesive energy of the material. It is not surprising that soft substances such as Zn, Cu, Ag, Cd, Au and Bi have large sputtering coefficients, while hard elements such as Cr, Mo and Hf have values an order of magnitude lower (4).

Since it is necessary to implant rather large concentrations of ions into metals, in general, to obtain a significant change in material properties, sputtering is a major consideration and this is particularly true if the metal is soft, or the angle of incidence is far from normal, as can be the case for complex shapes. The erosion rate due to sputtering can then limit severely the incident dose that is retained because a dynamic equilibrium can be established in which target material is sputtered away at such a rate that the implanted profile is steadily consumed. Under these conditions the peak of the implanted ion (and damage) distribution lies at the surface. Two ways of overcoming this constraint are (i) to find some ion species which will produce the required effect in lower concentration (which may also have economic benefits) and (ii) to induce transport inwards of the implanted material in association with a point-defect flux (see below). A third method is to allow or induce the presence in the critical outer layers of metal of some low-mass hard species (such as carbon) that will thereby raise the cohesive energy and lower the sputtering coefficient. Carbon frequently accumulates at ion bombarded surfaces as a result of the 'cracking' or ionic decomposition of adsorbed hydrocarbon originating in pumping fluid or lubricant. Some, at least, of the observed variation in sputtering coefficient with atomic number of the target material is due to variation in the chemical affinity for carbon, and the ability to produce a hard carbide: thus the low coefficients for Ti, Cr, V, Nb, Ta and W can be explained in part by this factor. It is one which can be exploited in the choice of material for ion assisted coatings (see below).

Sputtering coefficients tend to increase with temperature, especially as one approaches the melting temperature T_m since the cohesive energy diminishes. Ion sources operating on solid materials often make use of sputtering to release atoms from an electrode surface, and hence it is beneficial to arrange the thermal conditions, i.e. support conduction, so that these electrodes operate at a high temperature, but not so high that melting occurs.

Sputtering rates vary from grain to grain in a polycrystalline metal, and may be particularly low at local inclusions e.g. of carbide. A surface texturing can therefore occur during sputtering, and this becomes further modified by the angular dependence of sputtering, so that cones and ridges may develop after prolonged irradiation (Fig. 3). In an alloy there will generally be different sputtering rates for the various constituents. For some time it was believed that these effects could be explained simply in terms of the mass of the atoms involved, i.e. the energy partition during collisional interactions. However, differential sputtering in the outer atomic layers merely results in a change in composition which would balance the erosion rates. As Kelly and Harrison (5) have shown, the dominating factor is continuous segregation of an alloy constituent from beneath the surface, to replenish the concentration of the more easily sputtered species. This diffusion may be Gibbsian or it may be a process of radiation enhanced diffusion driven by the flux of point defects produced by bombardment. Clearly, both the phenomena outlined in this paragraph will interfere with analytical techniques such as SIMS which rely upon ion sputtering to profile the material surface, and for this reason methods of taper sectioning combined with the use of a scanned microbeam are becoming more popular.

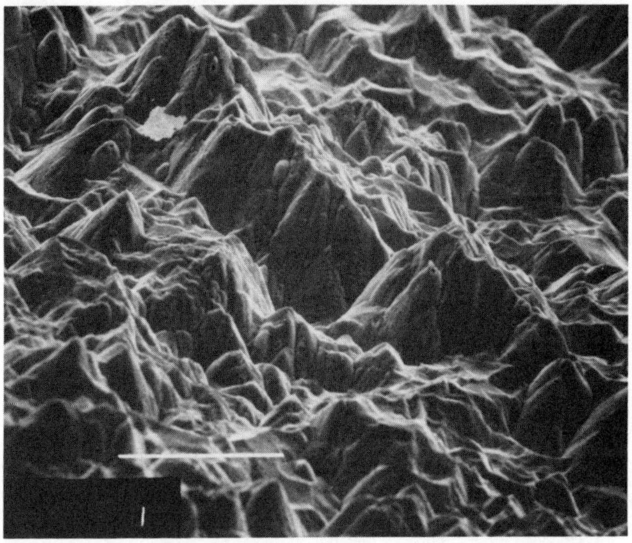

Fig. 3 - The surface of polycrystalline copper after intense sputtering with 2 keV Ar^+ ions. The marker denotes 1 μm.

INTERACTION BETWEEN POINT DEFECTS AND IMPURITIES

As we have seen, ion implantation can create large numbers of defects and these may interact strongly with the implanted atoms, thereby altering their lattice location or spatial distribution and so affecting the properties of the alloy. There are many possible interactions to consider, between impurities and vacancies or interstitials, between the point defects themselves, and subsequently between defect clusters and impurities. Models are available for estimating the (mainly electronic) interaction potentials by a second-order perturbation theory. Thus Takai et al. (6) have calculated many of the binding energies for 16 different impurities in the HCP metals Mg, Zn and Cd. The binding between vacancies and impurity atoms increases with the size of the atom, but there is a marked dependence upon electronegativity, so that a repulsion exists between vacancies and large atoms of low electronegativity, such as Sn, Pb or Bi. The interaction between these species is such that they will tend to cluster together, the strongest binding occurring if all the impurity atoms lie on the basal plane. The interstitial-substitutional impurity interaction potential is, to a first approximation, the inverse of the vacancy-impurity potential, but there is a tendency, e.g. for Cu or Mn in irradiated Al, to form mixed dumbbells consisting of an impurity and a host atom on adjacent interstitial sites.

The thermal migration of such impurity-defect complexes may be appreciably greater than that of substitutional impurity atoms in an unirradiated lattice. Diffusion of substitutional impurities commonly takes place via adjacent thermally-generated vacancy production, but the availability of such sites into which the impurity can hop is much enhanced by irradiation. The classic form of radiation-enhanced diffusion is one that is driven by the concentration gradient of the impurity but at a rate that is elevated by the non-equilibrium defect concentration. Ion bombardment produces defects inhomogeneously, however, and creates powerful fluxes of both vacancies and interstitials, the latter being mobile even at very low temperatures. If there is a binding between impurities and defects then there will be a net migration which is driven by the gradient or divergence of the defect concentration: this effect has sometimes been called a 'vacancy wind'. In the absence of such binding there can also be an induced transport of impurity atoms if, for example, the rate at which they can hop into a neighbouring vacancy site is greater than that for the corresponding host atoms. Then the impurities will progressively accumulate towards the peak of the vacancy generation profile, by what is effectively the inverse of the Kirkendall process (7,8). As in that situation of mutual interdiffusion, the strength of the effect depends upon the difference between the equilibrium thermal diffusivities of host and impurity, and it requires a species that is able to diffuse rapidly but does not bind to defects.

The redistribution of implanted atoms by enhanced diffusion can be very pronounced, especially at elevated temperatures at which vacancies are mobile (typically above 0.5 T_m). Watkins (9) has observed a number of such cases during the ion implantation of titanium at temperatures above 500°C. Figure 4 shows the interesting double peak which can develop if some implanted atoms are trapped locally e.g. by some other impurity or by clustering together, while other, isolated atoms are driven inwards by a defect flux (generally created at about 0.7 R_p). Depending upon the impurity-impurity interaction, the redistributed implanted atoms may (or may not) cluster together and the precipitate boundary so formed may well act as a sink for vacancies. In this way there would be a net flux of host atoms outwards and foreign atoms inwards. Watkins was able to demonstrate that Fe implanted above 500°C does aggregate into clusters since these were sizeable enough to be ferromagnetic.

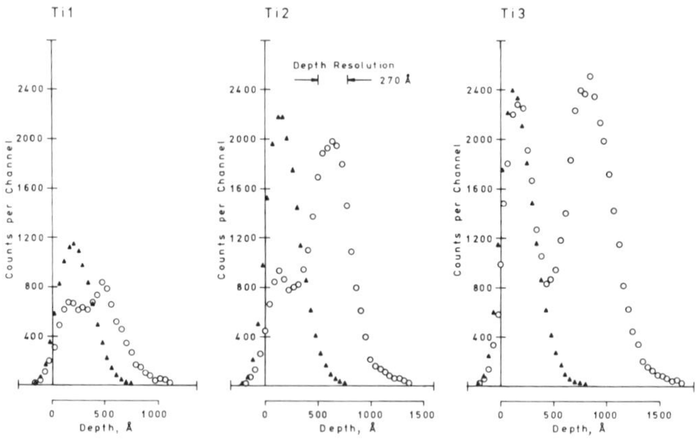

Fig. 4 - The depth distribution of Cu^+ ions implanted into titanium at 500°C showing the development of a second, deeper peak caused by radiation enhanced diffusion (after Watkins (9)).

Depending upon the ion flux and temperature, second phase precipitates in an irradiated alloy may either be dispersed or may undergo Ostwald ripening by which the larger ones grow at the expense of smaller. The factors governing such behaviour have been discussed by Nelson (10) with particular reference to the γ' precipitates of Ni_3Al present in Ni-Al alloys. Ordered precipitates can be first disordered and then dissolved, the dispersion coming about mainly by radiation-enhanced diffusion from the interface.

ION IMPLANTATION METALLURGY

Ion implantation has become so widely used as a means of preparing metals for scientific research that the above term has been developed (11). The chief virtue is that, unlike conventional alloying or powder metallurgy, chosen atoms can be introduced in an atomically dispersed manner and to concentrations that are not governed by equilibrium solubilities or diffusion rates. Furthermore, this addition can be carried out at low temperatures without, generally, altering the grain structure of the host material. The implanted atoms lie in a well-defined zone and it is possible to measure their subsequent migration using modern analytical methods, such as ion backscattering, with a depth resolution of 5 nm or better.

As an example of the usefulness of this, Myers et al. (12) have investigated beryllium implanted with copper to concentrations above the limit of solid solubility. Upon annealing some copper precipitates as Be_3Cu while the rest diffuses into the bulk. The concentration at the boundary between the two-phase and one-phase regions gives the solid solubility of copper at that temperature, while the concentration profile could be fit with a complementary error function to give the diffusivity. In this way data on diffusion rates was extended by five orders of magnitude. In general, such techniques provide valuable information regarding the low temperature behavior of metals and alloys.

Some combinations of metal and additive can, under appropriate conditions, form an amorphous or glassy structure and this is interesting both from the scientific and technological standpoint. Thus combinations of transition metals and metalloids, or transition metals and rare earths or noble metals and refractory metals will form amorphous alloys if subjected to rapid quenching from the melt or under ion bombardment. Large differences in atom size or electronegativity limit stable alloy formation and any tendency towards anisotropic bonding favours the glassy state. In one of the earliest examples Ali et al. (13) showed that P^+ implantation of the transition metals Fe, Co and Ni would produce a thin amorphous alloy at metalloid concentrations between 10 and 20 atomic per cent. Transmission electron diffraction showed that the structure resembled that in rapidly quenched metals. Steel implanted with sufficiently high doses of titanium and carbon (above about 20 atomic per cent) also becomes amorphous (14) and the surface then possesses interesting low-friction properties. After heat treatment micro-crystallinity is restored, and the friction increases, although wear rates remain low. The reasons for this are not yet known.

The precise lattice location of implanted atoms can be determined by the technique of channelled Rutherford backscattering (RBS). If the probing beam, generally of 2 MeV $^4He^+$ ions, is directed along major crystal axes the yield of particles scattered in a backward direction can be related to the position of the impurity with respect to the host lattice. This is because the channelling phenomenon (15) steers incident ions towards the center of the open channel and reduces the scattering probability (cross-section) for substitutional atoms, while increasing it for interstitial species. Thus Figure 5 shows the variation with angle of the backscattered yield of alpha particles from tantalum and lanthanum ions implanted into single

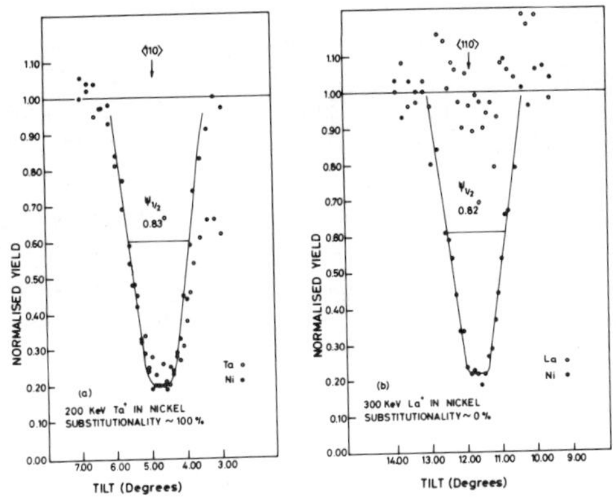

Fig. 5 - The angular distribution of the back-scattered yield of ^4He ions from Ni single crystals ion implanted with (a) Ta^+ at 200 keV and (b) La^+ at 300 keV. The dip for Ta matches that for Ni, showing it is entirely sub-stitutional, while La gives no dip, showing that it occupies random sites in the lattice (after Sood (16)).

crystal nickel (16). Tantalum occupies sub-stitutional sites, while the larger lanthanum atoms are randomly distributed. Sood (17) has shown that atoms up to about 40 per cent larger than the host can be accommodated substitution-ally by ion implantation: the normal limits for differences in atomic radius and electro-negativity in alloy formation are, according to the Hume-Rothery rules, only ± 15 per cent. Figure 6 shows a variety of species which, when introduced into Fe by ion implantation, are mainly substitutional. Smaller species, however, such as boron, remain interstitial.

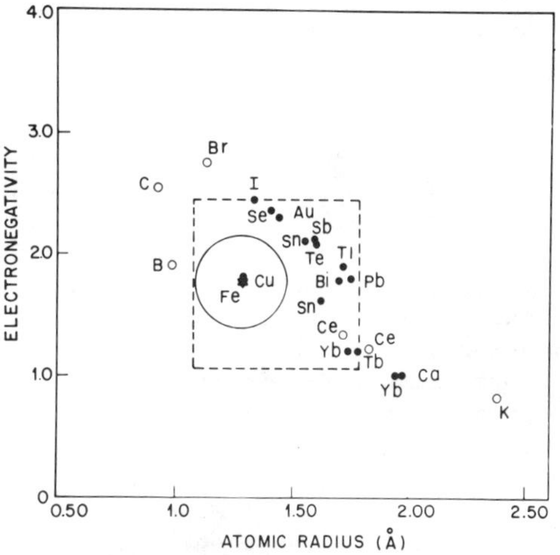

Fig. 6 - Atoms that are mainly substitutional (closed circles) or mainly interstitial (open circles) when implanted into single crystal iron, plotted as a fraction of their size and electronegativity. The circle around Fe corresponds to the Hume-Rothery rule (after Sood (17)).

One possibility is that the vacancy-rich regions near to the ion trajectory produce an elastic strain which aids the accommodation of oversized substitutional atoms: the vacancies may thus act locally as a third constituent of the alloy. As a result of local disorder the electron:atom ratios will be disturbed, and it appears that this relaxes the normal constraint on electronegativity differences. Thermal annealing restores equilibrium, and insoluble implanted atoms tend to precipitate and to migrate towards the surface, often in association with point defects, for which the surface acts as a sink. The eventual fate of these atoms is often to appear as second-phase crystallites at the surface.

Hydrogen provides an example of a species which is normally highly insoluble in metals such as iron. As a small interstitial species its mobility is very high, and it is located mainly at traps such as dislocations (in cold-worked material) or at interfaces such as grain boundaries and boundaries with inclusions such as MnS or TiC (18). In proton-bombarded material the irradiation induces defects which can trap hydrogen. Thus, Myers et al. (19) were able to show that hydrogen is associated with vacancies in implanted Fe, and that the trapping activation energy is 0.53 eV, i.e. about the same as that between hydrogen and dislocations. These observations are relevant to the problems of the trapping and retention of tritium in fusion reactors and other systems.

The role of trapping centers for highly mobile interstitial species such as H, B, C, N in metals is important since, generally, the solid solubility of these atoms at low temperatures is very small. If, however, there is a source of these elements at the metal surface and an available trapping site near to the surface, then it is observed that remarkable quantities of material can be introduced. Thus, Singer and colleagues (19) have observed that if iron or steel is implanted with high doses of Ti^+ ions, carbon from the atmosphere of the work-chamber can diffuse inwards and form small TiC precipitates. The rate of in-diffusion of carbon increases as the dose of titanium rises, because the trap density goes up. Furthermore, hydrogen is available, perhaps by the radiolytic decomposition of absorbed hydrocarbon, and as Lee and Lee (18) have shown, TiC:Fe interfaces are powerful traps for hydrogen, the trapping enthalpy being 50 per cent higher than that at vacancies or dislocations in iron. As a result, large amounts of hydrogen are incorporated in association with carbon, but there is no reason to believe that these two species migrate together. Since TiC precipitates are increasingly used as strengthening agents in alloy composites, it is necessary to be aware that hydrogen embrittlement may be influenced also.

Antimony, together with Sn and As, is an increasingly common 'tramp' impurity in steels. It can segregate to grain boundaries and lead to de-cohesion. Myers et al. (20) have shown that Sb is trapped strongly at the interface between TiC or TaC precipitates in Fe. Here again, RBS analysis was used to demonstrate the process in specimens which were ion implanted with Sb^+ ions at different temperatures (Fig. 7). The

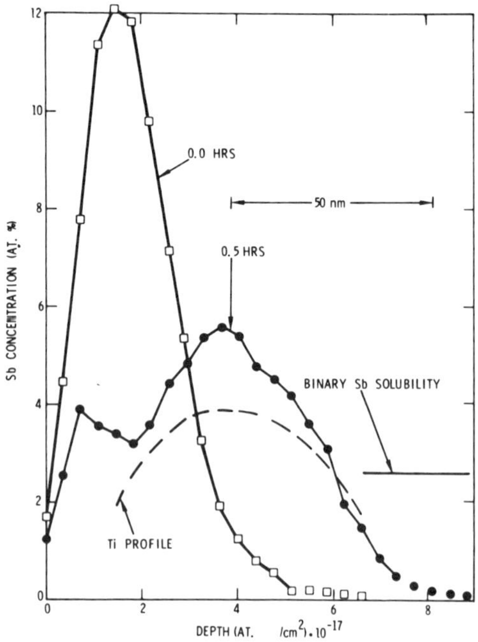

Fig. 7 - The thermally-induced redistribution of implanted antimony to decorate a deeper layer of TiC precipitates in iron, also introduced by ion implantation (Myers et al. (20)).

experimental data were consistent with a trapping enthalpy of 0.39 eV. It is possible that such trapping, in alloys containing relatively small amounts of tramp impurities, will scavenge the system and lessen the likelihood of grain boundary trapping and subsequent decohesion. If so, the same arguments may apply to alloys strengthened with TiC, such as 'FerroTic'(R) and to the problem of hydrogen embrittlement, mentioned above. If hydrogen is trapped with a trapping enthalpy greater than that of dislocations or grain boundaries, then it may no longer be available to assist crack growth. It is not yet clear whether it is possible to create stronger traps either by powder metallurgy or by ion implantation, but the advantages of doing so would be considerable.

ION BEAM MIXING

Because it is difficult to generate ion beams of certain refractory or noble metal species, there is interest in ion beam mixing as an alternative means of introducing foreign elements. A thin coating (up to about 200 nm thick) is deposited in vacuum and this so-called 'bilayer' is ion bombarded with the intention of incorporating the coating into the substrate by various processes of interdiffusion and mixing.

Collisional processes of direct recoil (of struck atoms) or of cascade mixing, which occurs during bombardment with heavy ions that are capable of displaying all the atoms along their trajectory, provide athermal mechanisms for mixing. Collisional redistribution of atoms, can be modelled mathematically and contributes an irreducible amount of intermixing between any two systems, formed either as a 'bilayer' of material deposited on substrate or as a 'marker' layer of atoms sandwiched between two regions of host material. Calculations by Sigmund and Gras-Marti (21) show, however, that in most experimental cases the degree of mixing is greater than would be expected, and the mixing parameter (mean square deviation of the redistribution function) does not depend smoothly on the atomic weight.

Such observations show that thermally-activated diffusive processes, taking place on a far longer timescale, account for the major part of the mixing in many cases. The various kinds of radiation-enhanced diffusion and inverse Kirkendall processes described above are believed to be responsible. At relatively low temperatures the migration of interstitials can induce transport (always in the same direction) while at higher temperatures vacancy-assisted migration occurs.

The dominant factor controlling the rate of intermixing appears to be the equilibrium diffusivity of the species involved, but the heat of formation of compounds determines whether a definite phase is produced. It is remarkable that the violent and non-equilibrium bombardment of a bilayer can often result in a specific intermetallic phase being produced throughout a layer perhaps 200 nm thick. The composition and structure of the mixed phase depend on its heat of formation, the diffusivities of the constituents within that phase and possibly the relationship between the structure of the phase and that of the original layers.

There will usually be present at the interface between deposited films and the substrate a thin layer of oxide and adsorbed contaminants. Normally such a layer is a cause of poor coating adhesion and it can act as a diffusion barrier for intermixing. Even a 2 nm film of SiO_2 on silicon provides a substantial barrier for thermal diffusion but fortunately the barrier properties of the film are disrupted by energetic ion bombardment. A flux of the order of 10^{16} ions cm^{-2} penetrating the interfacial layer is usually sufficient to allow subsequent thermal interdiffusion to proceed rapidly and uniformly. The mechanism by which the oxide is disrupted is not well established, but it appears likely that it involves the intense ionizing effect of a heavy ion passing through a thin film of insulator. The breaking of bonds and the local stresses generated seem effective in creating pathways for cationic diffusion.

A similar effect has been observed by Goode (22) in a rather unusual example of ion beam mixing between a solid and a gaseous phase. He showed that stainless steel subjected to ion bombardment in the presence of residual molecular oxygen will develop an oxide film even at temperatures as low as 50°C. The process is thermally activated and the film thickness increases parabolically with time, suggesting that it is governed by diffusion through the thickening film, which is found to consist of very fine-grained Fe_2O_3. It is concluded that the normally very slow diffusion of Fe through this oxide is enhanced greatly by ion bombardment (with $\sim 10^{13}$ N^+ ions cm^{-2} sec^{-1} at 100 keV) and that chemical interaction takes place with oxygen at the oxide surface.

ION BEAM ENHANCED DEPOSITION AND ION ASSISTED COATINGS

It has been known for many years that ion bombardment can be used to modify the properties of deposited coatings. Christian Weissmantel (23) has been a pioneer of this approach, using mainly ion energies of 0.5 - 2.0 keV to bombard metal surfaces during deposition or while exposed to organic vapor, e.g. butane. Adsorbed hydrocarbon becomes 'cracked' to produce a hydrogen-containing film of carbon, called i-carbon (i standing for ion) in which some of the bonding is tetrahedral as in diamond. Likewise boron (24,25) under ion bombardment with nitrogen develops a proportion of tetrahedrally-bonded BN. These films are hard, inert and semitransparent, and therefore have a number of practical applications.

There are two techniques available: the ion bombardment can be carried out, generally with low energy ions, continuously during film deposition, or the bombardment or implantation, generally with more energetic reactive ion species such as N^+, can be performed after the film has been laid down. The results are likely to be different in terms of structure, adhesion and homogeneity of composition. More energetic light ions such as N^+ will lose a significant proportion of their energy in the form of ionizing processes and insofar as the coating develops insulating properties there is the possibility of radiation annealing of the structure. The desired non-equilibrium high temperature/high pressure phases analogous to diamond are more likely to be produced at lower ion energies (< 10 keV) perhaps around the peak of the energy loss rate in the form of elastic collisions. The effective high temperatures are brought about collisionally and because large amounts of additional material are injected the effective pressures and compressive stresses are high. For example, Xe^+ ions implanted into metals form 'bubbles' which are in fact solid due to the local pressure generated by creating a small bubble in a material with high cohesive energy.

Sequential deposition and ion bombardment of titanium using N_2^+ ions at about 30 keV has been carried out by Kant et al. (26) to develop wear resistant films of TiN (containing some TiC and TiO_2) on steel substrates. By carrying out the process in a residual atmosphere of N_2 it was possible to induce some enhanced reaction at the surface of the film and thereby increase the net take-up of nitrogen. Either because of the fine micro-structure, or due to the presence of residual metal, these films were deformable and could be crushed by several microns into the substrate steel without decohesion. Because such films are only 100 - 300 nm thick, they are smooth and free from the nodules which develop in CVD or PVD coatings of TiN. The low temperature at which they can be formed makes them suitable for precision components such as bearings, molds, etc. Dearnaley (27) has shown that the best treatment in terms of microhardness and abrasion resistance may be to implant first with nitrogen, then deposit the coating (B,Ti,Hf etc.) and bombard first with energetic ions in order to induce interfacial mixing, after which the ion energy is programmed downwards to implant into the film.

Holleck (28) has argued that it is advantageous to deposit coatings in which there are several distinct phases, with interphase boundaries which are able to absorb energy and arrest crack formation. This is particularly desirable in hard coatings which are likely to be brittle. It is possible, in the process described above, to lay down alternate films, perhaps of boron and titanium, and to control the stoichiometry so that the optimum ratio of nitrogen:cation species is obtained. The heats of formation and bond strengths between the various species Ti, B and N involved are all high and this will favour the creation of a fine dispersion of stable precipitates.

In summary, the chief virtues of using energetic ion bombardment, in vacuum, to assist coating properties are (i) the non-equilibrium nature of the process, enabling unusual high temperature phases to be produced; (ii) the controllability of the process in terms of the penetration depth, and concentration and uniformity of the implanted atoms, and (iii) the low temperatures involved. The development of powerful sources, at least for gaseous ions, capable of 0.1 - 1.0 A of beam current, it is now possible to envisage economic methods of treating large areas, as long as the more conventional coating process can be sufficiently well controlled!

STRENGTHENING MECHANISMS

Ion implantation is being used more widely as a sophisticated method of improving the microhardness and wear resistance of metals and alloys. Here it is possible to give only a brief summary of the basic mechanisms by which implanted atoms can strengthen the material. It must be recognized that wear is, however, a highly complex process involving surface chemistry, lubrication, the secondary action of wear debris, and many other controversial factors.

In terms of the classical ideas of physical metallurgy, strengthening is achieved by impeding the movement of dislocations, brought about during plastic deformation. This is done by creating obstacles, i.e. points at which energy is absorbed in the movement of a dislocation. The simplest obstacles are foreign atoms, either interstitial or substitutional. Interstitial atoms, such as nitrogen or carbon, are able to segregate at dislocations (if they are mobile) and so create an 'atmosphere' which resists perturbation. A dispersion of second phase particles will resist deformation, particularly if they are hard: they should not be too large, however, or the interface can crack. The skill of medieval swordmakers lay in being able to produce a dense dispersion of fine carbide or carbonitride particles in steel by forging at relatively low temperatures. Being a low temperature process, ion implantation is ideal for introducing a very fine dispersion of second-phase precipitates. Dispersion strengthening, rather than solid solution hardening, is the chief mechanism when reactive ions such as nitrogen are implanted into metals such as titanium or chromium.

The high dose ion implantation of Ti^+ or Ta^+ into iron or steel does not, by itself, produce a significant improvement in wear resistance, but in association with comparable concentration of carbon or nitrogen the surface becomes highly wear-resistant (29). As was mentioned above, the structure becomes amorphous when more than about 20 atomic per cent of Ti and C are present in steel (14), but the wear resistance remains high after thermal treatment sufficient to restore microcrystallinity. It is therefore reasonable to attribute the strengthening to the fine dispersion of TiC precipitates.

Fleischer (30) identified a more efficient mechanism of strengthening involving an oversized substitutional atom adjacent to an interstial impurity, which is trapped there by the strain surrounding its large neighbour. Such asymmetric defects or dipoles (Fleischer called them 'tetragonal defects') interact with the shear stress of a nearby dislocation and act as energy-absorbing obstacles. As Fleischer pointed out, the introduction of large numbers of greatly oversized atoms is normally impossible, but he suggested that "a beam of energetic particles" might be used. This is what can now be done by ion implantation, and Dearnaley et al. (31) have reported that yttrium (which is 40 per cent greater in radius than iron) will improve wear resistance in steels even at low doses of $\sim 10^{16}$ ions cm^{-2}. Nitrogen or carbon must be present as well, but the quantities of these and of yttrium required are small: a dose of about 10^{16} Y^+ ions cm^{-2} is sufficient, corresponding to only about 1 atomic per cent of additive. It was shown (31) that at a temperature of $200^{\circ}C$ or less nitrogen will move to and decorate implanted yttrium, a process which serves to retain the mobile interstitial to higher temperatures than would otherwise be the case. Yttrium atoms have a size which should just enable them to occupy substitutional sites in iron, according to Sood's empirical rule, illustrated in Figure 6 above. Its diffusivity in iron is very low, and it is therefore unlikely to form aggregates. The obstacles impeding dislocation movement are therefore likely to be atomically dispersed, which is a requirement for Fleischer's dipoles to be effective. This application of ion implantation makes good use of the non-equilibrium nature of the process.

Structural factors are also important in a tribological situation since they influence the work-hardening mechanism that takes place as dislocations move and entangle with one another. The fcc structure has many equivalent slip planes and thus generally supports work hardening to a much greater extent than in hcp metals, where slip is confined more or less to the basal plane. This may explain why nitrogen ion implantation into cobalt-cemented tungsten carbide is more successful if carried out at elevated temperatures (> 400°C), because the cobalt binder undergoes a phase transition from hcp to fcc at around that temperature. Dearnaley et al. (32) have suggested that implanted nitrogen will lock the structure into the fcc phase and that the interaction between nitrogen interstitials and dislocations generated during wear will provide greater strengthening in this condition. The idea is that a dislocation entanglement involving multiple slip planes will have a greater resistance to stress.

Complex work-hardening processes occur during the running-in stage of wear and it is then that ion implanted impurities, close to the surface, can exert most effect. Under favorable conditions a reduced wear rate can persist to depths of wear that far exceed the depth of ion implantation. Steady conditions of slow attrition, such as those occurring during mild abrasive wear by fine particles of debris or other materials, are required for such persistance to be observed. It does not appear in accelerated wear tests or under severe conditions of adhesive or abrasive wear e.g. in rock drilling.

MECHANISMS OF CORROSION RESISTANCE

Ion implantation is capable of bringing about major changes in the corrosion rate of metals. Generally this comes about as a consequence of compositional changes and for this reason it is often possible to correlate the behavior of an ion implanted surface with that of a conventional alloy of similar composition.

Under conditions of aqueous corrosion, improvements in the rate of general corrosion are obtained either by reducing the anodic or cathodic processes. The electrochemical potential of the alloy can be altered, for example by implanting atoms such as Cr or Ti into iron, or by introducing them by ion beam mixing of a coating. It is also possible to implant atoms such as tantalum into iron, and to obtain good corrosion resistance: Ta has too large an atom to form an alloy with iron, but the comparison with chromium made using ion implanted specimens shows that tantalum gives a lower critical current density (Fig. 8).

Some implanted species, such as the noble metals Pt or Pd act by catalysing surface electrochemical reactions in which hydrogen is released in gaseous form. Even small amounts of

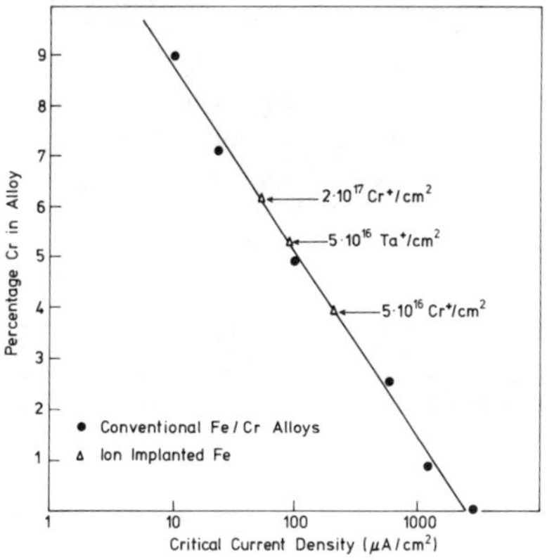

Fig. 8 - The critical current density (a measure of corrosion rate) for iron-chromium binary alloys in an acidic electrolyte, compared with that for ion implanted Fe with a similar surface composition. The data point for Ta^+ implanted iron shows that it is superior to Cr^+ (from Ashworth et al. (33)).

implanted catalyst can reduce markedly the corrosion of titanium in acid.

Localized corrosion causing pitting is a source of failure in highly-stressed bearings because it can initiate cracking. The ion implantation of Cr^+ is now being used to overcome the problem, probably by the formation of a protective film of chromium oxy-hydroxide. The effect is enhanced by chloride ions, which occur in a marine environment.

Thermal oxidation is another common form of corrosion and in this case conditions are more favorable for ion implantation because there is no longer the risk of galvanic attack caused by differences in electrochemical potential. Such effects occur if the thin implanted layer is scratched and an electrical path exists through the fluid medium. The initial rate of thermal oxidation is generally fast enough to consume the implanted atoms in oxide, so that any lasting effect they may have must come from an improvement in the protective quality of the oxide.

This can come about in various ways, such as:-

(i) the formation of a continuous barrier layer which is less permeable to anions or cations;
(ii) the relief of stress in the oxide, especially as it thickens, so that cracking or other structural changes which allow corrodent to penetrate are reduced;
(iii) improvement in the bonding of oxide scale to the metal surface so that it remains in place after thermal cycling (which causes stress by the mismatch of expansion coefficients).

Each of these has been observed or proposed as an explanation of the oxidation in some ion implanted metal. It is generally easier to block short-circuit diffusion pathways rather than providing a continuous diffusion barrier. Oversized cations such as yttrium or rare earths have a tendency to segregate at grain boundaries.

within an oxide film, during its growth. It is not yet clear just how the presence of these atoms will necessarily block the migration of other species. Further work on the modelling of ionic structures and migration along grain boundaries may resolve this question.

REFERENCES

1. Dearnaley, G. Proc. Seminar on Surface Alloying by Ion, Electron & Laser Beams, Toronto (1985) (to be published by ASM).
2. Lindhard, J., M. Scharff and H. E. Schiøtt, Kgl. Danske Vid. Selsk. Matt.-Fys. Medd. 33 no. 14 (1963).
3. Biersack, J. P. and L. G. Haggmark, Nucl. Instr. & Methods 174, 257 (1980).
4. Almén, O. and G. Bruce, Trans. 8th National Vacuum Symposium, Washington (1961) p. 245, Pergamon Press, New York.
5. Kelly, R. and D. E. Harrison, Mater. Sci. & Eng. 69, 449-456 (1985).
6. O. Takai, M. Doyama and Y. Hisamatsu, "Point Defects & Defect Interactions in Metals" p. 113, University of Tokyo Press, Tokyo (1982).
7. Dearnaley, G., Appl. Phys. Lett. 28, 244-245 (1976).
8. Marwick, A. D., J. Phys. F. 8, 1849 (1978).
9. Watkins, R. E. J., Radiation Effects 84, 27-44 (1985).
10. R. S. Nelson, "Applications of Ion Beams to Metals" ed. S. T. Picraux, E. P. EerNisse and F. L. Vook, p. 221, Plenum Publishing Corp., New York (1974).
11. C. M. Preece and J. K. Hirvonen, "Ion Implantation Metallurgy", Met. Soc. AIME, Warrendale, Pa. (1980).
12. Myers, S. M. and J. E. Smugeresky, Met. Trans. 8A, 609 (1977); Myers, S. M., S. T. Picraux and T. S. Prevender, Phys. Rev. B.9, 3953 (1974).
13. Ali, A., W.A. Grant and P. J. Grundy, Radiation Effects 34, 251 (1977).
14. Follstaedt, D. M., J. A. Knapp and S. T. Picraux, Appl. Phys. Lett. 37, 330 (1980).
15. D. V. Morgan (ed.) "Channelling" John Wiley & Sons, London (1973).
16. Sood, D. K. and G. Dearnaley, Radiation Effects 39, 157-162 (1978).
17. Sood, D. K., Phys. Lett. 68A, 469-470 (1978).
18. Lee, J. L. and J. Y. Lee, Metal Science 17, 426-432 (1983).
19. Singer, I. L., R. N. Bolster and C. A. Carosella, Thin Solid Films 73, 283 (1980).
20. Myers, S. M., D. M. Follstaedt and H. J. Rack Appl. Phys. Lett. 33, 396-8 (1978).
21. Sigmund, P. and A. Gras-Marti, Nucl. Instr. & Methods 182/183, 25 (1981).
22. Goode, P. D. and I. J. R. Baumvol, Nucl. Instr. & Methods 189, 161-168 (1981).
23. Weissmantel, C., J. Vac. Sci. & Technol. 18, 179 (1981).
24. Satou, M. and F. Fujimoto, Jap. J. Appl. Phys. 22, L171-L172 (1983).
25. Guzman, L., F. Marchetti, L. Calliari, I. Scotoni and F. Ferrari, Thin Solid Films 117, L63-L66 (1984).
26. Kant, R. A., B. D. Sartwell, I. L. Singer and R. G. Vardiman, Nucl. Instr. & Methods B7/8, 915-919 (1985).
27. Dearnaley, G., Mater. Sci. & Eng. 69, 139-147 (1985).
28. Holleck, H. Proc. 12th Int. Conf. on Metallurgical Coatings, Los Angeles (April 1985) to be published in Thin Solid Films.
29. Follstaedt, D. M., L. E. Pope, J. A. Knapp, S. T. Picraux and F. G. Yost, Thin Solid Films 107, 259 (1983).
30. R. L. Fleischer, "Strengthening of Metals" D. Peckner (ed.), Reinhold Publishing Corp., New York (1964), p. 138.
31. Dearnaley, G., P. D. Goode, F. J. Minter, A. T. Peacock and C. N. Waddell, Proc. 12th Int. Conf. on Metallurgical Coatings, Los Angeles (April 1985) to be published in Thin Solid Films.
32. Dearnaley, G., B. James, D. J. Mazey, F. J. Minter, P. K. Rol and V. Thompson, Proc. 11th Int. Plansee Seminar, Reutte (Metallwerk Plansee, Reutte, 1985) 627-638.
33. Ashworth, V., D. Baxter, W. A. Grant and R. P. M. Procter, Corros. Sci. 16, 775 (1976).

FORMATION OF HARD SURFACES BY THE LASER MELT-PARTICLE INJECTION PROCESS

Khershed P. Cooper
Geo-Centers, Inc.
Suitland, Maryland, USA

Abstract

A process for forming hard metal matrix-carbide composite surfaces using an oscillating laser beam is described. By this technique, injected layers up to 12 mm wide were produced with a single melt pass. Results showing the effect of sample translation rate, laser power and powder feed rate on injected layer characteristics of Ti-6Al-4V are presented. Microstructural analysis of Inconel 625 and Ti-6Al-4V alloys treated with WC and TiC is discussed in terms of phase distributions, microanalysis and the dependence on the degree of dissolution of the carbide phase into the melt. Hardness data from the various portions of the processed area are summarized and corelated with microstructural information. With particle injection, the hardness of Inconel 625 was increased from less than 20R_c to 56 R_c. Finally, results of co-efficient of friction (μ_k) measurements are given and discussed in terms of surface damage and type of wear. After 15 wear cycles, untreated Inconel 625 had a μ_k value of 0.7 and showed extensive plowing, while the carbide injected surface exhibited a value of 0.4 and showed minimal deformation.

THE LASER MELT-PARTICLE INJECTION process was developed to form hard, wear resistant surfaces on alloys (1,2). Extensive development work was carried out especially on light weight, high strength alloys, such as 5052 Al and Ti-6Al-4V, for which alternative surface hardening processes are not plentiful (1,3). By this process, ceramic particles, such as carbides, are blown into the melt pool formed by a high energy laser beam on a moving substrate. Subsequent solidification results in a composite metal-carbide layer exhibiting continuity of properties across the substrate-melt interface.

With a single melt pass and a stationary beam, layers 2-3 mm wide could be processed. Wider areas were processed by overlapping several melt passes with incremental advancement of the work-piece (2). Besides being a relatively slow process, the overlapped regions sometimes showed excessive thermal cracking and carbide dissolution. For example, in Ti-6Al-4V alloy injected with WC, complex resolidified carbides were found within the overlapped region (4). As a first step towards the commercialization of the process, a means to produce wide injected strips with a single melt pass was developed. For this purpose, the process was modified to include laser beam oscillations to form wider melt pools. With this technique, injected layers 5-10 mm wide were produced (5). Thus, wear resistant surfaces requiring contact with wide mating faces could be fabricated.

THE LASER MELT-PARTICLE INJECTION PROCESS

A schematic representation of the laser melt-particle injection process is given in Figure 1. A high energy oscillating laser beam forms a shallow, but wide melt pool on the translating sample surface. Ceramic particles, carried by a helium gas stream, are injected into the melt pool via a dual nozzle located nearby. The spacing between the nozzle outlets can be adjusted to distribute the particles uniformly over the width of the melt pool. The injected layer consists of hard, ceramic particles embedded in a metal matrix.

PROCESSING APPARATUS - The laser processing apparatus is described in Figure 2. It consists of a vacuum chamber which is usually backfilled with helium. This prevents oxidation and enhances heat transfer. The sample is clamped to a translating stage on the rotating table. Sample translation rates can be varied from 0.5 to 10 cm sec^{-1}. This device was used to make 6-8 cm long individual passes, as well as 4-5 cm wide areas by overlapping several passes (2).

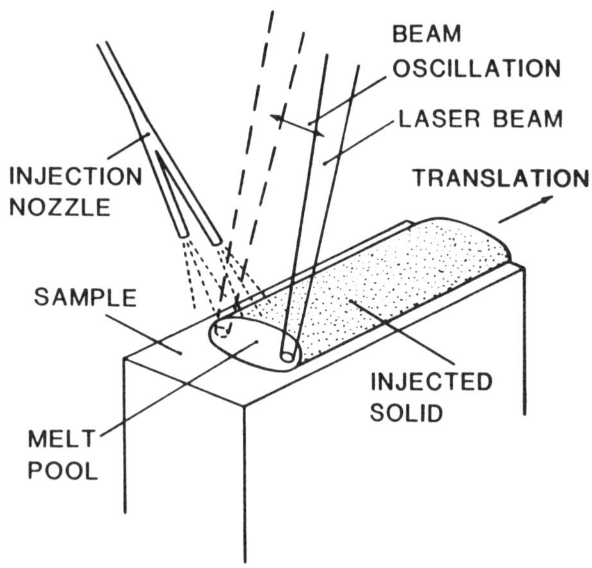

Fig. 1 - Schematic diagram of laser melt-particle injection processing with an oscillating beam.

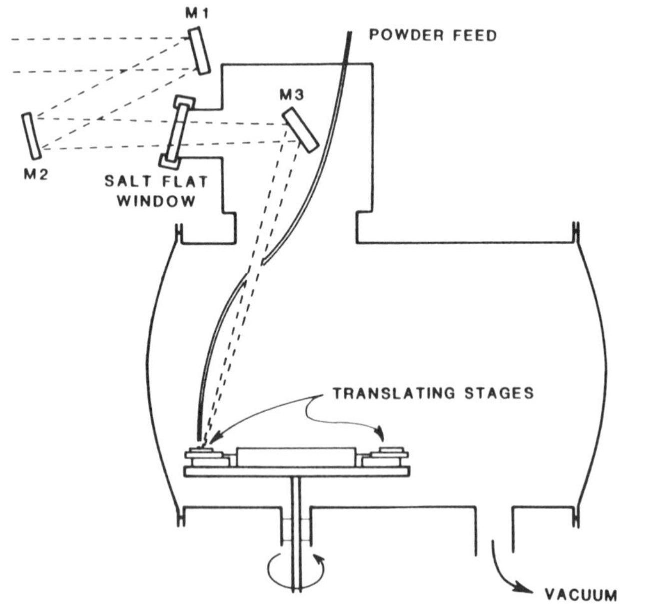

Fig. 2 - Sketch of the laser processing apparatus showing the laser optics.

A continuous wave CO_2 laser beam operating in the near gaussian mode was used at power levels of 4 to 12 kW. The collimated beam was made to oscillate by a mirror placed at a distance of about 1.5 m from the focussing mirror M2 (see Figure 2). An electromagnetic driver resonated the oscillating mirror at a frequency of 110-120 Hz. By adjusting the current to the driver, melt passes of desired widths could be achieved.

The laser beam enters the chamber through the salt window shown in Figure 2. Alternatively, a vacuum-sealed, pneumatically operated door is used when the chamber is back-filled to ambient pressure. The pneumatic door is kept open only for the duration of the process. This avoids problems due to overheating and cracking of the salt flat which can occur during long exposure to the laser beam. Mirror M3 directs the beam onto the sample at an angle that determines the radius of curvature of the melt pass.

Powder particles are metered into the carrier gas stream by a rotating auger. Irrespective of the density, the volume flow rates of the particles are constant for a given auger speed. Powder feed rates of 0.1-0.7 cm^3 sec^{-1} were used. The carrier gas flow was maintained with a pressure of 35-50 kPa. The distance between the nozzle and sample was between 1.8 and 2.2 cm. When too close, instabilities in the melt pool resulted. Further away, particle injection became ineffective. A photograph of the process during operation appears in Figure 3. Particles passing through the laser beam are heated to incandescence. The flowing particles improve the efficiency of laser-metal coupling.

Besides processing small sections for experimental work, the apparatus is capable of forming annular rings and cylinderical shapes. An example of an annular injected surface is given in Figure 4. The sample is an Inconel 625 alloy shaft seal with a 10 mm wide, 127 mm diameter annular layer injected with 45-75 μm size WC. After grinding the injected layer to a flat, the test piece was tested in sliding contact with a mating graphite ring in a test apparatus that simulates the wear conditions in a submarine shaft seal. Initial results showed a rate of wear substantially below the acceptable level with excellent water sealing capability (6).

Fig. 3 - Photograph of the injection process showing particle incandescence.

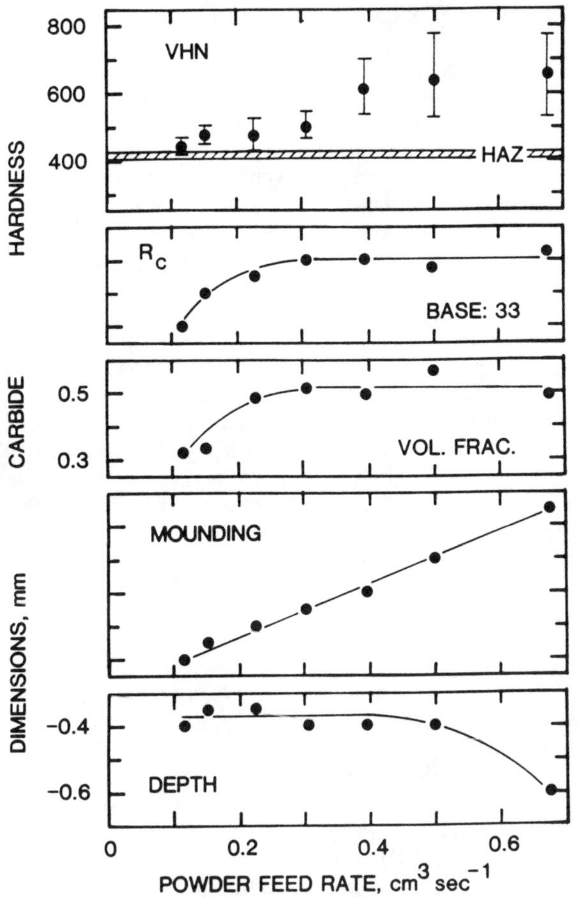

Fig. 8 - Effect of powder feed rate on injected layer characteristics.

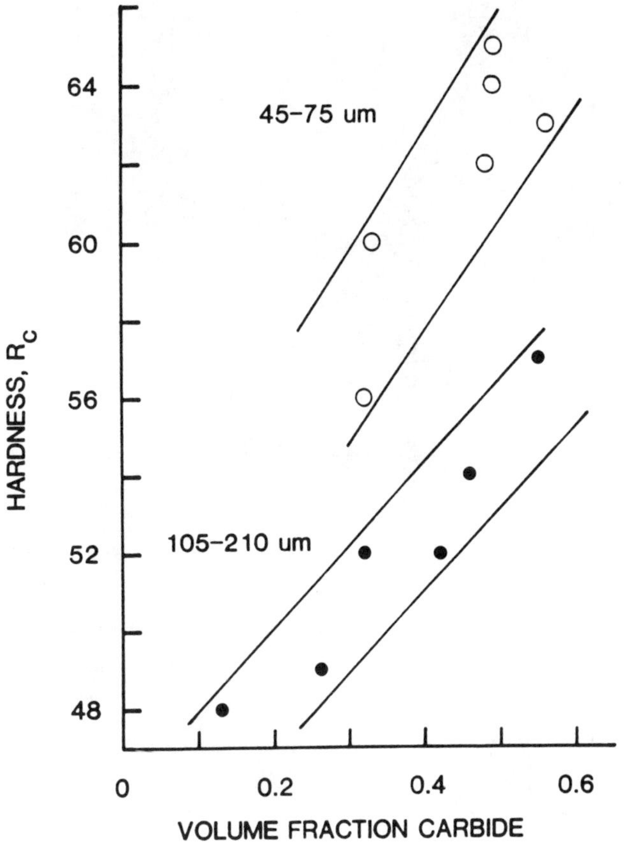

Fig. 9 - Linear trends of surface hardness with carbide volume fraction in Ti-6Al-4V injected with TiC.

sets of data correspond to the latter two studies described above. The scatter within each data set is due to the heterogeneous nature of the injected surface. Since the superficial (15N-brale) hardness indentor leaves an impression with a diameter equivalent to a few carbide particles in width, hardness values seem to depend on particle size. For the same carbide volume fraction, the indentor will intercept more finer (45-75 μm) particles than coarser (105-210 μm) particles. Hence, consistently higher hardness values (open circles) were obtained with the finer carbides as shown in Figure 9.

PROCESS CONTROL - In the previous sections, dimensional control of the injected layer with processing was demonstrated. The results are summarized in Table I. The choice of sample speed depends on the melt width desired and the ability to promote effective laser-metal coupling. Once laser-metal interaction is achieved, injected layer characteristics can be reasonably controlled with laser power and powder feed rate.

Deeper melts are obtained with increasing laser power levels and greater mounding with increasing powder feed rate. But there are upper limits to the two variables (5). Larger melt volumes resulted in reduced carbide volume fractions that were distributed nonuniformly. Excessive powder feed rates promoted surface instabilities and extreme carbide dissolution. It appears that the volume fraction of carbides can be increased with lower laser power levels and higher powder flow rates, but up to a limit. This limit is between 0.5 to 0.55 for TiC injected in Ti-6Al-4V alloy. As discussed earlier, a drawback of a high carbide content is its association with resolidified dendritic carbides within the matrix which can promote matrix cracking (7).

ALLOYS STUDIED

Ti-6Al-4V and Inconel 625 alloy specimens were injected with WC and TiC particles of 45-75 μm and 75-150 μm size distributions. Processing conditions were optimized to give consistent injected layer dimensions. Optimized processing parameters and injected layer characteristics for different alloy/carbide combinations are given in Table II. Approximately 5 mm wide strips were injected on edges of Ti-6Al-4V alloy coupons which were subsequently machined into impact wear test specimens. Results of these tests have been reported elsewhere (5). Since the carbide particles differ in size and density, different powder feed rates were necessary to obtain treated surfaces similar in dimensions and carbide volume fraction (see Table II).

Fig. 4 - An example of carbide injected wear test sample: Inconel 625 alloy shaft seal.

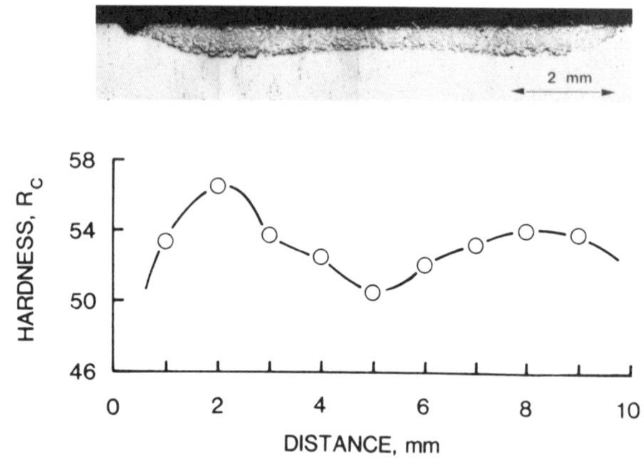

Fig. 5 - Cross-section of WC injected Inconel 625 and hardness profile across injected surface.

INJECTED SURFACE CHARACTERISTICS - A cross-sectional view of the injected surface obtained with beam oscillations is shown in Figure 5. This micrograph is of Inconel 625 alloy injected with 45-75 μm size WC and ground to a width of 10 mm. Variations in the penetration depth are between 0.4 to 0.8 mm and are due to the sinusoidal oscillations of the mirror. These variations are not a cause for concern since the minimum injected depth is much deeper than the observed depth of wear (6).

As seen in Figure 5, the carbide particles are uniformly distributed throughout the cross-section over a volume fraction of 0.45. Structural continuity across the melt/base metal interface is evident. Figure 5 also traces the hardness profile across the injected strip. Hardness variations appear to correspond to variations in the penetration depth. But these variations (50-57 R_C) are minor compared to the significant increase in hardness from that (<20 R_C) of the base metal.

CHOICE OF PROCESSING CONDITIONS

Proper selection of the processing parameters necessitated a detailed study of the process. The influences of sample translation rate, laser power and powder feed rate on the injected layer dimensions, carbide content and hardness were investigated. Layer dimensions, such as thickness, penetration depth (deepest point in the melt pool) and degree of mounding (highest point in the injected layer), were measured under the optical microscope using verniers. Depth and mounding were measured relative to the original surface. Variations in these quantities along the length of the melt pass were minor compared to variations due to processing conditions. Carbide volume fraction was estimated by standard quantitative metallography techniques. Hardness of the ground and polished surface was measured using a 15 kg superficial indentor on a Rockwell hardness tester. For comparison, the 15N scale was converted to the R_C scale. Due to the shallowness of the injected surface, readings from the standard (150 kg) R_C indentor tended to be influenced by the base metal. Also determined was the Vickers microhardness of the matrix using a 100 g load.

EFFECT OF SAMPLE TRANSLATION RATE - Since an oscillating beam interacts with the sample over a wider area, slower sample scanning rates below the beam were used to achieve laser power densities comparable to that with a stationary beam. Figure 6 demonstrates the influence of sample translation rate on injected layer thickness for different values of laser power and powder feed rate. For this experiment Ti-6Al-4V alloy samples were injected with 75-150 μm size WC.

Referring to Fig. 6, the sample scanned at the rate of 2.5 cm sec^{-1} (closed symbols) exhibited an injected layer thickness of 0.4 to 0.5 mm for laser power levels of 5 to 7 kW and powder feed rates of 0.3 to 0.5 cm^3 sec^{-1}. At higher powder feed rates, indicated by arrows at the end of the dashed lines, coupling between the laser beam and sample could not be achieved because most of the beam energy was absorbed by the powder stream. The end result was an unbonded deposited layer of sintered carbide particles over an unmelted sample surface. More effective laser-metal interaction was possible at the slower scanning rate of 1.5 cm sec^{-1}, open symbols in Fig. 6, with 7 to 8 kW laser power and moderate powder feed rates of 0.2 to 0.4 cm^3 sec^{-1}. With these parameters, uniform layers, 0.5-1.5 mm thick, were formed. No change in carbide fraction was observed for the two scanning rates studied.

EFFECT OF LASER POWER - The effect of laser power on injected layer characteristics is shown in Figure 7. In this study, Ti-6Al-4V alloy was injected with 105-210 μm size TiC and the powder feed rate was held constant. As expected, due to the progressive increase in heat input, deeper penetration occurred with increasing laser power.

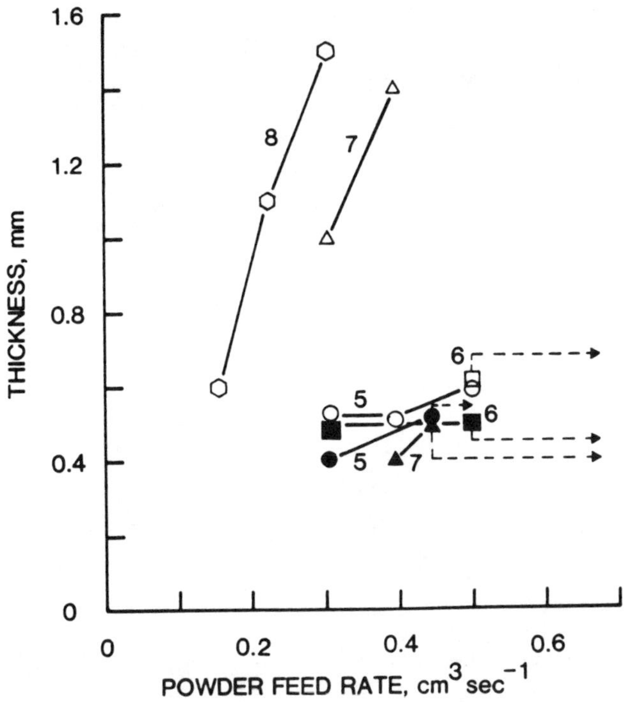

Fig. 6 - Variation of injected layer thickness as a function of powder feed rate and laser power (numerical values) for two sample scan rates: 2.5 cm sec^{-1} (closed symbols) and 1.5 cm sec^{-1} open symbols). Arrows indicated failure of laser-metal coupling.

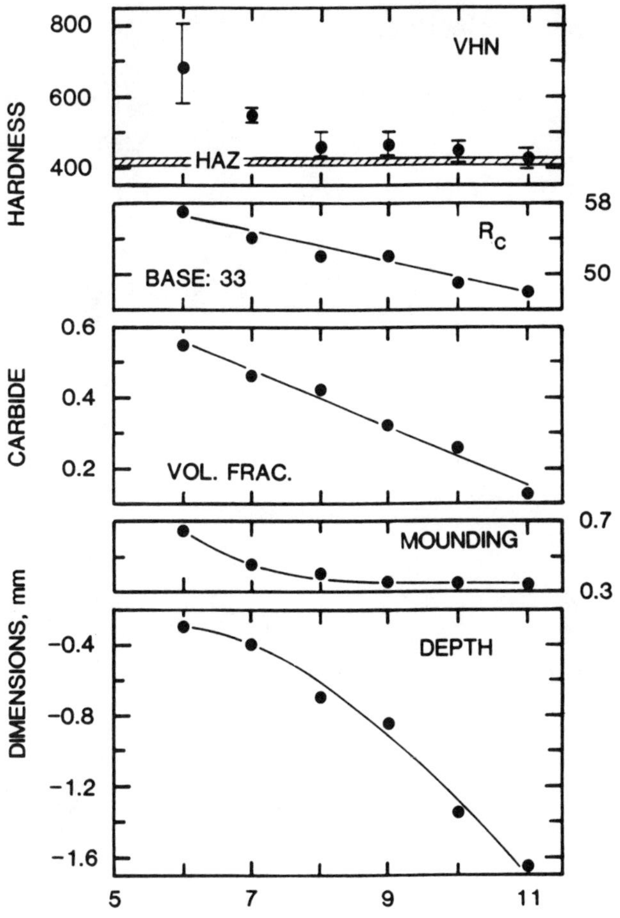

Fig. 7 - Effect of laser power on injected layer characteristics.

Since a constant amount of powder was fed into a progressively larger volume of melt, volume fraction carbide decreased significantly with laser power. The lower carbide volume fractions obtained with increased laser power had little influence on mounding, keeping it constant after an initial decrease (see Figure 7).

Referring to Figure 7, the reduction in Rockwell hardness seems to parallel the reduction in particle volume fraction with laser power. But the lowest value of 48 R_C with 0.15 vol. fraction carbide is significantly higher than the base metal value of 33 R_C. Matrix microhardness appears to vary little above that (405-425 VHN) of the HAZ except at the lower laser power levels. At low laser power levels, carbide volumes were high causing excessive dissolution of the carbide into the melt. Networks of resolidified carbides that form as a result of the dissolution process enhance the matrix hardness well above that of the HAZ (7). Carbide dissolution appears to be an uneven process as evidenced by the wide scatter in the microhardness values given in Fig. 7.

EFFECT OF POWDER FEED RATE - The effect of powder feed rate on injected layer characteristics is illustrated in Figure 8. For this study, 45-75 μm size TiC was injected in Ti-6Al-4V samples with a laser power of 8 kW. Except at the highest feed rate, penetration depth remains constant over a wide range of powder feed rates. This suggests that a constant volume of melt forms independent of powder feed rate. But, since a larger flux of particles enters the melt, the degree of mounding is observed to increase linearly. On the other hand, according to Figure 8, instead of increasing proportionately, the carbide volume fraction appears to level off after an initial increase with increasing powder feed rate.

In a similar fashion, the Rockwell hardness of the injected surface remains constant after an initial increase as shown in Figure 8. In spite of the carbide volume being constant over a wide range of powder feed rates, increasing volumes of resolidified carbides contribute to the progressive increase in the degree of mounding. The presence of increasing quantities of resolidified carbides is indicated by the increasing microhardness of the matrix given in Figure 8.

From the hardness data in Figures 7 and 8, it appears that although resolidified carbides enhance matrix microhardness, they have little influence on the macrohardness of the injected surface. Macrohardness appears to depend solely on the carbide volume fraction for a given particle size distribution. Approximate linear trends of Rockwell hardness with carbide content is graphically depicted in Figure 9. The two

Table I - Control of Injected Layer Characteristics in TiC Injected Ti-6Al-4V

	Increasing Sample Speed	Increasing Laser Power	Increasing Powder Feed Rate
Layer thickness	↓	↑	↑
Penetration depth	↓	↑	-,(↑)
Mounding	↓	(↓),-	↑
Vol. frac. carbide	-	↓	(↑),-
Resolidified carbides	-	(↓),0	↑
Hardness	-	↓	(↑),-

Minor trend given in brackets
- no change
0 none

Table II - Optimized Processing Conditions and Injected Layer Characteristics of Alloys Studied

Sample No.	Alloy	Carbide Type	Carbide Size (μm)	Laser Power (kW)	Sample Speed ($cm\ sec^{-1}$)	Powder Feed Rate ($cm^3 sec^{-1}$)	Layer Dimensions (mm) Width	Depth	Mounding	Vol Frac Carbide
1	Ti-6Al-4V	WC	45-75	8	1.5	0.115	5.2	0.45	0.40	0.55
2		WC	75-150	8	1.5	0.305	4.8	0.40	0.45	0.43
3		TiC	45-75	8	1.5	0.395	4.9	0.40	0.40	0.49
4	Inconel 625	WC	45-75	10	0.75	0.350	8.8	0.4-0.7	0.8	0.46
5		WC	75-150	10	0.75	0.445	8.6	0.2-0.4	0.7	0.42
6		TiC	45-75	10	0.75	0.500	9.5	0.4-0.7	1.1	0.48
7		TiC	75-150	10	0.50	0.500	10.7	0.7	0.8-1.3	0.52

Wider (9-12 mm) injected layers were formed in Inconel 625 alloy test pieces by increasing the amplitude of the laser beam oscillations and reducing the sample scan rate. These melt pass tests were a precursor to the fabrication of wear resistant surfaces on the shaft seal shown in Figure 4. Unlike the nearly lenticular shape of the narrower layers on the Ti-6Al-4V alloy coupons, injected layers on Inconel 625 alloy showed variations in thickness across the transverse section. Thickness variations were due to variations in penetration depth as shown in Figure 5 or due to variations in mounding (e.g. sample No. 7 in Table III). Variations in mounding, in the form of parallel ridges, were caused by surface tension effects and were observed when attempts were made to produce wider (>10 mm) melt pools injected with the lighter TiC particles. As explained earlier, the observed variations in depth are not considered a problem. Also, since the top surface of the injected layer is ground to a flat after processing, variations in mounding are not considered serious as long as the minimum mounding is sufficiently thick.

STRUCTURAL EVALUATION

Surface properties such as wear and corrosion will depend upon the degree of microstructural homogeneity and modification within the solidified melt pool. Particle injected surfaces contain several regions of interest. These are the HAZ, the melt-substrate interface, the injected particle, the particle-matrix interface, the matrix, and the dissolution products. The nature of each of these microstructural components will be discussed for the samples listed in Table II.

Table III - Microhardness Measurements in Various Portions of the Treated Surface of Samples Given in Table II

	Microhardness (VHN)			
			Matrix	
Sample No.	Base	HAZ	Lower	Upper
1	354	400	423	455
2	363	407	400	405
3	348	414	439	626
4	284	-	523	681
5	282	-	524	644
6	306	-	336	468
7	296	-	313	422

Metallographic samples were prepared using standard preparation techniques. Microstructures were evaluated under the SEM and phases were analyzed using Energy Dispersive Spectroscopy (EDS). For microanalysis, peak intensity ratios of the main elements in a given spectrum were compared. Microstructural information was supplemented with Vickers microhardness measurements using a 100 g load. By this means the extent of hardening in the various sections of the treated surface was determined. Microhardness values in the base metal, the HAZ and the matrix in the lower and upper regions of the injected layer are summarized in Table III.

MICROSTRUCTURAL ANALYSIS - Of the two alloys, an extensive HAZ was found only in laser-processed Ti-6Al-4V alloy. The HAZ resembled a structure quenched from the β region consisting of α' platelets in a β matrix. True to its design as a high temperature alloy, Inconel 625 alloy showed no structural change in the base metal. Examples of substrate microstructures are visible in Figure 10. An examination of the microstructures in Figure 10 clearly shows structural continuity across the substrate-layer interface. This supports the fact that the modified surface is an in situ composite and not a deposited coating.

Particle distribution within the injected layer is fairly uniform as a result of intimate mixing with the melt. Melt-particle interaction is expected to result in some amount of dissolution of the injected phase. The dissolution rate depends upon the alloy, the type of carbide, the particle size, the volume of carbide and the interaction kinetics. Since carbide volume fraction for the various samples in Table II varies over a narrow band (0.42-0.55), it would not contribute to the observed variations in the microstructures. A consequence of the dissolution process is the appearance of resolidified carbides in the interparticle matrix as reported by Ayers (7).

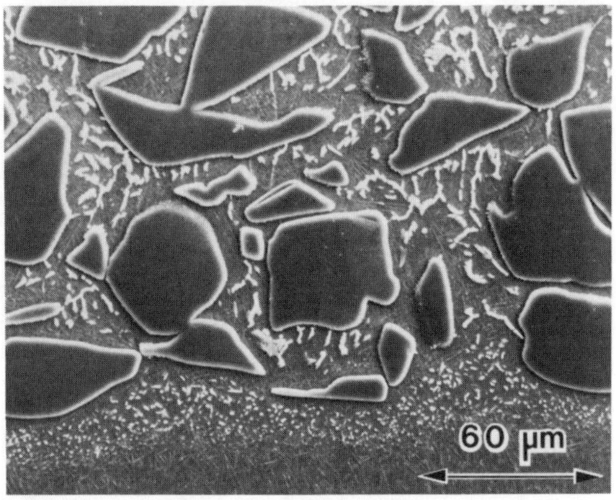

(a)

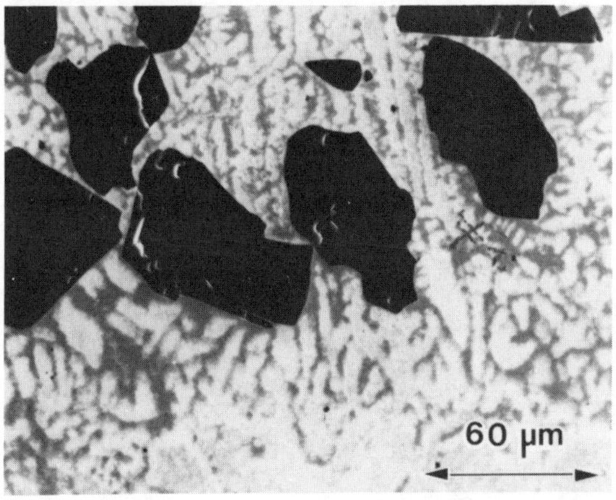

(b)

Fig. 10 - Microstructures in the lower portions of TiC injected layers. (a) Ti-6Al-4V. (b) Inconel 625.

When injected with TiC, resolidified carbides appear as a network of dendrites, examples of which are given in Figure 11. In WC injected Inconel 625, resolidified carbides appear as blocky dendrites as shown in Figure 12. Also shown in Figure 12 is a possible mechanism for carbide dissolution. The WC particle edges appear highly perturbed and seem to spherodize into globules. The globules act as nuclei for the resolidified carbide dendrites. A similar dissolution mechanism was also observed with TiC injection. From their morphology, it can be concluded that resolidified carbides precipitate from the melt as initial solidification products.

Resolidified carbides are usually confined to the upper regions of the melt pool where mixing is especially vigorous. The absence of resolidified carbides in the lower portions of the injected layer suggests reduced melt-particle interaction. In fact, in some cases, the matrix resembles the substrate in nature. As shown in Figure 10(a), in TiC injected Ti-6Al-4V, the matrix resembles the HAZ and is made up of α'

and phases. A similar martensitic matrix structure was reported by Flower et al (8) in Ti-6Al-4V that was laser surface alloyed with carbon. Shown in Figure 10(b), the matrix in TiC injected Inconel 625, consists of cored (Ni)-phase dendrites resembling the structure obtained upon simple laser remelting of the alloy surface.

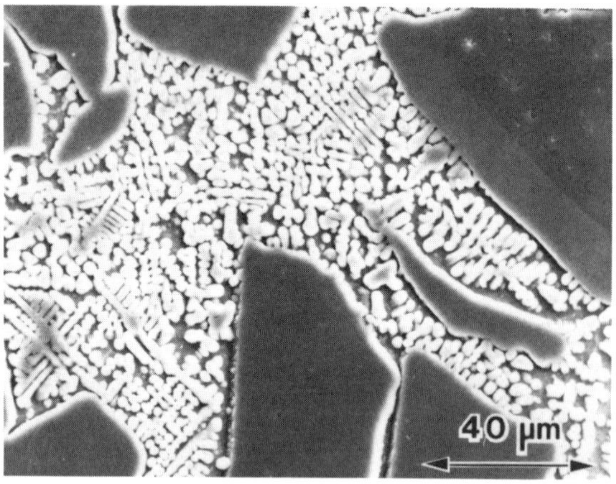

(a)

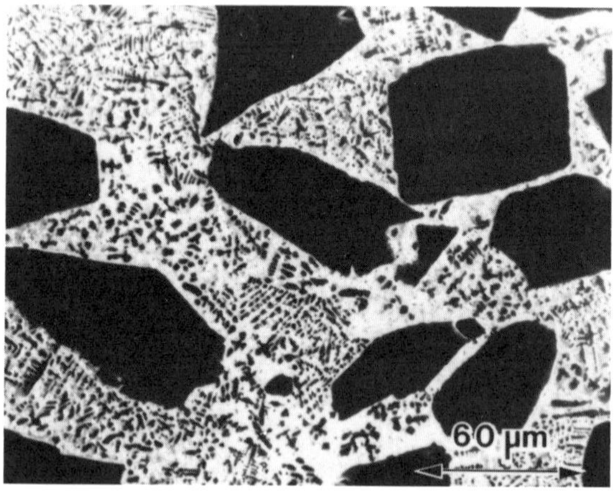

(b)

Fig. 11 - Network of resolidified carbides found in the upper regions of TiC injected layers. (a) Ti-6Al-4V. (b) Inconel 625.

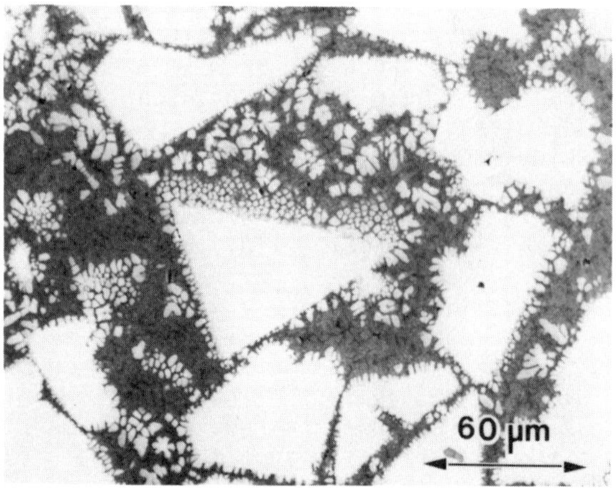

Fig. 12 - Blocky resolidified carbides found in WC injected Inconel 625.

Besides the dendritic resolidified carbides, intercellular (Figure 13(a)) and inter-dendritic (Figure 13(b)) solute-rich products were also found in the matrix. Unlike resolidified carbides, which are the primary solidification products, the solute-rich phases form towards the end of solidification. From its relationship to the primary carbide particles, shown in Figure 13, the solute-rich phase is probably a complex alloy carbide having formed after limited dissolution of the injected phase. Solute-rich phases were also found in association with resolidified carbides. The preceding microstructural observations were based on the evaluation of samples injected with the finer (45-75 µm) carbide particles. When injected with the coarser (75-150 µm) particles, the dissolution products in the microstructures were similar in nature but in reduced amounts.

MICROANALYSIS - With the presence of dissolution products in the melt leading to a variety of microstructures, compositional changes in the injected surface were expected. For this analysis, elemental peak intensity ratios from EDS measurements on the various phases found in the treated surface were obtained. This analysis is qualitative and only relative changes are compared. For these alloys, the following peaks were evaluated: $W(L\alpha_1)$, $Ti(K\alpha_1)$, $Cr(K\alpha_1)$, and $Ni(K\alpha_1)$. In WC injected Ti-6Al-4V, small amounts of W were found in the matrix (W/Ti=0.03) and along the carbide edges (W/Ti=0.07). This supports the contention that the intercellular phase may contain some W probably combined with Ti and C.

Alloying of Inconel 625 with W (W/Ni=0.25 and 0.14 in the lower and upper parts of the layer, respectively) in the WC injected sample was evident. Besides being part of the interdendritic solute-rich phase, some amount of W may be expected to form a solid-solution with the (Ni)-phase. This can be so because W has fairly high solubility (40 w/o W) in pure Ni (9). Resolidified carbides showed significant Ni (W/Ni=1.91) and Cr (W/Cr=1.17) pick-up, although Cr pick-up was in larger amounts than Ni (Cr/Ni=1.64 compared to 0.56 in the substrate). The edges of the injected WC showed a similar analysis. Accompanying the Cr enrichment of resolidified carbides, Cr depletion of the matrix in the upper regions was observed (Cr/Ni=0.40 compared with 0.56 for the substrate).

With TiC injection, the extent of alloying in Inconel 625 was not as significant as with WC injection. Maximum solubility of Ti in pure Ni is only 12 w/o (9). Negligible amounts of Ti (Ti/Ni=0.04) were found in the matrix in the lower portions of the injected layer, while some amount of Ti (Ti/Ni=0.15) was picked up by the matrix in the upper portions of the melt pool. The resolidified carbides contained small amounts of Ni (Ti/Ni=5.00) and Cr (Ti/Cr=7.69) with somewhat more Cr than Ni (Cr/Ni=0.66 compared with 0.58 in the substrate). Again resolidified carbides were accompanied by a slightly Cr-depleted matrix (Cr/Ni=0.54 compared to 0.58 in the substrate). Besides the solubility limits mentioned above, the relative ease with which WC dissolves in Ni compared to TiC can be readily understood by comparing their free energies of formation (ΔF). At 1500°C, ΔF for WC is -8.4 kcal $mole^{-1}$ compared to -39.5 kcal $mole^{-1}$ for TiC. Hence, it follows that WC dissociates more readily than TiC. This also explains the ease with which WC dissolved in Ti-6Al-4V.

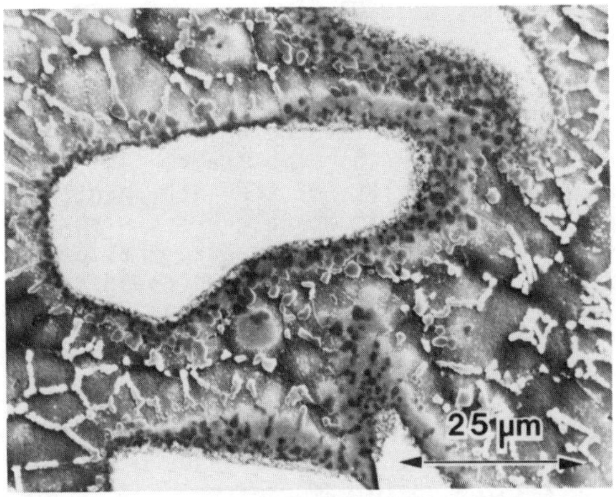

(a)

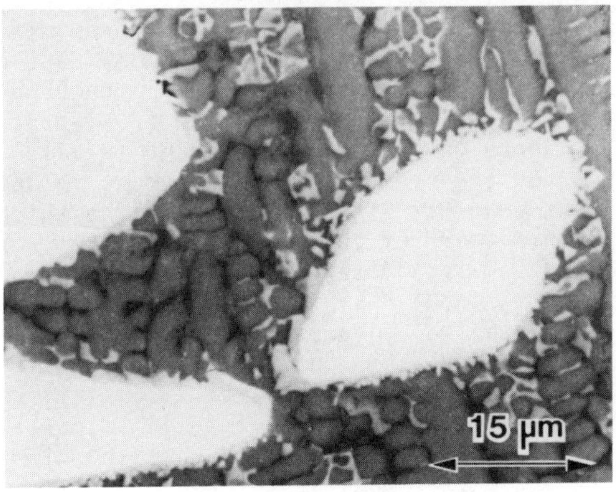

(b)

Fig. 13 - Morphologies of solute-rich dissolution products found in WC injected layers. (a) Ti-6Al-4V. (b) Inconel 625.

MICROHARDNESS - The presence of dissolution products in the matrix has a varying influence on its microhardness. Ayers (7) has identified several possible hardening mechanisms in injected surfaces. The hardening trend in the Ti-6Al-4V alloy surface injected with TiC is graphically illustrated in Figure 14. From a base metal value of about 340 VHN, the microhardness increased to 400 VHN in the HAZ. Within the injected layer, matrix hardness was 480-500 VHN in the lower regions and 600-700 VHN in the upper regions. Higher hardness values in the lower portions were probably due to the feathery intercellular phase found in the matrix shown in Figure 10(a). Resolidified carbides contributed to the vastly increased hardness in the upper portions of the injected layer.

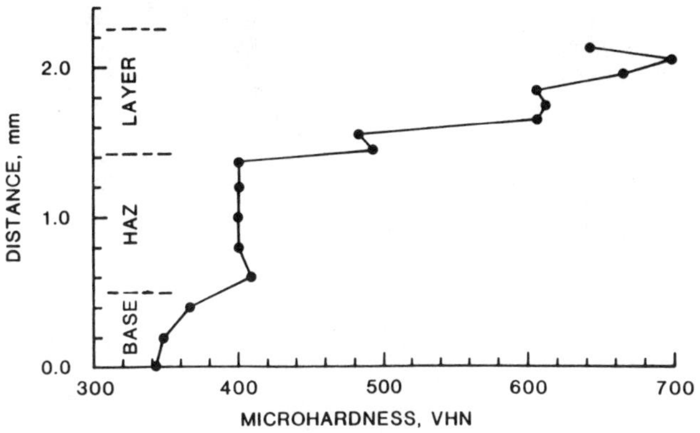

Fig. 14 - Microhardness variation across the modified surface in TiC injected Ti-6Al-4V.

Examination of Table III demonstrates varying degrees of hardening in the injected surface. In WC injected Ti-6Al-4V (samples 1 and 2), hardening of the matrix was negligible over that of the HAZ. WC dissolution in Ti-6Al-4V was not sufficient to form resolidified carbides but did form intercellular products as shown in Figure 13(a). These solute-rich phases had minimal influence on the matrix microhardness.

Hardening of the matrix above that of the substrate is significant in WC injected Inconel 625 (samples 4 and 5 in Table III). The presence of solute-rich interdendritic phases shown in Figure 13(b), and possible solid-solution strengthening of the (Ni)-phase by W contributed to the near doubling of the matrix microhardness over that of the substrate. In addition, the presence of resolidified carbides in the upper matrix, Figure 12, further boosted the microhardness to nearly 2.5 times that of the substrate. Similar magnitudes of hardness increase were not observed in Inconel 625 injected with TiC. Referring to Table III, in the lower portions of the injected layer, the matrix microhardness is nearly the same as that of the substrate. Neither dissolution products, Figure 10(b), nor alloying with Ti was observed in this portion of the injected layer. In the upper portions, resolidified carbides contributed to the 1.4-1.5 times increase in the microhardness of the matrix. The difference in hardening in the injected layer with WC and TiC is graphically illustrated in Figure 15 for Inconel 625. Microhardness relative to base metal of WC injected matrix is consistently higher across the thickness than that of the TiC injected matrix.

An examination of Table III reveals hardening trends to be similar for the two particle size distributions studied except that the degree of hardening of the matrix was consistently lower when injected with the coarser (75-150 µm) particles. Carbide dissolution rate depends to a small extent on particle size. The nature of the microhardness gradient through the thickness of the injected layer is important. Since nearly half of the injected layer is removed by grinding to prepare the sample for wear studies, the decreased hardness of the matrix after grinding may decrease its wear resistance.

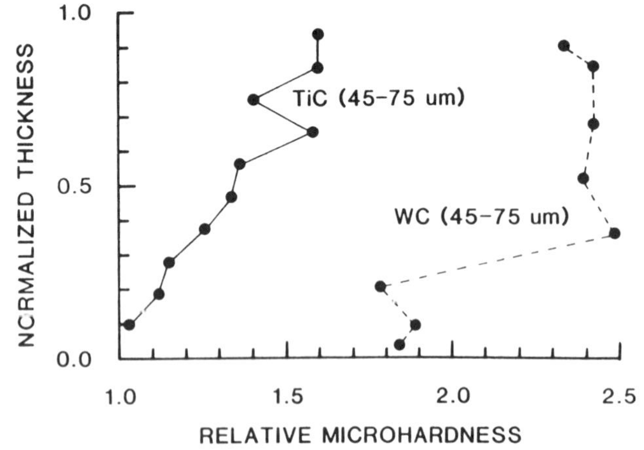

Fig. 15 - Variations of microhardness relative to base metal across the injected layer thickness in Inconel 625.

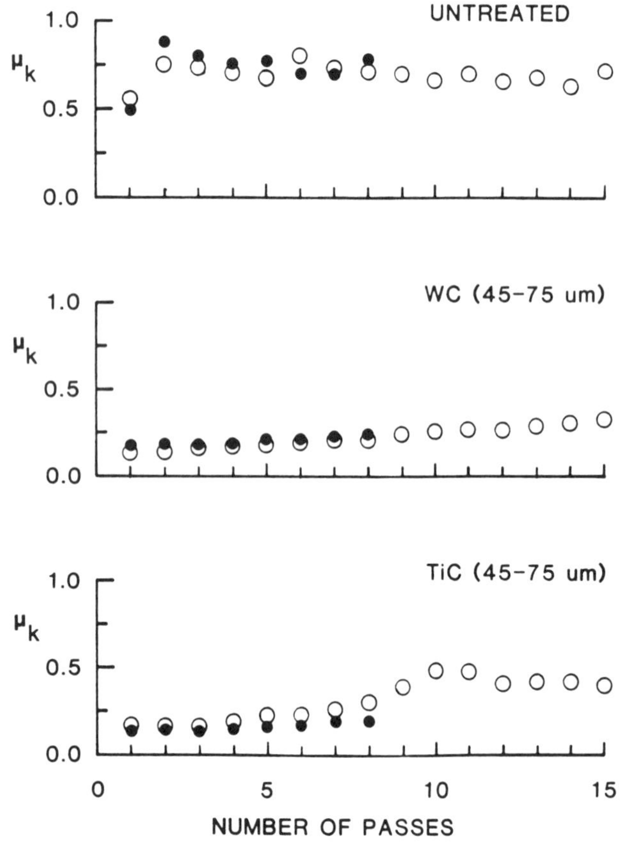

Fig. 16 - Co-efficient of friction (μ_k) measurements on untreated and injected Inconel 625 samples.

WEAR BEHAVIOR OF INCONEL 625

To assess the wear behavior of particle injected Inconel 625 alloy surfaces, kinetic co-efficient of friction measurements (μ_k) were made using a 1.27 cm diameter 52100 steel ball and a 1 kgf load. This is a standard test used to determine the wear behavior of shallow treated surfaces under sliding contact. The tests were conducted without lubrication at 25°C employing a slider velocity of 0.01 cm sec^{-1}. Sample preparation involved polishing the injected surface to a 3 µm diamond finish. Before testing, the slider and samples were washed in benzene and rinsed in 2-propanol. A total of 15 passes were made in the same track. In order to judge the repeatability of the test, additional 8 passes were made in a separate track with a fresh slider. Results of the μ_k measurements are plotted in Figure 16. In the untreated case, μ_k was 0.55 after the first pass, increased to 0.70 after the second pass and remained around 0.7 after 15 passes. When injected with 45-75 µm size WC, μ_k gradually increased from 0.15 to 0.33 at the end of 15 slides. The 45-75 µm size TiC injected surface yielded a μ_k value of 0.15 initially which rapidly increased to 0.40 after 9 passes and remained 0.4 after 15 passes.

Table IV summarizes the μ_k values after the first and fifteenth pass, the wear scar width and also gives hardness and carbide volume fraction for the different samples studied. Since the carbide content does not vary much (0.47-0.51), its influence on μ_k will be the same in all cases. For the treated samples, the μ_k values were less than 0.2 after the first slide which is less than a third of the μ_k value for the untreated sample. After 15 passes, μ_k is about 0.4 which is about half the value for the untreated sample. In a similar study, Ayers et al (10) reported a $_k$ value of 0.16 on the first slide on Ti-6Al-4V injected with TiC. This value gradually increased to 0.26 after 19 slides. The untreated Ti-6Al-4V alloy had a μ_k value of 0.4 for all passes. Hence, μ_k for the treated surface was 0.40 and 0.65 times that of the untreated surface after the first and the 19th slides, respectively. These relative values of μ_k for Ti-6Al-4V are comparable to those found for injected Inconel 625. From these studies it is concluded that substantial reduction in μ_k is achieved by particle injection thereby limiting wear due to sliding contact.

Examination of the wear scars on the sample surface under the optical microscope revealed the nature and extent of damage sustained by the sample surface. As shown in Figure 17(a), after 15 passes, the wear track on the untreated surface was 300 µm wide as a result of extensive plowing. By contrast, as shown in Figure 17(b), the damage done to the 45-75 µm WC injected surface, after the same number of passes is negligible. A few, shallow scars add up to a width of about 90 µm, while most of the contact surface appears unaffected. Closer examination of the wear track revealed mild abrasion of the

Table IV - Results of Co-efficient of Friction (μ_k) Measurements on Inconel 625

Sample	Carbide Type	Carbide Size (μm)	Carbide Vol. Fract.	Hardness R_c	Hardness VHN**	μ_k 1 Pass	μ_k 15 Passes	Scar Width* (μm)
A	Untreated	-	-	<20	292	0.55	0.72	300
B	WC	45-75	0.47	51	618	0.15	0.33	90
C	WC	75-150	0.51	54	605	0.19	0.41	100
D	TiC	45-75	0.50	49	516	0.15	0.41	80
E	TiC	75-150	0.48	43	406	0.17	0.44	150

*after 15 passes
**of matrix

matrix as shown in Figure 18(a), and deposition of wear debris in the form of thin films as shown in Figure 18(b). EDS analysis of the debris on a carbide particle indicated the presence of Fe which meant that material transfer occurred from the slider to the injected surface. This material transfer is probably responsible for the gradual increase in μ_k of the WC treated sample as shown in Figure 16.

(a)

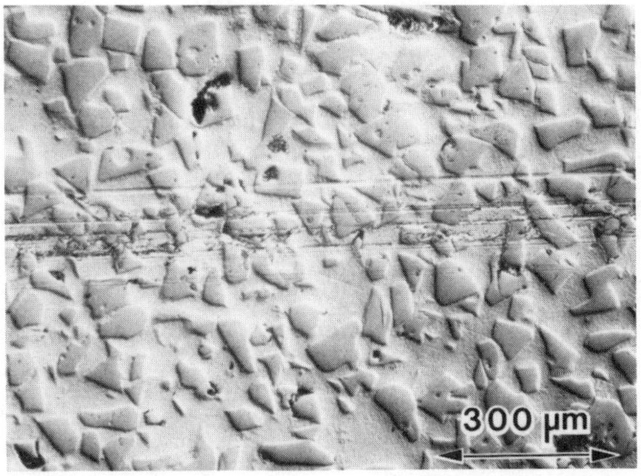

(b)

Fig. 17 - Wear scar characteristics showing significant reduction in wear due to particle injection. (a) Untreated. (b) Injected with 45-75 μm size WC.

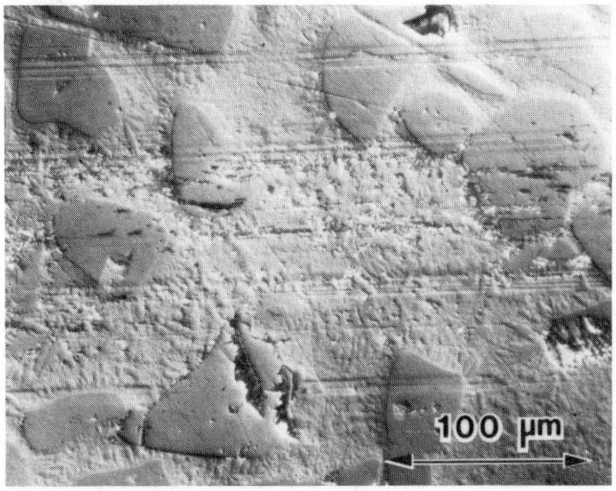

(a)

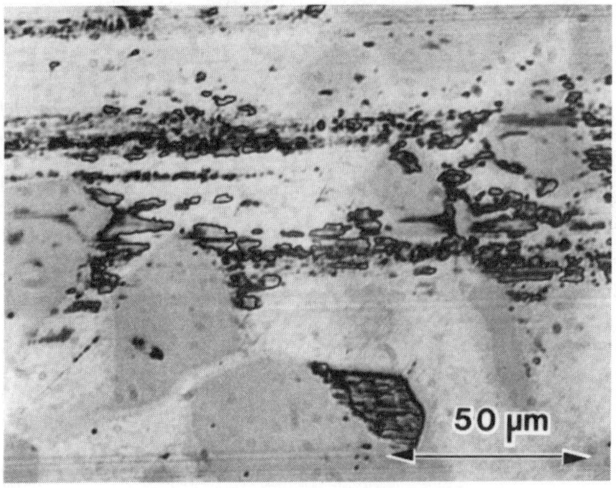

(b)

Fig. 18 - Wear modes in 45-75 μm size WC injected Inconel 625. (a) Mild abrasion of matrix. (b) Deposition of wear debris.

The wear track width on the 45-75 μm size TiC injected surface is 80 μm as shown in Figure 19 and is indicative of improved wear resistance. But, compared to the negligible wear of the harder (618 VHN) WC injected matrix, the softer (516 VHN) TiC injected matrix shows damage from plowing. The initiation of matrix wear may have contributed to the rapid rise in μ_k for this sample as shown in Figure 16. In either case, as mentioned earlier, considerable improvement in wear resistance is realized by particle injection. The wear track characteristics described above for the composite surface indicate that during the wear process, the slider rides over the protruding carbides, allowing only intermittent metal-to-metal contact. A harder matrix offers resistance to wear giving a lower μ_k value, while a softer matrix yields to the wear process giving a higher μ_k value. The shielding of the metal from wear by the hard carbides is responsible for the superior sliding wear resistance of injected surfaces.

Fig. 19 - Matrix damage due to sliding wear in 45-75 μm size TiC injected Inconel 625.

CONCLUSIONS

With an oscillating laser beam, wide (up to 12 mm) metal matrix-carbide composite layers were formed by the laser melt-particle injection process with a single melt pass. Reasonable control over the injected layer characteristics by a suitable choice of the processing conditions was demonstrated. In general, higher carbide contents in the injected layer were promoted by lower laser power levels and higher powder feed rates, but up to a limit. This limit was 50-55 volume percent for Ti-6Al-4V injected with TiC. Attempts to increase the carbide content further resulted in excessive carbide dissolution into the melt and matrix embrittlement due to resolidified carbides.

As a result of varying degrees of dissolution of the carbide phase in the melt, an assortment of solidification products was seen in the matrices of the various samples. Where dissolution was extensive, in the upper portion of the melt pool, dendritic carbides appeared in the matrix. Besides depleting the matrix of its key alloying elements, e.g. Cr in Inconel 625, resolidified carbides increased the microhardness of the matrix considerably. Where the dissolution process was less vigorous, in the lower portions of the melt pool, solute-rich products appeared in the inter-cellular regions of the matrix. While solute-rich phases due to WC may have contributed to the hardening of the matrix in Inconel 625, in Ti-6Al-4V matrix hardening from WC injection was minor. EDS analysis demonstrated that in most cases, the dissolution process was accompanied by some degree of matrix alloying of the carbide constituents. In the case of Inconel 625, alloying occurred more readily with W than with Ti.

Co-efficient of friction tests revealed that μ_k of Inconel 625 could be reduced to 0.4 by particle injection from a value of 0.70 for the untreated alloy. In the WC injected sample, wear debris was picked up from the steel slider and the harder matrix showed little wear. In the TiC injected sample, some amount of plowing occurred in the softer matrix. In either case, the improved wear resistance was a consequence of intermittent contact between slider and matrix due to the presence of the protruding carbides.

The effects of the two particle size distributions on the microstructure, the microhardness and the wear behavior were similar except that the influence of the coarser particles was on a reduced scale.

ACKNOWLEDGMENTS

The author would like to thank Dr. J. D. Ayers, Mr. W. P. Robey and P. Slebodnick of the Physical Metallurgy Branch and Dr. R. L. Stegman and Mr. R. E. Beigel of the Laser Materials Application Center at the Naval Research Laboratory for their assistance in the research reported here.

REFERENCES

1. Ayers, J. D., Thin Solid Films 84, 323-331 (1981).
2. Ayers, J. D., Schaefer, R. J., and Robey, W. P., J. Metals 33, 8, 19-23 (1981).
3. Ayers, J. D., Wear 97, 249-266 (1984).
4. Ayers, J. D. and Bolster, R. N., Wear 93, 193-205 (1984).
5. Cooper, K. P. and Ayers, J. D., Surface Engineering 1, 4 (1985).
6. Private Communication, David Taylor Naval Ship Research and Development Center, Annapolis, MD 21402.
7. Ayers, J. D., "Lasers in Metallurgy", p. 115, AIME, Warrendale, PA (1981).
8. Flower, H. M., Walker, A., and West, D.R.F., Scripta Met 9, 923-926 (1985).
9. Hansen, M. and Anderko, K., "Constitution of Binary Alloys," McGraw-Hill, NY (1958).
10. Ayers, J. D., Tucker, T. R., and Bowers, R. C., Scripta Met 14, 549-550 (1980).

INFLUENCE OF MACHINING ON THE MICROSTRUCTURE, MICROHARDNESS AND PRECIPITATION IN THE ALUMINUM ALLOY 7075

C. A. Brown, F. Dupont, B. Ilschner
Department of Materials
Swiss Federal Institute of Technology-Lausanne
Laussane, Switzerland

ABSTRACT

A 7075 aluminum alloy has been machined by turning in the solutionized and quenched condition and subsequently artificially aged. Microhardness as a function of depth from the machined surface shows the surface layers to be harder than the bulk material before peak aging of the bulk and softer after. A microstructural investigation by TEM of foils prepared perpendicular and parallel to the machined surface shows a fine cell structure very near the surface below which there is evidence of gross deformation which diminishes with depth. Precipitate free zones were not observed at depths of about 50 μm after artificial aging but were at 100 μm. There also appears to be an influence of machining on matrix and grain boundary precipitate sizes in the artificially aged material.

AS THE TOOL passes during conventional machining by chip removal the newly formed surface and sub-surface layers can be subjected to large deformations at a high rate with resulting heating. This thermomechanical process to which the surface region is subjected affects its microstructure and properties and the reaction of the surface layers to subsequent fabrication processes or to chemical and mechanical attacks.

The present work investigates the microstructure and microhardness of surface layers of commercial 7075 Al after machining the solutionized and quenched material and after subsequent artificial aging. The surfaces were created by turning using conventional machining conditions. The surface and sub-surface layers have been tested for microhardness and examined using scanning and transmission electron microscopy. Significantly, specimens perpendicular to the surface have been observed with the latter. Using these techniques it is possible to see the influence of machining on precipitation as a function of depth from the machined surface.

Haslum and Rubenstein (1970) present microhardness data for aluminum machined orthogonally at a slow speed (2.3 m/min) showing work hardening that monotonously increases as the surface is approached. Whereas a maximum at about 100 μm from the surface and a hardness at closest approach to the surface lower than the interior has been reported for 2024 aluminum; also orthogonally machined but at higher speeds (6 - 192 m/min) (Elkhabeery and Bailey 1984; Bailey and Elkhabeery 1984). Transmission electron microscopy on foils prepared at and parallel to an aluminum surface orthogonally machined at very slow speeds (< 1 m/min) reveals 1.0 - 1.5 μm diameter cells similar to those formed in the tertiary deformation zone of quick stop specimens (Kaneeda et al. 1983). Observations (TEM) have been made perpendicular to surfaces machined orthogonally (at similarly slow speeds) in brass (Turley 1971) and copper (Turley and Doyle 1983). Both observed long and narrow subgrains in a heavily deformed region or fragmented zone near the surfaces. The former reports strains as high as 7 diminishing to 2.3 at the lower boundary of this region (0.08 mm maximum depth with a - 35° rake angle). Several researchers have plated copper on worn copper surfaces and then thinned specimens perpendicular to the surface electro-chemically (Turley and Samuels 1981; Van Dijck 1976; and Heilmann et al. 1983). These researchers have observed fine cellular substructure near the surface and at greater depths layers of diminishing deformation similar to the machined surfaces. Ion milling has been used to prepare specimens perpendicular to a worn steel surface after vapor deposition of copper and nickel electroplating (Ives 1983).

The sensitivity of precipitation phenomena in the Al-Zn-Mg systems to quenching, aging and to the concentration and distribution of defects, especially in regard to precipitate free zones (PFZs) adjacent to the grain boundaries, is well known (Embury and Nicholson 1965; Lorimer and Nicholson 1966; Joshi et al. 1981). Plate-shaped, transition phase η' particles, some of which are heterogeneously nucleated on dislocation lines, are believed to be responsible for the maximum hardness in 7075 Al (Park and Ardell 1983). Overaging results in a coarsening of matrix (Park and Ardell 1983) and grain boundary precipitates, an increase in their spacing and a widening of PFZs (Narasimha Rao 1981). Additionally, intrinsic and extrinsic grain boundary defects have been observed to influence grain boundary precipitation in Al-Zn-Mg alloys (Gronsky and Furrer 1981; Vanderwalker and Vander Sande 1981).

The precipitation process in Al-Zn-Mg has been found to be sensitive to prior cold plastic deformation. The resulting defects provide nucleation sites and there is an accelerating effect on aging kinetics (Thompson et al. 1974; Baranovskiy et al. 1980; Papazian, 1981). Kawabata et al. (1984) have shown that ductility at cryogenic temperatures may be improved by cold working Al-Zn-Mg before aging. The improvement was thought to be due to more favorable precipitate distribution, irregular grain boundary shape and higher density of sub grain boundaries in the cold worked specimens.

There is evidence that reduction of the size or elimination of the PFZ will improve mechanical properties (Kawabata and Izumi 1976; Kawabata et al. 1984). And it has been shown that fatigue resistance in 7075 and Al-Zn alloys is improved with thermomechanical treatment by homogenizing the microstructure thus dispersing slip (Ostermann 1971; Lu and Weissmann 1978). As machining after solutionizing and quenching heats and deforms the surface layers of the specimen during a critical period for precipitate nucleation the microstructure of these surface layers after aging will be influenced and it may be possible to use this process to improve mechanical properties.

MATERIALS AND PROCEDURE

The composition of the as received 40 mm bars was: 4.7 Zn; 2.04 Mg; 1.58 Cu; 1.33 Fe, 0.24 Cr; 0.19 Mn; 0.13 Si. The bars were turned before heat treatment to 29 mm diameter (in order to remove microstructural gradients remaining from the fabrication process) and cut to 100 mm segments. Solutionizing was done in air at 460 °C for 115 minutes followed by quenching in water at 19 °C. Figures 1 and 2 show the initial microstructure after solutionizing and quenching, at a distance of approximately 12.5 mm from the center of the bar, i.e. the same as the final machined surface.

The specimens were machined to a final diameter of 25 mm in two equal passes by turning approximately 25 minutes after quenching. A tungsten carbide tool insert with a 0.8 mm nose radius, 45° angle of approach and 30° rake angle was used. The machining conditions were as follows: speed 140 m/min; feed 0.2 mm/rev; depth of cut 0.1 mm. Macrohardness was measured before and after turning.

Microhardness profiles were made on specimens not subjected to artificial aging within 10 hours of machining. Microstructural examination by TEM on the specimens not subjected to artifical aging was done in the order of 10 days after machining. Artificial aging was done in air at 118 °C ± 2 °C for 24 hrs, beginning approximately 1 hour after the quench (see Fig. 3).

Before microhardness testing specimens were electrolytically plated with a layer of nickel approximately 0.2 mm thick. A plane, inclined

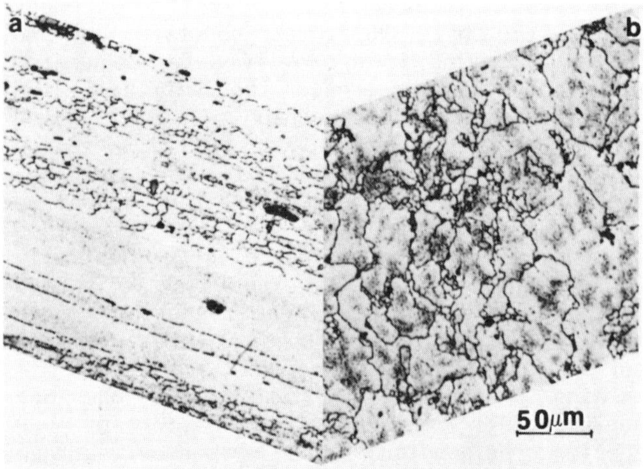

Fig. 1 Optical micrographs at the same depth as the machined surface : a) longitudinal, b) transverse.

Fig. 2 Microstructure after solutionizing, and quenching at the same depth as the machined surface: a) longitudinal, b) tranverse.

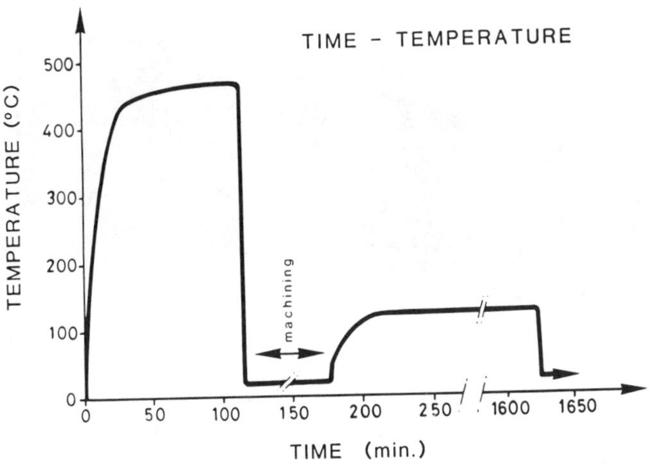

Fig. 3 Solutionizing, quenching, machining and aging sequences.

at less than 1° to the longitudinal axis of the bar, was polished using 1 μm diamond paste so as to minimize the depth of deformation due to polishing while preserving a clear distinction between the machined and polished surfaces. Vickers microhardness tests were done on the polished surface with a load of 110 ± 2 g. The depths of the tests from the surface were estimated from the tops of the asperities on the machined surface by a modification of the cord method (McNamara and Peters, 1985). Electron transparent foils for TEM examination were prepared parallel to the machined surface at several depths from the machined surface using a variation of the controlled electrochemical thinning method described by Straver et al. (1984). Foils perpendicular to the surface were prepared by first depositing 2 - 3 mm of nickel then cutting with EDM 0.3 mm thick, 3 mm diameter discs perpendicular to the surface which were then thinned electrochemically.

RESULTS AND DISCUSSION

The median cutting forces were 143 ± 3 N with arange of about 100 N. The mean chip thickness was 0.25 mm giving a chip thickness ratio of 0.51. Comparable values for machining in the peak aged condition are 0.18 mm and 0.70 respectively. The mean peak to valley surface roughness (Rz) was found to be 18 μm compared to a theoretical value of 5.1 μm (Armarego and Brown, 1969) and 7.29 μm when machined in the peak aged condition. Scanning electron microscopic examination of the chips and machined surface showed considerable evidence of built up edge deposits and tearing of the surface. The solutionized and quenched material does not optimize machinability.

The microhardness as a function of depth from the machined surface are shown in Figures 4 and 5 before and after aging respectively. On these plots 0 μm of depth is taken as the peaks of the asperities. The are no points in approximately the first 15 μm of depth because hardness tests were not performed on the asperities. It should be noted that these hardness values have not been corrected for the influence of the hardness gradient in the underlying layers (Tönshoff and Brinksmeier 1980). The monotonous decrease in hardness with distance from the surface is similar to that noted by Haslum and Rubenstein (1970) and is due at least in part to work hardening, steadily decreasing with depth. However, because of the metastability of the solutionized and quenched material, precipitation phenomena should be considered as contributing to the hardness gradient in Fig. 4. The aging conditions were chosen to bring the bulk material approximately to maximum hardness. Fig. 5 shows that the bulk material, apparently removed from the machining influence, has increased about 100 HV. However, after aging, the surface region is softer than the interior and there is

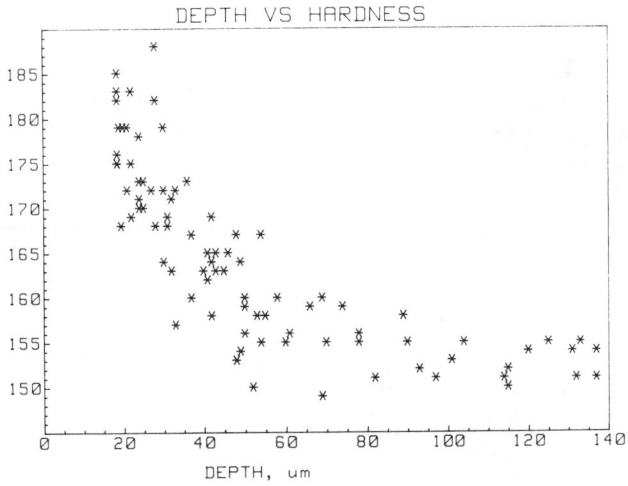

Fig. 4 Microhardness as a function of depth from the machined surface after solutionizing, quenching and machining.

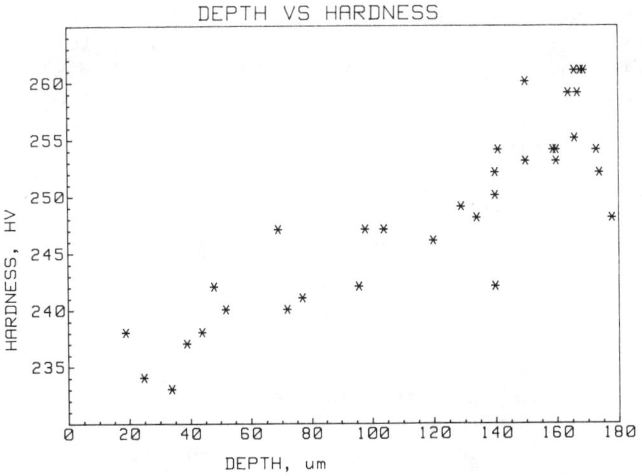

Fig. 5 Microhardness as a function of depth from the machined surface after solutionizing, quenching, machining and aging.

Fig. 6 TEM micrograph of a foil perpendicular to the machined surface, solutionized, quenched and machined then artificially aged.

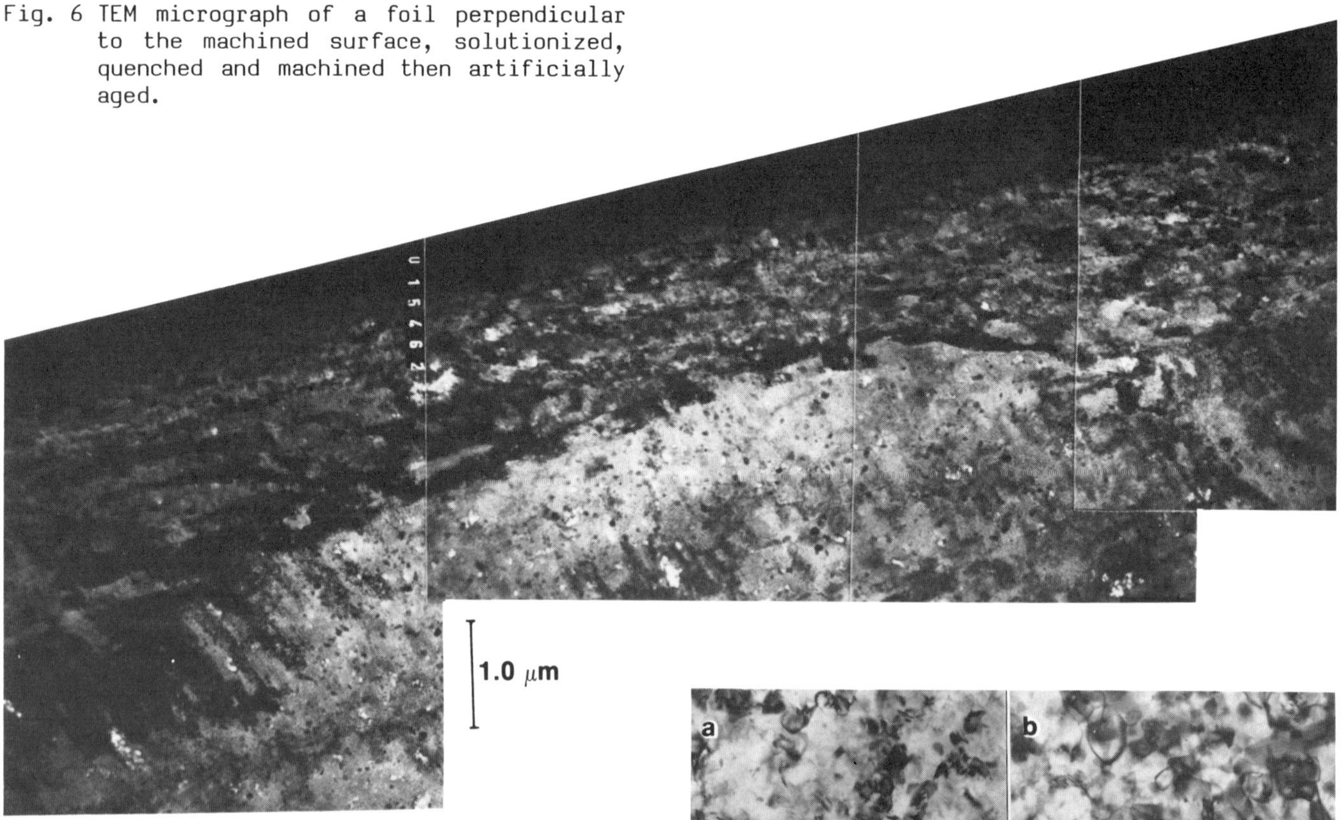

a steady increase in hardness with depth, indicating an interaction of the deformed structure resulting from machining with the precipitation process.

Microstructural gradients after aging as a function of depth from the machined surface are discernable in Fig. 6, a TEM micrograph of a foil prepared perpendicular to the machined surface and parallel to the cutting direction. The sequence of a fine cellular substructure near the surface, then elongated sub-grains and heavily deformed grains is typical of that observed in specimens prepared perpendicular to the surface, parallel as well as perpendicular to the cutting direction. The bands of dislocation that are evident in the lower left portion of Fig. 6 have previously been observed by Ryum (1969) in a deformed and aged Al-Zn-Mg alloy.

The fine cellular substructure near the surface is shown in Fig. 7 a&b, TEM micrographs of foils parallel to and virtually at the machined surface before and after aging. These fine cells, larger and more distinct after aging than before are thought to be the result of polygonization, typical after large deformation in high stacking fault energy material (Meyers and Chawla, 1984). Figures 8 - 10, TEM micrographs of foils prepared parallel to the machined surface, show the deformed microstructure as a function of depth in specimens not artifically aged after machining. At 6 μm from the surface (Fig. 8) large areas of dislocation

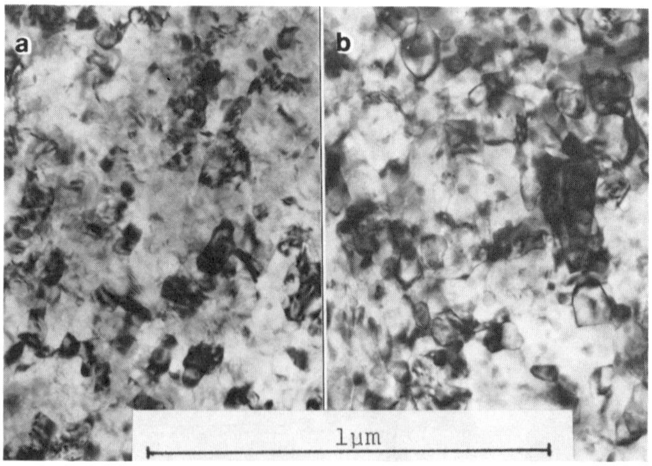

Fig. 7 TEM micrographs of a foil parallel to and at machined surface, a) before artificial aging, b) after artificial aging.

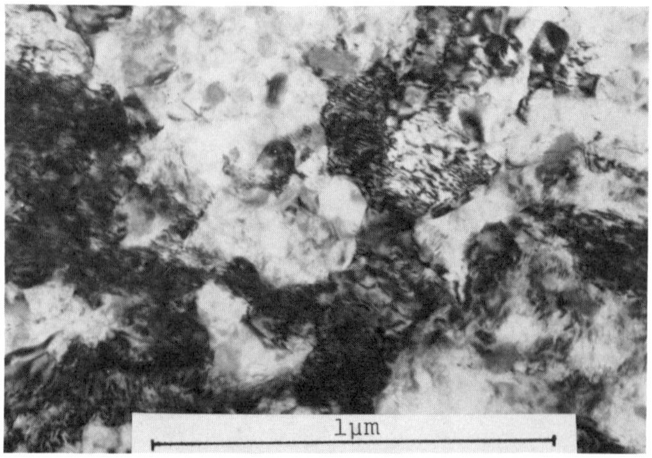

Fig. 8 TEM micrograph of a foil parallel to and about 6 μm below the machined surface, not artificially aged.

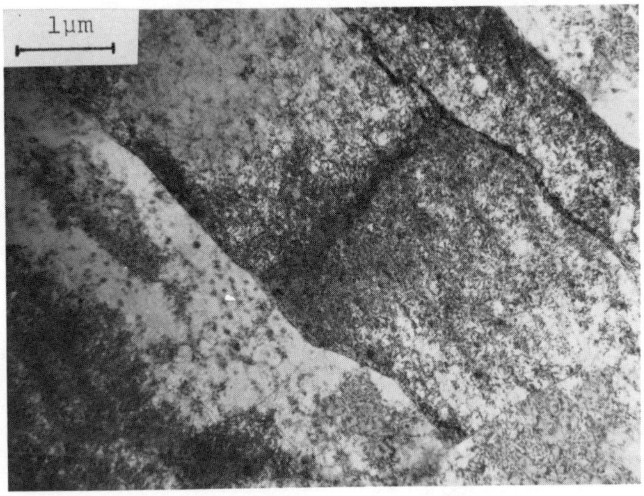

Fig. 9 TEM micrograph of a foil parallel to and about 20 µm below the machined surface, not artificially aged.

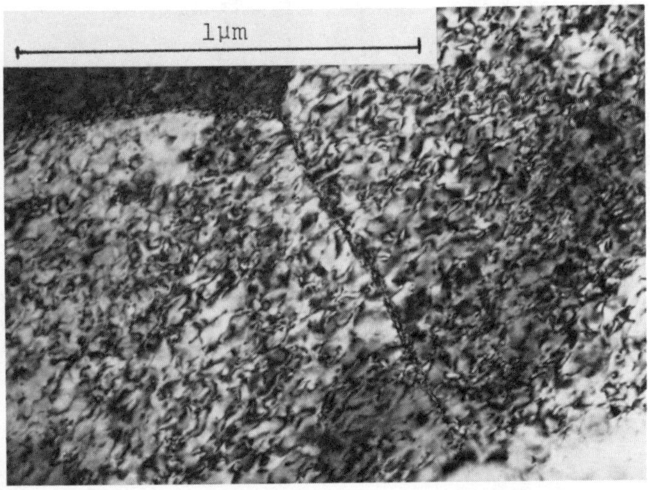

Fig. 10 TEM micrograph of a foil parallel to and about 50 µm below the machined surface, not artificially aged.

Fig. 11 TEM micrograph of a foil parallel to and about 50 µm below the machined surface, solutionized, quenched and machined then artificially aged.

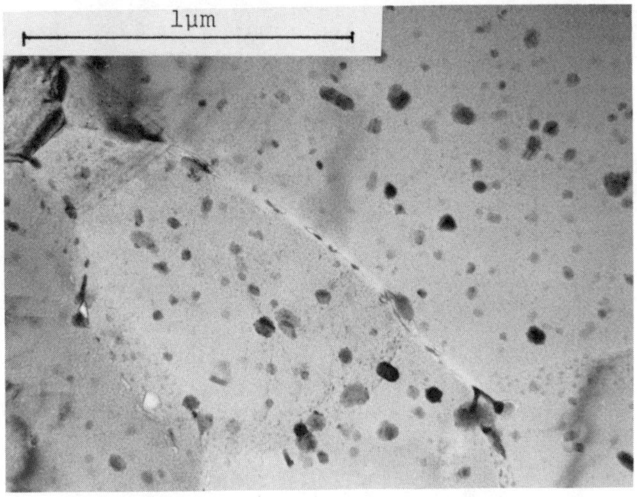

Fig. 12 TEM micrograph of a foil parallel to and about 100 µm below the machined surface, artificially aged.

tangles are evident.* Grain boundaries are discernable at a depth of 20 µm (Fig. 9) and there is a high dislocation density typical of large deformations. Still, at 50 µm from the machined surface in the unaged condition, (Fig. 10) there is considerable contrast due to local strain fields.

After artificial aging, precipitation is evident in the matrix (Figures 11 and 12). Typically, at about 50 µm from the surface (Fig. 11), there is no evidence of PFZs, grain and sub-grain boundary precipitates are small (upper right corner) and there is clear evidence of precipitation on dislocations (center). At 100 µm from the machined surface, PFZs are evident (Fig. 12) and grain and sub-grain boundary precipitates are generally larger than those at 50 µm.

The extent to which clustering of zinc and magnesium and formation of GP zones may be responsible for hardening of the surface layers (Fig. 4) beyond that due to work hardening is masked in the TEM by the high level of strain contrast which exist in these specimens.

Evidently the heating and deformation of the surface layers has influenced the precipitation. The injection of nucleation enhancing defects into the grain boundary region during the machining before artificial aging could be responsible for the lack of PFZs noted above (Fig. 11). This would be consistant with the observation of Embury and Nicholson (1965) regarding nucleation and those of Kawabata et al. (1984) in regards to cold working reducing PFZ width. The relatively lower hardness of the surface layers after artificial aging (Fig. 5) is believed to be due to overaging caused by cold working as discussed by Baranovskiy et

* Depths for TEM specimens prepared parallel to the surface are measured from the bottoms of surface valleys. Add 18 µm for comparing with microhardness depths.

al. (1980) and Papazian (1981). This proposition is supported by qualitative observations of relative coarsening of the matrix precipitates in the aged specimens as one approaches the surface (Figures 6, 11, 12).

ACKNOWLEDGEMENTS

The authors gratefully acknowledge the support of the Swiss National Science Foundation (project No. 2.371.084 FN), the helpful cooperation of the Interdepartmental Institute of Microscopy at the Swiss Federal Institute of Technology, Lausanne, and the Swiss Aluminium Corp. for providing material.

BIBLIOGRAPHY

Armarego, E.J.A.; Brown, R.H., "The Machining of Metals", p. 172, Prenice-Hall, New Jersey (1969).

Bailey, J.A.; Elkhabeery, M.M., J. Eng. Mater. Technol. Trans. ASME, 106, 161-166 (1980).

Baranovskiy, A.N.; Romanova, R.R.; Pushin, V.G.; Buynov, N.N., Phys. Met. Metallogr., 49/4, 159-167 (1980).

Elkhabeery, M.M.; Bailey J.A., J. Eng. Mater. Technol. Trans. ASME, 106, 152-160 (1964).

Embury, J.D.; Nicholson, R.B., Acta Metall, 13, 403-417 (1965).

Gronsky, R.; Furrer, P., Metall. Trans. A, 12A, 121-127 (1981).

Haslam, D.; Rubenstein, C., CIRP Ann., 18, 369-381 (1970).

Heilmann, P.; Clark, W.A.T.; Rigney, D.A., Acta Metall, 31/8, 1293-1305 (1983).

Ives, L.K., Wear, 86, 151-156 (1983).

Joshi, A., Shastry, C.R., Levy M., Effect of heat treatment on solute concentration at grain boundaries in 7075 Al-alloy, Metall. Trans. A, 12A, 1081-1088 (1981).

Kaneeda, T.; Ikawa, N.; Kawabe, H.; Tsuwa, H., Bull. Jpn. Soc. Precis. Eng., 17/1, 25-30, (1983).

Kawabata, T.; Izumi, O., Acta Metall, 24, 817-825 (1976).

Kawabata, T.; Suenaga, H.; Izumi, O., J. Mater. Sci., 19, 1007-1021 (1984).

Lorimer, G.W.; Nicholson, R.B., Acta Metall, 14, 1009-1013 (1966).

McNamara, M.M.; Peters, T.L., Met. Prog., 127/6, 41-48 (1985).

Meyers, M.A.; Chawla, K.K., "Mechanical Metallurgy Principles and Applications", p. 377, Prentice - Hall, New Jersey (1984).

Narasimha Rao, B.V., Metall. Trans. A, 12A, 1356-1359 (1981).

Ostermann, F., Metall. Trans. A., 2, 2897-2902 (1971).

Papazian, J.M., Mater. Sci. Eng., 51, 223-230 (1981).

Park, K.J.; Ardell A.J., Metall. Trans. A, 14A, 1917-1965 (1983).

Ryum, N., Acta Metall, 17, 821-830 (1969).

Straver, W.T.M.; Rozendaal, H.C.F., Mittemeijer, E.J., Metall. Trans. A., 15A, 627-637 (1984).

Thompson, D.S.; Levy, S.A.; Spangler, G.E., Aluminium, 50/10, 647-645 (1974).

Tönshoff, H.K.; Brinksmeier, E., CIRP Ann., 29/2, 519-539 (1980).

Turley, D.M., J. Inst. Met., 99, 271-276 (1971).

Turley, D.M.; Doyle, E.D., Strength of Metals and Alloys ICSMA 6, Gifkins, R.C. ed., Pergamon Press, Oxford, 553-558 (1983).

Turley, D.M.; Samuels, L.E., Metallography, 14, 275-294 (1981).

Vanderwalker D.M.; Vander Sande, J.B., Metall. Trans. A, 12A, 1787-1793 (1981).

Van Dijck, J.A.B., Wear, 42, 109-117 (1977).

AN ASSESSMENT OF THE ENERGY CONSERVATION POTENTIAL OF SURFACE MODIFICATION TECHNOLOGIES

Hien K. Le
DHR Incorporated
McLean, Virginia, USA

ABSTRACT

A model to assess the impact of the use of surface modified tools on energy consumption in the U.S. metalworking industries has been developed. The model takes into account the number of metalcutting and metalforming machines, the average horsepower and energy consumption of their main drive motor, the metalworking conditions such as tool speed and feed rates, and actual data on friction and wear reduction in metalworking with surface modified tools. The model estimates that the tribological sink in the U.S. metalworking industries amounts to 27 trillion Btu/year, of which about 7 trillion Btu might be saved had surface modified tools been used.

ENERGY LOSSES DUE to friction and wear (tribological losses) amounted to more than four quadrillion Btu/year in the United States alone in 1978. The awareness of these losses has spurred considerable research activities in the development of new technologies to control them. Of these, surface modification technologies such as ion implantation and electron beam and laser hardening are extremely promising, since rather than attempting to develop new materials, they involve imparting tribologically superior properties to the surface of a material without changing its bulk properties. Currently, the prime benefactor of surface modification technologies is the metalworking industry, which is making increasing use of surface modified tools such as drill bits and cutters. For the purpose of energy conservation R & D planning, it is highly relevant to be able estimate the impact of the use of these tools on the energy consumption of the metalworking sector. This paper presents such an attempt to estimate tribological losses in metalworking in the United States, and the potential energy savings that the use of surface modified metalworking tools might bring about.

THE MODEL

A flow chart of the model used to estimate tribological sinks (energy losses) and the energy savings which might be attributed to the use of surface modified tools in metalworking is shown in Figure 1. Two estimates contribute to the tribological sink. The first is the frictional sink, which is the sum of frictional energy losses at the machine drive system and at the tool-workpiece interface. The second is the wear sink which is the sum of the energies embodied in the alloys used to make the worn metalworking tools. The tribological sink is the sum of the two estimates. The energy savings due to surface modification are the tribological sink adjusted downward to take into account the increase in tool life and the reduction in friction.

THE FRICTIONAL TRIBOLOGICAL SINKS IN METALCUTTING - Industrial metalcutting is essentially a batch operation with productive cutting time estimated at about 20 % of the total working time (Ref. 1). Furthermore, metal cutting machines, whether operated manually or by computer, require specific settings for the type of material being cut. From these considerations, the first cooponent of frictional losses in metalcutting, the energy losses at the drive systems, can be estimated as follows:

$$Z_c = (1 - e)\, E_c \quad (hp - hr) \qquad (1)$$

where e is the efficiency of the spindle drive systems, E_c the annual metalcutting energy consumption:

$$E_c = \sum_{i,j} n_{i,j}\, P_{i,j}\, h \quad (hp - hr) \quad (2)$$

In Eq. (2), $P_{i,j}$ is the horsepower of the main motor drive of the metalcutting machine. To take into account the type of metalcutting machines and the alloys being cut, $P_{i,j}$ is written as:

$$P_{i,j} = \frac{1}{e} \sum_{i,j} P_{a,i} \frac{C_{a,i}}{\sum_a C_{a,i}} \quad (hp) \quad (3)$$

where:

$n_{i,j}$ = number of machines of type i in a given industry j

h = total hours worked/year

$P_{a,i}$ = power required at the tool point of machine i to perform a standard operation on an alloy a (hp)

$C_{a,j}$ = consumption of an alloy "a" in a given industry j (ton/year)

e = efficiency of the spindle drive system

s = percent of the total number of working time actually spent in productive metalcutting.

The second component of frictional energy losses in metalcutting is at the tool point. It is equal to the power used to overcome friction multiplied by the metalworking time:

$$I_c = \sum_{I,J} n_{I,J} \left[P_{i,j}\,(1 - e) - P_{wi,j}\right] h\, s \quad (4)$$

(hp-hr)

In equation (4), the new variable, $P_{wi,j}$ is the power used to cut the metal; $P_{i,j}(1-e)$ is the power delivered at the tool point; and their difference, the power used to overcome frictional forces.

An efficiency ratio u can now be defined as the ratio of the power used to overcome friction to the power used to work the metal:

$$u = (P_{i,j}\,(1 - e) - P_{wi,j})/P_{wi,j} \quad (5)$$

Then, I_c can be rewritten as:

$$I_c = \sum P_{ij} \frac{(1 - e)}{(u + 1)}\, u\, n_{ij}\, h\, s \quad (6)$$

(hp-hr)

Estimates for the efficiency ratio u in metalcutting are computed using the orthogonal cutting model (Ref. 2) This model is commonly used to describe force relations in most metalcutting operations such as those considered in this study. Using to this model, u is given by:

$$u = \frac{\sin\phi\cos\phi - (f_t/f_c)\,\sin^2\phi\,(\cot\phi + \tan(\phi - \alpha))}{((f_t/f_c)\cos\alpha + \sin\alpha)((\sin\phi/\cos(\phi-\alpha))} \quad (7)$$

where:

ϕ = shear angle, defined as:

$$\phi = \tan^{-1} \frac{b\cos\alpha}{1 - b\cos\alpha} \quad (8)$$

α = effective rake angle of the tool as measured in a plane perpendicular to its cutting edge, in degree;

f_t = thrust force, force component acting in a direction perpendicular to the surface generated and,

f_c = cutting force, force component acting in the direction of tool travel.

THE FRICTIONAL TRIBOLOGICAL SINK IN METALFORMING - In contrast to metalcutting operations and with few exceptions, industrial metalforming is a high production operation. This is true for Industry Group 34 of the Standard Industrial Classification, Fabricated Metal Products, which uses most of the metalforming machines in the United States. A high production rate implies that metalforming machines often run continuously in one, two, or even three shifts per day and at their peak rated power and capacity. Thus, there is no need to adjust the machine power requirement to take into account the metal being formed, as it was done in the metalcutting model.

The frictional energy losses E_f at the drive systems can then be estimated by the following equation:

$$E_f = \sum_{i,j} n_{i,j}\, P_{i,j}\, h \quad (hp - hr) \quad (9)$$

and the frictional energy losses at the tool-workpiece interface by the equation below:

$$I_f = \sum P_{ij} \frac{(1 - e)}{(u + 1)} u \, n_{ij} \, h \, s \qquad (10)$$

(hp-hr)

In the above equations, the symbols are as defined previously and the subscript c has been changed to f to reflect metalforming operations.

Unlike metalcutting, the efficiency ratios in metalforming cannot be described by a single mathematical model. Rather, an efficiency ratio must be derived for each metalforming operations.

In shearing of thin sheets, friction does not have a significant effect on the shearing load force (Ref. 3). For thick sheets, force relations similar to those found in metalcutting most likely will apply (Refs. 4, 5). Since most sheet metal shearing involves thin sheets, an efficiency ratio equal to zero can be used for shearing.

In punching, force relations are extremely complex to model. Experimental data (Refs. 6, 7) shows friction accounting for 10-30 percent of the force requirement in punching. Thus, the value of u would have the same range, i.e. 10-30 percent.

In bending and forming operations, such as rotary draw and roll bending, the frictional component of the load force is negligible because the tool-work piece contact is minimal. On the other hand, in compression bending, stretch forming and most tube forming operations (such as flaring, flanging, beading, curling expanding reducing and swaging), a certain amount of friction is desirable.(Ref. 8). Therefore, for these above operations, the efficiency ratios are equal to zero.

In forging and pressing, the efficiency ratio can be approximated by the product of the coefficient of friction and the height to width ratio of the work piece(Ref. 9). When the width of the work piece is equal to its height, the efficiency ratio is then equal to the coefficient of friction.

THE WEAR TRIBOLOGICAL SINK - The wear tribological sink is given by:

$$T_w = \sum_{a,j} q_a \, m_{a,j} \qquad (11)$$

where:

q_a = the embodied energy of an alloy a used to make metalworking tools, and

$m_{a,j}$ = the consumption of an alloy a for metalworking tools in a given industry j (ton/-year).

THE ENERGY SAVINGS DUE TO FRICTION REDUCTION IN METALWORKING - The net effect of using surface modified tools is to increase the efficiency ratio u, since less work is now required to overcome friction. The energy savings due to friction reduction S in both metalcutting and metalforming can thus be written as:

$$\Delta I = \sum_{i,j} n_{i,j} \, p_{ij} h \, s \, (1 - e) \frac{(u_2 - u_1)}{(u_2 + 1)(u_1 + 1)} \qquad (12)$$

where u_2 and u_1 are the efficiency ratios with and without surface modified metalworking tools, respectively.

THE ENERGY SAVINGS DUE TO WEAR REDUCTION IN METALWORKING - Reducing wear increases metalworking tool life. If the average increase in tool life is defined as v (in percent), the energy savings can be written as:

$$\Delta T_w = \sum_{a,j} q_a \, m_{a,j} \left(\frac{v}{v + 1}\right) \qquad (11)$$

THE DATA

The following data are needed to implement the above model: the numbers of metalworking machines in the United States, the rated horsepower of their main motor, typical metalworking conditions, friction coefficients in metalforming with and without surface modified tools, efficiency of spindle drive, and consumption of alloys in the metalworking industries.

METALWORKING MACHINES CONSIDERED - The number and type of metalcutting and metalforming machines considered in this study are shown in Tables 1 and 2 respectively. The 13th American Machinist Inventory of Metalworking Equipment is the primary source for these numbers (Ref. 10). The study considered 70 of all metalworking machines in the United States (except those associated with primary metals processing).

RATED HORSEPOWER OF THE MAIN DRIVE MOTOR OF METALWORKING MACHINES - A survey of metalworking machine manufacturers was conducted. Data was obtained on the horsepower of the main drive motor of the metalworking machines. The results of the survey are shown in Table 3.

ALLOY CONSUMPTION FOR METALWORKING TOOLS - These data are shown in Tables 4 and 5 and are derived from the U.S. Department of Commerce Census of Manufacturers (Ref. 11).

INCREASE IN TOOL LIFE WITH SURFACE MODIFICATION - With surface modification, tool life has been shown to increase anywhere from 3 to 6 times (Ref. 12) for metalforming tools such as dies, and 7 to 25 times for metalcutting tools such as TiN coated drill bits, taps, hobs, and end mill cutters (Ref. 13, 14).

FRICTION REDUCTION WITH SURFACE MODIFIED METALWORKING TOOLS - A seven percent decrease in the coefficient of friction has been found when going from carbide dies to sapphire dies in drawing experiments (Ref. 15). A reduction of the same order of magnitude would be expected with surface modified dies. Since the metalforming efficiency ratio u is proportional to the coefficient of friction, it would decrease by the same amount when the coefficient of friction is decreased. In metalcutting, when surface modified tools such as TiN coated drill bits are used instead or carbide ones, torque and thrust are lowered anywhere from 25 to 43 percent (Ref. 13, 14). This means that the metalcutting efficiency ratio u would also decrease by the same amount.

OTHER DATA - The efficiency e of the spindle drive system of metalworking machines is taken to be about twenty percent (Ref. 2). The embodied energy in steel is estimated at 19.2 million Btu/ton (Ref. 16). Metalforming coefficients of friction in most cases vary from 0.05 to 0.10 (Ref. 17)

CALCULATIONS

Calculation of the tribological sinks was conducted on an IBM Personal Computer using the Lotus Development Corporation's 1-2-3 spreadsheet software.

CALCULATION OF THE METALCUTTING TRIBOLOGICAL SINKS - To model the metalcutting frictional tribological sink, a large spreadsheet consisting of three modules was constructed and linked together. The first module (Table 6) calculates the required spindle drive motor horsepower for standard cutting of carbon steels, alloy steels, aluminum and aluminum alloys, and copper and copper alloys. The second module uses the horse power data calculated in the first module and data on alloys consumption in the metalworking industries to compute the Material Averaged Motor Horsepower, $P_{i,j}$ (defined in Eq. 3). Results of calculations in the second module are shown in Table 7. The third module uses the results of module 2 to calculate the tribological sinks and energy savings attributable to the use of surface modified metalworking tools.

CALCULATION OF THE METALFORMING TRIBOLOGICAL SINKS - The tribological sinks in metalforming are calculated using a single module having a structure similar to the third module of the metalcutting spreadsheet.

RESULTS AND DISCUSSION

The tribological sinks in metalworking are shown shown in Table 8. The total frictional sink is 27.9 billion BTu/year while the wear sink is about 7.7 billion Btu/year.

The energy savings achievable when surface modified tools are used in metalworking are shown in Table 9. About 7.3 billion Btu/year can be saved, of which 1.8 would be from friction reduction and 5.5 from wear reduction.

A breakdown of the sinks and savings by major industry and machine type is shown in figures 2 and 3.

The estimates obtained are probably conservative. The model considered only about 72 percent of all metalworking machines in the United States when estimating the tribological sinks. The basis for eliminating a particular machine type from the study was absence of severe frictional losses at the tool point and amenability of the tools to surface modification. Because of this, tool point sink estimates are probably correct while the drive sink estimates are low. The same rationale applies to the estimated energy savings. Thus, the effect of having low friction components, such as surface modified bearings in the drive systems of metalworking machines, would be to increase the energy savings estimated.

Estimates of the wear sink do not include actual wear sink produced by companies that manufacture their own tools. Furthermore, the estimates were derived using a low value for embodied energy for steel (19.7 M Btu/ton vs. 37.4 M Btu/ton for steel produced from the ore). If these had been included, the wear sink would be bigger.

By changing the various model parameters within their allowable ranges and recalculating the energy savings and tribological sinks using the spreadsheet facility, various ranges of estimates were obtained. However, these estimates do not vary significantly from the results presented.

SUMMARY AND CONCLUSIONS

According to the model developed in this paper, the tribological losses in the metalworking industries of the United States amount to 27 billion Btu/year. Of these, about 7.4 billion Btu/Year might have been saved had surface modified tools been used extensively.

ACKNOWLEDGMENT

This work was funded by The U. S. Department of Energy's Energy Conversion and Utilization Technology Division and monitored by Pacific Northwest Laboratory / Battelle Memorial Institute through Contract DE-AC06-76RLO 1830.

REFERENCES

1. Lyle, C. S. "A Proven Tooling System for Flexible Turning Machines", Paper presented at the 2nd Biennal International Machine Tool Technical Conference, Septemebr 5-13, 1984, Chicago, Illinois.

2. Drozda, T. J. ed. "Tool and Manufacturing Engineers Handbook -- Volume 1: Machining", 4th Edition. Society of Manufacturing Engineers, Dearborn, Michigan (1983)

3. Shey, J. A. "Tribology in Metalworking, Friction, Lubrication and Wear", American Society For Metals, Metal Park, Ohio (1983)

4. Rowe, G. W. "Lubrication and Lubricant", Ed. E.R. Braithwaite. Elsener Publishing Company, Amsterdam, New York (1967)

5. Shey, J. A. " Metal Deformation Processes, Friction and Lubrication", 2nd ed. Marcel Dekker, Inc. New York (1972)

6. Freeman. R. and W. Leeming. "Ironing of Thin Walled metal Cup: The Distribution of Punch Load," BISRA Report No. MW/E/53 (1953), cited by Hosford, W.F. in "Metalforming: Mechanics and Metallurgy", P(rentice hall Publisher, Englewood Cliffs, New Jersey (1978)

7. Blazynsky, T.Z. "Metalforming Tool Profiles and Flow", J.Wiley and Son, New York (1976)

8. Kervitch R. J. and R. K. Springborn "Cold Bending and Forming of Tubes and Other Sections", American Society of Tools and Manufacturing engineeers, Dearborn, Michigan (1966)

9. Harris J. B., "Mechanical Working of Metal: Theory and Practice", Pergammon Press, Oxford, New York (1983)

10. American Machinist , " The 13Th American Machinist Inventory of Metalworking Equipment". McGraw-Hill Inc. New York (1983)

11. U. S. Bureau of The Census, "1982 Census of Manufacturers", MC82-84 (Part 2), Government Printing Office, Washington, D.C. (1983)

12. Kirvonen, J. K. " Industrial Application of Ion Implantation", Paper presented at the Materials Research meeting, Boston, Massachusetts, November 1893.

13. Hatschek, R. L. American Machinist Special Report 752, March 1983.

14. Henderer, V. "Performance of Titanium High Speed Steel Drills", Paper presented at the 11 North American Metalworking Research Conference, University of Wisconsin, Madison, Wis., May 24-26, 1983.

15. Wright, P. K. and R. S. Rao, " Friction Reduction in Machining and Forming", Paper presented at the 11 North American Metalworking Research Conference, University of Wisconsin, Madison, Wis., May 24-26, 1983.

16. Elliot, J. F. " Energy for Steel Production-Prospects and Options", In "Energy Use and Conservation in the Metal Industry", Y. A. Chang, W. M. Danvers and J. M. Cigan Eds. The Metallurgical Society of the AIME, New York, New York (1975)

17. Schey, J. A. "Tribology in Metalworking, Friction, Lubrication and Wear", American Society for Metals, Metals Park, Ohio (1983)

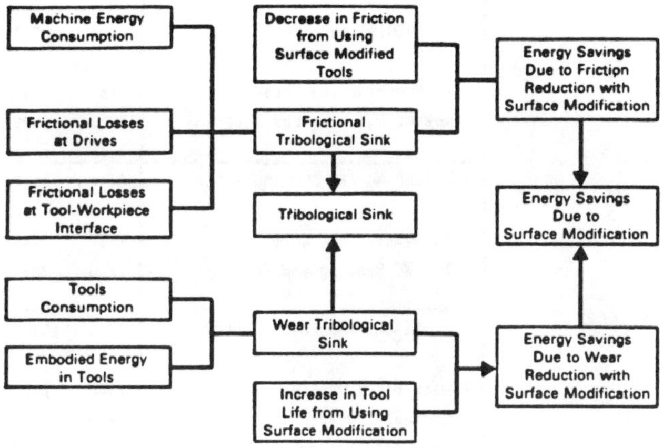

FIGURE 1. Energy Savings Estimation Model

Machine Type	Industry Group[1] SIC 25	SIC 33	SIC 34	SIC 35	SIC 36	SIC 37	SIC 38	SIC 39	TOTAL
1. TURNING MACHINE GROUP	982	20904	85327	191914	42330	59868	24123	8834	435208
- NC Turning Machines	46	842	4348	18592	2677	4922	1538	387	33352
- Non-NC Turning Machines	811	17390	75235	136490	34652	39818	20362	7569	332327
- NC Boring Machines	25	271	500	2935	182	907	106	138	5064
- Non-NC Boring Machines	74	2087	3592	19980	3065	10094	877	693	40462
- Machining Centers	26	350	1652	13917	2644	4127	1240	47	24003
2. DRILLING MACHINE GROUP	2606	18528	68675	125146	44179	40497	21686	7971	329288
- NC Drilling Machines	58	261	1121	4205	1150	718	227	253	7993
- Non-NC Drilling Machines	2271	14873	54602	109924	37116	36191	19200	7276	281453
- Tapping Machines	257	2620	8085	7989	5121	2007	1669	391	28139
- Threading Machines	20	774	4867	3028	792	1581	590	51	11703
3. DRILLING MACHINE GROUP	1181	9610	39471	128121	27206	42089	15317	6247	269242
- NC Milling Machines	45	433	1676	7876	1983	2145	1132	639	15929
- Non-NC Milling Machines	1099	8658	36102	103072	23862	27199	12907	5580	218479
- Gear Cutting Machines	37	519	1693	17173	1361	12745	1278	28	34834
4. BROACHING MACHINE GROUP	33	652	2624	6825	821	5197	233	120	16505
5. SAWING MACHINE GROUP	2259	10458	40115	56807	18911	21371	8587	6999	169107

Table 1: **Metalcutting Machine Tools Considered in this Study**

Machine Type	Industry Group[1] SIC 25	SIC 33	SIC 34	SIC 35	SIC 36	SIC 37	SIC 38	SIC 39	TOTAL
1. PUNCHING AND SHEARING GROUP	1854	2924	26139	19571	9565	7222	3296	2015	72586
- NC Punching and Shearing	166	95	2490	1290	1013	373	368	428	6223
- Non-NC Punching and Shearing	1688	2829	23649	18281	8552	6849	2928	1587	66363
2. BENDING AND FORMING GROUP	4487	3177	30196	19945	8432	9591	2569	1473	79870
- NC Bending and Forming	197	145	989	493	247	428	71	15	2585
- Non-NC Bending and Forming	4290	3032	29207	19452	8185	9163	2498	1458	77285
3. PRESSES AND FORGES GROUP	9040	16380	114356	46479	45114	32316	11570	14279	289534
- Mechanical Presses	8508	7187	89051	29289	29544	19010	6376	10008	198973
- Hydraulic Presses	283	6422	10494	13300	9131	9712	2530	1420	53292
- Pneumatic Presses	159	567	3512	2309	5592	1557	2563	1397	17656
- Forges	90	2204	11299	1581	847	2037	101	1454	19613

Table 2: **Metalforming Machine Tools Considered in this Study**

(1) SIC Codes:

25: Furniture and Fixtures
33: Primary Metal Industries
34: Fabricated Metal Products
35: Machinery, Except Electrical
36: Electric and Electronic Equipment
37: Transportation Equipment
38: Instruments and Related Products
39: Miscellaneous Manufacturing Industries

MACHINE TYPE	DRIVE HORSEPOWER L (hp)	H (hp)	A (hp)
1. METALCUTTING MACHINES			
1.1 NC TURNING MACHINES			
-Horizontal under 9" chuck	7.5	20	13.75
-Horizontal 9" to under 13" chuck	30	50	40
-Horizontal 13" to under 20" chuck	20	50	35
-Horizontal 20" chuck and larger	40	60	50
-Vertical turn & bore (VTM, VBM)	75	150	112.5
1.2 NON NC TURNING MACHINES			
-Eng & tlrm (not tracer) over 8" to 16" swing	4	7.5	5.75
-Eng & tlrm (not tracer) over 16" swing	10	30	20
-Tracer lathes (all sizes)	10	20	15
1.3 NC BORING MACHINES			
-Horizontal bore, drill, mill (bar machining)	8	28	18
1.4 NON NC BORING MACHINES			
-Hor. bore, drill, mill (bar mach) table & planer type	8	10	9
-Hor. bore, drill, mill (bar mach) floor type	50	70	60
-Precison, horizontal and vertical	87	120	103.5
1.6 NON NC DRILLING MACHINES			
-Radial	5	16	10.5
1.7 NC MACHINING CENTERS			
-Automatic tool change: vertical Y-axis 26" or less	8	20	14
-Automatic tool change: vertical Y-axis over 26"	7.5	25	16.25
-Automatic tool change: horizontal Y-axis under 27"	7.5	25	16.25
-Automatic tool change: horizontal Y-axis 27" or over	20	30	25
1.8 NC MILLING CENTERS			
-Profiling and dupplicating	15	25	20
1.9 NON NC MILLING MACHINES			
-General purpose, knee or bed: vertical	3	10	6.5
-General purpose, knee or bed: horizontal	7.5	10	8.75
-Automatic and manufacturing		22.5	11.25
1.10 GEAR CUTTING AND FINISHING MACHINES			
-Gear hobbers	24	37	30.5
-Gear shapers	3	12	7.5
1.11 NC GRINDING MACHINES (ALL TYPES)	1.5	30	15.75
1.12 NON NC GRINDING MACHINES			
-External: plain center type	20	25	22.5
-External: universal center type	7.5	13	10.25
-External: centerless (including shoe type)	15	50	32.5
-External: chucking	5	15	10
-Surface: reciprocating, vertical and horizontal, power	5	19	12
-Tool and cutter	1	1.5	1.25
-Bench, floor and snag	0.33	5	2.665
-Disk grinders, single and double spindle	3	20	11.5
-Abrasive belts (except finishing)	5	100	52.5
1.13 HONING, LAPPING, POLISHING MACHINES			
-Polishing stands (incl. abrasive belts-not grinding)	1	10	5.5
1.17 CUTOFF AND SAWING MACHINES			
-Abrasive wheels	10	60	35
-Band saws, countour sawing and filing	1	10	5.5
2. METALFORMING MACHINES			
2.2 NC PUNCHING & SHEARING MACHINES			
-Punching machines (incl. comb. punching & shearing)	0.5	15	7.75
-Plate and sheet shears: hydraulic	3	25	14
2.4 NON NC BENDING AND FORMING MACHINES (POWER)			
-Press brakes: mechanical	3	20	11.5
-Press brakes: hydraulic and pneumatic	3	10	6.5
-Bending rolls, sheet and plate	2	25	13.5
-Bending rolls, angles, bars, shapes, pipes	3.5	100	51.75
-Rotary bending and forming machines	0.75	1.25	1
-Ram and press bending machines	20	25	22.5
2.5 MECHANICAL PRESSES-POWER (NOT FORGES)			
-Open back (OBI) &gap: 51 tons and over	15	30	22.5
-Vertical straight side or arch:single point	5.5	50	27.75
-Two point: 300 tons or under	10	40	25
-Two point: 301 tons or over	30	40	35
2.6 HYDRAULIC PRESSES (NOT FORGES)			
-Vertical, straight side or column	7.5	75	41.25
-C-Frame or gap:15 tons or under	3	20	11.5
-C-frame or gap: over 15 tons	4	40	22
-Vertical, double action	10	20	15
2.8 COIL PROCESSING SYSTEMS	25	40	32.5
2.12 THREAD ROLLING MACHINES	3	20	11.5

Note: HP=spindle horsepower, hp. L=low, H=high, A=average

Table 3: Spindle Drive Horsepower of Selected Machine Tools

Alloy	1977(a) All Forming Accessories	1977(a) Forming Dies(b)	1983(c) Forming Dies	1977(a) All Cutting Accessories	1977(a) Cutting Tools(d)	1983(e) Cutting Tools
Total	212.50	127.25	272.61	92.60	56.66	121.39
Carbon Steel	167.00	100.00	214.24	43.20	26.43	56.63
Alloy Steel	42.30	25.33	54.26	48.70	29.80	63.84
Stainless Steel	3.20	1.92	4.11	0.70	0.43	0.92

(a) Ref. 17.
(b) Calculated as 59.88% of all forming accessories.
(c) Calculated using an annual growth rate of 13.54%.
(d) Calculated as 61.19% of all cutting accessories.
(e) Calculated using an annual growth rate of 9.33%.

Table 4: Estimated Alloy Consumption for Metalworking Tools (1000 tons)

Materials and Processing	SIC 25 (%)	SIC 33 (%)	SIC 34 (%)	SIC 35 (%)	SIC 36 (%)	SIC 37 (%)	SIC 38 (%)	SIC 39 (%)
1. Carbon Steels								
Cutting	88.42	0.0	82.67	74.63	72.39	83.07	22.79	7.33
Forming	87.09	0.0	76.45	75.03	67.71	81.30	22.39	4.50
2. Alloy Steels								
Cutting	2.67	0.0	8.67	13.34	7.45	8.92	58.21	92.67
Forming	2.63	0.0	7.97	12.49	6.96	8.04	57.18	56.97
3. Copper and Copper Alloys								
Cutting	0.0	64.33	4.64	6.27	11.62	3.31	11.93	0.0
Forming	0.0	88.44	3.92	5.06	14.34	3.33	13.50	0.0
4. Aluminum and Aluminum Alloys								
Cutting	8.91	35.67	4.02	5.76	8.54	4.69	7.06	0.0
Forming	10.28	11.56	11.66	7.42	10.99	7.33	6.94	38.53
Total Cutting:	1	1	1	1	1	1	1	1
Total Forming:	1	1	1	1	1	1	1	1

Table 5: Industrial Metalworking of Major Alloys, by Percentage, 1977

MATERIALS & PROCESS	TOOL MATERIAL	DEPTH	FEED			SPEED	METAL REMOVAL RATE	UNIT POWER	REQUIRED SPINDLE	REQUIRED MOTOR
		d (in)	fr (ipr)	fm (ipm)	ft (ipt)	Vc (fpm)	Q (in3/min)	P (hp)	HPs POWER (HP)	HPm POWER (HP)
1. TURNING										
-Carbon steels	HSS	0.25	0.02	n.a	n.a	120	7.20	1.50	10.80	13.50
-Alloy steels	HSS	0.25	0.02	n.a	n.a	100	6.00	1.30	7.80	9.75
-Al & Al alloys	HSS	0.25	0.03	n.a	n.a	800	72.00	0.25	18.00	22.50
-Cu & Cu alloys	HSS	0.25	0.02	n.a	n.a	300	18.00	1.00	18.00	22.50
2. FACE MILLING (milling width: 5 in.; 10 teeth cutter; mill cutter dia.= 5in)										
-Carbon steels	HSS	0.15	n.a	n.a	0.012	200	3.60	1.40	5.04	6.30
-Alloy steels	HSS	0.15	n.a	n.a	0.007	175	1.84	1.40	2.57	3.22
-Al & Al alloys	HSS	0.15	n.a	n.a	0.01	250	3.75	0.32	1.20	1.50
-Cu & Cu alloys	HSS	0.15	n.a	n.a	0.01	250	3.75	1.00	3.75	4.69
3. DRILLING (1/2 in. diameter drill)										
-Carbon steels	HSS	n.a	0.01	n.a	n.a	80	2.40	1.30	3.12	3.90
-Alloy steels	HSS	n.a	0.002	n.a	n.a	45	0.27	1.50	0.41	0.51
-Al & Al alloys	HSS	n.a	0.012	n.a	n.a	200	7.20	0.20	1.44	1.80
-Cu & Cu alloys	HSS	n.a	0.01	n.a	n.a	140	4.20	0.80	3.36	4.20
4. BROACHING (8 tooth broach; width of cut W = 1 in.)										
-Carbon steels	HSS	0.04	n.a	n.a	0.005	40	19.20	0.50	9.60	12.00
-Alloy steels	HSS	0.032	n.a	n.a	0.004	30	11.52	1.00	11.52	14.40
-Al & Al alloys	HSS	0.048	n.a	n.a	0.006	40	23.04	0.20	4.61	5.76
-Cu & Cu alloys	HSS	0.04	n.a	n.a	0.005	25	12.00	0.40	4.80	6.00

NOTES: A) Metal removal rate and horsepower equations:

- Turning : Q = 12 x d x fr x Vc
- Milling : Q = w x d x fm = w x d x ft x n x rpm = w x d x ft x n x 3.82
- Drilling : Q = 3.1416 x D x fr x 3.82 x Vc / 4
- broaching : Q = 12 x Vc x ft x n = 12 x Vc x dt
- Required power: HPs = Q x P (at spindle)
- Required power: HPm = Q x P / E (at motor)
- Drilling rpm : 3.82 x Vc / D

where: Q = metal removal rate, cubic inches per minute
d = depth of cut, inches
dt = total stock removed per stroke, inch. For Broaching, dt = chip load
fr = feed, inches per revolution
Vc = cutting speed, feet per minute
W = width of cut, inches
fm = feed rate, inches per minute. For drilling, fm = fr x rpm = fr x 3.8
ft = feed rate, inches per tooth
n = number of teeth in cutter
rpm = revolution per minute of work or cutter
D = diameter of drill or milling cutter, inches
HPs = required horsepower at spindle, hp
HPm = required horsepower at motor, hp
E = efficiency of spindle drive, about 80%

P = unit power, horsepower per cubic inch per minute

B) HSS = high speed steel

Table 6: Cutting Conditions and Derived Motor Horsepower

PROCESS	MATERIALS AVERAGED MOTOR HORSEPOWER BY INDUSTRY (hp)[1]							
	SIC 25	SIC 33	SIC 34	SIC 35	SIC 36	SIC 37	SIC 38	SIC 39
1. Turning	14.20	22.50	13.95	14.08	15.04	13.89	13.02	10.02
2. Drilling	3.64	3.34	3.54	3.35	3.50	3.51	1.81	0.76
3. Milling	5.79	3.55	5.76	5.51	5.47	5.75	3.97	3.44
4. Broaching	11.56	5.91	11.68	11.58	10.95	11.72	12.24	14.22

Table 7: Materials Averaged Motor Horsepower, by Operation and Industry

Operation	Frictional Sink Drive	Frictional Sink Tool	Wear Sink	Total
Metalcutting	2,490	3,320	2,323	8,133
Metalforming	10,400	4,000	5,370	19,770
Total	12,980	7,320	7,693	27,903

Table 8: Tribological Sinks in Metalworking (billion Btu/year)

Operation	Reduction in the Friction Sink: Drive	Reduction in the Friction Sink: Tool	Reduction in the Wear Sink	Total
Metalcutting	0	739 (22%[a])	1,936 (83%)	2,675 (53%)
Metalforming	0	1,100 (25)	3,580 (67)	4,680 (24)
Total	0	1,839 (25)	5,536 (72)	7,355 (26)

(a) Percent of the sink.

Table 9: Energy Conservation with Surface Modification Technologies (billion Btu/year)

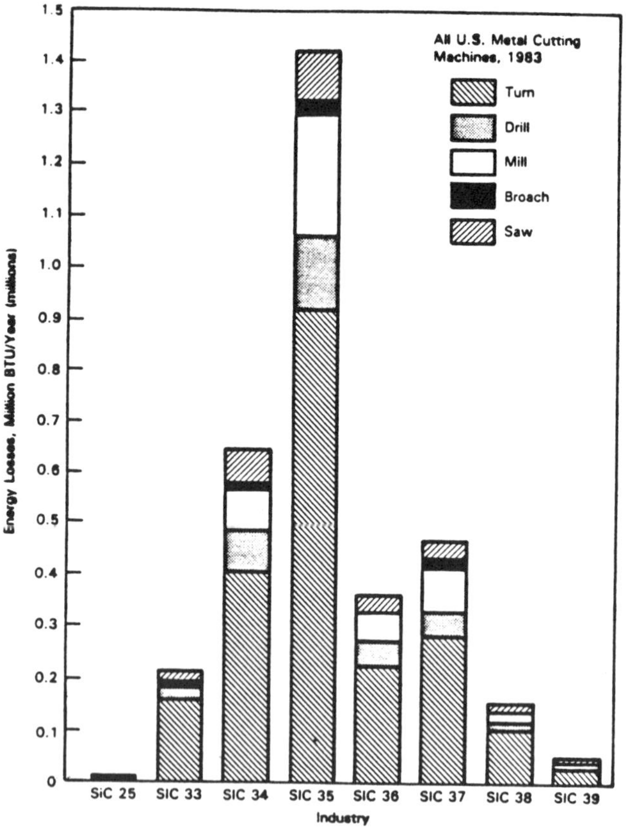

a) Frictional Losses at the Tool Point

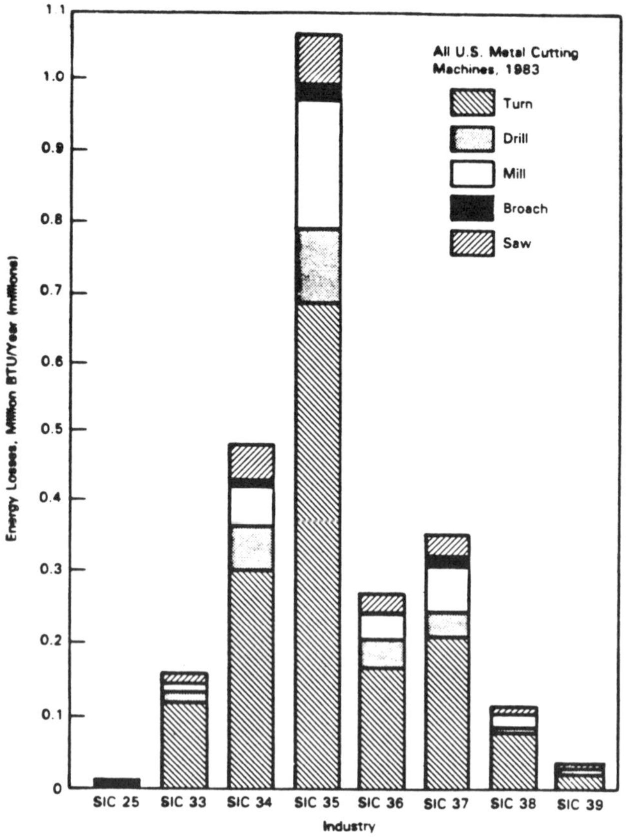

b) Frictional Losses at the Drive System

Figure 2: Tribological sinks and energy saving potential by machine type and industry in metalcutting

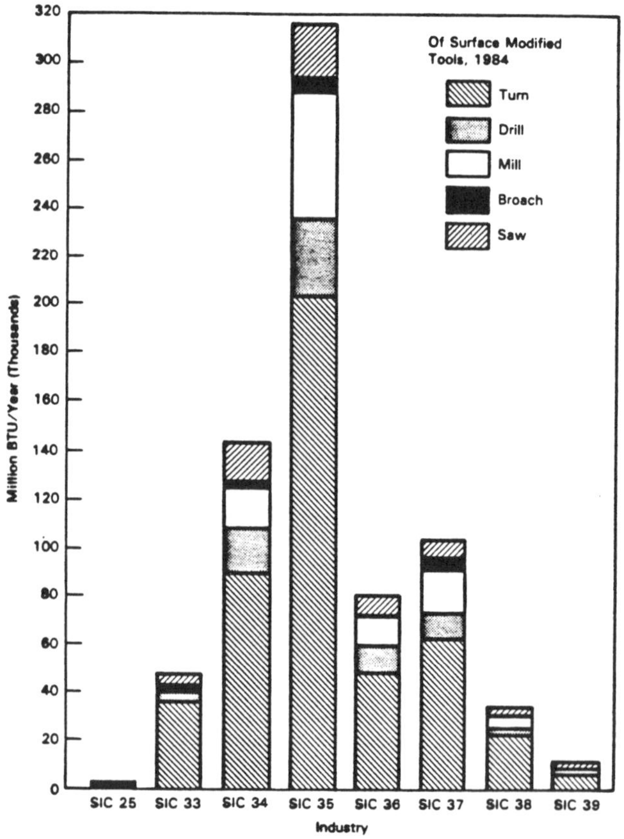

c) Metalcutting Energy Savings Potential

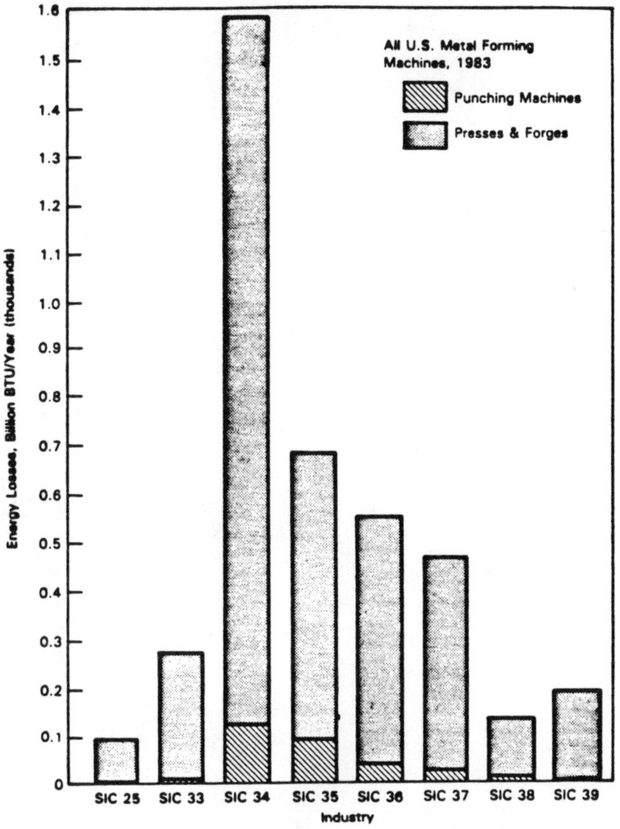

a) Frictional Losses at the Tool Point

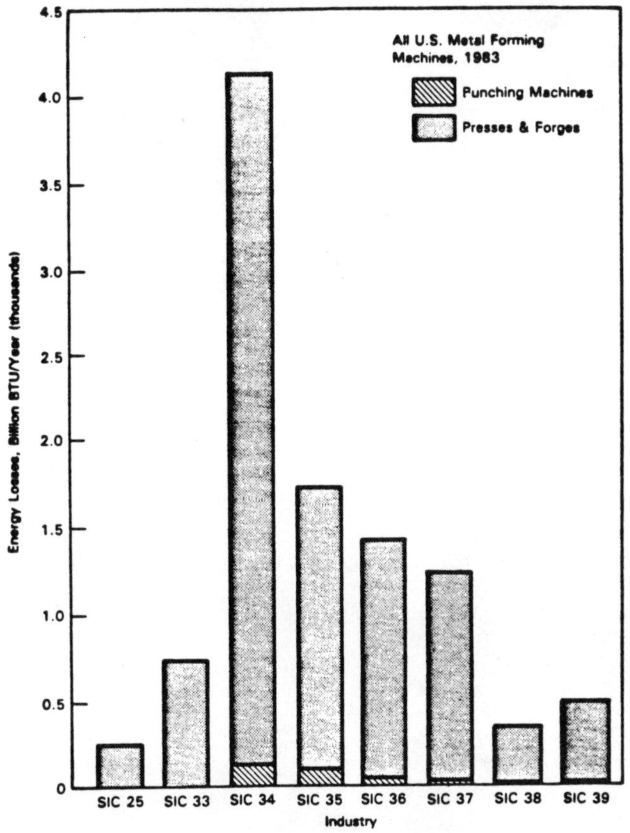

b) Frictional Losses at the Drive System

Figure 3: Tribological sinks and energy saving potential by machine type and industry in metalforming

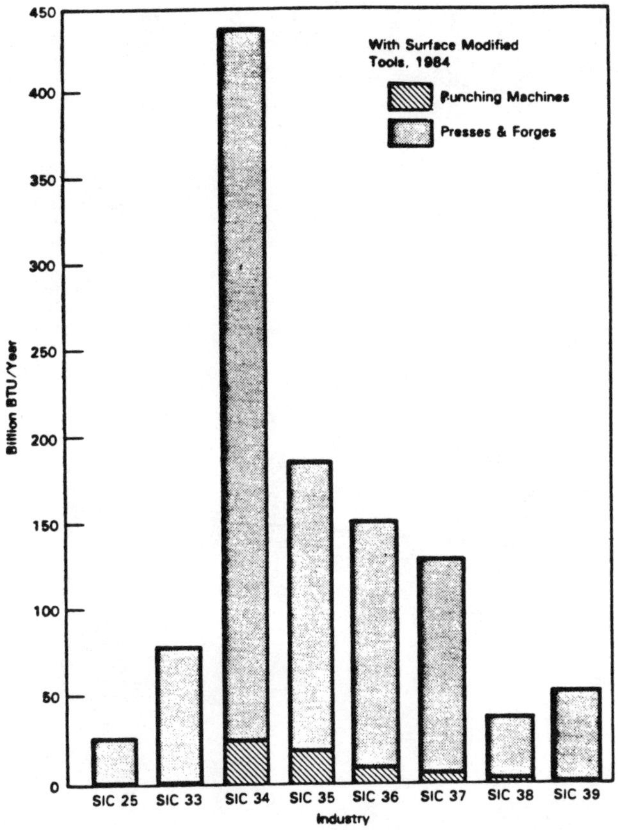

c) Metalforming Energy Savings Potential

SURFACE STRESS MODIFICATION FOR REPAIR OF STRESS-RELIEVED CARBON STEEL FABRICATIONS

Charles F. Jenkins
E. I. du Pont de Nemours & Co.
Savannah River Plant
Aiken, South Carolina USA

ABSTRACT

Pitting corrosion was observed in welded carbon steel tanks which had been stress relieved to improve resistance to stress corrosion cracking. Successful repair required the development of a cleaning and repair method which would not increase residual tensile stresses at the steel surface. An experimental study indicated that uniform residual tensile stresses of 27.6 - 34.5 megapascals (4,000-5,000 psi) remained in the plate after the original stress relief. Grinding, shot peening, and grit blasting were evaluated. In laboratory tests, metal grit blasting changed the average residual stresses at the surface to 34.5 - 69 megapascals (5,000-10,000 psi) compressive. Moreover, the depth of the compressive zone was substantial - perhaps 138 megapascals (20,000 psi) at 0.25 mm (0.010 inch). This is beneficial and will contribute to an increase in resistance to stress corrosion cracking. The repair technique employed on the tanks consisted of blast cleaning the entire surface to a uniform Class II finish with minimum change in the configuration of the pits.

STRESS CORROSION CRACKING (SCC) near butt welds in large diameter carbon steel waste tanks used for storage of alkaline nitrate solutions occurred largely as a result of residual stresses.[1-3] This led to a requirement for in situ stress relief of all future tanks.

A uniform heat treatment can be applied only to new tanks. Following their fabrication, the tanks are heat treated by raising the temperature to 1,100°F, as required for stress relief. Following this in situ field heat treatment, the tanks are encased in concrete, precluding general reheating for stress relief purposes.

The heat treatment effectively relieves welding stresses and reduces long range reaction stresses sufficiently to prevent SCC in the solutions of interest.[3] Any additional work on or in the tanks subsequent to the stress relief is monitored closely to ensure that no long-term, localized stresses are produced. For example, work inside the 85-foot-diameter tank proceeds only after the vessel floor is protected from weld splatter and the possibility of mechanical damage.

In one case, the tank bottom was covered with 1/2-inch plywood sheeting during this period. The floor and plywood were often wet and after several months pitting was noted in the floor plates. Some pits were about 1/8-inch deep (25% of the wall thickness). A program to clean, characterize pit size and distribution, and develop a repair method was initiated.

The cleaning procedure consisted of metal grit blasting of the bottom surfaces. This was followed by examination of the pits. Blasting will remove scale, etc., and impart residual stress to the previously stress-relieved tank.[4] These stresses are additive to the working stresses and need to be considered in the overall stress analysis of the vessels.[5] In addition, complications can be expected at locations where repairs to the tank surfaces are made subsequent to the stress relief heat treatment and/or the blast cleaning. The levels of stressing and the possibilities for modifying and controlling the stress by additional surface treatments were unknown. To answer the questions about stress levels and to develop an acceptable post-heat-treatment repair technique for use in the stress-relieved structures, a study of residual stresses commenced.

EXPERIMENTAL

The blind hole drilling method[6] was applied to laboratory samples in order to determine the residual stresses which exist after the various treatments. The method uses strain gages and is semidestructive, requiring drilling of a small shallow hole. This precludes its application directly in the tank. On the other hand, the method is applicable to nearly any surface and may be used wherever a strain gage rosette pattern can be attached and a hole can be drilled with precision.

Data were taken at 0.005-inch increments of metal removal. The depth of stress and the stress profile were determined by repeated application of the basic equations.[6] The subsurface stress distribution was determined because it was believed that this could be even more important than the surface stress.

The bulk of the samples were 10 inch x 10 inch plates, large enough to prevent planar relaxation in the samples for the particular situations being evaluated. A rectangular rosette strain gage pattern was used (Figure 1). Both smooth plate and artificially pitted plates (predrilled 1/8-inch holes, 1/8-inch deep) were studied. Most of the samples were made form 1/2-inch A-537 Class I carbon steel plate as used in tank fabrication. Heat treatment, grinding, shot peening, and grit blasting variations were evaluated.

The heat treatment employed was the same as used in the field - uniform heating and cooling at 100°F/hour to 1,100°F with a 1-hour hold at the higher temperature. Type C Almen intensity strips[7] were used to establish control and reliability for blasting in both the field and laboratory.

RESULTS AND DISCUSSION

AS-FABRICATED STRESS VALUES - Application of the drilling method on samples of the as-received steel plate indicated an average surface tensile stress of 5,500 psi (Table 1), with a subsurface stress distribution as shown in Figure 2. This is a typical distribution, as all samples which had undergone some form of working exhibited a peak stress beneath the surface substantially greater than that on the surface. The sample plate had been formed by cross-rolling at the mill.

Stress remaining in the plate after the stress relief heat treatment is also shown in Figure 2. The surface stress is only slightly less than that before heat treating. However, a significant reduction of stress beneath the surface occurs, resulting in the establishment of a nearly uniform distribution of less than 4,500 psi.

The anticipated residual stress is approximately 7,000 psi for 50,000 psi yield strength steel, given the prescribed heat treatment.[8]

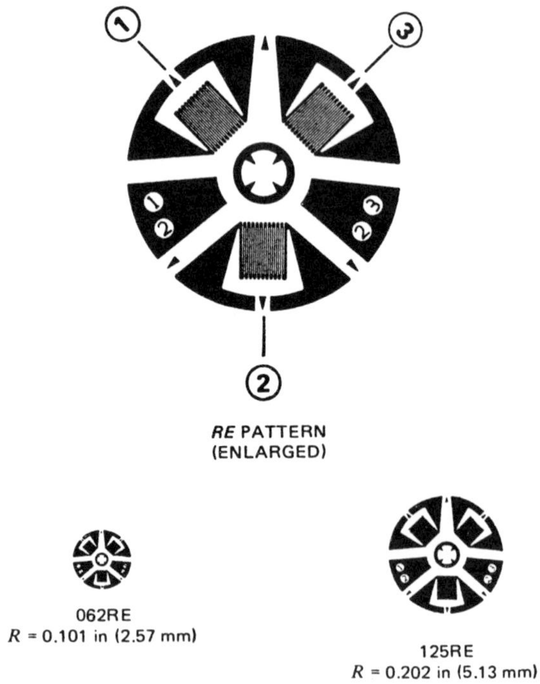

Fig. 1. Micro-measurements RE-pattern strain gage rosette for residual stress measurement (used in the present study)

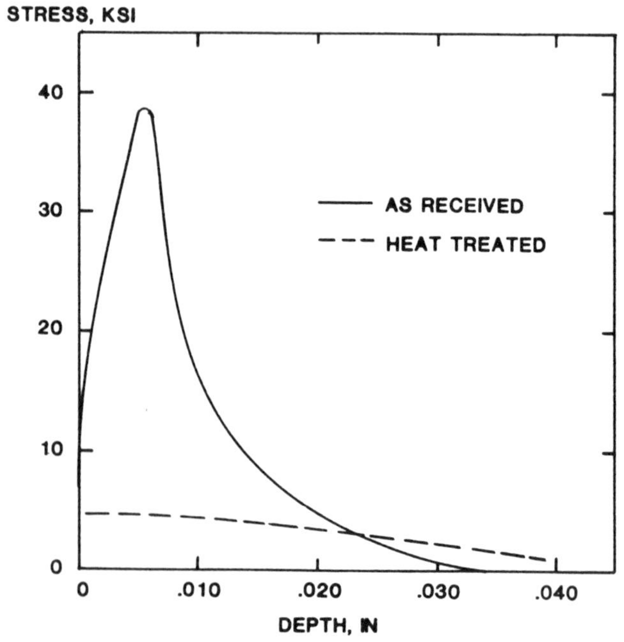

Fig. 2. Measured stress distribution in normalized and rolled ASTM A537 Class I steel plate before and after stress relief heat treatment

Table 1 - Residual Stresses in A537, Class 1 Carbon Steel Plate

Description	Average Surface Stress (psi)	Range of Maximum Stress, inches	Maximum Average Stress (psi)
As-Received	+5500	0.006	+38,000
Stress-Relieved	+4700	-	-
Class II Finish, Grit Blast	-1200	0.002 - 0.006	-53,000
Class I Finish, Grit Blast	-8000	0.002 - 0.006	-53,000
Class II Finish + Shot Peen	-10,000	0.002 - 0.006	-53,000
Abusively Ground	+19,000 to 29,000	0.002 - 0.010	+53,000
Abusive and Starr Blast	-5000 to +9000	0.002 - 0.006	-40,000 to -53,000
Abusive and Shot Peen	-5000	0.002 - 0.010	-53,000
Wet Grind (Nonabusive)	-2000 to +5000	0.005 - 0.006	+10,000 to +28,000
Wet Grind and Starr Blast	0 to -1000	0.002 - 0.005	-44,000
Wet Grind and Metal Grit Blast	-19,000	0.002 - 0.010	-53,000
Wet Grind and Shot Peen	-15,000 to -19,000	0.002 - 0.010	-53,000
Stone Grind/Spray Cool	+15,000	0.010	+38,000
Stone Grind + Shot Peen	-15,000	0.005	-53,000

Notes:

- Steel Grit: GL-40, 45-55 R_c
- Shot: Grade 230, 45-55 R_c; Blast: 200% Coverage
- Starr Blast: Silica-Zirconia Powder

STRESSES AND APPEARANCE AFTER GENERAL CLEANING - Effects of grit blasting are shown in Table 1 and Figure 3. The samples were prepared during actual cleaning operations and are representative of the tank bottom.

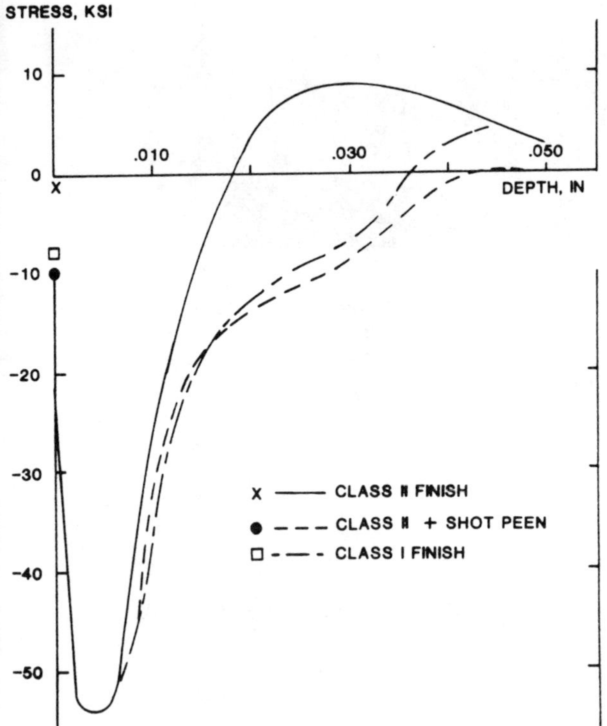

Fig. 3. Measured stress distributions in metal grit blasted steel plate

Surface stresses are compressive and range from -1,000 to -10,000 psi after cleaning. However, large compressive stresses are developed to significant depths by the blasting. Near-subsurface stress distributions for the commercial or Class II finish and for the "white" metal or Class I finish are similar, with yield strength being reached. The additional working of the metal beyond the Class II finish provided by Class I blasting or shot peening changes the stress distribution somewhat. Similar stress profiles have been reported for other materials after grit blasting or shot peening.[9]

Grit blasting is a metal removal technique as well as a cleaning method for metal surfaces. The erosion rate was not studied, but it will vary with the blast conditions. One data point obtained for zero degree impingement of the grit on the surface resulted in metal removal at a rate of 0.001 to 0.002 inch per second.

Grit blast cleaning was chosen because of its ability to remove scale. Photomicrographs in Figures 4 and 5 illustrate scale removal and surface roughening due to grit blasting. Care must be exercised to minimize development of surface flaws such as folds, entrapped grit, etc. The depth of subsurface plastic deformation due to grit blasting was observed to vary from 0.0005-inch to greater than 0.0040 inch (Figure 6). This correlates well with the stress measurements (Figure 3).

Grit blasting of pits roughens pit surfaces to the same extent as the plate surface (Figure 7). Compressive stresses in the pit bottoms will be similar to those measured for the plate. Moreover, edges will become rounded by the abrasive blasting, further reducing stress concentration effects. It is evident that most small flaws will not require removal after grit blasting.

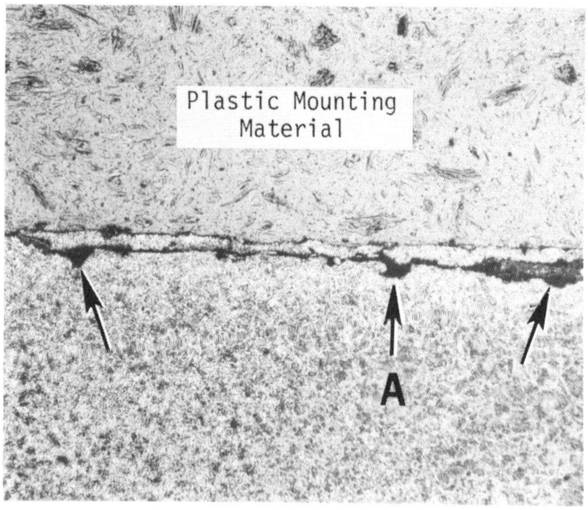

Fig. 4. Sectional view of carbon steel surface with iron oxide scale and other surface flaws. Magnification 50 X. The need for removing the outer layer is evident in this photomicrograph illustrating hidden incipient pits and crevices (arrows) beneath the outer layer. Point A is apt to be folded down during blasting.

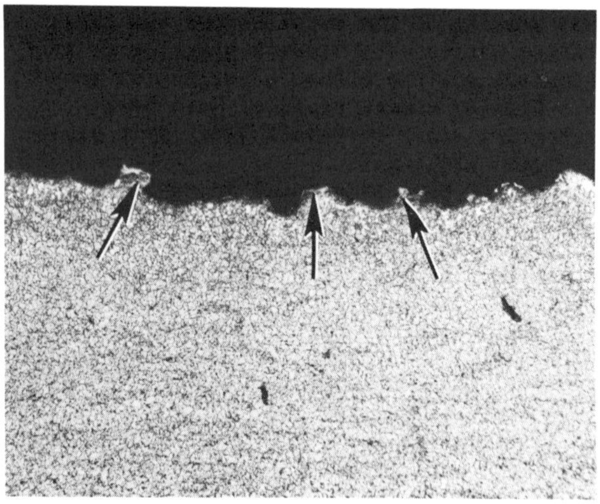

Fig. 5. Cross sectional view of grit blasted surface from which a thick (>0.005-inch), relatively non-adhering scale had been removed. Note the obvious surface roughness. Subsurface deformation was mild in this case. Marked points could fold over with additional blasting. Magnification 100 X.

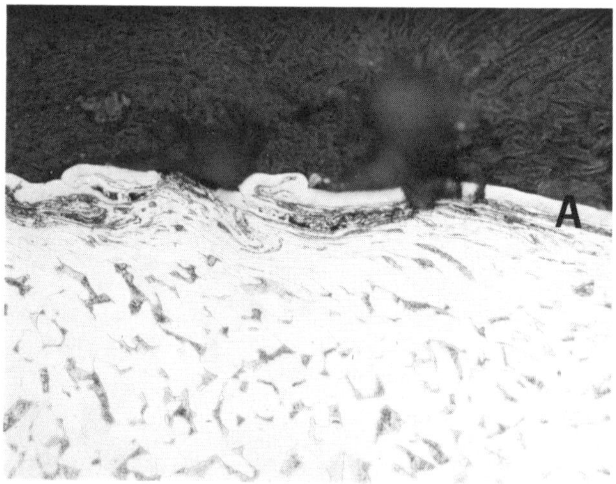

Fig. 6. Photomicrograph illustrating subsurface plastic deformation caused by grit blasting. Magnification 200 X. Nickel-plating was utilized for edge preservation. Depth of the distorted layer exceeds 0.004 inch in places.

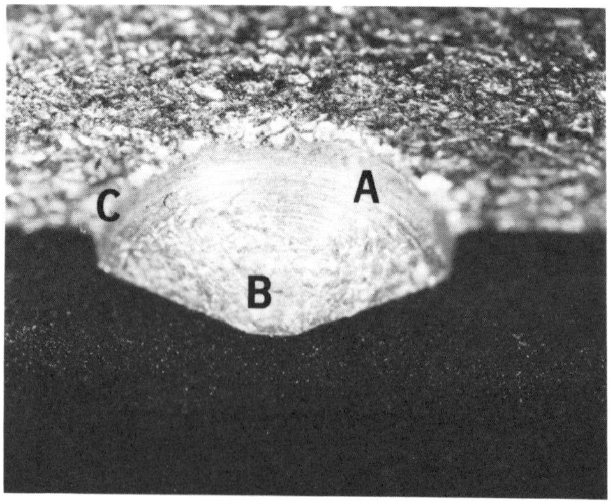

Fig. 7. Microscopic sectional view of drilled 1/8-inch pit (0.0125 inch deep) in grit blasted steel plate. Note drill marks (A) on vertical wall where blasting has not altered the surface. The bottom (B) of the pit has been roughened to the same extent as the top surface by the blasting operation. Also, the edge (C) of the pit has been blunted by the blasting.

REPAIR METHODS STUDY - Residual stresses resulting from welding preclude filling a pit or flaw by a welding method unless some form of post-weld heat treatment is used. The large tensile stresses which are retained in the weld heat-affected zones have led to stress- corrosion cracking during service.

Stress relief heat treatment is designed to reduce residual stresses to the 4,000 to 7,000 psi level discussed previously. After the stress relief heat treatment, care is used to avoid any localized stressing of the tank walls and bottom.

Alternative repair methods include grinding of the pits to reduce their stress intensifying character. Basically this means enlarging the diameter without increasing the depth. The stress concentration factor and the stress intensity vary with pit geometry. Deeper, narrower pits produce greater concentrations of stress and result in lower allowable loads. Where deep pits occur, it should be beneficial to widen the affected area. However, the decision to repair by grinding to increase the aspect ratio of a pit must be made on an individual basis.

Residual stresses for a harshly ground steel surface are contained in Table 1, and a typical average stress profile is shown in Figure 8. Surface temperatures exceeded 400°F during grinding. Surface stress is high, +29,000 psi, and yielding occurs deep in the material. The advantage of shot peening the ground surface is also shown in Figure 8, where a compressive surface stress and subsurface deformation are in evidence.

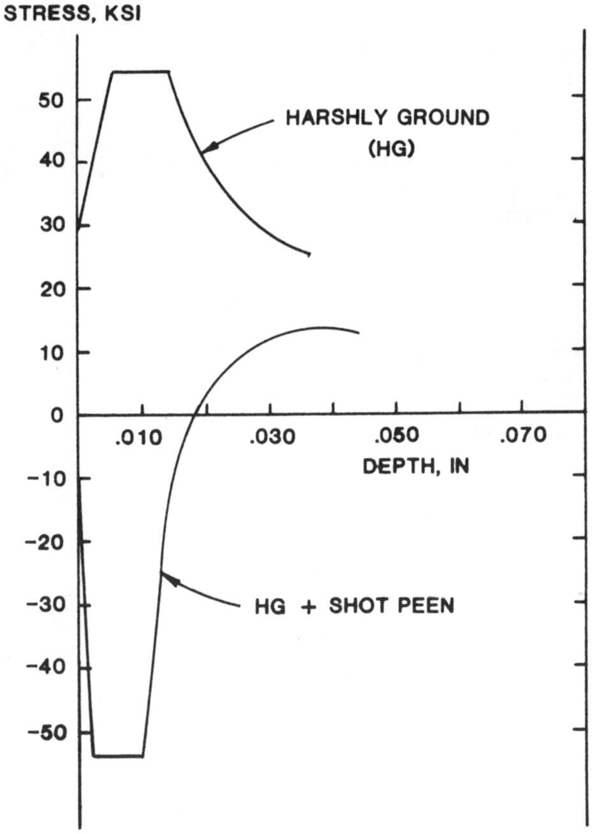

Fig. 8. Measured stress distributions in steel plate after abusive grinding and after shot peening

Other methods of grinding were evaluated. Wet-grinding, a slow non-abusive method which resulted in a very little surface temperature change, nevertheless caused an increase in subsurface tensile stress to about +29,000 psi (Figure 9). Subsequent blast finishing provided substantial compressive stresses and completely suppressed the tensile stresses resulting from grinding (Figure 9). Note that metal grit blasting induces a stress profile similar to that resulting from shot peening.

A more rapid method of nonabusive grinding utilizes an oxyacetylene nozzle for air/water spray cooling. With a coarse stone wheel, temperatures were maintained below 100°F.

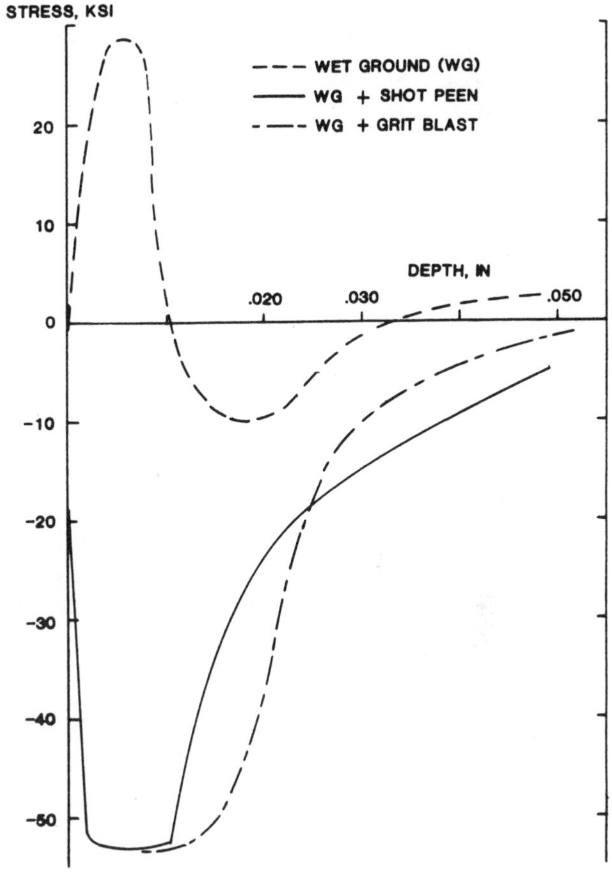

Fig. 9. A typical stress profile in a plate section following relatively gentle grinding. Effects of subsequent grit blasting or shot peening are also illustrated.

The spray cooling, stone grinding method was applied to samples which had not been heat treated (Table 1). An increase in surface stress to +15,000 psi tensile occurred, but the subsurface maximum stress was unchanged at +38,000 psi. Thus, it appears that this method allows rapid grinding without altering subsurface stresses significantly. Shot peening of the spray cooled/stone ground steel resulted in average compressive stresses of

-15,000 psi on the surface with a stress distribution similar to those noted in Figures 3, 8, and 9.

It is evident that metal grit blasting of the carbon steel surfaces yields residual stresses similar to those obtained by shot peening. Some variations can be expected with shot size or grit size, or if pressure or angle at the nozzle is changed.[4] However, metal grit blasting will suffice for post-grind finishing, and shot blasting is unnecessary. Almen intensities for both peening and blasting operations were 4-5C.

In evaluating the various repair methods, grinding, grit blasting, and shot peening were always applied over an area about 4 inches in diameter centered on a simulated 1/8-inch pit (1/8-inch drill bit, 1/8-inch deep). The outer part of the 4-inch area was a transition region receiving some metal work but not to the same depth or extent as at the center pit. Measurements in the transition zone indicate stresses intermediate between the center and the area outside the 4-inch zone.

The repair procedure recommended for removal of any flaw in the heat-treated structures is contained in Table 2. When applying the technique, it is important to create a gradual finish transition to avoid sudden changes of stress in the material. Thus, any ground area should be feathered. The grit blast finish should also be feathered.

CONCLUSION

Cleaning and repair of flaws in a stress-relieved structure should be finished by metal grit blasting. This will leave the surfaces in a state of compressive residual stress which is highly resistant to SCC. Stresses beneath the surface will also be compressive to substantial depths, and can approach the yield strength in the material. Consideration of grit blasting as a normal means of improving resistance to SCC is indicated.

ACKNOWLEDGEMENT

The information contained in this article was developed during the course of work under Contract No. DE-AC09-76SR00001 with the U.S. Department of Energy.

Table 2 - Post Heat Treatment Repair Procedure for Pits or Other Flaws in Stress Relieved Structures

The procedure recommended for modification or removal of pits, gouges, or other flaws includes spray cool wet-grinding to "dish-out" an area of approximately 4 inches diameter followed by grit blasting to finish the ground surface.

Wet Grind

- Coarse No. 24 stone wheel, air-driven tool.
- Oxyacetylene torch nozzle with compressed air and cold water feed lines attached.

Adjust the spray to provide an atomized water flow with water accumulation kept at low amounts. Apply spray continually to the surface being ground. Contain the water (puddle) and occasionally wipe-up. Wipe dry when grinding is complete.

Feather the grinding toward the outer regions of the 4-inch area.

Finishing

- GL40 steel grit, R_c 45-55.

Apply a Class I finish in the center of the ground area and feather the grit blast beyond the 4-inch area. The Almen intensity in the center should be 4+ to 6 using "C"-type test strips.

REFERENCES

1. L. P. Costas, M. L. Holzworth, and W. C. Rion. Stress Corrosion Cracking of Carbon Steel in Simulated Waste Solution. DP-1023, E. I. du Pont de Nemours and Company, Savannah River Laboratory, Aiken, SC (1966).

2. M. L. Holzworth, R. M. Girdler, L. P. Costas, and W. C. Rion. "How to Prevent Stress Corrosion Cracking of Radioactive Waste Storage Tanks." Material Performance, 7, 36 (1968).

3. R. S. Ondrejcin, S. P. Rideout, J. A. Donovan. "Control of Stress Corrosion Cracking in Storage Tanks Containing Radioactive Waste." Nuclear Technology, 44, 297-306 (1979).

4. S. S. Birley and A. Owens. "Surface Stresses Induced by Grit Blasting." Anti-Corrosion (July 1979).

5. J. A. Donovan. Material Aspects of SRP Waste Storage - Corrosion and Mechanical Failure. DP-1476, E. I. du Pont de Nemours and Company, Savannah River Laboratory, Aiken, SC (November 1977).

6. "Measurement of Residual Stresses by the Blind; Hole Drilling Method." Photolastic Division, Vishay Technical Data Bulletin T-403.

7. Shot Peening Applications. Metal Improvement Company, Inc. (1980).

8. AWS Welding Handbook, 7th ed, Vol 1, p 272. American Welding Society (1976).

9. S. S. Birley, M. J. Morton, and J. F. Alder. "Residual Stresses in Welded Joints and the Effects of Mechanical Surface Treatments," Residual Stresses in Welded Construction and Their Effects. International Conference, London, England (November 1977).

THE MICROSTRUCTURE OF 55 W/O Al-Zn-Si (GALVAUME) HOT-DIP COATINGS

J. H. Selverian and M. R. Notis
Lehigh University Bethlehem, PA, USA

A. R. Marder
Bethlehem Steel Corporation
Bethlehem, PA, USA

Abstract

Light metallography, x-ray diffraction, SEM, EMPA, and STEM were used to study the microstructural characteristics of 55 w/o Al-Zn-Si (Galvalume) hot-dip coatings on steel. Processing variables of coating thickness and dip time were studied and a thin foil sample preparation technique involving ion beam thinning for the STEM evaluation of coating cross-sections was developed. The spangled coating surface was comprised of a fine dendritic network, and correlations between dendrite arm spacing and spangle size with coating thickness have been made. The overlay, or solidified bath, consists of grains of Al-rich dendrites along with Zn-rich interdendritic zones. EMPA in conjunction with TEM imaging showed that these Zn-rich interdendritic regions had a heavy concentration of precipitates. Selective etching of the overlay revealed a forest of Si particles which grew out from the alloy layer. The alloy layer was studied using the EMPA and STEM and it was found that up to five intermetallic phases were present after long dip times.

GALVALUME SHEET* coating is a hot dipped coating applied to steel for improved atmospheric corrosion. The commercial coating consists of nominally 55 Al, 43.4 Zn and 1.6 Si (all in w/o). Silicon is added to the bath to prevent a very strong and rapid exothermic reaction between the Al-Zn bath and the steel sheet[1]. During the coating process a diffusion layer forms between the bath and the steel. This region is called the alloy layer, and consists of Fe-Al and Fe-Al-Si compounds[2,3]. Another feature of the Galvalume coating is that Si particles are formed in the overlay and result from the supersaturation of the Al-Zn bath upon cooling.

The studies of Cleary[4] and Zoccola et al[5] concentrated on the corrosion behavior of the coating, while Hart[3] and Friel[2] completed some of the more recent work on determining the composition of the alloy layer. Cleary found that as the secondary dendrite arm spacing (DAS) and the Al dendrite size decreased and the density of the Si particles increased, the corrosion resistance increased. Zoccola et al investigated the atmospheric corrosion behavior of a wide range of Al-Zn alloys, which led to an Al/Zn ratio in Galvalume of 1.2. Hart studied the constitution of the alloy layer in a rapidly cooled commercial panel using Auger electron microscopy, and found that the approximate composition was 55.8% Al, 33.7% Fe, 6.7% Si, 3.8% Zn. Friel found two alloy layers in a commercial panel using EMPA, where the compositions were 53.1% Al, 25% Fe, 8.9% Si, 13% Zn and 56.6% Al, 36.7% Fe, 3% Si, 3.7% Zn.

EXPERIMENTAL PROCEDURE

Three bath compositions were used, 55% Al with 0.7, 1.7, or 6.88% Si (balance Zn). All baths were saturated with Fe prior to dipping. The temperature of the baths was held at 610°C. The panels were degreased, pickled, fluxed, and dipped for the following times: 4, 25, 100, 256, 400, 900, and 1800 seconds. Following dipping, the panels were water quenched to freeze in the structure of the alloy layer as it existed at temperature. In order to determine the effect of coating thickness on structure, preheated panels were dipped in a 610°C bath for 2.2 seconds, wiped with air knives to obtain coatings of nomnially 2, 13, 22, and 41 microns, and cooled at a rate of 22°C/sec. Standard microstructural techniques were used for preparation of the samples. DAS measurements were made according to previous techniques[4]. Spangle size was measured on the coated sheet surface at a very low magnification by the intercept method. The average alpha Al dendrite cross section, alpha size , was also measured by the intercept method. In this case, the number of

*GALVALUME SHEET is a registered trademark of the Bethlehem Steel Corporation for steel sheet coated with 55% Al-Zn alloy.

boundary intercepts on a line perpendicular to the coating/substrate interface was measured and fifty lines were used for each sample.

A JEOL 733 microprobe was used for quantitative chemical analysis. The operating conditions were an accelerating voltage of 18 kV and a beam current of 10 nA. X-ray diffraction using a Diano 8535A diffractometer and Cr-K_{α} radiation was also used for phase identification. A sample preparation technique for viewing the coating on thin sheet was developed, and is described elsewhere[6]. Table 1 outlines the essential steps in this technique.

1. Cut panel into 3mm X 20mm strips.
2. Glue strips together to from a sandwich.
3. Mount the sandwich vertically in a metal tube with an epoxy mounting compound.
4. Cut 300 micron slices from the mount on a diamond saw.
5. Grind individual slices down to 50-75 microns on SiC paper.
6. Separate the individual strips and mount them on a slotted TEM grid.
7. Ion beam thin to electron transparency.

Table 1. Essential steps in preparing STEM specimens from thin coated sheet Figure 1. Macrophoto of a panel of commercially produced Galvalume sheet.

RESULTS AND DISCUSSION

MICROSTRUCTURAL CHARACTERISTICS - A macrophoto of a commercially produced Galvalume panel is shown in Figure 1. The obvious visible feature of this panel is its spangled appearence. The spangles consist of Al-rich dendrites and are analogous to grains. It was found that the spangle size is a direct function of the coating thickness, Figure 2. Another characteristic of these coatings is the dendrite arm spacing (DAS), shown in Figure 3. The secondary DAS is plotted as a function of coating thickness in Figure 4. In all cases increased coating thickness increased spangle size and DAS.

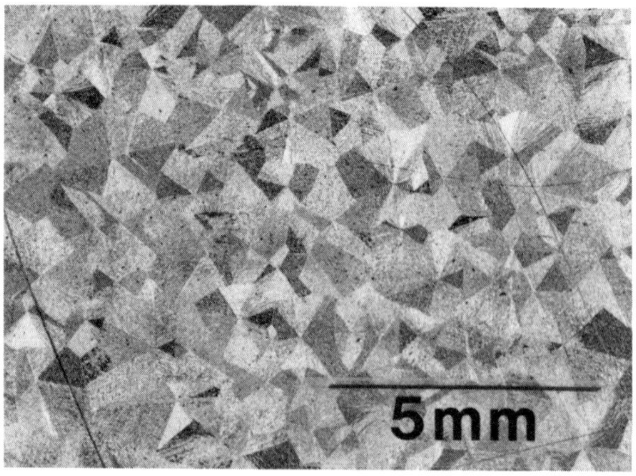

Figure 1. Macrophoto of a panel of commercial Galvalume showing spangled appearance.

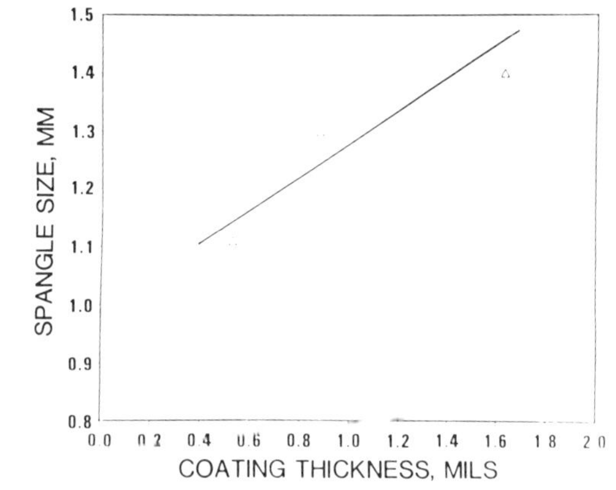

Figure 2. Plot of spangle size verses coating thickness.

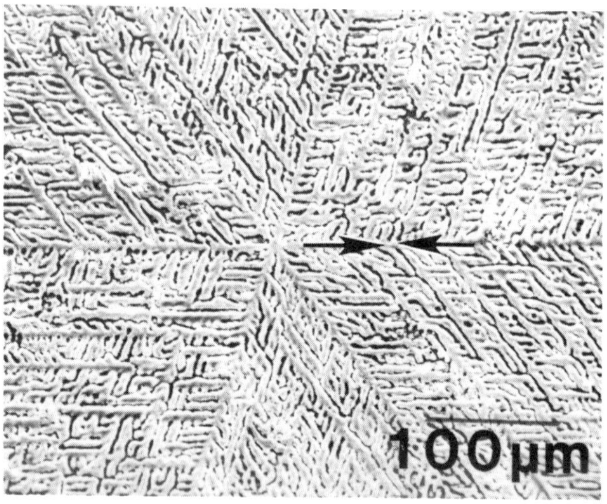

Figure 3. SEM micrograph showing the center of an Al dendrite. The DAS is defined as the length between the two arrows,approximately 12 microns.

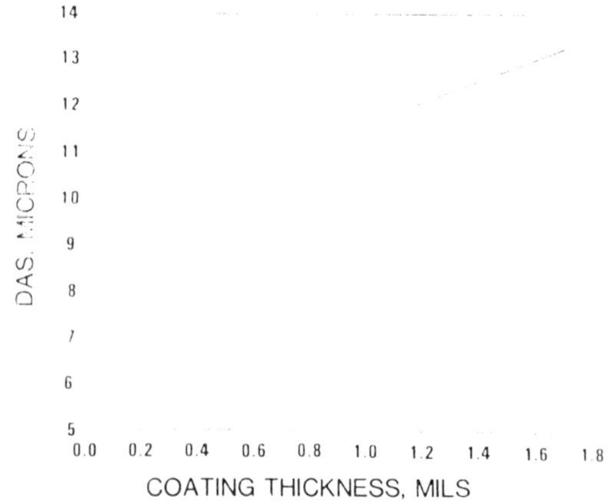

Figure 4. Plot of secondary dendrite arm spacing (DAS) verses coating thickness.

A micrograph of the cross-section of a commercial panel is shown in Figure 5. Three features of the coating are defined in this figure: the alpha Al dendrite size (A), the Zn-rich interdendritic regions (B), and the Si particles (C). These three features define the part of the coating referred to as the overlay (or solidified bath).

The effect of coating thickness on the alpha Al dendrite size is described in Figure 6. The alpha Al dendrite size decreases slowly in the range of coating thicknesses from 40 to approximately 8 microns. In this range the main effect of the coating thickness is to increase the cooling rate. Faster cooling rates increase supercooling of the liquid, and more dendrite nuclei form, thus decreasing the alpha Al dendrite size. It should be noted that the alpha Al dendrites cannot be larger than the coating thickness. Once the coating thickness is the same as the alpha Al dendrite size (dotted line in Figure 6), the size of the alpha Al dendrites are determined by the coating thickness. Therefore, at larger coating thicknesses the cooling rate is the controlling factor in determining the alpha Al dendrite size for a given panel, however at small thicknesses, $<$ 8 microns, the alpha Al dendrite size is determined by the coating thickness.

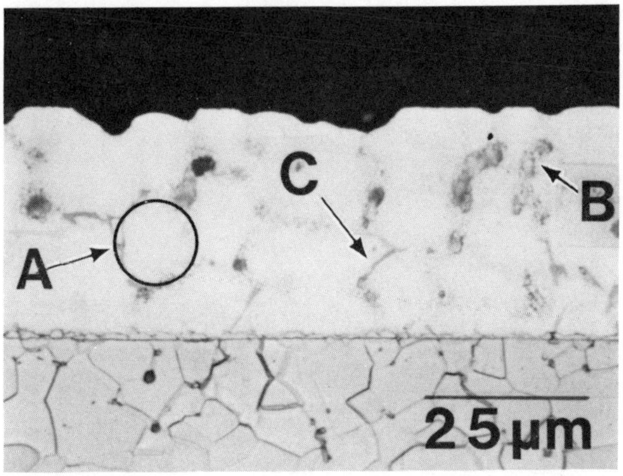

Figure 5. Microstructure of a commercial panel. A defines the alpha Al dendrite size, B is the Zn-rich interdendritic regions, and C is a Si particle, 1% nital etch.

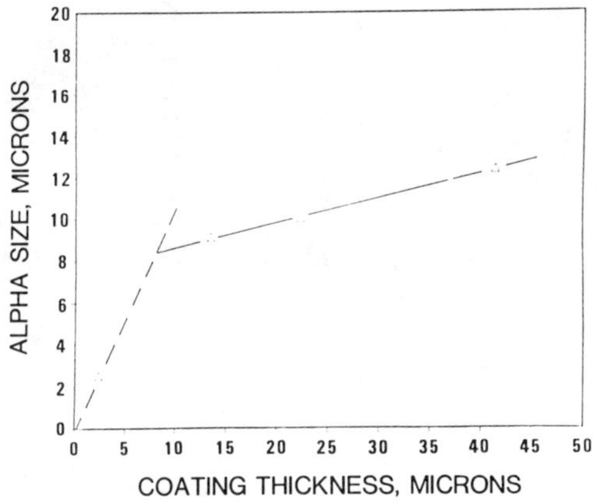

Figure 6. Plot of alpha Al dendrite size verses coating thickness. Below a coating thickness of approximately 8 microns the alpha Al dendrite size equals the coating thickness.

Cleary[4] studied the effects of cooling rate on the DAS. His results show that as the cooling rate increased, the DAS decreased, Figure 7. Concurrent with a decrease in DAS, Cleary noted that the density of Si particles in the overlay increased, Figure 8. It appears that by increasing the cooling rate, more Si particles are formed which then act to refine the structure by either physically constraining the growth of the alpha Al dendrites, or by providing numerous heterogenous nucleation sites for the dendrites.

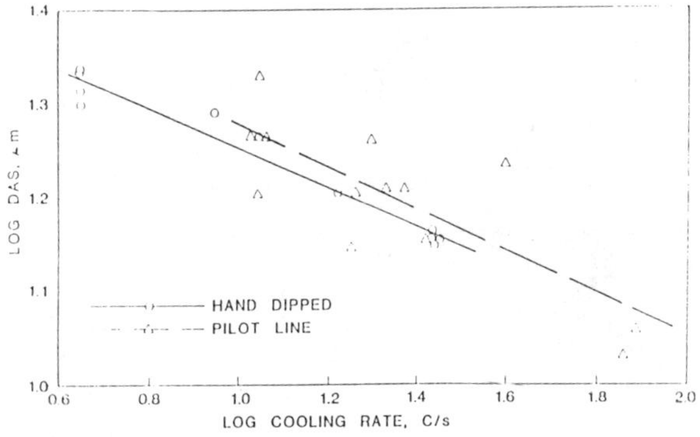

Figure 7. Plot of log DAS versus log cooling rate (ref. 4).

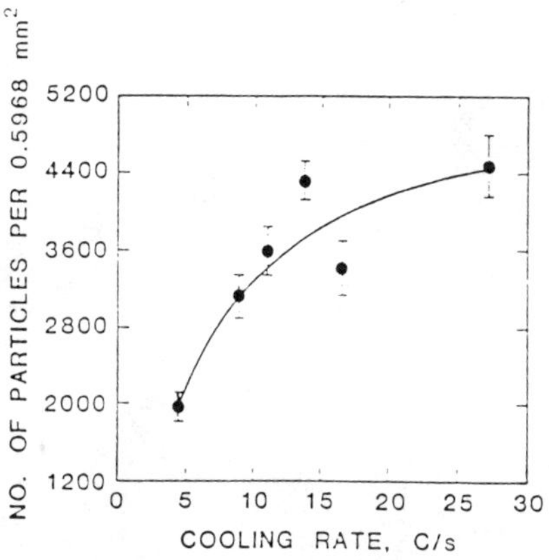

Figure 8. Plot of Si particles per area versus cooling rate (ref. 4).

A comparision of panels with identical coating thickness and cooling rates but different Si contents shows the effect of Si in refining the microstructure. Figure 9a shows a panel dipped in a 0.7% Si bath and then water quenched: Figure 9b shows a panel dipped in a 6.88% Si bath and then water quenched. In both the 0.7% Si bath (Figure 9a) and the 6.88% Si bath (Figure 9b) the coating thickness of the panels is approximately 45 microns. It is apparent from these figures that the panel with the higher Si content has a finer structure than the low Si panel. The morphologies of the Si particles in these two panels are shown in Figures 10a and b after selectively etching away the Al dendrites and Zn-rich interdendritic regions. It is obvious from these figures that the number of Si particles increases as the Si content of the bath increases. The Si particles appear to nucleate at the alloy layer/bath interface and grow into the bath.

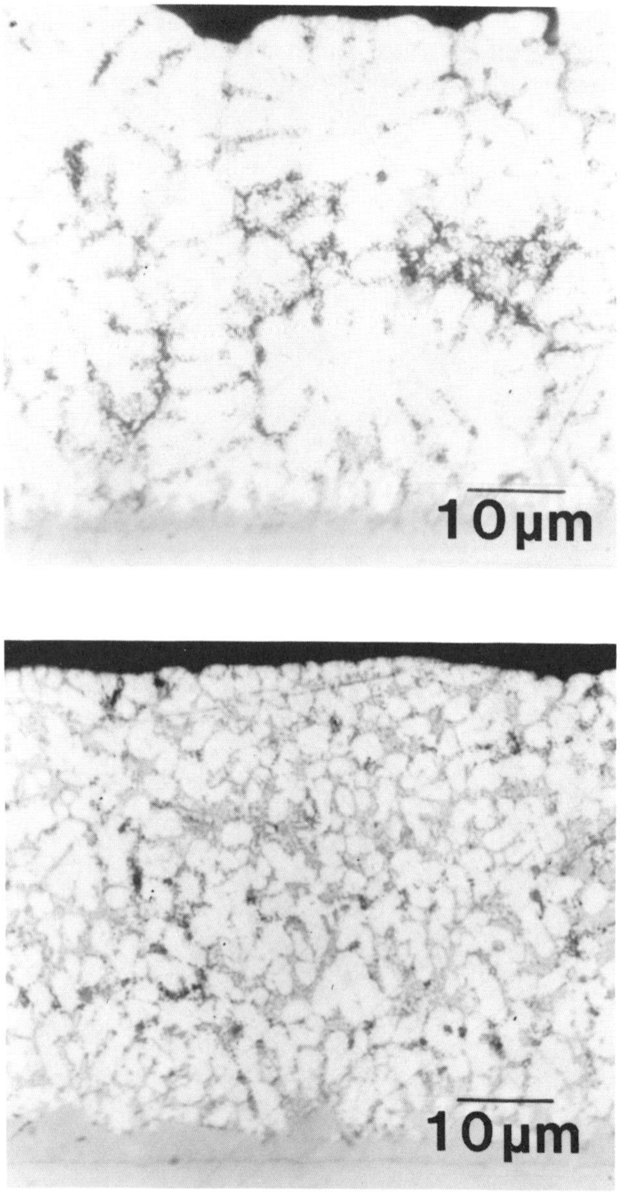

Figure 9. a) Overlay of a water quenched panel dipped in the 0.7% Si bath for 4 seconds, b) same as a) except dipped in a 6.88% Si bath.

The effects of the different variables on the microstructure of the overlay can be summarized as follows. The faster the cooling rate: the smaller the alpha Al dendrite size, the smaller the DAS, and the higher the Si particle number density. Also, as the Si content of the bath is increased a refinement of the overlay occurs, i.e., the alpha Al dendrite size decreases.

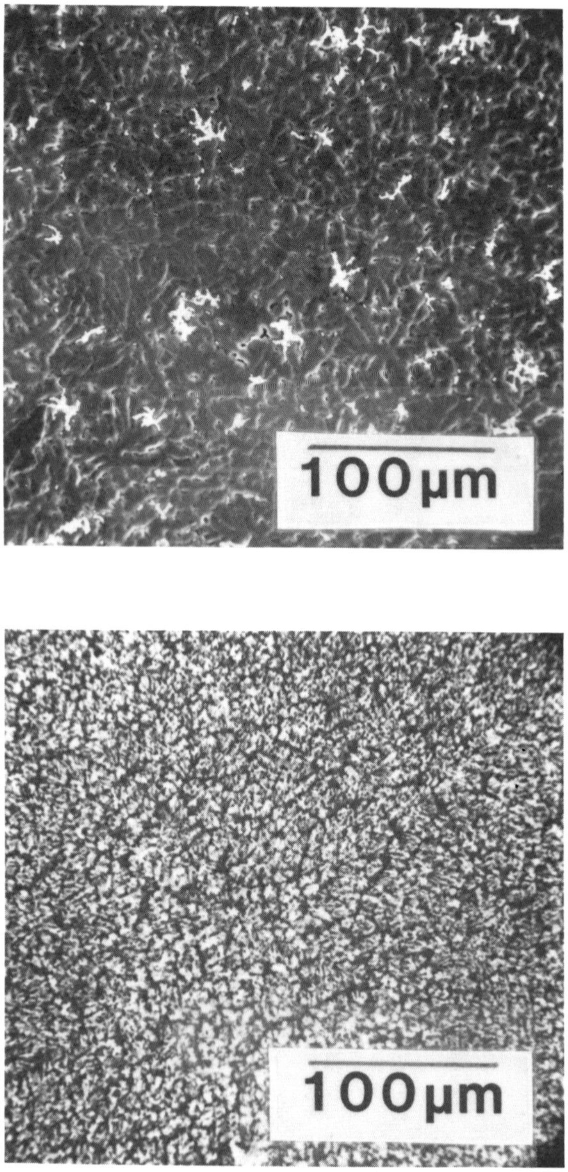

Figure 10. a) Si particles in the panel from Figure 9a, b) Si particles in the panel from Figure 9b, Electrolytically etched in 20 V/O perchloric - ethanol at 30 V. (SEM micrographs)

STEM INVESTIGATION - The results from a STEM-EDS x-ray trace across an Al dendrite from a panel dipped in the 1.7% Si bath for 4 seconds, are shown in Table 2. The scan was taken from one Zn-rich interdendritic region, through the Al dendrite, to another Zn-rich interdendritic region, Figure 11. The results show that the core of the Al dendrite consists of approximately 80% Al and 18% Zn. An interesting feature is that a Si gradient exists across the dendrite, which varies from approximately 1.8% Si in the interdendritic regions to 0% Si in the center of the dendrite.

Point	Compositions (w/o) Al	Fe	Si	Zn
1	56.7	3.5	1.0	38.7
2	77.8	2.7	1.8	17.7
3	79.7	2.2	0.9	17.2
4	79.7	2.1	0.4	17.7
5	80.5	2.0	0.0	17.5
6	79.5	2.7	0.0	17.8
7	79.3	2.8	0.0	17.8
8	77.9	2.6	1.5	18.0
9	75.8	2.6	1.5	20.0
10	73.8	2.5	1.3	22.3
11	68.8	2.8	1.7	26.6
12	64.6	3.1	1.0	31.3
% relative error	±6.5	±3.0	±6.0	±3.0

Table 2. Results of scan through Al dendrite.

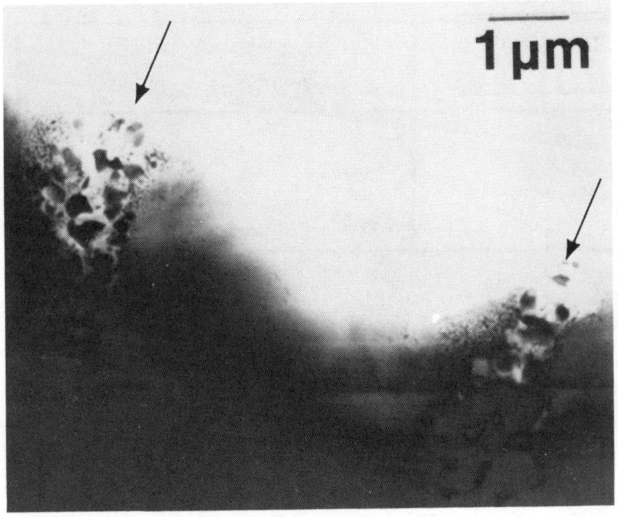

Figure 11. An Al dendrite in the 1.7% Si bath. The arrows define the Zn-rich interdendritic regions. (STEM micrograph)

The composition of the Al dendrite compares favorably to the composition expected from the binary Al-Zn equilibrium phase diagram[7]. The composition of the first solid to solidify at 610°C is approximately 85% Al and 15% Zn. Also, the maximum solid solubility of Si in Al is approximately 1.6% Si[8] which compares favorably to the experimental value of 1.8% Si.

The Zn-rich interdendritic region is heavily populated by precipitates, Figure 12. The large precipitates in the center of the figure are most likely the metastable R phase[7] and the smaller precipitates are probably coarsened Guinier-Preston (G.P.) zones. Figure 13 shows a twinned Si particle, the twin can be seen as a band running through the Si particle. These twins are evidence of the twin reentrant corner growth mechanism[9], which occurs at high growth rates.

Figure 14 a-c show the microstructures representative of the 4, 900, and 1800 second dip times in the 1.7% Si bath, respectively. The alloy layer in the panel dipped for 4 seconds was studied in STEM, Figure 15. The average composition of this layer, obtained in STEM, is 52.2 ± 5% Al, 31.1 ±1% Fe, 10.8 ± 1% Si and 5.5 ± 0.5%

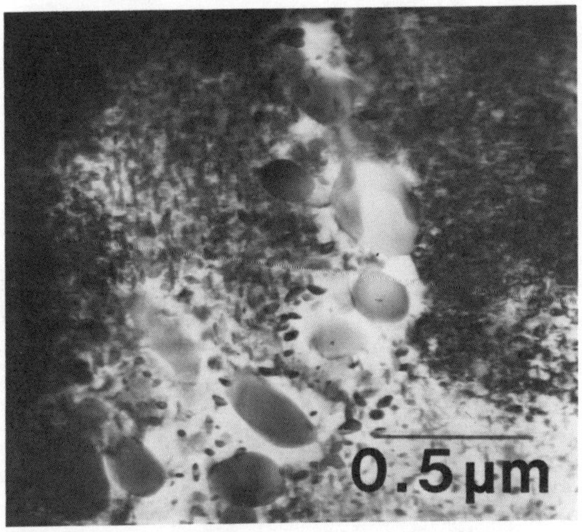

Figure 12. The Zn-rich region in the 1.7% Si bath. The large precipitates are the metastable R phase and the smaller precipitates are probably coarsened G.P. zones. (TEM micrograph).

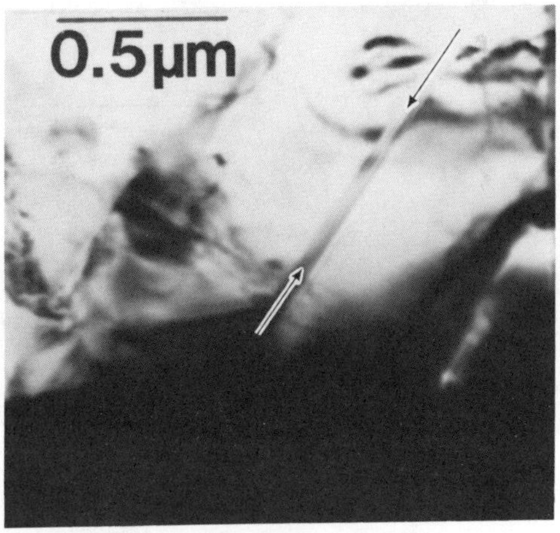

Figure 13. Twins can be seen in the Si particle as a dark band (arrows). (TEM micrograph)

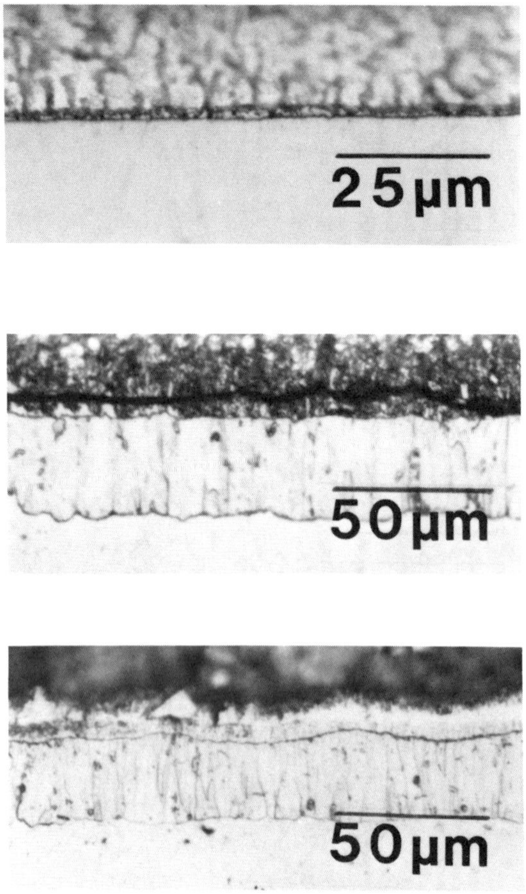

Figure 14. a) Microstructure of the panel dipped for 4 seconds in the 1.7% Si bath, b) dipped for 900 seconds, c) dipped for 1800 seconds. Keller's etch.

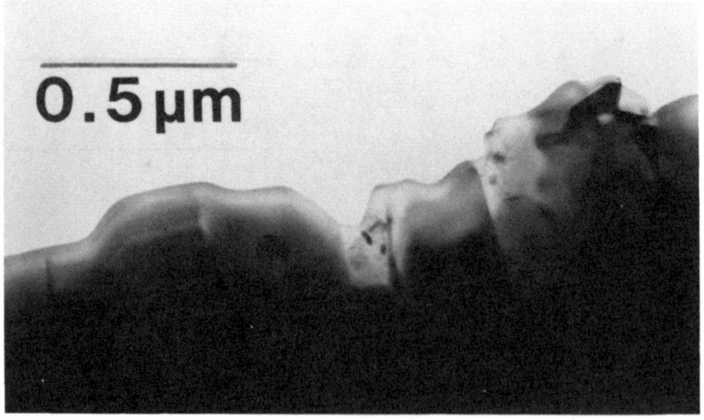

Figure 15. Microstructure of the alloy layer in the 1.7% Si bath. The black region is the Fe panel, The gray region is the alloy layer (T_{5C} phase), and the small precipitates are probably G.P. zones in the overlay which exist in the discontinuities in the alloy layer. (TEM micrograph).

Zn. The alloy layers have been identified as a function of dip time by x-ray diffraction and EMPA. The results of these investigations are summarized in Table 3. The phases present in the panel dipped for 1800 seconds are Fe_2Al_5, T_1, $FeAl_3$, T_{5C}, and T_{5H}, where T_{5C} is the cubic form of T_5 and T_{5H} is the hexagonal form of T_5, and the compositions of these phases determined in this study are given in Table 4. These phase designations refer to those used by Rivlin and Raynor[10]. The diffusion path for the 11000 second panel is shown on the Fe-Al-Si ternary phase diagram in Figure 16. This diagram is an approximation, because the presence of Zn has been ignored and the compositions of the Al, Fe, and Si were normalized to 100%. The diffusion path traces the following sequence:

1. Fe panel
2. Fe_2Al_5: $Fe_2Al_5 + T_1$
3. $FeAl_3$; $FeAl_3 + T_1$
4. $T_{5C} + T_{5H}$
5. Al-Zn-Si bath

It is believed that the T_1 phase precipitated out within the Fe_2Al_5 and $FeAl_3$ phases during cooling.

Dip time (s)	Phases present in the 1.7 w/o Si bath
4	T_{5C}
900	Fe_2Al_5, T_1, $FeAl_3$, T_{5C}
1800	Fe_2Al_5, T_1, $FeAl_3$, T_{5C}, T_{5H}

Table 3. Phases present in the alloy layer of the 1.7% Si bath as a function of time.

Phase	Composition (w/o)			
	Al	Fe	Si	Zn
Fe_2Al_5	51.26	43.79	2.92	2.03
$FeAl_3$	57.94	38.18	2.26	1.67
T_1	25.14	54.65	20.02	0.19
T_{5H}	57.66	33.06	8.50	1.71
T_{5C}	58.21	29.73	6.70	6.96

Table 4. Compositions of phases found in this study.

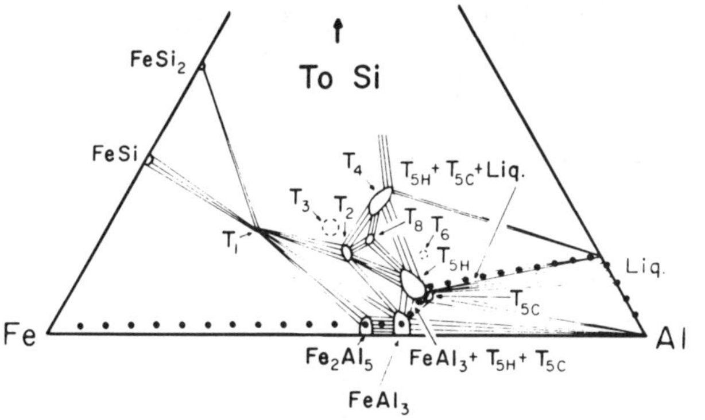

Figure 16. Diffusion path in the panel dipped for 1800 seconds in the 1.7% Si bath.

The results of the STEM investigation of the panel dipped for 4 seconds in the 1.7 w/o Si bath can be summarized as follows. The alpha Al dendrite shows a composition of approximately 80 w/o Al and 18 w/o Zn at the center. A Si gradient exists across the dendrite which goes from 0 w/o at the center to approximately 1.8 w/o in the Zn-zich interdendritic regions. The alloy layer consists of the T_{5C} phase (52.2 $\pm$ 5% Al, 31.1 $\pm$ 1% Fe, 10.8 $\pm$ 1% Si, and 5.5 $\pm$ 0.5% Zn). The Zn-rich interdentric regions were heavily populated by coarsened G.P. zones and precipitates of the metastable R phase.

CONCLUSIONS

1. The Si particles present in the overlay are responsible for the refinement of the microstructure. The refinement is caused by an increased number of Si particles which physically constrain the Al dendrite growth and/or act as heterogenous sites for dendrite nucleation.
2. The overlay consists of Al dendrites which contain approximately 18% Zn and 0-1.8% Si in solid solution. The composition of the dendrite as determined by STEM is in good agreement with the binary Al-Zn phase diagram. These dendrites are surrounded by Zn-rich interdendritic regions which have a high concentration of Zn-rich precipitates. Also found in these regions are Si particles which are twinned and are composed of essentially pure Si.
3. The alloy layer in a water quenched panel dipped for 4 seconds in the 1.7% Si bath consists of the T_{5C} phase (52.2% Al, 31.3% Fe, 10.8% Si and 5.5% Zn). In a panel dipped for 1800 seconds in the 1.7% Si bath the following layer sequence is observed;

- Fe panel
- Fe_2Al_5 ; $Fe_2Al_5 + T_1$
- $FeAl_3$; $FeAl_3 + T_1$
- $T_{5C} + T_{5H}$
- **Al-Zn-Si bath**

ACKNOWLEDGMENT

We would like to thank the Bethlehem Steel Corperation for their support of this project and to the many people who provided help along the way.

REFERENCES

1. Borzillo, A.R., Horton, J.B., U.S. patent # 3343930, September 26, 1967
2. Friel, J., Homer Research Labs, Bethlehem Steel Corp., Private Communication
3. Hart,R., Homer Research Labs, Bethlehem Steel Corp., Private Communication
4. Cleary, H.J., The Microstructure and Corrosion Resistance of 55%Al-Zn Coatings on Sheet Steel,Microstructural Science v. 12 Northwood, D.O., White, W.E., Van der Voort, G.F.,ASM, Metals Park, OH., p. 103,(1985)
5. Zoccola, J.C.,Townsend, H.E.,Borzillo, A.R., Horton, J.B., Atmospheric Corrosion Behavior of Al-Zn Alloy Coated Steel, Atmospheric Factors Affecting the Corrosion of Engineering Materials, ASTM STP 646, Coburn, S.K., Ed., ASTM, p. 165,(1978)
6. Selverian, J.S.,Notis, M.R., Marder, A.R., To be submitted to Metallography
7. Murray, J.L., The Al-Zn System,Bull. Alloy Phase Diagrams, Vol. 4, No. 1, p. 55,(1983)
8. Murray, J.L., McAlister, A.J., The Al-Si System,Bull. of Alloy Phase Diagrams, Vol. 5, No. 1,p. 74,(1984)
9. Hamilton, D.R., Seidensticker, R.G., Propagation Mechanism of Germanium Dendrites,J. Appl. Phy., Vol. 31, No. 7,p. 1165, 1960
10. Rivlin, V.G.,Raynor, G.V.,Critical Evaluation of Constitution of Al-Fe-Si System,Inter. Metals Reviews, No. 3,p. 133,(1981)

THE WEAR OF STAINLESS STEEL AND ITS RELATION TO SURFACE MODIFICATION

H. Hong and Robert F. Hochman
School of Materials Engineering
Georgia Institute of Technology
Atlanta, GA 30332-0245, USA

T.F.J. Quinn
Division of Engineering, Sci. & Mater.
United States International University
Europe Campus
The Avenue, Bushey
Herts. WD2 2LN, England

ABSTRACT

The wear and friction behavior of unmodified, TiN plated, and nitrogen implanted 316 stainless steels were investigated using pin-on-disk machine. The wear tests were carried out at relatively slow sliding speeds without lubrication.

The wear behavior of unmodified 316 stainless steel showed the lack of any load transition which was observed at high sliding speeds. TiN plated and N implanted 316 stainless steels showed an improvement of wear resistance.

THERE HAS BEEN several researchers [1-4] who reported that oxidation plays an important role during the wear process of steels. However, 316 stainless steels show transitional behavior from mild wear (mainly oxidational wear) to mixed wear. Some previous work [4] indicated that a transition occurs around 10 Newtons at sliding velocities of 3.0 m/s and 3.5 m/s. In this study, relatively low sliding speeds were used to study the effect of sliding speed on the transition load. In addition, TiN ion plating and nitrogen ion implantation were carried out in an attempt to overcome the poor wear resistance of 316 stainless steels during sliding. TiN has been used as a hard coating on hardened substrate steels to improve their wear properties [5-10]. However, some researchers [11,12] have reported the feasibility of hard coatings on a soft substrate and this reason was carried out with the objective of studying whether or not TiN could be used on relatively soft 316 stainless steels. Nitrogen implantation has also been used as a wear reducing method for many steels [13-17]. However, previous researchers [13-16] have been limited to mainly lubricated conditions or extremely low loads. In this study, the effect of nitrogen implantation on 316 stainless steel was investigated under dry wear.

EXPERIMENTAL PROCEDURES

The wear tests were carried out at sliding speeds of 0.25 m/s and 0.92 m/s using pin-on-disk testing equipment. The pin-on-disk machine used has been detailed elsewhere [18]. The load is applied with dead weights and the friction force at the pin was measured continuously throughout the test using strain gages. The volume of wear could be calculated from the decrease in pin height as measured by a linear differential voltage transducer (LVDT).

The 316 stainless steel was used to make pins and disks for the wear test. The pins were 12 mm thick and 50.8 mm diameter. Pins and disks were ground to a surface finish of approximately 0.05 μm Centre Line Average (CLA). The specimens were degreased using detergent, washed in water, then in acetone, and dried. After fine polishing, a number of disks were subjected to either TiN ion plating or nitrogen ion implantation. Pins and disks were accurately weighed before and after the test, using a balance sensitive to a few micrograms. A stream of compressed air was used to remove all loose wear debris.

The TiN coating was performed using an ULVAC hollow cathode source system. The substrate was sputter cleaned before plating. The substrate temperature was measured throughout the plating operation and remained at or below 150° C. The thickness of the TiN plating was 1 μm. The bias voltage was 500 V and the argon pressure was 9×10^{-4} Torr.

Nitrogen implantation was performed at an energy of 100 kev and a dose was 2.5×10^{17} ions/m^2. During implantation, substrate temperature was maintained at approximately room temperature.

A Phillips diffractometer (CuKα radiation) was used to find the crystallographic orientation and phases present in the TiN coatings. A Knoop microhardness tester was used to measure the hardness of the TiN coated disks and nitrogen implanted disks. Scanning Electron Microscopy (SEM) was used to examine worn surface topographies of pins and disks. Auger Electron Spectroscopy was used to determine the depth profile of nitrogen after the implantation.

To compare the wear rate of uncoated and TiN coated disks, the Maximum Wear Depth (MWD), which is defined as the distance from the original surface to the deepest part of the profile, was measured since there is a weight difference between the TiN and 316 stainless steels. In this study, a mean maximum wear depth ($\overline{MWD}$) was used. It was obtained by averaging the MWD of four traverses of the profilometer stylus across the wear track.

RESULTS

Figure 1 shows the wear pattern (wear rate vs. load) for sliding speeds of 0.25 m/s and 0.92 m/s. The wear rate at the slower sliding speed of 0.25 m/s shows a higher wear rate than that of 0.92 m/s sliding speed. Note that there is not a transition behavior from mild wear to mixed wear as reported by several researchers [2-4] for high sliding speeds.

Figure 2 shows the plot of the friction coefficient versus load for both speeds. The friction coefficient was measured continuously throughout the wear test and only the equilibrium friction coefficient was used. The friction coefficient at the sliding speed of 0.25 m/s is higher than that for the 0.92 m/s sliding speed.

Scanning Electron Micrographs (SEM) of the worn pin surface topographies show different wear behavior (see Figure 3). The evidence of oxide formation can be seen on the pins at the sliding speed of 0.92 m/s.

Figures 4 and 5 are plots of mean maximum wear depth ($\overline{MWD}$) vs. load for uncoated and TiN coated 316 stainless steel disks. The TiN coated 316 steel disks show a smaller $\overline{MWD}$ than that of uncoated 316 stainless steel for both sliding speeds tested.

Figures 6 and 7 are plots of friction coefficient versus sliding distance for both sliding speeds. Initially, TiN coated 316 stainless steel shows a much lower friction coefficient than that of uncoated 316 stainless steel.

The ratio of damaged track width to total track width as a function of load for the sliding speed of 0.25 m/s is shown in Figure 8. The TiN coated 316 stainless steel shows much less damage following testing at the same sliding distance. Figure 9 shows Scanning Electron Micrographs of surface topographies of the TiN coated 316 stainless steel disk.

Figure 10 is a plot of the wear volume for nitrogen implanted and unimplanted 316 stainless steel disks. The nitrogen implanted disks definitely show less wear volume under the same testing conditions. Friction coefficients of nitrogen implanted pins were compared with that of unimplanted pins (see Figure 11). Here again, the nitrogen implanted pins show lower friction coefficient than that of unimplanted pins.

Figure 12 shows the depth profile of nitrogen after the implantation. Auger depth profile shows a Gaussian distribution and the maximum concentration was approximately 14 atomic percent.

DISCUSSION

WEAR AND FRICTION BEHAVIOR OF UNCOATED 316 STAINLESS STEEL - The wear behavior of 316 stainless steel at relatively slow sliding speeds showed a different behavior than that reported for high sliding speeds as indicated by the lack of any load transition in Figure 1.

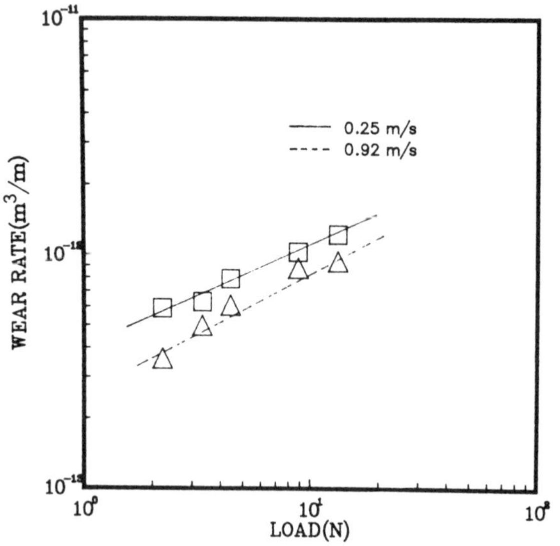

Figure 1. The wear pattern of 316 stainless steel pins on 316 disks at 0.25 m/s and 0.92 m/s.

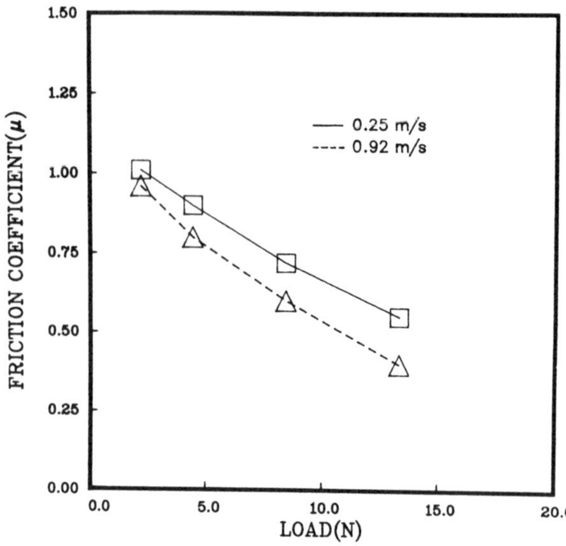

Figure 2. The friction coefficient of 316 stainless steel pins on 316 disks at 0.25 m/s and 0.92 m/s.

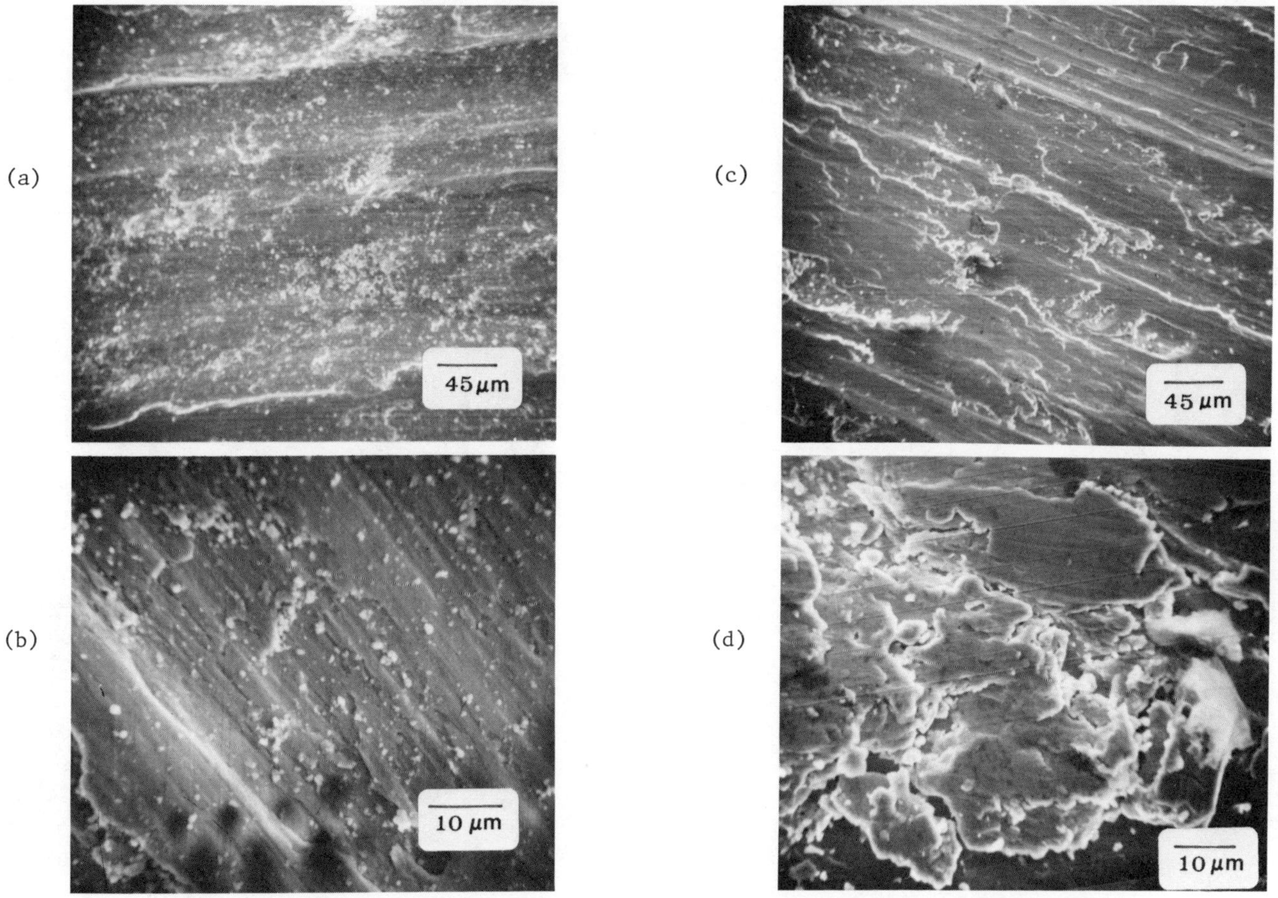

Figure 3. Scanning Electron Micrographs of worn surface of 316 stainless steel pins on 316 disks: (a) and (b) at 0.25 m/s and 8.9 N, (c) and (d) at 0.92 m/s and 8.9 N.

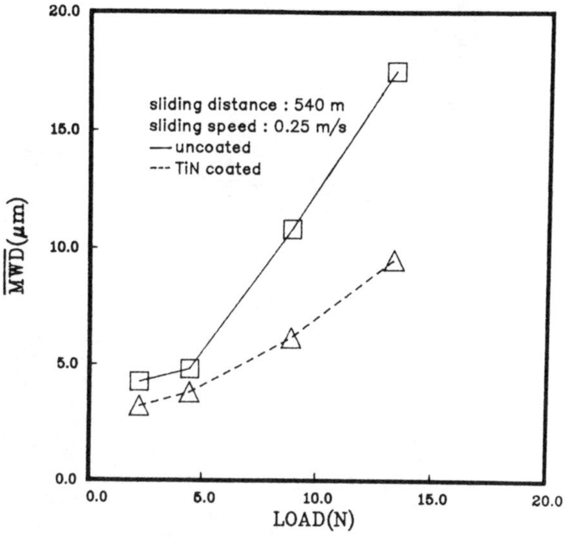

Figure 4. Mean maximum wear depth ($\overline{MWD}$) versus load for uncoated and TiN coated 316 stainless steel disks on 316 pins at 0.25 m/s.

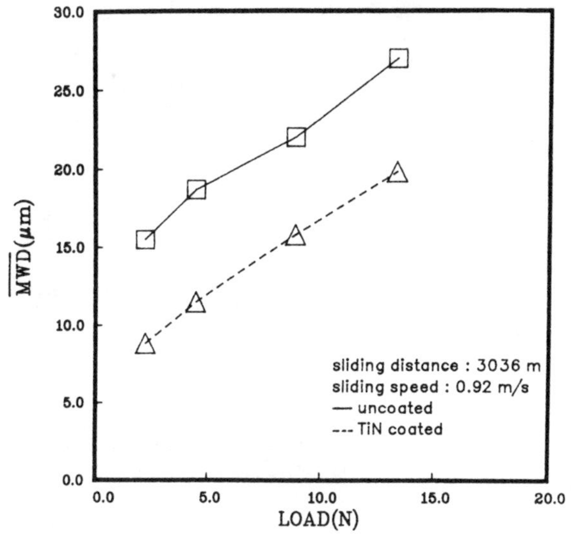

Figure 5. Mean maximum wear depth ($\overline{MWD}$) versus load for uncoated and TiN coated 316 stainless steel disks on 316 pins at 0.92 m/s.

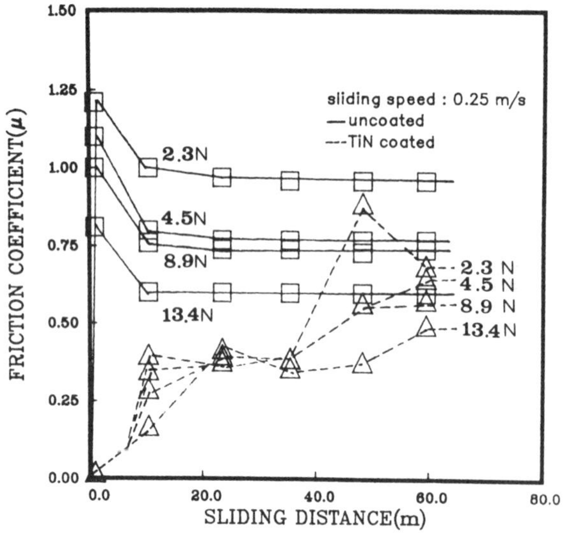

Figure 6. Friction coefficient versus sliding distance for 316 stainless steel pins on uncoated and TiN coated 316 disks at 0.25 m/s.

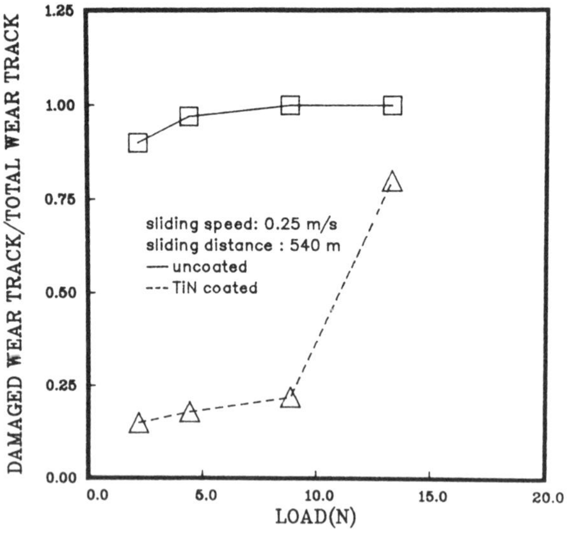

Figure 8. The ratio of damaged track width to total track width for uncoated and TiN coated 316 stainless steel disks on 316 pins at 0.25 m/s.

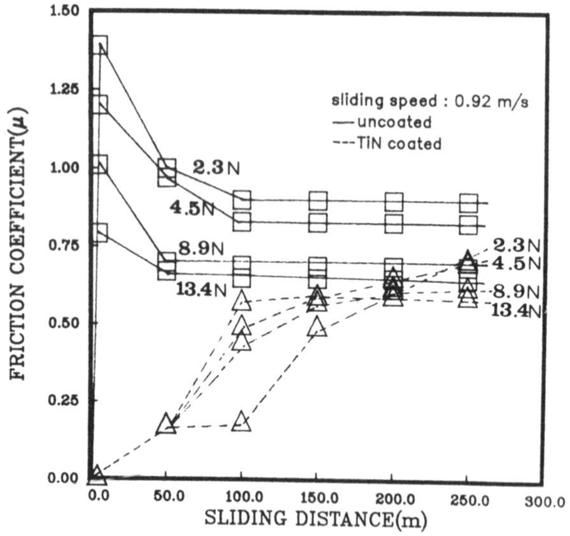

Figure 7. Friction coefficient versus sliding distance for 316 stainless steel pins on uncoated and TiN coated 316 disks at 0.92 m/s.

(a)

(b)

Figure 9. Scanning Electron micrographs of the wear track of TiN coated 316 stainless steel disk at: (a) low and (b) high magnification.

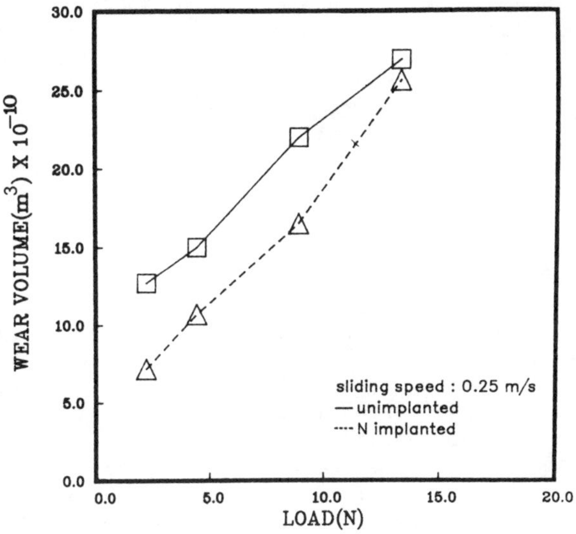

Figure 10. The wear volume of nitrogen implanted and unimplanted 316 stainless steel disks on 316 pins at 0.25 m/s.

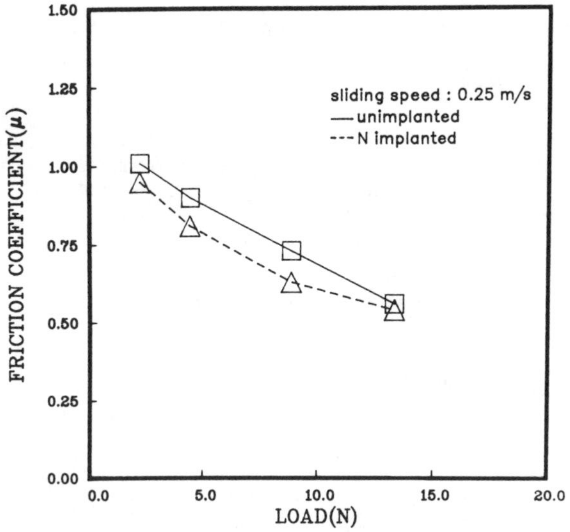

Figure 11. The friction coefficient of 316 stainless steel pins on nitrogen implanted and unimplanted 316 disks at 0.25 m/s.

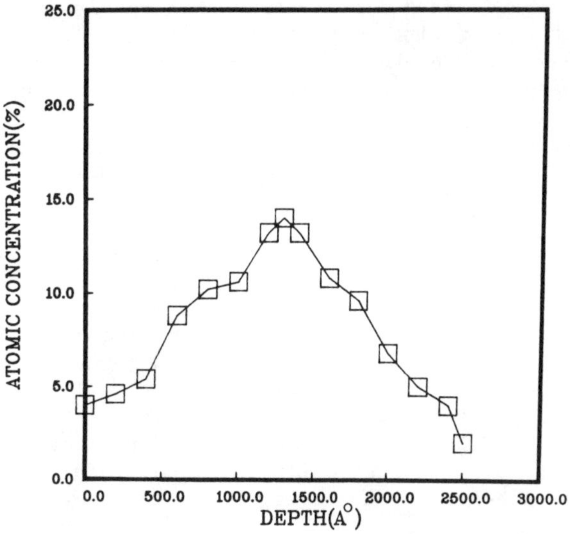

Figure 12. AES depth concentration profile of nitrogen after implantation.

Scanning electron micrographs show a prow formation with grooves in the flat on the worn surface of pin (see Figure 3) which is in general agreement with the observation of Smith [1] from tests conducted at very low sliding speed. The worn surface also shows an extensive oxide formation probably due to increased frictional heating. Therefore, it is possible that prow formation is the dominating mechanism of wear at very low sliding speed. As the sliding speed increases, one obtains both prow formation and oxide formation. At high sliding speeds, oxidation of the surface prevails. It is interesting to note that oxide formed at the sliding speed of 0.92 m/s was removed by delamination (see Figure 13). This oxide removal mechanism was not observed at high sliding speeds [4].

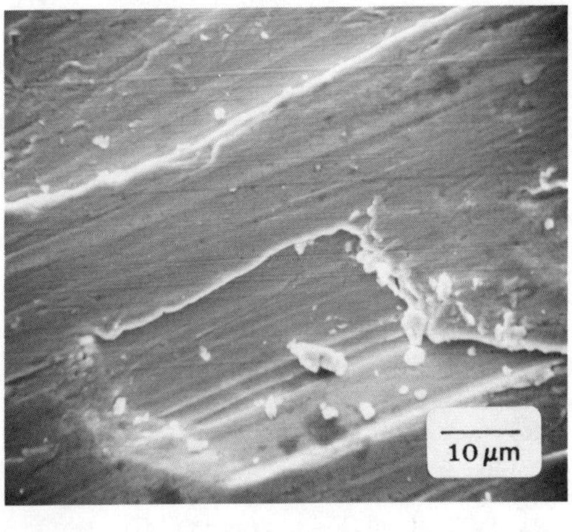

Figure 13. Scanning Electron Micrograph showing oxide removal by delamination.

THE PROPERTIES OF TiN COATING - The titanium nitride coating had a golden yellow color and x-ray diffraction (XRD) results showed that the coating was essentially a single TiN phase. XRD peaks from the TiN coating also showed a preferred (111) surface orientation in the (111) plane. These results are in good agreement with those of Kobayashi and Doi [6]. They also reported that (111) plane showed a maximum wear resistance.

Hardness results using a load of 2 grams showed approximately 2000 kg/mm^2. The adhesion of TiN coating is a controlling factor in improving the wear resistance on 316 stainless steel. To determine the adhesion of coating, an impact tester was used to bend TiN coated disks so that the substrate deformed plastically and the interface between TiN coating and substrate was subjected to high stress. Figure 14 shows the coating at test well above the yield strength of the substrate.

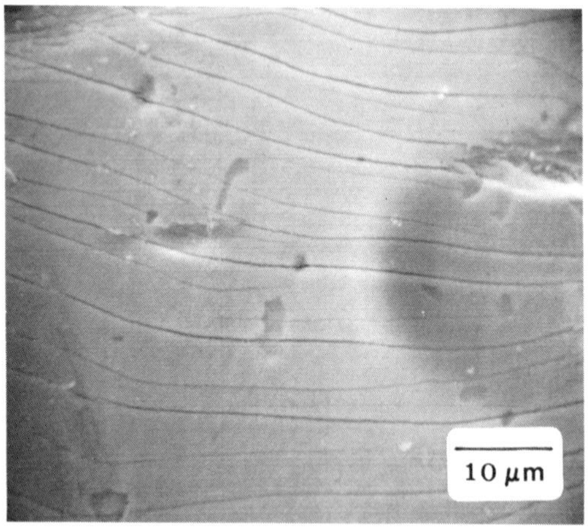

Figure 14. Scanning Electron Micrograph showing the surface of TiN coated 316 stainless steel disk after plastic deformation.

THE WEAR AND FRICTIONAL BEHAVIOR OF TiN COATED 316 STAINLESS STEELS - TiN ion plated 316 stainless steel show a major improvement in wear resistance. At the sliding speed of 0.25 m/s and a sliding distance of 540 m, the MWD was about half of the MWD of uncoated specimen and the ratio of damaged track width to total track width was very low for TiN coated 316 stainless steel. The overall wear resistance of the TiN coated 316 was increased up to six times. Even at high sliding speed (0.92 m/s) and longer sliding distance (3036 m), the wear resistance still improved about 50%.

The friction coefficient for coated specimens was significantly less than for uncoated ones. However, during the wear test, the friction coefficient of the coated specimens gradually approached the uncoated values. This was possibly caused by a penetration of TiN coating in spite of the excellent adhesion properties of the film.

Scanning electron micrographs of worn surface of TiN coated 316 stainless steel disk show that there was penetration of coating during the wear test. However, even in the damaged wear track, a TiN film can still be seen.

THE WEAR AND FRICTIONAL BEHAVIOR OF NITROGEN IMPLANTED 316 STAINLESS STEELS - Nitrogen implanted 316 stainless steel disks show less wear volume especially at low loads. At the load of 2.3 N, the wear volume was about half of unimplanted disks. The improvement in wear resistance can be explained by the increase in surface hardness after nitrogen implantation. The ratio of the hardness of the implanted 316 stainless disk to that of the unimplanted 316 stainless disk was 1.27. In addition, nitrogen is an austenite stabilizer, it can prevent or delay the martensitic transformation of 316 stainless steel in the running-in period (severe wear) observed by previous researcher [1,4]. The beneficial effect of nitrogen implantation decreased rapidly by increasing the load because the MWD of the nitrogen implanted 316 stainless steel disks are far beyond the maximum penetration depth of nitrogen.

A slightly lower friction coefficient was observed for the nitrogen implanted 316 stainless steels following test at relatively long sliding distance. This can be most likely due to the migration of nitrogen during the wear test observed by other researchers [13,15,16]. As contact between the asperities generate localized heat the nitrogen diffuses away from the contact area.

CONCLUSIONS

1. The wear behavior of 316 and surface modified 316 stainless steel varied according to the sliding speed. Both a prow formation and adherent oxide formation was observed at the sliding speed of 0.92 m/s and 0.25 m/s. There was not a transition load in the range of 2-15 N.

2. At the sliding speed of 0.92 m/s, adherent oxide was removed by delamination.

3. The TiN coating was essentially single phase TiN with a (111) preferred orientation and a good adhesion to the substrate.

4. The feasibility of hard coating on soft substrate was established. Even though there was a penetration of the TiN coating due to the severe test conditions, the wear resistance improved up to 6 times compared to that of uncoated 316 stainless steel. The friction coefficient was always initially very low for coated samples and increased to the uncoated values within a relatively short time.

5. Nitrogen ion implantation improved the wear resistance of 316 stainless steels especially

at low test loads. The friction coefficient was slightly lower for implanted samples than for unimplanted 316 stainless steels.

ACKNOWLEDGEMENTS

The authors wish to acknowledge the partial support of National Aeronautics and Space Administration (NASA)-Marshall Space Flight Center under the contract NAS 8-350-48.

Thanks are due to Dr. J. K. Hirvonen of Spire Corporation for nitrogen implantation, Dr. K. Moriyama of ULVAC for TiN plating, and Dr. W. B. Carter of Georgia Tech for AES analysis.

REFERENCES

1. Smith, A.F., Wear, 96, 301-18 (1984).

2. Allen, C.B. and T.F.J. Quinn, Proc. ASME-ASLE Lubrication Conf., San Diego, CA, October 1984, to be published.

3. Quinn, T.F.J., Tribol. Int., 16, 257-71, 305-12 (1983).

4. Hong, H.S., R.F. Hochman, and T.F.J. Quinn, manuscript in preparation for publication.

5. Buhl, R., H.K Pulker, and E. Moll, Thin Solid Films, 80, 265-70 (1981).

6. Kobayashi, M. and Y. Doi, Thin Solid Films, 54, 67-74 (1978).

7. Sirvoi, E.H., M. Sulonen, and H. Sundquist, Thin Solid Films, 96, 93-101 (1982).

8. Jamal, T., R. Nimmagadda, and R.F. Bunshah, Thin Solid Films, 73, 245-54 (1980),

9. Kloos, K.H., E. Bronszeit, M. Gabriel, and H.J. Schroder, Thin Solid Films, 96, 67-77 (1982).

10. Gabriel, H.M. and K.H. Kloos, Thin Solid Films, 118, 243-54 (1984).

11. Bair, S., S. Ramalingam, and W.O. Winer, Wear, 60, 413-19 (1980).

12. Ramalingam, S., and W.O. Winer, Thin Solid Films, 73, 267-74 (1980).

13. Yost, F.G., S.T. Picraux, D.M. Follstaedt, L.E. Pope, and J.A. Knapp, Thin Solid Films, 107, 287-95 (1983).

14. Hirvonen, J.K., J. Vac. Sci. Technol., 15, 1662-68 (1978).

15. Bolster, R.N., and I.K. Singer, ASLE Trans., 24, 526-32 (1980).

16. Pollock, J.T.A., M.J. Kenny, and P.J.K. Paterson, Mat. Res. Soc. Symp. Proc., 27, 691-97 (1984).

17. Oliver, W.C., R. Hutchings, and J.B. Pethica, Met. Trans. A, 15, 2221-29 (1984).

18. Ramalingam, S., W.O. Winer, and S. Bair, Thin Solid Films, 84, 273-79 (1981).